Truelove Lowland, Devon Island, Canada: A High Arctic Ecosystem

Truelove Lowland, Devon Island, Canada: A High Arctic Ecosystem

edited by L.C. Bliss

The University of Alberta Press 1977

First published by
the University of Alberta Press
Edmonton, Alberta
Canada 1977

Canadian Cataloguing in Publication Data
Main entry under title:
Truelove Lowland, Devon Island, Canada

"This book is the summary of 33 research projects conducted under the auspices of the International Biological Programme."
Includes bibliography and index.
ISBN 0-88864-014-5

1. Tundra ecology—Northwest Territories—Devon Island.
2. Devon Island, N.W.T. I. Bliss, Lawrence C., 1929-
II. International Biological Programme.
QH541.5.T8T78 574.5'264 C77-002072-0

Printed by Printing Services of
The University of Alberta, Edmonton, Alberta, Canada.

This book is dedicated to the memories of
Professors Thorvald Sørensen and B.N. Gorodkov
who pioneered arctic biological research in
Greenland and the U.S.S.R.

Contents

Preface

The International Biological Programme was initiated in the late 1950s to study biological production in relation to human welfare, to advance scientific knowledge, and to promote international scientific cooperation. Approximately 50 countries participated in detailed planning and preparation from 1964 to 1967, research from 1967 to 1972, and syntheses from 1972 to 1975.

The Canadian Committee for the International Biological Programme (C.C.I.B.P.) sponsored two terrestrial productivity projects including this one. Planning for this study took place from 1969 to 1970, field research from 1970 to 1974, and synthesis in 1974 and 1975. Due to the late initiation of this project and the previous full commitment of C.C.I.B.P. funds, special approval by the committee was obtained to have the National Research Council of Canada provide basic core funding. Additional funds were obtained from Canadian government departments (Environment, Indian and Northern Affairs, Energy Mines and Resources) and from 24 petroleum companies and consortia.

The Devon Island Project was one of 14 major studies within the I.B.P.-P.T. Tundra Biome Programme, a programme that included four arctic tundra sites. This one was the most northern.

This book contains summary papers of all of the field and laboratory studies conducted over the five year period. A preliminary book was published in 1972 (Bliss 1972) and synthesis papers of the project were published in the books entitled Structure and Function of Tundra Ecosystems (Rosswall and Heal 1975) and Canadian Contribution to the International Biological Programme (Cameron and Billingsley 1975).

May 1976

L.C. Bliss
Project Director

References

Bliss, L.C. (ed.) 1972. Devon Island I.B.P. Project, High Arctic Ecosystem. Dept. Botany, Univ. Alberta Printing Services. Edmonton, 413 pp.

Cameron, T.W.M. and L.W. Billingsley. (eds.). 1975. Energy Flow—Its Biological Dimensions. Roy. Soc. Canada. Ottawa. 319 pp.

Rosswall, T. and O.W. Heal (eds.). 1975. Structure and Function of Tundra Ecosystems. Ecol. Bull. No. 20. Swed. Nat. Sci. Res. Council. Stockholm. 450 pp.

Acknowledgments

A project of this magnitude could not be carried out without the support of many people and organizations. Unlike the other Canadian I.B.P. Projects, this one was funded from several sources. Major funding was provided by the following federal and provincial agencies and universities: National Research Council of Canada's C.C.I.B.P. and its Division of Building Research; Environment Canada and its Canadian Wildlife Service; Arctic Land Use Research programme and the Environmental-Social Program Northern Pipelines Committee of the Department of Indian and Northern Affairs; Polar Continental Shelf Project of the Department of Energy, Mines and Resources; Alberta Environment; and The University of Manitoba Northern Studies Committee. The Arctic Institute of North America managed and provided part of the Base Camp facilities. The University of Alberta, The University of Calgary, Laurentian University, and The University of Manitoba provided teaching assistantships for some of the graduate students. Several researchers used their N.R.C.C. operating grants.

The Executive of the Arctic Petroleum Operators Association with the following 24 member companies and consortia provided financial support: Atlantic Richfield Canada Ltd., Arctic Gas Study Ltd., British Petroleum Oil and Gas Ltd., Canadian Arctic Gas Study Ltd., Canadian Geothermal Oil Ltd., Canadian Southern Petroleum Ltd., Canadian Superior Ltd., Chevron Canada Ltd., Canada Cities Service Ltd., Elf Oil Exploration and Production Canada Ltd., Gulf Oil Canada Ltd., Hudson's Bay Oil and Gas Co. Ltd., Imperial Oil Ltd., King Resources Co., Mackenzie Valley Pipe Line Research Ltd., Mobil Oil Canada Ltd., Northwest Projects Study Group Ltd., Panarctic Oils Ltd., Pan Canadian Petroleum Ltd., Shell Oil Canada Ltd., Sun Oil Co., Tenneco Oil and Minerals Exploration Ltd., Texaco Canada Ltd., and Union Oil Canada Ltd.

Special thanks are due to King Resources Co. (1970), Sun Oil Co. (1971) and Imperial Oil Ltd. (1972, 1973) for logistic support between Edmonton and Resolute and to the Polar Continental Project which provided most of the logistic support from Resolute to Devon Island. Without this support the project could not have functioned in such a remote area.

Panarctic Oils, Elf Oil, and Sun Oil provided most of the inter-island logistic support for the surface disturbance studies. Environmental instruments were loaned by the Atmospheric Environment Service of Environment Canada.

The University of Alberta and its Department of Botany absorbed many of the associated administrative costs and this is gratefully acknowledged.

The Provincial Soil and Feed Testing Laboratory provided nutrient analyses for plant material and for soil samples of various project researchers.

Special aerial photography was used for development of the plant community and soil maps. K. Arnold of the Inland Waters Branch, Environment Canada, and T.J. Blachut and M. Van Wijk of Photogrammetry Branch, N.R.C., deserve special acknowledgement. The maps were prepared by G.A. Lester and staff of the Cartographic Branch, Department of Geography, The University of Alberta.

The project operated through a Steering Committee whose members came from universities and the petroleum industry. Corresponding members from government agencies, the C.C.I.B.P. office, and industry also aided in major decisions. Their help is gratefully acknowledged. Administrative details are contained in Appendix I.

I especially want to thank the many people who worked on the project: the undergraduate field assistants who gained insight into research, the graduate students and Post-Doctoral Fellows who conducted much of the research and whose papers are included along with those of faculty members. Last but not least, are those people in Edmonton and elsewhere who contributed in many support aspects, yet who did not have the thrill of experiencing an arctic summer. Only through the dedication, cooperation, and enthusiasm of these many people could such a project have been successfully completed.

To Mrs. B. Walker a special thank you for her excellent cooking which helped maintain high moral, and another to Ward Elcock and Mike Hoyer of the Arctic Institute of North America for their fine camp management.

All manuscripts received one to three reviews and as a result the chapters of this book have been subjected to the same rigors as is customary for scientific journals. The editor gratefully acknowledges the editorial help provided by the following people:

G.O. Batzli
T. Beck
W.D. Billings
I.M. Brodo
R.T. Coupland
K.R. Everett
P. Flanagan
D.R. Flook
M.M.R. Freeman
L.J. Fritschen
K.D. Hage
H. Halvorson
F.K. Hare
O.W. Heal
M. Hickman
C. Jonkel
J. Kalff
K.A. Kershaw
R. Knowles
A.H. Laycock
R.E. Longton
S.F. MacLean, Jr.
J.R. Mackay
W.J. Maher
P.C. Miller
H. Nichols
E.A. Paul
J. Phillipson
F.A. Pitelka
R.E. Redmann
D.E. Reichle
G. Rempel
C.E. Schweger
C.W. Slaughter
K. Van Cleve
G.C. West
R.G. White
J.B. Whittaker
D.T. Wicklow
J.R. Willard

The total costs of typesetting, printing, and binding this book have been provided by the Arctic Petroleum Operators Association, Arctic Land Use Research Program of the Department of Indian and Northern Affairs, Environment Canada, and the President's Fund, The University of Alberta.

Truelove Lowland, Devon Island, Canada: A High Arctic Ecosystem

Introduction

L.C. Bliss

Reported here is a research summary of the structure and function of a high arctic ecosystem. The study formed a part of the Canadian contribution to the International Biological Programme and was one of 14 major studies within the I.B.P. Tundra Biome programme.

Location

Truelove Lowland (43 km^2), on the northeastern coast of Devon Island (Fig. 1) (75° 33' N, 84° 40' W), and named after an early whaling ship, is the westernmost in a series of four lowlands (Skogn 13 km^2, Sparbo-Hardy 86 km^2, and Sverdrup 26 km^2) that support a greater biological diversity than the surrounding lands. Most of Devon Island (54,000 km^2), the fifth largest island within the Canadian Arctic Archipelago (1,424,000 km^2), is a Polar Desert (Fig. 2). The High Arctic, with its sparse vegetation cover and limited wildlife, contains relatively few areas (6%) such as Truelove Lowland, especially within the Queen Elizabeth Islands (1%).

The area was chosen because of its:

1. biological diversity;
2. reasonable logistic distance (315 km) from Resolute Bay, Cornwallis Island;
3. existing research camp established by the Arctic Institute of North America in 1960;
4. background of scientific data; and
5. representativeness of lowlands, especially in the eastern High Arctic, that contain herds of muskox.

Objectives

This study was undertaken as a part of the Canadian contribution to the International Biological Programme. It was one of 14 major ecosystem studies conducted within the I.B.P. Tundra Biome. Within this group of cooperative projects, it was the only one conducted within the High Arctic, out of a total of four major arctic studies.

The major objectives of this study were to:

1. determine population numbers and standing crop (weight, calories,

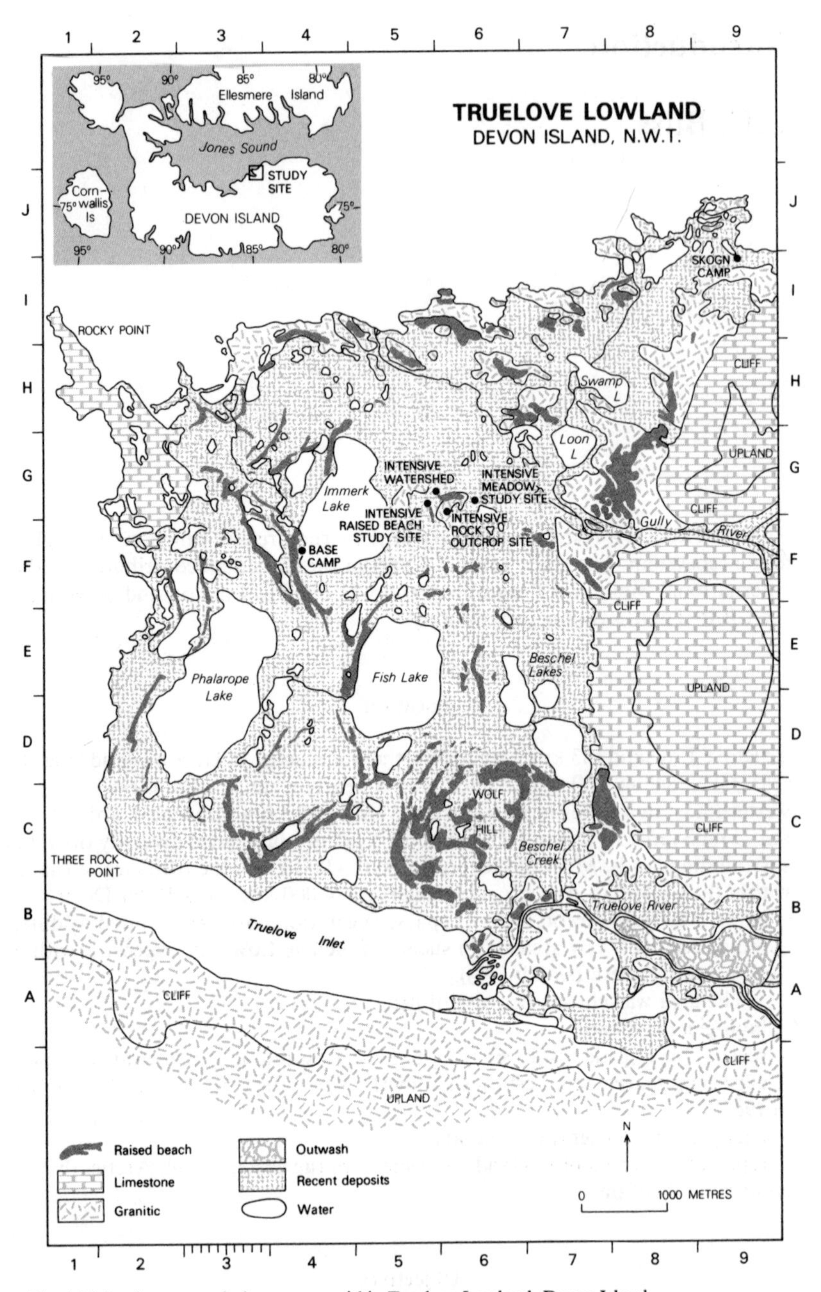

Fig. 1 Major features and place names within Truelove Lowland, Devon Island.

nitrogen) for the major biological components of the Lowland and to express these data on a land area basis by major habitats (topography — plant community);

2. determine the rates of energy flow through the total system and its two major habitat subcomponents (sedge-moss meadows and raised beach systems);

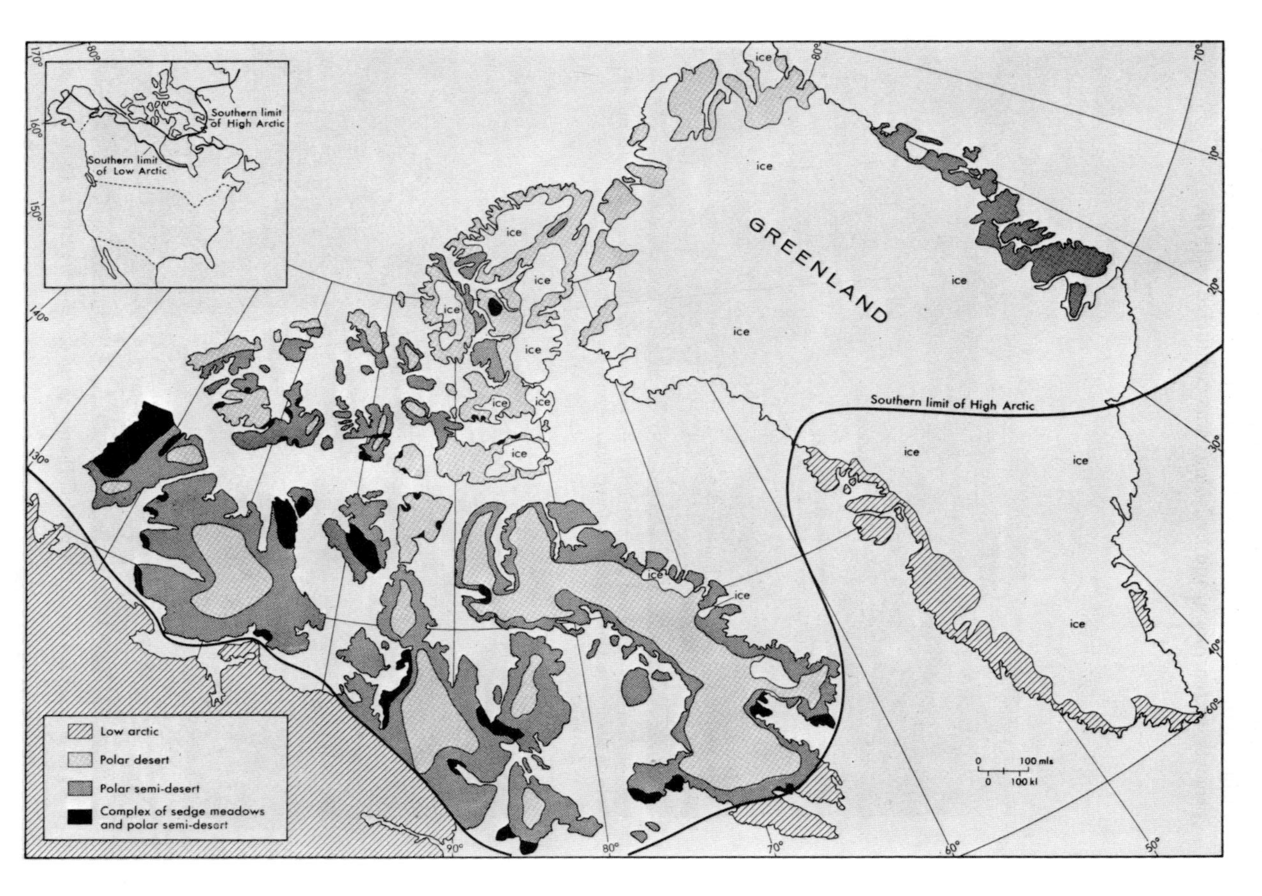

Fig. 2 The High Arctic of North America, including Greenland. Delineation of the vegetation for Greenland was provided by T.W. Böcher, and for the District of Keewatin by J. Svoboda and R.W. Wein.

3. determine efficiency of the system in capturing and utilizing energy at different trophic levels;
4. determine the environmental and biological limiting factors for the growth and development of important plant and animal species; and
5. develop static and dynamic models of high arctic ecosystem function and the function of its component parts.

Additionally the role of Inuit (Eskimo) as harvesters within northern systems, the hydrologic role of rivers, streams and lakes, and the biological components of lakes were studied.

Major research was conducted in the summers of 1970 through 1973 including the 1972-73 winter. A more minimal program of several biological projects, meteorology, and hydrology was conducted in 1974.

Fig. 3 View of the lowland and dolomite cliffs to the east from the plateau beyond the Truelove River. Note the granitic rock outcrops in the foreground beyond the shadow.

Fig. 4 Polar desert at 400 m on the plateau. The non-sorted polygons are 2 to 3 m across. Total plant cover is $<2\%$.

Site description

Truelove Lowland (43 km^2) and Valley (3.6 km^2) form a compact unit, isolated on the north, west, and part of the south by a 24 km shoreline. To the east and south, steep cliffs rise 300 m to the barren plateau (Fig. 3). The plateau (Fig. 4), covered with sedimentary rocks of probably Cambrian age, rises gradually to the ice cap 17 to 20 km inland from Truelove Inlet. The ice cap begins at an elevation of *ca.* 450 m and reaches a maximum ice thickness of 500 to 700 m. Maximum elevation on the ice cap is 2,080 m. Much of the island was deglaciated 8,700 years B.P. (Andrews 1970).

Most of the Lowland is covered with Pleistocene age deposits that overlay a Precambrian complex of granulites and granitic gneisses. These rocks outcrop along the north coast, the Gully River, and the Truelove River (Fig. 1). Dolomites of probable Cambrian age occur at Rocky Point and form the cliffs (>200 m) to the east of the Lowland where they rest on a 2 to 3 m section of sandstone that overlays the granites (Krupicka this volume).

These coastal lowlands resulted from postglacial rebound following ice retreat. Radiocarbon dating has shown that the upper marine limit lies at 76 m, dating to *ca.* 9,450 B.P. (Barr 1971). Total uplift is 107 m, corrected for sea rise. For the first 2,000 years, uplift averaged *ca.* 3.2 m 100 $years^{-1}$ and *ca.* 0.6 m 100 $years^{-1}$ over the last 7,450 years (based upon Barr 1971). At the present, uplift about equals eustatic sea level rise. With uplift, lagoons were cut off by off-shore bars which formed beach ridges, resulting in the formation of shallow lakes. Some of these have filled in to form meadows (Fig. 5). Of the three larger lakes, Phalarope Lake (155 ha), Fish Lake (101 ha), and Immerk Lake (96 ha), the latter two along with the smaller Loon Lake (16 ha), are deep enough (7 to 8.5 m, mean depth 2.9 to 3.2 m) to support arctic char *(Salvelinus alpinus).* There are 15 other lakes of medium size and many small water bodies (Fig. 1). Lakes and ponds cover 22% of the Lowland (Table 1).

The raised beaches occur in a sequence of more than 20 "steps" across the Lowland. They effectively block the drainage of water and, because of their elevated nature (1 to 5+ m) and coarse textured gravels and sands, provide much drier habitats for plants and animals. These raised beaches are often 30

Fig. 5 Hummocky sedge-moss meadow in which *Carex stans* dominates with lesser amounts of *Eriophorum angustifolium*, and *Arctagrostis latifolia*.

Table 1. Major topographic-plant community units within Truelove Lowland.

Unit	Area (ha)	Lowland (%)
Salt water marsh	22	0.5
Lichen barren on limestone pavement	180	4.2
Cushion plant-lichen on raised beaches	215	5.0
on limestone pavement	26	0.6
on rock outcrops	52	1.2
Cushion plant-moss on raised beaches	274	6.4
separate from raised beaches	298	6.9
Dwarf shrub heath on rock outcrops	533	12.4
Hummocky sedge meadow	883	20.6
Frost-boil sedge meadow	796	18.4
Wet sedge meadow	88	2.1
Lakes and ponds	933	21.7
Total	4300	100.0

Fig. 6 Raised beach ridge dominated by *Dryas integrifolia*, with lesser amounts of *Carex nardina* on crest and *C. rupestris*, *Saxifraga oppositifolia*, and *Salix artica*, on slope, Intensive Study Site. Note insect emergent trap background and meteorological station beyond.

to 100 m wide and 500 to 1000+ m in length (Fig. 6). Stranded whale bone and shells have permitted radiocarbon dating; beaches near Rocky Point being 2,900± 85 years B.P., the Base Camp is located on one 6,100± 125 years B.P., and our Intensive Study Site was on a raised beach *ca*. 7,500 years B.P. (Barr 1971). Typically, lakes or wet sedge-moss lands occur upslope of these ridges; downslope better drained sedge lands occur. Drainage across the Lowland is impeded by the sequence of raised beach ridges and periodic rock outcrops (Fig. 7); only three small creeks drain the lakes.

The Truelove River, which originates at the ice cap, *ca*. 30 km inland, provides a minor influence in the Truelove Valley and Lowland because of its entrenchment. Gully River flows from the plateau to the east through Loon and Swamp Lakes and provides water for meadows in that portion of the Lowland (Fig. 1).

The Lowland has been mapped for soils and plant communities (see separate chapters). Maps (in pocket) have been prepared, based upon

Fig. 7 Dwarf shrub heath dominated by *Cassiope tetragona* within the Rock Outcrop Intensive Study Site.

photography flown for this purpose (see Appendix 2). Although more units were mapped, seven major topographic-plant community units with several subdivisions have been recognized (Table 1). Most of the biological data are presented in relation to the land area of one or more of these units. Of these units, hummocky sedge-moss meadows (20.6%), raised beach ridges (11.4%), and rock outcrops (12.4%) were studied most intensively.

General research plan

The major objective of this study was to determine energy flow through the lowland system and its major habitat subsystems. Therefore it was necessary to determine the major bioenergetic components and then to locate representative soil-plant-animal (topographic) assemblages for study. Through a systematic inspection of a series of meadows and raised beaches, a representative of each was selected and designated as the Intensive Study Site (Fig. 1). These sites were within 700 m of each other; between them a rock outcrop was selected for more limited meteorological and plant production research and an adjacent small sedge-moss meadow-watershed for part of the hydrology study.

Within the meadow and raised beach intensive sites, a 5×5 m grid network was established and divided into areas for:

1. destructive sampling;
2. non-destructive sampling; and
3. observation only (control) (Figs. 8 and 9).

Since the Lowland consists of a mosaic pattern of sedge-moss meadows, cushion plant communities on raised beaches, and dwarf shrub heath in the rock outcrops, a number of raised beaches, meadows, and rock outcrops (generally 3 to 8) were selected for additional research (Extensive Study Sites). Thus the data collected were more representative of the biological diversity occurring throughout the Lowland than would have been the case if only the intensive sites had been studied. The Intensive and Extensive Study Sites included the following research:

Intensive study site research	
Meteorology	Nitrogen Fixation
Soils	Invertebrates (general)
Hydrology	Collembola
Plant Production	Nematodes
Photosynthesis	Microbiology and Decomposition
Evapotranspiration	Surface Manipulation

Extensive study site research	
Meteorology	Lemming
Soils	Muskox
River Hydrology	Arctic Hare
Plant Production	Weasel, Fox
Nitrogen Fixation	Muskox Dung Decomposition
Invertebrates (general)	Limnology
Birds	

Research on the mobile animals (birds and mammals) was conducted wherever the animals could be found and observed in the Lowland and Truelove Valley.

In addition, the utilization of wildlife by Grise Fiord Inuit (Eskimo) was studied on north Devon Island and southern and central Ellesmere Island. The surface disturbance studies, associated with petroleum exploration, were conducted in the western Queen Elizabeth Islands, although surface disturbance studies were also conducted on Truelove Lowland.

Site representativeness

This Lowland and comparable areas with sedge-moss or grass-moss and cushion plant communities occupy only 1% of the Queen Elizabeth Islands (*ca.* 418,000 km^2). Most of the land is a Polar Semi-desert of cushion plant-moss/lichen communities as on the raised beach ridges in Truelove Lowland or the herb-moss communities of the western islands (see Bliss and Svoboda 1977). Such lands occupy *ca.* 25% of these northern islands. Polar Deserts with <2 to 5% plant cover comprise an additional 49% and ice caps the balance (25%) (see Fig. 2). A mosaic pattern of lowland sedge-moss and cushion plant communities comprise a larger percentage (7%) of the southern arctic islands and mainland, yet they cover only a small portion of the total lands of the High Arctic (6%) compared with Polar Semi-deserts (45%) and Polar Deserts (41%). The remainder of the area is ice (8%).

These lowlands, usually with considerable amounts of ponded water, are more closely related to the Low Arctic of the arctic coastal plain of Alaska, areas in the Mackenzie Delta region, and areas on the Taimyr Peninsula, U.S.S.R. in terms of ecosystem structure and function, than to broad areas in the High Arctic. Communities of dwarf shrub heath are limited to warmer, snow protected slopes throughout the High Arctic, and in all landscapes are a minor component.

It is these few oases of more complete plant cover and scattered lakes that provide an adequate food base for muskox, lemming, waterfowl, and associated predators which permits this level of biological diversity and

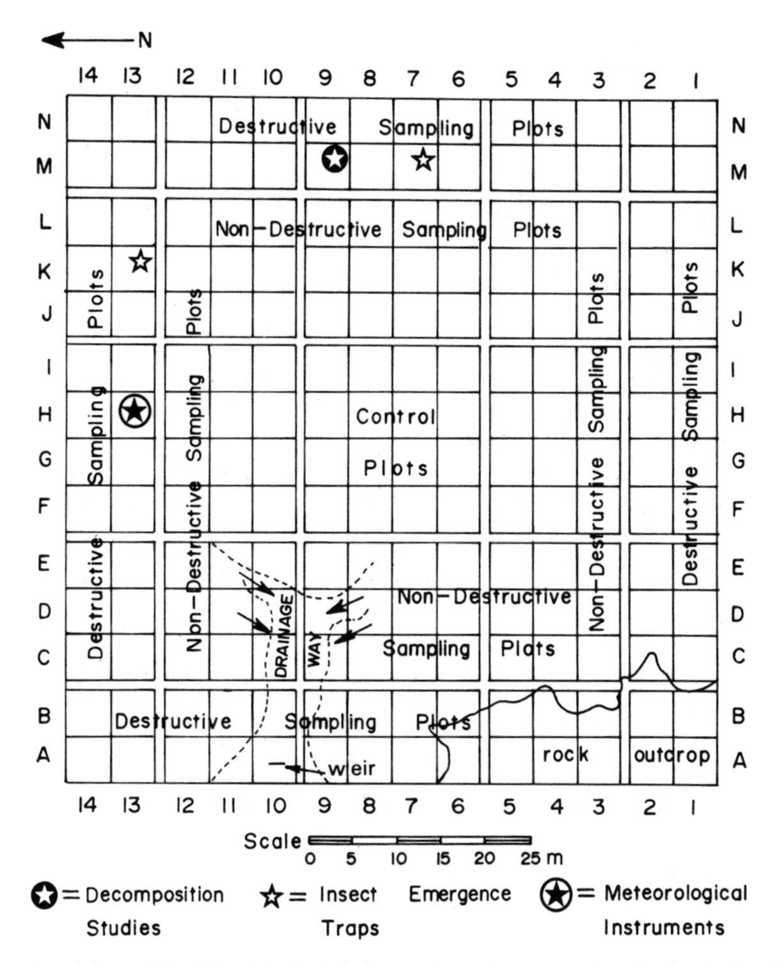

Fig. 8 Map of the Intensive Study Site hummocky sedge-moss meadow showing the location of studies.

productivity at these latitudes. At least some of these oases have higher solar radiation and the warming effect of Föhn winds which gives them higher temperatures in summer than surrounding lands. Table 2 shows that Eureka, Ellesmere Island, and Truelove Lowland have accumulative degree day values much more like those of Barrow, Alaska than to other locations in the Queen Elizabeth Islands.

The Truelove Lowland flora includes 96 species of vascular plants, 132 species of mosses, 30 hepatics (Vitt and Pakarinen this volume), 182 lichen species (Richardson and Finegan this volume), and 92 species of fungi (see Appendices 3-6). Of the nine terrestrial high arctic mammals (including polar bear), Peary's caribou *(Rangifer tarandus pearyi)* and brown lemming *(Lemmus trimucronatus)* are the only mammals absent. Caribou were present in the recent past, having become extinct following Inuit hunting. There are 17 to 19 species of birds that nest in the Lowland, far more than in the Polar Desert or Semi-desert (see Appendix 8). While the invertebrate data are not complete, a minimum of 150 species of insects in 7 orders occur along with *ca.*

Table 2. Accumulated degree days above 0°C for arctic locations 1970-1973.

Location	Latitude °N	Accumulated Degree Days 1970	1971	1972	1973	Mean	Mean Number Days Above 0°C
Eureka, Ellesmere Isl.	80	292	420	252	338	326	81
Isachsen, Ellef Ringnes Isl.	79	132	220	80	188	155	60
Truelove Lowland, Devon Isl.	76	295	292	172	354	278	75
Resolute, Cornwallis Isl.	75	259	292	131	307	247	66
Sachs Harbour, Banks Isl.	72	520	463	364	591	485	93
Barrow, Alaska	71	230	201	402	362	269	78
Tuktoyaktuk, N.W.T.	70	821	968	630	1207	932	114

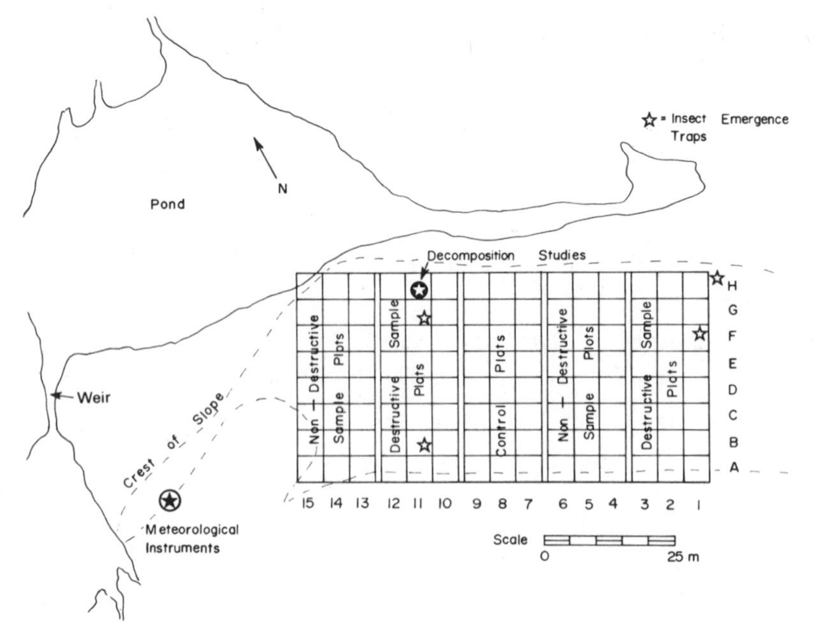

Fig. 9 Map of the Intensive Study Site raised beach ridge (crest) showing the location of studies.

50 species of Nematoda-Enchytraeidae and Protozoa (Ryan 1977) (see Appendix 7).

These biological components and others to be discussed elsewhere in this volume, show that the diversity and productivity of such lowlands are unusual for these latitudes and that special environmental conditions permit the development of these oases. By studying an ecosystem of high productivity, we have maximized the amount of biological information possible and the information necessary to manage plant-animal systems at these latitudes.

References

Andrews, J.T. 1970. A geographical study of postglacial uplift, with particular reference to Arctic Canada. Inst. Brit. Geogr., Spec. Pub. 2. 156 pp.

Barr, W. 1971. Postglacial isostatic movement in northeastern Devon Island: a reappraisal. Arctic 24: 249-268.

Bliss, L.C. and J. Svoboda. 1977. Plant communities and plant production in the western Queen Elizabeth Islands, N.W.T. (in preparation).

Ryan, J.K. 1977. Comparison of invertebrate species lists from IBP tundra sites. *In:* Tundra: Comparative Analysis of Ecosystems. J.J. Moore and S. Jonsson (eds.). Cambridge Univ. Press (in press).

Abiotic components

Permafrost investigations on Truelove Lowland

R.J.E. Brown

Introduction

Permafrost, or perennially frozen ground, occurs everywhere beneath the land surface of Truelove Lowland, which is situated in the northern part of the continuous zone (Brown 1967). The presence of thick continuous permafrost throughout the Canadian Arctic Archipelago has been known for a long time, but until recently investigations have been limited to a few scattered locations (Brown 1970, Brown 1972a). During the past few years, observations have increased considerably, especially in the Queen Elizabeth Islands, with the onset of petroleum explorations (Judge 1973). The Canadian Tundra Biome Study of the I.B.P. provided an opportunity to obtain information on permafrost conditions in the major terrain types at Truelove Lowland. No permafrost investigations had been carried out previously on Devon Island. A preliminary indication of permafrost conditions was available from observations at Resolute, N.W.T., 320 km to the southwest on adjacent Cornwallis Island. At this location permafrost was found to be about 400 m thick with a temperature of −12.5°C at the depth of zero annual amplitude (Cook 1958, Misener 1955).

Methods

A preliminary reconnaissance was made in August 1970 to survey surface features associated with permafrost, to obtain measurements of thickness of the permafrost active layer in the major terrain units in the Lowland, and to select sites for the installation of thermocouple cables. A drilling program was carried out in June and July 1971 to obtain information on the perennially frozen soils and bedrock and to install thermocouple cables for monitoring ground temperatures in the permafrost. A total of ten holes (nine in the Lowland and one on the adjacent upland plateau to the east) were drilled in six of the major terrain units to varying depths down to 9 m using a two-man portable Packsack 10 hp diamond drill (Fig. 1, Table 1). Holes in the gravelly soils were limited in depth to less than 3 m by the lack of sufficient power to drill casing. Holes in bedrock were deeper, but limited to depths of 8 to 9 m because of the very cold ground temperatures which caused freezing problems in the drilling. The restriction of logistical support to light aircraft at this remote location in the High Arctic prevented bringing in a sufficiently heavy duty drill rig for the installation of deeper holes.

Table 1. Soil and rock profiles at thermocouple cable sites, Truelove Lowland, Devon Island, N.W.T.

A.* *Beach Ridge—South Slope* (38 m above sea level)	
0 to 3 cm	Organic material
3 cm to bottom of hole (1.45 m)	Well graded sandy gravel (beach material)
A. *Beach Ridge—Top* (38 m above sea level)	
0 to bottom of hole (1.20 m)	Well graded sandy gravel (beach material)
A. *Beach Ridge—North Slope* (38 m above sea level)	
0 to 5 cm	Organic material
5 cm to bottom of hole (0.75 m)	Well graded sandy gravel (beach material)
B. *Ice-Wedge Polygon—Centre* (30 m above sea level)	
0 to 40 cm	Peat
40 cm to bottom of hole (1.35 m)	Ice with scattered peat stringers
B. *Ice-Wedge Polygon—Trench* (30 m above sea level)	
0 to 8 cm	Peat
8 cm to bottom of hole (1.65 m)	Ice
C. *Tundra-Meadow* (53 m above sea level)	
0 to 1.5 m	Peat
1.5 m to 2.4 m	Stony clayey silt (till) with granite boulders up to 30 cm diameter
2.4 m to bottom of hole (2.80 m)	Medium sand
D. *Limestone-Coast* (1 m above sea level)	
0 to 4.5 m	Pitted (vesicular) limestone (porosity—3%; thermal conductivity**—11 to 12, 4.62 to 5.04)
4.5 m to bottom of hole (8.70 m)	Mottled brownish green grey dolomitic limestone (porosity—7 to 13%; thermal conductivity—10 to 11, 4.20 to 4.62)
E. *Limestone-Inland* (9 m above sea level)	
0 to 3.9 m	Pitted (vesicular) limestone (porosity—3%; thermal conductivity—11 to 12, 4.62 to 5.04)
3.9 m to bottom of hole (6.90 m)	Mottled brownish green grey dolomitic limestone (porosity—7 to 13%; thermal conductivity—10 to 11, 4.20 to 4.62)
F. *Granite Gneiss* (75 m above sea level)	
0 to bottom of hole (6.45 m)	Granite-gneiss (porosity—0.3%; thermal conductivity—6, 2.52)
G. *Upland Plateau* (306 m above sea level)	
0 to 1.5 m	Stoney clayey silt
1.5 m to bottom of hole (7.90 m)	Fissured sandstone and dolomite (porosity—3 to 10%; thermal conductivity—7.4 to 8.6, 3.11 to 3.71)

*See map (Fig. 1).
**First values in CGS units—m cal $cm^{-1}sec^{-1}C^{-1}$. Second values in SI units—watts per metre kelvin ($Wm^{-1}k$). Thermal conductivity determination by Geothermal Studies Section, Seismology Division, Earth Physics Branch, Department of Energy, Mines and Resources, Ottawa, Ontario.

The first ground temperature measurements were taken in July 1971 and monthly readings were continued until the end of the field season in October when the Base Camp operation was discontinued for the winter (Brown 1972b, Brown 1973). Monthly observations were resumed in May 1972 and continued through the winter of 1972-73 to August 1973 when the program was terminated. The depth of snowcover was also noted monthly at a snow stake at each site during the observation periods.

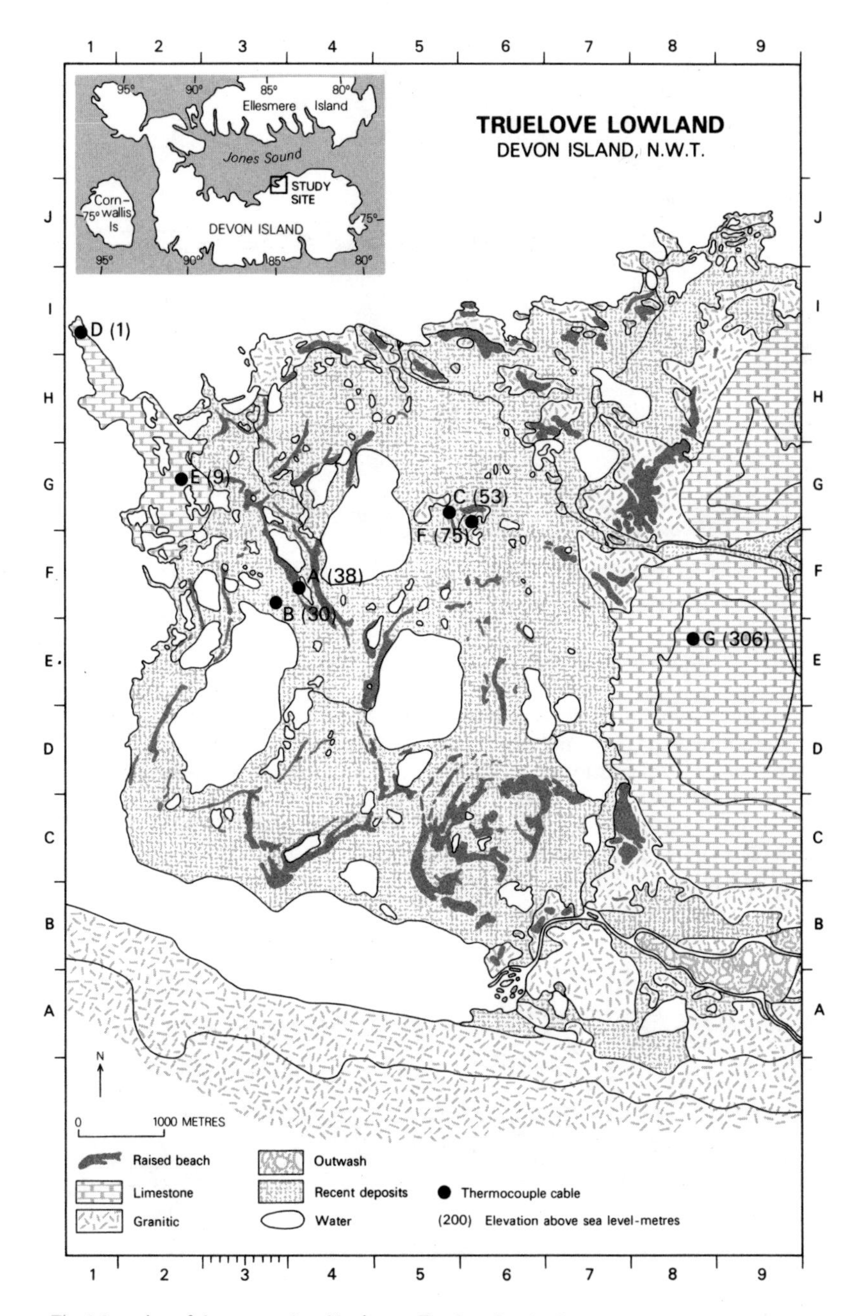

Fig. 1 Location of thermocouple cable sites on Truelove Lowland.

Fig. 2 Beach ridge thermocouple cable sites—north slope on left, top in centre, south slope on right, July 1971.

Terrain characteristics of thermocouple cable sites

Three thermocouple cables which have sensors at 15 cm intervals were installed on the south slope, top, and north slope of a beach ridge (designated as the Beach Ridge Site) southwest of Base Camp similar to the Intensive Raised Beach Site (Fig.2). Two cables, also with sensors at 15 cm intervals, were placed in a nearby area of high centre ice-wedge polygons (Ice-Wedge Polygon Site), one cable in the centre of a polygon, and the other in a trench (see Fig. 5 Muc and Bliss this volume). One cable which has sensors at 0.75 and 1.5 m

Fig. 3 Limestone-inland thermocouple site, July 1971.

Fig. 4 Upland plateau thermocouple cable site, July 1971.

intervals was installed in a hummocky sedge-moss meadow adjacent to the Intensive Meadow Site. Four thermocouple cables with the same sensor intervals were installed in exposed bedrock. Two of these were located in the limestone area in the northwest section of the Lowland, one at the coast at Rocky Point (Limestone-Coast Site), and the other 0.5 km inland (Limestone-Inland Site) (Fig. 3). One hole was drilled in a granite gneiss outcrop (designated as the Granite Gneiss Site) near the Intensive Meadow Site. The fourth was installed on the 300 m high upland plateau surface (Upland Plateau Site) east of the Lowland (Fig. 4). Data on the ground (soil/rock) profiles, depth of thermocouple cables, elevation above sea level, and rock thermal conductivity values are given in Table 1.

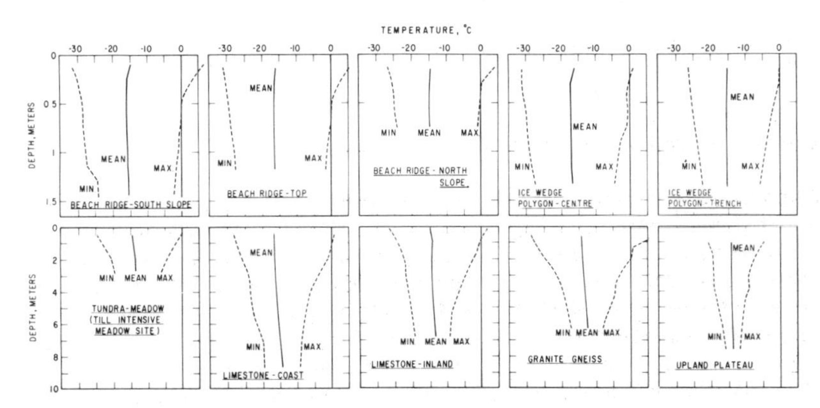

Fig. 5 Ground temperature envelopes of thermocouple cable sites.

Table 2. Active layer (depth of thaw) at thermocouple cable sites, Truelove Lowland, Devon Island, N.W.T.

Site	*Active layer (cm) 1971*	*Depth of thaw (cm) 1972*		*Depth of thaw (cm) *1973*
		July	*August*	*July*
Beach Ridge—South Slope (A)[1]	110	49	40-70	88
Beach Ridge—Top (A)	91	52	46-76	not available
Beach Ridge—North Slope (A)	61	34	46	55
Ice-Wedge Polygon—Centre (B)	30	21	16-30	46
Ice-Wedge Polygon—Trench (B)	30	12	16	46
Tundra-Meadow (C)	70	30	37	not available
Limestone-Coast (D)	122	58	46-122	91
Limestone-Inland (E)	140	55	16-91	116
Granite Gneiss (F)	223	146	137	168
Upland Plateau (G)	<30	<30	30	not available

*August 1973 not available.
[1]See map.

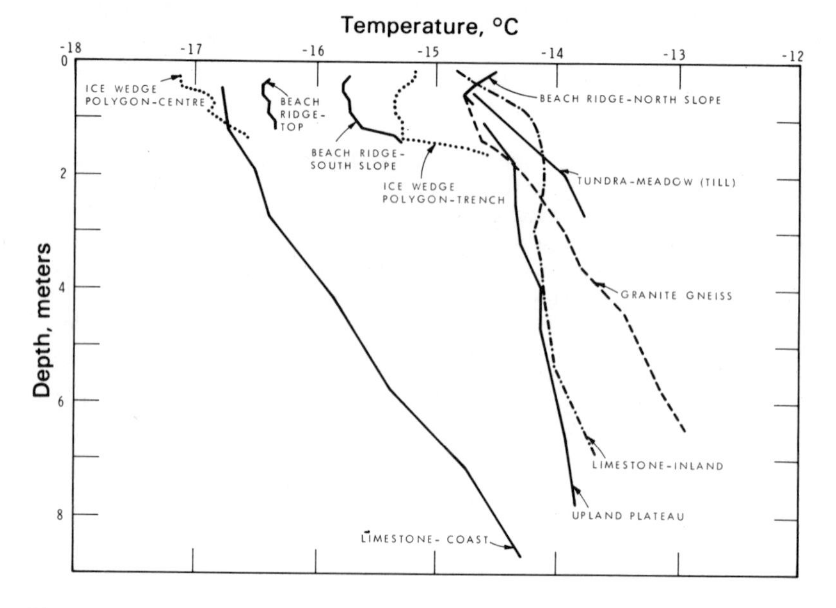

Fig. 6 Mean annual ground temperature graphs at thermocouple cable sites.

Ground temperature measurements

Although ground temperature measurements were taken monthly for a 16 month period from May 1972 to August 1973 inclusive, some unavoidable gaps and readings of questionable accuracy in the summer of 1973 necessitated restricting the data used for calculating mean annual ground temperature values to the period of May 1972 to May 1973 inclusive. Within this period, values for missing readings were interpolated as the arithmetical mean between the two adjacent monthly observations. Mean annual temperature values were obtained for each thermocouple junction for 12 monthly readings for the two periods of May 1972 to April 1973 and June 1972 to May 1973. The average of

Table 3. Snowcover (cm) at thermocouple cable sites (May 1972-May 1973), Truelove Lowland, Devon Island, N.W.T.

Site	*May*	*Jun*	*Jul*	*Aug*	*Sep*	*Oct*	*Nov*	*Dec*	*Jan*	*Feb*	*Mar*	*Apr*	*May*
Beach Ridge—South Slope (A)[1]	20	23	—	—	—	15	18	15	15	13	18	18	13
Beach Ridge—Top (A)	30	25	—	—	—	10	18	13	15	10	18	18	15
Beach Ridge—North Slope (A)	53	48	—	—	—	20	28	28	30	30	36	36	30
Ice-Wedge Polygon—Centre (B)	10	10	—	—	—	10	15	3	5	10	8	10	—
Ice-Wedge Polygon—Trench (B)	84	86	—	—	—	53	58	61	61	61	71	69	69
Tundra-Meadow (C)	46	46	—	—	—	10	*15	20	25	30	30	30	30
Limestone-Coast (D)	28*	33*	—	—	—	10	*18	25	15	23	30	30	33
Limestone-Inland (E)	28	33	—	—	—	23	*24	25	25	25	28	30	41
Granite Gneiss (F)	15	—	—	—	—	10	*7	3	15	13	23	5	10
Upland Plateau (G)	38	36	—	—	—	10	28	*27	25	25	30	36	30

*Interpolated Values.
[1]See map.

Fig. 7 Polygonal crack (snow covered) and peat wedge in active layer in Base Camp beach ridge, August 1970.

Fig. 8 Polygonal cracks in limestone near limestone-inland site, July 1971.

these two annual means and the maximum and minimum temperatures of each depth were used to plot the ground temperature envelope for each site shown in Fig. 5. The same mean annual ground temperature values for all sites are plotted on one graph to provide a visual comparison of them (Fig. 6). Active layer and depth of thaw determinations at all sites for the three summers of 1971, 1972, and 1973 are listed in Table 2. The depth of snowcover at each site is shown in Table 3.

Fig. 9 Turf hummocks near ice-wedge polygon site, August 1970.

Results and Discussion

Surface features characteristic of high latitude continuous permafrost conditions are prevalent in the Lowland. Polygons and polygonal cracks due to thermal contraction occur in the beach ridges (Fig. 7), meadows, and even the weathered limestone where the surface is fragmented (Fig. 8). Surface features associated with severe frost action in the active layer of fine-grained soils are also widespread (Washburn 1973). They include nonsorted circles (stony earth circles), stone nets, and turf hummocks (Fig. 9).

Although it is risky to generalize on the permafrost ground temperatures on the basis of such a short observation period, some broad patterns appear to exist. The Granite Gneiss Site had the highest mean ground temperatures for the one-year observation period of the exposed bedrock areas and the Limestone-Coast Site the lowest. The Upland Plateau Site had the smallest gradient in the study area. The Granite Gneiss and Limestone Sites had comparable ground temperature envelopes, the former being slightly larger, but the Upland Plateau displayed much smaller amplitudes of maximum to minimum temperatures. The Tundra-Meadow Site (hummocky sedge-moss meadow) was the warmest of the other sites and in fact it was warmer than the Granite Gneiss Site on the basis of the mean annual ground temperatures. There was wide variation among the three Beach Ridge Sites depending on slope orientation. The same was true of the Ice-Wedge Polygon Site, the centre being colder than the trench. At all sites the mean annual ground temperatures were slightly closer to the minimum temperatures in the envelopes than to the maximum.

Thickness of the active layer and depth of thaw varied greatly among the sites and from one summer to another (Table 3). Difficulties with temperature observations in the summers of 1972 and 1973 prevented an accurate assessment of active layer thickness similar to 1971. The general patterns indicated, however, that the Granite Gneiss Site had consistently the thickest active layer (depth of thaw), and the Upland Plateau had the thinnest of the rock sites. For the other sites, the thickness of the active layer at the Beach Ridge Sites was similar to the Limestone Sites; the Tundra-Meadow Site had a thinner active layer, and the peat at the Ice-Wedge Polygon Sites the least, being similar to the Upland Plateau Site. The ground began to thaw from the surface in mid-to-late June as soon as the snow melted. By mid-September all thawed ground above the permafrost table was refrozen. Ground thawing was deepest in 1971, almost as deep in 1973, and much shallower in 1972.

The depth of snowcover was quite variable among the sites. The Granite Gneiss Site had the thinnest snowcover of the rock sites and the Limestone-Inland the thickest. The centre of the Ice-Wedge Polygon Site had the least snowcover, and other exposed sites such as the Beach Ridge top and the Limestone-Coast also had less snowcover than the other sites.

Permafrost environmental relationships

Despite the short observation period, irregular readings, and shallowness of the thermocouple cables, some broad correlations are evident in the relationships between the permafrost and environmental factors.

Air temperature data were not available for Truelove Lowland at the time of writing, so observations at Resolute were used to provide a general idea of yearly comparisons. The summers of 1971 and 1973 were considerably warmer than 1972. Degree days of thawing, based on monthly mean air temperatures for June, July, and August, were 255 in 1971, 116 in 1972, and 272 in 1973. January and February mean monthly air temperatures in the winter of 1971-72 were nearly 3°C colder than in 1972-73, about −19.5°C in January and February 1972, and about −16.8°C in January and February 1973. Variations in the active layer (depth of thaw) were considerably less in 1972 than in the other two summers (Table 2). It was also found that for May and June 1972 the ground temperatures were 2-3°C colder than those of May and June 1973 as would be expected from colder air temperatures in the winter of 1971-72.

Snowcover is an important environmental factor throughout the permafrost region and its influence in Truelove Lowland was very marked during the observation period. Variations in mean annual ground temperatures from one site to another due to snowcover were particularly evident at the Beach Ridge Sites and Ice-Wedge Polygon Site. The north slope of the Beach Ridge had the highest mean annual ground temperatures, due apparently to its having twice the depth of snow as the top and south slope. This ameliorating influence of the winter snowcover more than compensated for the lower ground temperatures near the surface and lesser depth of thaw that would be expected for a north slope in summer. At the Ice-Wedge Polygon Site, winter ground temperatures were higher in the trench, where snow accumulated to considerable depth in contrast to the centre which was windblown. In summer, the temperatures in the trench were similar to those in the centre and the depth of the thaw was the same at both locations.

The type of soil and rock has an important influence on the thickness of the active layer and the thickness of the permafrost. At the Tundra-Meadow Site and Ice-Wedge Polygon Site the surface peat layer was responsible for the thin active layers due to the insulative properties of this earth material. The dolomite (limestone) was found to have a higher thermal conductivity (11-12 m cal $cm^{-1}sec^{-1}C^{-1}$ or 4.20 to 4.62 W $m^{-1}k^{-1}$) than the granite gneiss (6 m cal $cm^{-1}sec^{-1}C^{-1}$ or 2.52 W $m^{-1}k^{-1}$) (Brown 1973, Table 1). The decrease in amplitude of ground temperature with depth was more rapid for the Granite Gneiss Site as would be expected because of this factor.

Before the temperature observation program was initiated it was anticipated that the ground temperatures at the Upland Plateau Site would be lowest of all the sites because of its elevation, 300 m above Truelove Lowland. The Limestone-Coast proved, however, to have the lowest average annual ground temperature (−16.8°C) and the Upland Plateau the highest (−14.6°C). The slope of the former was much steeper than of the latter and the curves probably intersect a short distance below the bottom depth of the thermocouple cables. The proximity of the Limestone-Coast site to the cold coastal seawater probably affected the summer ground temperatures. Amelioration of winter ground temperatures were possibly reduced with the formation of thick sea ice. The Upland Plateau Site ground temperature graph had the smallest temperature change with depth (smallest temperature gradient) of all the sites, which may indicate the thickest permafrost is at this location.

Acknowledgments

The author wishes to thank Dr. L.C. Bliss for inviting him to participate in the Devon Island Project which afforded the opportunity of obtaining permafrost information in the Canadian Arctic Archipelago. It would not have been possible to carry out these investigations, especially the drilling program, at such a remote location without the support facilities existing there. The Devon Island Project provided invaluable assistance in shipping equipment to the site. Thanks are due both to the International Biological Programme (I.B.P.) and the Arctic Institute of North America (A.I.N.A.) for providing accommodation and support services. In this regard the assistance of Dr. G.M. Courtin, I.B.P. Field Director and Mr. Ward Elcock, A.I.N.A. Camp Manager is particularly appreciated.

Thanks are also due to Dr. Courtin for organizing the ground temperature observation program. Measurements were taken at different times by M. Claude Belcourt, Mr. B. Hubert, C.L. Labine, Miss Janet Marsh, Dr. R. Riewe and Dr. W. Speller. The author is indebted to these six individuals for persevering in the face of difficult field conditions and periodically malfunctioning equipment. A special acknowledgment is due to M. Belcourt who carried out the observation program during the winter of 1972-73.

The author is grateful to Dr. L.W. Gold, N.R.C., Professor J.R. Mackay, Department of Geography, University of British Columbia; and Mr. G. Rempel, Imperial Oil Limited, for helpful suggestions on the manuscript.

The author wishes to acknowledge his appreciation to Dr. Judge for preparing the following section on permafrost thickness.

References

Brown, R.J.E. 1967. Permafrost in Canada. Map NRC 9769. National Research Council Canada, Division of Building Research, Ottawa, and Geological Survey of Canada Map 1246A.

———. 1970. Permafrost in Canada — Its Influence on Northern Development. Univ. Toronto Press, Toronto. 234 pp.

———. 1972a. Permafrost in the Canadian Arctic Archipelago. Zeitschr. Geomorphol. Suppl. No. 13: pp. 102-120.

———. 1972b. Permafrost investigations on Devon Island, N.W.T. Canadian Tundra Biome Study Site (I.B.P.). 30-45. *In:* Devon Island

I.B.P. Project-High Arctic Ecosystem. L.C. Bliss (ed.) Univ. Alberta, Edmonton. 413 pp.
———. 1973. Influence of climatic and terrain factors on ground temperatures at three locations in the Permafrost Region of Canada. pp. 27-34. *In:* Permafrost-Proceedings of North American Contributions to Second International Conference. National Academy of Sciences. Washington. 783 pp.
Cook, F.A. 1958. Temperatures in permafrost at Resolute, N.W.T. Geogr. Bull., No. 12: 5-18.
Judge, A.S. 1973. The prediction of permafrost thickness. Can. Geotech. J. 10: 1-11.
Misener, A.D. 1955. Heat flow and depth of permafrost at Resolute Bay, Cornwallis Island, N.W.T., Canada. Trans. Am. Geophys. Union. 36: 1055-1066.
Washburn, A.L. 1973. Periglacial Processes and Environments. Edward Arnold. London, England. 320 pp.

Calculations of permafrost thickness

A.S. Judge

Earth Physics Branch, Department of Energy, Mines, and Resources, Ottawa, Ontario

The equilibrium state and its perturbations

The simplest model of the thermal condition of the earth's crust is to consider a semi-infinite homogeneous body with a constant surface temperature and a constant terrestrial heat flow from below. In this case the surface temperatures, and hence the permafrost thickness, are dependent only on the above factors and the thermal conductivity of the rocks present:

$$T_z = T_s + \frac{q}{k} . z \qquad (1)$$

where T_z represents the temperature at any depth z, T_s is the surface temperature, q the terrestrial heat flow and k the thermal conductivity. In a multilayer media this can be generalised by rewriting k as:

$$k = \frac{1}{z_n} \sum_{i=1}^{n} \frac{\Delta z_i}{k_i} \qquad (2)$$

With the exception of highly contorted, complexly faulted, or folded areas the spatial variation of rock lithology, and hence thermal conductivity and equilibrium terrestrial heat flow, is small. Surface temperatures may vary considerably across the half space, however, due to such terrain factors as variations in elevation, exposure and vegetation, and presence of nearby bodies of water. If each of these features is assumed to have been present and possessed its present temperatures for an infinite time, then the change to the subsurface temperature at a point from that for an infinite half space with constant surface temperature (assumed to be collar temperature) is proportional to the solid angle subtended by the feature Ω and its temperature difference from the collar temperature V, and is given by:

$$\Delta T_z = \sum_i \frac{\Omega i}{2\pi} . \Delta_i V \qquad (3)$$

Analytical results for Ω are available in polar coordinates, which is the simplest configuration to use when estimating or predicting temperatures measured in a borehole or below a point on the earth's surface. A simple polar grid is established following the method of Jeffreys (1940).

It is again reasonable to assume a constant equilibrium terrestrial heat flow and thermal conductivity for many temporal problems outside of young volcanic areas or areas with rapid sedimentation or erosion. Surface temperatures have changed considerably with time in response to fluctuations in the Pleistocene ice sheets, postglacial marine submergence and emergence, and changes in terrain features. These variations can be represented by a simple step function and becomes:

$$T = 1 - \text{erf}\,[\,Z_{/2(kt)}{}^{1/2}\,] \tag{4}$$

where κ is the thermal diffusivity of the medium.

Thermal parameters

Surface temperatures at the thermocouple cable sites on Truelove Lowland and adjacent upland are shown in Table 4. They vary from −16.8°C at the Limestone-Coast Site near sea level to −14.6°C at the Upland Plateau Site with an elevation of 306 m. Little difference is noted between the latter location and inland sites on the Lowland. Seawater temperatures were not measured at the Lowland but water temperatures at another location at the same latitude and where the sea remained largely ice-covered in the summer averaged −1.7°C. Temperatures in the lakes and ponds are not available either and so have been assumed to be slightly above 0°C, i.e., 1°C.

Measured average thermal conductivities range from 2.51 W $m^{-1}K^{-1}$ for the granite gneisses, 3.35 for the sandstones, and 4.5 for the dolomitic limestones. Local undisturbed equilibrium heat flow in that locality is unmeasured but following Judge (1973) probably varies between 62.7 mWm^{-2} and 83.6 and thus calculations have been carried out using both values. Lithological characteristics at each of the sites at which calculations were made are shown in Table 1.

Calculations

Assuming equilibrium conditions initially and the absence of any topographic features the distribution of permafrost would be Ze as shown in Table 5. Permafrost thickness ranges from 592 to 700 m for a heat flow of 62.7 mWm^{-2}. The variations in thickness are due only to the lithological variations at different sites.

At each of the sites a polar grid was drawn with radii drawn at 10° intervals and 30 concentric rings around the borehole collar to a maximum radius of 2500 m. The resulting topographic information derived at each intersection of radius and circle presented a digital map of the land surface consisting of 1080 points which was used to calculate topographic corrections. Zp in Table 5 shows the values of permafrost thickness corrected for the presence of topographic features appearing after the disappearance of glacial ice.

Finally and most important, changes of the surface with time must be taken into account. It is assumed using the data presented by Prest (1970) that at a time before 9000 years ago the glacial ice cover was complete across the area and that ice bottom temperatures were similar to present surface temperatures. This is reasonably consistent with the present measured ice

Table 4. Surface temperature, postglacial history, lithology of five rock thermocouple cable sites on Truelove Lowland, Devon Island, N.W.T.

	Surface temp. (°C)	*Time ice free—years*	*Time of emergence—years*	*Lithology-depth (m)*		
				Sandstone	*Limestone*	*Granite Gneiss*
Tundra-Meadow (C)	−14.8	9,000	6,500	—	—	near surface
Limestone—Coast (D)	−16.8	9,000	1,000	—	0-60	below 60
Limestone—Inland (E)	−14.9	9,000	3,000	—	0-30	below 30
Granite Gneiss (F)	−14.8	9,000	not submerged	—	—	from surface
Upland Plateau (G)	−14.6	9,000	not submerged	0-300	—	below 300

Table 5. Equilibrium permafrost thickness at five rock thermocouple cable sites on Truelove Lowland, Devon Island, N.W.T.

	Depth interval (m)	*Temperature gradient °C km⁻¹ (q=62.7 mWm⁻²)*	*Permafrost thickness Ze[1] (m)*	*Permafrost thickness Zp[2] (m)*	*Temperature gradient °C km⁻¹ (q=83.6 mWm⁻²)*	*Permafrost thickness Ze (m)*	*Permafrost thickness Zp (m)*
Tundra—Meadow (C)	from surface	25	592	565	33	448	408
Limestone—Coast (D)	0-60	14			18		
	below 60	25	700	615	33	536	425
Limestone—Inland (E)	0-30	14			18		
	below 30	25	610	590	33	469	435
Granite Gneiss (F)	from surface	25	592	588	33	448	442
Upland Plateau (G)	0-300	19			25		
	below 300	25	600	659	33	515	514

[1]No consideration of topographic features.
[2]Correction for topographic features.

Table 6. Probable permafrost thickness at five rock thermocouple cable sites on Truelove Lowland, Devon Island, N.W.T.

Site	*Permafrost thickness (m) (q=62.7)*	*Permafrost thickness (m) (q=83.6)*
Tundra—Meadow (C)	525	385
Limestone—Coast (D)	250	210
Limestone—Inland (E)	470	340
Granite Gneiss (F)	588	442
Upland Plateau (G)	659	514

bottom temperatures in the Arctic (Paterson 1968, Hansen and Langway 1966). About 9000 years ago the ice was removed, but because of isostatic changes in land elevation and eustatic changes in sea-level any land at present below an elevation of 70 m is assumed to have been submerged and then to have slowly emerged according to Walcott's (1972) pattern of emergence for Jones Sound. The permafrost thickness values calculated for this effect are shown in Table 6 and these values should approximate fairly closely the thicknesses to be observed in Truelove Lowland. It should be emphasized that this is a very simplified set of circumstances to simplify the mathematical calculations. Further complications such as latent heat, zoning of lake temperatures with depth, zoning of ground surfaces according to snow cover, vegetation cover, aspect, etc., could be included as well as more complex surface history situations. The only limitations on the sophistication of the models is the available experimental data and the cost of computing.

Conclusions

Permafrost thickness at Truelove Lowland may be as thin as 210 m in coastal areas (Limestone-Coast Site) and as thick as 659 m on the adjacent upland (Upland Plateau Site). The presence of a nearby relatively warm sea has a highly modifying influence on coastal areas increasing the subsurface temperature at the Limestone-Coast Site by 6°C at a depth of 90 to 170 m. Likewise the uplift history and the time since emergence from the sea is highly significant causing a further modification of temperature at the stated depths by 2.5 to 4.8°C and causing a maximum disturbance at 400 m of 7.3°C. In contrast the permafrost thickness at the Upland Plateau Site has probably been little modified by either of these influences.

References

Hansen, B.L. and C.C. Langway. 1966. Deep core drilling in ice and core analysis at Camp Century, Greenland 1961-1966. Antarctic J. 1:207-208.

Jeffreys, H. 1940. The disturbance of the temperature gradient in the earth's crust by inequalities of height. Month. Nat. Roy. Astron. Soc., Geophys. Suppl. 4: 309-312.

Judge, A.S. 1973. The prediction of permafrost thickness. Can. Geotech. J. 10: 1-11.

Paterson, W.S.B. 1968. A temperature profile through the Meighen Ice-Cap, arctic Canada. IUGG Commission Snow and Ice Rept. No. 79: 440-449.

Prest, V.K. 1970. Quaternary Geology of Canada. Chap. 12. pp. 674-764. *In:* Geology and Economic Minerals of Canada. 5th Ed. Geol. Surv. Can. 838 pp.

Walcott, R.I. 1972. Late Quaternary vertical movements in Eastern North America: quantitative evidence of glacio-isostatic rebound. Rev. Geophys. 10: 849-884.

Soils of Truelove Lowland and Plateau

B.D. Walker and T.W. Peters

Introduction

Much information concerning the soils of Soviet and North American Arctic regions have been collected since 1950. So-called genetic soil types (Great Soil Groups) have been recognized and qualitative pedogenic processes contemplated (Krieda 1958, Tedrow and Cantlon 1958, Ignatenko 1965, Rozov and Ivanova 1968). Pedologists now discuss genetic soil types and pedogenic gradients in terms of biogeographic or pedogenic zones that have been imposed upon the Arctic (Ivanova et al. 1961, Tedrow 1968, 1973; Aleksandrova 1970).

Yet until recently pedologists have not constructed relatively detailed soil maps, partly because of landscape phenomena produced by frost disturbance and manifest in the disruption of soil profiles and horizons. Terrain systematic approaches to mapping of soils have resulted in the use of extremely complex map units (Brown 1966, King 1969, Ignatenko 1973 — soil map in Matveyeva et al. 1974) or generalized map units (Cruickshank 1971) where groups of patterned ground types were recognized in delineating terrain (map) units but not in classifying the soils. Holowaychuk et al. (1966) were probably the first pedologists to recognize the single taxonomic class or pedon concept (Soil Survey Staff 1960) as applied to mapping of soils in patterned ground.

The prime objective of this paper is to depict local distribution of soils within Truelove Lowland. Mapping problems, alluded to above, were overcome through: (1) recognition of frost disturbance as a pedogenic factor; (2) adherence to the pedon concept (Soil Survey Staff 1960) as specified for northern soils (Canada Soil Survey Committee 1973); and (3) use of a tentative classification system for northern soils — the Cryosolic Order (C.S.S.C. 1973). Consequently mapping difficulties have been reduced to dependence upon complexities in parent material types and vegetation and drainage patterns. Such is the case on the Truelove Lowland sector of the study area.

Soil map units relate, via legend format, soil subgroup classes (C.S.S.C. 1973) with parent materials, drainage patterns, periglacial features, and plant community types. It is hoped that the mapped distribution of these units (based on soil series or complexes) will provide a land base for interpretation or extrapolation of other ecological data gathered within the framework of the Devon Island I.B.P. program.

A secondary objective was to characterize some of the major or significant soils of the study area. Profile descriptions and accompanying analytical data of sampled profiles, considered to be representative of certain map units, facilitate this aim.

Since drainage is considered to be the most important of environmental factors determining local pedogenic trends, soil-drainage relationships are briefly explored and correlated with plant communities.

Methods

Field survey techniques

The field work segment of this study was conducted during the summers of 1971 through 1973. Initial objectives were to become familiar with the landscape and its components and to develop concepts pertaining to terrain systematics.

Detailed study of the Raised Beach and Meadow Intensive Study Sites (Introduction), less intensive surveying of the remaining area, and a fairly comprehensive soil profile sampling program enabled the realization of these objectives. Sampling sites, selected because of their representativeness, were sampled on a horizon basis. Profile morphologies and site characteristics were described using the guidelines of C.S.S.C. (1970).

In an attempt to illustrate some of the critical drainage-soil relationships, five line transects were studied at various locations on Truelove Lowland. Along these transects slopes were measured with an Abney level, and plant community and soil changes were identified and noted.

Black-and-white air photos having a scale of *ca.* 1:5,000 were very useful in the field work and mapping. The final map (separate in back folder) is produced upon an orthophoto (controlled) mosaic having a scale of *ca.* 1:15,000.

All facets of this study contributed to the formulation of landscape systematics concepts and, ultimately, to the development of a soil map legend. The final phase of the field program involved the testing of the proposed legend plus ground truthing some of the air photo interpretations. As a result the legend was modified and refined.

Analytical methods

Sample preparation involved air drying followed by roller-grinding and collection of soil material passing a 2 mm sieve.

Chemical and physical analyses were carried out according to the routine procedures used by the Alberta Institute of Pedology. These involved determination of soil reaction (water-saturated paste method); total nitrogen (macro Kjeldahl method); calcium carbonate equivalent by the manometric method of Bascombe (1961) on samples having pH values of 7.0 or greater; total carbon (dry combustion using a Leco induction furnace); organic carbon (difference between total carbon and $CaCO_3 - C$); exchangeable hydrogen ion (displacement with barium acetate on those samples having pH values of 6.5 or less); exchangeable cations by displacement with neutral, 1*N* ammonium acetate and determination of Ca, Mg, Na, and K by atomic absorption spectrophotometry (on samples having pH values of 7.5 or less); oxalate extractable iron and aluminum (acid ammonium oxalate extraction of McKeague and Day 1966); and particle size distribution analysis by the pipette

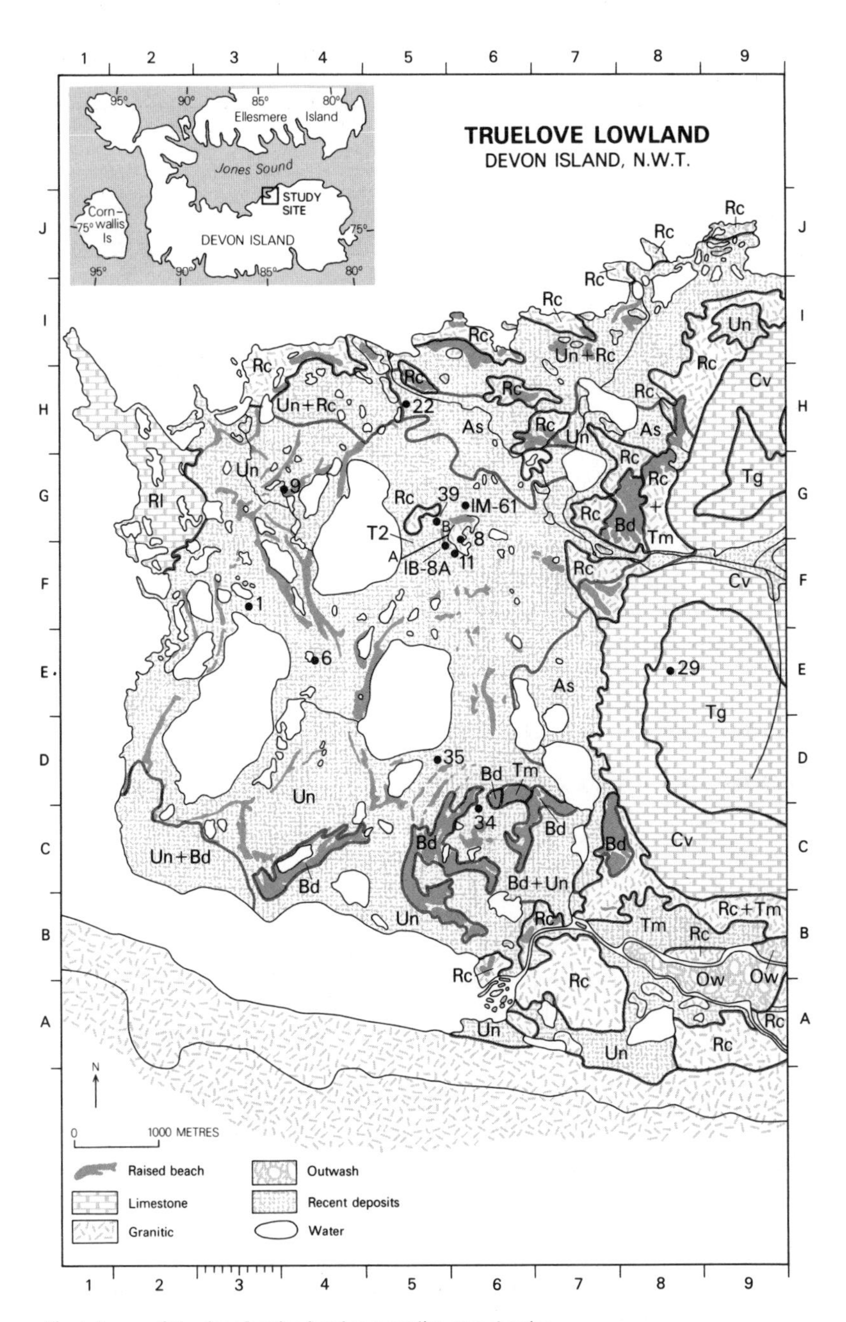

Fig. 1 A map of Truelove Lowland and surrounding area showing:
a. Locations of selected sampling sites identified by number, IB-8A (Intensive Beach — Plot 8A), and IM-61 (Intensive Meadow — Plot 6I).
b. Location of soil-topography transect 2 (T2).
c. General distribution of parent materials using the following legend: As — alluvial-lacustrine; Bd — beach materials (also dark shading showing raised beaches); Cv — colluvium; Ow — Outwash; Rc — crystalline bedrock (also shading pattern designating granitic areas); Rl — dolomite bedrock; Tg — glacial till of the upland; Tm — glacial till of the Lowland; Un — undifferentiated materials.

Table 1. General characteristics of mineral parent materials and exposed rock in the study area

Id. sym.	*Parent material*	*Texture*	*Reaction*	*Calcareous class*
As	Alluvial-lacustrine material	variable; medium to fine textured	medium acid to neutral	weakly to non-calcareous
Av 1	Alluvial material	medium to moderately fine textured	mildly alkaline	extremely calcareous
Av 2	Alluvial material	fragmental		
Bd 1	Beach material	coarse-skeletal	moderately alkaline	moderately to strongly calcareous
Bd 2	Beach material	coarse-skeletal	neutral to mildly alkaline	weakly to non-calcareous
Bd 3	Beach material	fragmental	*moderately alkaline	*extremely calcareous
Bd 4	Beach material	fragmental		
Cv	Colluvium	fragmental		
Ow 1	Outwash	fragmental		
Ow 2	Outwash	coarse-skeletal		
Rc 1	Crystalline rock -consolidated	(consolidated)		
Rc 2	Crystalline rock -unconsolidated	coarse-skeletal	neutral to mildly alkaline	weakly to non-calcareous
R 1	Limestone rock	(consolidated)		
Tg	Till	medium to moderately fine textured	mildly alkaline	extremely calcareous
Tm	Till	moderately coarse to coarse-skeletal	moderately alkaline	very strongly calcareous
Un	Undifferentiated	Moderately coarse textured	Mildly to moderately alkaline	strongly to very strongly calcareous

*Reaction and calcareous classes are of underlying coarse-skeletal sands and gravels.

method (organic matter and carbonates removed) or the hydrometer method (organic matter removed).

A number of samples were submitted for available nutrient analyses according to methods used by the Alberta Soil and Feed Testing Laboratory, Edmonton. These analyses included the determination of available nitrate-nitrogen (extracted by a $CuSO_4 - Ag_2SO_4$ solution); available phosphorus (extracted by an $NH_4F - H_2SO_4$ solution); available potassium (extracted by NH_4OAc); available sodium (extracted by NH_4OAc); and water soluble sulfate (extracted by a $CaCl_2 \cdot 2H_20$ solution).

Dom. lith. source	Coarse fragment description	Landform	Distribution
probably dolomite	Very few to none	Alluvial-lacustrine	As on Fig. 1
dolomite	Content increasing with depth; variable sizes and shapes.	Localized floodplains on other landforms	Most common near Gully River
dolomite	Rockland; angular and thin, flat gravels to boulders.	Alluvial cone	Base of eastern escarpment; coordinates E 7 & D 7, Fig. 1
dolomite	Randomly stratified, rounded & subrounded sands, gravels, and cobbles; excessively stony areas common.	Raised beach	Shaded areas & Bd on Fig. 1
crystalline rock	50-90%; rounded & subrounded gravels to stones; some stratification.	Veneer over crystalline bedrock	Numerous crystalline rock outcrops in area
dolomite	Rockland of angular gravels; sometimes overlying coarse-skeletal sands & gravels.*	Raised beach	Raised beaches on and near Rocky Point
variable	Rockland; angular & subrounded gravels, cobbles, and stones.	Present beach and tideland	West and south coastline
dolomite	Rockland; angular to thin, flat stones and boulders. Small areas with thin, flat gravels.	Escarpment (includes Gully floor)	Cv on Fig. 1
variable and angular	Rockland; rounded, subrounded, gravels to stones.	Active stream channel	River and stream beds
dolomite	50-90%; rounded and subrounded gravels and cobbles.	Rolling outwash plain	Ow on Fig. 1
		Crystalline rock outcrop	Rc and tone code on Fig. 1
crystalline	50-90%; angular gravels to stones.	Part of crystalline rock outcrop	<40% of crystalline rock
		Limestone plain	Rl on Fig. 1
dolomite	<20% angular gravels; occ. small islands of rockland.	Ground moraine	Plateau, Tg on Fig. 1
dolomite	20-90%; subrounded gravels to stones; both dolomite and crystalline stones common.	Lateral and end moraine	Tm on Fig. 1
dolomite	<20% rounded to subrounded gravels to cobbles; excessively stony areas common.	Undifferentiated plain	Un on Fig. 1

All results are reported on an oven dry basis and pertain to the fine earth (2 mm and less) fraction of the soil body.

Some samples were collected on a volume basis (core techniques) that facilitated the determination of bulk densities and coarse fragment contents (particles >2 mm). These results are reported on an oven dry basis and pertain to the sum of all fractions making up horizons sampled in such a manner.

Results

Physiography and environment

A general site description plus portrayal of bedrock geology, macro- and micro-climate, plant community patterns, and permafrost characteristics are more than adequately covered by other chapters in this book. However, adequate characterization of landforms or surficial deposits existing within the study area was not previously attempted nor was it included in the I.B.P. study program.

Parent materials

Of paramount significance to soil classification at lower hierarchial levels is the identification and separation of the surficial deposits occurring within an area.

Table 1 presents general characteristics and descriptions of mineral parent materials and landforms occurring in the study area. Fig. 1 depicts the general distributions of most of these materials.

Two types of organic deposits were also identified. Localized areas of the undifferentiated plain (Un, Fig. 1) exhibit sufficiently thick deposits of organic material that no mineral layers are encountered within the active layer. From an ecological standpoint, such organic deposits, regardless of their total depths of accumulation, must be considered as the parent materials of the organic soils which develop in them.

The most common of these organic deposits has developed essentially *in situ* and is derived, in the main, from sedges and *Hypnum* mosses. Reaction of this peat is normally slightly acid to neutral, however more acid pH values do show up in some stream-side microenvironments. Unrubbed fibre content is roughly 50% based on a few tests using the syringe method of C.S.S.C. (1974). Lenses and discontinuous layers of unwashed, fluvial sands can be occasionally found in this deposit.

Small fields of high-centered, ice-cored polygons, consisting of dry surface peat and ground ice mixed with organic materials, are located at what may be critical (to their development) places along streams or among lakes. Characteristic of the landform are the convex to flat-topped, ice-cored mounds, often 1.5 to 5 m across, raised some 0.5 to 1.5 m above narrow (1 m), steep-sided trenches which connect to form the polygonal pattern. The trenches are underlain by ice-wedges that are frequently exposed in places.

Most of the dry peat mantling these mounds is not unlike the wet peat described above. However, in some polygon fields there occurs an underlying, secondary organic deposit which resembles coprogenous or diatomaceous earth (C.S.S.C. 1970) but contains a considerable amount of algal cells and cell fragments. A similar, but more raw, material was observed along the shores of a number of lakes and ponds in the study area. This appears to be slightly decomposed bluegreen algae.

King (1969), Barr (1971), and Barrett (1972) examined, in greater detail, various aspects of glacial and postglacial geomorphic history, physiography, and postglacial isostatic movement of northeastern Devon Island.

Periglacial modification of surficial deposits

Periglacial features are widespread in the study area. Frost processes include congelifraction or frost shattering of rocks and exposed bedrock; congelifluction, manifest by minor gelifluction lobes and slopes; congeliturbation, revealed by the presence of circles, nets, and stripes; and ice-

Table 2: Periglacial features commonly associated with the various surficial deposits of the study area.

Deposit	*Commonly associated periglacial features*
Alluvial-lacustrine	Occasional ice-wedge fissures
Alluvial	None
Beach (All)	Dry—ice-wedge fissures, widely spaced and, therefore, generally not forming a polygonal pattern
Beach (Bd 1)	Imperfectly drained — circles and nets on level ground, stripes on slopes
Colluvium	Occasional "earth debris islands" (Barrett 1972)
Outwash (Ow 1)	None
Outwash (Ow 2)	Occasional ice-wedge fissures
Rock (Rc)	Frost shattering and occasional ice-wedge fissures
Rock (Rl)	Frost shattering
Till (Tg)	Whole morainal plain (plateau) modified by sorted stripes and mass movement (gelifluction) in the direction of lateral ground-water flow; most rubbly areas exhibit rock circles
Till (Tm)	Much of this deposit modified by nets and stripes
Undifferentiated	Much of the undifferentiated plain modified by circles and nets on level ground, stripes on slopes
Organic (wet)	Occasional ice-wedges fissures
Organic (dry)	High-centered, ice-cored, ice-wedge polygons

wedge fissure formation. The periglacial features, identified using the unified classification of Washburn (1956), are commonly associated with certain surficial deposits as tabulated in Table 2.

Soils

The tentative system for northern soils (C.S.S.C. 1973, Zoltai and Tarnocai 1974) is used for classifying, to the subgroup level, the soils of the study area. Taxonomic correlation at the subgroup level with three other commonly used North American systems is presented in Table 3. To better indicate local pedogenic trends and emphasize the biological importance of organic layers, 7th Approximation (Soil Survey Staff 1970) and, to a lesser extent, Cyrosolic Order (C.S.S.C. 1973) thickness rules were liberalized when applied to soils of the study area. Consequently the use of "Histic" is qualified by stating that the surface organic layer (assumed continuous in all cases) must occupy, in the vertical dimension, at least 25% of the total active layer thickness and be at least 5 cm thick. Soils lacking vertically continuous mineral horizons within the active layer are classified as organic soils (Pergelic Cryofibrists and Fibric Organo Cryosols). Permafrost consisting of frozen organic material was observed under such soils but few checks concerning depths of organic matter accumulation were made.

The complexity of the Lowland landscape and, in particular, surficial materials necessitated further soil separations within most subgroups. Consequently, several unnamed "soil series" (C.S.S.C. 1970) were identified. With one exception, soil series are separated on the basis of parent material diversity and can be identified on the soil map legend by comparing subgroup classification taxa, and parent material descriptions. The one exception involves distinguishing the Brunisolic Static Cryosol of map units 11 and 21 from the Brunisolic Static Cryosol of map unit 12 primarily on the basis of a dissimilar plant community-drainage (or moisture regime) relationship that has resulted in minor soil morphology differences.

The most striking distinction concerning soil variability and distribution within the study area is that the Lowland sector is characterized by complexity,

Table 3. Taxonomic correlation at the subgroup level of three classification systems commonly used for soils of the Arctic with the tentative system for northern soils (C.S.S.C. 1973) as applied to soils found in the study area

Tentative System for Northern Soils (C.S.S.C. 1973)	*Existing Canadian System (C.S.S.C. 1970, 1973)*	*Seventh Approximation (Soil Survey Staff 1970)*	*Great Soil Group* (Tedrow & Cantlon 1958, Tedrow* et al. *1958, Tedrow 1968)*
Brunisolic Turbic Cryosols	Turbic Eutric Brunisols	Pergelic Cryochrepts	
Regosolic Turbic Cryosols	Turbic Orthic Regosols	Pergelic Cryothents	
Gleysolic Turbic Cryosols	Turbic Rego Gleysols	Ruptic-Histic** Pergelic Cryaquepts; Pergelic Cryaquepts	Perhaps: Soil of Polar Desert-Tundra Interjacence***
Brunisolic Static Cryosols	Cryic Melanic Brunisols	Pergelic Cryochrepts	Arctic Brown (Miniature or Shallow sub-group); Arctic Brown; Shallow Arctic Brown
Regosolic Static Cryosols	Cryic Orthic Regosols	Pergelic Cryothents	
Lithic Regosolic Static Cryosols	Lithic Orthic Regosols	Lithic Cryothents	
Gleysolic Static Cryosols	Cryic Rego Gleysols; Cryic Rego Humic Gleysols	Histic** Pergelic Cryaquepts; Pergelic Cryaquept	Meadow Tundra
Fibric Organo Cryosols	Cryic Fibrisols	Pergelic Cryofibrists	Bog or Half Bog
Glacic Fibric Organo Cryosols	Cryic Fibrisols	Pergelic Cryofibrists	Bog

*Subgroup supplied where known.
**Organic layers not thick enough to meet the definition of a histic epipedon; therefore its use (histic) is qualified by stating that the organic layer (assumed continuous in all cases) must occupy, in the vertical dimension, at least 25% of the total active layer thickness and be at least 5 cm (2 inches) thick.
***Tedrow 1970.

whereas the upland (plateau) is typified by relative simplicity or uniformity. Of the 27 soil series plus miscellaneous land types identified, two belong to the interior plateau and one separates this upland from the Lowland proper.

Placed in regional perspective with respect to soil development, the study area occurs within the Polar Desert soil zone (Tedrow 1974). Yet on the Lowland sector vegetation patterns and, to a lesser extent, climate, excepting permafrost conditions, are not indicative of polar desert conditions but reflect features representative of sedge-grass tundra found in the Low Arctic (Bliss 1975). The adjacent upland section of the study area displays physiographic features more typically associated with the term "Polar Desert."

Soil map units

The complexity of the Truelove Lowland terrain in conjunction with the mapping scale used, necessitated rearrangement of the 19 "soil series" and 8 miscellaneous soils to form 24 map units, most of which represent unnamed soils separated at a level corresponding to soil series (C.S.S.C. 1970). However, eight map units represent soil complexes in which two "soil series," soil association in American terminology (Anon. 1973), or a "soil series" and a miscellaneous soil are so intimately intermixed geographically that it is practical to separate them only at scales larger than 1:5,000.

A second level of complexing was necessitated by landscape complexity, interpretation difficulties, and the scale of mapping. Consequently, many of the soil areas delimited on the map are identified by hyphenated map units (a maximum of three for any one soil area).

Map unit descriptions centre around conceptual pedons. In most cases, variation in map unit characteristics are also depicted. Sampled soil individuals, deemed most closely resembling the conceptual type pedons, are described (with accompanying analytical data) as part of the more important map unit portrayals.

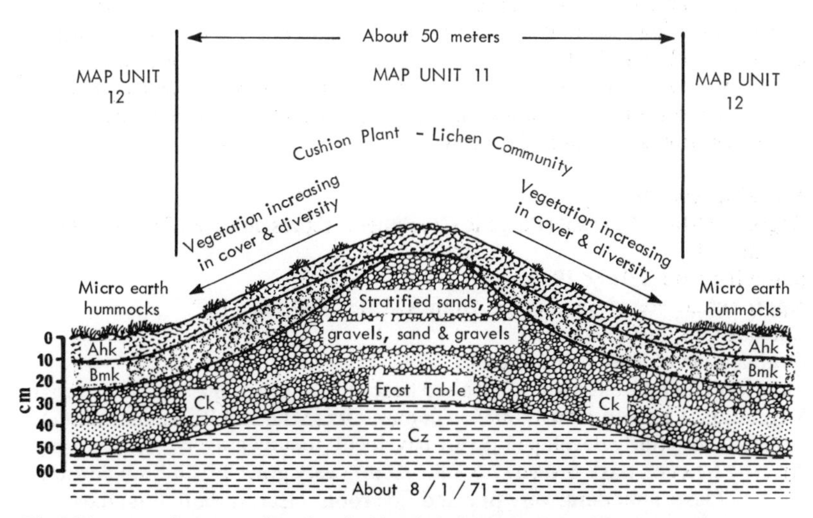

Fig. 2 Diagrammatic cross-section of a raised beach designated as belonging to map unit 11. Brunisolic Static Cryosols (Ahk, Bmk, Ck, Cz horizon sequence) dominate in this map unit. Regosolic Static Cryosols (Ahk, Ck, Cz horizon sequence) are subdominant.

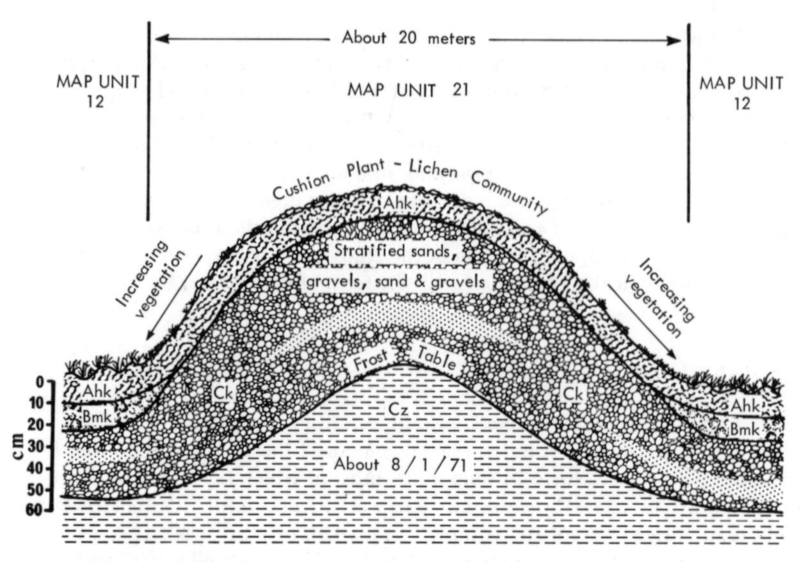

Fig. 3 Diagrammatic cross-section of a raised beach designated as belonging to map unit 21. Regosolic Static Cryosols (Ahk, Ck, Cz horizon sequence) dominate in this map unit. Brunisolic Static Cryosols (Ahk, Bmk, Ck, Cz horizon sequence) are subdominant.

Map Units 11 and 21
(Sampling Sites 39 and Intensive Raised Beach—Plot 8A)

Both these map units, each representing a soil complex, are associated with slope and crest positions of raised beach landforms containing Bd 1 beach material (Table 1). Cushion plant-lichen vegetation (Muc and Bliss this volume) and rapidly to well drained conditions are characteristic. The more commonly occurring soil map unit 11 is dominated by Brunisolic Static Cryosols with subdominant Regosolic Static Cryosols. Order of dominance is reversed in map unit 21.

Figs. 2 and 3 are diagrammatic cross-sections of two raised beaches, the first representing map unit 11 and the second, map unit 21. Regosolic Static Cryosols (Ahk, Ck, Cz horizon sequence) commonly occupy crest positions and show a range in development from profiles lacking Ah horizons (ochric epipedons) to profiles bordering on the types of development that may be classed as Brunisolic Static Cryosols.

Sampling Site 39

Regosolic Static Cryosol (Pergelic Cryorthent) representative of map units 11 and 21
Sampling date: 9 July, 1971
Approximate elevation and age: 28 m AMSL; 7500 years B.P.
Site position: Ridge crest
Drainage class: Rapidly drained
Landform and parent material: Raised beach (crest zone); Bd 1 beach material
Vegetation: Cushion plant — lichen community (Muc and Bliss this volume)

0-2 cm; poorly developed desert pavement, mainly gravels.

Ahk 2-13 cm; dark brown (7.5 YR 3/2 m*) very gravelly sand; single grain; loose; abundant, very fine and fine roots; clear, smooth boundary; 11 ± 3 cm thick; moderately alkaline.

*m = moist color.

Ck 13-82 cm; very gravelly sand to very gravelly loamy sand; single grain; loose; plentiful very fine roots to 33 cm; abrupt smooth boundary; 69 ± 8 cm thick; moderately alkaline.

Cz 82 plus cm; frozen gravels and sands.

Remarks: A color value is not provided for the Ck horizon because individual grain colors prevail and no uniform matrix color exists. Random stratification of sands and gravels occurs throughout the profile. The Ck horizon becomes moist at about 5 cm above the frost table.

In more favorable positions, on slopes and the broad flat tops of raised beaches, Brunisolic Static Cryosols are more likely to occur. Under increased plant cover (Svoboda this volume), these soils have Ah horizons with increased humus content and weak B horizons (cambic horizons) that have been enriched by small amounts of mobile humus forms. These soils are usually bounded, down slope, by similar soils displaying stronger development (map unit 12) and occurring under cushion plant-moss vegetation (Muc and Bliss this volume).

Intensive Raised Beach Site (Plot 8A)

Brunisolic Static Cryosol (Pergelic Cryochrept) representative of map units 11 and 21

Sampling date: 26 July, 1971

Approximate elevation and age: 28 m AMSL; 7500 years B.P.

Slope class and aspect: C; WSW

Drainage class: Well to rapidly drained

Landform and parent material: Raised beach (slope zone); Bd 1 beach material

Vegetation: Cushion plant — lichen community (Muc and Bliss this volume)

LF 2-0 cm; discontinuous litter layer occurring beneath cushion plants; contains some eolian sands.

Ahk 0-10 cm; very dark brown (10 YR 2/2 m) sand; single grain; soft; abundant, very fine and fine random roots; abrupt, smooth boundary; 10 ± 3 cm thick; mildly alkaline.

Bmk 10-26 cm; very dark grayish brown to dark brown (10 YR 3/2-3/3 m) very gravelly sand; single grain; loose; plentiful, very fine vertical roots; gradual, wavy boundary; 13 to 21 cm thick; moderately alkaline.

Ck 26-62 cm; very gravelly sand; single grain; loose; few, very fine roots to 35 cm; very few, very fine roots to 46 cm; abrupt, smooth boundary; 33 to 40 cm thick; moderately alkaline.

Remarks: A color value is not provided for the Ck horizon because individual grain colors prevail and no uniform matrix color exists. Sands and gravels are randomly stratified throughout the profile. The Ck horizon becomes quite moist at about 15 cm above the frost table. The B horizon of this individual has less organic matter than the modal pedon (*ca.* 1% org. carbon).

Raised beaches belonging to map unit 21 are usually narrower, have steeper but more rounded slopes and tops, and are elevated higher above adjacent land than those belonging to map unit 11 (Fig. 3 vs. Fig. 2). Small depressions, characterized by Brunisolic Static Cryosols similar to those of map unit 12, are infrequent inclusions associated with map unit 11. Likewise, ice-wedge fissures, generally sufficiently far apart that polygonal patterns are not formed, are considered as common inclusions in both map units. These ice-wedge fissures have developed, on a micro scale, a soil and vegetation sequence similar to that of "tri-zonal" (crest, slope, transition) raised beaches. Active layer depths range from 0.6 to 1.0 m in these two map units.

Table 4. Chemical and physical data for selected Brunisolic and Regosolic Static Cryosols

Horizon	Depth (cm)	pH H_2O	Org. Carbon %	Total N %	$CaCO_3$ Equiv. %	Exchangeable Cations (me/100g) H	Ca	Mg	Na	K	T.E.C. (me/100g)	NO_3-N
Site 39—Regosolic Static Cryosol in beach material (Bd 1)—representative of map units 11 and 21:												
Ahk	2-13	8.1	1.2	0.04	11.7	—	—	—	—	—	—	2
Ck	13-34	8.3	0.0	0.04	19.7	—	—	—	—	—	—	0
Ck	34-59	8.3	0.0	0.01	20.3	—	—	—	—	—	—	0
Ck	59-82	8.3	0.0	0.01	18.6	—	—	—	—	—	—	0
Intensive Raised Beach Study Site (Plot 8A)—Brunisolic Static Crysol in beach material (Bd 1)—representative of map units 11 and 21:												
Ahk	0-10	7.8	3.7	0.12	9.6	—	—	—	—	—	—	1
Bmk	10-26	8.0	Tr	0.03	10.1	—	—	—	—	—	—	0
Ck	26-46	8.1	0.0	0.01	13.6	—	—	—	—	—	—	0
Ck	46-62	8.1	0.0	0.01	18.6	—	—	—	—	—	—	0
Site 11—Brunisolic Static Crysol in beach material (Bd 1)—representative of map unit 12:												
Ahk	0.8	7.1	8.4	0.81	1.9	—	59.0	18.9	0.1	0.1	66.7	9
Bmk	8-23	7.7	1.4	0.10	11.7	—	—	—	—	—	—	1
Ck	23.61	7.8	0.0	0.02	20.2	—	—	—	—	—	—	1
Site 8—Lithic Regosolic Static Crysol in residual material (Rc 2)—representing subdominant soils of map unit 83:												
FH	5-0	6.2	24.0	1.64	—	17.8	69.5	21.1	0.1	0.7	102.4	23
Ah	0-5	6.9	1.5	0.12	—	—	7.1	2.9	Tr	0.1	9.7	2
Ah	5-31	7.4	4.1	0.27	1.8	—	21.3	7.4	0.1	0.1	23.5	2

*Visual estimate (% by volume).

Map Unit 12 (Sampling Site 11)

Map unit 12 occupies lower slope positions of raised beaches or all of those raised beaches having very low relief and is characterized by Brunisolic Static Cryosols developed in Bd 1 beach material (Table 1) under cushion plant-moss vegetation (Muc and Bliss this volume) and moderately well drained conditions. This map unit identifies a habitat type that is considered transitional between raised beach proper and meadow. It belongs to raised beach landforms and forms the third (bottom slope) zone of "tri-zonal" raised beaches (Fig. 2).

Soil characteristics range between those of Brunisolic Static Cryosols of map unit 11 and those of the gleyed mineral soils characterizing most meadows. "Snowpatch" habitats and their gleyed Brunisolic Static Cryosols are included.

The modal pedon of map unit 12 displays darker colored A and B horizons (increased humus contents) and stronger solum development than the most closely related "soil series," the Brunisolic Static Cryosol of map units 11 and 21. Soils having Ah horizons replaced by thicker FH or H (organic) horizons are included within the limits of map unit 12. Micro-earth hummocks (less than 10 cm high and 20 cm across) may occur on the soil surface; their presence appears related to specific vegetation types. Maximum active layer thicknesses range between 0.5 and 0.7 m.

Sampling Site 11

Brunisolic Static Cryosol (Pergelic Cryochrept) representative of map unit 12
Sampling date: 12 August, 1971
Approximate elevation and age: 28 m AMSL; 7500 B.P.
Slope class and aspect: C; W

Available Nutrients (ppm) P	K	Na	SO_4-S	Oxalate Fe %	Al %	% Coarse Fragments (>2mm)	Particle Size Distribtn. (% of <2mm fraction) Sand (2.0-0.05mm)	Silt (0.05-0.002mm)	Clay (<0.002 mm)	Bulk Density (g/cc)
1	8	2	3.8	0.05	0.05	65	88	10	2	1.6
1	6	1	1.0	0.02	0.00	77	84	16	0	1.6
1	5	2	1.0	0.01	0.00	75	93	6	1	1.7
1	5	0	0.5	0.01	0.00	58	95	5	0	1.6
1	12	1	0.5	0.10	0.03	17	92	6	2	1.3
1	5	1	0.0	0.06	0.00	57	97	2	1	1.6
1	3	1	0.0	0.04	0.00	50	91	9	0	1.6
1	5	2	0.5	0.08	0.00	64	90	8	1	1.6
0	34	7	3.2	0.22	0.07	45	78	8	14	0.7
0	10	1	2.3	0.10	0.02	51	86	9	5	1.4
2	7	3	1.0	0.07	0.00	56	86	12	2	1.7
5	217	23	5.4	0.52	0.07	—	—	—	—	—
2	24	3	2.0	0.25	0.09	<20*	73	21	6	—
1	23	5	2.6	0.39	0.14	>50*	60	28	12	—

Drainage class: Moderately well drained

Landforms and parent material: Raised beach (lower slope, transition zone); Bd 1 beach material

Vegetation: Cushion plant-moss community (Muc and Bliss this volume)

LFH 2-0 cm; black (10 YR 2/1 m) litter layer occurring beneath cushion plants and mosses.

Ahk 0-8 cm; black (10 YR 2/1 m) gravelly sandy loam; single grain; very friable; plentiful, very fine and fine roots; abrupt, smooth boundary; pH 7.1.

Bmk 8-23 cm; very dark grayish brown to dark brown (10 YR 3/2-3/3 m) very gravelly loamy sand; single grain; loose; plentiful, very fine and fine roots; clear, wavy boundary; pH 7.7.

Ck 23 plus cm; very gravelly sand; single grain; loose; very few, very fine roots to 36 cm; cobbles and stones increasing in quantity with depth; pH 7.8.

Remarks: The landform in which this site occurs is a long inclined slope of beach material lying against a large rock outcrop. This slope provides drainage of higher land to the east. Although the solum is considered typical of Brunisolic Static Cryosols in map unit 12, vegetation at this site is not characteristic of the cushion plant-moss community type in which it is grouped. A color value is not provided for the Ck horizon because individual grain colors prevail and no uniform matrix color exists. Sands and gravels are randomly stratified throughout but cobbles and stones increase in number with depth such that excavation was restricted to 61 cm. Free water was encountered at 53 cm.

Map Unit 13

Map unit 13 is dominated by Brunisolic Static Cryosols developed in Bd 2 beach material (Table 1) that thinly mantles portions of some rock outcrops. Although a type of profile development similar to the Brunisolic Static Cryosols of map unit 11 and 12 is considered to be dominant and modal, map unit 13 has much broader limits than those discussed above and includes soils with stronger development (analogous to the Brunisolic Static Cryosols of map unit 12) as well as Regosolic Static Cryosols.

Map Unit 22

Except for landform and parent material type, this map unit resembles, very closely, map unit 21. The dominant soil, Regosolic Static Cryosols, developed in coarse-skeletal, fluvial outwash (Ow 2), is generally weaker, having very weak or no Ah horizons. Brunisolic Static Cryosols are considered as minor inclusions rather than subdominant soils.

Map Units 31 and 71
(Sampling Sites 6, 9, and Intensive Meadow — Plot 6I)

Forming a complex mosaic over much of the Lowland are Gleysolic Static Cryosols of undifferentiated material (Un) and Fibric Organo Cryosols in fibrous organic material (wet). These soils, occurring under hummocky and wet sedge-moss meadow communities (Muc and Bliss this volume) are represented by two map units; 31, in which Gleysolic Static Cryosols are dominant, and 71, in which Fibric Organo Cryosols are dominant (See Fig. 6 for diagrammatic representations of these soil profiles).

Average maximum thaw depths of 0.3 to 0.4 m are common for Gleysolic Static Cryosols of these map units. The type pedon has a surface layer of fibrous, commonly minerotrophic, organic material that occupies 1/4 to 1/2 (excluding thickness of moss layers) of the active layer thickness. However, it is differences in peat layer thickness that account for much of the variability away from the characteristic, conceptual pedon. Dull or bluish colors indicative of gleying and commonly associated with reducing conditions are seldom found in the mineral portions of these soils.

Sampling Site 6

Gleysolic Static Cryosol (Histic Pergelic Cryaquept) representative of map units 31 and 71
Sampling date: 27 July, 1971
Approximate elevation and age: 10 M AMSL; <5700 years B.P.

Table 5. Chemical and physical data for selected Gleysolic Static Cryosols

Horizon	Depth (cm)	pH H_2O	Org. Carbon %	Total N %	$CaCO_3$ Equiv. %	H	Exchangeable Cations (me/100g) Ca	Mg	Na	K	T.E.C. (me/100g)	NO_3-N
Site 6—Gleysolic Static Cryosol in undifferentiated material (Un)—representative of map units 31 and 71:												
Of	23-0	6.2	26.1	2.24	—	13.8	49.2	15.1	0.1	0.2	75.9	0
Ckg	0-5	7.6	0	0.04	35.5	—	—	—	—	—	—	0
Site 22—Gleysolic Static Cryosol in alluvial-lacustrine material (As)—representative of map unit 32:												
Of	8-0	6.5	34.3	1.89	—	29.3	45.1	15.9	0.2	0.4	98.0	—
Ahg	0-5	5.7	5.8	0.45	—	8.5	9.9	3.7	0.1	0.2	26.7	—

*Estimate based on data collected from other sites.

Slope class and aspect: B; W
Drainage class: Very poorly drained
Landform and parent material: Undifferentiated plain; undifferentiated (Un) material
Vegetation: Hummocky sedge-moss meadow (Muc and Bliss this volume)
28-23 cm; moss layer
Of 23-0 cm; black (10 YR 2/1 m) fibrous organic matter; abundant, very fine vertical roots; abrupt, wavy boundary; 18 to 28 cm thick; pH 6.2.
Ckg 0-5 cm; dark gray to gray (2.5 Y 4.5/0 m) sandy loam; slightly sticky, non-plastic; few very fine and fine roots; abrupt, smooth boundary; 2 to 8 cm thick; pH 7.6.
Cz 5 plus cm; frozen mineral soil.
Remarks: Much variation in thickness of the Of horizon is due to hummock versus depression microtopography. Thickness of peat (Of horizon) in this soil individual does not represent the modal condition for Gleysolic Static Cryosols of map units 31 and 71 as it is 10 to 15 cm too thick.

Fibric Organo Cryosols, developed in fibrous, minerotrophic peat, exhibit average maximum thaw depths of between 0.2 and 0.3 m. Maximum and average thicknesses of peat were not investigated. However, since the two soils of map units 31 and 71 represent a continuum, it must be assumed that this property is variable across the map units. Other variations inherent in Fibric Organic Cryosols include color of peat (very dark brown and dark reddish brown) and reaction. Most common are the slightly acid to neutral Fibric Organo Cryosols but very strongly to medium acid members, normally associated with some stream-side habitats, also occur.

Intensive Meadow Site (Plot 6I)
Fibric Organo Cryosol (Pergelic Cryofibrist) representative of map units 31 and 71
Sampling date: 17 August, 1971
Approximate elevation and age: 35 m AMSL; <8000 B.P.
Slope class and aspect: B-C; W
Drainage class: Very poorly drained
Parent material: Slightly decomposed sedges and mosses
Vegetation: Hummocky sedge-moss meadow (Muc and Bliss this volume)
2-0 cm; moss layer
Of_1 0-8 cm; very dark brown (10 YR 2/2 m) slightly decomposed organic matter; abundant, very fine and fine random roots; abrupt, smooth boundary; 8±2 cm thick; slightly acid.

Available Nutrients (ppm) P	K	Na	SO_4-S	*Oxalate* Fe %	Al %	*% Coarse Fragments (>2mm)*	*Particle Size Distribtn. (% of <2mm fraction)* Sand (2.0-0.05mm)	Silt (0.05-0.002mm)	Clay (<0.002 mm)	*Bulk Density (g/cc)*
5	105	40	50+	0.54	0.13	—	—	—	—	0.2-0.3*
1	18	4	21.2	0.62	0.02	—	63	31	6	—
—	—	—	—	4.04	0.03	—	—	—	—	—
—	—	—	—	0.97	0.18	—	2	43	55	—

Table 6. Chemical and physical data for Selected Fibric Organo Cryosols

Horizon	Depth (cm)	pH H_2O	Org. Carbon %	Total N %	$CaCO_3$ Equiv. %	H	Exchangeable Cations (me/100g) Ca	Mg	Na	K	T.E.C. (me/100g)	NO_3-N
Intensive Meadow Study Site (Plot 61)—Fibric Organo Crysol in sedge-moss derived peat (wet)—representative of map units 31 and 71:												
Of_1	0-8	6.5	38.1	3.06	—	13.0	110.7	35.2	0.2	0.4	132.5	4
Of_2	8-31	6.2	42.2	2.68	—	22.3	93.5	23.4	0.2	0.2	127.9	6
Site 9—Fibric Organo Cryosol in sedge-moss derived peat (wet)—variant within map units 31 and 71:												
Of	0-3	5.9	38.7	1.69	—	24.6	31.8	4.5	0.4	0.8	83.0	5
Of	3-6	5.2	40.3	1.97	—	45.7	39.8	14.0	0.2	0.6	130.6	0
Of	6-9	5.0	36.4	2.01	—	54.5	21.3	0.8	0.3	0.4	104.6	0
Of	9-12	4.9	22.6	1.80	—	39.6	15.3	Tr	0.0	0.1	75.8	2
Of	12-15	5.0	29.1	1.85	—	45.8	19.0	4.2	0.0	0.1	79.6	0
Of	15-18	5.1	30.4	1.93	—	45.9	22.8	6.9	0.0	0.2	77.6	0
Site 1—Glacic Fibric Organo Cryosol in dry peat overlying ground ice—representative of map unit 72:												
Litter	2-0	6.0	34.6	2.23	—	15.8	50.1	34.7	0.8	2.2	98.1	200+
Of_1	0-13	5.8	27.8	2.00	—	17.6	44.2	27.9	0.6	0.3	88.2	88
Grnd. Ice**	76-130	6.2	37.3	2.40	—	17.6	64.4	38.9	0.7	0.4	103.3	0

*Estimate based on data collected for these and similar meadows by Muc (this volume).
**Results obtained from organic fraction only and not adjusted for high ice content of sample.

Of_2 8-31 cm; dark reddish brown (5 YR 2/2 m) slightly decomposed organic matter; abundant, very fine and fine random roots; abrupt, smooth boundary; 23±2 cm thick; slightly acid.

Oz 31 plus cm; frozen organic material.

Remarks: The Of_1 and Of_2 horizons are separated entirely on the basis of color.

Sampling Site 9

Fibric Organo Cryosol (Pergelic Cryofibrist) representing common variants in map units 31 and 71

Sampling date: 29 July, 1971

Approximate elevation and age: 11 m AMSL; 6000 years B.P.

Slope class and aspect: A-B; S

Drainage class: Very poorly drained

Parent material: Slightly decomposed sedges and mosses.

Vegetation: Wet sedge-moss meadow (Muc and Bliss this volume)

5-0 cm; moss layer

Of 0-18 cm; black to very dark brown (10 YR 2/1-2/2 m) slightly decomposed organic matter, abundant, very fine and fine random roots; abrupt, smooth boundary; 18±3 cm thick; very strongly to medium acid.

Oz 18 plus cm; frozen organic material.

Remarks: This small meadow occupies a narrow swale between two raised beaches. A stream flows through its centre and most of the meadow is generally under 2 to 5 cm of water throughout the summer.

Map Unit 32 (Sampling Site 22)

Flat to undulating alluvial-lacustrine plains dominated by Gleysolic Static Cryosols and hummocky to wet sedge-moss meadows (Muc and Bliss this volume) belong to soil map unit 32.

The type pedon is characterized by a surface organic horizon, with

								Particle Size Distribtn. (% of <2mm fraction)			
Available Nutrients (ppm)				Oxalate		% Coarse	Sand	Silt	Clay	Bulk
P	K	Na	SO_4-S	Fe %	Al %	Fragments (>2mm)	(2.0-0.05mm)	(0.05-0.002mm)	(<0.002 mm)	Density (m/cc)
0	129	21	5.4	0.92	0.15	—	—	—	—	0.2'*
0	47	25	8.0	0.26	0.13	—	—	—	—	0.2'*
10	393	35	6.8	11.08	0.03	—	—	—	—	0.1-0.2'*
10	210	35	21.2	3.34	0.05	—	—	—	—	0.1-0.2'*
10	108	30	46.0	2.72	0.09	—	—	—	—	0.1-0.2'*
5	61	23	44.0	3.14	0.13	—	—	—	—	0.1-0.2'*
10	78	30	38.7	2.34	0.16	—	—	—	—	0.1-0.2'*
10	78	35	44.0	3.20	0.18	—	—	—	—	0.1-0.2'*
140	1335	75	14.7	—	—	—	—	—	—	—
10	275	100	9.2	—	—	—	—	—	—	—
65	160	65	11.4	—	—	—	—	—	—	—

properties analogous to peat of the organic soils and Gleysolic Static Cryosols of map units 31 and 71 (see above discussion), overlying organo-mineral horizons thought to be accretionary (deposition of mineral material on mats of vegetation) rather than pedogenic. Average maximum thaw in these soils is estimated at between 0.2 and 0.3 m.

Color, texture, organic matter content, reaction, and $CaCO_3$ content of the organo-mineral horizon, designated as "Ahg" in most cases, account for most of the variation in the soils of map unit 32. Grayish brown to very dark grayish brown moist colors predominate but dark gray and black colors may be seen on occasion. Textures range from sandy loam to clay but clay loam and silty clay textures predominate. Organic carbon, ranging from 3.1 to 11.1% in Ahg horizons of the sampled profiles, varies among sites and within profiles. $CaCO_3$ content and pH show significant variation among sampling sites only. Medium acid to neutral and non-calcareous conditions prevail but, on occasion, weakly calcareous, mildly alkaline Ahg horizons occur. Areas adjacent to the escarpment show tendencies towards higher calcareous and reaction classes than areas farther away.

Sampling Site 22
Gleysolic Static Cryosol (Histic Pergelic Cryaquept) representative of map unit 32.
Sampling date: 31 July, 1972
Approximate elevation and age: 25 m AMSL; 7500 years B.P.
Slope class and aspect: A; level
Drainage class: Very poorly drained
Landform and parent material: Alluvial-lacustrine plain; alluvial-lacustrine (As) material
Vegetation: Wet sedge-moss meadow (Muc and Bliss this volume)
14-8 cm; moss layer
Of 8-0 cm; dark reddish brown (5 YR 2/2 m) with bands and patches of

very dark brown (10 YR 2/2 m) slightly decomposed organic matter; abundant, very fine to medium random roots; abrupt, smooth boundary; 8±2 cm thick; slightly acid.

Ahg 0-5 cm; grayish brown (2.5 Y 5/2 m) silty clay; amorphous; sticky, plastic; plentiful very fine and fine roots; abrupt, smooth boundary; 5±2 cm thick; medium acid.

Ahz 5 plus cm; frozen organo-mineral acid.

Remarks: Although this profile is very shallow, it must be pointed out that the summer of 1972 was abnormally cool and that this site was examined relatively early in the summer (probably three to four weeks prior to maximum thaw). Similar soils exist under the more commonly occurring hummocky sedge-moss meadow vegetation (Muc and Bliss this volume).

Map units 31, 32, and 71 have nearly identical plant communities and drainage features and occupy similar, often contiguous positions in the Lowland landscape. Soils of these units interfinger along designated soil boundaries and, consequently, add to map unit diversity.

Map Unit 33

In localized floodplains alluvium (Av 1) has been and continues to be deposited over other surficial materials. The poorly drained soils (Gleysolic Static Cryosols) of these small, melt-water discharge slopes lack surface organic horizons that are characteristic of other Gleysolic Static Cryosols (see discussion above).

Significant variation in map unit 33 involves depth of alluvium. During the survey, alluvium thicknesses ranging from 0.1 to greater than 0.6 m (frost encountered) were observed. Parameters such as pH (7.5-8.0), texture (loam and silt loam), calcareousness (35-50% $CaCO_3$ equivalent), and the nature of organo-mineral horizons (usually present as thin bands in these soils) vary slightly among sites.

Map Unit 34

Map unit 34, occupying lower portions of a lateral moraine, resembles very closely map unit 31 with the exception that map unit 34 lacks subdominant organic soils. Gleysolic Static Cryosols developed in till (Tm) are dominant and morphologically identical to those of map units 31 and 71. Map unit 34 is distinguished by steeper slopes and a more even distribution of subrounded stones. In addition, the parent material (Tm) tends towards slightly finer textures. Within the study area, map unit 34 occurs mainly on the south-facing lower slope of a lateral moraine within the Truelove River Valley.

Map Unit 41 (Sampling Site 34)

Relatively small raised beaches, forming slightly elevated, convex-shaped landforms in otherwise flat terrain and characterized by patterned ground features (circles and nets according to Washburn 1956) and cushion plant-moss (frost-boil) plant community (Muc and Bliss this volume) belong to map unit 41. Imperfectly drained Brunisolic Turbic Cryosols (Fig. 4), characteristic of this map unit, show considerable variability because of physical deformation and the unusual width of pedons — from sparsely vegetated frost-boil (or hummock) centre to densely vegetated trough centre, a distance ranging from 1 to 3 m. Morphologically, A-B horizon sequences usually occur, either vertically or obliquely, to frost under trough positions while hummocks, covering about 40% of the map unit, consist of C horizon (Fig. 4). Due to the confining microtopography, A horizons of troughs are usually organic (greater

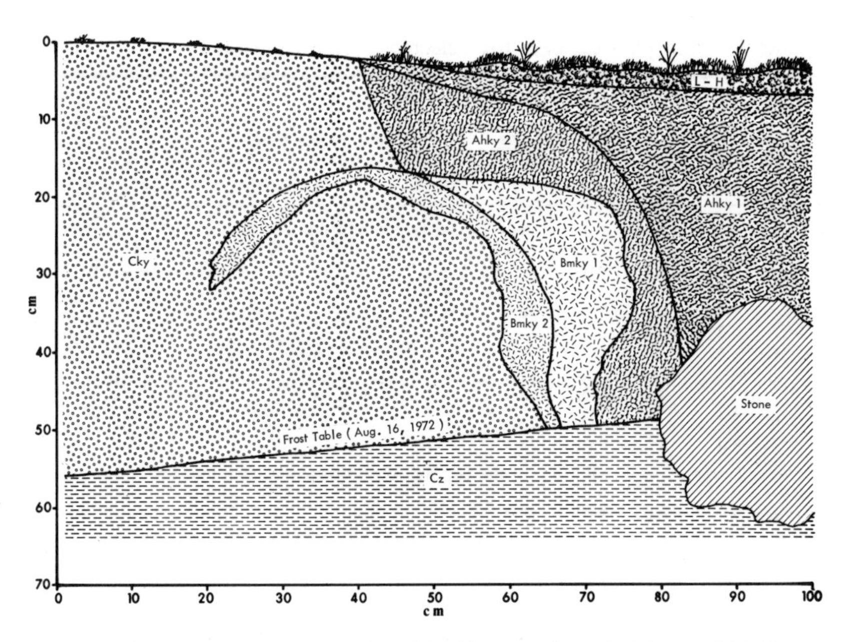

Fig. 4 Diagrammatic cross-section of a Brunisolic Turbic Cryosol in beach (Bd1) material. Diagram based on sampling site 34. This type of development is dominant among soils of map unit 41.

than 17% organic C). B horizons display considerable variability, ranging from weakly developed cones that display only slight humus accumulation and chemical alteration to those that may have received small amounts of mobile iron, based on color and oxalate-extractable Fe data (McKeague and Day 1966). Texturally, the soils of map unit 41 range from loamy sands to loams. Coarse fragments are not as abundant as in beach material (Bd 1) of more prominent raised beaches (map units 11, 12, and 21) but cobbles and stones are more frequent under trough positions indicating some degree of sorting by cryoturbation processes.

Sampling Site 34

Brunisolic Turbic Cryosol (Pergelic Cryochrept) representative of map unit 41
Sampling date: 16 August, 1972
Approximate elevation and age: 65 m AMSL; 8900 years B.P.
Slope class and aspect: A-B; W
Drainage class: Imperfectly drained
Landform and parent material: Raised beach; Bd 1 beach material
Vegetation: Cushion plant-moss with frost-boils (Muc and Bliss this volume)

LFH 2-0 cm; slightly to well decomposed litter occurring beneath cushion plants and mosses; 0 to 3 cm thick.

$Ahky_1$ 0-36 cm; black (10 YR 2/1 m) gravelly loamy sand (very high organic matter content); amorphous; very friable; plentiful, very fine and fine oblique and horizontal roots; estimated coarse fragments 40% (cobbles and stones concentrated near lower boundary); abrupt, broken boundary; 0 to 41 cm thick; mildly alkaline.

$Ahky_2$ 8-19 cm; very dark brown (10 YR 2/2 m) loamy sand; single grain; very friable; plentiful, very fine and fine oblique and horizontal roots; abrupt, broken boundary; 0 to 15 cm thick; mildly alkaline.

Table 7. Chemical and physical data for Selected Turbic Cryosols

Horizon	Depth (cm)	pH H_2O	Org. Carbon %	Total N %	$CaCO_3$ Equiv. %	H	Exchangeable Cations (me/100g) Ca	Mg	Na	K	T.E.C. (me/100g)	NO_3-N
Site 34—Brunisolic Turbic Cryosol in beach material (Bd 1)—representative of map unit 41:												
$Ahky_1$	0-36	7.4	16.5	1.40	3.9	—	71.2	18.7	0.2	0.3	87.6	—
$Ahky_2$	8-19	7.6	4.3	0.36	17.2	—	—	—	—	—	—	—
$Bmky_1$	20-29	7.9	—	—	24.0	—	—	—	—	—	—	—
$Bmky_2$	20-24	7.9	—	—	26.8	—	—	—	—	—	—	—
Cky	0-25	8.2	0.3	0.01	26.2	—	—	—	—	—	—	—
Cky	25-56	8.1	—	—	30.1	—	—	—	—	—	—	—
Site 35—Gleysolic Turbic Cryosol in undifferentiated material (Un)—representative of map unit 61:												
Ofy	18-0	7.2	25.9	2.21	1.3	—	63.7	21.2	0.3	0.3	86.7	—
Ckg	0-40	7.9	—	—	23.3	—	—	—	—	—	—	—
Ckgy	0-51	8.0	0.5	0.01	25.3	—	—	—	—	—	—	—
Site 29—Gleysolic Turbic Cryosol in till (Tg) of the plateau—representative of map unit 62:												
$Ckgy_1$	0-21	7.8	0.2	0.03	52.4	—	—	—	—	—	—	—
$Ckgy_2$	21-36	7.7	0.0	0.03	47.2	—	—	—	—	—	—	—

*Visual estimate (% by volume).

$Bmky_1$ 20-29 cm, dark brown (10 YR 3/3 m) loamy sand; single grain; friable; few, very fine and fine roots; abrupt, broken boundary; 0 to 10 cm thick; moderately alkaline.

$Bmky_2$ 20-24 cm; strong brown (7.5 YR 5/6 m) loamy sand; single grain; friable; very few, very fine and fine roots; clear, broken boundary; 0 to 5 cm thick; moderately alkaline.

Cky 0-56 cm; yellowish brown (10 YR 5/4 m) sandy loam; common, medium faint yellowish brown (10 YR 5/6 m) and dark yellowish brown (10 YR 4/4 m) mottles; friable; very few, micro and very fine roots to 25 cm; abrupt, broken boundary; 0 to 56 cm thick; moderately alkaline.

Cz 56 plus cm; frozen beach material; upper boundary at 45 to 56 cm from mineral soil surface.

Remarks: According to criteria specified in Canadian soil taxonomy (C.S.S.C. 1970) the $Ahky_1$ horizon is borderline between mineral and organic horizons. The typical pedon representing map unit 41 would have an FH or H horizon replacing the $Ahky_1$ horizon of this soil individual because surface organic horizons in trough positions are more prevalent. See Fig. 4 for diagrammatic representation of this profile.

Noted as inclusions in map unit 41 are those areas no longer undergoing active physical disruption (cryoturbation). In such situations, hummocks have recently become vegetated and, in many cases, normal (vertical) horizon development is as yet imperceptible.

Map Unit 42

Soil map unit 42 is associated with moderately to strongly sloping, stony beach (Bd 1) material and is dominated by imperfectly drained Brunisolic Turbic Cryosols and Regosolic Turbic Cryosols. Stripes (Washburn 1956) are the prevailing patterned ground feature. The Brunisolic Turbic Cryosols resemble those of map unit 41. Regosolic Static Cryosols lack B horizons under trough positions. These soils are probably less stable than those of map unit 41 due to greater slope gradients and stoniness.

Available Nutrients (ppm)				*Oxalate*		*% Coarse Fragments*	*Particle Size Distribtn. (% of <2mm fraction)*			*Bulk Density*
P	*K*	*Na*	*SO₄-S*	*Fe %*	*Al %*	*(>2mm)*	*Sand (2.0-0.05mm)*	*Silt (0.05-0.002mm)*	*Clay (<0.002 mm)*	*(g/cc)*
—	—	—	—	0.43	—	<40'*	—	—	—	—
—	—	—	—	0.37	0.01	<30'*	81	9	10	—
—	—	—	—	0.22	Tr	<20'*	84	10	6	—
—	—	—	—	0.70	0.02	<20'*	79	15	6	—
—	—	—	—	0.40	0.01	<20'*	65	26	9	—
—	—	—	—	0.53	0.01	<20'*	65	27	8	—
—	—	—	—	0.40	0.02	<20'*	—	—	—	—
—	—	—	—	0.43	0.01	<20'*	65	27	8	—
—	—	—	—	0.19	0.01	<10'*	65	26	9	—
—	—	—	—	0.14	Tr	<20'*	22	53	25	—
—	—	—	—	—	—	<20'*	20	48	32	—

Map Unit 51

Soil map unit 51 is associated with moderately to strongly sloping, very stony till (Tm) and is dominated by imperfectly to moderately well drained Regosolic Turbic and Brunisolic Turbic Cryosols. Very intense sorting during formation of striped patterned ground has, in many cases, resulted in sparsely vegetated troughs that are dominated by stones. Unstable conditions plus the lack of vegetation have produced weakly developed soils. Some elevated positions labelled as map unit 51 become quite dry during late summer.

Map Unit 61 (Sampling Site 35)

Map unit 61 occupies a large part of undifferentiated plains and is commonly associated with map units 31 and 71 with respect to topographic position and drainage. Gleysolic Turbic Cryosols (Fig. 5) are dominant and develop under a frost-boil sedge-moss meadow community (Muc and Bliss this volume). All patterned ground types (circles, nets, and stripes) are found in this map unit with stripes frequenting the steeper sloping areas.

Gleysolic Turbic Cryosols exhibit much less variability than Brunisolic Turbic Cryosols (see discussion under map unit 41) considering that pedon widths are about the same for both taxonomic types. Various degrees of sorting, manifest within profiles by concentration of gravels and cobbles below organic layers (Fig. 5), take place in Gleysolic Turbic Cryosols. Organic layer thicknesses are somewhat variable.

As with Gleysolic Static Cryosols in undifferentiated material, soils of map unit 61 do not exhibit strong gley coloration. However, some profiles have, at the mineral soil surface, thin bands of grayish colored mineral soil (1 cm thick in frost-boil positions but thickening to 5-10 cm under organic layers). Their position may reflect an input of humic material plus warmer temperatures.

Sampling Site 35

Gleysolic Turbic Cryosol (Ruptic-Histic Pergelic Cryaquept) representative of map unit 61

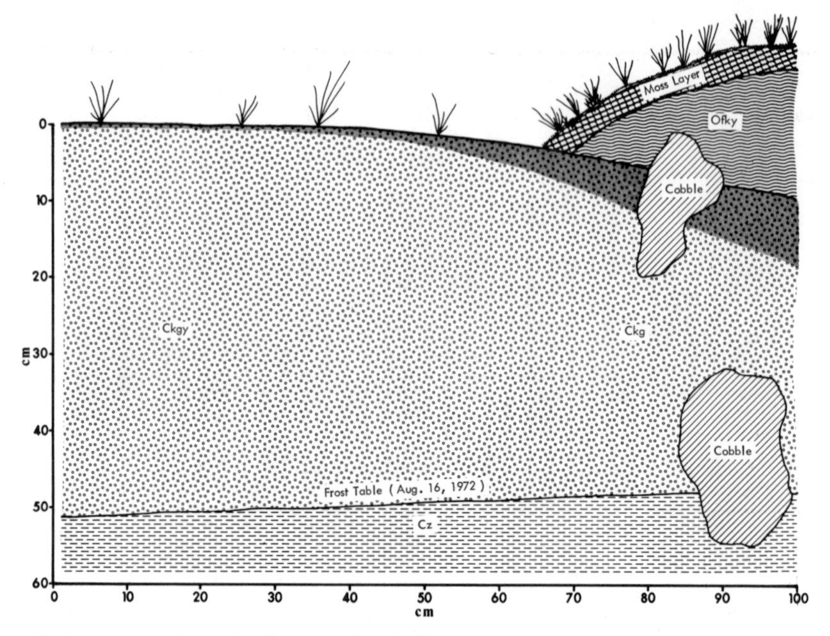

Fig. 5 Diagrammatic cross-section of a Gleysolic Turbic Cryosol in undifferentiated (Un) material. Diagram based on sampling site 35. This type of development is dominant among soils of map unit 61.

Sampling date: 16 August, 1972
Approximate elevation and age: 25 m AMSL; 7300 years B.P.
Slope class and aspect: A; level
Drainage class: Very poorly to poorly drained
Landform and parent material: Undifferentiated plain; undifferentiated (Un) material
Vegetation: Frost-boil sedge-moss meadow (Muc and Bliss this volume)
3-0 cm; moss layer; 0 to 4 cm thick

Ofy 18-0 cm; black to very dark brown (10 YR 2/1-2/2 m) slightly decomposed organic matter; plentiful fine vertical roots; some cobbles at lower boundary; abrupt, broken boundary; 0 to 18 cm thick; neutral.

Ckg 0-40 cm; light olive brown (2.5 Y 5/4 m) sandy loam; thin band of olive gray (5 Y 4/2 m) sandy loam beneath Ofy horizon; slightly sticky, nonplastic; contains some gravels and cobbles; abrupt, smooth boundary; 35 to 43 cm thick; moderately alkaline.

Ckgy 0-51 cm; light olive brown (2.5 Y 5/4 m) sandy loam; very thin band of olive gray (5 Y 4/2 m) sandy loam at the surface; slightly sticky, nonplastic; abrupt, smooth boundary; 48 to 51 cm thick; moderately alkaline.

Cz 51 plus cm; frozen mineral soil; frost at 35 to 51 cm below mineral soil surface.

Remarks: Refer to Fig. 5 for a diagrammatic representation of this profile. It is assumed that mineral material underlying the organic layer (Of horizon) is not frost-churned and is therefore labelled Ckg. Material forming the frost-boil is assumed to be churned and heaved and is therefore labelled Ckgy. However, boundaries separating these two horizons cannot be seen in any Gleysolic Turbic Cryosols. It is assumed that a diffuse boundary does exist but its shape and position are not known. A slight increase in coarse fragment content of the Ckg horizon relative to the Ckgy horizon indicates some degree of sorting. The frostboil is vegetated by sporadically occurring sedges and algae.

Map Unit 62 (Sampling Site 29)
Gleysolic Turbic Cryosols dominate the plateau portion of the study area. These soils, developed on fine-textured, calcareous till (Tm) of an extensive ground moraine, lack the surface organic layers of their lowland counterparts. Duller colors and more intense mechanical sorting of gravels are also characteristic features. Striped ground is the most common manifestation of cryoturbation with stripes being aligned downslope and parallel to the direction of groundwater flow, thus indicating a certain amount of gelifluction or mass flow.

Sampling Site 29
Gleysolic Turbic Cryosol (Pergelic Cryaquept) representative of map unit 62
Sampling date: 11 August, 1972
Approximate elevation: 330 m AMSL
Slope class and aspect: C; W
Drainage class: Poorly drained
Landform and parent material: Ground moraine; glacial till (Tg)
Vegetation: Moss-herb (Polar Desert) (Muc and Bliss this volume)
$Ckgy_1$ 0-21 cm; olive (5 Y 5/3 m) silt loam; very sticky, plastic; gradual, smooth boundary; 18 to 25 cm thick; mildly alkaline.
$Ckgy_2$ 21-36 cm; pale olive (5 Y 6/3 m) silty clay loam to clay loam; very sticky, plastic; abrupt, smooth boundary; 15±3 cm thick; mildly alkaline.
Cz 36 plus cm; frozen mineral soil.
$Ckgy_3$ 0-10 cm; olive (5 Y 5/3 m) gravelly silt loam; very sticky, plastic; clear, broken boundary; 0 to 15 cm thick.
Remarks: The $Ckgy_3$ horizon represents a cross-section of a vegetation-stone stripe (wedge or V-shaped in cross-section) that is oriented parallel to the slope. Separation was based on coarse fragment content which decreases with depth in the $Ckgy_3$ horizon but, in terms of the whole profile, indicates that mechanical sorting has taken place. These stripes, ranging from 5 to 15 cm in surface width, are depressed up to 5 cm below the centres of the slightly convex frost-boils.

Map Unit 72 (Sampling Site 1)
Soil map unit 72 encompasses those areas characterized by high-centred, ice-cored polygons (see landform and parent material description in Parent Material section). Glacic Fibric Organo Cryosols, developed in dry fibrous peat, occupy polygon tops. Frozen peat is underlain at variable depths, by ground ice containing bands of frozen peat with variable but fairly high contents of ice.

The available nutrient status of these soils is unusually high relative to all other soils of the study area. Although the exact significance of this trait has not been determined, it must be, in part, a reflection on the relatively warm, dry microclimate affecting these soils.

Sampling Site 1
Glacic Fibric Organo Cryosol (Pergelic Cryofibrist) representative of map unit 72

Sampling date: 26 June, 1971
Approximate elevation and age: 8 m AMSL; 2450 years B.P. (basal peat age according to Barr 1971)
Site position: Slightly rounded top of a polygon
Drainage class: Moderately well drained
Landform and parent material: High centred, ice-cored polygons; fibrous, dry peat overlying ground ice
Vegetation: Ice-wedge polygon (cushion plant) (Muc and Bliss this volume)
Litter 2-0 cm; moss and litter layer
Of 0-13 cm; black to very dark brown (10 YR 2/1-2/2 m) slightly decomposed organic matter; quite compact; few, very fine and fine roots; abrupt, smooth boundary; 13 ± 2 cm thick; medium acid.
Oz 13-38 cm; black to very dark brown (10 YR 2/1-2/2 m) frozen peat with a high ice content.
Grd Ice 38-140 cm; predominantly ice containing peat layers of variable thicknesses and organic matter contents.
Cz 140 plus cm; frozen gravels and stones.
Remarks: A small diameter, hand-operated ice corer was used to describe and sample this profile to the Cz layer.

Significant soil variation in map unit 72 occurs in interpolygon trenches where trench walls and bottoms are inhabited by wetter, more fibrous, very shallow organic soils.

Map Unit 81 (Sampling Site 32)
Raised beaches, 3300 years old and younger and composed of angular to thin, flat dolomitic gravels and cobbles (Bd 3), belong to map unit 81. This material is recognized as rubble land rather than soil, however. Peculiar Regosolic Static Cryosols, similar to profile 32, are designated as being subdominant in the map unit. Within the age range specified above, older raised beaches tend towards soils with this type of development.

Sampling Site 32
Regosolic Static Cryosol (Pergelic Cryorthent) representative of subdominant soils in map unit 81
Sampling date: 14 August, 1972
Approximate elevation and age: 4 m AMSL; 3300 years B.P.
Site position: Crest of a raised beach
Drainage class: Rapidly drained
Parent material: Rubbly beach material (Bd 3) overlying Bd 1 beach material
Vegetation: Cushion plant-lichen community on rubble (Muc and Bliss this volume)
Rubble 0-32 cm; angular to thin, flat fragments of gravel and cobble sizes; dolomite origin; estimated coarse fragments 90-100%; thin calcite films on undersides of many fragments; abrupt, smooth boundary.
$IICk_1$ 32-45 cm; very gravelly sandy loam; loose; clear, smooth boundary; 13 ± 5 cm thick; estimated coarse fragments 50-60%; $CaCO_3$ equivalent 86.0%; no organic C; pH 7.9.
$IICk_2$ 45-75 cm; very gravelly sand; loose; abrupt, smooth boundary; 30±7 cm thick; estimated coarse fragments 70-80%; $CaCO_3$ equivalent 55.9%; pH 8.2.
Cz 75 plus cm; frozen beach material.
Remarks: No colors are provided for the IICk horizons because individual grain colors prevail and no uniform matrix colors exist.

Map Unit 83 (Sampling Site 8)
Outcroppings of crystalline rock (Rc) form jagged, irregular-shaped knobs or buttes that are prominent, characteristic features of Truelove Lowland. These crystalline rock outcrops, excepting limited areas which are overlain by beach materials (map unit 13), belong to soil map unit 83 and are dominated by consolidated bedrock and lichen vegetation. Small, locally depressed niches characterized by dwarf shrub heath-moss vegetation (Muc and Bliss this volume) and Lithic Regosolic Static Cryosols developed in frost-shattered residual materials (Rc 2) occupy an estimated 15 to 40% of rock outcrop units. These soils, designated as subdominant in the map unit, vary with respect to depth of solum, depth of horizons, and organic matter content of horizons. Being located, however, in microclimatically favorable positions relative to most other soils of the study area, organic matter decomposition and incorporation of humus into mineral soil is probably comparatively rapid for such latitudes. B (cambic) horizons, produced by slight chemical alteration, were not found in such microsites but their occurrence could be expected where residual material is deeper and contains less coarse fragments.

Sampling Site 8
Lithic Regosolic Static Cryosol (Lithic Cryorthent) representative of subdominant soils in map unit 83
Sampling date: 28 July, 1971
Approximate elevation: 32 m AMSL
Site position and aspect: Small, level niche in the middle slope section of a steeply sloping rock outcrop; SSW
Drainage class: Moderately well to well drained
Landform and parent material: Crystalline rock outcrop; residual crystalline material (Rc 2) overlying bedrock (Rc 1)
Vegetation: Dwarf shrub heath-moss community (Muc and Bliss this volume)
L 7-5 cm; undecomposed litter.
FH 5-0 cm; black (10 YR 2/1 m) moderately to well decomposed organic matter; abundant, very fine and fine roots; abrupt, broken boundary; 0 to 8 cm thick; slightly acid.
Ah_1 0-5 cm; very dark grayish brown (10 YR 3/2 m) sandy loam; single grain; loose; plentiful, very fine and fine roots; clear, broken boundary; 0 to 10 cm thick; neutral.
Ah_2 5-31 cm; very dark brown to very dark grayish brown (10 YR 4/2-3/2 m) very gravelly sandy loam; single grain; loose; few, very fine roots; abrupt, wavy boundary; 15 to 28 cm thick; mildly alkaline.
R 31 plus cm; lithic contact composed of closely spaced stones and boulders.
Remarks: Within Ah horizons there is some horizontal variation with respect to colors and concentrations of organic matter and coarse fragments. The presence of a few dolomite erratics suggests that this deposit may in fact be glacial.

Map Units 82, 84, 85, 86, 87, and 88
The remaining miscellaneous land types do not have subordinate soils as do map units 81 and 83 but are dominated by rock land or rubble land (classed as Regosolic Static Cryosols).

Present-day marine beach and tideland areas belong to map unit 82. The material (Bd 4) forming these features is dominated, lithologically, by dolomite on and near Rocky Point to crystalline fragments adjacent to Truelove Inlet. All shapes and sizes of coarse fragments can be found.

Map unit 84 encompasses the dolomite plain (consolidated dolomite bedrock — Rl) that is Rocky Point. Shallow mineral gley soils and organic soils that occupy some local depressions, particularly around lakes and ponds, are designated as minor inclusions.

Rubble land associated with present-day stream channels belongs to map unit 85. Angular dolomitic cobbles and stones predominate in this outwash (Ow 1).

Large cones of rock debris, invading the Lowland from the steeply sloping eastern escarpment but having their apices above its base, are apparently alluvial (Av 2) in origin. Map unit 86 designates these alluvial cones. The rubble, angular to thin, flat gravels to stones, is predominantly dolomite in origin.

Stony colluvium (Cv) forms an escarpment which marks the eastern boundary of Truelove Lowland and rises some 250 to 300 m to a plateau which forms the main body of Devon Island. Map unit 87 incorporates the features of this escarpment. Lithologically, the angular to thin, flat rubble is dominated by dolomite with crystalline fragments increasing in number towards the escarpment base. Classed as extremely sloping with slope measurements averaging 35° (70%), well within the limits defining the angle of repose, the escarpment includes, amongst the rock rubble, occasional vertical prominences of consolidated dolomite bedrock and small "ledges" where fine earth mineral soils may occur.

Map unit 88 applies to those areas on the plateau, usually heights of land and steep, short slopes, characterized by thin, flat to angular rubble dominated by dolomitic stones. Rock circles (Washburn 1956) occur frequently in this material.

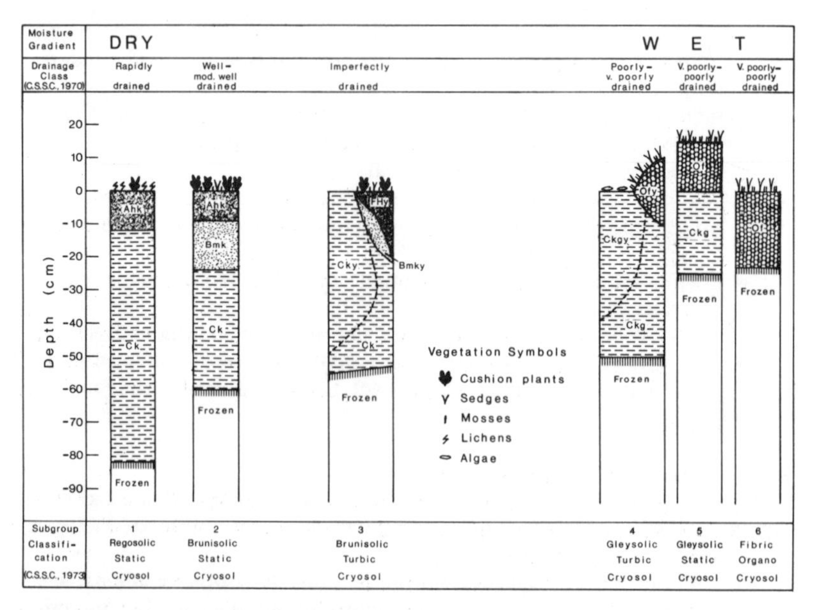

Fig. 6 Diagrammatic representations of 6 major types of soil development, arranged according to moisture gradient, occurring on Truelove Lowland. Each is classified to the subgroup level of abstraction (C.S.S.C. 1973).

Table 8. Correlation of plant communities and soil subgroups arranged according to moisture gradient

Moisture gradient	*Plant communities (*sensu *Muc 1975)*	*Soil classification to the Subgroup level (C.S.S.C., 1973)*
Dry	Cushion plant-lichen community on rubble	Miscellaneous soils Regosolic Static Cryosols >
	Cushion plant-lichen community	Brunisolic Static Cryosols > Regosolic Static Cryosols
	Dwarf shrub heath-moss community	Lithic Regosolic Static Cryosols
	Cushion plant-moss community	Brunisolic Static Cryosols
	Cushion plant-moss (frost-boil) community	Brunisolic Turbic Cryosols > Regosolic Turbic Cryosols
	Frost-boil sedge-moss meadow	Gleysolic Turbic Cryosols
Wet	Hummocky sedge-moss meadow	Gleysolic Static Cryosols > Fibric Organo Cryosols
	Wet sedge-moss meadow	Fibric Organo Cryosols > Gleysolic Static Cryosols

Soil-drainage-vegetation relationships

On the Lowland, relationships among soils, vegetation types (plant communities *sensu* Muc and Bliss this volume), and moisture status or drainage are quite predictable (Fig. 6, Table 8) until conditions of severely restricted drainage are encountered. Under such conditions, soil differences, from a classification viewpoint, remain considerable whereas plant community differences become more subtle (Fig. 6) than on drier sites.

Concepts regarding drainage and soil development must be re-evaluated as they apply to these northern environments. Firstly, most drainage is lateral rather than vertical because of the impervious frost layer at shallow depths. Secondly, topographic variance (relief) is subtle, yet produces complex patterns of drainage, soils, and vegetation (Fig. 7). Thus, subtle aspects of drainage other than rate of water supply versus rate of removal must be considered for partial explanation of the differences in profile development encountered, especially in wet sites.

Comparison of soils of selected Tundra Biome sites

Of the numerous international Tundra Biome sites, Barrow, Alaska, U.S.A. (71° 25' N, 156° 41' W); Agapa, Western Taimyr, U.S.S.R. (71° 25' N, 88° 53' E); and Tareja, Western Taimyr, U.S.S.R. (73° 12' N, 90° 54' E) express ranges of soil-forming environments most comparable with those of C.C.I.B.P.'s Devon Island, Canada (75° 33' N, 84° 40' W) site. Various levels of information concerning soils of the above four sites issue from Ignatenko (1973), Vasil'evskaia et al. (1973), Brown and Veum (1974), and Everett (1974).

Landscapes at Barrow, Agapa, and Tareja, all occurring in the Tundra soil zone (Tedrow 1973) or the Low Arctic (Bliss 1975), appear to be most severely and extensively affected by patterned ground phenomena — mainly various kinds of polygonal ground sometimes in association with circles and nets. Classification of cyclic pedons (repeating lateral horizon variations) has not been attempted at these sites, but rather profiles occupying different micro-

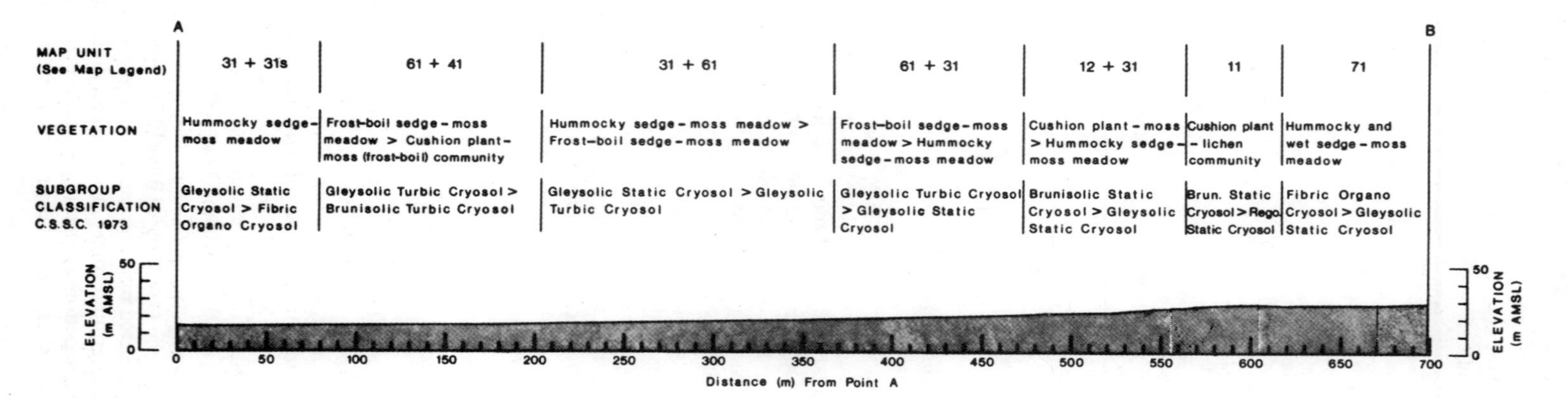

Fig. 7 A typical line transect (Transect No. 2) across meadow areas and a raised beach. Map unit, plant community, and soil changes along this transect are indicated.

positions of a polygon, circle, or net mesh are individually classified. It is quite possible that the C.S.S.C. (1973) system used in this paper will need further study and expansion to accommodate the soil variability of tundra polygons.

Arrangement of soils according to a qualitative moisture gradient (dry to wet) provides a means for comparing other general pedogenic trends among sites. Soils of wet locales (gleyed mineral soils and organic soils) are basically similar among the five sites, differing only in degree of gley coloration, reaction (pH), and depth of active layer. Organic and gleyed mineral soils of the Devon Island site generally lack pronounced gley colors and have slightly higher pH values (a reflection of local substrata) and shallower active layer depths than counterparts of the other sites.

At the Devon Island site, the "Turbic" group of soils exist under intermediate moisture conditions. The Soviet sites display "tundra weakly gleyed soils" that, in most cases, lack distinct organic or organo-mineral surface horizons but have gleyed B horizons. Barrow soils qualifying in the intermediate moisture range are quite similar to those of wet sites but occupy microtopographic elevations in the landscape and display better humified organic layers.

Soils developed under relatively well drained conditions better reflect the regional climates affecting those sites. Soil variations among the three sites belonging to the Tundra soil zone (Tedrow 1973) or Low Arctic (Bliss 1975) reflect local drainage peculiarities and, in some cases, age. Well drained soils have not been reported for the Barrow I.B.P. intensive sites since they occupy land of low absolute relief. At Tareja there occur "tundra illuvial-humus soils" characterized by weakly gleyed lower solums and humus-enriched, upper B horizons overlain by organo-mineral A horizons. Dry, immature "alluvial soils" with organo-mineral surface horizons occur at Agapa. Regosolic and Brunisolic Static Cryosols (see previous section) have developed in well drained locales at the Devon Island site. The former are similar in morphology to the "alluvial soils" of Agapa.

In summary it becomes apparent that soil-forming environments are more severe at Devon Island, particularly on the upland plateau sector where ahumic, non-peaty Gleysolic Turbic Cryosols, unique among the Tundra Biome sites evaluated, are present. Brunisolic Static Cryosols of raised beach slopes also tend to be unique among the Biome sites but morphologic counterparts have been reported for the Barrow area (Drew and Tedrow 1957).

Conclusion

Using generally accepted soil survey techniques, C.C.I.B.P.'s Tundra Biome study area on Devon Island was investigated for purposes of determining kinds and distribution of soils. This study area encompasses coastal lowland, not unlike sedge-grass tundra found in the Low Arctic (Bliss 1975), and interior plateau or upland, analogous to Polar Desert (Tedrow 1973) conditions.

Soil separations to the sub-group level of classification were accomplished using the tentative classification system for northern soils — the Cryosolic Order (C.S.S.C. 1973). By definition the Cryosolic Order recognizes, at the primary level of separation, the regional climate which, through its influence over other soil-forming factors, creates a peculiar kind of soil-forming environment. Using the pedon concept (Soil Survey Staff 1960), this system adequately handles cyclic features produced by congeliturbation. This physical process, a function of climate, drainage, and parent material, is considered pedogenic. Furthermore, conventional nomenclature (e.g. Brunisolic, Gleysolic), reflecting concepts gained from the study of temperature and boreal soils where climate is significantly different, is relegated to a proper level of abstraction (the sub-group tier) for arctic environments. Therefore, the use of this soil classification system (C.S.S.C. 1973) represents an advance in the recognition of soils whose development is severely restricted by cold climates and has reached equilibrium with present-day environments. It is felt that significant morphologic changes in most soils of the study area will not occur provided climate does not change significantly.

Within sub-groups finer separations approximating the soil series level of generalization (C.S.S.C. 1970) were made. The resulting units, or intimate combinations of them, formed the bases for developing map units used to depict the distribution of soils at a scale of 1:15,000 (see separate map in back folder).

Through observation, it is felt that drainage (or moisture regime) including various aspects of groundwater hydro dynamics, exerts the greatest influence with respect to local pedogenic trends. In this regard, soil-vegetation patterns are quite predictable except under the wettest conditions.

It can be stated that most soils of the study area, and in particular those of the Truelove Lowland sector, represent northern extensions of many soils found in temperate, boreal, subarctic and tundra biogeographic zones. Soil-forming processes common to all or some of these zones, plus the Devon Island site, namely, translocation of humus, oxidation, gleization, organic residue accumulation and decomposition, and cryoturbation, tend to decrease in intensity from south to north. In this regard the Gleysolic Turbic Cryosols of the plateau sector, typical of local "Polar Desert" conditions, are unique within the study area.

Acknowledgments

Support for this research was provided by the Canadian Committee for the International Biological Program. The authors also express appreciation to members of the Alberta Institute of Pedology, University of Alberta for advice and support. Special thanks are extended to W. McKean and A. Schwarzer (Alberta Institute of Pedology Soil Laboratory) for assistance with laboratory analyses and to the Alberta Soil and Feed Testing Laboratory, Edmonton, for available nutrient analyses.

Northern Forest Research Centre (Canadian Forestry Service, Environment Canada), Edmonton, provided space for manuscript and map preparation. Appreciation is extended to G. Lester (Department of Geography, University of Alberta) for final map preparation.

References

Aleksandrova, V.D. 1970. Vegetation and primary productivity in the Soviet Subarctic. 93-114. *In:* Productivity and Conservation in Northern Circumpolar Lands. W.A. Fuller and P.G. Kevan (eds.). Morges, Switzerland: IUCN N. Ser. No. 16. 344 pp.

Anonymous. 1973. Glossary of Soil Science terms. Soil Science Society of America. Madison, Wisconsin. 33 pp.

Barr, W. 1971. Postglacial isostatic movement in northeastern Devon Island: A reappraisal. Arctic 24: 249-268.

Barrett, P.E. 1972. Phytogeocoenoses of a Coastal Lowland Ecosystem, Devon Island, N.W.T. Ph.D. thesis. Univ. British Columbia, Vancouver. 292 pp.

Bascombe, C.L. 1961. A calcimeter for routine use on soil samples. Chemistry and Industry. Part II: 1826-1827.

Bliss, L.C. 1975. Tundra grasslands, herblands, and shrublands and the role of herbivores. Geoscience and Man 10: 51-79.

Brown, J. 1966. Soils of the Okpilak River Region, Alaska. CRREL Research Report 188. Hanover, New Hampshire. 49 pp.

———. and A.K. Veum. 1974. Soil properties of international Tundra Biome sites. 27-48. *In:* Soil Organisms and Decomposition in Tundra. A.J. Holding, O.W. Heal, S.F. MacLean, Jr., and P.W. Flanagan (eds.). Tundra Biome Steering Committee, Stockholm, Sweden. 398 pp.

Canada Soil Survey Committee (C.S.S.C.) 1970. The System of Soil Classification for Canada. Canada Department of Agriculture. Queen's Printer, Ottawa. 249 pp.

———. 1973. Report of the working group on northern soils: Tentative classification system for Cryosolic soils. 346-358. *In:* Proceedings of the Ninth Meeting of the Canada Soil Survey Committee. J.H. Day and P.G. Lajoie (eds.). Univ. Saskatchewan, Saskatoon. 358 pp.

———. 1974. The System of Soil Classification for Canada. Revised edition. Canada Department of Agriculture Publication 1455. Information Canada, Ottawa. 255 pp.

Cruickshank, J.G. 1971. Soils and terrain units around Resolute, Cornwallis Island. Arctic 24: 195-209.

Drew, J.V. and J.C.F. Tedrow. 1957. Pedology of an arctic brown profile near Point Barrow, Alaska. Soil Sci. Soc. Amer. Proc. 21: 336-339.

Everett, K.R. 1974. Principal soils and geomorphic units of the Barrow sites. U.S. Tundra Biome Data Rpt. 74-7. 45 pp.

Holowaychuk, N., J.H. Petro, H.R. Finney, R.S. Farnham, and P.L. Gersper. 1966. Soils of Ogotoruk Creek Watershed. 221-273. *In:* Environment of the Cape Thompson Region, Alaska. N.J. Wilmovsky and J.N. Wolfe (eds.). U.S. Atomic Energy Commission, Div. of Technical Information. Oak Ridge, Tenn. 1250 pp.

Ignatenko, I.V. 1965. The characteristics of soil formation in the different subzones of the eastern European tundra. Problems of the North, No. 8: 213-225.

———. 1973. Soils of the main types of tundra biocoenoses in western Taimyr. CRREL Draft Translation 408, Hanover, New Hampshire. 67 pp.

Ivanova, Y.N., N.N. Rozov, A.A. Yerokhina, N.A. Nogina, V.A. Nosin, and K.A. Ufimtseva. 1961. New materials on the general geography and classification of soils in the polar and boreal belts of Siberia. Sov. Soil Sci. No. 11: 1171-1181.

King, R.H. 1969. Periglaciation on Devon Island, N.W.T. Ph.D. Thesis. Univ. Saskatchewan, Saskatoon. 470 pp.

Kreida, N.A. 1958. Soils of the eastern European tundras. Sov. Soil Sci. No. 1: 51-56.

Matveyeva, N.V., T.G. Polozova, L.S. Blagodatskykh, and E.V. Dorogostaiskaya. 1974. A brief essay on the vegetation in the vicinity of the Taimyr Biogeocoenological Station. IBP Tundra Biome Translation. 56 pp.

McKeague, J.A. and J.H. Day. 1966. Dithionite- and oxolate-extractable Fe and Al as aids in differentiating various classes of soil. Can. J. Soil Sci. 46: 13-22.

Rozov, N.N. and E.N. Ivanova. 1968. Soil classification and nomenclature used in Soviet pedology, agriculture and forestry. 53-77. *In:* Approaches to Soil Classification. Soil Map of the World FAO/UNESCO Project, World Soil Resource Rpt. 32. FAO/UNESCO, Rome. 143 pp.

Soil Survey Staff, 1960. Soil Classification: A Comprehensive System, 7th Approximation. Soil Conserv. Service, USDA. 265 pp.

———. 1970. Selected chapters from the unedited text of the soil taxonomy of the National Cooperative Soil Survey. Soil Conserv. Service, USDA, Washington, D.C.

Tedrow, J.C.F. 1968. Pedogenic gradients of the polar regions. J. Soil Sci. 19: 197-204.

———. 1970. Soil investigations in Inglefield Land, Greenland. Medd. om Grønland. 188: No. 3. 1-93.

———. 1973. Polar soil classification and the periglacial problem. Builetyn Peryglacjalyny No. 22: 285-294.

———. and J.E. Cantlon. 1958. Concepts of soil formation and classification in arctic regions. Arctic 11: 166-179.

———. J.V. Drew, D.E. Hill, and L.A. Douglas. 1958. Major genetic soils of the Arctic Slope of Alaska. J. Soil Sci. 9: 33-45.

Vasil'evskaia, V.D., V.V. Ivanova, and L.G. Bogatyrev. 1973. Natural conditions and soils of "Agapa" station (Western Taimyr). IBP Tundra Biome Translation 8. 28 pp.

Washburn, A.L. 1956. Classification of patterned ground and review of suggested origins. Bull. Geol. Soc. Amer. 67: 823-866.

Zoltai, S.C. and C. Tarnocai. 1974. Soils and vegetation of hummocky terrain. Environmental-Social Program, Northern Pipelines, Rpt. No. 74-5. Ottawa. 86 pp.

Bedrock geology of the Truelove River area

J. Krupicka

Introduction

The Truelove River area on the northeastern coast of Devon Island is underlain by two geological units:
1. the Canadian Shield,
2. the Arctic Platform.
The crystalline rocks of the Shield outcrop in the Lowland, in the steep fault zone of the cliff south of the Truelove River, and in the lower part of the slope of the eastern plateau. The platform sediments build the eastern plateau and the Rocky Point, and form a shallow cover upon the basement rocks of the southern block. The general petrographic features are given in Fig. 1.

Results and discussion

The crystalline rocks of the Canadian Shield

The rocks are exposed as glacially rounded and striated outcrops in the lowlands region and in the Truelove River Valley, and as a steep continuous cliff in the vault wall south of the valley and the inlet. (Fig. 2)

In the hand specimen many of the rocks show little foliation. On outcrop surfaces, however, foliation is generally easily determined. Especially quartz-rich bands stand out on the weathered surface. Schistosity is almost non-existent; gneissosity is well developed in the migmatites. Lenticular and sill-like bodies of endemic granitic matter emphasize the general trend of foliation.

Foliated and more massive rock types are closely intermixed. Strike and dip measurements are possible in every larger outcrop. The average strike is N 10° W; the dip is invariably steep, in many cases vertical or sub-vertical. The data for the strike agree with those of Roots (in Fortier et al. 1963) who reported a strike of N 10° W for the rocks west of Cape Skogn, with a westerly dip of 60-80°. Christie (1969) noted for the north coast of Devon Island mainly E-W strikes, but also some SE trending strikes. Prest (1952) found in the area of Dundas Harbor on the southeastern coast of Devon Island the same general strike as in the Truelove River area.

The crystalline complex in the Truelove River area consists almost

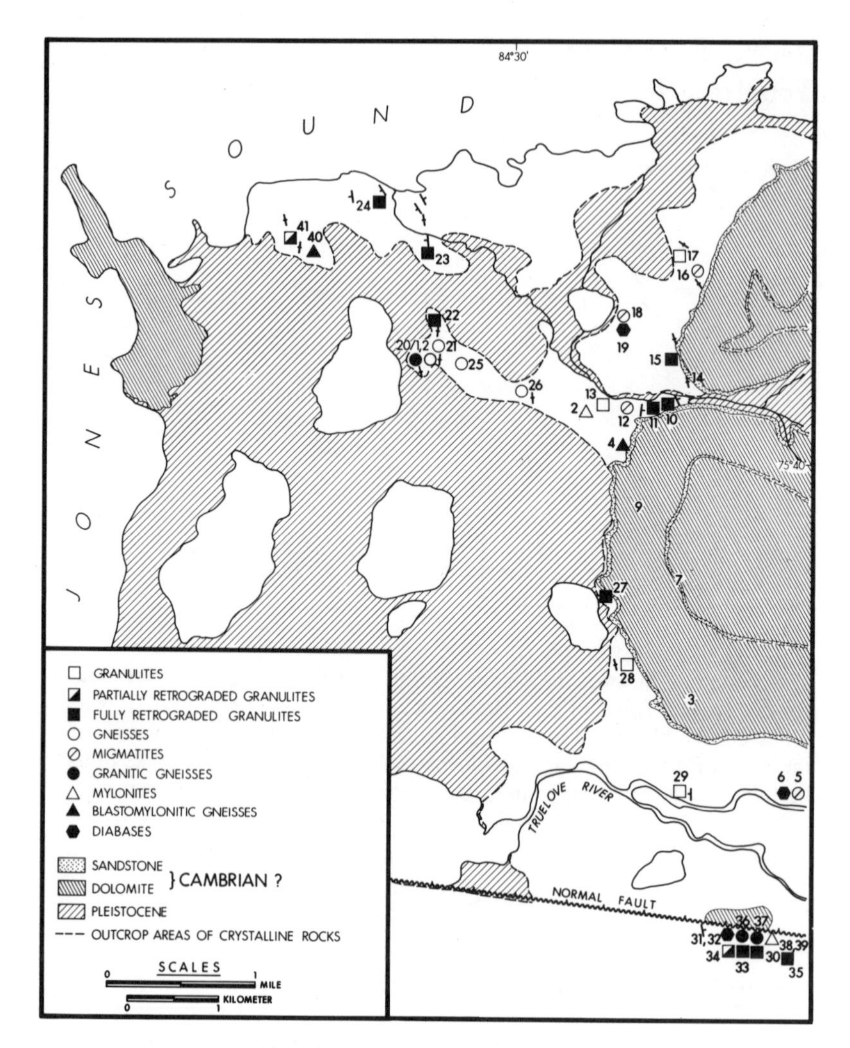

Fig. 1. Bedrock geology of the Truelove River area.

entirely of metamorphic rocks; the only unmetamorphosed crystalline rocks are younger diabases which cut the metamorphic rocks but do not penetrate their sedimentary cover. Granulites and gneisses directly derived from the granulites form more than half of the area. Both rocks, the highgrade true granulites and their retrograded amphibolite facies equivalents, occur intermixed throughout the complex.

No evidence has been found for the existence of younger supracrustal rocks that have undergone progressive metamorphism under amphibolite facies conditions. Nor has there been found evidence of a preceding progressive metamorphism leading to the formation of granulite facies rocks from adjacent lower-grade rocks.

The nearest granulite facies rock occurrences have been reported from eastern Ellesmere Island (Washington 1916), northern Baffin Island (Jackson

1969), Somerset Island (Blackadar 1967), and from Boothia Peninsula (Blackadar ibid., Brown et al. 1969).

Granulites are known just across Baffin Bay, in Inglefield Land in northwestern Greenland. The Tectonic/Geological Map of Greenland (1970) shows much of the area of Inglefield Land to be underlain by enderbitic gneisses (=granulites).

The observable metamorphic history of the Truelove River crystalline area begins with high-grade granulitic rocks, and develops as several stages of polymetamorphism under the relatively water-rich conditions of the amphibolite facies or, in some local cases in the granitic gneisses, even of the greenschist facies. Gneisses carrying pseudomorphs after hypersthene are linked by transitional types with biotite gneisses lacking identifiable alteration products of hypersthene, and these in turn are closely linked with granitic gneisses. Therefore it can be assumed that there exists a continuous retrogressive metamorphic series from granulites to granitic gneisses, and that the rocks of lower metamorphic grade than the granulites represent polymetamorphic products of original granulites.

Predominantly static, partial recrystallization led to the development of gneisses with abundant pseudomorphs after hypersthene. In addition, thorough mechanical reworking accelerated the process of recrystallization and granitization and resulted in the formation of microcline-rich granitic gneisses and blastomylonites. The spinel- and cordierite-bearing garnet migmatites, although without hypersthene, may represent an older period of migmatization, perhaps that occurring towards the end of the granulite facies metamorphism.

The granulites themselves suggest a sedimentary rather than igneous origin. They contain abundant, evidently primary (not introduced) quartz and a rather low proportion of dark minerals, which would not be expected were they derived from basic or intermediate igneous rocks. For an original acid igneous rock, the plagioclase is too calcic. The foliation of the granulites is not accompanied by the development of the typical platy quartz indicative of strong penetrative movement as is that of the blastomylonites. It probably is derived from original bedding rather than from mechanical reworking.

The main features of the retrogressive metamorphism are: the alteration and disappearance of hypersthene; the decrease in the amount of plagioclase and in its anorthite content; the decrease in the amount of antiperthite; the gradually rising degree of alteration of plagioclase; the growing separation and the increasing proportion of potassium feldspar accompanied in the later stages by the development of typical gridded microcline; the growing complexity of diablastic feldspar intergrowths in the middle stages of the process; the shift towards biotite and in the latest stages towards chlorite in the dark minerals; the tendency to ever more inequigranular and porphyroblastic texture; and the general tendency to more massive and granite-like rocks.

It is generally accepted that dynamic reworking accelerates adaptation to new equilibrium conditions and promotes a general tendency to granitization. The increase in the amount of potassium feldspar and the decrease in the An-content of plagioclase are especially pronounced in intensely reworked rocks, and the most granitoid rocks are in reality blastomylonites and recrystallized cataclasites.

A similar connection between strong deformation and granitization (microclinization) was stated by Berthelsen (1960) and by Windley (1969) for the crystalline rocks of southwestern Greenland, and has been found widespread in the basement rocks of western Canada (Burwash and Krupicka 1969).

The increasing amount of potassium feldspar originated in two ways. The first was by mobilization of the abundant antiperthitic exsolution material in the plagioclase of the granulites. This represented, in effect, a redistribution and individualization of the K-feldspar content already present in the rock, and may be considered as the more important source in the first stages of the retrogressive metamorphism. For the formation of the more advanced products of granitization, the granitic gneisses and the blastomylonites, an import of potassium must be postulated. It remains an open question whether this is merely a redistribution of the potassium content within the complex itself, or whether the source lies outside the complex. With the first explanation, of course, one would expect to find complementary K-impoverished rocks in the area. No such rocks have been found. The interpretation that the granulites themselves were a source (see Dawes 1970) meets with a fundamental objection: the granulites had lost their potassium long before the granitization of the complex started.

The textural relationships of the recrystallized cataclasites and mylonites show very clearly the younger nature of most of the potassium feldspar in them. Most notable are the K-porphyroblasts growing amidst granulated and altered plagioclase. This potassium feldspar appears mainly in the form of microcline, and is late deformational or post-deformational in age. Heier (1957) gives a temperature of about 500°C for the transition from monoclinic potassium feldspar to triclinic microline, which corresponds to conditions of the lower amphibolite facies.

Both types of granitization, the isochemical and the metasomatic, are assumed by Berthelsen (1960) to have taken place in the Tovqussap granulites. Windley (1969) also considers potash metasomatism to have been operative in the formation of the reworked microcline-rich belts in the granulite complex of southwestern Greenland. Dawes (1970) states that widespread metasomatism occurred during the granitization in the area adjacent to the Tassiussaq charnockites, and (ibid.) that the notable feature was the migration and diffusion of potash. Blackadar (1967) assumed widespread potash metasomatism in all pre-Helikian rocks of the northern Boothia Peninsula and southern Somerset Island.

No isotopic ages have been obtained from the crystalline rocks of Devon Island. For the Arctic part of the Canadian Shield north of Baffin Island, only one date has been published (Christie 1962): a K/Ar date for biotite in gneiss giving 1,760 m.y. A number of age determinations from northern Baffin Island also give Hudsonian ages, except for one whole-rock Rb/Sr date of 2,340 m.y. for a gneiss. Farther southeast, in central Baffin Island, another gneiss (granulite?) with a Rb/Sr date of 2,600 m yr appears among the Hudsonian dates. Stockwell (1961) includes the whole Canadian Shield of the Eastern Artic in the Churchill Province where the last strong metamorphism occurred during the Hudsonian Orogeny, approximately 1,800 m.y. ago.

Most granulites in the Canadian Shield have been firmly established as Archean rocks. Some occurrences of granulite facies rocks in the Churchill Province, however, have been considered as prograded equivalents of adjacent lower-grade metamorphic rocks (Heywood 1970, Jackson 1969). This interpretation could imply the assumption of a Proterozoic age for the granulites. Blackadar (1967) includes the granulites of Boothia Peninsula and Somerset Island among Aphebian rocks.

Without isotopic age dating it is futile to speculate whether the granulite facies metamorphism of the Devon Island rocks was Kenoran, or whether it may have been even older, corresponding to the 3,200 m.y. old granulites at Fiskenaesset. In either case there is little doubt that it was the Hudsonian metamorphism that left the last strong imprint on the original granulite complex.

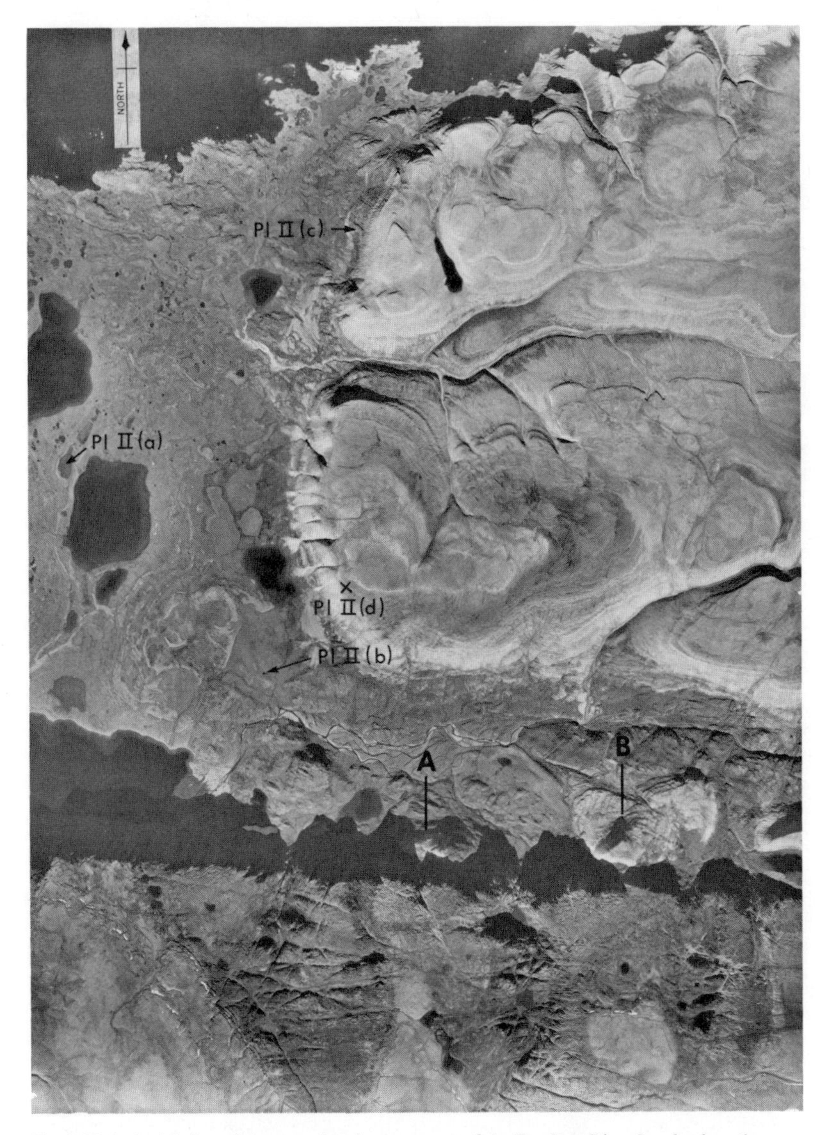

Fig. 2. Pl. I. Aerial view of the central and eastern part of the Truelove River Lowland on the northeastern coast of Devon Island. Note the sedimentary plateau to the east, and the basement fault-line cliff and plateau to the south. The northern (sedimentary) block subsided by approximately 300 m along an east-west normal fault, coinciding with the conspicuous cliff south of the river and the inlet. Sedimentary outliers (A, B and outliers to the east) belong to the subsided northern block. The arrows mark positions from which the photographs, Pl. II(a) to II(d) were taken. *(National Air Photo Library.)*

Pl. II(a)

Pl. II(b)

Fig. 3. Pl. II(a) View of the Truelove River Lowland, with cliffs of basement rocks and downfaulted sedimentary outliers (A, B) in the background. View southeast.

Pl. II(b) View southwestward toward the fault-line scarp on the south side of Truelove Inlet. Note the thin layer of Paleozoic sediments along the skyline, and the gentle westward dip of the unconformity. In the foreground are outcrops of crystalline rocks along the lower reaches of Truelove River.

Pl. II(c) Granulite outcrop. Behind the outcrop is an elevated beach, approximately 80 m above the present sea level. Although the beach is on crystalline bedrock, the gravel is derived almost exclusively from the dolomitic sediments outcropping higher on the slope. The beach is visible as a white strip on Pl.I.

Pl. II(d) Stromatolites in a slab of silty dolomite. Frost-heaved outcrops on the edge of the eastern plateau.

Pl. II(c)

Pl. II(d)

The sediments of the Arctic Platform

The boundary between the underlying crystalline complex of the Canadian Shield and the overlying unfolded and unmetamorphosed sedimentary cover is well marked in the field by the contrast between the light-colored sediments and the dark basement. The contact was followed continuously for miles. Many places were observed where it is not covered by talus and the outcrops are bare; without exception, however, the lowermost member of the sedimentary sequence had been peeled off from the hard, flat surface of the crystalline basement by frost action, and slabs of the basal sandstone lie on the surface of the basement. A few metres higher the sedimentary beds may be found lying in situ.

The boundary is a classical nonconformity: horizontal sedimentary strata are in contact with a crystalline metamorphic basement. Most of the contact is original and transgressional. Its flat surface [Pl. II(b)], dipping gently to the west, shows that peneplanation of the crystalline complex occurred before the deposition of the sediments. The photograph shows the eastern outskirts of the sediments of the Arctic Platform which, approximately 50 km to the west, reach sea level.

Along the big Truelove River cliff, however, the sedimentary rocks are faulted against the basement rocks [Pl. II(a)]. The northern block subsided along an east-west-trending normal fault, and the stratigraphic displacement is more than 300 m. The entire north-facing cliff comprises crystalline rocks. Erosional remnants of the sedimentary cover of the subsided, northern block are visible in plate I (A, B and other outliers to the east). The valley of Truelove River separates them from the main body of sediments. Plate II(a) shows the same features from the ground. The main fault lies north of the scarp which, however, does coincide with secondary fault planes. The steepness of the upper parts of the scarp, 70°-80°N, reflects the dip of the main fault plane.

The subsided northern block is tilted towards the south. The surface separating sediments and basement rises northwards [Pl. I, near the arrow marked Pl. II(c)]. In the area of Cape Skogn, 25 km farther north, a high coastal cliff again is formed by basement rocks.

The lowermost member of the sedimentary sequence is a 2- to 3-m thick, slightly dolomitic sandstone. It is usually bimodal. The average grain size of the matrix is approximately 0.2 mm; the matrix contains phenoclasts from 1 to 4 mm in diameter. Frequent grains of bluish quartz point to granulite as one of the main sources. The phenoclasts are subrounded and consist of quartz, "quartzite", and feldspar (potassium feldspar predominating over plagioclase). The "quartzite" grains are, in reality, recrystallized quartz from the stringers and lenses in the blastomylonitic and cataclastic gneisses. They have the typical serrate texture of a sedimentary metaquartzite but are identical down to minute details, with the recrystallized quartz crush from the sheared gneisses. Their presence in the basal sandstone is proof of the Precambrian age of the period of deformation and recrystallization of the basement rocks.

The 180 to 200 m of overlying sedimentary rocks are mainly dolomite. The sequence contains three horizons of massive dolomite forming conspicuous cliffs for long distances, separated by bedded silty and argillaceous dolomite and some dolomitic shale. The bedded sediments are almost always hidden under a thick cover of talus from which protrude the cliffs of massive dolomite. Above the uppermost horizon of the massive dolomite, on the edge of the plateau, lies a thin (approximately 2 m) bed of fine-grained quartzitic sandstone (orthoquartzite).

No outcrop was found on the plateau above the escarpment; all is covered by masses of debris from the underlying rock, disintegrated by frost action.

The bedded dolomite is a fine-grained, lutaceous rock. It shows numerous slump textures, and contains intercalations of intraformational conglomerates. The rock may be packed with ovoidal or, in places, ovoidal-to-elongated and curved phenoclasts suggestive of fossils. The only real organogenic fossil structures found in the whole sedimentary sequence were stromatolites [Pl. II(d)], which are rather common.

The sample of silty quartzo-feldspathic dolomite exhibiting slump textures has three layers: (1) a very fine grained iron-red colored layer, with an average grain size of less than 0.01 mm, composed of approximately 90% dolomite which with somewhat larger subangular grains of quartz and feldspar; (2) a complex, brecciated zone where slab-like fragments of layer (1) float in a larger-grained (0.1 mm) matrix composed of 50% dolomite and 50%

clastic quartz and feldspar; and (3) a more even-grained layer consisting of 70% dolomite and 25% quartz and feldspar with the remaining 5% comprised of vividly green glauconite, pleochroic biotite, and some muscovite.

About half of the felsic clastic minerals are feldspars, both microcline and plagioclase, markedly clear and unaltered. Their mainly angular character and freshness, as well as the preservation of biotite, indicate rapid disintegration of the parent rock and short transport, at least partly in the form of slumps.

The rock of the upper arenite bed is a dolomitic quartzitic sandstone (orthoquartzite) with abundant crossbedding. It is composed of 80.5% quartz, 19% dolomite, and 0.5% feldspar, mostly potassium feldspar. Texturally the rock is a very good example of orthoquartzite. Well-rounded and well-sorted grains of clastic quartz are cemented together by accretion rims growing in full optical continuity with the clastic grain. The accretion rims are marked by lines of minute (0.001 mm) unidentifiable grains. They are also of clastic origin, with subsequent partial recrystallization.

The sediments of the investigated area are assigned Cambrian and Ordovician ages on the new Geological Map of Canada (Douglas 1970). Christie (1967) and Kerr (1967) consider analogous sediments in east-central Ellesmere Island to be Early Cambrian. The stratigraphic sequence there appears to be similar to that in the Truelove River area. The basal clastic, unfossiliferous beds (Rensselaer Bay Formation) are overlain abruptly by fine-grained dolomite (Cape Leiper dolomite) which, in turn, is overlain by massive, crystalline, vuggy dolomite of the Cape Ingersoll Formation. In the Truelove River area, however, the alteration of bedded and massive dolomite is a repeated phenomenon, and starts immediately above the basal sandstone. Glenister (in Fortier et al. 1963) considers equivalent rocks in the Sverdrup Inlet area, approximately 50km to the west, to be "Lower Ordovician and (?) earlier." On the south coast of Devon Island, near Dundas Harbor, Prest (1952) also found a thin basal sandstone, which he classifies as Paleozoic, overlain by "limestone, sandy limestone and limy sandstone."

References

Berthelsen, A. 1960. Geology of Tovqussap Nuna. Medd. om Grønland, 123, no. 1.

Blackadar, R.G. 1967. Precambrian geology of Boothia Peninsula, Somerset Island, and Prince of Wales Island, District of Franklin. Geol. Surv. Can., Bull. 151.

Brown, R.L., I.W.D. Dalziel, and B.R. Rust, 1969. The structure, metamorphism and development of the Boothia Arch, Arctic Canada. Can. J. Earth Sci., 6: 525-543.

Burwash, R.A., and J. Krupicka, 1969. Cratonic reactivation in the Precambrian basement of western Canada. I. Deformation and chemistry. Can. J. Earth Sci., 6: 1381-1396.

Christie, R.L. 1962. Geology, southeast Ellesmere Island, District of Franklin; Geol. Surv. Can., Map 12-1962.

———. 1969. Eastern Devon Island and southeastern Ellesmere Island, District of Franklin. Geol. Surv. Can. Paper 69-1. Pt. A. p. 231-234.

Dawes, P.R. 1970. The plutonic history of the Tassiussaq area, South Greenland. Medd. om Grønland, 189: no. 3.

Douglas, R.J.W. 1970. Geology and economic minerals of Canada; Geol. Surv. Can., Econ. Geol. Rept. no. 1.

Fortier, Y.O., R.G. Blackadar, B.F. Glenister, H.R. Greiner, D.J. McLaren, N.J. McMillan, A.W. Norris, E.F. Roots, J.G. Souther, R. Thorsteinsson, and E.T. Tozer, 1963. Geology of the north-central part of the Arctic Archipelago, Northwest Territories (Operation Franklin). Geol. Surv. Can., Mem. 320.

Heier, K.S. 1957. Phase relations of potash feldspar in metamorphism; J. Geol. 65:468-479.

Heywood, W.W. 1970. Geological reconnaissance of Southampton Island, District of Keewatin; Geol. Surv. Can., Paper 70-1A, p. 144.

Jackson, G.D. 1969. Reconnaissance of north-central Baffin Island. Geol. Surv. Can., Paper 61-1, Pt. A, p. 171-176.

Kerr, J.W. 1967. Stratigraphy of central and eastern Ellesmere Island, Arctic Canada. Geol. Sur. Can., Paper 67-27, Pt. I.

Prest, V.K. 1952. Notes on the geology of parts of Ellesmere and Devon Islands, Northwest Territories. Geol. Surv. Can., Paper 52-32.

Stockwell, C.H. 1961. Structural provinces, orogenies, and time-classification of rocks of the Canadian Precambrian Shield; Geol. Surv. Can., Paper 61-17.

Washington, H.S. 1916. The charnockite series of igneous rocks. Am. J. Sci., 4th Ser. 41, p. 323-338.

Windley, B.F. 1969. Evolution of the early Precambrian basement complex of southern West Greenland; Geol. Assoc. Can., Spec. Paper no. 5: p. 155-161.

Microclimatological studies on Truelove Lowland

G.M. Courtin and C.L. Labine

Introduction

In spite of recent interest in polar regions there remains a paucity of literature on arctic microclimates. All studies are very recent and most of these have been confined to Low- or Mid-Arctic (Ahrnsbrak 1968, Wendler 1971, Romanova 1972, Rouse and Stewart 1972, Brazel and Outcalt 1973a, b; Weller and Holmgren 1974) or close to centres of populations in the High Arctic (Vowinckel 1966). Ohmura's work on Axel Heiberg and Smith's research on the Fosheim Peninsula, Ellesmere Island are the only other microclimatic research undertakings in the High Arctic (Ohmura 1970, Smith 1975).

With the onset of the I.B.P. more intense work was initiated with attempts at much longer time spans of data gathering (Weller and Cubley 1972, Skartveit et al. 1975). The Canadian program, initiated in 1970, included the first attempt at long range microclimatic measurement in the Canadian High Arctic. Furthermore, the studies on the Truelove Lowland were undertaken with the biota in mind and most data gathering was concentrated in the biologically active snow-free period and the months that immediately preceded and followed it. The program had one complete over-winter period of data collection.

Thus the studies on Truelove Lowland were microenvironmental in nature as opposed to being purely microclimatological and were initiated with three principal aims in mind:

1. The descriptive microclimatology of a high arctic coastal lowland including the spatial variation that exists over the 43 km^2 lowland, 4 km^2 valley, and adjacent plateau.
2. The evaluation of energy input, both as shortwave and net radiation, to the Lowland.
3. Assistance with environmental measurement requested by other researchers in the project, either because of a requirement for precision beyond that sought at the microclimatological stations or in locations not covered by the station network.

Description of the area

Macroclimate

The dominant factors which shape the arctic climate are: the character of the solar energy input, the nature of the immediate and adjacent surfaces, weather systems, and topography (McKay et al. 1970).

The macroclimate of the Canadian Arctic Archipelago, of which Devon Island is a part, is that of a high latitude continental landmass. Even though the islands are surrounded by ocean, the water is frozen for approximately 10 months of the year and provides only a limited amount of moisture to affect the precipitation pattern.

No long term weather records exist for Devon Island but records from adjacent stations (Table 1) indicate continentality with low precipitation and low mean annual temperature. It must be appreciated, however, that the stations tend to be located close to the coast where there is a modifying effect on temperature by the proximity of large masses of water in spite of a thick covering of sea ice. Temperatures inland may well be more extreme.

Polar regions are characterized by surface weather systems associated with the large-scale, cold-cored circumpolar vortex. This homogeneous arctic core of the westerly vortex in the Canadian Arctic is most intense and westerly flow is strongest in winter. During this time the vortex shows three centres: over the Canadian Archipelago, Kamchatka, and Novaya Zemlya. In summer the vortex weakens and contracts to a single mean centre near the pole. The circulation tends to be more zonal (i.e. from west to east) and moving cyclones of the arctic frontal belt often traverse the area, producing light precipitation typical of the season (Wilson 1967, Hare 1968, Hare and Hay 1974). This strengthening and weakening of the vortex is related to the fluctuating radiation regime. The long polar night and the equally long polar day serve to give a very steep temperature gradient from equator to pole in winter and a much shallower gradient in summer.

The relative infrequency of cylonic storms, the shallow pressure gradients in summer, the lack of a neighboring moisture source for much of the year, and the high latitude combine to give the High Arctic a cold desert climate with yearly water equivalent precipitation no higher than 180 mm, and sub-freezing mean annual temperatures (Table 1). Furthermore, mean annual winds are no greater than in temperate regions but periods of high wind appear to be associated with cylonic disturbances (Courtin in Bliss et al. 1973).

Physiography

The general geographic and physiographic characteristics of Truelove Lowland have been described earlier in this volume. For the purpose of the

Table 1. Annual mean climatic data for weather stations adjacent to Devon Island (Thompson 1967).

Station	*Latitude N*	*Longitude W*	*Temp (°C)*	*Mean Annual precip. (mm)*	*Wind (km hr⁻¹)*
Arctic Bay	73° 00'	85° 18'	-13.9	125	9.0
Resolute Bay	74° 41'	94° 54'	-16.2	130	18.7
Eureka	80° 00'	85° 56'	-16.4	67	12.2
Alert	82° 30'	62° 20'	-17.8	147	9.1

microenvironmental work three major lowland physiographic types were recognized: the Raised Beach Ridge, the Sedge-moss Meadow, and the Granite-gneiss Rock Outcrop. Also included was one upland type, the edge of the limestone plateau to the east of the lowland, which may well give an idea of the mean macroclimatic conditions that affect the general area and which do differ markedly from the conditions that prevail in the lowland below (Labine 1974).

Lowlands of the kind typified by Truelove Lowland and adjacent lowlands such as Skogn and Sparbo-Hardy are by no means common in the High Arctic. The floristic diversity found is characteristic of much lower arctic latitudes, and gives rise to the question of why such lush sites should be found in an area that is almost uniformly Polar Desert.

Methods

The mosaic of physiographic types, the effect of the steeply-slanted cliffs to south and east, the two west-trending valleys of the Gully and Truelove Rivers, the large number of lakes and ponds, and the proximity of the Arctic Ocean all contribute to affect the microclimate of the Lowland and to impart to it a mosaic of microclimates that are arranged both in time and space.

The realization that the microclimate of the Lowland was not nearly as uniform as first conceived, and that the biologists had a demand for data from a wide variety of sites and habitats, resulted in the number of weather stations being increased from five in 1970 to twelve in 1972 (Table 2, Fig. 1). Table 2 also indicates the period of time that each station operated. During the winter of 1972-73 and the following summer the number of stations was again reduced because of a shortage of manpower.

Limitations on the amount of instrumentation available, lack of personnel, and the enormous task of data reduction and synthesis dictated that environmental measurement should conform to the same pattern followed by almost all researchers on the project. This was to perform intensive studies on a limited number of sites and extensive studies on several others to give a measure of the variability to be expected from microhabitat to microhabitat and within the same general physiographic type from one part of the Lowland to another.

The list of instruments used at each station may be found in Table 3. The most intensively instrumented stations were those at the Beach Ridge (Figs. 2 and 3) and Meadow Intensive sites (Fig. 4). Moreover, the periodicity of measurements was very high and the degree of accuracy of measurement was greater. The Base Camp station was operated principally as a standard synoptic weather station (Fig. 5), to which was added vertical temperature and shortwave radiation measurements.

Measurements of shortwave incoming radiation and of the temperature profile were also made at the Cliff, Wolf Hill (Figs. 6 and 7) and Plateau stations (Fig. 8). The remaining stations each consisted solely of a precipitation gauge, a hygrothermograph, and an odometer-type 3-cup anemometer (Fig. 9).

Problems of instrumentation and measurement

In general the field of microclimatology has been developed to meet the demands of agriculture in temperate regions where the climate is not rigorous

Table 2. Micrometeorological stations used during the Devon Island Project and the time that each was in operation.

Site	*Summer 1970*	*Summer 1971*	*Summer 1972*	*Winter 1972-73*	*Summer 1973*
Base Camp	8 June-20 Sept.	31 May-20 Sept.	6 May-23 Aug.	24 Aug.-30 April	1 May-20 Aug.
Beach Ridge	22 June-30 Aug.	14 June-22 Aug.	6 May-23 Aug.	24 Aug.-5 May	6 May-23 Aug.
Hummocky Sedge-					
moss Meadow	6 July-30 Aug.	14 June-22 Aug.	6 May-23 Aug.	24 Aug.-5 May	6 May-23 Aug.
Rock Outcrop	30 June-30 Aug.	9 July-12 Aug.	20 July-7 Sept.		
Cliff	30 June-30 Aug.	14 June-22 Aug.	6 May-9 Oct.	10 Oct.-30 April	1 May-18 Aug.
Plateau		21 June-22 Aug.	6 May-7 Sept.		16 May-18 Aug.
Rocky Point		20 June-22 Aug.	6May-12 Oct.		15 June-18 Aug.
Wolf Hill		21 June-19 July	10 May-12 Oct.		
Wet Sedge-moss					
Meadow		20 July-8 Aug.	5 Aug.-18 Aug.		
Truelove Inlet			19 June-7 Sept.		
Skogn Exclosure		10 Aug.-22 Aug.	12 June-29 Aug.		
Truelove Valley			20 June-27 Sept.		

and line power is normally available. In the Arctic, however, the most simple microclimatic measurements become very difficult and sometimes impossible. At the start of the project the lack of sturdy, sophisticated equipment forced the authors to keep most instrumentation very simple. This, in turn, led to the use of clock-driven recorder-type instruments. Power-driven recorders and integrators were necessary, however, simply to increase the accuracy and intensity of measurement. This increase in the level of complexity meant that it became impossible to repair instruments on site. Serious instrument breakdown necessitated that the unit be sent south for repair. The minimum time in which one could reasonably expect to have such an instrument back in the field was six weeks, virtually the length of the snow-free period. The process of data gathering often became a gamble between high quality data or no data. A combination of long and sometimes rough handling of instruments in transportation, low temperatures, and high humidity tended to give the project a higher than normal rate of instrument breakdown that resulted in major gaps in data. Graphic recorders were also preferred, simply to avoid the problem of not being able to ascertain when a measuring sensor had failed until a copy of the uncoded data became available. To minimize disturbance of the ecosystem, generators were restricted to Base Camp and to periodic use at the intensive study sites for measurement of photosynthesis (Mayo et al. this volume). Other than at Base Camp then, all recorders and integrators were battery driven. The cool temperatures, even in summer, seriously affected both battery life and the efficiency of recorders having any sort of mechanical operation.

By using recorders the longest period that instrument failure could go undetected was a week, the period in which the stations were serviced. The problem with this approach was that data reduction was very time-consuming. An endeavor to code all data in the field was only partially successful because of many other demands on the researchers' time in keeping stations running efficiently and performing other associated duties.

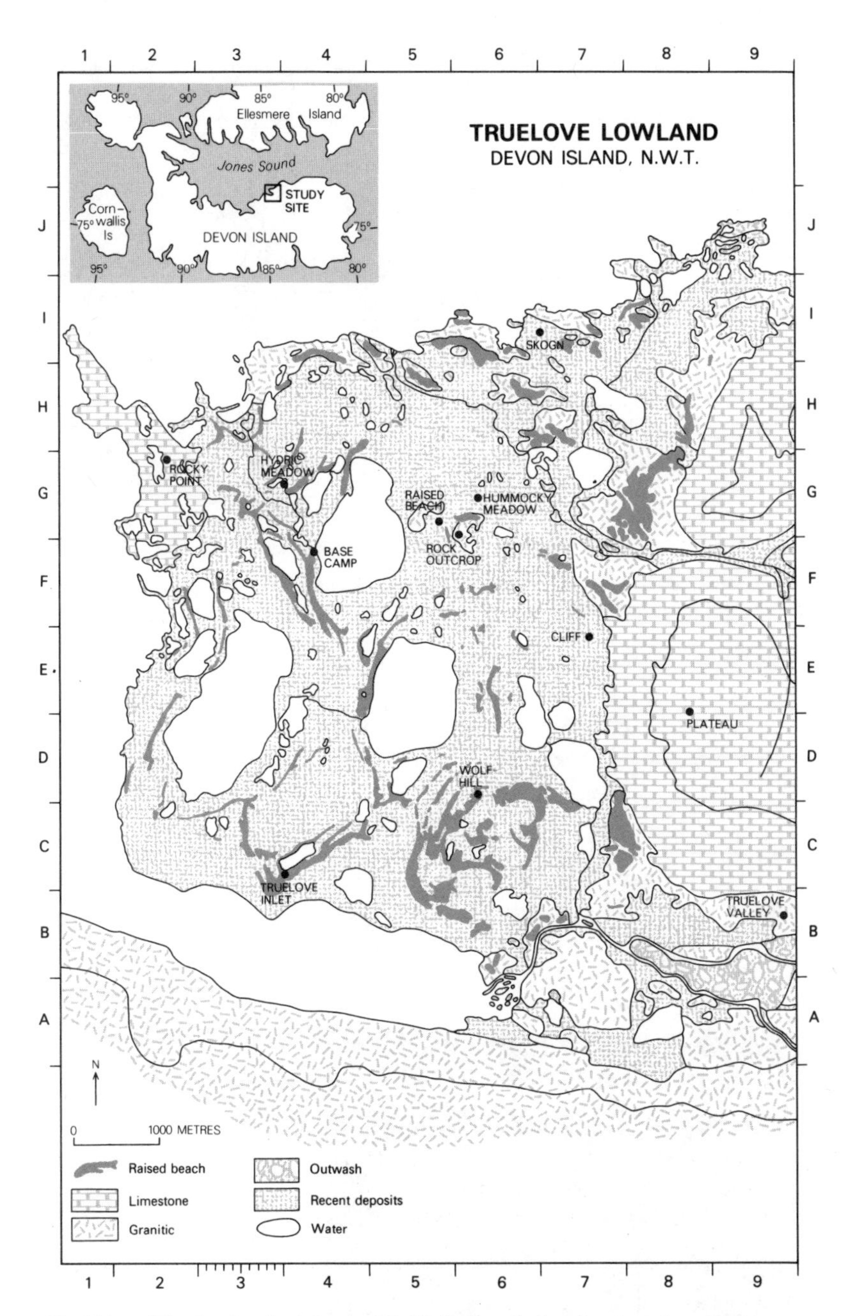

Fig. 1 Map of Truelove Lowland showing the distribution of microclimatic stations which were established between the years 1970 and 1973. All stations were in operation only in 1972.

Fig. 2 General view of the Raised Beach Ridge microclimatological site in the early spring (May), showing instrument mast, albedometer, and instrument shelter.

Fig. 3 Instrument tower at the Raised Beach Ridge microclimatological site to show deployment of sensors.

Fig. 4 Aerial view of Hummocky Sedge-moss Meadow microclimatological site.

Fig. 5 Anemometer tower (10 m) and Stevenson Screen, Base Camp.

Fig. 6 Wolf Hill extensive microclimatological site.

Fig. 7 Temperature profile mast showing self-aspirating shield and pyranograph in background, Wolf Hill extensive microclimatological site.

Fig. 8 Plateau extensive microclimatological site.

Fig. 9 Hydric meadow extensive microclimatological site.

Table 3. Distribution of instrumentation for the micrometeorological stations used during the Devon Island Project.

Parameter	*Base Camp*	*Beach Ridge*	*Hummocky Sedge-moss Meadow*	*Rock Outcrop*	*Cliff*	*Plateau*	*Rocky Point*	*Wolf Hill*	*Wet Sedge-moss Meadow*	*Truelove Inlet*	*Skogn Exclosure*	*Truelove Valley*
Radiation												
Pyranometer	X											
Net radiometer	X+	X	X	X*								
Albedometer	X+											
Pyranograph					X	X	X	X	X**		X**	
Temperature												
Grant Profile (20 probe)		X	X									
Grant Profile (14 probe)	X							X	X**		X**	
Grant Profile (7 probe)					X							
Lambrecht hygrothermograph		X	X									
Belfort hygrothermograph	X			X	X	X	X	X	X	X	X	X
Max. thermometer	X											
Min. thermometer	X											
Wind												
Standard anemometer + wind vane	X											
Sensitive anemometer		X	X									
Belfort recording anemometer		X+	X+			X						
Belfort anemometer				X	X	X	X	X	X	X	X	X
Precipitation												
Standard snow gauge	X											
Standard rain gauge	X											
Taylor rain gauge		X	X	X	X	X	X	X	X	X	X	X
Atmospheric Moisture												
Lambrecht hygrothermograph		X	X									
Belfort hygrothermograph	X			X	X	X	X	X	X	X	X	X
Soil Flux												
Soil flux plate		X	X									
Soil Moisture												
Gravimetric		X	X									

+Winter 1972-73.
*5 July-23 Aug., 1972 several locations.
**Two to three week period only.

Instrumentation

Radiation

Shortwave incoming radiation was measured with an Eppley Black and White Pyranometer Model 8-48 placed on top of the Stevenson Screen at the Base Camp station. The signal was either recorded continuously on a Leeds and Northrup Speedomax-W chart recorder or integrated over 15 min periods on a Lintronic Mark V volt-time integrator. Robitzsch bimetallic strip pyranographs (Belfort Instrument Co. Model 51850) were placed at the Cliff, Wolf Hill, and Plateau stations to give an idea of spatial variation across the network of stations.

Albedo was measured using a Kipp and Zonen Albedometer Model CM-4 over the beach ridge and meadow. Owing to a shortage of recorders only the lower sensor was used. The signal was measured either with a portable strip chart recorder Esterline Angus Model T-171-B or with a Lintronic integrator that integrated over 15 min periods.

Net radiation was measured only at the beach ridge and meadow. The sensor was a Funk-type net radiometer (Middleton and Co. CN-1). The signal was integrated over 15 min by a Lintronic integrator.

Temperature

Temperature was recorded in two ways, either in the form of a temperature profile from the atmosphere into the soil recorded hourly with a multipoint recorder, or at a height of 10 cm by the temperature sensing portion of a hygrothermograph. The hygrothermographs were placed in small screened shelters (Vogel and Johnson 1965). The multipoint recorders used varied in the number of channels available. All were Grant Model D recorders. At the Raised Beach Ridge and Hummocky Sedge-moss Meadow Intensive Study Sites 20 channels were recorded; the profile of probes being as follows: 200, 150, 100, 75, 50, 25, 15, 10, 5, 1, 0, −1, −2, −5, −10, −15, −25, −50, −100 cm. Fast response "B"-type thermistor probes were used above ground whereas slow response "C"-type probes were used in the soil. The thermistors were in stainless steel tubes 44 mm long and 3 mm in diameter. The probes at the surface and −1 cm were "E"-type probes about 1 mm in diameter and length. Owing to difficulties in measuring the actual surface temperature this smaller probe was buried *ca.* .1 cm below the surface.

At the Base Camp a profile was taken to 10 m and recorded on a 20-channel recorder. The spacing of the sensors was: 10, 7, 5, 2, 1, 0.5, 0.25, 0.10, 0, -0.02, -0.05, -0.10, -0.25, -0.50 m. All probes were of the "B" and "C" type.

Profiles were also taken from the Cliff and Wolf Hill stations using "B" and "C"-type probes. The spacing at the Cliff was: 100, 25, 10, 2, 0, −2, and −25 cm. At Wolf Hill the spacing was: 100, 50, 25, 10, 5, 2, 0, −1, −2, −5, −10, −25, and −50 cm.

All sensors above the soil surface were shielded from direct sunlight. In 1970 simple tubular shields made from aluminum foil were used, but were found to be inadequately ventilated owing to the low mean wind speeds experienced in the area. From 1971 on, a self-aspirating shield (Fig. 7) that worked on the Venturi principle (Courtin, in preparation) was used with greater success. Tests made on this shield indicate that temperatures were no more than 2°C above the actual temperature under all but the worst conditions. These were a combination of calm winds, cloudless sky in the middle of the day, and an even cover of snow on the ground beneath the shield. These conditions were rarely experienced but gave errors as great as 3°C, when compared to a very small unshielded thermocouple (0.025 mm diam).

Wind

Three types of 3-cup anemometers were used to measure wind. At Base Camp a standard Type 45B A.E.S. weather office anemometer and windvane were mounted at 10 m and the signal led to an anemograph that recorded each mile of wind run. Wind direction, divided into the eight principal compass points, was recorded simultaneously. At the beach ridge and meadow, Casella and Co. Sheppard-type sensitive anemometers were mounted with the cup wheel at 35 cm. These anemometers have a very low starting speed (*ca.* 0.1 m sec^{-1}) but will record winds as high as 15 m sec^{-1}. Each anemometer was connected to a

Sodeco PL 103 Counter/Printer that integrated continuously but was made to print every 10 min by means of a Geodyne 190 A time clock. The number of counts integrated was related to actual wind speed via a separate calibration curve for each instrument.

All other stations were equipped with Belfort Model 5-349 odometer-type anemometers that were read at irregular intervals by any researcher in the vicinity, but never less than once a week when the stations were serviced. The exception was the Plateau station where the same model anemometer was connected to a Sodeco counter/printer that recorded each mile of wind run, over 10 min time periods.

Precipitation

At the Base Camp weather station two standard weather office gauges were used. The first was a snow gauge 12.7 cm in diameter mounted in a Nipher shield, and the second was a 9 cm diameter rain gauge that was unshielded. At all other stations a circular Taylor Clear-Vu gauge (unshielded) with a diameter of 10.0 cm was used. In the latter a small quantity of mineral oil was placed to prevent evaporation of precipitation in between visits to the station. The Base Camp gauges were examined at least every 12 hr.

Atmospheric humidity

Hair element-type hygrothermographs were used at all stations. The instrument chosen was a Belfort Model 5-594 with the exception of the beach ridge and meadow where Lambrecht Model 252 C instruments were used both because they were available and because they are slightly more accurate. All instruments had weekly charts with the exception of that in the Truelove Valley which had a monthly chart. The Base Camp instrument was placed in the standard Stevenson Screen whereas the remainder were placed in similar but smaller screened shelters (Vogel and Johnson 1965) at ground level so as to better sample the atmospheric humidity in the zone where much of the tundra biota exists. Because the hair element is mounted vertically it sampled a zone approximately 10 to 20 cm above the ground. At Base Camp relative humidity and dew point were measured and recorded using a sling psychrometer wetted with distilled water. The humidity and temperature charts were changed each week and calibrated using either a sling psychrometer or an Atkins aspirating psychrometer Model 3F01-F46.

Winter data

During the winter of 1972-73 microclimatological measurements were continued. Owing to the small size of the overwintering team, certain changes were made in instrumentation and the number of stations that were kept in operation (Table 3).

In summary, these were the principal changes that were made. Four stations were kept in operation: Base Camp, Beach Ridge, Sedge-moss Meadow, and Cliff; all radiation sensors were taken to the Base Camp beach ridge on the premise that since the ground was covered with snow it made little difference upon which beach ridge the sensors were located and that it was easier to service them if they were close to camp; the Casella anemometers were removed and replaced with Belfort anemometer/Sodeco counter-printer systems; the cups were placed at 2 m to minimize interference from blowing snow; the moisture pen was removed from the Cliff hygrothermograph because it interfered with the operation of the temperature pen when temperatures dropped below −15°C; the Grant multipoint recorder was disconnected since it would not operate at temperatures much below freezing; and precipitation gauges other than those at Base Camp were removed since measurement of snow in the Arctic is always of dubious accuracy because of blowing snow and requires the use of special shielding. The Nipher shield used at Base Camp is presently being tested exhaustively and has been found to be more accurate than any other shield tested (Goodison 1975).

Results and Discussion

Radiation

Fig. 10 summarizes the incoming shortwave radiation data as five-day means collected over four summer field seasons and one winter. Continuous data exist from early May 1972 to mid-August 1973. The summer field season data indicate the large fluctuations that occur from one season to another and throughout any given season. Even though there is considerable seasonal scatter, it is evident from the mean line drawn through the data that the shortwave incoming radiation is strongly skewed. The reason is attributed to increase in cyclonic activity with reduced zonal circulation, increase in cloud

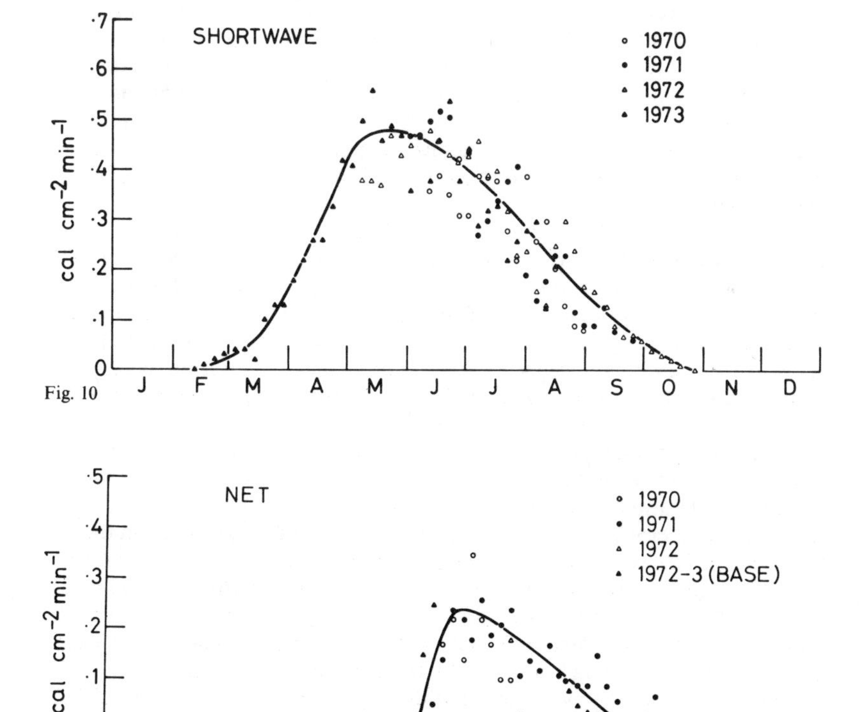

Figs. 10 and 11 Total shortwave incoming (10) and net allwave radiation (11) (Raised Beach Ridge and Base Camp) for the years 1970 to 1973. The solid line represents a "mean" curve through the available data, Truelove Lowland.

Table 4: Dates of ice break-up (sea); ice melt and refreeze (lakes) for the 1970-1973 field seasons.

	Date of break-up or ice melt		*Date of refreeze*		*No. of days ice-free*
Sea Ice	28 August	1970			
	27 July	1971			
	15 August	1972			
	20 July	1973			
Immerk Lake	30 August	1970	2 September	1970	2
	18 July	1971	30 August	1971	42
	50% Thaw	1972		1972	0
	20 July	1973	12 August	1973	21

that accompanies snowmelt over the land, and breakup of the sea ice. Examples of marked reductions in radiation that were caused by frontal storms were late June and late August 1970, and early August 1971.

Several shorter term trends are also of note. Researchers arriving early in the field season have often reported a period of clear weather followed by deterioration in late May and early June. This trend is seen for 1973 when a period of calm and clear weather was experienced. At this time Resolute Bay, 350 km to the west southwest, was experiencing severe storm conditions with high winds and blowing snow that reduced visibility at times to zero.

We suggest that this period of fine weather is occasioned by the same phenomenon that brings fair, mild weather to the subarctic in April and May. Bryson (1966) attributed this to a broad diffluent stream of arctic air during that period. Whilst this stream reaches its greatest extent in subarctic latitudes, its source is the High Arctic and the diffluent flow is present already over Devon Island.

There appears to be a reduction in radiation in all four years that coincides with the main period of snowmelt. This may be seen in 1971 and 1973 when snowmelt occurred at the end of June and in 1970 and 1972 when the snow did not melt until early July (Table 4). A similar reduction is seen in early August and may result from the break-up of the sea ice and thaw of the larger lakes. A recovery in late August is thought to be due to a return to cooler temperatures and therefore less humidity. There then follows a steady decline into late October and the end of the "solar" year.

Fig. 11 is a presentation of beach ridge net radiation but the record is not as complete as for shortwave incoming radiation owing to instrument failure. Continuous data were collected from late August 1972 to mid-June 1973 over the Base Camp beach ridge. The summer field season data are all from the Beach Ridge Intensive Study Site. Seasonal fluctuations during the "summer" months do not closely correspond to those of incoming shortwave as might be expected. The reason is thought to be the long wave component of incoming radiation that is propagated in the lower 100 m of the atmosphere (Vowinckel and Orvig 1965) and which is sensed by the polyethylene shielded net radiometer, but not the pyranometer with its glass shield. This matter is fully discussed by Addison (this volume).

The seasonal curve for net radiation is skewed, like that of shortwave incoming radiation. In this case, however, the reason is not only an increase in cloudiness with the onset of snowmelt but the marked and rapid changes in albedo that occur with exposure of the dark colored ground.

Differences in net radiation between two of the major physiographic types in the Lowland, the Beach Ridge and the Hummocky Sedge-moss Meadow,

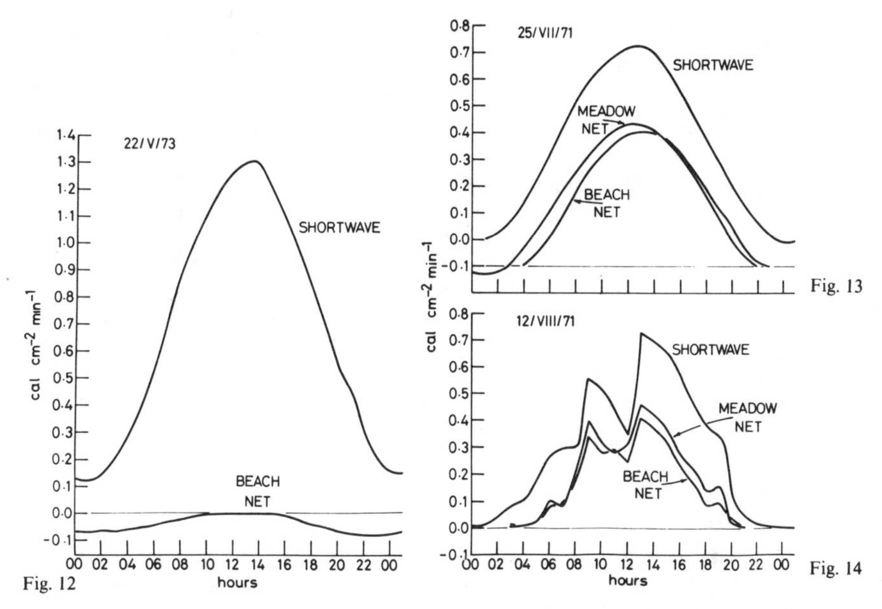

Figs. 12, 13, and 14 A comparison of total incoming (Base Camp) and net radiation (Raised Beach Ridge and Hummocky Sedge-moss Meadow) for three selected days, Truelove Lowland.

can be calculated only for a two-month period in 1971. The mean value for the beach ridge was 0.156 ly min^{-1} whereas that for the meadow was 0.136 ly min^{-1}, a reduction of 13%. One might have expected a higher net radiation from the meadow when considering the light color of the beach ridge with the almost complete vegetation cover of the meadow. It is assumed that the site, being saturated to the surface, or even inundated, forms a highly reflective surface, thus accounting for the lower radiation values (Geiger 1965).

To indicate the variation in diurnal fluctuation in both shortwave incoming and net radiation, data are presented from early, middle, and late in the period when the sun is above the horizon (Figs. 12, 13, and 14). The 22 May, 1973 was perfectly cloudless (Fig. 12) and demonstrates the very high level of incoming radiation; a level that is only equalled in certain alpine regions where the levels are elevated owing to the reduction in both the density of the atmosphere and the amount of pollutants. Mooney et al. (1965) reported midday values in early August as high as 1.76 ly min^{-1} for a site at 2,774 m in the White Mountains of California. In contrast the fluctuation in net radiation is very small and furthermore remains very slightly negative even at solar noon.

Fig. 13 is for 25 July, 1971, and represents a clear day in the middle of the plant growing season. Not only is the curve for shortwave incoming radiation given, but also that for both beach ridge and meadow net radiation. Although "night time" values of incoming radiation are similar to those shown for May, the peak for the day is much lower, .825 ly min^{-1} as against 1.3 ly min^{-1}. Net radiation, for both the beach ridge and meadow, however, are much higher, about 60% of incoming at the peak of the day and only become slightly negative for 2 hr each side of midnight (not shown). The minimum value at midnight is −0.5 ly min^{-1}. The net radiation curves for the meadow and beach ridge are very similar for that particular day but are out of phase by about 1 hr. This is not due to an error in timing but no ready explanation can be offered.

By 12 August (Fig. 14), clear days are very rare because of the abundance

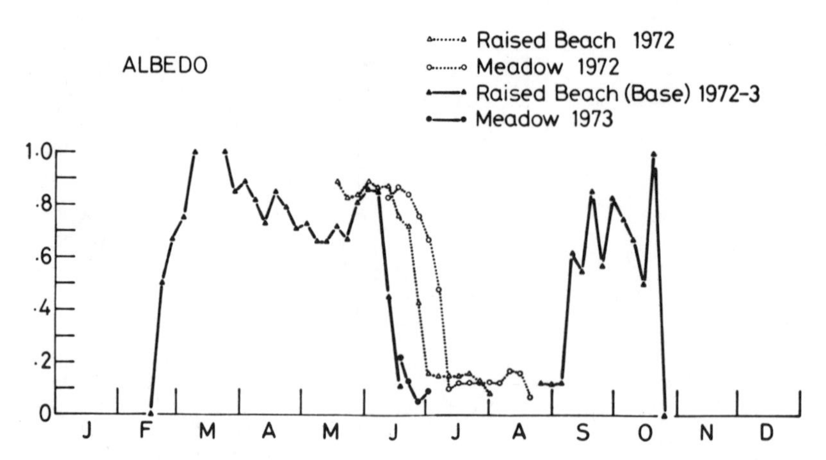

Fig. 15 Reflected shortwave radiation (albedo) for two surfaces (Raised Beach Ridge and Hummocky Sedge-moss Meadow) for the years 1972-73.

of open water that brings about a corresponding increase in the amount of cloud. The trends, however, remain the same. Net radiation values are about 60% those of incoming shortwave radiation. Meadow values are again slightly higher than those of the beach ridge and in phase, peak values are lower than three weeks previously by about $-.1$ ly min^{-1} and the period of negative "night time" values increased from 3 to 5 hr.

The change in albedo with time gives a very clear indication of the solar regime for the Lowland and the onset of the snow-free season. Fig. 15 covers the winter period from late August 1972 to mid-June 1973 over the Base Camp beach ridge and the 1972 field seasons for beach ridge and meadow. A short period of meadow data for June 1973 also is shown.

Albedo decreases dramatically with the loss of the sun in late October and rapidly increases again with the return of the sun in mid-February. An albedo of 1.0 indicates that all shortwave radiation received is also being reflected. Published values of albedo for snow-covered surfaces are usually less than 1.0. Geiger (1965) cited albedos of 0.75 to 0.95 and 0.40 to 0.70 for new and old snow respectively. It is possible that the measured value of 1.0 could be due to the very low sun angle experienced at high latitudes because albedo usually increases with decreasing solar altitude (Wilson 1967).

The fluctuating but generally decreasing pattern experienced in April and May reflects new falls of snow and periods when snow was being reworked by the wind and darkened by the accretion of wind-blown dust and organic debris that tends to be removed from exposed ridge crests. A period of fresh snow in late May and early June once again increased albedo before snowmelt reduced it to summer values of less than 0.2. The short period of sedge meadow data for 1973 shows that snowmelt was only about five days behind that at the Base Camp beach ridge. It should be noted, however, that at the Base Camp, the albedometer was a little below the ridge crest where snow tends to be deeper in winter and to take longer to melt in the spring. The curves for 1972 show that snowmelt was much later (the latest of the four study years) and that the beach ridge became snow-free about 10 days ahead of the meadow.

Snowmelt and the associated decrease in albedo normally occurred over a period of 10 to 14 days. In 1970 however most of the snow melted in an 8 hr period of Föhn winds. This had a dramatic effect on the biota, especially lemmings, as discussed by Fuller et al. (this volume).

The snow-free values of albedo for both gravels (beach ridge) and sedge vegetation (meadow) compare favorably with values of 0.3 to 0.6 for sandy soil, and 0.12 to 0.30 for meadow and fields given by Geiger (1965). Independent measurements of the same sites (Addison this volume) give values of 0.14 for dry, beach ridge sites and 0.11 for wet, meadow sites.

The progressive drop to values of 0.08 for the beach ridge and 0.07 for the meadow are changes brought about by moisture changes at the sites. Summer values of 0.15 to 0.24 were measured by Ahrnsbrak (1968) over various lichen covered surfaces and Weller and Cubley (1972) found that the albedo of a wet meadow increased from .15 to 0.20 when the sites dried out. The 1972 values for September and October rise and fall with each snowfall and the melt that followed it.

The mean curve drawn through the summer beach ridge and the winter Base Camp data for both shortwave incoming and net radiation (Figs. 10 and 11) not only indicates the seasonal relationship between the two but also permits calculation of the annual energy regime. Shortwave incoming radiation increases rapidly from the time that the sun reappears in early February to a peak in late May. Again, the reason that the peak does not occur at the solstice is that incidence of cloud increases with the amount of ablation and the consequent increase in atmospheric humidity. Net radiation, on the other hand, remains slightly negative until the ground becomes snow-free (in this case the beach ridge crest) after which it increases rapidly to a peak in early July, and then declines gradually, paralleling the curve for shortwave incoming radiation. Net values become negative again in early October, whereas shortwave values do not drop to zero before the end of the month.

Integration of the area under each curve gives the mean annual radiation received. The value for shortwave was 87.3 kly yr^{-1} whereas that for net radiation was 17.8 kly yr^{-1}. These measured values are in marked contrast to calculated values for Resolute Bay of 76.9 kly yr^{-1} for shortwave radiation and 2.5 kly yr^{-1} for net radiation (Hare and Hay 1974). Furthermore, the same authors present annual net radiation isopleths which show that values in the Low Arctic may vary from 11 kly yr^{-1} (Barrow, Alaska) to 25 kly yr^{-1} (Hudson Bay Lowland) and only reaches zero at the latitude of northern Ellesmere Island. That is to say, a positive radiation balance exists over virtually the entire terrestrial arctic. By interpolation, the net radiation balance for Devon Island should be about 3 kly rather than 17.8 kly as presented above. This value is of a magnitude cited by Hare and Hay for the mid- to low-arctic. Even supposing that the measured value is not absolute because of measurement error and the short span of measurement, it must be assumed that the magnitude is correct. This elevated radiation regime, together with plentiful water, helps to account for the lush flora of Truelove Lowland that is more commonly found at a much lower latitude. A possible reason for this anomaly will be discussed later.

Temperature

Whilst the pattern of heating over the Lowland is largely a function of solar radiation, the seasonal changes are also affected by the proximity of the Lowland to the Arctic Ocean and to the surrounding topography. A series of isotherm maps designed to indicate these changes is shown in Fig. 16. The 5-day mean data for each isotherm map were selected to span a summer season from the period prior to spring thaw through to the following freeze-up. The 1972 data were chosen because the network of stations was most complete.

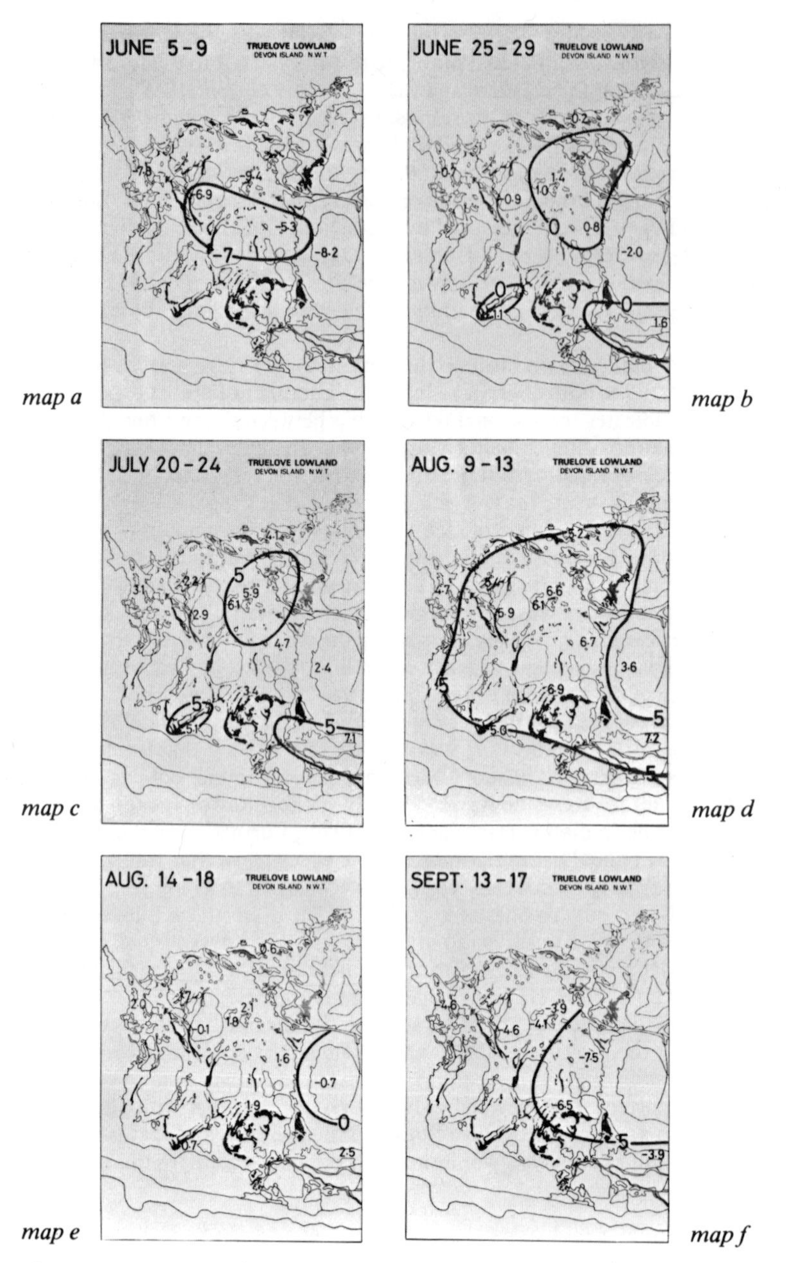

Fig. 16 Isotherm maps showing temperature distribution (°C) for Truelove Lowland, for selected 5 day mean periods, 1972.

Map *a* shows that in early June temperatures in the Lowland were well below freezing with the centre of the Lowland being somewhat warmer. The highest temperatures were found against the west-facing limestone cliffs. The exposed faces of these cliffs are largely blown free of snow and furthermore are more or less normally exposed to the rays of the afternoon sun. The effect of their heating is felt along the base of the cliffs.

By the end of June (map *b*), although the Lowland was still largely snow-covered, temperatures were close to 0°C. Three areas were above freezing, the mouth of the Truelove Valley and two "islands" within the Lowland. The smaller island is coincident with a large beach ridge that is mostly snow-free and the other probably owes its existence to the heating of the cliffs mentioned previously and also perhaps to a tongue of warm air issuing from the valley of the Gully River in the same way as that from the Truelove Valley. Unfortunately, the sparse network of stations did not permit this possibility to be verified. It is thought that the heating of the valley walls (canyon effect) results in localized heating of the air within the valley. Most of this warmed air must, of necessity, rise; but as the phenomenon intensifies, the mass of air pushes out in all directions, thus forming a tongue of warmer air at the mouth of the valley.

The pattern of heating was maintained throughout July (map *c*). The warmest portions of the Lowland remained as for late June but the entire Lowland was snow-free and above freezing.

The overall Lowland reached its highest temperatures in mid-August (map *d*) just before the return of freezing temperatures. There is, therefore, a considerable temperature lag behind the summer solstice which occurs before snowmelt. The coolest areas at this time were the limestone upland and coastal areas. Whilst the same relative pattern was maintained, the change that occurred within the next five days was dramatic (map *e*). The plateau was below 0°C and coastal areas, with the exception of Rocky Point, were close to freezing. The Truelove Valley was again the warmest area.

A month later (map *f*) the entire area was well below freezing with those stations furthest from the coast being the coldest. This probably resulted from a moderating effect of bodies of water such as the ocean and the larger lakes, reradiation, and possibly katabatic drainage from the Devon Icecap *ca*. 20 km to the south and east (Holmgren 1971).

The year 1972 was used for the isotherm maps because those data were the most complete available. It was, however, the coldest of the four years studied. Ten-day means for 1971 to 1973 from a single site (the beach ridge) are given in Table 5. Not only was 1972 cold but it was also the shortest frost-free season (69 days). In contrast, 1973 was the longest frost-free season (99 days) and also the warmest; 1971 and 1972 may be considered average in the context of data available. The length of the frost-free season is not entirely dependent upon latitude, for whereas a more temperate arctic site such as Abisko, Sweden (lat. 68° 22' N) has a longer growing season, *ca*. 145 days (Skartveit et al. 1975); the growing season at Barrow, Alaska (lat. 71° 15' N) is only *ca*. 40 days (Barry, pers. comm.).

The pattern of temperature for a given site changes greatly from year to year. Although one finds the normal progression of temperatures from cold to warm and back to cold within a season, marked changes do occur such as the cold period in mid-August 1973. The warmest 10-day period in 1970 at Base Camp was in early July whereas it occurred a month later in 1973.

The two intensive study sites, the beach ridge and the hummocky sedge meadow, whilst being spatially close, differ greatly in their exposure, amount of snowcover, soil type, vegetation, and drainage. Figs. 17 and 18 show the

Table 5. Comparison of 10-day mean temperature (°C) at the Beach Ridge (height, 150 cm) for June through mid-September, 1970-1973.

Year	*31 May- 9 June*	*10 June- 19 June*	*20 June- 29 June*	*30 June- 9 July*	*10 July- 19 July*	*20 July- 29 July*	*30 July- 8 Aug.*	*9 Aug.- 18 Aug.*	*19 Aug.- 28 Aug.*
1970	—	—	—	6.4	5.5	4.1	5.0	3.5	2.5
1971	—	—	6.9	5.8	6.9	6.1	4.6	0.9	3.8
1972	-5.5	-2.9	1.3	2.4	4.4	4.0	3.6	3.4	0.4
1973	-1.6	-0.1	4.0	9.0	10.5	9.3	13.6	10.6	—

Year	*29 Aug.- 7 Sept.*	*8 Sept.- 17 Sept.*
1970	—	—
1971	-1.6	-2.4
1972	-2.6	-2.8
1973	—	—

temperature-time relationship that exists at three heights for these two sites over the course of the year 1972-73. Several characteristic differences exist between the temperature regimes that reflect the physiographic and vegetational differences. Mean monthly maximum temperatures at the beach ridge site reached 8°C, whereas those at the meadow only reached 2°C. Owing to the exposure at the beach ridge there was little snowcover and temperatures rose above freezing in late June but only 10 days later at the meadow where snow accumulates to 30 cm or more. Minimum temperatures were lower at the beach ridge than at the meadow by 7°C. In summary, the beach ridge has a greater yearly amplitude of temperature than the meadow because of increased exposure.

The wet, peat soil (−50 cm) of the meadow not only shows the above characteristic of lesser amplitude but also a soil temperature curve that lags air and surface temperature by almost two months, reaching a maximum of −2°C in September. This is in contrast to a mire with permafrost in northern Sweden with milder winter temperatures; where the active layer was more than twice as deep, surface refreeze occurred in late September, and refreeze at −50 cm did not take place until December (Skartveit et al. 1975). There was no such lag at the beach ridge and, furthermore, mean monthly temperatures were above freezing in July and August at this depth. The cold wet soil also affected surface temperatures so that the highest temperatures in summer were air temperatures, whereas at the dry, well-drained, gravelly beach ridge, maximum temperatures were realized at the surface. In winter, the deep meadow snow insulated the surface, and air temperatures were lowest. At the beach ridge, where snow was almost absent, reradiative processes were more extreme and the coldest temperatures were experienced at the surface.

The tautochrones presented in Figs. 19 to 22 attempt to clarify further the relationship of temperatures from air to soil. Each point on a tautochrone is a five-day mean. The tautochrones in Figs. 19 and 20 are for the beginning, middle, and the end of the growing season and for mid-winter for the beach ridge and the hummocky sedge-moss meadow. The data of Figs. 21 and 22 are less complete and are presented to give perspective on another beach ridge and meadow.

The mid-winter profile for the beach ridge (Fig. 19) shows soil

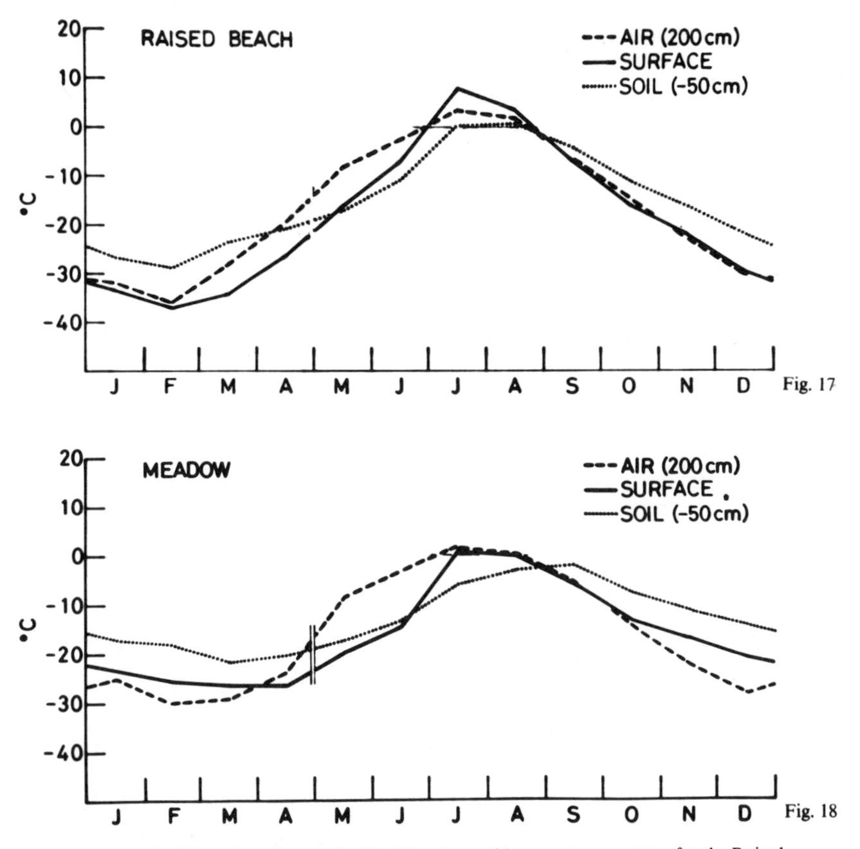

Figs. 17 and 18 Air (200 cm), surface, and soil (−50 cm) monthly mean temperatures for the Raised Beach Ridge and Hummocky Sedge-moss Meadow, May 1972 through April 1973, Truelove Lowland.

temperatures that were only slightly higher than air temperatures. There was a very sharp inversion at the surface that should have had a marked, if less severe, effect on the sensors immediately above the surface. This did not occur and the only explanation that can be offered is that since the probes were not immediately above each other to avoid interference, snow may have built up around these probes thus insulating them. The insulating properties of snow are clearly seen for the same mid-winter tautochrone at the meadow where temperatures gradually increased from the air (−31°C) to the soil (−19°C) through *ca.* 30 cm of snow. The same phenomenon was found in late June, but at this time the air was warmer than the soil. The lower temperature found immediately at the surface may have been associated with *pukak* (Pruitt 1970) that forms during the period of refreeze in the autumn. The late-June tautochrone for the beach ridge was affected by *ca.* 5 cm of snow and the elevated temperatures above the snow are thought to have been caused by advection of warm air from adjacent snow-free areas.

The tautochrone for mid-July shows soil temperatures that were lower than the air with a maximum temperature at the surface for both the beach ridge and the meadow. This surface heating was more marked at the beach ridge than at the meadow. The active layer at the meadow had only reached 8 cm whereas at the beach ridge the active layer was 60 cm.

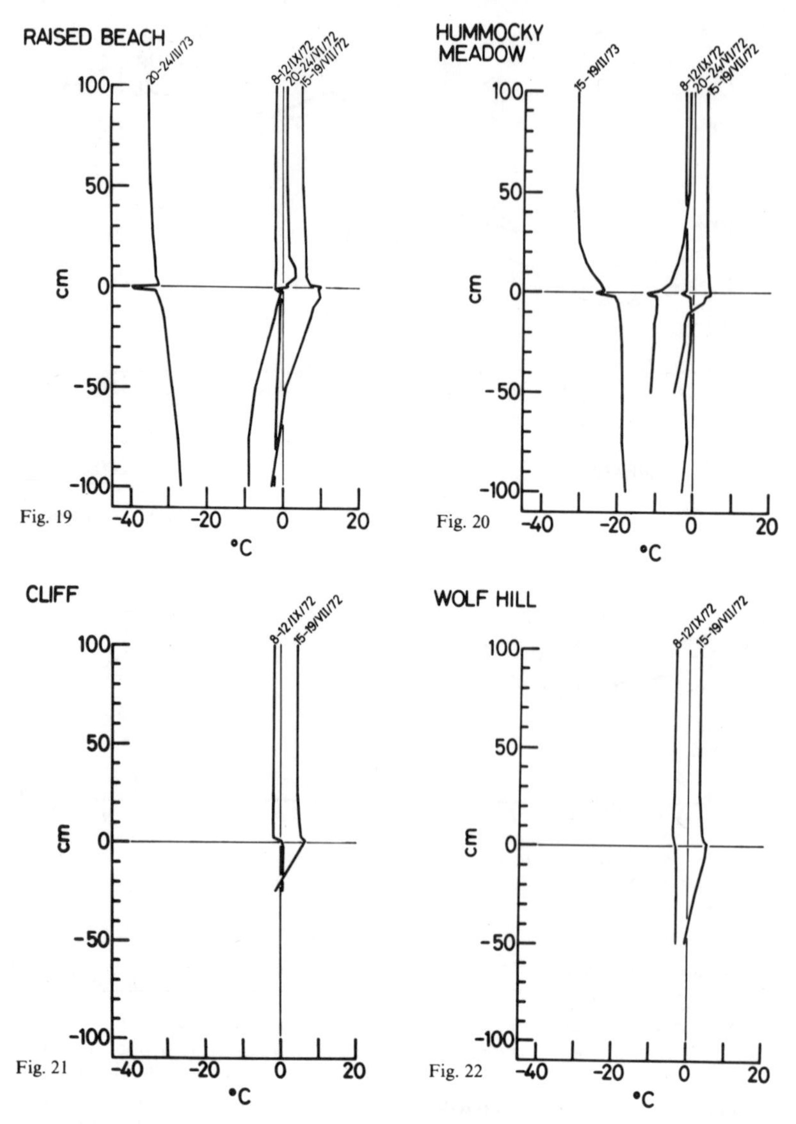

Figs. 19, 20, 21 and 22. Temperature profiles at four microclimatological stations on Truelove Lowland for selected 5 day mean periods during the years 1972 and 1973.

The period of refreeze that usually begins in early September is marked by isothermal profiles. These are almost identical (−2°C) for both sites.

The mid-July profile for Wolf Hill, a complex of beach ridges, was very similar to that at the beach ridge site. The air temperature at Wolf Hill was 2°C lower than at the beach ridge and the soil was not quite as deeply thawed (40 cm versus 60 cm). The greater elevation in surface temperature at the beach ridge was probably caused by the use of a smaller probe that more closely measured the actual surface temperature. The tautochrone for September was again almost isothermal and showed a slight inversion from 5 to 100 cm. The two profiles at the cliff meadow site closely parallel those at the hummocky

Table 6. Maximum depth of soil active layer (cm) at five stations at Truelove Lowland, 1970-1973.

Station	*1970*	*1971*	*1972*	*1973*
Base Camp	—	—	50	>50
Beach Ridge	75	<90	80	>100
Hummocky Sedge-moss Meadow	<50	25	12	50
Cliff	—	—	<40	>25
Wolf Hill	—	—	>50	—

sedge-moss meadow. The cliff soil active layer in July was deeper (20 cm) than that at the hummocky sedge meadow (8 cm), possibly reflecting the history of warmer early season temperatures (Fig. 7). In September, the cliff soil was still thawed, in contrast to that at the meadow, perhaps reflecting the deeper active layer developed during the summer.

It has been seen that soil temperature and depth of the active layer vary greatly from site to site. Table 6 serves to compare depth of the active layer at all sites where data were taken for the four years of the project. How much thaw takes place is clearly a function of radiation and temperature since the greatest thawing took place in 1973, the warmest year, and the least took place in 1972, the coolest year. Again, the well-drained gravelly sites thawed to a much greater depth than did the wet peaty soils with their greater insulation and greater thermal inertia. Base Camp data were taken off the beach ridge crest where there is more vegetation cover and more organic matter in the soil. Although still a mineral soil, the thaw was reduced almost by half that of the beach ridge site where measurements were made on the crest of the ridge.

Precipitation

Precipitation in the Lowland is very variable both from month to month and from year to year. Total precipitation data for the Base Camp station are presented in Table 7. Although the data are incomplete, some inferences may be drawn. Precipitation in the Arctic tends to peak in late summer with a smaller peak in early spring following the very dry mid-winter (Thompson 1967). Weather station data for the High Arctic indicates that the maximum rainfall occurs either in July or August (Rae 1951) but greater variation was

Table 7. Monthly water equivalent precipitation (mm) for Base Camp station, Truelove Lowland for years 1970-1973. Bracketed figures indicate the amount of precipitation that fell as rain.

Year	*Jan.*	*Feb.*	*Mar.*	*Apr.*	*May*	*June*	*July*	*Aug.*	*Sept.*	*Oct.*	*Nov.*	*Dec.*
1970						25.8	9.2	16.0	18.4*			
						(3.3)	(6.3)	(5.1)	(1.6)			
1971						00.0	14.2	34.0	12.7	3.0		
							(14.2)	(31.2)	(1.0)			
1972					1.14	7.61	4.4	9.5	3.5	18.3	25.9	10.1
						(0.3)	(4.4)	(9.5)				
1973	6.4	6.6	17.8	5.4	0.0	12.2	39.9	22.1				
						(12.2)	(39.9)	(21.0)				

*Data for 3 weeks.

experienced at the study site. Maximum precipitation, in successive years, was received in June 1970, August 1971, November 1972, and July 1973. There was evidence of a spring peak in March 1973. Practically all the July and August precipitation was received as rain, but June was much more variable in this respect. Precipitation in May was very light as is true for other parts of the Arctic Archipelago (Rae 1951, Thompson 1967).

The total precipitation from May 1972 to April 1973 was 116 mm and may give an indication of the minimum amount of annual precipitation that can be expected. Most of the precipitation fell as snow (102 mm) and the remaining 14 mm as rain. The total for Barrow, Alaska for the same period was 137 mm (Barry, pers. comm.) whereas the mean precipitation for the years 1971-1973 at Abisko, Sweden was 363 mm (Rydén, pers. comm.).

Devon Island precipitation is low, but we may well have underestimated it, because "trace" precipitation is common and does not form part of the measured total (Rydén, this volume). There were also difficulties in estimating winter snow water equivalent (Rydén this volume). The biota also benefits directly from condensation due to fog, and indirectly from moisture advected from the open sea which condenses as rime when it moves over a rapidly cooling land surface in the autumn. Rydén (this volume) has estimated that annual precipitation averages 185 mm on the Lowland.

The low precipitation of 1972 and the high precipitation of 1973 at Truelove Lowland can be coupled with years that were correspondingly cool and warm, whereas 1970 and 1971 were intermediate as regards both precipitation and temperature. It is presumed that this is simply a function of the lower amount of water held in cool air versus warm air, thus making less moisture available to be precipitated. In 1972 the melt was also very late because of the cool conditions, and the peak precipitation was not experienced until November. Whether these two phenomena are related is not known. By comparison, the peak at Barrow, Alaska was also later than normal, occurring in September and October (Barry, pers. comm.).

Table 8 places the summer precipitation regime of Truelove Lowland into the context of other high arctic sites. The Lowland lies intermediate between Arctic Bay and Alert, both as regards latitude and precipitation. Eureka has much lower, and Resolute Bay higher, precipitation than the other sites. While

Table 8. Summer precipitation (mm) at Truelove Lowland (4 year mean) and at adjacent High Arctic weather stations (long-term means).

Station	*Latitude*	*Longitude*	*June*	*July*	*August*
Truelove	75° 33'	84° 40'	11.4	16.9	20.4
Alert	82° 30'	62° 20'	13.5	18.0	27.4
Arctic Bay	73° 00'	85° 18'	7.6	19.3	24.1
Eureka	80° 00'	85° 56'	3.8	12.9	9.1
Resolute Bay	74° 41'	94° 54'	12.4	26.4	30.5

Table 9. Total precipitation (mm) for meteorological stations on Truelove Lowland during the field seasons, 1970-1973.

Year	*Period*	*Base Camp*	*Beach Ridge*	*Hummocky Sedge-moss Meadow*	*Rock Outcrop*	*Cliff*	*Plateau*	*Rocky Point*	*Wolf Hill*
1970	30 June-13 Aug.	6.7	13.8	7.0	17.1	17.6			
1971	5 July-23 Aug.	39.5	39.4	41.6		42.1	40.7		
1972	5 June-18 Aug.	22.7	21.8	31.6		20.1	34.4	20.9	27.0
1973	1 May-13 Aug.	87.7	86.7	100.3		90.2	87.2	76.4	

the period from which the Truelove means are drawn is short, the July and August maxima experienced at the other stations is also borne out. Although the network of stations at Truelove Lowland lies within a circle 6 km in diameter, there is considerable variation from station to station. This may be due in part to the use of unshielded precipitation gauges, with the exception of the Base Camp snow gauge. This may account for the higher precipitation at the hummocky sedge meadow in three of the four years (Table 9).

Rocky Point, the most westerly station, situated on the coast, tends to have lower values than other stations and suggests that the land itself triggers precipitation from moisture that is carried in on the prevailing westerly wind (Table 9).

Atmospheric humidity

Table 10 lists monthly mean data for all stations operated in the Lowland in 1972. The data are given simply for completeness but little can be inferred with respect to changes that occur either from station to station or from month to month. The atmosphere above the Lowland is uniformly moist owing to the abundance of water surfaces in the form of wet meadows, ponds, lakes, and the Arctic Ocean that lies to the westward, in the direction of the prevailing wind. Even in May, when there is no open water, moisture enters the atmosphere through sublimation of snow by radiation that is already very high (Fig. 12).

On a year to year basis (Table 11) no strong pattern of change in atmospheric humidity emerges with the exception of 1972, where the rather lower values (10% to 20%) may be linked with cooler conditions (Table 5) and lower precipitation (Table 7).

Table 10. Relative humidity (%) from May through September for all Truelove Lowland stations, 1972.

Station	*May*	*June*	*July*	*August*	*September*
Base Camp	83	82	84	88	79
Beach Ridge	81	83	81	73	—
Hummocky Sedge-moss Meadow	87	84	81	83	—
Rock Outcrop	—	—	—	84	—
Cliff	92	90	86	84	85
Plateau	98	97	81	86	—
Rocky Point	100	99	94	90	81
Wolf Hill	98	97	81	88	94
Hydric Moss Meadow	—	—	79	87	—
Truelove Inlet	—	—	80	87	—
Skogn	—	67	61	78	—
Truelove Valley	—	—	75	77	76

Wind

As a general rule, wind speeds in the High Arctic are no greater or less than those in temperate latitudes (Rae 1951). Peak monthly mean values in Truelove Lowland (Table 12) are experienced during the summer months and are in accord with data from other stations in the Canadian Archipelago (Thompson 1967).

The lower wind speed values are shown for the beach ridge and hummocky sedge-moss meadow versus those at Base Camp may be attributed

Table 11. Comparison of 10-day mean relative humidity (%) at the Beach Ridge, June through mid-September, 1970-1973.

Year	*31 May- 9 June*	*10 June- 19 June*	*20 June- 29 June*	*30 June- 9 July*	*10 July- 19 July*	*20 July- 29 July*	*30 July- 8 Aug.*	*9 Aug.- 18 Aug.*	*19 Aug.- 28 Aug.*	*29 Aug.- 7 Sept.*	*8 Sept.- 17 Sept.*
1970	—	—	88	83	85	90	88	89	94	—	—
1971	—	—	94	92	90	92	—	86	86	—	—
1972	—	86	80	78	70	—	72	68	78	77	—
1973	86	88	81	90	83	88	86	88	—	—	—

largely to the height at which wind was measured. The Base Camp anemometer was at a standard height of 10 m whereas those at the intensive study sites were at 0.5 m in the summer and 2 m in the winter. A comparison of the wind data from the 10 m anemometer with that gathered at other sites can be made using the standard logarithmic profile of wind (Lowry 1969, Monteith 1973):

$$U = u^{*}/k \ln[(z-d)/z_{o}] \qquad (1)$$

and using a roughness parameter z_o=0.1 cm and a zero displacement d=1.0 cm, values given by Lowry (1969) for mown grass. These are assumed to be a reasonable estimate of the gravelly, sparsely-vegetated beach ridge surface. It is recognized further that the equation is correct only under neutral stability conditions, but it is felt that days with strong instability are rare.

Solution of the equation gives values at a height of 2 m that are 75% that of values at 10 m. These corrected values still exceed those of the beach ridge for the majority of monthly data given.

Generally, lower wind speeds were experienced at the meadow than at the beach ridge, owing to the former site lying in a depression and therefore being somewhat sheltered.

Table 12. Monthly mean wind speed (m sec^{-1}) for the year 1972-1973 at the three stations on Truelove Lowland. Anemometer height: Base Camp, 10 m.; Beach Ridge and Meadow, 0.5-1.0 m.

Year	*Station*	*Jan.*	*Feb.*	*Mar.*	*Apr.*	*May*	*June*	*July*	*Aug.*	*Sept.*	*Oct.*	*Nov.*	*Dec.*
1972	Base Camp						3.69	3.10	3.10	2.30	1.88	1.40	1.90
	Beach Ridge						2.53	2.20	2.10	1.30	1.59	1.18	1.22
	Meadow						1.92	1.60	1.40	1.10	1.65	1.10	1.66
1973	Base Camp	1.83	1.20	1.15	1.50	2.09	3.24	3.88					
	Beach Ridge	1.40	1.06	1.13	1.26	1.61	2.20	2.68					
	Meadow	1.14	0.94	1.05	1.59	1.53	2.63	1.32					

Table 13. Two-year mean wind speed (m sec^{-1}) for five-day periods during the 1971 and 1972 field season, at five stations on Truelove Lowland and adjacent Plateau.

Station	*29 June*	*4 July*	*9 July*	*14 July*	*19 July*	*24 July*	*29 July*	*3 Aug.*	*8 Aug.*	*13 Aug.*	*18 Aug.*	*23 Aug.*
Base Camp	2.90	2.82	3.68	2.65	2.44	2.53	3.52	1.88	4.76	5.24	4.01	2.90
Beach Ridge	2.02	1.98	2.60	2.05	1.99	1.74	2.24	1.40	3.05	3.45	2.52	2.82
Hummocky Sedge-moss Meadow	1.76	1.56	2.10	1.76	1.48	1.38	1.82	1.05	2.09	2.24	1.87	1.35
Cliff	1.55	1.60	1.82	1.86	1.66	1.45	1.55	0.98	2.73	3.60	1.35	1.12
Plateau	1.07	1.67	1.83	1.84	2.73	1.52	1.94	1.48	3.84	3.67	1.73	2.30

Besides topographic changes within the Lowland, there is a pattern of wind speed that corresponds with distance in from the western coast. Rocky Point data for 1972 only (not shown) were higher than the Base Camp, Beach Ridge, Meadow, or Cliff for the same year, indicating that wind diminishes either because of frictional effects of the topography or proximity to the cliffs.

The 1971-1972 means shown in Table 13 clearly indicate this trend from Base Camp to Cliff. The Base Camp data have been corrected using equation (1) and the same assumptions as above.

It has been suggested by Labine (1974) that the wind, as measured on the plateau, gives a good estimate of the wind field that lies above the Lowland. It should be noted, however, that whereas most values are comparable or a little lower on the plateau (Table 13), during periods of higher plateau wind these exceed wind speeds recorded at Base Camp. Wilson (1973) states that winds that flow over high land reach their greatest speed at the base of the obstruction on the leeward side. It is presumed that this is the situation that exists for Truelove Lowland.

Characteristics of the overall wind pattern of the Lowland is the sporadic occurrence of Föhn winds (Table 14) which are always from the south, often very violent, and which frequently bring a temperature increase of 5°C or more and a marked reduction in relative humidity. Besides these violent wind storms there are more frequent periods, lasting often no more than a few hours, where the wind is again from the south (but often very light), where temperature is increased and relative humidity is decreased. The violent southerly winds appear to be associated with cyclonic storms that pass over Devon Island whereas the origin or cause of the lighter winds is not yet clear.

Table 14. Characteristics of strong Föhn winds experienced during the I.B.P. study period.

Date	*Duration (hr)*	*Mean Speed (m sec^{-1})*	*Direction*	*Temp. increase C°*	*Relative humidity decrease (%)*
26 June 70	9	18	S	6	22
31 Aug. 70	34	23	S	9	44
4 Aug. 71	6	19	S	5	5
21 Oct. 71	16	11	S	6	12
8 Aug. 72	32	15	S	12	25

Soil moisture

Table 15 gives mean values of soil moisture at the two intensive study sites. In both sites the soil moisture at 12-15 cm, chosen to correspond to the middle of the rooting zone, is drier than that at 0-3 cm. Values at the beach ridge were quite constant during the field season, whereas the meadow soils lost moisture progressively during the season, but never dried below saturation. The high values measured for the beach ridge hollow are a function of shelter, giving rise to more complete vegetation cover and a greater abundance of organic matter, rather than finer texture.

Table 15. Mean soil moisture (% O.D.W.) at two depths for the period 8 July-31 Aug., 1972 at the Beach Ridge and Hummocky Sedge Meadow.

	Beach Ridge			*Meadow*		
Depth	*Crest (top)*	*Crest (side)*	*Hollow*	*Hummock (top)*	*Hummock (side)*	*Hollow*
0-3 cm	12	14	166	645	652	975
12-15 cm	8	7	37	436	425	487

Comparison of plateau and lowland

The incidence of a lush, vegetated lowland in the midst of a polar desert climate and physiography, suggests that a comparison should be made to see how different the two areas are from one another. Truelove Lowland, and the plateau above it and to the east, are under the same regional synoptic weather pattern and yet the microclimate of each is different. A detailed study of the microclimate of the plateau was made in 1973 (Labine 1974). Table 16 seeks to show the differences in four environmental parameters for varying periods of time over three field seasons at the Base Camp and Plateau stations to represent lowland and polar desert microclimate respectively. It should be noted that whereas the observation periods for 1972 and 1973 were reasonably long and of the same duration, that for 1971 was fairly short (44 days).

In 1971 and 1972 incoming shortwave radiation was only slightly lower on the plateau than at the Base Camp but in 1973 the reduction was very marked

Table 16. Comparison of incoming shortwave radiation, temperature, precipitation and wind speed between the Base Camp and Plateau sites; 1971-1972.

Year	*Period*	*Site*	*Radiation ($ly\ min^{-1}$)*	*Temperature (C°)*	*Precipitation (mm)*	*Wind speed ($m\ sec^{-1}$)*
1971	10 July-	Base Camp	0.27	4.2**	36.6	2.5++
	23 Aug.	Plateau	0.25	4.3+	38.7	2.6
1972	6 May-	Base Camp	0.35	-1.6	28.7	2.2
	7 Sept.	Plateau	0.33	-3.3	37.9	2.3*
1973	6 May-	Base Camp	0.40	-0.5	82.2	2.2
	13 Aug.	Plateau	0.28	-1.5	86.8	3.2

* 11 May-7 Sept.
** Temperature recorded at 1.5 m Beach Ridge.
\+ Temperature recorded at 0.15 m.
++ 10 m wind speed adjusted to 1 m.

(*ca*. 25%). These changes in radiation are thought to be a function of the periodic warm southerly flow of wind mentioned previously that tends to dissipate the low stratus decks commonly found during the summer months. The height of the plateau (300 m) above the Lowland often means that clouds that were classified as stratus over the Lowland appeared as fog over the plateau.

On the basis of the 1973 growing season (July, August) and only using the data of Labine (1974), incoming radiation was less on the plateau than on the Lowland, .27 versus .20 ly min^{-1}, whereas the reverse was true of net radiation, .09 versus .07 ly min^{-1}. In terms of total incoming radiation the plateau net radiation was 45%, whereas that of the lowland was only 26%. Labine accounts for this on the basis of greater incoming longwave radiation owing to the lower cloud height (Vowinckel and Orvig 1965) and lower outgoing radiation because of cooler surface temperatures and additional surface moisture that lower albedo (Geiger 1965).

Mean temperature was lower on the plateau during the two years with long periods of observations, but the amplitude was greater (more continental) than in the Lowland.

In general, snowmelt was two weeks later and refreeze was at least one week earlier than in the Lowland, giving a much shorter growing season. For a brief period, early to mid-July to early August, air temperatures were higher on the plateau than at Base Camp. Soil moisture was much higher on the plateau than on the most comparable lowland site, the beach ridge, with only surface drying occurring. The result was that the 6°C growing season mean at −10 cm was 5 to 6°C lower than that of the lowland beach ridge soil.

The slight increase in precipitation on the plateau in all years indicates not only additional cloud but also that either the cooler mean temperatures bring about greater precipitation or that at least some clouds borne inland from the west yield orogenic precipitation.

Plateau winds were often slightly lower than those of the lowland Base Camp but sufficient periods of high plateau winds were recorded to show higher mean winds at the latter site.

Labine (1974) found, when considering the energy budget of both lowland beach ridge and plateau, that both had similar Bowen Ratios and that the values were characteristic of dry areas. This apparent disparity for the poorly drained plateau was the result of surface drying at that site. The soil heat flux term was much higher for the plateau (40% of net radiation) than for the beach ridge (6% of net radiation). Yet beach ridge soil temperatures were warmer. This indicates that most of the energy absorbed by the plateau went into melting the active layer and little energy was available for sensible or latent heat fluxes. The plateau soils warmed up very little although thermal conductivity was high (aided by the high water content) and thermal diffusivity was low, as opposed to the Beach Ridge where the conductivity was low and diffusivity was high.

In summary, the Plateau has a harsher climate than the Lowland by virtue of a shorter growing season and colder soils. Even though air temperatures are higher in mid season and net radiation generally is higher, the extra solar energy is used to heat the very cold soil and contributes little to improving the surface conditions for plant growth.

On a broader scale the plateau falls within the definition of Polar Desert both from the standpoint of vegetation (Svoboda this volume) and summer temperature (Bliss et al. 1973). Aleksandrova (1970) defined Polar Desert, in the Soviet sector of the Arctic, as lying north of the 2°C July mean isotherm. Only where there is an Atlantic influence is Polar Desert found south of the isotherm, whereas in more conventional areas, it is found only where temperatures are lower than 2°C. July mean temperatures for the plateau for 1971, 1972, and 1973 were 6.0, and 4.0°C, respectively with a mean of 3.8°C. This indicates that Aleksandrova's value of 2°C is too low for the southern limit of continental high arctic polar deserts. It may well be that the occurrence of open water or partially open water in the Archipelago during the summer generally elevates temperatures sufficiently to account for the difference and gives rise to a similar extension of the southerly limit to that found by Aleksandrova with maritime influence.

Implications

Truelove Lowland is essentially an oasis in the High Arctic Polar Desert. The phenomenon, however, is not unique. Adjacent lowlands to the Truelove: Sparbo-Hardy, Sverdrup, Skogn, and Newman-Smith, show similar characteristics if not so well developed. The Fosheim Peninsula, Lake Hazen, and Tanquary Fiord on Ellesmere Island are also atypical for their latitude. Corbet (1969) has sought to explain the situation at Lake Hazen in terms of solar altitude. He states that with increasing latitude there is a reduction in the diurnal fluctuation of sun altitude that, in turn, reduces the diurnal range in temperature causing a net amelioration in climate. This situation may exist to some degree at the latitude of north Devon Island. The high yearly net radiation balance recorded, however, must be accounted for at a latitude that has a normal climatic regime with a high incidence of cloud. It is clear from the data reported that the overall climate of Truelove Lowland is more favorable than that of adjacent stations such as Resolute Bay. The reason is probably a function of the Lowland's topographic position with respect to the cliffs to the south and east, and the Devon Island Icecap (height 1800 m), to the south and east. Airflow from any direction between southwest and east must cross this high land and while returning to sea level, is heated dry adiabatically. This flow can result in the dissipation of cloud decks whose persistance would reduce the influx of radiant energy. It is not necessary for these winds to reach ground level in the Lowland to dissipate clouds and the sheltering effect of the cliffs may well often prevent them from doing so. Nevertheless, it has been noted that short periods of southerly winds are quite frequent, especially in the late summer (September), and are often accompanied by a slight increase in temperature and slight decrease in relative humidity. This seasonal bias in the case of Truelove Lowland not only has the effect of reducing cloud but also may help to extend the growing season longer than would otherwise be the case.

Combined with the strong "summer month" differences discussed above, the data from the one winter period (1972-1973) indicate that Resolute Bay emits 50% more radiation than does the Lowland (net radiation, Resolute −.023 vs Truelove −.0115 ly min^{-1}, October 1972 to March 1973 mean). Not only does the Lowland receive more energy when the sun is above the horizon but it loses less energy during the dark period. It is suggested that the lower winter radiation flux may be influenced by the increased cyclonic activity (McKay et al. 1970), the nearby ice-free North Water area, and the resulting greater snowfall on Truelove (winter ppt totals 1972-1973 Truelove 88.6 mm vs Resolute 30.9). The data span, however, is much too short to allow full understanding of the processes involved.

Snowmelt and the associated decrease in albedo normally occurred over a period of 10 to 14 days. In 1970, however, most of the snow melted in an 8 hour period of Föhn winds. This had a dramatic effect on the biota, especially lemmings, as discussed by Fuller et al. (this volume).

Barry and Jackson (1969) suggest that downfiord winds at Tanquary Fiord have a similarly ameliorating effect on climate and give a marked temperature increase and precipitation decrease over that at Eureka, situated *ca.* 200 km to the southwest.

More profound analyses of the Devon Island data are being undertaken to better understand the influence that Föhn winds may play on a local scale in the High Arctic.

Truelove Lowland is characterized not only by a higher insolation and warmer temperatures but also by abundant soil moisture in the meadows that represent a major portion of the terrestrial vegetated area. The cause and effect of vegetation in this regard is not clearly distinguished. The establishment of vegetation in ponds trapped behind emergent beach ridges has given rise to the accumulation of peat that has decomposed only slowly owing to the cold climate. The peat has a very high water holding capacity and high insulative qualities. Depth of thaw is thus minimized and radiant energy is used to melt ice rather than evaporate moisture. The topographic relationship of beach ridge to meadow leads to the trapping of snow in the meadow depressions, thus reducing the length of the thaw season and providing further moisture. Sedges, in their turn, thrive in this environment and continue year by year to increase the depth of peat and maintain the reservoir of water. The net result is a longer, warmer growing season with water being possibly limiting to meadow vegetation only in the physiological sense.

Much of the Lowland is made up of meadows and beach ridges with the outcrops only being a dominant physiographic unit along the northern coast and at Rocky Point. Beach ridges are an environmental contrast to the meadows in spite of their proximity. The porous nature of the gravels reduces the amount of water held in the soil which results in a much deeper active layer. Drought and exposure, however, combine to greatly reduce vegetation cover. The climate of exposed sites on the rock outcrops appears to be similar to that of the beach ridges, but there is no measureable mean climatic regime for an outcrop that has any validity because their dissected nature leads to a spectrum of microsites that vary from bare rock, through beach ridge, to pockets of meadow. All degrees of exposure and shelter from wind and radiation are possible.

The progressive change in temperature in the Lowland throughout the summer season appears to be more closely coupled to the physiography that surrounds the Lowland rather than to the physiographic units of beach ridge, meadow, and rock outcrop within the Lowland. The heating that occurs furthest inland along the limestone cliffs may be due in part to the heating of the cliff face and in part to the influence of the Gully and Truelove Valleys. Hubert (this volume) indicates that the calving activity he witnessed along the limestone cliffs and in the Truelove Valley may well be the result of the warmer spring conditions existing in these areas. In May 1973 the authors noticed that muskox were feeding in a line parallel to the Truelove Inlet where warming prior to snowmelt has been observed (Fig. 16, map *b*). The feeding may have been either by chance or because the snow was not as deep as elsewhere. It may also have resulted from warmer conditions that would soften the snow and make the digging of feeding craters easier.

Summary

The present research has formed part of a base line study on a high arctic ecosystem. It has been shown that it is possible for thermal oases to exist at high latitudes and that these can have a fundamental effect upon the biota.

Oases may be formed as a result of strong physiographic controls over microclimate, such as: high cliffs, proximity to large bodies of water, and ice caps.

The localized nature of the Truelove Lowland oasis is very strictly defined since the adjacent plateau is true polar desert characteristic of the latitude.

Even within the oasis, physiographic controls give rise, not only to different plant communities, but also to a variety of microclimates.

The study has been achieved using modest meteorological instrumentation under most extreme conditions. The data presented, therefore, are incomplete and have limitations both with respect to accuracy and the degree to which they may be interpreted.

The study as a whole has served not only to document the nature of high arctic microclimate but also to point to the necessity for further work in the acquisition of quantitative and qualitative data, and in the field of instrumentation that is better designed to cope with the rigors of the climate.

Acknowledgments

Almost all the field personnel gave assistance at one time or another during the project, especially with respect to reading anemometers and reporting non-functioning instruments. Their names are too numerous to list, but their efforts were much appreciated. Special thanks are accorded to J. Marsh, P.A. Addison, and Glen Pierce who were directly associated with the field research and synthesis; and to Claude Belcourt who maintained the stations during the winter 1972-1973. Specific aid to the project was given by E.I. Mukammal and L. Wiggins of the Atmospheric Environment Service in the form of funds, equipment loan, and most helpful advice. The Service's research division gave welcome and immediate assistance when trouble was experienced in the field. The very large quantity of data collected was processed partially at The University of Alberta and partially at Laurentian University. We wish to give thanks especially to R. Goodwin, D.W.A. Whitfield, P.A. Addison, H. Dow, S. Allaire, and T. Lalonde for their help with this crucial phase of the work.

The field work was made possible by grants from National Research Council (A-5071) and I.B.P.-N.R.C.

References

Ahrnsbrak, W.F. 1968. Summertime radiation balance and energy budget of the Canadian Tundra. Technical Report No. 37 Dept. Meteorol., Univ. Wisc., Madison, Wisconsin. 50 pp.

Aleksandrova, V.D. 1970. Vegetation and primary productivity in the Soviet Subarctic. pp. 93-114. *In:* Productivity and Conservation in Northern Circumpolar Lands. Fuller, W.A. and P.G. Kevan (eds.). Morges, Switzerland. IUCN New Ser. No. 15. 344 pp.

Barry, R.G. Pers. comm. Univ. Colorado, Institute of Arctic and Alpine Research, Boulder, Colorado.

———. and C.J. Jackson. 1969. Summer weather conditions at Tanquary Fiord, N.W.T. 1963-67. Arctic Alp. Res. 1: 169-180.

Bliss, L.C., G.M. Courtin, D.L. Pattie, R.R. Riewe, D.W.A. Whitfield, and P. Widden. 1973. Arctic tundra ecosystems. Ann. Rev. Ecol. 4: 359-399.

Brazel, A.J. and S.J. Outcalt. 1973a. The observation and simulation of diurnal evaporation contrast in an Alaskan alpine pass. J. Appl. Meteor. 12: 1134-1143.

———. 1973b. The observation and simulation of diurnal surface thermal contrast in an Alaskan alpine pass. Arch. Met. Geoph. Biokl., Ser. B. 21: 157-174.

Bryson, R.A. 1966. Airmasses, streamlines and the boreal forest. Geogr. Bull. 8: 228-269.

Corbet, P.S. 1969. Terrestrial microclimate amelioration at high latitudes. Science 166: 865-866.

Dorsey, H.G. 1951. Arctic meteorology pp. 942-951. *In:* Compendium of Meteorology. Malone, T.F. (ed.). Am. Meteorol. Soc. Boston. 1334 pp.

Gieger, R. 1965. The Climate Near the Ground. Harvard Univ. Press, Cambridge, Mass. 611 pp.

Goodison, B. 1975. Snow Studies at Cold Creek. Testing and Evaluation of Gauges Used in Canada for Fresh Snowfall Measurement. I.H.D. Can. Hydrol. Symp. Aug. 1975, Winnipeg.

Hare, F.K. 1968. The Arctic. Quart. J. Roy. Meteorol. Soc. 94: 439-459.

———. and J.E. Hay. 1974. The climate of Canada and Alaska. pp. 49-192. *In:* Climates of North America. *In:* World Survey of Climatology, Volume 11. Bryson, R.A. and F.K. Hare (eds.). Elsevier Scientific Publishing Co., Amsterdam, 420 pp.

Holmgren, B. 1971. Climate and energy exchange on a sub-polar ice cap in summer. Meteorologiska Institutionen, Upsala. Universitet Meddelande Nr. 107.

Labine, C.L. 1974. Measurement and Computer Simulation of Microclimate Differences Between a Polar Desert Plateau and a Nearby Coastal Lowland. M.Sc. Thesis. Univ. Guelph, Guelph, Ontario. 85 pp.

Lowry, W.P. 1969. Weather and Life. An Introduction to Biometeorology. Academic Press Inc., N.Y. 305 pp.

McKay, G.A., B.F. Findlay and H.A. Thompson. 1970. A climate perspective of tundra areas. pp. 10-33 (see Aleksandrova 1970).

Monteith, J.L. 1973. Principles of Environmental Physics. American Elsevier Pub. Co. Inc. N.Y. 241 pp.

Mooney, H.A., R.D. Hillier and W.D. Billings. 1965. Transpiration rates of alpine plants in the Sierre Nevada of California. Am. Midl. Natur. 74:374-386.

Ohmura, A. 1970. Some climatological notes on the expedition area. pp. 5-13. *In:* Report of McGill University Axel Heiberg Expedition. Mueller, F. (ed.). McGill Univ. Montreal.

Pruitt Jr., W.O. 1970. Some ecological aspects of snow. pp. 83-99. *In:* Ecology of the Subarctic Regions. Helsinki Symposium. UNESCO, Paris. 364 pp.

Rae, R.W. 1951. Climate of the Canadian Arctic Archipelago. Meteorol. Div. Can. Dept. Transp. Toronto. 89 pp.

Romanova, E.N. 1972. Microclimate of tundras in the vicinity of the Taimyr field station. International Tundra Biome Trans. No. 7.

Rouse, W.R. and R.B. Stewart. 1972. A simple model for determining evaporation from high-latitude sites. J. Appl. Met. 11: 1063-1070.

Rydén, B.E. Pers. comm. Hydrology Div., Dept. of Physical Geography, Univ. Upsala, Upsala, Sweden.

Skartveit, A., B.E. Rydén, and L. Karenlampi. 1975. Climate and hydrology of some Fennoscandian tundra ecosystems. *In:* Ecological Studies. Analysis and Synthesis, Vol. 16. Fennoscandian Tundra Ecosystems, Part I. Plants and Microorganisms pp. 41-53. Wielgolaski, F.E. (ed). Springer Verlag, New York, 366 pp.

Smith, M. 1975. Numerical Simulation of Microclimate and Active Layer Regimes in a High Arctic Environment. Ind. North. Aff. Publ. No. QS-8039-000-EE-A1 Ottawa. 29 pp.

Thompson, H.A. 1967. The Climate of the Canadian Arctic. Meteorol. Branch, Air Services, Dept. of Transp., Toronto.

Vogel, T.C. and P.L. Johnson. 1965. Evaluation of an economic instrument shelter for micrometeorological studies. Forest Sci. 11: 434-435.

Vowinckel, E. 1966. The surface heat budgets at Ottawa and Resolute, N.W.T. Arctic Meteorology Research Group. Dept. of Met., McGill Univ. Publication in Met. No. 85.

————. and S. Orvig. 1965. Radiation balance of the troposphere and of the earth atmosphere system in the arctic. Archiv. Fur. Met. Geo. und Biokl. Ser. B. Band 13.

Weller, G. and S. Cubley. 1972. The microclimates of the arctic tundra. pp. 5-12. *In:* Proc. 1972 Tundra Biome Symp. Bowen, S. (ed.). CRREL. Hanover, N.H. 211 pp.

————. and B. Holmgren. 1974. The microclimates of the arctic tundra. J. Appl. Met. 13: 854-862.

Wendler, G. 1971. An estimate of the heat balance of a valley and hill station in central Alaska. J. Appl. Met. 10: 684-693.

Wilson, C. 1967. Cold region science and engineering. Monogr. 1-A3a. U.S. Army. CRREL. Hanover, New Hampshire.

Wilson, H.P. 1973. Arctic Operational Meteorology. Arctic Weather Central, Edmonton. Queen's Printer. 348 pp.

Hydrology of Truelove Lowland

B.E. Rydén

Introduction

The water budget of high arctic and many high mountain areas is characteristic of extremely cold hydrological environments. They have the shortest season with liquid precipitation, accounting for less than 30% of the annual total. They receive high values of global solar radiation which allows rapid snowmelt, but the summer solstice is often past before the land is snow-free. Melt water movement downward is prevented by permafrost. The occurrence of glaciers in some areas increases dramatically the supply of water to downstream parts of basins during summer. After snowmelt, most streams carry a limited amount of water that comes from thawing of the active layer and from ground water storage.

According to Cook (1967) and McCann et al. (1972), most small catchments within the High Arctic yield 80% to 90% of the annual runoff (predominantly snowmelt) over a period of 2-3 weeks in late June and early July. Spring is very short and is characterized by a rapid decrease in snow cover; air temperature at the vegetation level rises rapidly (Courtin and Labine this volume). Lakes and ponds occur in great numbers on flat coastal plains in the Arctic (MacKay and Løken 1974); their detention storage often modifies runoff on lowlands.

The objectives of this study were to determine: (1) the water budget of the Intensive Study Sites and the Truelove Lowland; (2) the hydrological regime of rivers and streams connected to the Lowland, in particular the dynamics of their discharge during snowmelt, and (3) the pattern of evaporation from meadows and lakes.

Methods

Instrumental and measurement problems

In an area of hydro-climatological extremes and where water budget variables have a small magnitude, both precipitation and runoff amounts may show great year to year differences. Thus data of a few years duration may not be representative. Moreover, investigations in such a remote area are associated with various instrument problems including measurement of precipitation (Courtin and Labine this volume).

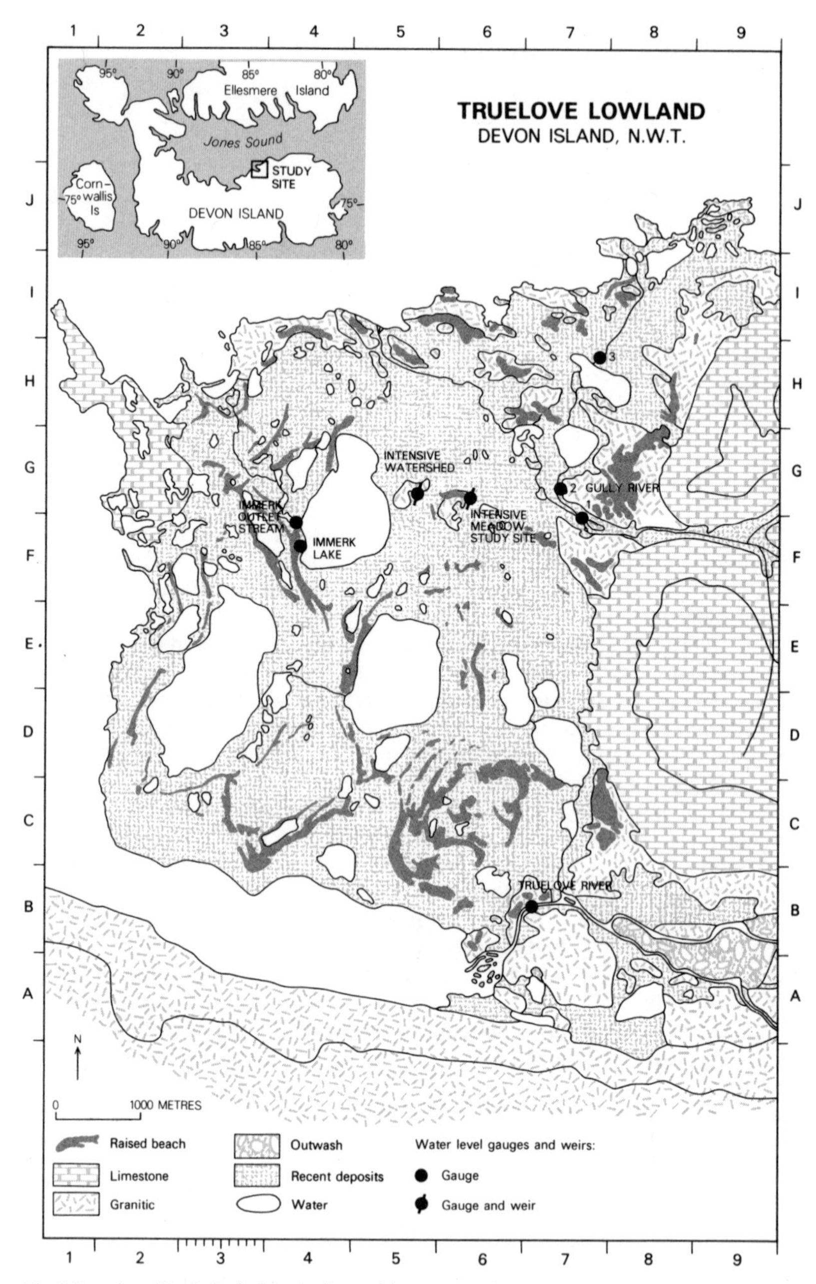

Fig. 1 Location of hydrological study sites and instrumentation over a three year period.

At times alternating freezing and thawing make runoff studies very difficult. Few measurements have been made of snowmelt; estimates of change in soil water content and groundwater storage data are only rough approximations due to lack of an acceptable field method. The hydrology study and its methods were mainly chosen to achieve a general knowledge of the water budget of Truelove Lowland.

Instrumentation

The hydrological variables were recorded in the Raised Beach and Hummocky Sedge-Moss Meadow Intensive Study Sites, at the Base Camp meteorological station and at the outlet of Immerk Lake (Fig. 1). The following is a description of the instrumentation and methods used during the years 1972 and 1973 (Holecek and Vosahlo 1974) and 1974.

Water level

Recording gauges of the float type (Ott, Leupold and Stevens, Munro) were mounted on stilling wells at the outlet of Immerk Lake (1972 and 1973) and at the main outlets of the Intensive Study Sites (1972-1974). Weirs (wooden or galvanized iron sheets) were installed by Holecek and Vosahlo at these three locations with stilling wells at the two outlets.

In 1974 the stilling well at the outlet of Immerk Lake was moved to the upper weir of the Intensive Meadow and a recording gauge installed. Thus three recording gauges (Munro) were operated in 1974 in the Lowland, i.e. in the main outlets of the Intensive Sites and in the upper part of the Meadow. This upper stream was considered to be the main inflow into the Intensive Meadow and was visible a large part of each summer as a narrow stream fed by a small lake several hundred meters to the east.

The water level of Immerk Lake was observed in 1974 with a staff gauge levelled to a bench mark datum set up on the arctic stone north of the Base Camp. Reported variations of level and area of the lake in previous years are not based upon any datum levels (Minns this volume). Comparisons will be qualitative. The Gully River, which drains part of the plateau east of the Lowland, and the Truelove River, having its drainage area southwest of the Lowland, were monitored in 1972 and 1973 (Holecek and Vosahlo 1974). Recording water level gauges (Leupold and Stevens) were placed at three control locations of Gully River (Fig. 1). Another recording water level gauge (Leupold and Stevens, pressure type) was installed in a straight, deep canyon of Truelove River.

For further details on instrumentation in 1972 and 1973 see Holecek and Vosahlo (1974). It is inferred that stage-discharge relationships were established at the hydrometric stations in Gully River and Truelove River, although the above report does not present judgments on the accuracy of these relationships or mention the various low and high water stages covered by the velocity measurements (Ott current meter and Revolution counter). It is known (Vosahlo, pers. comm.) that some problem arose from ice in the control sections during early melt runoff.

Stream discharge

Discharge of the streams was mainly determined by V-notch weirs (1972-1974) having a 90°-notch of which the relationship between discharge Q and water level h above the notch level is given by the expression:

$$Q = \frac{8}{15} \mu \sqrt{2 g \cdot h^{5/2}} \qquad (1)$$

where μ is the discharge coefficient and g gravity. The notch level, h=0, should be related by levelling to a certain reading on the staff gauge. The equation presumes that the dam formed by the weir is wide enough to retard flow. Thus initial velocity for the overflow is zero and gravity alone brings the water over the notch of the weir. Upon inspection in 1973 the author observed that in the dams generated by the weirs this was probably the case only at low water stage.

The error thus resulted in underestimates of discharge. Some leakage occurred below and around the walls of the dam due to the shallow permafrost table. During the spring flood in 1974 the leakage of the meadow weir was estimated to be about 15-20% on certain occasions.

The discharge coefficient $\mu < 1$, expressing the losses caused by friction and contraction of the flow over the weir, decreases the real discharge compared to the theoretical by some 30% at sharp-crested weirs and by 30-50% at badly shaped weirs. If the overflow jet is only partly ventilated, ordinary readings will overestimate the discharge. Hydrometric calibrations of the weirs were not done in 1973. The error in determining the discharge and disregard of the discharge coefficient compensate to some extent the above-mentioned underestimations.

In an early phase of the spring flood, snowmelt is very unstable and cannot be reliably recorded by float gauges. For this purpose portable weirs of two different sizes were used, having maximum capacities of 1.5 l sec^{-1} and 32 l sec^{-1}, respectively. The weirs, constructed by the author, were wooden with a triangular orifice of 90° angle of galvanized iron with the wooden dam wall mounted to a large plastic sheet, 0.3 m^2 and 4.5 m^2, respectively (Fig. 2). As an example of their use, it was found that an early part of snowmelt passed across the raised beach instead of flowing via the "normal" path through the stream bed. The portable weirs were roughly calibrated in the field and later in the laboratory, giving a discharge coefficient equal to 0.7 and 0.5 for the 32 l and the 1.5 l weir, respectively. Leakage of a portable weir of this type can be prevented by adjusting the large plastic sheet along the bottom and sides of the channel (Fig. 2).

The discharge of very small streams or rills was surveyed simply by the use of plastic cans or buckets, preferably those having a weak, adjustable shape. The can was pressed to the bottom of the rill covering a distinct section and allowed to be filled with water during a certain number of seconds. The volume was measured by graduated cylinders, thus resulting in discharge data.

Water stage gauges and weirs were levelled to bench marks in the vicinity of each gauge. This was done at least twice in 1974, i.e. at one early and one late period of the runoff season.

Characteristics of water

Water temperature was measured upon every visit to the Intensive Sites, using a mercury thermometer with a large sensor (Rydén 1973a). The accuracy was about ±0.02°C.

During the snowmelt period of 1973 the transport of litter by melt water was also studied. A net with 0.9 mm mesh was set up covering the cross-section of a small stream downstream of the Intensive Watershed Site. The net was changed at intervals of 1 to 9 days. The samples were weighed, allowing determination of total discharge and specific transport during the sampling periods.

Snow depth

Snow depth was measured from 1972 to 1974 with a 100 cm steel rod graduated in cm. Some error was introduced because the rod normally does not penetrate the ice layer at the bottom of the snow cover. Such ice may form early in the winter but increases often in late spring when internal drainage of melt water refreezes at the snow pack bottom. When the snow course measurements are of greatest interest, the bottom ice may be at its maximum, i.e. 2-5 cm, and thus the error of water equivalent will be appreciable.

Fig. 2 Portable weir on 4 July, 1974 as seen from the downstream side looking east into the Intensive Watershed.

Water equivalent

The water equivalents of the snow cover were derived in 1972 to 1974 from extensive snow depth measurements over the entire Intensive Watershed and a limited number of snowpack density measurements (Mount Rose snow sampler 1972-1974, Leupold and Stevens snow-sampling tube 1974). Vertical variations of density within the snowpack were estimated several times in 1974 by use of snow sample cylinders of 5 or 12 cm height (volumes 203 and 1150 cm^3, respectively). This type of density measurement permits more accurate laboratory rather than field weighing. The inability to remove ice at the bottom of the snow core may result in a 5-15% underestimate of water equivalent in the Intensive Watershed.

Precipitation and soil water content

For methods used in measuring these parameters, see Courtin and Labine, and Addison (all this volume) respectively. Other meteorological parameters, applied in the present paper, were also measured and reported by these authors.

Evaporation

In 1972 and 1973 evaporation was recorded with a Class A Pan and recording evaporimeter, E-801 (Weather Measure Corp.), the latter operating on the principle that a wetted filter paper evaporates readily when heated by air or radiation. At moderate rains the record is reversed and the amount of rain can be evaluated from the evaporation chart (Holecek and Vosahlo 1974). It is questionable whether the evaporimeter reproduces conditions in nature and the interpretation of these data is difficult.

Results

Snowmelt and runoff

Water budget variables have been estimated within three different types of drainage basins connected to Truelove Lowland. Runoff records have been obtained from:

1. The Intensive Watershed and the Intensive Hummocky Sedge-Moss Meadow (1972-1974) (0.12 and 0.4 km^2).
2. The Gully River that drains part of the plateau, 300 m.a.s.l., and the Lowland northeast of the intensive study sites (1972-1973) (22.4 km^2).
3. The Truelove River, that drains part of the Devon Island ice cap, part of the plateau, and the Truelove River Valley southeast of the Lowland (1972-1973) (574 km^2).

The analyses of runoff regimes are thus based on records of two summers from the Gully and Truelove Rivers and three summers from the Intensive Watershed. The stream types listed show the diurnal discharge variation typical of a glacial or snowmelt regime. However, the Intensive Watershed yields its flood peak in mid-June to early July, Gully River in late June to early July, and Truelove River during late June to mid-July (Fig. 11). Depletion is rapid in the lowland streams, resulting in a negligible discharge from the Intensive Watershed in late July, when the other two streams still yield more than their summer mean.

Intensive Watershed

This and other lowland watersheds remain frozen until mid- or late June. The main snowmelt occurs during no more than two weeks; a period when lake ice is more or less intact. As the ground is frozen with thawing in only the upper few centimeters as summer progresses, snowmelt runoff occurs as overland flow. The melt runoff shows a high degree of irregularity. The runoff pattern changes from day to day following the fluctuation in solar radiation and air temperature. Snow is seldom a hinderance because ripe snow offers little visible resistance to water flow. A shallow and unstable overland flow of meltwater soon establishes itself. Hydrograms for the Intensive Watershed in 1972 and 1973 are presented in Figs. 3 and 4, and for the Intensive Meadow Study site in Fig. 5.

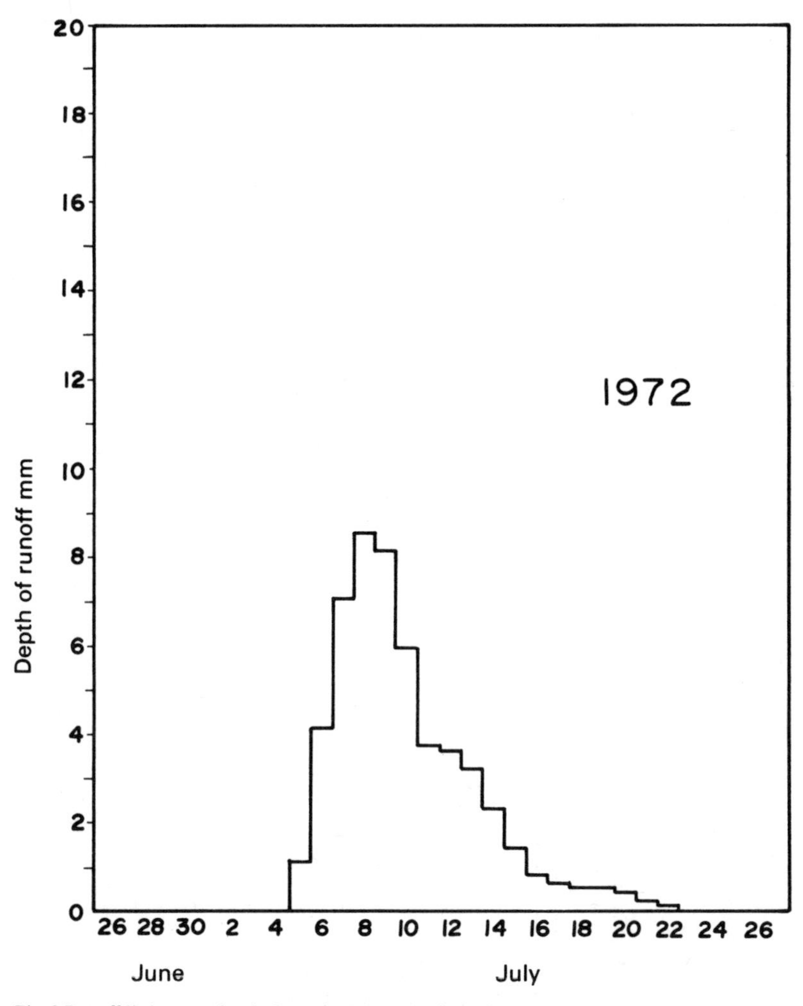

Fig. 3 Runoff hydrogram for the Intensive Watershed in 1972.

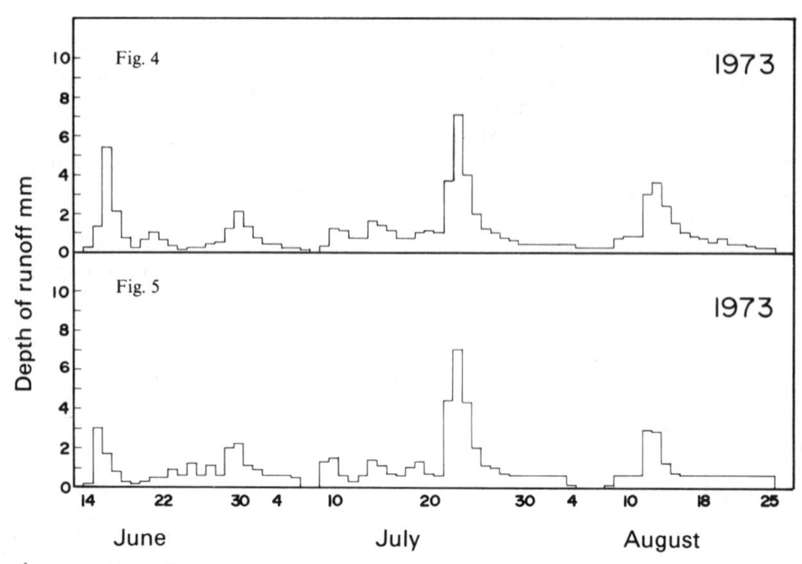

Figs. 4 and 5 Runoff hydrograms for the Intensive Watershed and the Intensive Meadow study site in 1973.

The irregular runoff is well illustrated by the spring outflow in 1974 from the pond at the Intensive Raised Beach Site. The outlet of the Intensive Watershed was blocked by solid ice that was higher than the raised beach around it. Consequently, the accumulating melt water on the ice-covered pond started to overflow the (early bare) raised beach and a series of small streams established themselves for three or four days across the beach. When the damming ice in the more permanent stream bed at the weir had melted enough, there was a sudden shift in outflow (Fig. 6). In a few hours the crossing streams dried and the outflow was established through the stream bed at the weir. During this period it was impossible to receive information from the recording water level gauges and thus the portable weirs were used. The discharge

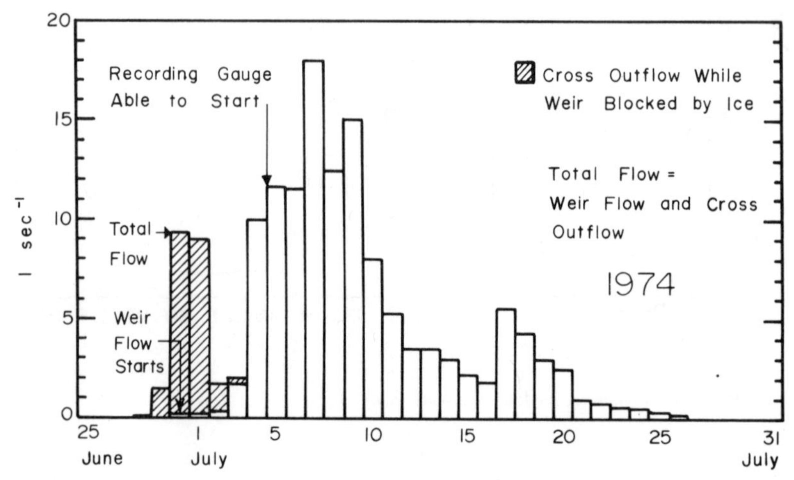

Fig. 6 Runoff hydrogram for the Intensive Watershed in 1974.

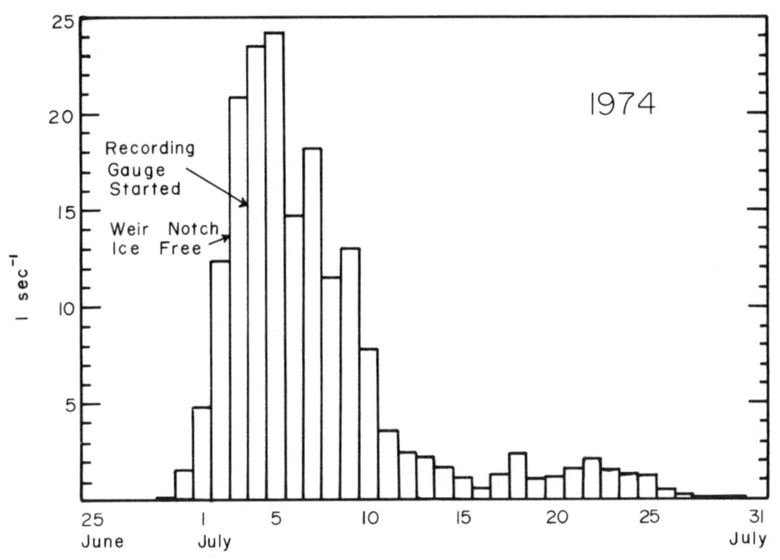

Fig. 7 Runoff hydrogram for the Intensive Meadow study site in 1974.

measured represented a considerable part of the spring flood. Taken together, these streams yielded about 9.0 l sec^{-1} which corresponds to a specific flow of 75 l sec^{-1} km^{-2} or 6.5 mm day^{-1}. The spring peak amounted to 14.0 l sec^{-1} (116 l sec^{-1} km^{-2} or 10.2 mm day^{-1}) and occurred a few days after the shift to the permanent stream bed when the recording gauge was installed (Fig. 6).

After the snowmelt peak there was a steady depletion, only interrupted by a few rains of frontal or instability origin. Cumulonimbus clouds were observed only twice in 1974. Most rains, however, were of low intensity, yielding small total amounts of precipitation (<0.5 mm) which are hardly visible in the hydrograms. A daily period of runoff was evident in 1974 until about 70% of the lowland snow cover had melted, or about the first week of August. Although the Meadow Site was wet all summer, the outflow from this basin decreased greatly by mid-July (Fig. 7). Discharge may cease long before soil refreezing occurs in late August or early September.

During the snowmelt there was considerable lateral movement of water across many sedge-moss meadows. In the Intensive Watershed Site the rate of maximum surface flow reached 100 l sec^{-1} km^{-2} in 1972 and 63 l sec^{-1} km^{-2} in 1973 (Holecek and Vosahlo 1974) whereas in 1974, with a 50-60% increase in snow depth, the rate extended to 116 l sec^{-1} km^{-2}.

Gully River

The Gully River drainage basin (22.4 km^2) upstream from the upper measurement station is nearly barren, almost exclusively composed of the branched system of deep narrow ravines and the plateau itself (Polar Desert).

The hydrologic regime of Gully River is similar to that of a high mountain river. During most of the warm season, Gully River shows a pronounced diurnal variation owing to melting snow. There was a snowmelt flood in mid-July 1972 and in late June in 1973. The flow rate in summer shows a high response to rain (Figs. 8, 9 and 10). As summer progresses the rate of discharge decreases. The occasions of recorded seasonal maximum flow are presented in Table 1.

In 1972 there was a deviation of 20 mm between measured runoff (138

Table 1. Maximum specific flow recorded in three different stream types, Devon Island.

Basin	*Date*	*Specific Discharge mm day^{-1}*	*Cause of flood*
Intensive Watershed 0.12 km^2	8 July 1972*	8.6	Snowmelt
	16 June 1972	5.4	Snowmelt
	23 July 1973	7.1	Rain
	10 July 1974	10.2	Snowmelt
Gully River 22.4 km^2	17 July 1972	23.6	Snowmelt
	27 June 1973	27.2	Snowmelt
Truelove River	18 July 1972	5.3	Snowmelt
	10 Aug 1972	25.5	Föhn
574 km^2, total; <400 km^2, contributing area	29 June 1973	9.7	Snowmelt
	22 July 1973	58.0	Rain
	12 Aug 1973	35.0	Rain

*Data from Holecek and Vosahlo 1974, except those of 1974.

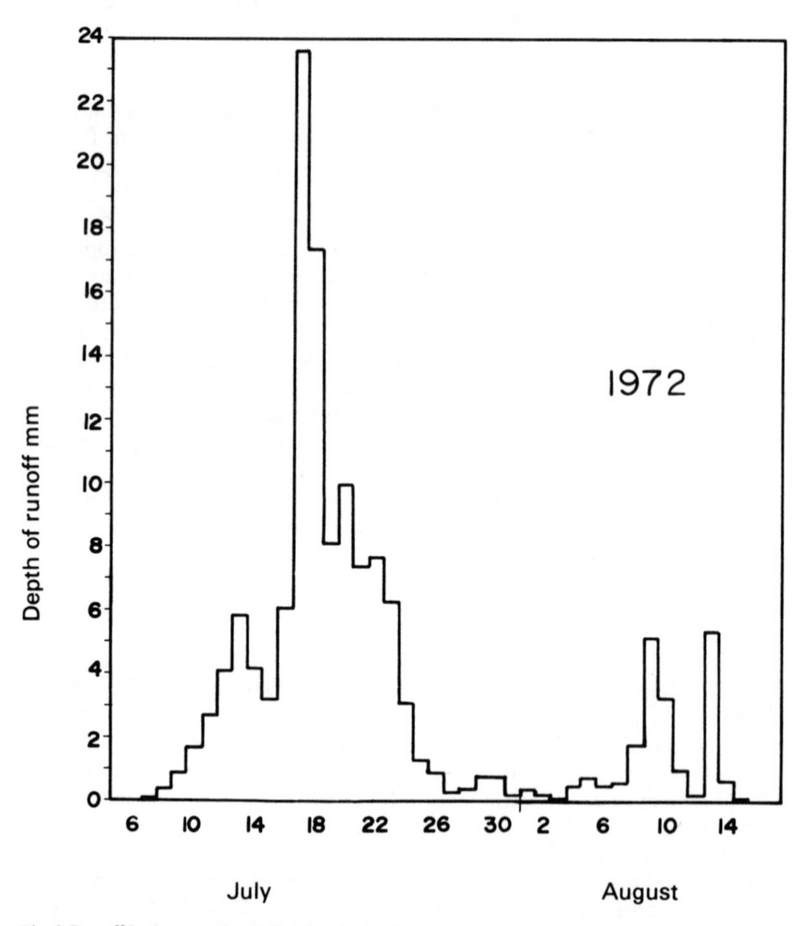

Fig. 8 Runoff hydrogram for Gully River in 1972.

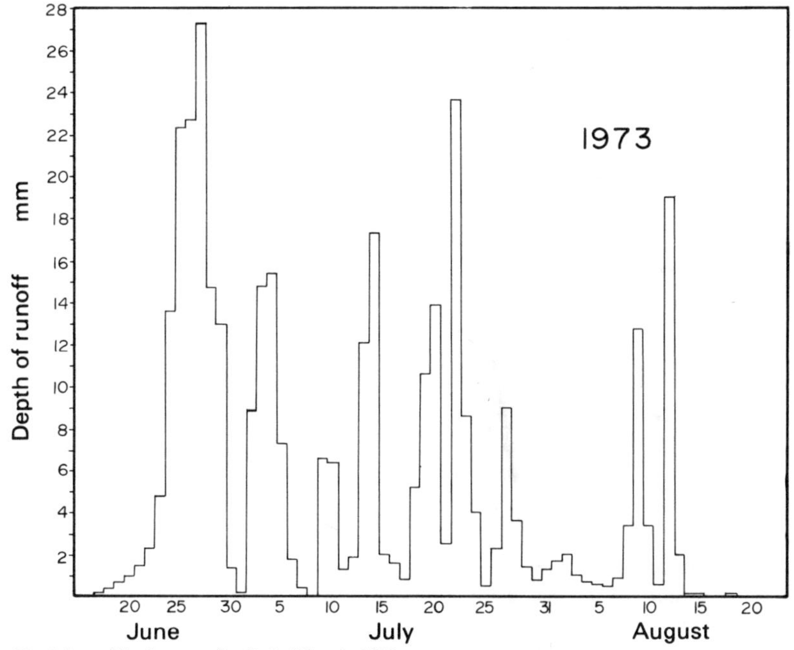

Fig. 9 Runoff hydrogram for Gully River in 1973

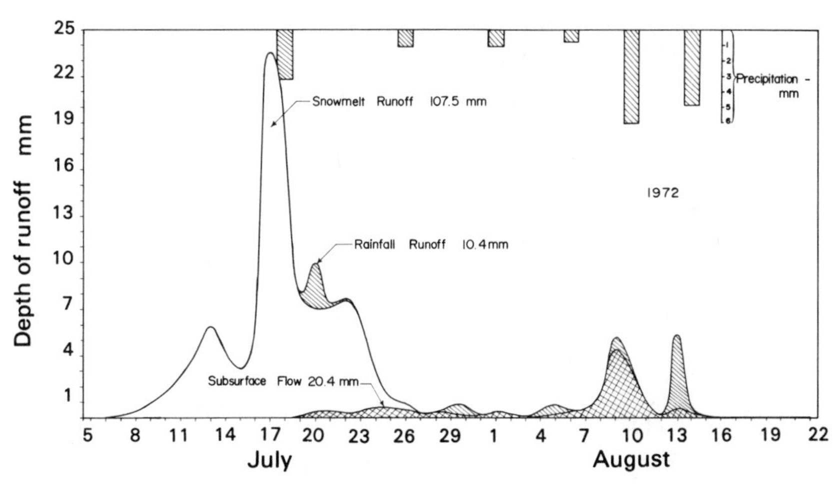

Fig. 10 Analysis of runoff in 1972, including the components of snowfall, rainfall, and meltwater from the active layer.

mm) and the calculated sum of snowmelt and rainfall runoff (118 mm) (Holecek and Vosahlo 1974). The difference is within the limits of the accuracy of winter precipitation measurements on Devon Island. Part of this deviation may also be regarded as active layer thaw, part or all of which is replaced each year by storing late seasonal rain or early snow that melts when the soil is thawed.

In the gullies of the catchment, wind blown snow accumulates, creating a precipitation storage that is probably of greater magnitude than that of the plateau portion of the basin. Total runoff in 1973 amounted to 361 mm in the Gully River (Holecek and Vosahlo 1974). However, the lowland precipitation was 174 mm and the water equivalent of the plateau snow cover was about 15

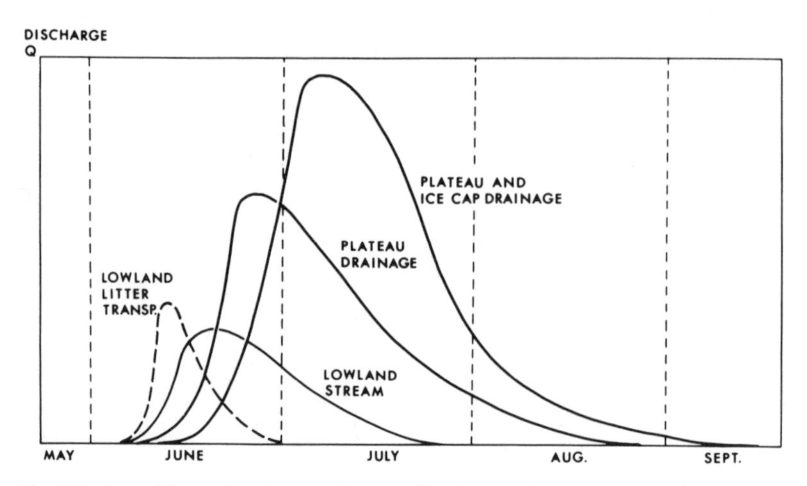

Fig. 11 Regimes of Devon Island. Generalized runoff hydrograms for the three types of water systems. The regimes are derived from a maximum of three years of record.

mm. The difference is too large to be accounted for by evaporation, active layer thaw and/or by increasing plateau precipitation by 10% to 20% due to increased elevation. The discrepancy could not be explained in that way, even if the runoff figure was based on a hydrologic year (from autumn to autumn). Possible explanations may be that there was additional snow storage in the deep ravines (where snow course measurements are extremely difficult to carry out) or that a very warm summer had melted snow drifts from previous years. These examples emphasize the difficulties involved in estimating water budget variables of some arctic basins.

Truelove River

The Truelove River has a complex hydrological regime with the glacial regime of the upper basin having a greater effect than the snowmelt regime of the lower basin. The drainage basin, defined by its topographic water divide is 574 km^2, but only part of this basin contributes water to drainage at any one time owing to the changing level of the 0°C isotherm (Holecek and Vosahlo 1974). The edge of the glacier is approximately 750 m.a.s.l. whereas the upper limit of effective catchment was about 1200 m.a.s.l., defined by the average level of the 0° isotherm as obtained from the radio soundings at Resolute Bay, 320 km to the west. The contributing part of the basin was estimated as 361 km^2 (1972) and 370 km^2 (1973) (Holecek and Vosahlo 1974). Maximum height of the glacier is 3200 m.a.s.l. The basin is nearly barren except for the Truelove Valley.

The first spring flood in the Truelove River occurred in late June 1973 and in mid-July in the two other years. As is often seen in the Arctic, this first warm period is followed by a cooler period which rapidly causes a decrease in runoff. Truelove River reflects one of the general characteristics of glaciers by delaying the melt runoff until later in the summer. The storing processes involved in this delay have been explained by Stenborg (1970), emphasizing the glaciological paradox: more energy being absorbed in June-July than is indicated by the relatively small discharge of melt water. The glacier produces, on the other hand, more melt water during July-August than the energy income should allow.

The diurnal discharge variation is very pronounced in the Truelove River, with the minimum occurring in afternoon and maximum around midnight throughout the summer. The flow rate also shows a high response to rain, partly owing to topography since water flows to the valley bottom along steep hillsides, partly owing to very little plant cover and to soils which are near saturation below 2-5 cm. The high response is also a result of an increased precipitation release with higher altitudes (Rydén 1973a). The orogenic effects are, most probably, well developed because the upper parts of the plateau and ice cap are exposed to all wind directions (Holmgren 1971).

One Föhn (chinook), 10 August 1972, resulted in a runoff peak of 25 mm day^{-1} (Holecek, pers. comm.). However, this Föhn did not cause the greatest flood recorded during the two years. Rainfalls in 1973 resulted in considerably greater discharges (Table 1). The exceptional storm on 22-23 July, 1973, to which the Lowland streams and the rivers showed marked response (Figs. 5, 10), was examined by Cogley and McCann (1976).

Summary of snowmelt and runoff

The main characteristics of the streams affecting the Lowland are their rapid snowmelt caused by considerable radiation income and drastically decreased albedo of the snowcover (Courtin and Labine this volume). A daily period of discharge is evident in all investigated streams, most pronounced however in the Truelove River. The snowmelt peaks occur from mid-June to mid-July in the lowland streams, Gully River, and Truelove River. Figure 11 shows the generalized pattern of seasonal variation in discharge, of the three regimes. In addition to patterns shown in the diagram there is a tendency for autumn floods to occur owing to increased frequency of rain at that time, quite similar to mountainous rivers in northern regions. For comparison, the litter transport by a lowland stream is also illustrated in Fig. 11; maximum transport occurs earlier than the discharge peak because litter is removed from a considerably larger area during the early non-channeled surface runoff. The response to rainfall is small in the Lowland streams, considerably greater in the rivers draining the plateau and ice cap.

Hydrologic year

When defining the environment of plants in the Arctic, the generally considered seasons (winter, spring, summer, and autumn) are unsatisfactory from the climatic viewpoint. Four seasons cannot be readily applied to arctic climates; partly because the coldest periods are found often during "spring," partly because three seasons (spring, summer and autumn) are crowded into a period of 3 to 4 months (Thompson 1965).

This is also true of arctic hydrology. Applying the periods suggested by Thompson to hydrological conditions of Truelove Lowland and ordering the periods chronologically within the hydrological (water) year, one arrives at the following surface regime.

September to November

Starting in the fall the depletion of stored water in the basin ends, some soil moisture recharge occurs, and the accumulation of snow and ice for next spring begins. In this scheme the ecosystem is first subject to freezing, beginning at the ground surface and slowly continuing through the thaw (active) layer. Raised beaches have only a thin cover of snow (Brown this volume), whereas heavy accumulation is common in the rock outcrops, slopes of raised beaches, ravines, and small valleys. Sea ice gradually forms, although in Jones Sound open water is found in October-November while complete

Table 2. Hydrological seasons of Truelove Lowland

Period of Winter formation	*The frozen, steady state period*	*Period of metamorphosis*	*Summer season*
Sept Oct Nov	Dec Jan Feb Mar April	May June	July August
Main part of yearly snowfall	Mostly darkness until Feb.	Polar day	Maritime influence
Active layer freezes	Bays and channels covered	Increased cloudiness and snowfall	Ground snowfree, partly dry
Lake and sea ice form	Air temperatures below $-15°C$	Frequent shifts between temp. below and above freezing	Channel flow
Streams and rivers cease	Very small amounts of precipitation	Crusts and ripe snow	Evaporation appreciable
	Permafrost and active layer form continuously frozen soil	Ablation of snow	Precipitation, mainly rain, but small amounts
	Snow cover persistent	Snowmelt runoff starts	Active layer developed
		Rivers start in late June	Temperature $+5 - +8°C$
		Ground mainly frozen except few cm	Flow along permafrost
		Wet ground	

freeze-over at Resolute Bay occurs in mid-October (Allen and Cudburd 1971). Water in the soil, streams, and lakes freezes and thus becomes non-movable. The cover of winter snow is formed. This period is the formation of winter (Table 2).

December to April

A long period of steady state follows, during which the polar night dominates. Bays and channels are covered with ice. Snow in the Lowland is partly redistributed by strong winds but little precipitation is added. Fog seldom occurs, partly owing to the very low moisture content of the cold arctic masses (Met. Branch, Dept. Transport, Canada 1970). For most processes the period may be regarded as stagnant.

May and June

In May the polar day is quite apparent with high insolation, increased cloudiness, and snowfall over the Lowland. A second maximum in precipitation can occur. There are frequent shifts in temperature around the freezing point. Snowmelt starts, resulting in wet or ice-covered ground. The streams do not normally break up until late June (Adams 1966, Stenborg 1970), but non-channeled flow is frequent and forms the first water source to the ecosystem. Soil thaw begins and the upper layer soon fills with water; this may be termed the period of metamorphosis (Table 2).

July and August

Two months can be regarded as the summer season when these islands are under maritime influence. The ground is snow-free and water discharge is channeled. Precipitation and evaporation balance on small amounts, rainfalls are frequent but of low intensity and amount, drizzle and dew are most probably sources of equal or greater importance to the ecosystem for its use of water. Fog is frequent over the sea and moves over the land forming stratus clouds. Water, gradually released when the soil thaws, is probably used by plants earlier than the infiltrated melt water; thus the degree of recharged moisture content at the time for the start of winter becomes important to root systems and soil organisms.

Water budget

Only after several years of field measurements are water budget variables satisfactorily known. Different aspects of the Truelove Lowland climate have been investigated during 5 to 6 summer periods and two over-wintering periods, i.e. 1961-62 (Dahlgren 1974) and 1972-73 (Courtin and Labine this volume). In addition, the hydrological investigations are not complete, nor were they planned to investigate primarily the water budget. Nevertheless, estimations of the interaction and balance of the hydrological variables permit judgments on water availability to vegetation and a key to some important plant-water relationships.

Solid precipitation

In the Arctic where liquid precipitation shows low intensity and small amounts, snow is the dominant influence upon the hydrological cycle. In remote areas the periods of field measurements of water budget variables are mostly restricted in time and make it necessary to estimate, for instance, the precipitation input to the budget. Although the snow course measurements were done earlier than anything else in the field program, the snowpack has often been subject to redistribution across water divides and appreciable evaporation has begun.

Snow course measurements (Table 3) are available for three years for the Intensive Watershed and the adjacent lowland. All these measurements have been done in early or mid-June, a period when accumulation of snow is at its maximum, evaporation losses probably are minor and visible snowmelt has not started.

Evaporation loss prior to snow courses (ablation)

For the years 1972 and 1973 results of the snow studies were presented by Holecek and Vosahlo (1974). The measurements in 1974 by the author are presented below. The field methods in these years were the same. The earlier authors have estimated the losses from the snowpack due to sublimation, although the applied formula(e) were not presented and thus it is not possible

Table 3. Estimation of water equivalent of the snowpack (mm).

Year	*Water equiv.*	*Evap. loss*	*Total*	*Corr. as to ice*	*Estimated total water equiv.*	*Average snow density*
20 June, 1972	98	26	124	+10	134	—
4 June, 1973	78	24	102	+ 8	110	—
15 June, 1974	168	15	183	+ 4	187	0.375

to make completely comparable estimates as to the corresponding loss in 1974.

Snow evaporation takes place when energy is available for the process, viz. not until one month or more after the end of the polar night. Potential sublimation can be calculated according to Eagleson (1970) if temperature is disregarded:

$$Y = Q_i\,(1-A) \cdot 1/q_s \cdot dt \qquad \text{cm} \qquad (2)$$

where Y is depth of sublimation loss in cm (or the depth of condensate), Q_i the incoming short wave radiation, cal cm^{-2} per time unit, A is albedo of the snow surface, q_s the latent heat of sublimation, (680 cal g^{-1}), and dt the time interval of integration. The actual season, April to mid-June, is a period of few radiation records in the Lowland. For the actual purpose it would be appropriate to add radiation data from 1961-62 (Dahlgren 1974), which also covers one winter period (Table 4).

The calculations show that possible ablation during March averaged 0.04 mm day^{-1}, April 0.16 mm, May 0.35 mm, and first part of June 2.65 mm day^{-1}. Thus, 56.6 mm may equal the potential loss prior to snow course measurements.

Obviously, part of the ablation of 2.65 mm day^{-1} in June is snowmelt (Table 4). However, it is difficult to distinguish between the amount of energy used for evaporation and that used for snowmelt. Advice on the solution of the problem can be derived from one particular spring situation. The snow course measurements in 1974 gave an average snow cover in the Intensive Watershed equal to 45 cm, having an average density of 0.38 g cm^{-3} just before rapid melt. As such a snowpack melted in 12 days, daily ablation rates equalled about 14 mm of water equivalent. The necessary energy for this process is 112 ly day^{-1}. The above calculated energy budget (Table 4) shows that early June 180 ly day^{-1} are available of the net radiation after heating snowpack and ground. Consequently, 68 ly day^{-1} may be used for evaporation, which corresponds to 1.0 mm day^{-1}. Although approximate, the figures give the order of magnitude of the processes, melting and evaporation of the snow. The above calculated potential loss of 2.65 mm day^{-1} in June involves snow evaporation of 1.0 mm day^{-1}. Thus 23 mm should correspond to the normal loss from the snowpack of the Intensive Watershed prior to snow course measurements in mid-June.

Another approach to the estimation of snow evaporation (Konstantinov 1963) makes it possible to calculate seasonal evaporation with data on air temperature and air humidity (vapor pressure) measured at standard observation level (200 cm). The approach which is partly theoretical, partly experimental, and statistical, is suitable for estimates over periods of a week to

Table 4. Ablation calculated from energy budget.

Years measured	*Months*	*R (ly day^{-1})*	*A*	*R_N*	*LE**	*Ablation (mm day^{-1})*
1962,** 1973+·	1-31 March	109	0.87	14	3	0.04
1962, 1973	1-30 April	369	0.85	55	11	0.16
1962, 1972, 1973	1-31 May	603	0.80	121	24	0.35
1961-62,** 1971-73+	1-15 June	600	0.50	300	180	2.65

* LE represents the energy that remains for evaporation after heating the snowpack and the ground: the order of magnitude of the relationship between R_N and LE follows the energy budget calculations for Barrow, Alaska, U.S.A. (Weller and Holmgren 1974).

** From Dahlgren 1974

\+ From Courtin and Labine this volume

Table 5. Estimation of evaporation from snow according to Konstantinov (1963).

Month	*Mean Air temp (°C)*	*Rel hum (%)*	*Vapour pressure (mb)*	Konstantinov method. $\Delta T°$	*Corr T°C*	Δe	*Corr e mb*	*Evap. (mm day^{-1})*
				Resolute Bay (20 year mean)				
March	−31.5	—	0.325	+ 14	−18	+0.8	1.13	0 or −0.08
April	−23.1	—	0.775	+ 13	−10	+ 1.3	2.1	0.03
May	−10.6	—	2.00	+ 8	− 2.5	+4.2	6.2	0.20
June	− 0.3	—	5.4	+ 2	+ 1.7	+2.5	7.9	0.80
				Base Camp, Truelove Lowland				
April								
1973	−15.4	72	1.15	+ 11.5	−3.9	+2.0	3.15	0.17
May								
1972	−10.5	83	2.07	+ 8.0	−2.5	+4.2	6.27	0.10
1973	−10.1	80	2.06	+ 8.0	−2.1	+4.2	6.26	0.16
						Average		0.13
June								
1970	+ 1.7	86	5.94	+ 2.0	+ 3.7	+2.4	8.34	1.00
1971	+ 3.5	88	6.91	+ 2.0	+ 5.5	+2.3	9.21	1.30
1972	− 2.5	82	4.07	+ 3.5	+ 1.0	+2.7	6.77	0.72
1973	−10.1	80	2.06	+ 5.0	−5.1	+2.6	4.66	0.09
						Average		0.78

a season. The results are checked with relatively long term experimental data, mostly from pans and lysimeters for both vegetation and snow. The final evaporation from soil, water, and snow can be read from a diagram.

Although this technique shows the lowest degree of accuracy in the range of temperature below −5°C and some steps in it may be subject to improvements, there are great advantages in using this technique. It is evident that at (air) temperature below −20°C, snow evaporation is negligible. Maximum evaporation at temperatures around and below −15°C equals 0.05 mm per day.

Meteorological variables are needed for the calculation over the late part of the "stagnation period" (Table 2), when few records from Truelove Lowland exist (see introduction to Water Budget). Since 20 year means are available, Resolute may serve as a reference. Table 5 presents air temperature and vapor pressure at Resolute, and Truelove Lowland, followed by the calculation steps suggested by Konstantinov. The flux of water from the snowpack is negligible in March, or there may be a sublimation gain. The following months show an increasing evaporation towards 1.0 mm day^{-1}. The few years of data from Truelove Lowland confirm the evaporation values derived from the 20 year series of Resolute, i.e. March 0, April 0.03, May 0.20 and June 0.80 mm day^{-1}. Moreover, the order of magnitude is the same as that using the energy budget approach. June evaporation differs, but this month is apparently a sensitive period for the snowpack. An average of the results of the energy and environmental approaches shows an evaporation loss from the snowpack in June of 0.9 mm day^{-1}.

In summary, evaporation from the snowpack, prior to snow course measurements at 15 June, should equal 20 mm.

Snow course measurements

A summary of the snow course measurements is presented in Table 3. The variations are great, but this may be quite natural. Indeed, the series is too short to establish a water budget term. However, the water equivalent of snow in 1972 and 1973 seems to be small compared with the 20 year mean of Resolute showing 80 mm. Precipitation increases eastwards over the Queen Elizabeth Islands; maps based on 20 year records show an increase from Resolute to Truelove Lowland (320 km to the east) of 20-50% of the Resolute value (Potter 1965).

In the snow course measurements a minor systematic error may have decreased the water equivalent, since snow samplers do not measure the ice layer at the bottom of the snowpack (see Methods). The error may amount to 10% of the water equivalent, inserted into Table 3. It seems appropriate to consider that solid precipitation over Truelove Lowland has a magnitude of 140-145 mm year^{-1}. This corresponds to a mean snowpack depth of about 40 cm.

Liquid precipitation

The part of annual precipitation that was rain equalled 12% and 43% in 1972 and 1973 respectively (Courtin and Labine this volume). Long term averages at Resolute also show that the liquid part (rain, drizzle, dew) of yearly precipitation is the smaller, 57 of 136 mm, or 42%. The variations are great, however (23 mm of 132 mm in 1972 and 88 of 167 mm in 1973) (Met. Branch, Dept. Transport, 1970, and partly unpublished). There is a tendency for decreasing rainfall when snowfall increases; the reverse is less true, i.e. there is no evident decrease in snowfall when yearly rainfall increases.

Rainfall is extremely variable from year to year over the Canadian Arctic (Thompson 1965) and the few years of records at Truelove Lowland confirm that picture. A summary of the various rainfall records over the Lowland is given in Table 6. Note that the values refer to recording periods of different lengths. However, there is a tendency of increasing precipitation towards the

Table 6. Liquid precipitation over the Truelove Lowland 1970 through 1974.

Site	*1970*	*1971*	*1972*	*1973*	*1974*	*Average*
Base Camp Stn.	16.3	46.5	15.2	72.8	30.8	36 mm
Intensive Watershed	—	—	35.4	53.3	39.0	43 mm

Table 7. Liquid precipitation simultanously recorded at three Devon Island sites in 1973, 1 July - 13 August.

Site	*Rainfall amount (mm)*
Base Camp Station	57.2
Intensive Watershed	40.7
Plateau, 300 m.a.s.l.	59.9

Data sources: Base Camp—Courtin and Labine (this volume), Intensive Watershed—Holecek and Vosahlo (1974), the plateau—Labine (1974). Instruments are standard gauges except for Intensive Watershed where the evaporimeter E-801 was operating.

interior of the Lowland (Table 6) and an increase with elevation (Table 7), both trends valid only for liquid precipitation.

If all values are lumped together the arithmetic mean could serve as an areal mean of yearly liquid precipitation over the Lowland, which equals 38.7 mm; considering the error sources of precipitation measurements it is assumed that 40 mm is closer to the "true" value.

Instrumental errors

All readings of precipitation, in particular those of exposed sites, are impaired by errors, the magnitude of which depends on shape of orifice and windshield, exposure, level, windspeed, type of precipitation, etc. All of them, however, result in a catch less than the real amount of precipitation (WMO 1970).

From experiences in exposed areas (Kirigin 1972, Rydén 1973a) it is concluded that unshielded gauges can lead to an error of at least 4% on the Lowland and 10% on plateau.

Moreover, the pattern of rainfall in high arctic areas seems to be that of frequent rain or drizzle but of low intensity and small amounts. There are problems of measuring "traces of rain" and getting all the water out of a gauge — thus a 4-6% error may arise, corresponding to 1-3 mm per summer.

Summary of precipitation

Characteristic snow depths have been established for a few distinct surface types of the Lowland. Based on the distribution of major topographic plant community units, Table 8 shows an estimate of an overall snow cover for Truelove Lowland. The resulting mean snow water equivalent for the entire Lowland is 142 mm.

The mean liquid part of precipitation, 39 mm, corrected according to errors discussed above, would equal 42 mm. Thus the total annual precipitation over the Lowland based on data from 5 years only (Table 6), should equal 185 mm.

Runoff

The regime and special characteristics of snowmelt and runoff in the three studied watersheds and their surroundings were described earlier. For water

Table 8. Snow accumulation on the Lowland.

Area (ha)	*Type of area*	*S Average snowdepth (cm)*	*V Snow volume ($cm^3 \times 10^{-8}$)*
293	Raised ridges, crest and upper slope	20	5860
274	Transition zones	100	27400
298	Slopes related to raised ridges	60	17800
533	Rock outcrops (2/3 of the area accumulates)	100	35500
180	Lichen barren	15	2700
1904	Meadow, ponds, and lakes	45	85680
	Total		174940

If snow density (d) averaged 0.35 g cm^{-3}, mean snow water equivalent for the entire Lowland is:

$V \times d / A = 142.4$ mm

budget estimates there is an additional need for an integration over the different years, particularly as to the Intensive Watershed.

Summer runoff during the years 1972 and 1973 was reported by Holecek and Vosahlo (1974) to equal 54 and 74 mm, respectively. The corresponding value for 1974 was 101 mm, a winter of more snow. Other reasons for the great differences, as described in Methods, result from missing non-channel flow of early snowmelt, and leakage through the weir. Both errors underestimated the total yield. To compensate for the error, we apply the relationship:

$$R(\text{"true"}) = R(\text{recorded}) + \Delta R(\text{non-chan.}) + \Delta R(\text{leak.}) \quad \text{mm} \qquad (3)$$

During 1974 that part of snowmelt that could not be recorded with water level gauge at the weir was *ca.* 15% of the total runoff from the Intensive Watershed. The snowmelt start in 1972 and 1974 show great similarities. Hence the missing discharge volume of 1972 is assumed to have the same portion, or 15%, whereas in 1973, with a slower beginning of runoff, the term ΔR (non-channel) is assumed to be one-half the portion of 1972. Leakage may have averaged about 7% in both years. According to (3) we thus arrive at 66 (instead of 54) mm in 1972, and 85 in 1973. The average of three years discharge, 84 mm, corresponds to a period of 42 days within the Intensive Watershed (drainage area 0.12 km^2). Specific annual discharge then equals 2.7 l s^{-1} km^{-2}.

Table 9. Total discharge from Intensive Watershed.

Year	*Dates*	*no. days*	*Discharge (mm)*	*Specific discharge* ($l\ sec^{-1}\ km^{-2}$)
1972	1 July (5)* — 23 July	23	66	2.1
1973	12 June (14)* — 25 Aug	74	85	2.7
1974	28 June — 26 July	28	101	3.2

*Dates for start of recording discharge.

Evapotranspiration

Mass balance method

An annual estimate of evapotranspiration (E) is provided with fairly good accuracy through the mass balance method, i.e. as the residual in a water balance equation:

$$P = R + E \pm \Delta S \qquad (4)$$

once the yearly runoff (R) and precipitation (P) are known from measurements in the field and the storage term $\pm \Delta S$ can be disregarded. Applying the results above, assuming that the averages found are of a long-term character, omitting the storage term the yearly evapotranspiration amounts to 185−84=101 mm. The snow evaporation was earlier estimated on a long term basis to be 20 mm. Thus the "normal" summer losses of water to the atmosphere from soil, plants, and open water surfaces should equal 81 mm.

Energy balance approach

An energy balance approach to the assessment of evapotranspitation is based on the relationship:

$$R_N - G - LE - H=0 \qquad ly\ min^{-1} \tag{5}$$

where R_N is the net radiation at the surface in ly min^{-1}, G is the flux of heat into the ground. LE is the latent heat of evaporation/condensation and H the sensible heat, the turbulent heat flux to the atmosphere. Purely biological, chemical, and physiological processes are disregarded. A requisite tacitly understood, is a continuous supply of water. The latent heat of evaporation is determined through:

$$L_T=596.9 - 0.580T \qquad cal\ g^{-1} \tag{6}$$

from basic thermodynamics. L_T is the latent heat of evaporation at the temperature T°C. In the terrain under consideration the air temperature range is roughly 0 to + 10°C during the growing season. For all estimations of evaporation at the present purpose L_T is averaged to 594 cal g^{-1}.

It has been found at high latitude bogs in the U.S.S.R. that, other factors being equal, the dependence of evaporation upon the energy flux is nearly linear, or:

$$E=K \times R_N \tag{7}$$

where E is the evaporation in mm hr^{-1}, R_N the net radiation is cal cm^{-2} hr^{-1} and K is an empirically determined coefficient (Romanov 1962). Once the relationship between E and R_N is found, it should be possible to estimate seasonal values. The coefficient has the dimension mm cm^2 cal^{-1}.

The relationships between the term LE and the other terms of equation (5) were investigated on the Lowland during selected days, i.e. sunny days and cloudy respectively (Addison 1973). From these results the following relationships analogous to (7) are derived, valid for raised beaches:

$$LE=0.154\ R_N \qquad Cal\ cm^{-2}\ min^{-1}\ \text{at clear sky} \tag{8a}$$
$$LE=0.271\ R_N \qquad Cal\ cm^{-2}\ min^{-1}\ \text{at overcast} \tag{8b}$$

The relationships (8) refer to two 48 hr periods in July 1972, one with clear sky conditions and the other 10 tenths of the sky covered. The actual values of LE, 0.051 and 0.026 cal cm^{-2} min^{-1}, respectively, result in daily evapotranspiration values 1.24 and 0.63 mm day^{-1}, respectively, on raised beaches.

Applying available data on average cloudiness and assuming the evaporating period being 60 days, the seasonal water loss from vegetation and soil at raised beaches is 74 mm. An approach of the same kind as (8) for meadow type of surfaces shows:

$$LE=0.512\ R_N \qquad Cal\ cm^{-2}\ min^{-1}\ \text{at clear sky} \tag{9a}$$
$$LE=0.545\ R_N \qquad Cal\ cm^{-2}\ min^{-1}\ \text{at overcast conditions} \tag{9b}$$

Actual values of LE equal 0.131 and 0.139 cal cm^{-2} min^{-1} which corresponds to daily rates of evapotranspiration of 3.20 and 3.39 mm day^{-1}, respectively. A calculation, similar to the previous, gives 185 mm which represents the water loss to the atmosphere from vegetation and soil during summer period of hummocky sedge-moss meadows. The figure means potential evapotranspiration from a sub area having the densest vegetation cover in the Lowland. A more realistic value for the whole area will come much closer to the mass balance result of 81 mm per year.

Yearly evapotranspiration

The estimations above consider raised beaches and meadows respectively. The whole area of the Intensive Watershed contains rock outcrop to a portion of 11.2%, meadows 63.1%, and raised beaches 25.7%. Consequently the evapotranspiration E is composed of three parts and the weighted mean for 1972 was 67.6 mm and for 1973, 119.7 mm. The corresponding mean daily values of evapotranspiration were 1.47 mm and 1.58 mm, respectively.

The flux of water to the atmosphere from lakes and ponds also must be considered in addition to the energy budget calculations above. Holecek and Vosahlo (1974) reported a water evaporation of 4 and 8 mm respectively, during the summers of 1972 and 1973. For 1971 as to standing water Addison (this volume) found an evaporation of 1.6 mm day^{-1}, which is 96 mm over a growing season of 60 days.

Applying pertinent net radiation (Courtin and Labine this volume) and assuming an albedo of the ponds and lakes to equal 50% when partly ice-covered (June), and 5% when ice-free (mid-July and August), the open water surfaces produce an evaporation of approximately 60 mm. One-half of this water loss to the atmosphere occurs during the melting phase until a water temperature maximum of 5° to 8°C is achieved in late August, the other one-half (30 mm) during 2-3 weeks before ice forms in September.

Albedo

In the calculations of evapotranspiration, the key to the order of magnitude seems to be the albedo. Bare ground conditions similar to those of Truelove Lowland have been reported by Gavrilova (1963) to have values of 15% to 18% for dry tundra with brownish remains of grass from the previous year.

The albedo values obtained at Devon Island in 1962 (Dahlgren 1974) refer to the hummocky, relatively dry meadow (the instrument location being 50 m W of the later erected meteorological tower). His result, a summer albedo of 15%, is valid for meadow and heath, while fine-grained and dry raised beaches with light lichens may show 20% as a maximum. The albedo variation (Courtin and Labine this volume) was 70% to 90% from March to May. During June there was a rapid decrease to 15% during the snow-free period. It is concluded that errors in albedo affect estimates of evapotranspiration for the wettest month, June, and that calculations thus may deviate by 10-20 mm.

Summary of evapotranspiration

From the different approaches above in estimating evapotranspiration and potential evapotranspiration the following is recalled. The mass balance method which considers the whole area, resulted in 81 mm of annual water loss. The energy budget for raised beaches and hummocky sedge meadows respectively, gave 75 and 185 mm under assumptions of generalized conditions over the growing season. In particular, energy budget calculations for the years

1972 and 1973 taking into account the distribution of the main surface types, viz. meadows, beaches, and rock outcrops, showed *ca.* 70 mm (1972) and 120 mm (1973).

In summary, evapotranspiration rates for Truelove Lowland and the Intensive Watershed averaged *ca.* 90 mm from the end of main snowmelt period through the end of the growing season.

Change in storage

As to exchange, there can be: (1) intensive water exchange, characterized by folded mountain regions; (2) weak water exchange of lowland areas; and (3) the intermediate changes of the highlands (MacKay and Løken 1974).

Truelove Lowland belongs to the second zone but one should consider also the third zone because of the cliff, slope, and foothill areas belonging to the plateau. With respect to drainage, the soils in the Intensive Watershed can be divided into two groups according to Rieger (1974): the poorly drained soils that are usually or always saturated and where the active layer thickness is equal to or less than 30 cm (i.e. the hummocky sedge-moss meadows), and the well-drained soils with dry permafrost, and active layer of threefold depth and having moisture contents below field capacity (i.e. the raised beaches).

The change of storage can be determined as a residual when other variables are known. In the water balance equation (4) the storage terms ΔS can be divided into several parts, representing all these items of water which take part in a complete hydrological cycle. Part of the water could be removed or stored to an unknown later year and enter into the water cycle of another year. Examples of natural storage are groundwater, soil water, lakes and ponds, snow and ice bodies, or:

$$\Delta S = \pm \Delta S_G \pm \Delta S_S \pm \Delta S_L \pm \Delta S_I \qquad (10)$$

The change of groundwater storage is unknown in this case because no groundwater level observations have been made. The change of soil water storage (see Methods) is possible to estimate, at least roughly, from measurements of soil water content at the beginning and end of the growing season (Courtin and Labine this volume). The analyses are based on weekly soil moisture content data, expressed as percent of oven dry weight, from the summers of 1970 through 1973 and volume-based moisture data from 1974, including studies on soil-water relationships.

At the sedge meadow the seasonal variation of soil moisture content shows a maximum value at snowmelt, which was the start of sampling, with a decreasing trend throughout the growing season. Owing to rainfall, intermittent peaks are superimposed on this trend. The soil moisture content is at annual minimum in August and recharges a little prior to freezing in August-September. The main recharge occurs during snowmelt in spring, when much of the surface also is subject to super-saturation. Results on the water exchange of a meadow profile and others are presented in Table 10.

A similar pattern of variation, but having a lower moisture content, is found on the raised beaches. The response to rain is considerable although the coarse material should allow for rapid percolation through the whole active layer. The lower 10-12 cm above the frost table are saturated with water (Walker and Peters this volume). The average porosity of the active layer is great enough for a storage of more than 10 years of normal summer precipitation. The annual exchange of soil water is approximately 30-50 mm, which equals rainfall in summer.

Table 10. Available water in the active layer.

Site	*Thaw depth max. (cm)*	*Bulk density ($g\ cm^{-3}$)*	*Density of solids ($g\ cm^{-3}$)*	*Water content % by volume*	*Field capacity (%)*	*Seasonal min (mm)*	*Seasonal max (mm)*	*Available soil water (mm)*
Raised beach	110[2]	1.3-1.7[6]	2.6	2-15[3]	15	22	165	176
Transition zone	40[2]	0.9-1.1		10-40[4]	54	20	136	205
Wet sedge hummocky meadow	25-30[1]	0.5-0.34[1, 5]	0.20-0.06	50-90	82	135	260	240
Plateau station	30[5]	1.6	2.6	10-54	48	30	162	120

Sources of data:
1. Muc, 2. Svoboda (both this volume); 3. Courtin and Labine (this volume and unpublished data); 4. Addison (pers. comm.); 5. Bliss (1975); 6. Walker and Peters (this volume).

Regarding plateau soils (excluding highly porous stonefields), and soil in the sedge meadows, one normally finds them saturated a few centimeters below the surface in late summer and more or less over-saturated after snowmelt.

The distribution of water below the surface is strongly affected by permafrost and the development of a relatively thin active layer in most of Truelove Lowland. Percolation is constrained by the permafrost table and the saturated layers above this level. Lateral movement can occur on the surface and along the permafrost table (Rydén 1973b).

Stored water gradually becomes available to root systems as the thaw proceeds downwards during the growing season (Rydén 1976). The soil thaw has been found to depend strongly on both precipitation and the varying insulating properties of the soil owing to different degrees of saturation and on microtopography (Rydén and Kostov 1977). Soils thaw slowly, but a thin layer of *ca.* 1 cm is thawed almost immediately after removal of the snow. In general 60% to 70% of the active layer thaw occurs by mid-season (Svoboda this volume). The active layer of Truelove Lowland soils reaches maximum thickness and maximum soil moisture capacity in late July (Muc this volume, Bliss 1975; compare with MacKay and Løken 1974) (Table 10).

During the growing season some of the soil moisture of the root zone is used because moisture flux through evapotranspiration is considerable (0.5 to 3.5 mm day^{-1}) (see section on Evapotranspiration). Not until autumn rain does the upward flux decrease enough to give way to some recharge of the active layer, especially in the coarse textured soils of the beach ridges.

Immerk Lake shows maximum water level during ice-melt in late June or early July, after which there is a steady decrease through the end of August. A certain response to rain is found, such as moderate peaks that last a day or two after rainfall. The difference between the extreme levels represents approximately the storage of water in the lake during the winter and the precipitation added until ice breaks up. During the years of 1972 and 1973 (Minns this volume) and in 1974 the level differences (6, 15, and 39 cm) correspond to a lake storage of 15, 9 and 89 mm. Although the two smaller figures are unrealistically small, the figure for 1974 corresponds fairly well to the winter precipitation of the lake.

The possible change of storage from year to year in the ponds of the Intensive Watershed has not been determined because of the difficulties involved. Although they represent a minor areal part of the watershed, the magnitude of the exchangable storage may equal the amount of summer precipitation.

The fourth type of storage is definitely the greatest and plays an important role in the long term water balance, i.e., the snow and ice component. In both severe and ordinary winters, a considerable amount accumulates in depressions and gullies that are partly shaded from solar radiation. It is shown (see Snowmelt and Runoff) that snowdrifts (plateau) of earlier years may melt during an unusually warm summer. Water is thereby added to the discharge from the basin in an amount that cannot be explained by means of the parameters of that year. This component is most probably responsible for the greatest part of the questioned storage term of 217 mm in 1973 (Holecek and Vosahlo 1974).

Fluvial transport

Some miscellaneous parameters are highly dependent on water occurrence but do not belong to the hydrological cycle other than in a chemical sense. Among these parameters is the transport of organic material by turbulent flow of melt water (Table 11). Although much suspended finer material passed through the mesh of the litter weir, it depicts the transport properties of flowing water. The shallow overland flow of melt water in the unstable non-channeled stage (see Runoff under Methods) is turbulent enough to carry a load of plant material, muskox dung, etc. The peak of litter transport was somewhat earlier than the water flood itself. The depletion of litter transport is much more rapid than that of sediment (see summary of Snowmelt and Runoff). The transport mechanism of shallow, turbulent water, mixed with ice particles is able to redistribute rather than remove litter, largely because of the resistance provided by standing vegetation (Muc this volume).

Table 11. Transport of organic material by turbulent flow of melt water July 1974.

Time period Start	End	Days	hrs	*Litter weight* (g)	*Discharge mean*	*integrated* ($l s\ m^{-3}$)	*Specific transport* ($g\ m^{-3}$)
1 July 2200 -	3 July 1600	1	18	22.3	6.5	990	0.020
3 July 1600 -	4 July 2245	1	07	37.15	6.4	710	0.050
4 July 2245 -	10 July 1645	5	18	22.61	6.8	3400	0.007
10 July 1645 -	19 July 1615	9	00	11.63	9.2	7200	0.002
19 July 1615 -	26 July 1230	6	20	1.60	1.0	590	0.003

Summary of water budget

An attempt to synthesize the mass budget terms of three years was shown in the section on Evapotranspiration. The water budget terms equalled:

$$\begin{array}{ccc} P & R & E \\ 185 = & 84 + & 101 \end{array} \quad \text{mm} \qquad (11)$$

The water loss term E consists of two parts, separated in time, viz. snow evaporation and evapotranspiration, during the growing season. Total snow evaporation equalled 20 mm, and 81 mm remained for evapotranspiration.

Evapotranspiration, calculated by different energy budget methods, averaged 90 mm, and the open water surfaces similarly lost *ca.* 60 mm over the

Table 12. Water budget (mm) of Truelove Lowland.

Year	Precipitation Snow	Precipitation Rain	Precipitation Total	Runoff	Winter: snow evaporation	Growing season: evapotranspiration	Change in storage
1972	132	36	168	66	32	69	Plus 1
1973	111	73	184	83	41	110	Minus 50
1974	169	35	204	101	34	65	Plus 4

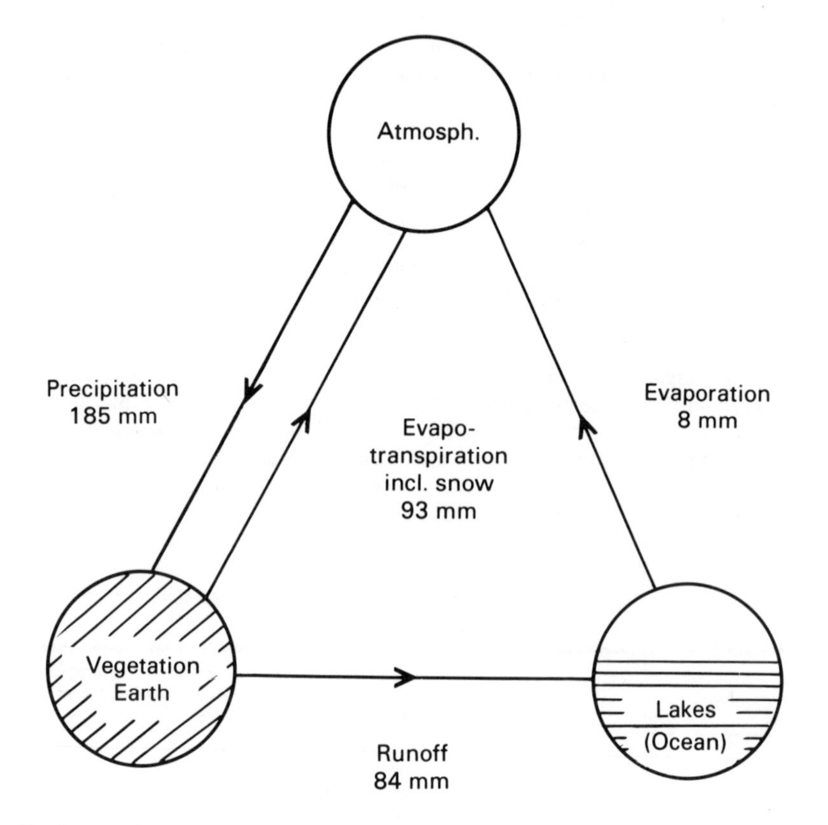

Fig. 12 Generalized water budget of Truelove Lowland.

growing season. Considering the areal distribution of evapotranspiring surfaces (plants and bare soil, 63%), evaporating surfaces (open water, 12%) and non-evaporating surfaces (rock outcrop, 25%), respectively, (Holecek and Vosahlo 1974), the water losses from the Intensive Watershed averaged:

E=	E (snow) +	E(open water) +	ET(plants and soil)	
101	20	8	73 mm	(12)

The expressions (11) and (12) represent the estimated total water balance on a long term basis for Truelove Lowland, in particular the 0.12 km^2 area of the Intensive Watershed.

Discussion and Summary

As with all research, some areas require special studies. Certain aspects of soil water occurrence and movement are not well known, in particular the recharge processes and the magnitude of storage available in the active layer. The role of a permafrost table in soil water movement could be studied under laboratory conditions. Vertical redistribution of soil moisture profiles of the active layer occur most probably during the periods when the ground is partly frozen: this process affects the moisture around the root systems and also changes the conditions for infiltration and percolation of snowmelt. The measurement of arctic precipitation seems to need a special recording gauge, i.e. one which is able to record occasions of dew, drizzle, and rainfall as traces that most probably are of greater use for high arctic plants than any pattern of greater rains (Cook 1960).

The problems encountered in the treatment of evapotranspiration were grouped into four parts, i.e. the water loss from snow and ice, from water surfaces, from bare soil, and from vegetation. Gradients probably exist, in part from the coast towards the interior of the Lowland, from the Lowland to higher elevations (Rydén 1973a) as expressed by level differences within the Lowland, and between the Lowland and the plateau. In order to quantify the water balance in this area, estimates of water loss from the plateau are needed. A carefully observed soil lysimeter on the plateau (Labine, unpublished data) showed in the present author's analysis, an average evapotranspiration over the summer season of 0.2 to 0.3 mm day^{-1}, whereas the energy balance approach gave about 0.96 mm day^{-1}. The latter, real or not, is three times greater than the measured value. Hydrogeological and aerodynamical reasons explain quantitatively that the measured values are too low. The estimated value, on the other hand, represents potential evapotranspiration and should thus be higher than the real evapotranspiration. It is evident, however, that the satisfactory establishment of a water budget needs a series longer than five years and that estimates must be checked against an independent approach.

Snowmelt runoff shows similarities with that of the arctic coastal lowlands in Alaska (Carlson and Kane 1975). In the Alaskan arctic coastal tundra (Bunnell et al. 1975), estimated annual values are 175 mm precipitation (measured), 103 mm runoff (estimated), and 72 mm evapotranspiration (estimated). As expected, the input to the hydrological system is of the same magnitude (185 mm for Truelove Lowland). The other two variables are, however, of reversed size (84 mm and 101 mm for runoff and evapotranspiration respectively). The greater evapotranspiration is most probably related to the radiation income which is considerably higher over the Truelove Lowland (Courtin and Labine this volume).

Water availability can be described in terms of hydrological seasons. The formation of winter (September to late November) is the freezing period, the period when the land becomes snow covered. The period of steady state (December to late April) is the long unbroken winter when snow is redistributed but when little precipitation is added. The metamorphosis period (May-June) is characterized by high insolation, increased cloudiness, and snowfall. Non-channeled flow is frequent by mid-June and the ground begins to thaw. The summer season (July-August) is the snowfree season and the period when water discharge is channeled. Rainfall is frequent but of low intensity and amount.

The freeze-thaw cycle is important in the water budget. During snowmelt in the meadows, the frozen ground forces the melt-water to flow above the surface, regardless of whether or not the soil is fully recharged in the layers below. Thawing to an appreciable depth does not occur until after snowmelt; this gives rise to two different periods of water movement, horizontal until snowmelt is complete and vertical during the remainder of the summer season. On the raised beaches with rapid soil thaw during snowmelt, some recharge occurs.

Ice cover on streams and lakes develops in September and lasts until June-July, although some ice in deeper lakes may remain all summer (see Courtin and Labine this volume).

The rivers of the Lowland, as well as the small streams, show a regime which is dominated by snowmelt. In the Intensive Watershed, the short period of non-frozen conditions comprises two different parts. An intensive melt period of 1 to 2 weeks which yields about 80-90% of the annual total discharge, followed by a period of steady depletion with moderate response to rainfall. Late summer conditions show negligible surface runoff.

On Truelove Lowland, runoff pattern is modified by the addition of water from the 300 m high plateau and from its deeply eroded ravines. A considerable amount of meltwater originates from winter precipitation accumulated in these ravines. In particular, a very warm summer can release water from old snowdrifts. The runoff thus contributed to the Lowland originates as snowmelt, active layer thaw, and rainfall. Meadows on the northeastern border of the Lowland are thereby provided with water throughout the summer. Runoff from the remainder of the Lowland receives no contribution from the plateau.

Summer runoff in the Lowland is characterized by a weak to moderate daily variation; in Gully and Truelove Rivers by a very pronounced daily variation in the discharge.

The snowmelt peaks in the three streams (Fig. 11) occur first in the Intensive Watershed, later in Gully, and latest in Truelove River, which is in agreement with the increasing height extension of their drainage basins. The response to rainfall also increases in the same direction. Knowledge of precipitation patterns on the polar desert plateau is badly needed.

The water budget within Truelove Lowland is not accurately defined on the basis of project data. The recorded seasons were few and the instrumentation not planned for water balance purposes. Thus, the calculated "long term water budget" is approximate and may guide future research:

$$P = R + E \qquad (13)$$

$$185 = 84 + 101 \text{ mm}$$

where P is precipitation, R is runoff, and E represents the total loss from soil, plants, snow, and free water surfaces. Dividing E into its components, one arrives at 20 mm snow evaporation, 8 mm open water evaporation, and 73 mm evapotranspiration. The budget terms thus defined may illustrate the water flow through this high arctic ecosystem (Fig. 12).

Evapotranspiration exhibits its maximum of about 3 mm day^{-1} during a period shortly after snowmelt and decreases to less than 1 min day^{-1} in late August. The total for the non-frozen period may amount to more than 150 mm during favorable summers.

Precipitation often shows two maxima during this year, both of them yielding solid precipitation. The first appears in the winter formation period and constitutes the winter snow cover. The winter minimum is pronounced and reflects the extremely low water content of air masses during the cold polar night. A second maximum may occur during the metamorphosis period yielding snow and in the end of this period mixed snow and rain. Summer precipitation is of low intensity and shows small amounts but is very frequent. This particular pattern of liquid precipitation is most probably of great importance for plant growth in the ecosystem. Of the total annual precipitation (185 mm), roughly 30% is liquid. The variation in amount of the portion of solid vs. liquid is large, which is the usual pattern of precipitation in the High Arctic.

Acknowledgments

Hydrological studies were performed by Mr. George Holecek and Mr. Milan Vosahlo, during the summer periods of 1972 and 1973. Their installations in the Intensive Study Sites and their basic analyses were necessary prerequisites for the 1974 study. Parallel to this were the meso- and micrometeorological studies, the results of which are frequently applied in the hydrology synthesis.

Unpublished data and results have kindly been given to me by Mr. Holecek and Mr. Vosahlo, Dr. G.M. Courtin, Mr. C.L. Labine, Dr. P.A. Addison, Dr. Jan Addison, Dr. K.L. Bell, Mr. R. Schulten, and Dr. L.C. Bliss. Valuable discussions with all of them and with Mr. A. Skartveit, Bergen, Norway, have resulted in this synthesis.

The financial support from Alberta Environment and the National Research Council of Sweden is gratefully acknowledged. Without the initiating efforts and support — on the Scandinavian side from Dr. Sten Johnsson, Upsala, and Dr. L.C. Bliss on the Canadian side — this paper would never have been written.

References

Adams, W.P. 1966. Ablation and run-off on the White Glacier, Axel Heiberg Island, Canadian Arctic Archipelago. McGill Univ., Montreal, Canada. 77 pp.

Addison, P.A. 1973. Studies on Evapotranspiration and Energy Budgets in the High Arctic: A Comparison of Hydric and Xeric Micro-environments on Devon Island, N.W.T. M.Sc. thesis. Laurentian Univ., Sudbury, Ontario. 119 pp.

Allen, W.T.R. and B.S.V. Cudburd. 1971. Freeze-up and break-up dates of water bodies in Canada. Canadian Meteorological Service. CLI-1-71. Toronto. 144 pp.

Anonymous 1970. Climate of the Canadian Arctic. Met. Branch, Dept. of Transport, Canada. Ottawa 1970. 71 pp.

Bliss, L.C. 1975. Devon Island. pp. 17-60. *In:* Structure and Function of Tundra Ecosystems. T. Rosswall and O.W. Heal (eds.). Ecol. Bull. Vol. 20. Stockholm: Swed. National Res. Coun. 450 pp.

Bunnel, F.L., S.F. Maclean, Jr. and J. Brown. 1975. Barrow, Alaska. pp. 73-124 (see Bliss 1975).

Carlson, R.F. and D.L. Kane. 1975. Hydrology of Alaska's Arctic. pp. 367-373. *In:* Climate of the Arctic. Proceedings of 24th Alaska Sci. Conf., Fairbanks. 436 pp.

Cogley, J.G. and S.B. McCann. 1976. An exceptional storm and its effects in the Canadian High Arctic. Arctic Alp. Res. 8: 105-110.

Cook, F.A. 1960. Rainfall measurements at Resolute, N.W.T. Revue Canadienne de Geographie, 14: 45-50.

———. 1967. Fluvial Processes in the High Arctic. Geogr. Bull. 9:262-8.
Dahlgren, L. 1974. Solar radiation climate near sea level in the Canadian Arctic Archipelago. Arctic Institute of North America Devon Island Expedition 1961-1962. Met. inst., Uppsala Univ., Medd. Nr. 121. 120 pp.
Eagleson, P.S. 1970. Dynamic Hydrology. McGraw-Hill, New York. 462 pp.
Gavrilova, M.K. 1963. Radiation climate of the arctic. Israel Progr. for Scient. Transl., Jerusalem 1966. 80 pp.
Holecek, G. and M. Vosahlo. 1974. Hydrology. Progress Report, Devon Island Project. 36 pp.
Holmgren, B. 1971. Climate and energy exchange on a sub-polar ice cap in summer. Arctic Institute of North America Devon Island Expedition 1961-1963. Met. Inst., Upsala Univ., Medd. Nr. 107-112. 391 pp.
Kirigin, B. 1972. A contribution to the problem of precipitation measurements in mountainous areas. pp. 1-12. *In:* Distribution of Precipitation in Mountainous Areas, WMO OMM No. 326. 587 pp.
Konstantinov, A.R. 1963. Evaporation in Nature. Transl. from Russian. Jerusalem 1966. 523 pp.
Labine, C.L. 1974. Measurement and Computer Simulations of Micro-climate Differences Between a Polar Desert Plateau and a Nearby Coastal Lowland. M.Sc. thesis, Univ. Guelph, Guelph, Canada. 85 pp.
McCann, S.B., P.J. Howarth, and J.G. Cogley. 1972. Fluvial processes in a periglacial environment. Queen Elizabeth Islands, N.W.T., Canada. Inst. Brit. Geogr. 55: 69-82.
MacKay, D.K., and O.H. Løken. 1974. Arctic hydrology. pp. 111-132. *In:* Arctic and Alpine Environments. J.D. Ives and R.G. Barry. (eds.). London: Methuen. 999 pp.
Potter, J.G. 1965. Snow Cover. Climatological Studies No. 3. Met. Branch, Dept. of Transport, Canada. Toronto 69 pp.
Rieger, S. 1974. Arctic soils. pp. 749-769. *In:* Arctic and Alpine Environments. J.D. Ives and R.G. Barry. (eds.). London: Methuen. 999 pp.
Romanov, V.V. 1962. Evaporation from Bogs in the European Territory of the U.S.S.R. Transl. from Russian. Jerusalem 1968. 183 pp.
Rydén, B.E. 1973a. Precipitation and Evaporation in a High Mountain Basin and Their Relationships with Altitude. Thesis, Hydrological Div., Dept. of Phys. Geogr., Univ. Uppsala. 228 pp.
———. 1973b. Markvattenrorelse i torv-ett experiment. (Soil water movement in tundra peat — an experiment.) pp. 59-64. In IHD field symposium on "Bog hydrology. Mathematical models in hydrology.", Trondheim 1973. (In Swedish) 211 pp.
———. 1976. Water availability to some arctic ecosystems, Nordic Hydrology. 6: No. 2. pp. 101-110.
———. and L. Kostov. 1977. The thaw-freeze cycle of tundra. pp. 16-48. *In:* Ecology of a Subarctic Mire. M. Sonesson (ed.). Ecol. Bull. No. 22. Stockholm: Natural Science Research Council. 440 pp.
Stenborg, T. 1970. The delay of runoff from a glacier basin. Geogr. Ann., Vol. 52, Ser. A, No. 1. 30 pp.
Thompson, H.A. 1965. The Climate of the Canadian Arctic. Met. Branch, Dept. of Transport. 32 pp.
Weller, G. and B. Holmgren. 1974. The microclimate of the arctic tundra. J. Appl. Meteorol. 13: 854-862.
WMO. 1970. Guide to Hydrometeorological Practices. 2nd ed. WMO — No. 168. TP. 82. Geneva 1970. 330 pp.

Vegetation history and plant communities

Palynological analysis of a peat from Truelove Lowland

Vlasta Jankovska and L.C. Bliss

Introduction

Because of logistic difficulties, the generally shallow nature of peats, and the problems of interpreting the meager pollen profiles, the recent history of vegetation in the High Arctic is little known (Nichols 1974). Peat depths in most sedge-moss lowlands are only 10 to 50 cm and in uplands seldom more than 1 to 3 cm. Peats more than 1 to 2 m in depth seldom occur, thus greatly reducing potential areas that can be analyzed for past vegetation-climatic conditions.

The age of basal peats was determined at three sites in the Lowlands by Barr (1971). They ranged in age from 6,900± 115 years B.P. in the Truelove Valley, 4,300± 95 years B.P. for peats along Beschel Creek, and 2,450± 90 years B.P. in ice-centre polygons 1 km SW of Base Camp. The latter site was cored for samples.

The objective of the study was to determine pollen and plant material content throughout the limited depth of the profile and to compare these results with those of other high arctic areas.

Methods

Core samples, 170 cm in length were obtained with a manual core drill from the centre of an ice-centre polygon. The cores (25 mm diameter) were deep frozen and therefore it was possible to rinse them in clean water before placing the 10 cm lengths in clean plastic bags. Due to high ice content (30% to 70% by volume) the samples were allowed to melt and the water evaporate prior to storage.

In the laboratory, samples for pollen analyses were processed according to the Erdtman (1943) method of acetalysis. Inorganic admixtures were dissolved in hydrofluoric acid. For each sample, all pollen was counted on three preparations of 20×20 mm areas; one additional replicate was checked cursorily. Pollen frequency is based upon absolute counts (Jorgensen 1967). Macroscopic analyses were made of the moss and vascular plant material and volume estimates made.

Results and Discussion

Macroscopic analysis

Throughout this profile moss peat and identifiable moss leaves predominated (Fig. 1). Mosses, rich in species diversity, were so well preserved in the permafrost, that chlorophyll was often present in leaf tissue, even in the deepest layers (150 to 170 cm). The most common genera included *Polytrichum*, *Mnium*, *Messia* cf. *uliginosa*, *Tomenthypnum nitens*, *Bryum*, *Calliergon*, and *Drepanocladus*.

Leaves of *Salix arctica* were found throughout the profile and *Dryas integrifolia* leaves were found below 135 cm. No seeds or fruits were found, which coincides with known low rates of seed production and relatively high rates of seed harvest by lemming and passerines. Lemming reach their highest numbers in this ice-centre polygonal area (Fuller et al. this volume).

The decomposed sediment appears to be partially decomposed masses of *Nostoc* which occur near the shore in Phalarope Lake. This material washes ashore and is found incorporated in moss peat in developing polygons near the lake. There is a progression of polygons with decreasing height but of comparable diameter as one approaches the lake.

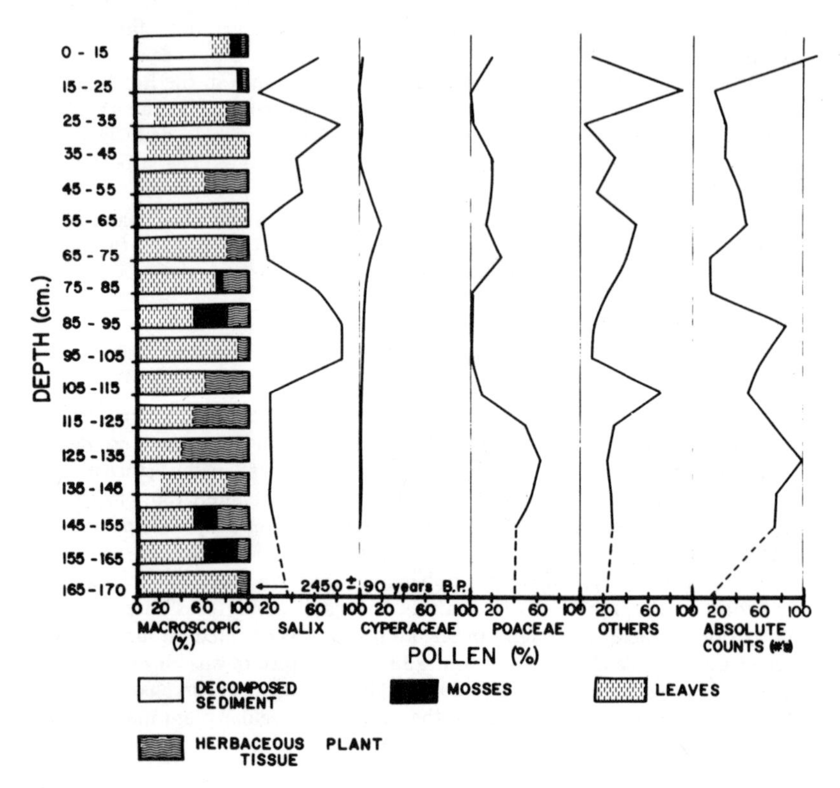

Fig. 1 Pollen and macroscopic plant material diagram from an ice-centre polygon, Truelove Lowland.

Pollen analysis

The absolute amount of pollen is low throughout the profile, averaging only 54 grains per three 20×20 mm sample areas. This corresponds with low pollen production by high arctic plants and makes interpretation of past climatic and vegetational patterns difficult. Only 16 groups of plants accounted for pollen diversity. Of these, *Salix arctica* and *Poaceae* accounted for most of it (Fig. 1). There were small amounts of *Betula*, *Pinus*, and *Alnus* pollen present but they represent only distant transport.

The pollen diagram shows some shift in species importance, an increase in *Salix* and a reduction in graminoids at the 85 to 105 cm depth.

Comparisons with other profiles

At this latitude, so far north of lands (1700 to 1800 km) where climatic and vegetational oscillation has been documented (Nichols 1967, 1970, 1974), and where peats are generally shallow and younger, it is difficult to draw conclusions.

On Axel Heiberg (79° 25' N) a peat profile, 2.7 in depth with a basal ^{14}C date of 4,160±100 B.P. consisted of *Carex* and *Drepanocladus* peat interspersed with waterlain sand. *Salix* and graminoid pollen occurred throughout the profile, though *Carex* pollen predominated, in amounts much greater than on Truelove Lowland. The pollen was mostly of local origin and changes in taxa were difficult to interpret in terms of regional climatic and vegetational change (Hegg 1963). He suggested that between 4,000 and 3,000 B.P. local climate was more favorable than today.

At Lake Klareso, North Greenland (82° 19' N), Fredskild (1969) reported a *Salix arctica* zone began at *ca*. 4,800 B.P. with a climatic maximum from 4,000 to 3,600 B.P. Driftwood in Inuit hearths dated between 4,000 and 3,600 B.P. with no driftwood younger than 2,720±100 B.P. There were numerous remains of muskox, indicating that a considerable plant cover existed then compared with the meagre plant cover and year-round frozen fiords of today. The limited pollen record dated at *ca.* 2,100 B.P. indicated a cold dry climate that has persisted to the present (Fredskild 1969). Our pollen diagram shows little shift other than a *Salix* maximum at about 1 m depth. This may coincide with a warming trend noted near the forest-tundra border at 1,800 to 1,600 B.P. (see Nichols 1974).

The slow accumulation of peat in Truelove Lowland agrees with that reported by Fredskild (1967) at Sermermuit, Jakobshavn, West Greenland (69° 12' N), Peary Land in North Greenland (Fredskild 1969), and by Iversen (1954) at Godthab Fiord, Greenland (65° N) and that a cold dry climate has prevailed for the past 1,000+ years.

Further interpretation of past climatic and vegetational changes in this portion of the High Arctic must await analyses of more profiles, additional ^{14}C dating, more knowledge of modern pollen rain from various plant communities, and permafrost-vegetation dynamics.

References

Barr, W. 1971. Postglacial isostatic movement in northeastern Devon Island. A reappraisal. Arctic 24: 249-268.

Erdtman, G. 1943. An Introduction to Pollen Analysis. Chronica Botanica Co., Waltham, Mass. 239 pp.

Fredskild, B. 1967. Palaeobotanical investigations at Sermermiut, Jakobshavn, West Greenland. Medd. om Grønland 178(4). 54 pp.

———. 1969. A postglacial standard pollendiagram from Peary Land, north Greenland. Pollen et Spores. 11: 573-583.

Hegg, O. 1963. Palynological studies of a peat deposit in front of the Thompson Glacier. Axel Heiberg Island Res. Rept. 217-219. McGill Univ., Montreal.

Iversen, J. 1953. Origin of the flora of Western Greenland in light of pollen analysis. Oikos 4: 85-103.

Jorgensen, S. 1967. A method of absolute pollen counting. New Phytologist 66: 489-493.

Nichols, H. 1967. The postglacial history of vegetation and climate at Ennadai Lake, Keewatin and Lynn Lake, Manitoba. Eiszeitalter und Gegenwart. 18: 176-197.

———. 1970. Late Quaternary pollen diagrams from the Canadian Arctic Barren Grounds at Pelly Lake, Northern Keewatin, N.W.T. Arctic Alp. Res. 2: 43-61.

———. 1974. Arctic North American palaeoecology: the recent history of vegetation and climate deduced from pollen analysis. pp. 637-667. *In:* Arctic and Alpine Environments. J.D. Ives and R.G. Barry (eds.). Methuen and Co. Ltd., London. 999 pp.

Plant communities of Truelove Lowland

M. Muc and L.C. Bliss

Introduction

One of the most distinctive features of this Lowland is its richness of plant species and mosaic pattern of plant communities. Most vascular plant and cryptogam species recorded for the Queen Elizabeth Islands have been found here. Plant cover and plant production are also much greater than usual at this latitude.

Within the Queen Elizabeth Islands, few detailed plant community studies have been conducted. Beschel (1970) described general patterns of plant communities on Axel Heiberg Island, and adjacent areas on Ellesmere and Devon Islands. Floristic lists and general descriptions of plants in relation to various habitats include areas of northern Ellesmere Island (Bruggemann and Calder 1953, Savile 1964), Ellef Ringnes Island (Savile 1961), Prince Patrick Island (MacDonald 1952-53, Tedrow et al. 1968), and Polunin (1948) described various plant communities of southern Ellesmere Island and Devon Island. The most detailed vegetation study in this Lowland was that of Barrett (1972). He recognized nine plant associations (phytogeocoenoses) grouped into seven Alliances using Braun-Blanquet methodology.

The objectives of this study were to:

1. map the major habitats with associated plant communities onto air photos (1:5000) and transcribe this information onto a meterically corrected stereo-orthophoto map (plant community and soil map in folder); and
2. describe in general terms the structure (physiognomy) and floristic composition of the plant communities.

Methods

The entire Lowland was covered on foot and all recognizable units of habitat and or plant community patterns were mapped in detail on the air photos. Plant community-topographic units were based upon uniform physiognomy of the vascular plant species. These units formed distinct patterns on the photos. Additional areas within the Lowland were specifically checked and mapped where other plant communities of limited extent were known to occur and where patterns on the photos could not be easily interpreted. Using this "ground truth," the photo-mosaic (orthophoto) map was completely mapped (see Appendix 2).

During this mapping project, plant communities were not quantitatively sampled. The species listed are those of relatively common occurrence in relation to their relative cover (D—dominant, C—common, M—minor). Detailed community analyses for vascular plants are presented for the three meadow types (Muc this volume), the two beach ridge cushion plant types (Svoboda this volume), and for all plants in the rock outcrop communities (Bliss et al. this volume). Detailed information on bryophyte communities are found in Vitt and Pakarinen (this volume) and for lichen communities in Richardson and Finegan (this volume). Data from these studies were utilized in preparing the species lists for the communities discussed below. As used here, habitat refers to landscape units that often contain two or more types of plant communities (e.g. cushion plant-lichen and cushion plant-moss communities on the raised beaches).

The map was then planimetered to provide the data in Table 1 in the Introduction. Because of their small areal extent, several plant communities reported here (hummocky graminoid and graminoid-moss meadows, ice-wedge polygons, herb-moss snowbed and unvegetated slopes and coastal shoreline) were not measured and included in Table 1.

Results

Major habitat — plant community types

Of the 18 units mapped, including the Polar Desert on the plateau to the east, only six are of major extent. Five of these types are depicted in Figs. 1 and 2 in terms of their general topographic relationships.

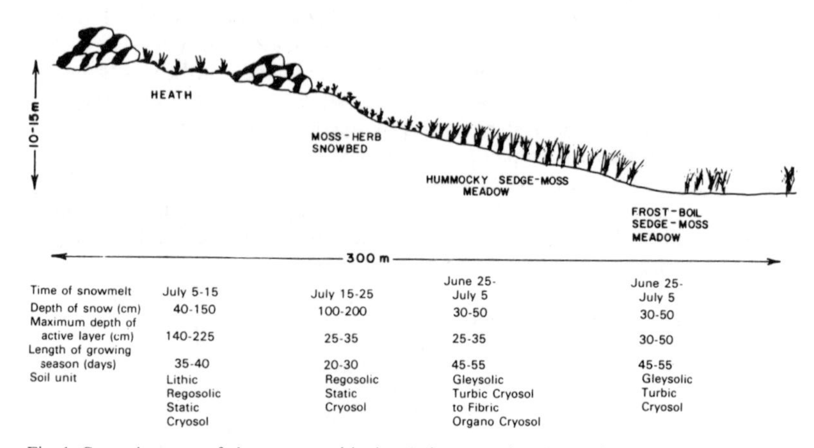

Time of snowmelt	July 5-15	July 15-25	June 25-July 5	June 25-July 5
Depth of snow (cm)	40-150	100-200	30-50	30-50
Maximum depth of active layer (cm)	140-225	25-35	25-35	30-50
Length of growing season (days)	35-40	20-30	45-55	45-55
Soil unit	Lithic Regosolic Static Cryosol	Regosolic Static Cryosol	Gleysolic Turbic Cryosol to Fibric Organo Cryosol	Gleysolic Turbic Cryosol

Fig. 1. General pattern of plant communities in relation to a raised beach ridge and an adjacent filled-in lake basin.

Hummocky Sedge-moss meadow

This, the most widespread community in the Lowland (20.5%), is most common toward the coast and along the face of the cliff to the east (see map). These communities have a distinct hummocky microrelief, the result of the differential growth of bryophytes. *Carex stans* is the dominant species with lesser amounts of *Eriophorum angustifolium*, *Arctagrostis latifolia*, and *Carex*

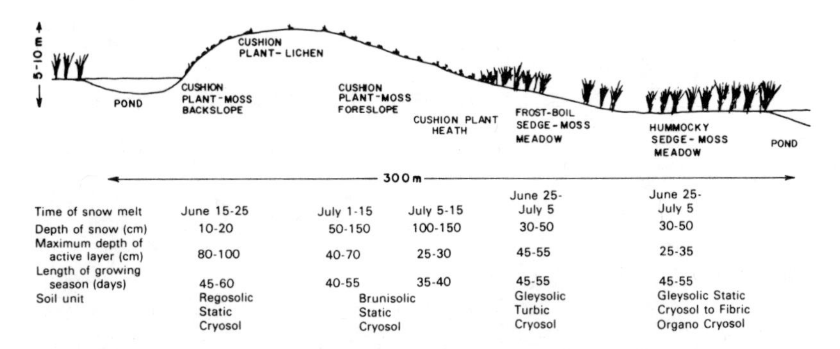

	Cushion plant-moss backslope	Cushion plant-moss foreslope	Cushion plant heath	Frost-boil sedge-moss meadow	Hummocky sedge-moss meadow
Time of snow melt	June 15-25	July 1-15	July 5-15	June 25-July 5	June 25-July 5
Depth of snow (cm)	10-20	50-150	100-150	30-50	30-50
Maximum depth of active layer (cm)	80-100	40-70	25-30	45-55	25-35
Length of growing season (days)	45-60	40-55	35-40	45-55	45-55
Soil unit	Regosolic Static Cryosol	Brunisolic Static Cryosol		Gleysolic Turbic Cryosol	Gleysolic Static Cryosol to Fibric Organo Cryosol

Fig. 2. General pattern of plant communities in relation to a granitic rock outcrop and adjacent slope and meadow.

membranacea (Fig. 5, Introduction). *Salix arctica* often occurs in hummocks that are elevated above the wet hollow surfaces. Lichen cover is extremely low (<1%), composed of *Xanthoria elegans*, *Cladonia pyxidata*, and *Lecanora epibryon*.

Vascular Plants		Bryophytes	
Carex stans	D	*Cinclidium arcticum*	D
Eriophorum angustifolium	C	*Drepanocladus revolvens*	D
Salix arctica	C	*Drepanocladus brevifolius*	C
Arctagrostis latifolia	C	*Riccardia pingius*	C
Carex membranacea	C	*Campylium arcticum*	C
Polygonum viviparum	C	*Distichium capillaceum*	C
Carex misandra	C	*Orthothècium chryseum*	C
Saxifraga hirculus	C	*Meesia triquetra*	C

Frost-boil sedge-moss meadow

These meadows are the second most abundant habitat type in the Lowland (18.4%). They are most prominent in the central and eastern part on older land surfaces (see map). Frost-boils, a characteristic feature of these meadows, average 49% of the surface. Some have bare soil due to cryoturbation but most are relatively inactive and covered with blue-green algae, some mosses, and scattered sedges (Fig. 3). Lichens *(Xanthora elegans*, *Solorina saccuta*, and *Mycoblastus sanguinarius)* are minor (<1%) as they are in all meadow types. *Campylium arcticum* reaches its highest cover in these meadows.

Vascular Plants		Bryophytes	
Eriophorum triste	D	*Drepanocladus revolvens*	D
Carex membranacea	D	*Campylium arcticum*	C
Carex stans	C	*Orthothecium chryseum*	C
Salix arctica	C	*Ditrichum flexicaule*	C
Arctagrostis latifolia	C	*Cinclidium arcticum*	C
Carex misandra	C	*Bryum pseudotriquetrum*	C
Polygonum viviparum	C	*Catoscopium nigritum*	C
Juncus biglumis	C	*Distichium capillaceum*	C
		Seligeria polaris	C

Fig. 3. Frost-boil sedge-moss meadow showing the pattern of *Carex membranacea*, *C. stans* and *Eriophorum triste* in relation to the 1 m soil boils.

Cushion plant—lichen

Cushion plant-lichen communities occur on the crests and upper slopes of raised beaches (5.0%), (Fig. 6, Introduction), comparable sites within rock outcrops (1.2%), and over much of the limestone pavement on and near Rocky Point (0.6% of the Lowland). These habitats have little winter snow cover and the soils are well-drained and warm in summer. Vascular plants provide *ca.* 20% cover, lichens and mosses 40% to 45% cover, and bare sand and gravel the balance (35% to 40%) on the raised beaches.

Vascular Plants

Dryas integrifolia	C
Saxifraga oppositifolia	C
Carex nardina	C
Salix arctica	C

Lichens

Alectoria pubescens	D
Thamnolia subliformis	C
Lecanora epibryon	C
Umbilicaria arctica	C
Hypogymnia subfusca	C
Rhizocarpon geographicum	C

Bryophytes

Encalypta rhaptocarpa	C
Distichium capillaceum	C
Mnium lycopodioides	C
Hypnum bambergeri	C
Myurella julacea	C
Ditrichum flexicaule	C

Cushion plant—moss

Communities dominated by *Dryas*, *Carex rupestris*, *Saxifraga oppositifolia*, and numerous mosses occupy the mid- and lower slopes of the raised beaches

Fig. 4. Slope of a raised beach showing cushion plant-moss community (dark area) and hummocky sedge-moss meadow beyond.

(6.4%) (Fig. 4) and extensive areas (6.9% of Lowland) separate from the raised beaches. These habitats have finer textured soils, more organic matter, higher soil moisture levels and greater moss cover than the communities on the lower slopes of the raised beaches. At the bottom of the beach ridge slopes where snow melts last, areas of *Cassiope tetragona* occur; a depauperate dwarf shrub heath-moss community.

Plant cover approaches 100% in these habitats with bryophytes and lichens contributing 30% to 50% of the total.

Vascular Plants		Bryophytes	
Dryas integrifolia	D	*Tortella arctica*	D
Carex rupestris	C	*Oncophorus wahlenbergii*	D
Cassiope tetragona	C	*Rhacomitrium sudeticum*	D
Saxifraga oppositifolia	C		
Carex misandra	C		
Salix arctica	C		

Lichens	
Cetraria cucullata	C
Cetraria nivalis	C
Thamnolia subliformis	C
Lecanora epibryum	C
Physcia muscigena	C

Dwarf shrub heath—moss

Communities dominated by *Cassiope tetragona* occur in the granitic rock outcrops (Fig. 7, Introduction) and on warm exposures in Truelove Valley. This is the most abundant community within the habitat diverse rock outcrops.

Pockets of cushion plant-lichen occur on sections of beach ridge, poorly drained areas contain sedge-moss meadow, and patches of graminoid-moss (high-dry) and herb-moss snowbed (moist) occupy special habitats. Rock outcrops (minus limestone pavement) occupy 12.4% of the Lowland. Mosses, especially *Rhacomitrium lanuginosum* are important among rock where there is little soil. The mats of *Cassiope* and associated herbs are confined to areas of deeper soil with higher moisture content. Floristically these are the richest communities on the Lowland.

Vascular Plants		Bryophytes	
Cassiope tetragona	D	*Rhacomitrium lanuginosum*	D
Dryas integrifolia	C	*Ditrichum flexicaule*	C
Saxifraga oppositifolia	C	*Hylocomium splendens*	C
Salix arctica	C	*Polytrichum juniperinum*	C
Carex misandra	C		

Lichens		on rocks	
Cetraria nivalis	C	*Rhizocarpon*	C
Thamnolia vermicularis	C	*geographicum*	C
Dactylina arctica	C	*Alectoria ochroleuca*	C
		Alectoria nigricans	C

Moss-Herb (Polar Desert)

The portion of the plateau to the east of the Lowland has a higher plant cover (5% to 7%), especially mosses, than do areas farther inland and to the south toward the ice cap (<2%) (Fig. 4, Introduction). Most plants, especially mosses, are concentrated in the shallow troughs of polygons and between or under frost shattered rock fragments.

Vascular Plants		Bryophytes	
Saxifraga oppositifolia	M	*Hypnum bambergii*	D
Saxifraga cernua	M	*Ditrichum flexicaule*	C
Saxifraga caespitosa	M	*Distichium capillaceum*	C
Papaver radicatum	M	*Mnium lycopodioides*	M
Cerastium alpinum	M		
Draba corymbosa	M		
Minuartia rubella	M		
Phippsia algida	M		

Lichens	
Polyblastia bryophila	M
Lecanora epibryon	M
Alectoria nigricans	M
Pertusaria dactylina	M
Rhizocarpon chioneum (on rocks)	M

Minor habitat — plant community types

Wet sedge-moss meadow

Meadows with a high water table are restricted (2.1%) to the outer edges of the Lowland along streams and pond margins. *Carex stans* is the only abundant vascular plant providing a relatively homogenous plant cover. Mosses, abundant in total cover and lower in species richness, help create a uniform surface. Lichens are excluded by the excessive moisture levels.

Vascular Plants		Bryophytes	
Carex stans	D	*Drepanocladus revolvens*	D
Arctagrostis latifolia	C	*Meesia triquetra*	D
Saxifraga hirculus	C	*Cinclidium arcticum*	D
Hierochloe pauciflora	C	*Calliergon giganteum*	D
Eriophorum angustifolium	C	*Drepanocladus brevifolius*	C
		Distichium capillaceum	C
		Campylium arcticum	C

Hummocky graminoid meadow

This is a minor type, found near the coast at both the northwest and southwest portion of the Lowland in more recently emerged areas (see map). Mosses are quite common but lichens are a minor component of the flora.

Vascular Plants		Bryophytes	
Arctagrostis latifolia	D	*Drepanocladus revolvens*	D
Dupontia fischeri	D	*Meesia triquetra*	D
Hierochloe pauciflora	C	*Cinclidium arcticum*	D
Carex stans	C	*Calliergon giganteum*	D
Salix arctica	C	*Drepanocladus brevifolius*	C
		Distichium capillaceum	C
		Campylium arcticum	C

Graminoid-moss meadow

These are small relatively dry meadows, mostly near the eastern cliff face. Small areas dominated by *Luzula confusa* and *Hierochloe alpina* occur on high, dry sites within the rock outcrops (Bliss et al. this volume).

Vascular Plants	
Arctagrostis latifolia	C
Carex membranacea	C
Eriophorum triste	C
Polygonum viviparum	C
Salix arctica	C
Dryas integrifolia	C

Herb-moss snowbed

This community occupies only small, scattered areas on the Lowland, generally in the lee of limestone ridges and granitic rock outcrops. Snow remains until mid- to late July most years. Mosses provide more cover than do vascular plants.

Vascular Plants

Phippsia algida	C
Papaver radicatum	C
Saxifraga cernua	C
Oxyria digyna	C
Cerastium regelii	C
Luzula nivalis	C
Ranunculus sulphureus	C

Bryophytes

Bryum cryophilum	D
Bryum pseudotriquetrum	D
Cinclidium arcticum	D
Catoscopium nigritum	D
Philonotis fontana var. pumila	C
Distichium capillaceum	C
Distichium flexicaule	C
Orthothecium chryseum	C

Lichens

Cetraria nivalis	C
Cetraria cucullata	C
Thamnolia subliformis	C

Ice-wedge polygons (raised-centre)

Ice-wedge polygons occupy a very small percentage of the Lowland (<0.5%), yet this is important habitat for *Dicrostonyx*. The raised-centre polygons east of Beschel Creek are dominated by *Carex stans* and *C. membranacea*, as are small areas of polygons in the northern part of the Lowland. The large raised-centre polygons southwest of the Base Camp are dominated by a cushion plant-moss community. The latter is described here (Fig. 5).

Fig. 5. Cushion plant-moss community with *Dryas* (flowers) and *Alopecurus alpinus* dominating in the raised centre ice-wedge polygons 1 km southwest of the Base Camp. Cores for pollen analyses were taken from this site. Two sites for measurement of permafrost temperature are also visible.

Vascular Plants

Dryas integrifolia	D
Salix arctica	D
Alopecurus alpinus	D
Carex misandra	C
Polygonum viviparum	C
Pedicularis hirsuta	C
Arctagrostis latifolia	C
Luzula confusa	C
Oxyria digyna	C

Bryophytes

Ditrichum flexicaule	C
Aulacomnium turgidum	C
Pogonatum alpinum	C
Psilopilum caifolium	R
Leptobryum pyriforme	R
Stegonia latifolia	R
Ceratodon purpureus	R

Lichens

Cladonia poccillum	C
Thamnolia subliformis	C
Cetraria nivalis	C
Lecanora epibryon	C

Tidal salt marsh

This narrow coastal fringe community, 10 to 30 m wide, is best developed along the west coast. *Puccinellia phryganodes* forms a solid turf in which other species occur widely spaced (Fig. 6). Lichens and mosses are very minor components.

Vascular plants

Puccinellia phryganodes	D
Carex ursina	C
Stellaria humifusa	C
Cochlearia officinalis ssp. *groenlandica*	C

Fig. 6. Tidal salt marsh community dominated by *Puccinellia phryganodes*.

Unvegetated coastal shoreline

On the coastal areas, primarily along the northern coastline where sea ice scours the surface, only occasional plants of *Cochlearia officinalis* occur.

Unvegetated scree slopes

The scree slopes along the west-facing slope to the plateau have small pockets of soil with occasional plants of *Saxifraga oppositifolia*, *S. cernua*, *Draba carymbosa*, *Papaver radicatum*, *Salix arctica*, and numerous species of bryophytes.

Discussion

The most common habitat-plant-community types of this Lowland are represented elsewhere in the High Arctic. Hummocky sedge-moss and frost-boil sedge-moss meadows with essentially 100% plant cover occur at Lake Hazen (Savile 1964), Van Hauen Pass area (Brassard and Longton 1970) and the Fosheim Peninsula, Ellesmere Island; Dundas Harbour on the south coast of Devon Island (Polunin 1948); local areas on Cornwallis Island (Thornsteinson 1958); and Bracebridge and Goodsir Inlets on Bathurst Island. Small lowlands with mostly grasses *(Dupontia fischeri*, *Pleuropogon sabinei*, and *Alopecurus alpinus)* occur on Prince Patrick and Melville islands. The southern islands of Prince of Wales, Victoria, and Banks have more extensive lowlands dominated by *Carex*, *Eriophorum*, *Dupontia*, *Alopecurus* and *Arctagrostis* (Bliss and Svoboda 1977a). These sedge-dominated communities, occurring only where drainage is impeded, are ecologically related to those of the Low Arctic (Bliss 1975).

The cushion plant-lichen or cushion plant-moss communities of the raised beaches, slopes below rock outcrops, and boulder areas (southwestern area) are types found elsewhere in the northern islands (Lake Hazen and Fosheim Peninsula, Ellesmere Island, numerous areas of Axel Heiberg, Bathurst, and Melville islands) (Savile 1964, Brassard and Longton 1970, Beschel 1970, Bliss and Svoboda 1977b). The southern arctic islands have vast areas dominated by these cushion plants (Bliss and Svoboda 1977a). These communities, with *Dryas integrifolia* dominating, form one of the most important components of the Polar Semi-desert (Bliss 1975).

Dwarf shrub heath-moss, dominated by *Cassiope tetragona*, occurs in local warm habitats as far north as Lake Hazen, Ellesmere Island (81° 49' N), (Savile 1964). Localized heath communities occur throughout the southern arctic islands and on Melville Island, but nowhere is the type common.

The coastal salt marsh with *Puccinella phryganodes* is a diminutive form of a type found throughout much of the Low and High Arctic (Polunin 1948, Jeffries 1977).

Floristically Truelove Lowland is rich in species compared with most lands at this latitude. Barrett and Teeri (1973) reported 93 species of vascular plants from the Lowland. To this we have added *Carex ursina*, *Stellaria humifusa*, *Luzula confusa* and *Pucinellia phryganodes*. The known vascular flora of Devon Island is 117 species (Barrett and Teeri 1973). Northern Ellesmere Island has 143 species (Brassard and Beschel 1968), 137 are reported on Axel Heiberg, and 72 species on Cornwallis Island.

The bryophyte flora is equally rich, having 132 species of moss and *ca*. 40 species of liverworts (Vitt 1975). Brassard (1971) reported 151 moss species from northern Ellesmere Island and Kuc (1973) 131 species from the Expedition Area of Axel Heiberg Island.

Barrett and Thomson (1975) reported 182 species of lichens from the Lowland. This list included 15 species provided by Richardson and Finegan (this volume). For further details on the floristics of plant groups other than algae, see appendices 3-6.

In summary this Lowland is a biological oasis in terms of total plant cover, diversity of plant communities, and floristic richness within lands that are mostly Polar Desert. The diversity and productivity of invertebrates and vertebrates, discussed elsewhere in this book, result from this rich food and diverse habitat base.

Acknowledgments

Financial support for this study was provided by an N.R.C. grant (A-4879) to the junior author. Space Optic Ltd. of Ottawa prepared the stereo-orthophotographs used for the final "photographic map" of the plant communities. Mr. J. Saastamoinen conducted the field survey, T.J. Blachut directed the work and M. Van Wijk planned the photography and aerial triangulation, and coordinated the orthophoto production; all of the Photogrammetry Branch of the National Research Council of Canada.

References

Barnett, D.M. and D.L. Forbes. 1973. Terrain performance, Melville Island, District of Franklin. Can. Geol. Surv., Rept, of Activities. Paper No. 73-1, part A. pp. 182-192.

Barrett, P.E. 1972. Phytogeocoenoses of a Coastal Lowland Ecosystem, Devon Island, N.W.T. Ph.D. thesis, Univ. British Columbia, Vancouver. 292 pp.

———. and J.A. Teeri. 1973. Vascular plants of the Truelove Inlet Region, Devon Island. Arctic 26: 58-67.

———. and J.W. Thomson. 1975. Lichens from a high arctic coastal lowland, Devon Island, N.W.T. Bryologist 78: 160-167.

Beschel, R.E. 1970. The diversity of tundra vegetation. pp. 85-92. *In:* Productivity and Conservation in Northern Circumpolar Lands. W.A. Fuller and P.G. Kevan (eds.). IUCN New Ser. No. 16. Morges, Switzerland. 344 pp.

Bliss, L.C. 1975. Tundra grasslands, herblands, and shrublands and the role of herbivores. Geoscience and Man 10: 51-79.

———. and J. Svoboda 1977a. Plant communities and plant production on Banks and Victoria Islands, N.W.T. (in preparation).

———. and———. 1977b. Plant communities and plant production in the western Queen Elizabeth Islands, N.W.T. (in preparation).

Brassard, G.R. 1971. The mosses of northern Ellesmere Island, Arctic Canada. II. Annotated list of the taxa. Bryologist 74: 282-311.

———. and R.E. Beschel. 1968. The vascular flora of Tanquary Fiord, Northern Ellesmere Island, N.W.T. Can. Field-Nat. 82: 103-113.

———. and R.E. Longton. 1970. The flora and vegetation of Van Hauen Pass, Northwestern Ellesmere Island. Can. Field-Nat. 84: 357-364.

Bruggeman P.E. and J.A. Calder. 1953. Botanical investigation in northeast Ellesmere Island, 1951. Can. Field-Nat. 67: 157-174.

Jefferies, R.L. 1977. Plant communities of muddy shores of arctic North America. J. Ecol. (In press.)

Kuc, M. 1973. Bryogeography of expedition area, Axel Heiberg Island, N.W.T., Canada. Bryophytorum Bibliotheca 2: 1-120.

MacDonald, S.D. 1952-53. Report on biological investigations at Mould Bay, Prince Patrick Island, N.W.T. Nat. Museum Can., Ann. Rept. Bull. No. 132: 215-238.

Polunin, N. 1948. Botany of the Canadian Eastern Arctic. Part III. Vegetation and Ecology. Nat. Museum Can. Bull. 104. Ottawa. 304 pp.

Savile, D.B.O. 1961. The botany of the Northwestern Queen Elizabeth Islands. Can. J. Bot. 39: 909-942.

————. 1964. General ecology and vascular plants of the Hazen Camp area. Arctic 17: 237-258.

Tedrow, J.C.F., P.F. Bruggemann, and G.F. Walton. 1968. Soils of Prince Patrick Island. Arctic Inst. N.A., Tech. Rept. ONR — 352:4. 82 pp.

Thornsteinson, R. 1958. Cornwallis and Little Cornwallis Islands, District of Franklin. Can. Geol. Surv. Memoir No. 294. 134 pp.

Vitt, D.H. 1975. A key and annotated synopsis of the mosses of the northern lowlands of Devon Island, N.W.T., Canada. Can. J. Bot. 53: 2158-2197.

Primary producers

Ecology and primary production of Sedge-moss Meadow communities, Truelove Lowland

Michael Muc

Introduction

Although lowlands with sedge and grass dominated meadows cover less than 2% of the landscape in the Queen Elizabeth Islands (Babb and Bliss 1974), they constitute the most productive landscape unit in the High Arctic. Nutrient deficiencies, low temperatures, and a relatively short growing season all contribute to a relatively low standing crop and annual production in this vegetation type, compared with more temperate communities. A distinctive feature of the meadow communities is their retention of the majority of vascular biomass belowground (Dennis and Johnson 1970) and maintenance of a relatively uniform growth rate over the growing season (Tieszen 1972, Muc 1973).

The objectives of this study were to determine plant communities, to determine aboveground and belowground net primary production within the three meadow community types present, to determine aboveground plant growth rates, and to relate carbohydrate, nutrient, and chlorophyll content to net production.

Study area

In Truelove Lowland, sedge-moss meadows cover approximately 40% of the total and 50% of the terrestrial lowland system. In a phytosociological study Barrett (1972) described ten plant associations, while the I.B.P. project worked only with seven community types based on dominant species and growth form (Bliss 1975). Muc (1973) identified three distinctive lowland sedge-moss meadow community types. The hummocky sedge-moss meadow community type is predominant (50% of all lowland meadows), frost-boil sedge-moss meadows the next most common (45%), and wet sedge-moss meadows (5%) least common. The following studies and comparisons are based on the vascular plant components of these meadow types, although mosses are a very important constituent.

Twenty lowland sedge-moss meadow stands (12 hummocky, 5 frost-boil, and 3 wet) were studied on the bases of their floristic and physiographic features (Fig. 1). The hummocky meadow at site 1 was studied intensely during the growing seasons of 1970-1973. In addition, comparative aboveground and belowground standing crop, primary production, phenology, and related data

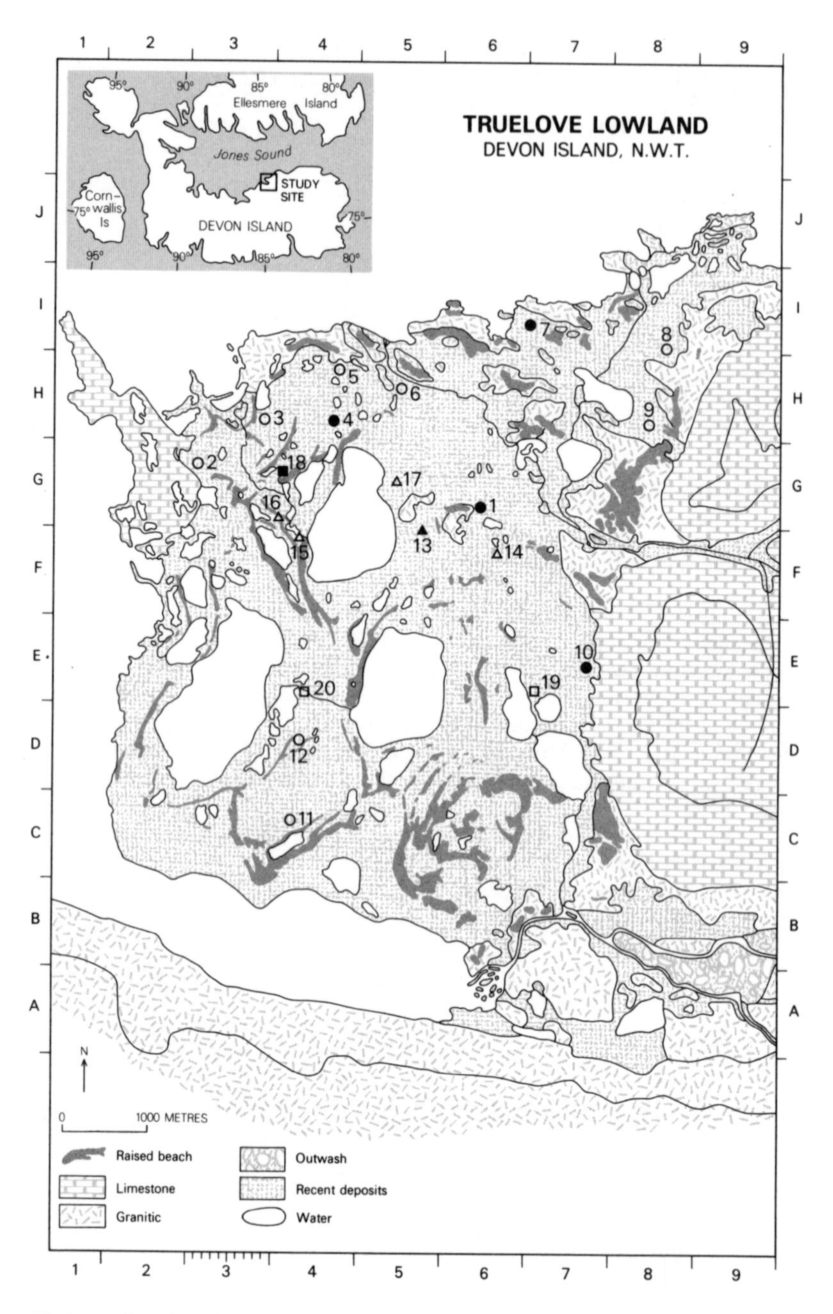

Fig. 1 Location of the 20 lowland sedge-moss meadows (hummocky O frost-boil △ wet □) at which phytosociological (open) and net primary production (closed) data were collected.

were collected from the frost-boil (site 13) and wet (site 18) meadow for three summers. Similar standing crop and primary production data were collected from three other hummocky meadows (sites 4, 7, 10) during 1971.

Methods

Environmental measurements

Snow depth and density measurements were taken in 1971 and 1972 from five permanent snow stations, located in a stratified random pattern, at the intensive hummocky meadow (site 1) and the wet meadow (site 18). At three-day intervals, from early June through to snowmelt, snow density measurements were taken with a U.S. Cold Regions Research and Engineering Laboratory snow kit. Snow depths (average of five measurements per station) were taken with a meter-long steel (1 cm diameter) probe.

Seasonal soil thaw (active layer) depth measurements were taken at each of the meadows sampled for standing crop (1970, 1971) and from 40 locations along a permanent 10 m transect in the meadows sampled in 1972. In 1970 and 1971, measurements were taken in conjunction with plant biomass harvests. In 1972, sampling was at regular three-day intervals. The 1973 thaw depth was measured on a single date (5 August), approximately at the time of maximum seasonal thaw, from 25 random locations at each of the meadows sampled for plant biomass.

Meadow physiographic features (hummocks, hollows, frost-boils) were measured quantitatively on a percentage cover basis by means of ten parallel 25 m transects at sites 1 and 13. Similar measurements in the remaining meadow stands were based on 10 to 16, 20×50 cm quadrat samples.

In 1971, weekly gravimetric soil moisture measurements were taken at sites 1, 13, and 18. Triplicate soil cores (65 cm diameter) were taken from each meadow on a weekly basis. The cores were divided into 5 cm segments, oven dried at 105°C for 24 hours, and soil moisture (% dry weight) and bulk densities (g cc^{-1} dry soil) calculated. Soil pH readings were taken from triplicate 17 July, 1971 samples; a 1:2 soil to water ratio was used.

Two 150×120 cm framed shelters, covered with heavy sheet plastic, served (1972) as field greenhouses at the intensive hummocky study meadow (site 1). The greenhouse roof was suspended 15 cm above the plant canopy and the greenhouse totally enclosed on all sides to eliminate any wind effect. Biological measurements within the greenhouses were restricted to *Carex stans* and consisted of photosynthetic leaf area, leaf length increment, tissue chlorophyll content, and flowering activity. Aboveground production was calculated from a single peak season harvest consisting of duplicate samples from within the greenhouses. These were compared with production estimates made on vegetation outside the greenhouses.

Plant communities

Early to mid-August floristic descriptions of the 20 lowland meadow stands were made at the time of peak floral development. Data were collected from sixteen 20×50 cm quadrats in meadows sampled for biomass and from ten similar quadrats in the remaining stands. Within each quadrat, actual foliar plant cover (%), vegetative and flowering tiller density, microhabitat area (hummocks, hollows, frost-boils) as percent of total area, and active layer (cm) depth was recorded. Prominence values (Beals 1960) were calculated for each of the vascular species and the values adjusted with a division of 10 to provide a maximum value of 100. The prominence values were used in an ordination program (Beals 1960) which had been revised and extended by M. Easton and D. Precht of the University of Alberta Computing Services Centre.

Aboveground biomass and net primary production

The harvest method was used in collecting primary production data. At site 1, sixteen 20×20 cm samples were harvested at five dates during the 1970 growing season. Sampling was modified to consist of ten 20×50 cm harvested quadrats at each of the six harvest dates in 1971, 1972. At sites 4, 7, 10, 13, and 18, only six 20×50 cm quadrats were harvested at each of the 4-5 harvest dates. Sites 4, 7, and 10 were only harvested in 1971. A single (5 August) harvesting was made at sites 1, 13, and 18 in 1973 and net production calculated on a standing crop:net production basis using the 1971 data.

In each of the harvested samples, all of the foliar vascular plant material was clipped at the moss surface. In addition, three 20×20 cm sod blocks (10 cm deep) were collected at each harvest date for stem base, rhizome, and root biomass information. The collected materials were sorted into live and dead, shoot, stem bases, rhizomes, and roots and dried at 85°C for 24 hr prior to weighing.

Net primary production was calculated as the weight difference between the lowest and highest seasonal biomass values for each live plant component. A correction for pre-peak dieback and post-peak growth was applied. Stem bases within the moss layer were included in the aboveground totals. Net primary production totals were based on separate calculations for each of the plant groups (graminoids, forbs, woody plants). This insured a more accurate estimate of net primary production by accounting for any asynchronous seasonal growth. Data for the frost-boil meadow were expressed on the basis of vegetated area (100% vegetation) and on total area (51% vegetation in total area). The average hummocky meadow data given were based on values from sites 1, 4, 7, and 10. Similar data based only on studies from the intensive hummocky meadow (site 1) were reported in an earlier paper (Muc 1973).

In this paper, biomass and standing crop are used interchangeably for live or dead plant material. Productivity refers to the rate of production (g day^{-1}) over the calculated meadow growing season.

Belowground biomass and net primary production

Belowground data are based on root material collected from soil cores (6.5 cm diameter). Sampling extended for the full depth of thaw and consisted of 16 cores at each of five harvest dates in 1970 at site 1 and six cores at sites 1, 13, and 18 in 1972 and 1973. In 1971 root standing crop data were calculated from three 20×20 cm sod blocks (10 cm deep) collected at the time of the aboveground sampling.

Roots were visually separated into live (turgid, whitish-yellow) and dead (flaccid, black) categories. Tetrazolium chloride root viability tests (Jensen 1962) showed 15% of the "dead" roots and all of the live roots to be viable. Subsamples taken from all harvested root material at site 1 over the course of the 1972 growing season were visually sorted into "live" and "dead" components (corrected for the viable component in the "dead" root portion) in order to establish a seasonal pattern of live and dead root proportions on the total root biomass. Seasonal root increment and growth pattern in the upper 10 cm of soil were studied in 20 transplants of *Carex stans* in which actual root growth was monitored over the 1971 season.

Graminoids contribute 95% of meadow root biomass. Extensive rooting depths and gradual seasonal soil thaw created difficulties in collecting root material and estimating seasonal net root primary production. The belowground graminoid net production data used in this paper were based on a linear regression analysis of root biomass data for the intensive study meadow (site 1) during 1972. These data were the best biomass estimates available because they were based on weekly sampling, seasonal estimates of live:dead root ratios, and knowledge of the vertical root distribution in the soil profile. The rooting depth was taken to 20 cm to maintain maximum data continuity. The calculated seasonal graminoid root production estimate, with 80% confidence limits was 5.06±6.03% of the initial total root biomass. An aboveground (green plus nongreen) to belowground net production ratio was calculated to be 1:2.73. This ratio was then used in calculating root production for the other meadow sites. These production estimates were based on the total seasonal aboveground graminoid production at each of the sites. The belowground graminoid production estimates were added on to the forb and woody plant root production estimates (based on seasonal belowground biomass change) to produce a total belowground net primary production estimate. Impetus for this method of estimating root production resulted from preliminary modelling of the graminoid data (Whitfield and Goodwin this volume) which initially showed an overestimate of belowground production.

Undoubtedly such a calculation of belowground production must be considered an underestimate. It does not account for potential pre- and post-sampling growth, nor does it take into account any potential growth below 20 cm. It seems, however, a reasonable approach to use, since seasonal aboveground production varies by less than 5%, inter-meadow standing crops and composition are relatively similar, and seasonal carbon fixation rates by the graminoids are similar.

Litter and organic matter

Litter was defined as unattached aboveground vascular plant material. Litter standing crop data were based on material collected from each of the quadrats (20×20 cm in 1970 and 20×50 cm thereafter) harvested for primary production. Litter redistribution by spring meltwaters was studied with the use of twenty-five 20×10 cm litter traps (1 mm^2 mesh) randomly distributed throughout meltwater rivulets at the intensive hummocky meadow (site 1) during 1972 and 1973. Each trap was secured to the frozen soil with long nails within a flowing rivulet. The litter collection opening of each trap was 15 cm in diameter. Litter was cleaned from each trap after 24 hr. Traps were relocated when water ceased flowing through them. The meadow surface area over which rivulet litter transport potentially occurred was calculated to be 1 or 3 m^2. Litter in ten 20×50 cm plots was collected from beneath the snow at site 1 for use in determining composition. Litter samples from hummocks and hollows were sorted into categories: graminoids, forbs, woody plants and mosses.

Root-free soil organic matter collections were made from soil root core washings which were passed through a 1 mm^2 mesh. Samples were collected from both hummock and hollow microhabitats at sites 1 and 13 and from the general meadow area at site 18.

Chlorophyll and leaf area index

In 1971, community chlorophyll measurements (mg m^{-2}) were taken weekly. Triplicate 20×50 cm samples were collected and the clipped plant material sorted by species. A 1 g freshweight subsample of photosynthetic tissue of each species was analysed for its chlorophyll content (mg g^{-1}).

Chlorophyll was extracted with 80% acetone and optical densities of the extracts read at wavelengths of 675 and 663μ. Chlorophyll content was calculated with Arnon's (1949) formulae.

Leaf area measurements for the individual species were based on similar 1 g freshweight subsamples which were pressed and subsequently measured on a leaf area photometer. Graminoid leaf area measurements were found (by graph paper measurements) to be underestimated (15%) as a result of tissue overlap. All reported leaf area measurements were corrected accordingly.

Plant constituents

Caloric and chemical analyses of meadow species were made on ground tissue material collected in the primary production harvests. An automatic 1200 Adiabatic Oxygen Bomb Calorimeter was used for the caloric analyses. Chemical analyses for protein, nitrogen, potassium, phosphorus, and calcium were made by the Alberta Soil and Feed Testing Laboratory in Edmonton.

Carbohydrate

Three sod blocks (20×20×10 cm deep) were collected weekly at site 1 to obtain aboveground and belowground plant material for carbohydrate extraction (*Carex membranacea* in 1970, *C. stans* in 1971, 1972). Two subsnowsurface collections of *C. stans* were taken at sites 1 and 18 on 15 May and 16 June, 1972. The washed and sorted shoots, roots and rhizomes were immediately fixed in an alkaline, boiling solution of 80% ethyl alcohol and sealed with wax in glass jars. Subsequent analyses of these samples for sugar and starch were made with Nelson's (1944) molybdate colorimetric technique. In the *C. stans* collected in 1972, plants were sorted according to their phenological stage of development and then divided into the above-named components as well as into bud and flowering portions. Mature plants were considered to be those tillers with more than two seasons of shoot growth. Reproductive (sexual and asexual) and non-reproductive tillers, as well as pre-tillering plants were also separated.

Plant growth and phenology

During each growing season, plant growth data were collected from 10 tagged plants of the dominant meadow species. Each plant was tagged with a black thread around its stem base, at the moss surface, and at two- or three-day intervals leaf measurements were taken from this base point. The data provided an accurate estimate of the amount of leaf tip browning before, and shoot growth after, the time of peak standing crop. Estimates of sub-snowsurface shoot growth in *C. stans* were made by: (1) tagging plants at the end of the 1971 growing season and remeasuring them immediately after snowmelt in 1972; (2) digging the plants out of the snow and tagging them in early spring (May 1972) and remeasuring them after snowmelt in 1972; and (3) by exposing them from a 35 cm snowcover, tagging, and taking shoot increment measurements every four days (1972).

The phenological data were gathered during the course of the community analysis in 1971 and in conjunction with biomass harvesting in 1972. Additional phenological observations were made during the course of the phenometric studies. Nine entire *Carex stans* and four entire *C. membranacea* belowground plant systems were dug up. Measurements were taken of inter-tiller rhizome lengths and live and dead, as well as immature and mature tiller distribution along the plant system.

Results

Environment

Late spring snow depths (at sites 1 and 18) were approximately 52 and 46 cm respectively in June 1971 but only 39 cm at the same time in 1972. Complete snowmelt did not occur at either site until 25 June, 1971 and 6 July, 1972. Early spring (May) snow densities at both sites averaged 0.26 g cm^{-3} and increased to 0.43 g cm^{-3} prior to snowmelt.

With the exception of sites 10 and 13, soil thaw depth variation in the various lowland meadows was less than ±5 cm. The frost-boil meadow (site 13) and the plateau base located hummocky meadow (site 10) with a gleysolic soil had thaw depths almost twice the depth of the other meadows (Fig. 2). The pronounced early August thaw depth increase at site 10 was associated with increased soil drainage. Seventy to 80% of maximum seasonal thaw is

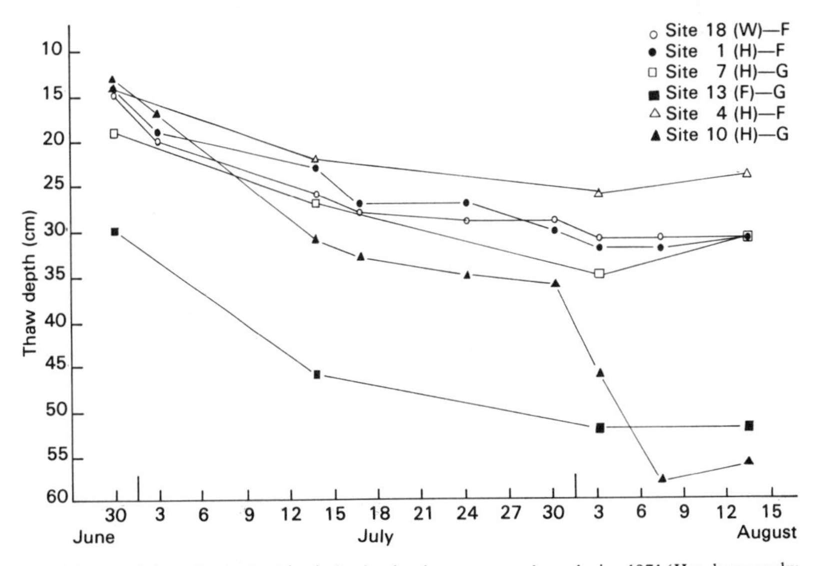

Fig. 2 Seasonal thaw depths (cm) in six lowland sedge-moss meadows during 1971 (H = hummocky, F = frostboil, W = wet) associated meadow soils (F = Fibric Cryosol, G = Gleysolic Cryosol).

developed three weeks after snowmelt. Peak thaw depths were found by early to mid-August. Refreezing at the permafrost table began by mid- to late August. Thaw depths were approximately 10 cm deeper in 1971 and 1973 which were earlier and milder seasons than in 1970 and 1972 which were later and cooler.

Hummocky meadows possessed a pronounced hummock and hollow microtopography with surface area cover of 61% and 39% respectively. The hummocks, 5-15 cm high and 15-90 cm in diameter, are the moss-derived type described by Raup (1965). Frost-boil meadows appeared to develop on lacustrine deposits. Mineral frost-boils covered 40% of their surface area and the remaining area was covered by hummocks (35%) and hollows (23%). The hummocks were of a slightly smaller dimension than those in the hummocky meadows. The majority of the frost-boils were covered with a 1 cm crust of organic material derived from algae and mosses. Vascular plant cover on the frost-boils was 10-15% of their surface area. Wet meadows developed in areas of impeded drainage (stream and pond margins) which remained inundated for the major portion of the growing season. Such wet habitats produced lush and uniform moss growth, resulting in a relatively uniform meadow topography.

The hummocky and wet meadows (sites 1 and 18) contained Fibric Organo Cryosols with low average soil bulk densities (0.19 and 0.15 g cm^{-3}, respectively). The soil pH was acidic (6.0 and 5.8 respectively). Seasonal soil moisture levels remained relatively uniform, averaging 340% and 450% respectively. The soils within the frost-boil meadow (site 13) were Gleysolic Static Cryosols under heavy vegetation and Gleysolic Turbic Cryosols under frost-boils (Walker and Peters this volume). Average soil bulk densities were 0.68 and 1.23 g cm^{-3} and soil pH was 7.0 and 7.7 respectively. The frost-boils were relatively dry (65% soil moisture); soils under a dense vegetation cover were wetter (285%). Gleysolic Static Cryosols were also found in hummocky meadows and 60% of the hummocky meadow stands studied possessed these soils.

Average greenhouse canopy air temperatures were 7°C greater than those outside the greenhouse. A supersaturated relative humidity was developed inside the greenhouses as a result of effective wind elimination.

Plant communities

A total of 34 vascular plant species was found in the lowland sedge-moss meadow communities (Table I). Average meadow species composition ranged from 18 to 21 species. The vascular plant cover of frost-boil meadows was primarily *Eriophorum triste*, *Carex membranacea*, *C. stans*, *C. misandra* and *Salix arctica*. *Saxifraga oppositifolia* and *Dryas integrifolia* were two prominent beach ridge species which extend their range into these meadows. *Draba alpina* and *Braya purpurescens* were exclusive to this meadow type. Frost-boils were generally sparsely vegetated (10-15%) by such species as *Eriophorum triste*, *Carex membranacea*, and *C. misandra*.

The vascular plant cover of hummocky meadows was mainly *Carex stans*, *Eriophorum angustifolium*, *Salix arctica*, *C. membranacea*, and *Arctagrostis latifolia*. Hummocks had a mixed cover of woody, forb, and graminoid species, while hollows were chiefly vegetated by *C. stans*.

In the wet meadows, *Carex stans* provided the major portion of the vascular plant cover. The remaining cover was contributed by *Eriophorum angustifolium*, *Arctagrostis latifolia*, and *Hierochloe pauciflora*. *Cardamine pratensis*, *Pleuropogon sabinei*, and *Ranunculus hyperboreus* were exclusive to these meadows.

Table 1. Mean prominence values for species, vascular plant cover, moss cover, bare ground plus rock cover, and active layer depth (cm) for the three lowland sedge-moss meadow community types.

Species	*Sedge-moss meadow type*		
	Frost-boil	Hummocky	Wet
Equisetum arvense	1.1	0.3	0.1
E. variegatum	0.1	0.1	0.1
Arctagrostis latifolia	6.1	4.2	3.2
Hierochloe pauciflora	1.0	1.1	3.0
Dupontia fischeri	1.5	—	0.2
Poa arctica	—	0.1	1.7
Pleuropogon sabinei	—	—	0.1
Carex stans	10.1	39.6	45.6
C. artrofusca	0.1	0.9	—
C. membranacea	22.3	7.4	—
C. misandra	8.2	1.2	—
Eriophorum triste	27.0	*	—
E. angustifolium	—	22.7	3.7
E. scheuchzeri	—	*	*
Juncus biglumis	1.9	1.2	0.1
Luzula confusa	0.1	0.2	—
Salix arctica	9.6	11.5	0.2
Polygonum viviparum	3.4	6.5	—
Melandrium apetalum	0.1	0.1	—
Stellaria longipes	0.1	0.1	0.5
Ranunculus sulphureus	0.1	0.1	—
R. hyperboreus	—	—	0.1
Cardamine pratensis	—	—	0.4
Braya purpurescens	0.1	—	—
Draba alpina	0.1	—	—
D. lactea	—	0.2	0.1
Eutremia edwardsii	0.2	0.1	—
Saxifraga hirculus	0.2	1.2	2.1
S. oppositifolia	1.4	0.1	—
S. cernua	—	0.1	0.7
S. foliolosa	—	0.2	0.2
Dryas integrifolia	0.4	0.1	—
Pedicularis sudetica	0.5	0.9	0.1
P. hirsuta	0.1	*	*
Mean No. species	20	21	18
Vascular plant cover (%)	58	86	77
Moss cover (%)	82	98	100
Bare groünd plus rock cover (%)	18	2	0
Active layer depth (cm)	49	29	31

*Species indistinguishable from others in the same genus but less prominent on the basis of flowering rates in the community.

Mosses were a major plant component within the lowland sedge-moss meadows. Details on moss composition are to be found in Vitt and Pakarinen (this volume). Lichens were an insignificant factor (<1% plant cover) in the hummocky and frost-boil meadows. They were entirely absent in the wet meadows. Details on meadow lichen composition may be found in Barrett (1972).

Ordination of the 20 lowland sedge-moss meadow stands sampled (Fig. 3) revealed a strong meadow community type separation along a soil moisture

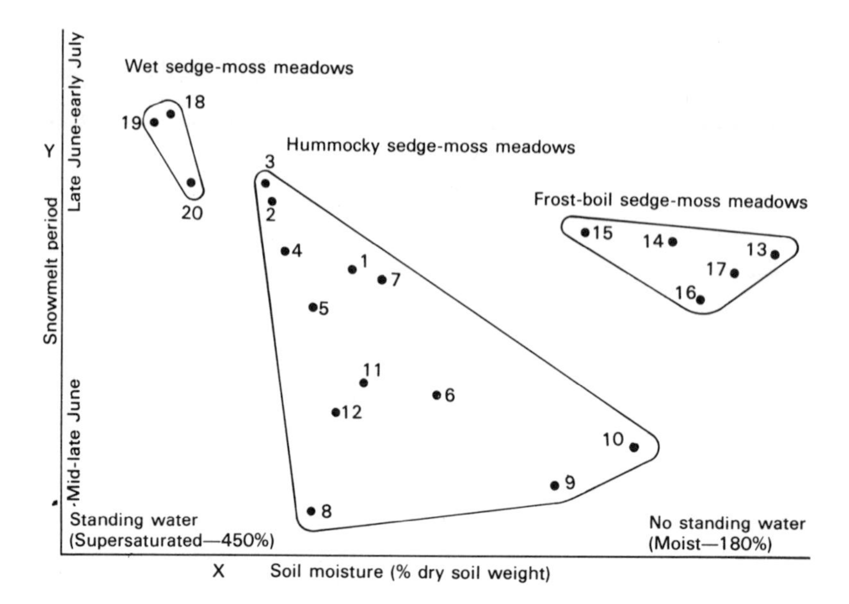

Fig. 3 Ordination of the 20 lowland sedge-moss meadow sites.

gradient and a less distinctive separation along a snowmelt gradient. Wet meadows were late melting and possessed standing water. The frost-boil meadows, although later melting, had no standing water present. Hummocky meadows ranged from early melting, moist soil stands to later melting stands with supersaturated soils.

Aboveground biomass and net primary production

In all of the meadows, graminoids contributed approximately 85% of the total biomass, woody plants 10%, and forbs the remaining 5%. The only exception was site 18, where monocot biomass constitutes 95% of the total biomass, and forbs the remaining 5%. Due to the relatively short growing season (50 day average), peak biomass development in all three plant groups was usually synchronous. In cases of asynchronous biomass development, forb and woody plant standing crops peaked a week or so earlier than did graminoids. Maximum total aboveground standing crops developed by early August in earlier, milder seasons (1971, 1973), and by mid-August in later, cooler seasons (1970, 1972).

Average maximum aboveground standing crops were 170 g m^{-2} cm in the frost-boil meadow, 258 g m^{-2} in the hummocky meadows, and 203 g m^{-2} in the wet meadow (Table 2). Variation in the aboveground standing crop between years was less than 10%; variation in the dead plant component being only slightly greater than that of the live plant component (Table 3).

Live shoot material constituted approximately 35% of the total aboveground biomass (Fig. 4). Green and non-green tissue provided almost equal proportions to the live plant component. In the hummocky meadow (site 1), non-green shoot material had considerably more biomass (12 g m^{-2} than in either of the other meadows. At the plateau base hummocky meadow (site 10), the non-green stem base biomass was only equal to 50% of the green shoot

Table 2. Average peak season vascular plant standing crops (g m^{-2}) of various sedge-moss meadow components in a frost-boil (site 13), hummocky (sites 1, 4, 7, 10), and wet (site 18) sedge-moss meadow and an average lowland meadow.*

	Sedge-moss meadow community type				
	Frost-boil		*Hummocky*	*Wet*	*Average*
Component	*Total*** *area*	*Vegetated area*			
Aboveground					
Green	27.1	53.3	38.3	40.0	33.4
Non-green	30.1	59.0	43.3	39.7	37.2
Total live	57.2	112.3	81.6	79.7	70.6
Attached dead	101.7	199.4	164.5	117.7	134.0
Litter***	10.7	20.9	12.1	5.3	11.1
Total dead	112.4	220.3	176.6	123.0	145.1
Grand total	169.6	332.6	258.2	202.7	215.7
Live:dead	1:2.0	1:2.0	1:2.2	1:1.5	1:2.1
Belowground (25 cm)					
Live rhizomes	17.3	33.9	54.8	50.0	37.7
Live roots	328.1	643.3	827.0	672.3	594.8
Total live	345.4	677.2	881.8	727.3	632.5
Dead rhizomes	3.1	6.1	9.0	13.7	6.6
Dead roots	302.1	592.9	742.5	625.8	538.6
Total dead	305.5	599.0	751.5	639.5	545.2
Grand total	650.9	1276.2	1633.3	1366.8	1177.7
Total aboveground plus belowground biomass	820.5	1608.8	1891.5	1569.5	1393.4
Ratios:					
Live:dead	1:0.9	1:0.9	1:0.9	1:0.9	1:0.9
Total live above: live below	1:6.0	1:6.0	1:10.8	1:9.1	1:9.0
Total dead above: dead below	1:2.7	1:2.7	1:4.2	1:5.2	1:3.8
Total above: below	1:3.8	1:3.8	1:6.3	1:6.7	1:5.5

* Based on lowland surface cover of each meadow type — frost-boil (45%), hummocky (50%), and wet (5%).

** Total area data based on 51% vegetation cover and vegetated data based on 100% vegetation cover.

*** Peak litter standing crops were from early season harvests.

Table 3. Seasonal peak aboveground and belowground live, dead (including litter) and total standing crops (g m^{-2}) in a frost-boil, (site 13), hummocky (site 1), and wet (site 18) sedge-moss meadows 1970-1973.

	1970			*1971*			*1972*			*1973*			
Meadow	*Live*	*Dead*	*Total*	*Live*	*Dead*	*Total*	*Live*	*Dead*	*Total*	*Live*	*Dead*	*Total*	*Average*
Aboveground													
Frost-boil	—	—	—	55	106	161	57	121	178	59	112	171	170
Hummocky	80	211	291	91	194	285	87	193	280	84	191	275	283
Wet	—	—	—	79	132	211	76	120	196	84	121	205	204
Belowground													
Frost-boil	—	—	—	336	298	634	370	329	699	331	290	621	651
Hummocky	997	848	1845	1073	947	2020	1186	1019	2205	1110	956	2066	2034
Wet	—	—	—	698	610	1308	683	599	1282	801	710	1511	1367

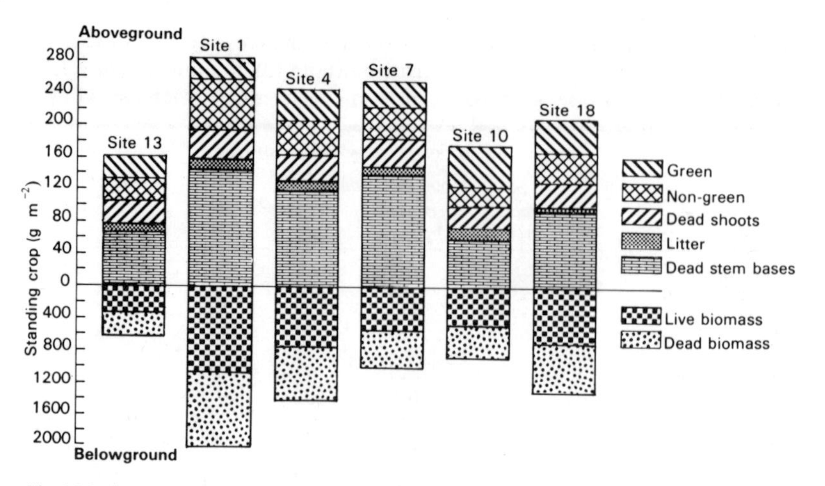

Fig. 4 Maximum aboveground and belowground standing crops in the frost-boil (site 13), hummocky (sites 1, 4, 7, 10), and wet (site 18) sedge-moss meadows, sampled for primary production in 1971.

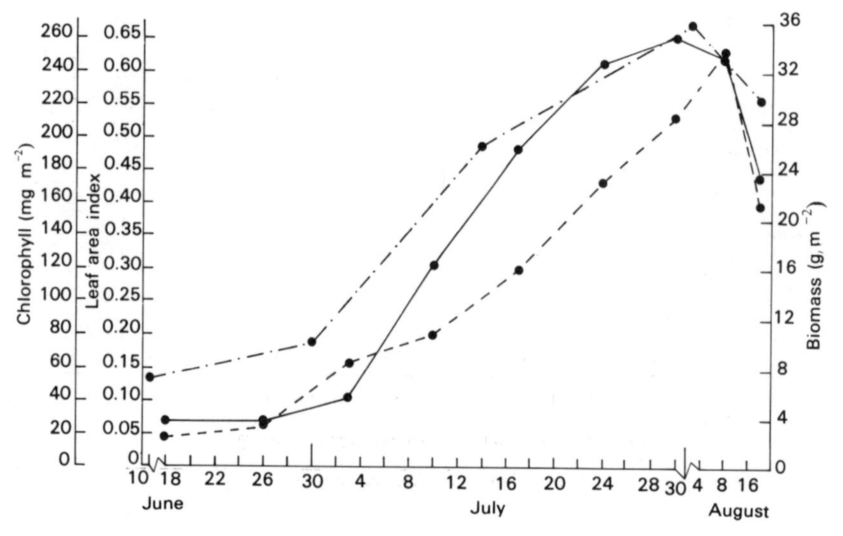

Fig. 5 Seasonal development of community chlorophyll (mg m^{-2}), photosynthetic leaf area index, and green shoot biomass (g m^2) in a hummocky meadow (site 1) during 1971.

biomass of the aboveground standing crop, approximately 80% consisted of stem base material in the moss mat. The remaining biomass portion was of dead shoot material above the moss canopy. At the plateau base meadow (site 10), dead stem base biomass comprised 68% of the total aboveground dead standing crop.

Seasonal green shoot weight increment paralleled seasonal community shoot chlorophyll and photosynthetic leaf area development (Fig. 5). A similar synchronous development was also found between shoot length increment and biomass increment, in both an earlier, milder (1971) and later, cooler (1972) season (Fig. 6).

Annual net production of the aboveground biomass averaged 28 g m^{-2} at

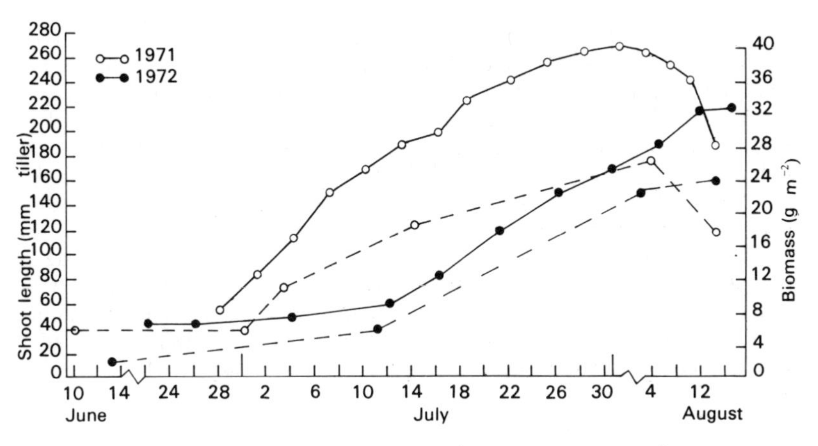

Fig. 6 Seasonal monocot green shoot length (mm plant^{-1}) and standing crop (g m^{-2}) in a hummocky meadow.

Table 4. Average aboveground net primary production [g m^{-2} and seasonal productivity* (g m^{-2} day^{-1})] for vascular plants in a frost-boil (site 13), hummocky (sites 1, 4, 7, 10) and a wet (site 18) hummocky sedge-moss meadow. In addition, an estimate of net primary production and productivity for an average lowland sedge-moss meadow* is included (± S.E.) (1970-1973).

Plant Group	*Component*	*Frost boil*** *Total area*	*Frost boil*** *Vegetated*	*Hummocky*	*Wet*	*Average***
Monocots	Green	15.8±0.6	31.0±1.1	23.1±1.6	37.7±0.6	20.6
	Non-green	3.3±0.3	6.5±0.7	7.0±1.4	6.3±0.3	5.3
	Rhizomes	2.8±0.4	5.6±0.9	7.2±0.9	7.5±0.9	5.2
	Roots***	52.2±2.4	102.4±4.8	82.6±5.4	120.5±6.5	70.8
Forbs	Green	1.3±0.3	2.5±0.6	3.6±0.6	1.4±0.2	2.5
	Roots	0.5±0.1	1.1±0.1	1.2±0.2	1.0±0.2	0.9
Woody plants	Green	4.5±1.7	8.8±1.9	5.3±0.6	—	4.7
	Non-green	2.8±0.3	5.5±0.5	2.5±0.5	—	2.5
	Roots	3.3±0.6	6.5±1.2	5.5±0.6	—	4.2
	Net aboveground production	27.7±1.4	54.3±2.7	41.7±2.0	45.4±0.05	35.6
	Net belowground production	58.9±2.3	115.4±4.5	96.5±5.6	128.7±1.0	81.1
	Total net production	86.6±3.4	169.7±6.7	138.2±7.6	165.1±8.8	116.7
	Ratio above:below	1:2.1	1:2.1	1:2.3	1:2.8	1:2.3
	Aboveground**** primary productivity	0.54±0.1	1.06±0.1	0.79±0.1	0.88±0.1	0.68
	Belowground primary productivity	1.16±0.1	2.27±0.2	1.84±0.1	2.51±0.2	1.57
	Net primary productivity	1.70±0.2	3.33±0.4	2.63±0.2	3.39±0.2	2.25

* Average meadow based on composite of contributions proportional to lowland area cover — hummocky (50%), frost-boil (45%), wet (5%).

** "Total area" data based on 51% vegetation cover; "vegetated" data based on 100% vegetation cover.

*** Calculated to a depth of 20 cm from the moss surface.

**** Based on a growing season of : 50 days/1970; 55/1971; 45/1972; 55/1973.

the frost-boil meadow, 42 g m^{-2} in the hummocky meadow and 45 g m^{-2} in the wet meadow (Table 4). Of the hummocky meadows, the intensive hummocky meadow (site 1) had the highest production with 47 g m^{-2} and the coastal hummocky meadow (site 7) the lowest with 30 g m^{-2} (Fig. 7). Seasonal productivity averaged 0.54 g m^{-2} day^{-1} in the frost-boil meadow, 0.79 g m^{-2} day^{-1} in the hummocky meadows and 0.88 g m^{-2} day^{-1} in the wet meadow. At least 80% of the aboveground production was of green shoot material. The highest proportions of green shoot production were found on sites 7 and 10. In the frost-boil and hummocky meadows, graminoids contributed 70% of the total aboveground production, woody plants 20%, and forbs 10%. In the wet meadow, 95% of the total aboveground production was by graminoids and the remaining 5% by forbs. Aboveground production was only 5% greater in an earlier, milder season as compared to a later, cooler season.

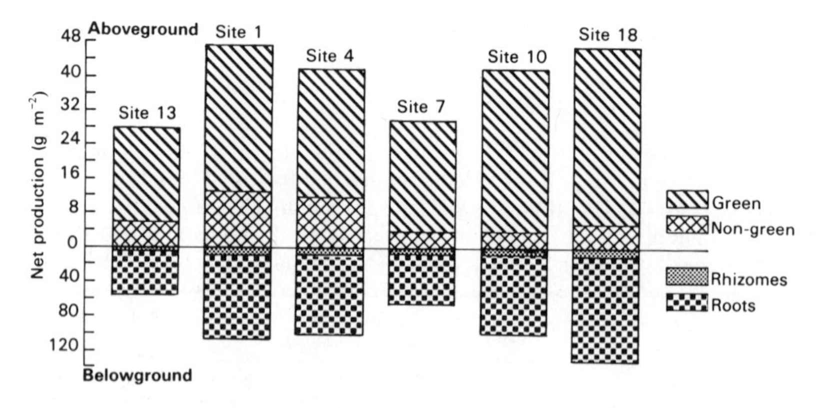

Fig. 7 Net aboveground and belowground primary production (g m^{-2}) in the frost-boil (site 13), hummocky (sites 1, 4, 7, 10), and wet (site 18) meadows sampled for primary production in 1971.

Belowground biomass and net primary production

The data show that about 65% of the root biomass occurred in the upper 15 cm of thawed soil (Table 5). No pronounced difference was found in the vertical root distribution beneath hummock and hollow microhabitats.

The average seasonal live root biomass was a relatively constant percentage of the total root biomass (52±0.7%). Average peak belowground biomass was lowest in the frost-boil meadow with 650.9 g m^{-2} and highest in the hummocky meadows, with 1,633.3 g cm^{-2} (Table 2). Of the hummocky meadows the intensive study meadow (site 1) possessed the greatest belowground biomass with 2,020 g m^{-2}. The lowest biomass was found in the plateau base hummocky meadow, (site 10) with 886 g m^{-2} (Fig. 4). Less than 5% of the total belowground biomass was of rhizome material. Slightly more than 5% of the total root biomass was of forb and woody plant roots.

Year to year variation in the belowground standing crop was less than 10% (Table 3). A correlation coefficient of $r=0.69\pm0.04$ for aboveground and belowground biomass at the intensive hummocky study meadow (site 1) indicated that less than half of the variance in belowground biomass is attributable to variance in the aboveground biomass. In the frost-boil and hummocky meadow, the greatest belowground biomass developed in a later,

Table 5. Vertical distribution and standard error (±) (5 cm intervals) of root (alive+dead) biomass (g m^{-2}) under hummocky, hollow, and overall meadow habitats (site 1, August 1973) (n=18).

Depth (cm)		*Meadow habitat*		
		Hollow	*Hummock*	*Average*
Moss	(0-5)	5.9±1.7	4.2±1.3	5.1±3.4
Soil	(5-10)	30.6±5.5	32.5±2.1	31.5±3.8
	(10-15)	27.3±3.0	29.0±5.1	28.2±4.2
	(15-20)	22.4±3.8	20.4±3.8	21.4±4.2
	(20-25)	13.8±5.9	13.9±3.8	13.8±5.1

cooler growing season. In the wet meadow, maximum belowground standing crops were found in earlier, milder seasons (Table 3).

Seasonal root increment of *Carex stans* transplants equalled 24% of their initial biomass. The major increment (80%) went into the formation of new and lateral roots and the remaining 20% into elongation of old roots. Seasonal net belowground production averaged 57 g m^{-2} in the frost-boil meadow, 97 g m^{-2} in the hummocky meadows, and 129 g m^{-2} in the wet meadow (Table 4). Based on biomass data, approximately 5% of the belowground production was in rhizome material and another 5% in forb and woody plant roots. Average belowground productivity was 1.16 g m^{-2} day^{-1} at the frost-boil meadow, 1.85 g m^{-2} day^{-1} at the hummocky meadows, and 2.51 g m^{-2} day^{-1} at the wet meadow.

Total meadow biomass and net primary production

The hummocky meadows had the highest average total biomass (1,892 g m^{-2}) and the frost-boil meadow the lowest (821 g m^{-2}) (Table 2). Of the hummocky meadows, the intensive meadow (site 1) had the highest total biomass (2,305 g m^{-2}) (Fig. 4). The wet meadow had the greatest total net primary production (165 g m^{-2}) and the frost-boil meadow the least (87 g m^{-2}) (Table 4). Total net primary production in the hummocky meadows averaged 138 g m and was highest in the intensive (site 1) meadow (147 g m^{-2}).

Litter and organic matter

The largest litter standing crops were found in the hummocky meadow (site 1) and the lowest in the wet meadow (site 18) (Table 6). Over the growing season, a 50% reduction was found in litter weight. The major portion of the litter developed through snow compaction of weathered and dead attached tissue; similar tissue also was added to the litter component during spring freeze-thaw periods. Litter traps revealed that *ca.* 4% to 11% of the meadow litter was redistributed within the meadow site by spring meltwaters but only *ca.* 0.1% (Rydén, pers. comm.) of the total litter was removed from the site, during spring runoff. Hummocks had almost 2.5 times as much litter as hollows. The majority of the hummock litter was of woody plant material (49%) while in the hollows, the majority (84%) was of graminoid tissue.

Soil organic matter content was 28,600 g m^{-2} (23 cm depth) in the hummocky meadow (site 1) and 13,400 and 10,400 g m^{-2} at the frost-boil (site 13) and wet meadow (site 18) respectively. At the former site, hummocks were underlain by almost twice the amount of organic matter found under hollows. Organic matter content under hummock and hollows at the frost-boil meadow

Table 6: Average (±S.E.) monthly litter standing crop in a frost-boil (site 13), hummocky (site 1), and wet (site 18) sedge-moss meadow, 1970-72 (n=6-16).

Site	Season	June*	July	August	September
13	1971	7.9±1.3	4.4±0.8	2.5±0.4	—
	1972	9.8±1.3	6.1±0.6	4.4±0.9	—
	Mean	8.9	5.3	3.5	—
1	1970	9.6±1.9	5.6±1.8	4.0±1.6	—
	1971	14.6±2.5	10.9±2.3	3.1±0.6	3.3±0.7
	1972	10.9±1.7	7.3±0.9	6.7±1.3	—
	Mean	11.7	7.9	4.6	—
18	1970	6.4±1.1	3.0±0.4	2.3±0.4	—
	1971	5.8±1.1	5.0±1.5	3.2±1.2	—
	Mean	6.1	4.0	2.8	—

*Sub-snow-surface samples.

(site 13) were approximately equal. In the hummocky and wet meadows (sites 1 and 18) there was an almost two-fold increase in organic matter content with depth, while in the frost-boil (site 13) meadow, the middle layer (8-15 cm) had almost four times the organic content of the upper and lower layers.

Chlorophyll and leaf area index

Chlorophyll content of meadow plant tissue ranged during the growing season from 3.1 to 11.1 mg g^{-1} dry wt. The lowest chlorophyll levels were found in the woody plants and the highest in the forbs. Comparison of *Carex stans* growing in all three meadow types showed frost-boil meadows (site 13) to have the highest chlorophyll levels (10-12 mg g^{-1}) and wet meadow plants (site 18) the lowest (6-8 mg g^{-1}) in early August. Under field equal greenhouse conditions, chlorophyll content of *Carex stans* was 27% greater than that in plants outside the greenhouse. This increase was equal to the 26% increase in net aboveground production within greenhouse plots.

Overwintering meadow plants retained an average of 20 mg m^{-2} of chlorophyll and less than 10% of this deteriorated after snowmelt. The chlorophyll content of woody and forb species peaked approximately two weeks later than in the graminoids. Peak community chlorophyll levels of 262 mg m^{-2} at the hummocky meadow (site 1) and 283 mg m^{-2} at the wet meadow (site 18) developed by late July. The peak (161 mg m^{-2}) did not come until early August at the frost-boil meadow (site 13). Graminoid tissues constitute 85-90% of the chlorophyll at the hummocky and frost-boil meadows (sites 1 and 13) and almost 100% at the wet meadow (site 18). These plant component contributions were proportional to those found for net primary production. A seasonal comparison of community chlorophyll content and green shoot biomass content at the intensive hummocky meadow (site 1) during 1971 showed a parallel development (Fig. 5).

Photosynthetic leaf area indices showed a relatively linear seasonal increment at all three meadows with peak values being developed by early August 1971. The same pattern of leaf area increment closely followed that of community chlorophyll and biomass increment (Fig. 5). Under greenhouse conditions, the leaf area index of *Carex stans* increased by 30% over that found in plants outside the greenhouse. This trend was similar to that found in the comparison of chlorophyll content and primary production.

Hummock *C. stans* had larger individual plant leaf areas than did hollow inhabiting plants. The highest individual plant leaf areas (8 cm^2) were found in wet meadow plants (site 18) and the lowest individual plant leaf areas (3 cm^2) were found in frost-boil plants (site 13).

Plant constituents

Inter-seasonal and inter-site caloric values of sedge-moss meadow species varied by less than 2% of the mean. The highest caloric levels were recorded in woody plants and the lowest in graminoids and Pteridophytes (Table 7). Caloric content of attached dead shoot material was only slightly lower (2%) than that of live shoots. On a total plant comparison, dead plants had slightly higher (4%) caloric content than did live plants. Belowground plant material was slightly lower (3%) in its caloric content than aboveground tissue. Average seasonal non-ash free caloric content of meadow litter was 4,616 cal g^{-1} and showed an overall seasonal decrease of only 2%. Litter ash content averaged 5% (2-7% range) in the various plant tissues, being highest in the forbs.

Table 7: Average seasonal (1970) non-ashfree caloric values (cal g^{-1} dry tissue weight) and ash content (%) for the major hummocky sedge-moss (site 1) meadow plant groups.

Plant portion	*Monocots*	*Forbs*	*Woody plants*	*Pteridophytes*
Live shoots	4785	4965	5100	4586
	(4.0)	(5.2)	(3.8)	(6.6)
Dead shoots	4777	4743	5002	4507
	(4.5)	(3.6)	(3.9)	(6.4)
Stem bases	4789	—	5106	—
	(4.1)	—	(1.9)	—
Roots	4670	4703	5106	—
	(3.3)	(4.4)	(4.0)	—
Rhizomes	4650	—	—	—
	(2.7)	—	—	—
Total live plant	4734	4804	5071	4547
	(3.7)	(4.4)	(3.4)	(6.5)
Total dead plant	4901	—	5282	—
	(5.6)	—	(3.7)	—

Comparison of similar plant constituents in *Carex stans* plants from the intensive meadow (site 1) showed inter-seasonal variation to be less than 20% of the mean; intra-seasonal variation was about 15%. Peak chemical levels developed by mid-season in aboveground (live and dead) tissue and in soil organic matter, but at the corresponding time interval, belowground (live and dead) tissue chemical levels were at their lowest (Table 8). Nitrogen and protein levels in aboveground live tissue were twice those of similar dead tissue, while phosphorus and potassium contents were three and six times greater, respectively, in similar comparisons. Aboveground dead biomass was four to seven times greater than the live tissue component, thereby multiplying the overall effect of greater chemical levels in the tissue. Belowground tissue chemical levels were comparable to those of aboveground dead tissue except for calcium. Total chemical content of the belowground component was three to five times greater than that of the aboveground component by virtue of the fact that there was three to seven times more biomass belowground.

Table 8. Chemical (% oven dry tissue weight) plant constituents of *Carex stans* tissue and soil organic matter at the intensive hummocky sedge-moss (site 1) meadow (1970).

Portion	*Date*	*Protein*	*N*	*Ca*	*P*	*K*
Aboveground live	1-7	17.4	2.78	0.20	0.31	1.72
	22-7	21.5	3.43	0.27	0.36	2.15
	17-8	18.5	2.96	0.24	0.33	1.78
	Mean	19.1	3.05	0.23	0.33	1.88
Aboveground dead	1 7	8.8	1.40	0.78	0.09	0.20
	22-7	10.9	1.74	0.84	0.13	0.39
	17-8	9.2	1.48	0.79	0.10	0.29
	Mean	9.6	1.54	0.80	0.10	0.29
Belowground live & dead	1 7	8.8	1.40	0.55	0.14	0.34
	22-7	7.7	1.23	0.36	0.10	0.38
	17-8	8.1	1.29	0.28	0.19	0.49
	Mean	8.2	1.30	0.39	0.14	0.40
Total dead plants	1 7	10.1	1.62	1.11	0.10	0.13
	22-7	9.2	1.47	0.78	0.10	0.17
	17-8	8.3	1.33	0.62	0.13	0.16
	Mean	9.2	1.47	0.83	0.11	0.15
Total live plants	1-7	11.7	1.86	0.51	0.18	0.75
	22-7	13.4	2.13	0.49	0.20	0.97
	17-8	11.6	1.91	0.44	0.21	0.85
	Mean	12.2	1.96	0.48	0.20	0.85
Soil organic matter	1-7	16.3	2.61	2.06	0.14	0.10
	22-7	16.1	2.58	1.90	0.15	0.15
	17-8	16.4	2.62	1.74	0.13	0.12
	Mean	16.3	2.60	1.90	0.14	0.12

Carbohydrate

Belowground carbohydrate reserves in *Carex membranacea* and *C. stans* showed an early season depletion and a late season buildup (Fig. 8). In both species, pre-snowmelt carbohydrate content was approximately 17% of the total plant weight and increased to 24% plant weight by the peak of the season. Early season carbohydrates were primarily in the form of sugars and by the peak of the season, sugar and starch levels were approximately equal. Aboveground tissue was predominantly (90%) sugar throughout the season. Oligosaccharides constituted the main type of sugar (98%). Monosaccharide levels in shoots were two times greater than in belowground tissue. Throughout the growing season, aboveground carbohydrate levels of mature plants exceeded those of immature plants and overall carbohydrate levels of reproductive plants were greater than those of vegetative plants.

Plant growth and phenology

The average duration of the potential lowland sedge-moss meadow growing season (period from time of 50% meadow snowmelt to the time of the first substantial snowfall) was 50 days. It ranged from 45 days in a later, cooler season (1970, 1972) to 55 days in an earlier, milder season (1971, 1973). Initiation of the growing season varied by as much as two weeks, commencing

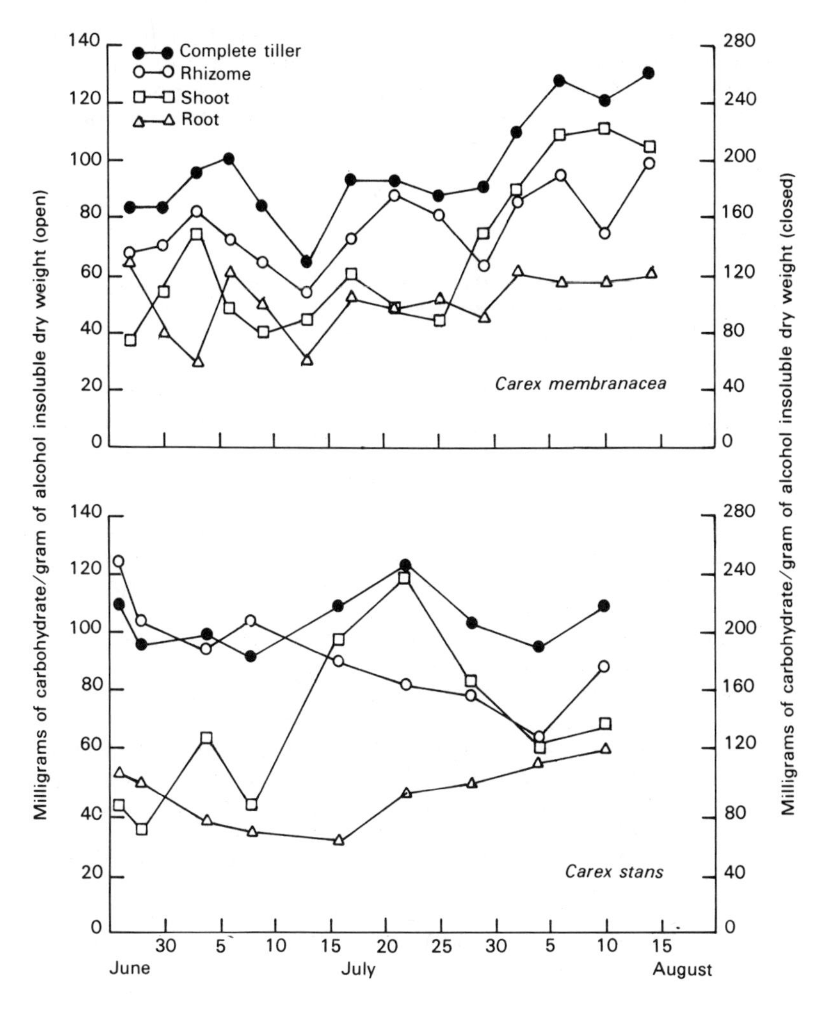

Fig. 8 Seasonal changes in the carbohydrate content of various *Carex membranacea* (1971) and *C. stans* (1971) tissue.

on 25 June, 1971 and not until 6 July, 1972. The belowground growing season was calculated to be of the same duration but lagged by three weeks. Shoot growth for graminoids averaged 30 to 35 days. Pronounced shoot senescence (dieback) coincided with the onset of seed set. In the earlier, milder seasons (1971, 1973), sedge shoot growth increased by 20% at the intensive study meadow (site 1) and by 75% at the wet meadow (site 18), compared to that in later, cooler seasons (1970, 1972). Shoot length increment in *Carex stans* closely followed the pattern of seasonal graminoid green shoot biomass increment (Fig. 6).

Inter-site comparison showed wet meadow (site 18) sedge shoot increment to be 35% greater than at the intensive study meadow (site 1). Intra-site comparisons at the intensive study meadow showed hummock sedges to have only 5% more shoot increment than those growing in hollows. At the frost-boil meadow (site 13), a similar comparison revealed the difference to be 30%.

Salix arctica and *Pedicularis sudetica* developed their full leaf complement two to three weeks ahead of the graminoids and retained a larger photosynthetic leaf area over a greater part of the growing season.

Both *Carex stans* and *C. membranacea* retained from 10-15% of their photosynthetic tissue overwinter. Tagged and snow covered *C. stans* plants at the intensive study meadow (site 1) showed an average pre-snowmelt shoot increment of 6-8 mm. Similar plants which had been exposed from under a 35 cm snow cover, in advance of regular spring snowmelt, did not show any shoot growth until after normal snowmelt occurred.

Overwintering *C. stans* plants retained one fully and three partially photosynthetic leaves. The three green leaves showed some growth over the growing season but eventually died back completely. The initial fully photosynthetic leaf showed only partial dieback after its seasonal growth. Over the growing season, three new leaves were initiated and grew. The two earliest formed leaves died back partially by the end of the season, the leaf initiated last remained fully photosynthetic overwinter. The life span of a *C. stans* leaf was two years but the leaves remained attached for another three years before breaking off and becoming litter.

Dieback (browning) of leaves prior to development of peak aboveground biomass, and growth after peak biomass development, provided an average underestimate of 4% and 3% respectively in shoot production. Forb and woody plant shoot material died back completely each season.

Slightly more than 50% of the tillers on *C. stans* and *C. membranacea* were dead. Of the remaining live tillers, 75% were mature plants and 25% were tiller buds. The suggested average life span of sedge tillers is slightly more than five years. Lowland sedge-moss meadow stem densities ranged from 1,100 stems m^{-2} in the frost-boil meadow (site 13) to 2,200 stems m^{-2} at the intensive study meadow (site 1). There were 1,300 stems m^{-2} at the wet meadow (site 18).

Graminoids, exclusive of *Luzula confusa*, had an average seasonal flowering rate of 5% of their population, while that of forb species was approximately 10% (Table 9). *Luzula confusa* had the highest flowering rate (35%). Flowering levels in the intensive study meadow (site 1) were approximately twice those in either the frost-boil or wet meadows (sites 13 and 18). Flowering after an earlier, milder season (1971) was 13% greater than after a later, cooler (1970) season. Following a later, cooler season, flowering rates

Table 9. Flowering percentage in various species at the frost-boil (site 13), hummocky (site 1), and wet (site 18) sedge-moss meadows (1971, 1972).

Species	*Frost-boil*		*Hummocky*		*Wet*	
	1971	*1972*	*1971*	*1972*	*1971*	*1972*
Carex stans	2.2	5.0	6.2	7.1	7.9	8.5
C. membranacea	3.5	3.9	2.8	3.9	—	—
C. misandra	5.7	8.3	2.4	6.3	—	—
Eriophorum triste	0.3	1.8	—	—	—	—
E. angustifolium	—	—	0.5	1.3	1.2	1.4
Arctagrostis latifolia	5.9	5.3	3.9	3.0	2.9	3.1
Poa arctica	—	—	—	—	2.9	2.4
Hierochloe pauciflora	3.3	1.0	8.7	10.0	1.8	1.9
Juncus biglumis	4.5	5.6	12.0	13.9	—	—
Luzula confusa	—	—	37.5	33.1	—	—
Polygonum viviparum	8.5	9.4	11.9	13.5	—	—
Saxifraga hirculus	—	—	2.6	1.3	—	—
S. foliolosa	—	—	8.5	11.9	—	—
Pedicularis sudetica	—	—	11.7	17.1	—	—
Mean	4.2	5.0	9.1	10.2	3.3	3.5

were depressed by as much as 80%. Within the field greenhouses, sedge flowering levels were 110% greater than those of sedges growing outside the greenhouses. In the following season, plants growing within the sites of the former greenhouses had flowering levels 20-65% greater than those in the immediate vicinity. In both cases, flowering levels were considerably lower (80%) than the year before.

Discussion

Physical environment

Snowcover in the range of 39-49 cm provided an effective insulation for the meadow vegetation and was the major source of spring meltwater flowing through the meadows. Snow, with densities as high as 0.43 g cm^{-3} developed prior to snowmelt and created a situation where miniature ice greenhouses could develop around individual plants. The importance of these ice greenhouses to sub-snow-surface plant growth will be discussed later.

The average maximum meadow soil thaw depths, in the range of 27-56 cm, were similar to those found in comparable low arctic communities (Pavlova 1969, Brown et al. 1970). Mineral soils in the frost-boil meadows thawed deeper as a result of a lower vegetation cover, better soil drainage and more compact soils. The pronounced early August increase in soil thaw of the plateau base hummocky meadow was attributed to increased soil drainage and accompanying soil warming. A milder season (1971, 1973) resulted in a 10 cm increase in soil thaw, compared to a cooler season (1970, 1972). This was in large part due to higher levels of incoming solar radiation in the early part of the season (Courtin and Labine this volume).

Frost-boil meadows bore a striking overall resemblance to the hummocky meadows but differed primarily in possessing prominent frost-boils. These frost-boils were generally poorly vegetated as a result of active cryoturbation. In a large number of these meadows, frost-boils possessed a thin, shallow (*ca.*/ 5 cm thick) organic layer and a thin cover of vascular plants. Frost-boil surface area cover in comparable low arctic meadows (Johnson et al. 1966, Aleksandrova 1970) was less than that found in the lowland meadows. This might reflect more active soil conditions within the high arctic environment. The hummock and hollow physiography in the hummocky meadows, as well as in heavily vegetated areas of the frost-boil meadows, provided two contrasting plant microhabitats. Hollows had reduced soil thaw depths, were colder, and possessed poorly aerated, supersaturated soils while the hummocks developed deeper soil thaw, were warmer, and had better aerated, warmer soil conditions. The presence of such differing microhabitats produced a distinctive plant distribution pattern (monocot concentration in hollows and a pronounced concentration of forbs and woody plants, in addition to graminoids, on the hummocks) and a slightly more vigorous plant growth on the hummocks. The relatively uniform habitat of the wet meadows caused a decrease in species diversity but did create a more uniform plant distribution.

The Fibric Organic Cryosols had a more extensive vegetation cover and a more effective insulation against soil thaw. Shallower soil thaw and lower soil temperatures within these meadows may have contributed to soil nutrient deficiencies by limiting the availability of nutrients as well as reducing their effective uptake by plants. The gleysolic soils of the frost-boil meadows, as well as the majority of the hummocky meadows had less organic material. The resulting decrease in insulation, coupled with higher soil density, resulted in deeper soil thaw. Deeper thaw and warmer soil temperatures did not increase soil nutrient availability since these soils were nutritionally poorer than the organic soils (Walker and Peters this volume).

Plant communities

Although hummocky and frost-boil meadows had similar land areas (50% and 45% respectively) the former meadow type predominates in the more recently emerged (north and west coast) portions of the Lowland as well as in the area along the base of the plateau. Hummocky meadows developed in areas of greater soil stability. The frost-boil meadows occupy the older (central and southern) portions of the Lowland in regions of heavy calcareous sediment accumulation. These areas appeared to be characterized by greater soil instability. Wet meadows had a limited distribution (5%) because of their dependence on an inundated type of habitat. Their origin was probably derived from partial habitat modification through flooding. This led to a reduction in species composition and a more uniform distribution of plants tolerant of flooded conditions.

Vascular plant composition in the three sedge-moss meadow community types was similar. The greatest species richness was found in the hummocky meadows which possessed a more intermediate type of habitat.

Similar meadow species similarities were observed for bryophyte components of the lowland meadows (Vitt and Parkarinen this volume). Of the species restricted to specific community types, Porsild (1964) described *Pleuropogon sabinei* and *Ranunculus hyperboreus* as being strongly associated with inundated habitats and *Eriophorum triste* and *Eutremia edwardsii* as being strongly associated with calcareous (alkaline) habitats.

The hummocky meadow type is a widespread tundra community (Polunin 1948, Savile 1964). The frost-boil meadow type is similarly a common tundra community (Porsild 1955, Johnson et al. 1966). The wet meadow type is most comparable to low arctic marsh communities (Beschel 1970). Although the meadows are a prominent component of Truelove Lowland, they are an insignificant component of the Queen Elizabeth Islands (Babb and Bliss 1974).

Although no quantitative data establishing lowland inter-meadow relationships are available, Sigafoos (1951) suggested that drainage of tundra marsh communities (wet meadows?) would lead to their replacement by more mesic types (hummocky meadows?). Similarly, Raup (1965) and Tikhomirov (1966) suggested that frost-boil communities were degraded forms of more mesic types. This latter meadow derivation could not be applied to the lowland meadows since it has been shown that the frost-boil meadows are on older surfaces than the hummocky sedge-moss meadows. Their development and maintenance appears dependent upon active, fine-textured soils with only a shallow to non-existant soil organic development. Furthermore, older frost-boil meadows took on the surface appearance of hummocky meadows when the inactive frost-boils were covered with a layer of mosses and vascular plants. In the lowland meadows, the suggested pattern of meadow development is one of frost-boil and hummocky meadows developing mainly independently but with a possible eventual modification (through maturity) of the former type into hummocky meadows. Wet meadows appear to be a modified (through flooding) form of hummocky meadows.

Biomass and net primary production

The moist habitat within these meadows was primarily responsible for the high proportion of graminoid biomass (85-100%). Monocot contribution was less pronounced in some other low arctic meadows (Andreev 1966, Aleksandrova 1970) and this might help explain the higher biomass found in these southern communities. The relatively uniform annual aboveground biomass and net primary production within the lowland meadows was attributed to a damping effect of seasonal climatic variation and low grazing pressure. Pronounced seasonal variations in aboveground biomass and production are generally associated with fluctuations in microtine grazing (Dennis and Johnson 1970). When compared to other communities of Truelove Lowland, average aboveground standing crops of 159 to 246 g m^{-2} in the meadows were comparable to the dwarf shrub heath type (Bliss et al. this volume) but lower than the cushion plant type where dead *Dryas* shoots comprised a large portion of the standing crop (Svoboda this volume).

The large belowground biomass was consistent with that reported from low arctic meadows (Dennis and Johnson 1970, Aleksandrova 1970). The majority of the root biomass was within the upper 15 cm of the soil surface, where soils were warmer and better aerated. The more optimal soil conditions within hummock microhabitats resulted in their root biomass being 35% higher in frost-boil and 95% higher in hummocky meadows, compared to hollow microhabitats. The consistent 1:0.9 live to dead root ratio throughout the growing season suggested a highly synchronized pattern of root growth and death rates through the season.

Litter averaged 20% of the biomass above the moss mat. This supports the five year shoot turnover rate calculated from phenological data. The importance of plant composition and microhabitat in determining litter biomass were shown by the fact that the hummocks, with primarily woody plants and forbs, had three times as much litter as did the hollows which were completely dominated by graminoids.

Primary production increased along a moisture gradient (frost-boil to wet meadows). Dennis and Johnson (1970) found similar community production gradients at Barrow, Alaska. Shoot production within the field greenhouses was 26% greater than that found under natural conditions. Higher temperatures (mean 7°C) were probably the most important factor in increased net production in the plastic greenhouses. Other factors were high humidity and reduced wind. Similar increased production within field greenhouses was reported for beach ridge cushion plants (Svoboda this volume).

Aboveground production of 28-45 g m^{-2} was approximately 50% that of comparable low arctic sedge communities (Tieszen 1972, Haag 1974). This was to be expected since these communities are the high arctic extension of low arctic meadow communities. Aboveground production accounted for approximately 30% of the total production, in contrast to beach ridge cushion plant communities where it was 80% of the total (Svoboda this volume).

Belowground production ranged from 58-129 g m^{-2}, a 5% annual increment of the belowground standing crop. Vassiljevskaya and Grishna (1972) reported annual belowground increments of 7% in West Taimyr tundra sites and Dennis and Johnson (1970) reported 17% increment at Barrow, Alaska. Roots which thawed later in the season had a shorter growth period and grew under lower soil temperature and oxygen regimes. Billings et al. (1973) have shown that low arctic soil temperatures decreased root growth.

Chlorophyll and leaf area index

Some 90% of the overwintering green tissue survived after snowmelt. The overwintering green tissue was considered to be important in the 2-6 mm of shoot length increment found to occur under the snow canopy. Presumably the plants were able to photosynthetize within the ice greenhouses. Tieszen (1972) reported that overwintering green tissue is minimal at Barrow and that it rapidly deteriorated after spring melt.

The chlorophyll content of 3-11 mg g^{-1} dry wt. was 3 to 5 times greater than in the beach ridge cushion plants (Svoboda this volume) and slightly higher than in low arctic graminoids (Tieszen 1972). Peak community chlorophyll levels of 161-283 mg m^{-2} were *ca*. 50% those reported for Barrow (Tieszen 1972) and a Norwegian alpine wet meadow (Berg et al. 1973).

Plant height, density, and leaf area index values were lower than those reported for Barrow (Caldwell et al. 1974). Peak photosynthetic leaf areas coincided with green shoot biomass peaks. A similar pattern was found within the plastic greenhouses and indicated that green tissue leaf area was closely correlated with green tissue biomass.

Carbohydrate

Seasonal changes in sedge carbohydrate reserves paralleled those found in low arctic meadow communities (McCown and Tieszen 1972). Spring reserve depletions were found to be 40-50% of maximum values, considerably less than the reported 70-80% depletion rate in temperate graminoids (Cook et al. 1962). The lower reserve depletion level in arctic species would serve as an adoptive mechanism in protecting against a serious reserve depletion which could not be compensated for during an unusually short and cold growing season. Depletion of sedge carbohydrate reserves continued until *ca*. 75-80% of aboveground growth was completed. This was consistent with the estimate for short growth for arctic species proposed by Billings and Mooney (1968).

Winter carbohydrate reserves were mostly in the form of sugars; summer reserves mostly as starch. In both cases, oligosaccharides were more abundant than monosaccharides, a condition also found in beach ridge cushion plants (Svoboda 1973) and in Norwegian wet meadow plants (Berg et al. 1973). Seasonal changes in sedge carbohydrate constituents were consistent with those described (Parker 1963) for plants that developed cold hardiness.

Vegetative sedge tillers concentrated their carbohydrate reserves belowground and used these for asexual reproduction. Increased short carbohydrate levels in flowering tillers indicated a depletion of belowground reserves to facilitate flowering. The relatively low level of flowering in meadow sedges may have resulted from insufficient carbohydrates available for floral reproduction. Flowering was highest in plants with greater carbohydrate reserves, e.g. those of the wet meadow, in plants after a relatively warm summer in which greater reserve accumulations could be developed, and in plants under the plastic greenhouse.

Plant constituents

Average caloric content of meadow species was consistent with that for low arctic sites (Tieszen 1970) and Norwegian alpine locations (Wielgolaski and

Kjelvik 1973). The higher caloric content of woody species was accounted for by their higher lipid levels (Hadley and Bliss 1964) and lignin content (Svoboda this volume). The slightly higher caloric values of dead versus live tissue could be explained by carbon concentration as a result of the loss of mobile elements (Malone and Swartout 1969) and slow decomposition rates (Widden this volume).

Meadow forbs had higher protein and nitrogen levels than did woody and graminoid species. Plant tissue nitrogen and protein levels were comparable with those of low arctic wet sedge meadow plants (Haag 1974) but were slightly lower than those in Norwegian alpine meadow plants (Wielgolaski and Kjelvik 1973). Live plant tissue had almost twice the nitrogen and protein content of dead tissue, indicating rapid nitrogen mineralization after tissue death. The low nitrogen levels of belowground tissue were attributable to the predominantly carbohydrate content of these tissues and suggested lower decomposition rates for the belowground tissue (Alexander 1967). Phosphorus levels varied less than those for nitrogen, a situation similar to that in low arctic meadow plants (Haag 1974). Phosphorus and potassium were two highly mobile tissue elements, while calcium was highly immobile and increased proportionally in dead tissue. Nitrogen, phosphorus, and potassium were therefore readily returned to the meadow nutrient pool through decomposition and leaching. Highest tissue nutrient levels were recorded in the latter part of July, approximately two weeks prior to peak standing crop development. After the peak standing crop, shoot nutrient content decreased by approximately 10% due to possible translocation belowground and reduced nutrient uptake.

Phenology and plant growth

Peak shoot growth and biomass content were reached 30-35 days after spring growth initiation. Plant shoot growth appeared to be "biologically fixed" in its development. Warmer summers resulted in an earlier start and end to the growth period. Total green shoot increment was, however, greater in such a summer. Meadow forbs and woody species developed their full leaf compliment two to three weeks earlier than the graminoids. This allowed them to make maximum use of the higher radiation levels in early July. Although graminoid leaf senescence and seed set initiation in early August coincided, low flowering rates (5%) ruled out any possible physiological interactions between the two processes.

The combination of high spring snow densities and freeze-thaw conditions may result in "ice greenhouse" formation around individual plants in a manner similar to that reported (Bell 1973) for *Kobresia bellardii* in the alpine tundra. These, along with the winter carryover of green tissue, would facilitate potential sub-snowsurface growth. Aleksandrova (1970) and Kovakina (1958) reported the occurrence of sub-snowsurface growth in other arctic species.

Carex stans produced one to two fewer leaves per shoot than the closely related *C. aquatilis* in the Low Arctic. The average life span of *C. stans* (5 years) was less than that of *C. aquatilis* which Shaver and Billings (1975) reported to be 7 years. A shorter life span would provide a more rapid nutrient turnover and would also, through translocation of nutrients between older and younger tillers (Allessio and Tieszen 1973), allow this rhizomotous species to produce numerous tillers. Tiller densities of 1,000-2,000 stems m^{-2} were comparable to the 1,600-2,000 stems m^{-2} reported for Eurasian meadows (Smirnov and Tokmakova 1972) but lower than the 4,800 stems m^{-2} at Barrow (Dennis and Tieszen 1972).

Meadow flowering levels were low but were comparable to those found at Barrow (Dennis and Tieszen 1972). Flowering in a given summer was dependent on growth conditions (climatic) of the previous year and may have been related to differences in seasonal carbohydrate reserve build-up.

Conclusions

Sedge-moss meadows, the dominant feature of Truelove Lowland, result from a topography that slows the flow of spring meltwater, keeping large areas wet to moist all summer. The communities are low in species diversity, yet high in plant production compared with other lowland plant communities. Community structure and function, including production, are quite similar to other low arctic wet sedge communities. Most of the plant biomass and production occur belowground and the rather large carbohydrate reserves permit rapid shoot growth as soon as snow melts. Most plant growth takes place over a 30-35 day period almost regardless of the summer climatic variability, indicating as with the beach ridge cushion plants, that they are biologically programmed in terms of their developmental pattern. Vegetative reproduction dominates and the relatively rapid rate of litter turnover ensures a more continual recycling of nutrients.

Acknowledgments

Many thanks to my supervisor, Dr. L.C. Bliss, for his direction in the fieldwork and comments on the interpretations of these results. Grateful appreciation to my field assistants over the course of the study — Ken Orich, Gabor Pal, Jr., Leslee Muc and Lisa Casselman. An added note of thanks to Mrs. Kveta Svoboda for her care and concern in related laboratory analyses.

Financial support for this study was provided in part by an N.R.C. grant (A-4879) to L.C. Bliss.

References

Aleksandrova, V.D. 1970 The vegetation of the tundra zones in the U.S.S.R. and data about its productivity. pp. 93-111. *In:* Productivity and Conservation in Northern Circumpolar Lands. W.A. Fuller and P.G. Kevan (eds.). IUCN N Ser. No. 16. Morges, Switzerland. 344 pp.

Alexander, M. 1967. Introduction to Soil Microbiology. John Wiley & Sons, Inc. New York. 472 pp.

Allessio, M.L. and L.L. Tieszen. 1973. Patterns of translocation and allocation of ^{14}C– photoassimilate *in situ* studies with *Dupontia fischeri* R. Br., Barrow, Alaska. pp. 219-229. *In:* Primary Production and Production Processes, Tundra Biome. L.C. Bliss and F.E. Wielgolaski (eds.). Tundra Biome Steering Committee. Edmonton. 256 pp.

Andreev, V.N. 1966. Peculiarities of zonal distribution of the aboveground and underground phytomass on the East European Far North. Bot. Zhur. 51: 1401-1411.

Arnon, D.J. 1949. Copper enzymes in isolated chloroplasts. Polyphenoloxidase in *Beta vulgaris*. Plant Physiol. 24: 1-15.

Babb, T.A. and L.C. Bliss. 1974. Susceptibility to environmental impact in the Queen Elizabeth Islands. Arctic 27: 234-237.

Barrett, P. 1972. Phytogeocoenoses of a Coastal Lowland Ecosystem, Devon Island, N.W.T. Ph.D. thesis. Univ. British Columbia, Vancouver. 292 pp.

Beals, E. 1960. Forest bird communities in the Apostle Islands of Wisconsin. Wilson Bull. 72: 156-181.

Bell, K. 1973. Autecology of *Kobresia bellardii*. Why winter snow accumulation pattern affects local distribution. Ph.D. thesis. Univ. Alberta, Edmonton. 167 pp.

Berg, A.O. Skre, R.E. Wielgolaski and S. Kjelvik. 1973. Leaf areas and angles, chlorophyll and reserve carbon in alpine and subalpine plant communities, Hardangervidda, Norway. 239-254. (See Allessio and Tieszen 1973.)

Beschel, R.E. 1970. The diversity of tundra vegetation. 85-92 (see Aleksandrova 1970).

Billings, W.D. and H.A. Mooney. 1968. The ecology of arctic and alpine plants. Biol. Rev. 43: 481-529.

———. G.R. Shaver, and A.W. Trent. 1973. Temperature effects of growth and respiration of roots and rhizomes in tundra graminoids. pp. 57-63 (see Allessio and Tieszen 1973).

Bliss, L.C. 1975. Devon Island, Canada. pp. 17-60. *In:* Structure and Function of Tundra Ecosystems. T. Rosswall and O.W. Heal (eds.). Ecol. Bull. No. 20. Swed. Nat. Sci. Res. Council, Stockholm, 450 pp.

Brown, J., H. Coulonmbe, and F. Pitelka. 1970. Structure and function of the tundra ecosystem at Barrow, Alaska. pp. 41-71 (see Aleksandrova 1970).

Caldwell, M.M., L.L. Tieszen, and M. Farred. 1974. The canopy structure of tundra plant communities at Barrow, Alaska and Niwot Ridge, Colorado. Arct. Alp. Res. 6: 151-159.

Cook, W.C. H.H. Biswell, R.T. Clair, E.H. Reid, L.A. Stoddart, and M.L. Upchurch 1962. Basic problems and techniques in range research. Pub. No. 890. National Research Council, Washington.

Dennis, J.G. and P.L. Johnson. 1970. Shoot and rhizome-root standing crops of tundra vegetation at Barrow, Alaska. Arct. Alp. Res. 2: 253-266.

———. and L.L. Tieszen. 1972. Seasonal course of dry matter and chlorophyll by species at Barrow, Alaska. pp. 16-21. *In:* Proceedings 1972 Tundra Biome Symposium. S. Bowen (ed.). CRREL, Hanover, New Hampshire. 211 pp.

Haag, R.W. 1974. Nutrient limitations to plant production in two tundra communities. Can. J. Bot. 53: 103-116.

Hadley, B. and L.C. Bliss. 1964. Energy relations of alpine plants on Mt. Washington, New Hampshire. Ecol. Monogr. 34: 331-357.

Jensen, W.A. 1962. Botanical Histochemical Principles and Practice. W.H. Freeman Co., San Francisco. 408 pp.

Johnson, A.W., L.A. Viereck, R.E. Johnson, and H. Melchior. 1966. Vegetation and flora. pp. 277-354. *In:* Environment of the Cape Thompson Region, Alaska. J.E. Wolfe and N.J. Wilimovsky (ed.). U.S. Atomic Energy Commission. Washington. 1250 pp.

Kovakina, V.A. 1958. Biochemical characteristics of some wintergreen plants of the far north. Bot. Zhur. 43: 1326-1332.

Malone, C.R. and M.B. Swartout. 1969. Size, mass, and caloric content of particulate organic matter in old field and forest soils. Ecology 50: 395-399.

McCown, B.H. and L.L. Tieszen. 1972. A comparative investigation of periodic trends in carbohydrate and lipid level in arctic and alpine plants. pp. 40-45. (See Dennis and Tieszen 1972.)

Muc, M. 1973. Primary production of plant communities of the Truelove Lowland, Devon Island, Canada — sedge meadows. pp. 3-14 (see Allessio and Tieszen 1973).
Nelson, N. 1944. A photometric adaption of the Somogyi method for the determination of glucose. J. Biol. Chem. 153: 375-380.
Parker, J. 1963. Cold resistance in woody plants. Bot. Rev. 29: 123-201.
Pavlova, E.B. 1969. Vegetal mass of the tundras of Western Taimyr. J. Moscow University No. 5: 62-67.
Polunin, N. 1948. Botany of the Canadian Eastern Arctic Part III. Vegetation and Ecology. National Museum of Canada. Bull. 104. Ottawa. 304 pp.
Porsild, A.E. 1964. Illustrated flora of the Canadian Arctic Archipelago. Second edn. National Museum of Canada, Bull. No. 146. 218 pp.
Raup, H.M. 1965. The structure and development of turf hummocks in the Mesters Vig District, Northeast Greenland. Medd. om Grønl. 166: 1-112.
Savile, D.B.O. 1964. General ecology and vascular plants of the Hazen Camp area. Arctic 17: 237-258.
Shaver, G.R. and W.D. Billings. 1975. Root production and root turnover in a wet tundra ecosystem, Barrow, Alaska. Ecology 56: 401-409.
Sigafoos, R.S. 1951. Soil instability in tundra vegetation. Ohio J. Sci. 51: 281-298.
Smirnov, V.S. and S.G. Tokmakova. 1972. Influence of consumers on natural phytocoenosis production variation. pp. 122-126. *In:* International Meeting on the Biological Productivity of Tundra. F.E. Wielgolaski and T. Rosswall (eds.). Tundra Biome Steering Committee, Stockholm, Sweden. 320 pp.
Svoboda, J. 1973. Primary production of plant communities of the Truelove Lowland, Devon Island, Canada — beach ridges pp. 15-26. (see Allessio and Tieszen 1973).
Tieszen, L.L. 1972. The seasonal course of aboveground production and chlorophyll distribution in a wet arctic tundra at Barrow, Alaska. Arct. Alp. Res. 4: 307-324.
Tikhomirov, B.A. 1966. The study of tundra biogeocoenoses. *In:* Programma i metodika biogeotsenologicheskikh issledovaniy Moscow, Izd-vo Nauka.
Vassiljeskaya, V.D. and L.A. Grishna. 1972. Organic carbon reserves in conjugate eluvial accumulative landscape of West Taimyr (station. Agapa). pp. 215-218 (see Smirnov and Tokmakova 1972).
Wielgolaski, F.E. and S. Kjelvik. 1973. Mineral elements and energy of plants at Hardangervidda, Norway. pp. 231-238 (see Allessio and Tieszen 1973).

Ecology and primary production of Raised Beach communities, Truelove Lowland

J. Svoboda

Introduction

Raised beaches have a special ecological significance for Truelove Lowland and can be classified as a part of the Polar Semi-desert. In terms of species diversity and plant production the plant communities are depauperate, yet they are richer than the true Polar Desert. Although this vegetation type, dominated by cushion plants, is of rather limited extent in the Queen Elizabeth Islands, it covers vast areas of the southern arctic islands.

The objectives of this study were: (1) to describe the plant communities and estimate standing crop and net plant production; and (2) to contribute to the understanding of primary production processes and survival strategy in cushion plants by studying phenology, growth, chemical components, and plant structure.

The field study was conducted during the summers of 1970 through 1973.

Study area

Raised beaches and other moderately to well drained gravel areas with a cushion plant vegetation represent *ca*. 20% of the Lowland. Three zones can be recognized (Fig. 1): dry crests and slopes, covering 293 ha; and the mesic gentle lower slopes which represent the ecological transition zone to the sedge meadows, covering 572 ha.

These raised beaches have been formed by ice push and tidal activity. Their stepwise sequence across the Lowland followed the glacio-isostatic

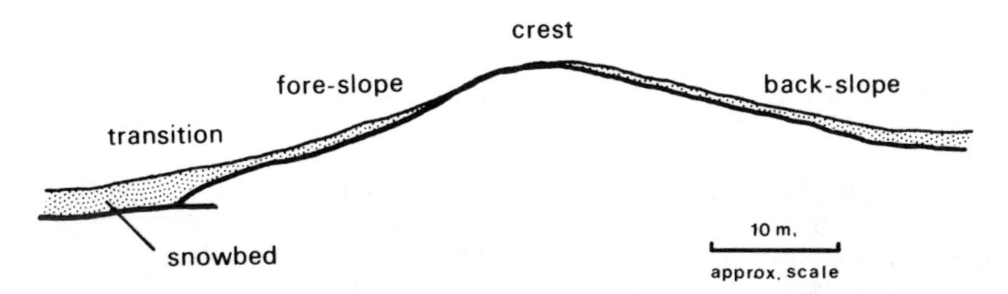

Fig. 1 Profile of a raised beach. Meadows typically occur beyond the fore-slope and back-slope.

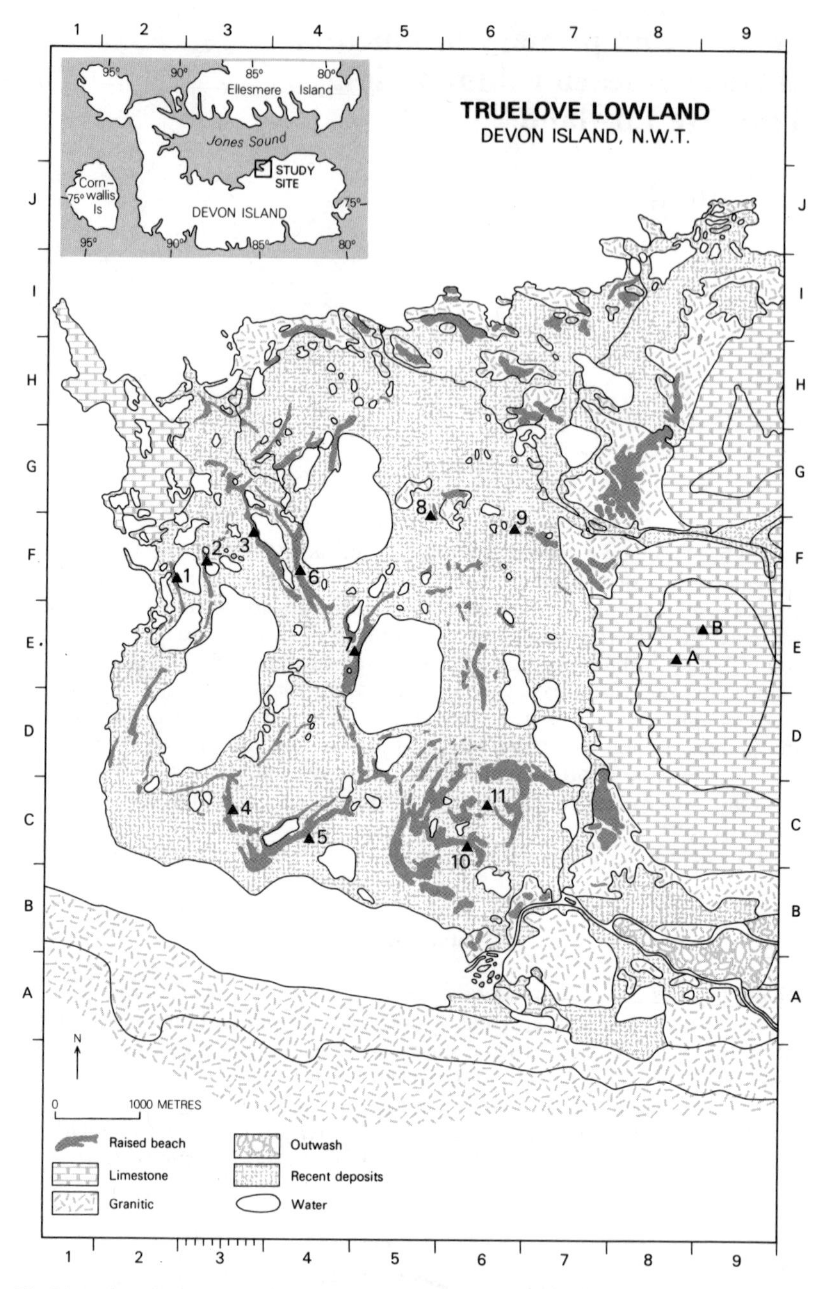

Fig. 2 Location of study sites on the raised beaches and the plateau, Truelove Lowland (△).

recovery some 9450 B.P. Post-glacial uplift (recovery) is computed as the amount of glacial emergence at specific times less the eustatic sea level at those times measured from the present sea level (Andrews et al. 1971). In Truelove Lowland the highest marine limits lie at 76 m.a.s.l. (King 1969, Barr 1971). Post-glacial uplift is still in progress at Devon and new beaches are forming, though its present rate is slow, 0.36 m/ 100 years (Andrews 1970). Much of the Lowland is covered by a Pleistocene mixture of crystalline, sandstone, and dolomitic components, which vary in degree of weathering. Raised beaches are formed of sand (there is little very fine grained sand) and gravel and their lithology reflects the nearest substrate. In cross-section a raised beach is a rounded convex body of coarse material (Fig. 1). At the base of each ridge there is a gentle concave slope of varying width. Because of its special ecological position (environmentally and floristically) between the dry ridge and moist meadow, this part was studied separately and named "transition."

Deep (30 to 100+ cm) ice-wedge cracks run along and across beach ridges and divide them into rectangles and polygons. Ice-wedges on the old uplift ridges appear to be inactive.

Raised beaches are one of the dominant topographic features. They shield the adjacent meadows from wind, increase meadow snow cover by acting as natural barriers, and subsequently increase summer moisture in meadows by restricting lateral drainage. Less pronounced smaller terraces lie parallel to the main axis of the ridge, and usually on its long fore slope. These contribute to microtopographic variability.

Methods

In addition to the intensive study an extensive study was conducted on eleven raised beaches in the Lowland and at two areas of the Plateau (Fig. 2).

Microclimate

To supplement the data of Courtin and Labine (this volume), temperature measurements (thermocouples) were made twice a week on three permanently established microclimatic stations, one in each of the beach ridge zones. Plant temperatures were measured by thermocouples introduced into the centre of a plant clump. Active layer depth (1 m intervals) was measured using a calibrated (1 cm) steel rod along a transect across the intensive study site every 2-3 days during the thaw period and once a week during refreeze. Soil moisture samples (n=4, gravimetric method) were collected weekly during the entire season from the 15 cm depth.

Plant communities and phenology

Community data from the intensive study site — Intensive Raised Beach Site (I.R.B.), were obtained by analysis of 60, 1×1 m quadrats, collected in a stratified random manner, in each of the three zones: crest, slope, and transition. In the extensive survey of 11 other raised beaches, 30 (50×50 cm) quadrats, 10 in each zone, were laid out. Percent cover, frequency, and number of individual plants per species were recorded.

Phenological measurements on cushion plants were performed mainly on the I.R.B. In 1972 *Dryas integrifolia* and *Saxifraga oppositifolia* were labeled (50 plants of each species in the crest and transition zones) and three phenological stages recorded: flower bud swell, flowering, and fruit development throughout the entire growing season. Clump area of labeled plants was measured and flowering was correlated with clump size and relative clump age. In order to obtain information on leaf expansion and shoot growth, in 1971 and 1972, shoots of some species were regularly collected, pressed as herbarium specimens or preserved in acetic acid-alcohol for later determination.

Vascular plant standing crop and production

Standing crop and net production were sampled by stratified random harvesting in 5×5 m subplots (destructive sampling areas, see Fig. 9 in the Introductory chapter). Sixteen samples (20×50 cm) were harvested on five separate dates in 1970 and 12 samples (100×100 cm) on five dates in 1971. In 1972 harvesting (two dates) was on an individual plant per species basis; 10 to 35 individuals per species per zone were harvested along with the root system after snowmelt (harvest 1), and at the peak of growth for a particular species (harvest 2). The area of each plant clump was measured and the standing crop and production values per m^2 were calculated using cover information per species for each zone. One set of samples was collected in 1973 using the same method as the previous year.

All material was sorted to species, then divided into roots and shoots which were further subdivided into standing live ("brown" shoots and "green" shoots) and standing dead. All plant material was ovendried at 80°C.

Due to a high variation in standing crop between harvests, calculation of seasonal production could not be based on the simple total weight differences between harvests. Production estimates (aboveground) were based upon only the green shoot-leaf increment. Annual increment of stem dry weight for *Dryas* was estimated by a laborious method of correlating shoot aging with shoot length. Annual stem dry weight increment was calculated to be *ca*. 21% of the net seasonal "green" shoot production. Net annual root production was assumed to be equal to stem (stripped of leaves) production. For *Dryas* the stem:root ratio was 1.8:1(±0.3). For more detail on methods see Svoboda (1974).

Active (living) and inactive (presumably dead) roots of *Dryas* were distinguished using radioactive phosphorus in the field (Svoboda and Bliss 1974).

Assuming that the strategy of the cushion species dominating these communities in the same habitat is fairly similar, the methods used for *Dryas* were applied to the other species in calculating community production. On slopes *Dryas* accounts for 78% of the standing crop and 68% of the net annual vascular plant production.

Leaf Area Index (L.A.I.)

Chlorophyll was extracted from the green parts of *Dryas*, *Saxifraga oppositifolia* and *Carex nardina* according to Arnon (1949) and determined spectrophotometrically (n=5) using a Spectronic 20 (Baush and Lomb). The amount of chlorophyll was calculated using Arnon's formula.

Carbohydrates

Three replicate samples of selected species were periodically collected, separated into roots and shoots, and put in jars with hot 80% ethyl alcohol in which a pinch of Na_2CO_3 was added to prevent spontaneous hydrolysis of non-reducing sugars. The jars were then heated (water bath, 15 min), sealed and stored for laboratory analyses. Before analyses the samples were dried, weighed, and ground to pass a 20 mesh screen, and further extracted on Soxhlet extractors for 6 hr. The insoluble residues were dried and stored for starch analyses.

Reducing sugars, starch, and total carbohydrates were determined using Nelson's (1944) colorimetric adaptation of the Somogyi spectrophotometric technique. A Spectronic 20 (Baush and Lomb) was used for colorimetric determination of reducing sugars. Carbohydrate data are presented in mg g^{-1} of insoluble residue and in mg g^{-1} of original dry weight. For more detail on methods see Svoboda (1974).

Lipid, protein, fibre, nutrient constituents

In the determination of lipids, the method of Bliss (1962) was followed. The percentage of volatile non-lipid compounds was assumed to be a minor fraction of the petroleum ether extract and was not determined. Analyses for protein, fibre and mineral content were made on dried material harvested on the slope of the I.R.B. three times during the 1970 growing season. Samples were analysed by using the standard analytical methods of the Association of Official Agricultural Chemists (1955).

Caloric content

Caloric content (n=2) was determined from the dried material harvested in 1970, using an automatic 1200 Adiabatic Oxygen Bomb Calorimeter (Parr Instrument Co. Inc.).

Leaf angle measurement

A special device designed by Whitfield and Svoboda for leaf angle measurements of very small leaves (Svoboda 1974) was used on three *Dryas* clumps (2 crest, 1 transition zone). Leaf angle to the horizontal plane was read directly on the loosely hinged protractor; leaf compass orientation was read on the "clock" established around the clump (12 hr=N). Plants were divided into four 90° compass segments and 150 leaves were randomly measured in each segment of each clump.

Results and Discussion

Microclimate

The Lowland is underlain by deep and cold permafrost. On the crest the active layer reaches a maximum of 80 to 100 cm in August but is shallower (40 to 70 cm) on slopes and even more shallow (25 cm) in the transition zone to a meadow. After snowmelt it takes about three days for the active layer depth to reach 25 cm during which time the soils warm only slowly. The root zone (*ca.* 25 cm depth) is influenced by the slowly lowering permafrost table for a much longer period.

Favorable temperatures exist only near the surface. Plant clump temperatures are similar to ambient in the first 3 cm but significantly higher than the air at 25 cm on a sunny day (15° to 20°C). Leaf temperature of these cushion plants may however exceed ambient by 20° to 30°C (Addison this volume). Surface temperatures were regularly 2° to 5° higher than on the slope because of solar angle on the crest. After mid-July, temperatures decreased steadily. Compared with the Intensive Meadow Site, raised beach average temperatures were almost 4°C higher near the ground (Courtin and Labine this volume).

Wind velocities increase with the progress of the season, 1.15 in early June to 3.54 m sec^{-1} in late August at 25 cm (Courtin and Labine). In winter snow is blown from the crest and deposited in the lee side of theridge. In June snow disappears rapidly from the crests and slopes but persists for 10 to 14 days in the snowbeds of the transitional zone. After snowmelt, crests are water saturated for a few hours and the transition zones for several days. At the end of the season, soil moisture in the transition zone was 3X that of the slope and the crest zones.

Precipitation was low each summer with much of it falling as fine mist. Summer rain is the main source of water for the raised beach plants after snowmelt. After this water has drained and evaporated, plants are subjected to considerable stresses (Addison this volume, Teeri 1973). On the contrary relative humidity is very high (*ca.* 90%) during the entire season.

Plant communities

Only about 15 species of vascular plants out of 95 were found on the raised beaches. Plant communities vary continually from the most densely covered transition zone to the almost barren crest. Since three physical zones could be

Table 1. Vascular plant cover (%) and number of individual plants (clumps) per m^2 (±SE) on the Intensive Raised Beach Site (I.R.B.) in Truelove Lowland (n=60).

	Cushion plant-lichen				*Cushion plant-moss*	
	Crest		*Slope*		*Transition*	
Species	*%*	*No.*	*%*	*No.*	*%*	*No.*
Woody						
Dryas integrifolia	7.55±1.3	5.75±0.8	18.68±1.8	12.68±1.9	27.80±1.7	25.40±2.5
Cassiope tetragona	—	—	—	—	7.60±3.2	7.07±2.4
Salix arctica	2.17±0.5	1.80±0.3	0.86±0.2	1.36±0.3	1.88±0.6	2.06±0.6
Monocots						
Carex nardina	5.80±0.6	14.35±1.5	3.96±0.5	9.52±1.2	0.20±0.0	0.20±0.0
Carex misandra	—	—	0.48±0.2	2.60±1.4	2.08±0.6	11.70±3.7
Carex rupestris	—	—	4.52±1.8	13.76±4.6	13.00±2.8	52.53±7.1
Forbs						
Saxifraga oppositifolia	4.05±0.3	37.60±2.9	6.28±0.3	51.00±3.7	4.60±0.6	28.00±5.2
Silene acaulis	—	—	0.05±0.0	0.66±0.0	0.20±0.0	0.66±0.0
Arenaria rubella	0.4 ±0.0	1.20±0.4	0.03±0.0	2.48±0.8		
Cerastium alpinum	0.06±0.0	0.70±0.3	0.01±0.0	0.16±0.0		
Draba spp.	—	—	0.02±0.0	1.32±0.5	0.00±0.0	0.06±0.0
Melandrium affine	0.00±0.0	0.20±0.0	0.00±0.0	0.13±0.0	0.06±0.0	0.26±0.0
Oxyria digyna	—	—	—	—	0.00±0.0	0.40±0.0
Papaver radicatum	—	—	—	—	0.00±0.0	0.13±0.0
Pedicularis lanata	0.05±0.0	0.10±0.0	0.06±0.0	0.40±0.1	0.22±0.0	0.60±0.0
Polygonum viviparum	0.00±0.0	0.05±0.0	0.00±0.0	0.13±0.0	0.66±0.0	1.60±0.0
Total for community	19.72±1.5	61.75±3.4	34.95±3.0	96.20±6.2	58.30±4.2	130.60±8.4

recognized, each zone was sampled for plant community composition.

1. *The crest zone*, with only 10% to 15% total plant cover, is dominated by a cushion plant-lichen community of *Saxifraga oppositifolia*, *Carex nardina* and *Salix arctica*. *Dryas integrifolia* is missing on the driest crests. Lichens, mostly crustose species, provide the greatest cover (Tables 1-3).

2. *The slope zone*, also covered by a cushion plant-lichen community, is

Table 2. Percentage frequency of main moss species, and percentage cover of main lichen species on I.R.B. (Mosses compiled after Vitt and Pakarinen*, lichens after Richardson and Finegan*.)

Moss species	*Cushion plant-lichen Crest and slope*	*Cushion plant-moss Transition*
	Frequency	
Distichium capillaceum	80	90
Encalypta rhaptocarpa	60	40
Mnium riparium	50	80
Bryum sp.	50	20
Myurella julacea	40	40
Isopterygium pulchellum	20	20
Dicranoweisia crispula	20	—
Didymodon asperifolius	20	—
Myurella tenerrima	20	—
Ditrichum flexicaule	—	70
Tomenthypnum nitens	—	40
Drepanocladus intermedius	—	30
Blepharostoma trichophyllum	—	30
Aulacomnium turgidum	—	30
Hylocomnium splendens	—	30
Pogonatum alpinum	—	20
Drepanocladus uncinatus	—	20
Orthothecium chryscum	—	20

Lichen species	*Cushion plant-lichen Crest*	*Slopes*	*Cushion plant-moss Transition*
Crustose		Percentage cover	
Rhizocarpon geographicum	3.3	1.9	—
Lecanora dispersa	1.0	1.2	—
Lecanora epibrion	3.4	18.0	1.5
Foliose			
Umbilicaria arctica	4.4	—	—
Umbilicaria lyngei	1.2	—	—
Parmelia separata	0.6	—	—
Hypogymnia subfusca	1.7	—	—
Physcia muscigena	—	—	1.9
Fructicose			
Cetraria cuculata	—	0.5	3.6
Cetraria islandica	—	1.0	—
Cetraria nivalis	0.5	1.7	0.5
Alectoria chalybeiformis	0.6	—	—
Alectoria nigricans	2.2	1.2	—
Alectoria ochroleuca	2.9	0.9	—
Alectoria pubescens	11.4	—	—
Thamnolia vermicularis	4.4	6.7	5.5
Total	37.6	33.1	13.0

Moss species with frequency values less than 20% and lichens with cover values less than 0.5% are not included. (Frequency values are based on 40 cm^2 sample plots.)

*See chapters this volume.

Table 3. Cover (%) of vascular plants, mosses and lichens on two raised beaches in Truelove Lowland (Compiled by Richardson, Svoboda, and Vitt.)

	Intensive	*Phalarope*
Crest zone		
Vascular plants	19.7	7.4
Mosses	2.0	3.0
Lichens	58.8	54.0
Bare	19.5	35.6
Slope zone		
Vascular plants	34.9	38.0
Mosses	7.4	8.1
Lichens	50.1	44.7
Bare	7.6	9.2
Transition zone		
Vascular plants	58.3	42.2
Mosses	22.3	58.8
Lichens	25.4	22.0
Bare	0	0

dominated by *Dryas integrifolia* with lesser amounts of *Saxifraga oppositifolia*, *Carex rupestris* and *C. nardina* (Table 1). Foliose and fruticose lichens are more abundant (Tables 2-3).
3. *The transition zone*, covered by a cushion plant-moss community, is dominated by *D. integrifolia*, *Carex misandra* and *Cassiope tetragona*. The lichen cover is less but there are more mosses (Tables 1-3).

Species of minor importance sampled but not included in Table 1 are: *Alopecurus alpinus*, *Draba alpina*, *D. corymbosa*, *D. subcapitata*, *Luzula confusa*, *L. nivalis*, *Pedicularis hirsuta*, *Saxifraga tricuspidata*, and *Tofieldia coccinea*. These species contribute less than 1% of the total vascular plant cover.

The extensive study data show a correlation between plant cover, standing crop, and age and elevation of the raised beaches (Fig. 3) as was also pointed

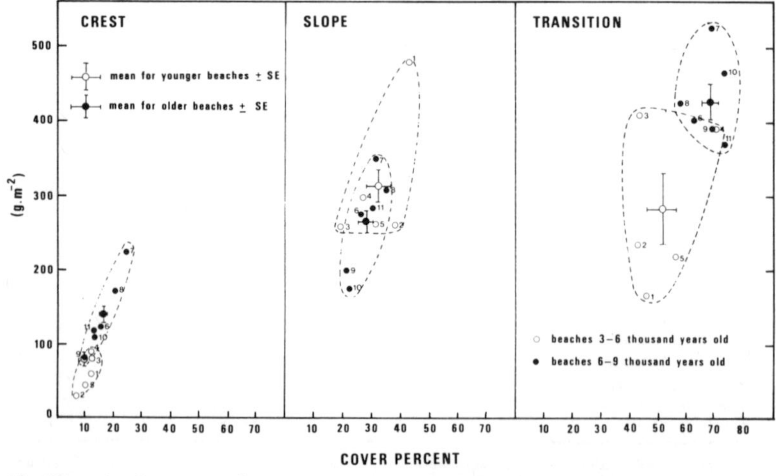

Fig. 3 Vascular plant community cover versus aboveground standing crop on eleven raised beaches in 1971.

out by Acock (1940) on Spitsbergen. The community and standing crop data on 10 lowland raised beaches (Table 4) show clearly the dominance of cushion species *Dryas integrifolia* and *Saxifraga oppositifolia* which are accompanied by tufted sedges (*Carex nardina* and *C. misandra*) and *Salix arctica*. *Cassiope tetragona* is abundant in the transition zone where snow lies longest.

Vascular plant cover is generally less than 2% on the plateau. Most plants and especially mosses occur where surface rock is common, especially in sorted polygons (Table 5).

Table 4. Vascular plant species, their average cover, number of individuals (density), dry weight, and aboveground standing crop on 10 raised beaches in Truelove Lowland.

Species name	*Cover±SE*	*Density±SE*	*Individual plant size (cm)*	*Individual plant weight (g)*	*Aboveground standing crop ($g\ m^{-2} \pm SE$)*
		Crest zone			
Dryas integrifolia	3.00±0.07	3.48±0.09	86.21	9.45	32.89± 0.73
Saxifraga oppositifolia	4.46±0.03	32.80±0.19	13.61	1.09	35.91± 0.24
Carex nardina	2.38±0.04	5.44±0.08	43.79	3.30	17.93± 0.27
Carex misandra	0.10±0.00	0.36±0.01	18.78	0.24	0.09± 0.01
Salix arctica	2.45±0.04	2.24±0.03	109.26	3.38	8.70± 0.16
Draba sp.	0.10±0.00	1.44±0.03	5.29	0.43	0.62± 0.03
Saxifraga tricuspidata	0.28±0.02	0.20±0.01	140.00	ND	T —
12 additional species					
		Slope zone			
Dryas integrifolia	17.65±0.15	11.20±0.07	157.61	20.30	227.36± 1.98
Saxifraga oppositifolia	5.80±0.04	38.80±0.24	14.96	1.17	45.57± 0.33
Carex nardina	2.00±0.03	5.52±0.08	36.23	1.33	7.34± 0.13
Carex misandra	0.26±0.01	1.68±0.05	15.62	0.23	0.39± 0.02
Carex rupestris	1.31±0.08	4.88±0.27	26.84	0.06	0.32± 0.01
Salix arctica	1.63±0.04	2.48±0.03	65.94	1.86	4.61± 0.12
Silene acaulis	0.10±0.00	1.04±0.03	9.86	1.32	1.38± 0.05
Pedicularis sp.	0.10±0.00	0.60±0.23	17.52	0.18	0.11± 0.01
Polygonum viviparum	0.15±0.00	1.44±0.04	10.42	0.02	0.03± 0.00
7 additional species					
		Transition zone			
Dryas integrifolia	15.73±0.15	11.68±0.09	134.65	20.02	223.90± 2.23
Saxifraga oppositifolia	10.11±0.08	41.04±0.31	24.63	0.73	30.12± 0.23
Carex misandra	6.52±0.06	18.64±0.14	34.98	1.09	20.35± 0.21
Carex rupestris	3.29±0.10	9.20±0.13	35.81	0.17	1.61± 0.06
Salix arctica	7.26±0.08	10.52±0.09	69.04	2.85	30.04± 0.37
Cassiope tetragona	7.87±0.17	8.20±0.17	95.98	3.81	31.21± 0.69
Alopecurus alpinus	3.75±0.08	11.52±0.24	32.57	0.07	0.78± 0.02
Arenaria rubella	0.10±0.00	2.32±0.05	4.24	0.16	0.36± 0.01
Carex subspathacea	1.12±0.03	3.20±0.21	35.16	0.07	0.21± 0.02
Carex stans	0.18±0.01	0.36±0.02	50.00	ND	T —
Equisetum arvense	0.11±0.01	1.64±0.16	7.01	ND	T —
Juncus biglumis	0.26±0.01	0.92±0.04	27.99	0.09	0.08± 0.00
Oxyria digyna	0.26±0.01	3.16±0.08	8.39	ND	T —
Pedicularis sp.	0.50±0.01	3.72±0.07	13.45	0.25	0.95± 0.02
Poa arctica	0.10±0.00	0.36±0.02	20.17	20.91	7.53± 0.39
Polygonum viviparum	3.42±0.07	27.08±0.45	12.62	0.04	1.11± 0.02
Papaver radicatum	0.10±0.00	1.28±0.04	7.67	0.60	0.77± 0.03
7 additional species					
Average for raised beach					
Crest Zone	12.97±0.50	49.01±1.60	—	—	97.07± 6.06
Slope Zone	29.07±0.83	69.02±2.20	—	—	287.30± 9.28
Transition Zone	60.77±1.39	156.01±5.00	—	—	359.44±12.79

T - trace
ND - No data

Table 5. Plant cover and standing crop (±SE, n=10) of the Polar Desert Plateau east of Truelove Lowland, Devon Island, N.W.T., 1970 and 1971.

	Area A (1970)			Area B (1971)		
Species	*Cover %*	*g m^{-2}*	*% of total*	*Cover %*	*g m^{-2}*	*% of total*
Arenaria rubella	—	0.02± 0.0	0.1	0.31±0.1	0.13±0.1	0.3
Cerastium alpinum	0.45±0.1	4.04± 1.5	17.1	0.91±0.2	1.17±0.5	2.6
Draba alpina	—	—	—	0.04±0.0	0.19±0.1	0.4
Papaver radicatum	0.40±0.2	1.02± 0.2	4.3	0.50±0.1	0.99±0.2	2.1
Phippsia algida	0.30±0.2	0.68± 0.6	2.9	0.86±0.5	0.54±0.3	1.2
Salix arctica	—	—	—	0.03±0.0	1.25±0.2	2.7
Saxifraga oppositifolia	3.90±1.1	15.67± 2.8	66.3	4.79±1.4	39.59±13.2	86.1
Saxifraga caespitosa	0.20±0.2	0.88± 0.9	3.7	0.55±0.2	1.39±0.6	3.0
Saxifraga cernua	0.15±0.2	0.38± 0.3	1.6	0.06±0.0	0.05±0.0	0.1
Stellaria longipes	0.12±0.2	0.92± 0.3	3.9	0.20±0.2	0.45±0.2	1.0
Total vascular plants	5.82±1.2	23.62± 3.8	—	7.98±1.6	52.9 ±16.0	—
Total moss	—	266.0 ±45.5	—	5.11±1.3	232.0 ±33.4	—
					4.30 unidentified roots	
					2.82 unidentified standing dead	

Ratios:

Vascular plants portions:
Total shoots:total roots 7.7:1
Live shoots:total roots 3.5:1
Live brown:green 1.8:1

Production (g m^{-2})	Area A	Area B
Vascular plants aboveground	2.0	4.5
Total	2.4	5.3

At Spitsbergen Rønning (1965) described the cushion plant-lichen community as *Nardino-Dryadetum*. In Truelove Lowland Barrett (1972) applied this association to the crest and slope communities but distinguished the transition community as a *Tetragono-Dryadetum* association.

Truelove Lowland is a mosaic of lakes, sedge meadows, rock outcrop areas, and raised beaches. The beaches, which are covered by cushion plants, represent the Polar Semi-desert type of vegetation that Bliss (1975) compared with Polar Desert on the upland plateau. This vegetation is much more depauperate in species diversity (only 8 to 10 vascular species) and very low in cover and standing crop. It fulfills the specifications of a Polar Desert outlined by Aleksandrova (1970). The vegetation is strictly High Arctic in composition, low and discontinuous in cover, restricted in vertical structure, and has a poorly developed root system.

Phenology and plant growth

After snowmelt, plants on the crest zone develop rapidly before the soil dries. In the transition zone (snowbed situation) plants are snow-free 2 to 3 weeks later and show different patterns of growth and development. *Saxifraga oppositifolia* flowers 3 to 5 days after snowmelt. A −2°C plant temperature is sufficient to break dormancy and 0°C to stimulate flowering in this species when grown in a growth chamber. Other species of the community flower 5 to 20 days later, the latest being sedges (Table 6). Phenological events in the crest and transition zones differ by more than two weeks and fruiting is usually not completed in plants from late snowbeds. In the late and cool summer of 1972 many *Dryas* flowers in the transition zone froze while in full bloom in late August while in the previous two years flowering was completed by the end of July and seed set in mid-August (Fig. 4).

A similar phenological pattern occurred in plants on "early" and "late"

Dryas Integrifolia

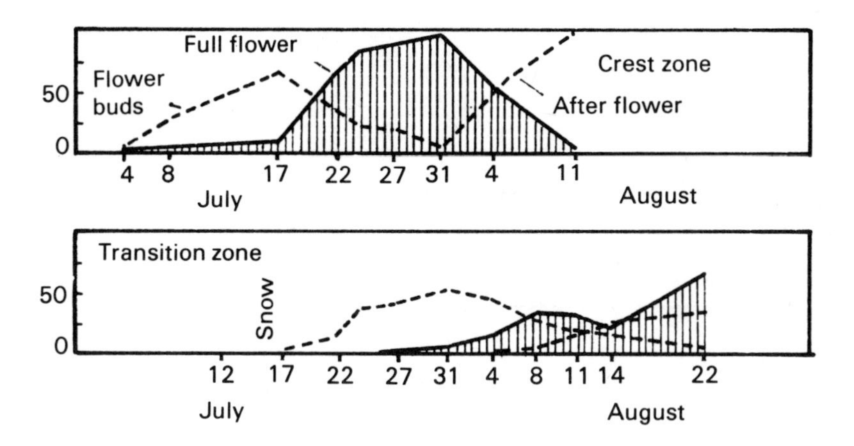

Saxifraga Oppositifolia

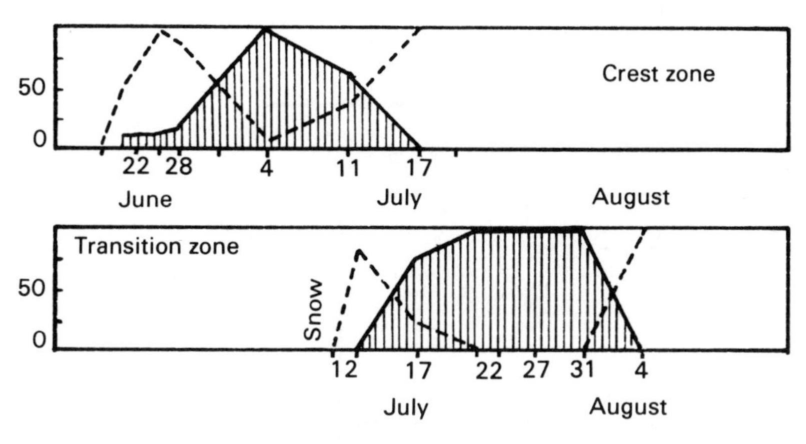

Fig. 4 Pattern of flowering of *Dryas integrifolia* and *Saxifraga oppositifolia* in two zones of I.R.B. in 1972, expressed as percentage of total flowering.

melting raised beaches. On 20 July 1972 when the I.R.B. (elev. 28 m) showed the first spots of snow-free ground, the crest of the Base Camp raised beach (No. 6, elev. 15 m) was 5 to 15% free of snow with *ca.* 5% of the *S. oppositifolia* population in flower, and the Air Strip raised beach (No. 3, elev. 12 m) was 50% snow-free with *ca.* 70% of *S. oppositifolia* in flower initiation and 30% in full flower.

Flowering pattern and microhabitat requirements differ in *Dryas* and *S. oppositifolia*. The mesic transition zone is the optimal habitat for *Dryas*, while *Saxifraga* flourishes best on drier slopes and the xeric crests. Morphological diversity resulting in typical crest and transition ecoforms occurs in the two species (Teeri 1972, Svoboda 1974). In the transition zone *Dryas* clumps are hemispherical and produce more leaves and flowers, while on crests *Dryas* is flat and prostrate, and most of the plant surface is dead. In the transition zone,

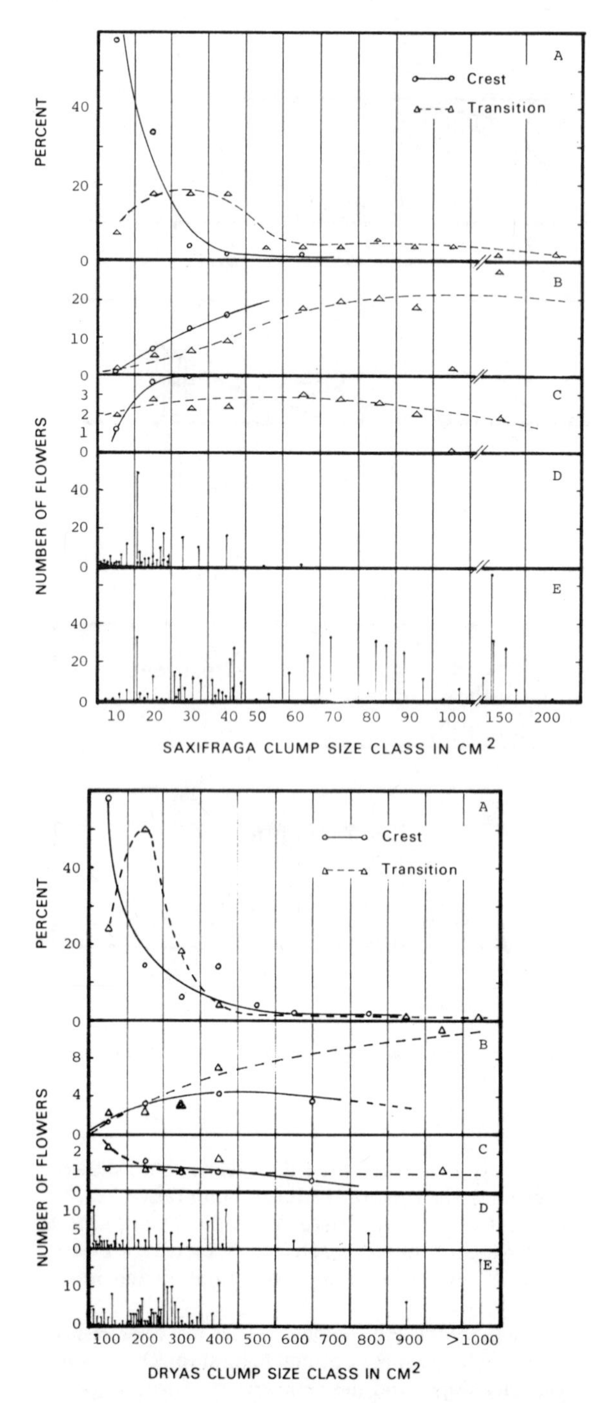

Fig. 5 Size distribution of *Dryas integrifolia* and *Saxifraga oppositifolia* clumps in the crest and transition zones (A); average number of flowers per clump in the different size classes (B); average number of flowers per 100 cm^2 (*D. integrifolia*) and 10 cm^2 (*S. oppositifolia*) in each class (C); actual variation of flower numbers per clump in relation to clump size in the crest (D); and the transition zone (E).

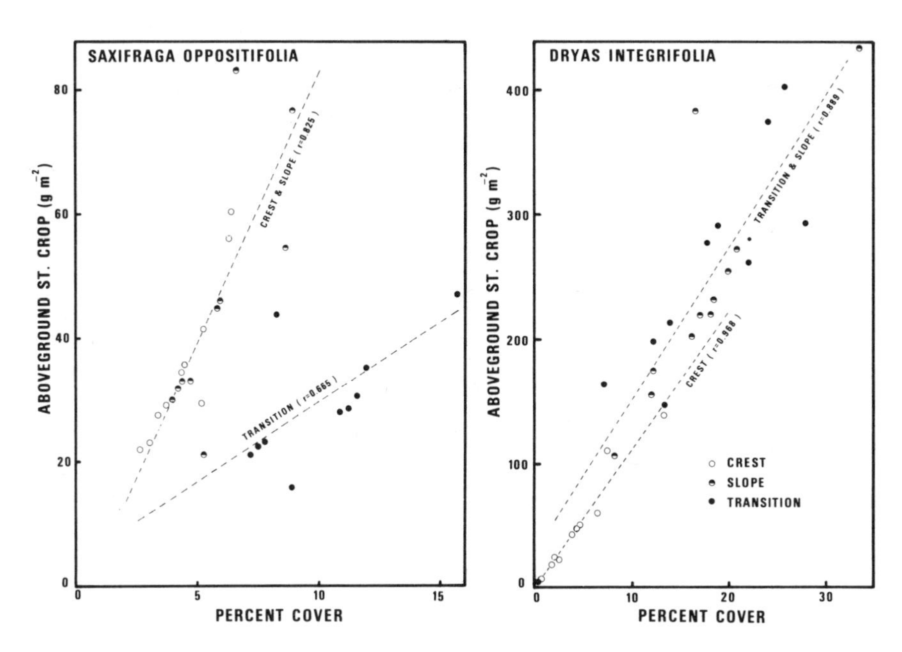

Figs. 6 and 7 Aboveground standing crop of *Saxifraga oppositifolia* (6) and *Dryas integrifolia* (7) versus species over in different raised beach zones. The correlation coefficient is significant at the 5% level.

average numbers of flowers increase almost linearly with clump size, which is not the case of *Dryas* on the crest, see Fig. 5(A).

On crests, *Saxifraga* forms compact cushions, 10 to 20 cm^2 with more loosely creeping shoots (clumps 20 to 40 cm^2) and fewer flowers in the transition zone, see Fig. 5(B). Consequently standing crop of *Saxifraga* shows a high correlation with cover in the crest and slope but a much lower correlation in the transition, while the correlations are high in *Dryas* in all zones (Fig. 6). The greater importance of *Saxifraga* on the top of ridges and *Dryas* in the lower and moister sites helps to explain why *Dryas* is an important species in the Polar Semi-deserts while *S. oppositifolia* extends in limited amounts into the Polar Desert (Table 5).

There is very little annual growth in these arctic cushion plants, in clumped sedges such as *C. nardina*, or in low shrubs such as *C. tetragona*. *Dryas* shoots carry 5 green leaves on the average producing 2.5 new leaves per growing season. Leaves are active for two seasons. After snowmelt old leaves are reddish-brown turning gradually green over 10 days. The process is reversed in mid-August with new leaves. Only leaves produced in a given year turn green again the second season. Linear growth takes place for *ca*. 4 weeks.

Saxifraga oppositifolia produces 7 to 8 leaves per year and they also remain functional for two growing seasons. In contrast to *Dryas* they do not redden in autumn. As with *Dryas*, the dead leaves remain on the shoot for many years, buffering and protecting the internal clump environment against temperature and moisture fluctuations.

Elongation of a single *Carex nardina* leaf lasts for 4 to 6 years before it is replaced. Leaves elongate over the entire season but die back to the sheath at the end of August. *Carex misandra* also forms dense but less compact tufts with leaves following a similar growth pattern to *C. nardina*.

Table 6. Phenology of the dominant species on the Intensive Raised Beach Site, Truelove Lowland (1971)

		Growing season on the Exp. raised beach ridge in weeks								
Zone	*Species*	*13-19 June*	*20-26 June*	*27 June-3 July*	*4-10 July*	*11-17 July*	*18-24 July*	*25-31 July*	*1-7 August*	*8-14 August*
Crest	*Saxifraga oppositifolia*	snow	part. free of snow, some flowers	flowering	flowering	seed formation	seed ripening	seed ripening	mature seed	new flower initiation
Crest	*Carex nardina*	snow	brown clumps	green leaf bases elongate	leaf elongation	inflorescence formation	some flowering	full flowering, leaf die back	seed formation. leaf die back	seed ripening intens., leaf die back
Slope	*Salix arctica*	snow	free of snow, dormant shoots	leaf buds opening	leaf devel-opment, sporadic flowers	leaf growth, full flowers	leaf growth, some flowers	seed formation and ripening	some leaves yellow, seed ripening	many leaves yellow, seed mature
Slope	*Pedicularis lanata*	snow	grey, dead shoots only	new shoot initiated	sporadic flowers	full flowering	flowering	some flowers, seed formation	seed formation and ripening	seed ripening
Slope	*Dryas integrifolia*	snow	brown clumps	brown-red tips	red-green tips	green leaves	flower initiation, full flowering	flowering, seed formation	seed formation and seed ripening	mature seed, leaf reddening
Slope	*Silene acaulis*	snow	grey clumps	greening clumps	new leaf full growth	some flowers	full flowering	flowering	some flowers	seed formation
Transition	*Cassiope tetragona*	snow	snow	brown-green tips	greening tips	some flowers	flowering	flowering	some flowers, seed formation	seed formation
Transition	*Carex misandra*	snow	part. free of snow	greening clumps	leaf growth	inflorescence formation	flowering	flowering	some flowers	seed formation and ripening

Salix arctica shows one of the widest ranges of tolerance to habitat factors of any arctic species. In Truelove Lowland its range extends from the most xeric sites on crests to hummocky tops in sedge-moss meadows, in rock outcrops as well as on the edge of the plateau. Plants show marked differences in size and vigor according to habitat. On the I.R.B. *S. arctica* develops 3-6 leaves on each terminal shoot per year. Catkins expand immediately after snowmelt and before leafing. Leaves develop slowly reaching full size in *ca.* 3 weeks. Shoot elongation averaged 7.7±.4 mm (n=87), leaf length varied between 17 to 25 mm (blade and petiole), and 9 to 12 mm in width.

There was no evidence of winter growth in the species examined by the author and Muc (this volume) on Devon Island.

Limited systematic data on the phenology of high arctic plants are available from Northeastern Greenland (Sørensen 1941), Bylot Island (Drury 1962), the Lake Hazen area on Ellesmere Island (Savile 1964) and some other sites within the Russian Arctic; Shamurin (1966) correlated flowering intensity and duration with habitat factors.

On Spitsbergen, Acock (1940) and on Truelove Lowland, Terri (1972) recorded the pattern of flowering in relation to microsites. On Ellesmere Island, Hocking (1968) related flowering duration and intensity to nectar production.

Growing season

Raised beach vegetation is heterogenous. Plant cover, standing crop, and species diversity decrease from slope to crest, with variation due to irregular microtopography. It was difficult to estimate seasonal changes in standing crop or production by sampling on an area basis, because of this heterogeneity and the very low rates of plant growth. Reasonable estimates were obtained by analysis at the species level and extrapolation to community level.

The growing season in the High Arctic is very short. The "potential" growing season, (snowmelt to mean temperature below freezing) is longer than the "actual" season (breaking of dormancy to 50% leaf coloration). Our phenological and growth rate measurements show that most of the plants are conservative and complete their seasonal cycle in 50 to 60 days. Their actual growing season is relatively fixed and constant compared to year-to-year variation in the potential growing season. The actual growing season varies slightly between species and shifts between zones. On the Base Camp raised beach the length of the potential and actual growing season varied in 1970 to 1972 as follows:

	Potential	*Actual*
1970	79 days (18 June - 4 Sept)	60 days (20 June - 18 Aug)
1971	80 days (13 June - 31 Aug)	55 days (15 June - 8 Aug)
1972	45 days (5 July - 18 Aug)	45 days (5 July - 18 Aug)

Standing crop

The gradual change in community composition and production of its main components (vascular plants, mosses, lichens) from the transition zone to the crest is related to differences in elevation, drainage, and exposure. Standing crop values of the vascular plants (based on the weighted average for the Intensive Raised Beach) represented *ca.* 80%; moss and lichens share almost equally the remaining 20% (Table 7). Mosses are abundant mainly in the moisture transition zone while lichens are more common on the slope and crest. Further data on moss and lichen distribution are presented in Table 2, and in Vitt and Pakarinen (this volume) and Richardson and Finegan (this volume).

Dryas integrifolia is the major contributor to the vascular plant standing crop in all three zones and it is followed by *Saxifraga oppositifolia*, *Carex nardina* and *Salix arctica*. *Cassiope tetragona* is prominent in the transition zone. Other species, mostly minute herbs, account for less than 1% of the standing crop (Table 8). On the crest and slope (cushion plant-lichen community) green shoots represent a small fraction of the standing crop (*ca.* 4% at the peak of growth). The non-photosynthetic stems (live brown) are 16%, roots 13%, and attached (standing) dead 68% of the total based on a three year average. In the transition zone (cushion plant-moss community), green shoots account for a larger percentage (Table 8). The high proportion of standing dead is a very prominent feature of the cushion plant community and its ecological significance has been mentioned earlier. The average I.R.B. dead:live aboveground ratio is 1.7:1 (maximum 3.3:1 on the slope, 1.6:1 in the transition, and 2.6:1 on the crest). The high proportion of standing dead on the slope and crest is probably related to the dry conditions and slow decomposition. The proportion of non-productive parts (aboveground brown:green, and total aboveground:green) increases upslope (Table 8). The increasing load of non-photosynthetic components lowers the overall performance and may reach a critical survival value for some species at different positions on the topographic gradient.

On the raised beach the total vascular standing crop averaged 400 g m^{-2} (Table 9). In addition there is 700 to 800 g m^{-2} of fibrous humus in the first 20 cm of soil, derived from living roots and representing an underground litter of unidentified root fragments. It has undoubtedly accumulated for hundreds of years owing to slow decomposition.

Standing crop data from 11 raised beaches sampled in the extensive program (Fig. 2) do not differ markedly from the I.R.B. On crests the aboveground standing crop ranged from 30 to 225 g m^{-2} (mean 97 g m^{-2}); on slope and transition zones it was 200 to 480 g m^{-2} (mean 287 g m^{-2}) and 165 to 530 g m^{-2} (mean 359 g m^{-2}) respectively (Fig. 3).

The data on individual plants (Table 4) also indicate the average size and individual dryweight of vascular plants in zones, the largest and heaviest being again *Dryas integrifolia*.

There is a direct relationship between age and elevation of raised beaches

Table 7. Cover and standing crop of plant community components on the I.R.B. in harvest 4 (20 July, 1970).

Standing crop components	*Cover* %	*Standing Crop* g m^{-2}	*Standing Crop* % *of total*
Vascular plants	41.2	596.6	80.1
Mosses (Mainly in the transition)	16.0*	77.3	10.4
Lichens (Mainly on crest and slope)	25.9**	70.8	9.5
Total	83.1	744.7	100.0

* Estimated value.
** Taken from Table 2.

Table 8. Contribution of main vascular species to the standing crop (by zones) of the Intensive Raised Beach Site (I.R.B.), harvested at the peak growth of the individual species (27 July to 11 August, 1972). Data are expressed as g m^{-2}.

Species	Cover %	Attached roots	Standing Crop green	Live brown	Standing dead	Shoots total	Total st. crop	% of total
Crest:								
Dryas integrifolia	7.55	12.7	3.6	13.4	93.5	110.5	123.2	47.52
Saxifraga oppositifolia	4.05	8.0	2.1	15.0	28.4	45.5	53.5	20.64
Carex nardina	5.80	15.7	0.4	4.4	24.6	29.4	45.1	17.39
Salix arctica	2.17	18.9	0.1	13.6	3.2	16.9	35.8	13.80
Other species	0.15	0.1	0.1	1.4	0.1	1.6	1.7	0.65
Total	19.72	55.4	6.3	47.8	149.8	203.9	259.3	100.00
Percentage of total		21.3	2.4	18.4	57.8	78.6	100.0	
Slope:								
Dryas integrifolia	16.68	34.7	20.4	43.0	320.5	383.9	418.6	77.72
Saxifraga oppositifolia	6.28	8.4	6.4	29.6	33.0	69.0	77.4	14.37
Carex nardina	3.96	7.7	1.5	3.6	12.3	17.4	25.1	4.66
Carex misandra	0.48	0.3	0.1	0.2	1.5	1.8	2.1	0.39
Carex rupestris	4.52	0.7	0.1	0.1	0.6	0.8	1.5	0.28
Salix arctica	0.86	6.3	0.4	4.0	2.4	6.8	13.1	2.43
Other species	0.17	0.1	0.4	0.3	—	0.7	0.8	0.15
Total	34.95	58.2	29.3	80.8	370.3	480.4	538.6	100.00
Percent of total		10.8	5.4	15.0	68.7	89.2	100.0	
Transition:								
Dryas integrifolia	27.80	23.0	36.6	49.4	210.8	294.8	317.8	70.54
Saxifraga oppositifolia	4.60	3.0	4.4	14.6	9.5	28.5	31.5	6.99
Carex misandra	2.08	2.5	1.4	1.7	9.3	12.4	14.9	3.30
Carex rupestris	13.00	9.9	2.2	2.0	3.6	7.8	17.7	3.93
Salix arctica	1.88	4.1	0.7	6.0	0.9	7.6	11.7	2.60
Cassiope tetragona	7.60	9.7	13.2	20.4	10.5	44.1	53.8	11.94
Other species	1.14	0.8	1.0	0.6	0.7	2.3	3.1	0.70
Total	58.30	53.0	59.5	94.7	245.3	397.5	450.5	100.00
Percentage of total		11.7	13.2	21.0	54.4	88.2	100.0	

Table 9. Mean vascular plant standing crop for the Intensive Raised Beach (weighted values averaging three zones, in the proportion 1:2:2 crest, slope and transition respectively) in 1970-1973, (g m^{-2}±SE).

	1970	1971	1972	1973	Mean*
Aboveground					
Green	29.3± 1.7	19.7± 2.6	24.8±7.9	27.9	25.4
Live brown	100.9±14.8	95.8± 2.7	70.6±4.5	107.7	93.7
Standing dead	169.0±12.3	179.1±19.6	255.2±20.9	NE	201.1
Total abvgr.	299.2	294.6	350.6		314.8
Belowground					
Attached roots	89.2± 3.2	62.7± 2.3	54.8±1.4	NE	68.9
Grand total	388.4	357.3	405.4		383.7
Ratios					
Aboveground live: Attached roots	1.5± 0.2	1.8± 0.1	2.0±0.2		1.7
Aboveground dead: Aboveground live	1.3± 0.1	1.5± 0.2	3.2±0.4		2.0
Abvgr.live brown: Abvgr.green	3.4± 0.4	5.1± 0.7	3.1±0.8		3.8
Total abvgr: Attached roots	3.6± 0.3	4.8± 0.3	7.5±0.4		5.3
Total non-green: Green	12.2± 0.9	18.2± 1.9	16.6±4.4		15.6

*Mean values of 5 harvests in 1970 and 1971, 2 harvests in 1972, and 1 harvest in 1973.
NE = not estimated.

and their plant cover and total standing crop. Relatively young raised beaches (3000 to 6000 years) have less cover and standing crop than those 6000 to 9000 years old. This correlation is pronounced mainly in crest and transition zones where both cover and standing crop values are significantly higher on older beaches compared to the younger ones. On slopes these differences are not so evident (Fig. 3).

On the east plateau, standing crop is only approximately 1/10 that of a Semi-desert raised beach, averaging 23 to 53 g m^{-2}. *Saxifraga oppositifolia* dominates here and it is accompanied by *Papaver radicatum*, *Cerastium alpinum* and tufts of *Phippsia algida*. Plants have tiny thread-like roots which results in a high shoot:root ratio (7.7:1). Mosses are abundant in proportion to vascular plants (Table 5). There are almost no lichens on the rocky ground surface except for big exposed rocks and large erratics. Both vascular plants and mosses grow between and under rocks where finer sand and silt accumulate.

Production and productivity

High arctic cushion species have a very low annual growth rate and therefore a very low production.

Net production on the I.R.B. decreased upslope in an approximate ratio of 3:2:1 for the *transition-slope-crest* zones respectively. Annual net production varied little from year to year (Table 10). Annual production is *ca.* 3% to 5% of the total standing crop, indicating turnover rates and average plant age, assuming that the vegetation is mature and in equilibrium with the environment.

Woody species, mainly *Dryas*, contribute over 50% of the total production followed by *S. oppositifolia* and *Carex* spp., the cover value for sedge is almost twice that of forbs. Cushion herbs are very compact on crests where they develop larger amounts of shoot terminals per unit plant area than in the transition zone.

Green shoots account for *ca.* 71%, with brown stems and roots the remaining 29% of net production. The 29% represents much of the perennial growth increment because most of the green shoots are leaves which will die quickly and appear as standing dead (the dead:live ratio is 1.5-3:1).

Estimates for root production range from 3 to 5 g m^{-2} for the total raised beach but are much lower for the crest. The autoradiographic technique (^{32}P) used to determine active roots of *Dryas* in the field (Svoboda and Bliss 1974) showed that the majority of attached roots are functional and alive. Only a limited number of roots are produced in these species each year, yet they are functional for a number of years.

In general climatic severity in terms of soil and atmospheric moisture stress increase from the base to the crest of each raised beach system. Aboveground alive standing crop and net annual production increase along this gradient from crest to transition zone (Fig. 8). Plant production is a larger increment (6.7%) of total standing crop on the transition zone than on the slope and crest (3.7%).

There is little difference in net vascular plant production between young and older raised beaches, although significant differences occur between the three zones in each age group (Table 11).

Table 10. Net annual production of vascular plants on the Intensive Raised Beach Site (I.R.B.) in three consecutive years (g m^{-2}).

		1970	g m^{-2}			1971	g m^{-2}			1972	g m^{-2}			1973	g m^{-2}		
Species	*cover %*	*Shoots*	*Roots*	*Total*	*% of Total*	*Shoots*	*Roots*	*Total*	*% of Total*	*Shoots*	*Roots*	*Total*	*% of Total*	*Shoots*	*Roots*	*Total*	*% of Total*
Crest:																	
Woody	9.72		Not estimated			3.2	0.5	3.7	56.74	8.4	1.4	9.8	65.35	3.8	0.6	4.5	35.2
Monocots	5.80		on a			0.6	0.1	0.7	11.21	1.3	0.2	1.6	10.52	3.0	0.5	3.5	27.3
Forbs	4.20		zonal basis			1.8	0.3	2.1	32.04	3.1	0.5	3.6	24.13	3.9	0.8	4.7	37.4
Total	19.72					5.6	0.9	6.5	100.00	12.8	2.1	15.0	100.00	10.7	2.0	12.7	100.0
Slope:																	
Woody	19.54					8.8	1.4	10.3	47.90	13.6	2.3	15.9	71.67	8.1	1.4	9.5	50.7
Monocots	8.96					0.6	0.1	0.7	3.60	1.4	0.2	1.6	7.45	4.3	0.7	5.0	26.8
Forbs	6.45					8.5	1.4	9.9	48.50	3.9	0.7	4.6	20.88	3.5	0.7	4.2	22.4
Total	34.95					17.9	3.0	20.9	100.00	18.9	3.2	22.1	100.00	15.9	2.8	18.7	100.0
Transition:																	
Woody	37.28					17.8	3.0	20.8	69.30	22.1	3.8	25.9	76.40	13.3	2.3	15.7	56.2
Monocots	15.28					2.8	0.5	3.3	10.96	1.7	0.2	1.9	5.80	6.8	1.1	7.9	28.4
Forbs	5.74					5.0	0.8	5.9	19.74	5.1	0.9	6.0	17.80	3.8	0.5	4.3	15.3
Total	58.30					25.6	4.3	30.0	100.00	28.9	4.9	33.8	100.00	23.9	3.9	27.9	100.0
Total Beach Ridge:																	
Woody	24.7	13.0	2.2	15.2	68.02	11.3	1.9	13.2	58.3	16.0	2.7	18.7	72.3	9.3	1.6	10.9	51.7
Monocots	10.8	1.5	0.2	1.7	7.99	1.5	0.3	1.7	8.0	1.5	0.2	1.7	7.4	5.0	0.8	5.8	27.7
Forbs	5.7	4.6	0.8	5.4	23.98	5.7	0.9	6.7	33.7	4.2	0.7	4.9	20.3	3.7	0.6	4.3	20.5
Total	41.2	19.1	3.2	22.3	100.00	18.5	3.1	21.6	100.0	21.7	3.6	25.3	100.00	18.0	3.0	21.0	100.0

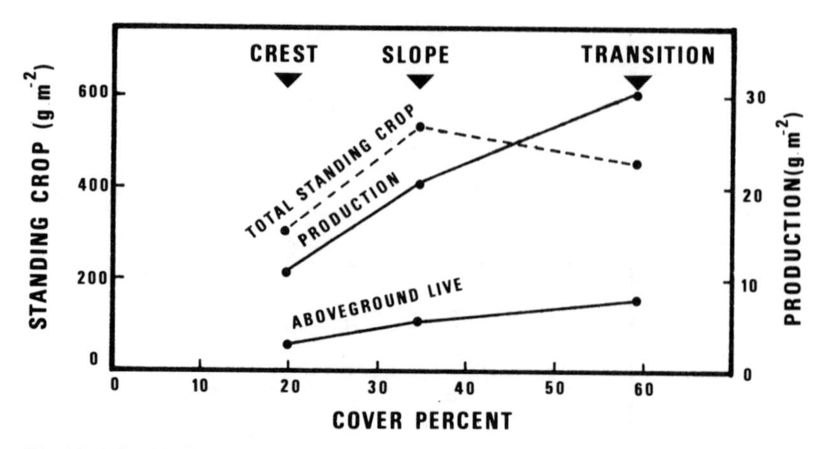

Fig. 8 Relationship between cover, standing crop, and production in three zones on I.R.B. Production figures are a three-year average.

On the plateau net vascular plant production was estimated to be 2 to 5 g m^{-2}, (10 to 15 g m^{-2} if moss production is included). Inland from the plateau margin where there is only 1% to 3% plant cover, net production probably averages only 0.5 g m^{-2}.

Net productivity, calculated on the basis of an actual growing season, was 0.37 in 1970 (60 days), 0.39 in 1971 (55 days), and 0.56 g m^{-2} day^{-1} in 1972 (45 days). Productivity in 1972 appears high for such a short growing season during that cool but wetter summer. The substantial part of annual production occurred during the first half of the summer with only sedges adding to shoot growth in August.

Raised beaches represent only a minor part of Truelove Lowland but they carry almost twice as much aboveground vascular plant standing crop per unit area as the meadows, three times as much as rock outcrops, and ten times as much as on the plateau. This results from a large component of dead material and the woody shoots of *Dryas*.

Table 11. Net annual production of vascular plants on eleven raised beaches in 1972.

No.	Name	*Raised Beach* Elev *(m)*	*Raised Beach* Age *(years BP)*	*g m⁻²* *Crest*	*Slope*	*Transition*	*Total**
1.	Sea	3.3	3283	11.1	20.5	45.0	28.4
2.	Phalarope	5.8	4000	2.5	11.6	25.2	15.2
3.	Air strip	11.9	5706	6.2	25.0	52.1	32.1
4.	North Inlet	12.8	5900	5.7	13.4	41.4	23.0
5.	Inlet	14.9	6200	4.1	12.5	23.6	15.3
6.	Base Camp	14.9	6209	5.6	13.2	51.8	27.1
Mean for the younger beaches		—	—	5.9	16.0	39.9	23.5
7.	Fish Lake	22.8	7068	5.3	16.8	54.5	29.6
8.	Experimental	28.0	7500	15.0	22.1	33.8	25.4
9.	Gully	36.9	8000	4.6	10.4	64.9	31.0
10.	Lower Wolf Hill	64.6	9000	9.1	10.2	48.8	25.4
11.	Wolf Hill	79.2	9500	4.9	18.1	40.7	24.5
Mean for the older beaches		—	—	7.8	18.5	48.5	28.4
Mean for all beaches		—	—	6.9	17.3	44.2	26.0

*Weighted value in the ratio 1:2:2 (*crest, slope, transition*).

Shoot production is only *ca*. 50% of that of the meadows, and root production is only an insignificant fraction compared to 120 to 185 g m^{-2} in the meadows (Muc this volume).

Between year variation in net production is small, demonstrating the conservative character of these plants. Plant growth and flowering in particular seem to be determined by climatic conditions of the previous year when the shoot and flower primordia were formed.

There is some variation in cover and standing crop among the raised beaches studied, but in general they are typical of the Polar Semi-desert, a minor area in extent on Devon Island which is mostly Polar Desert, but a type more prominent in the western Arctic Islands sharing the same latitude.

Though most of the literature on *Dryas* or cushion plant dominated communities deals only with community description, there are some data available on standing crop and net annual production. At Disco Island, Greenland (69°N), at the *Dryas integrifolia* fell-field community (20 vascular species, 25% cover), vascular standing crop was 114 g m^{-2} with 33 g m^{-2} lichens, and production of 42.5 and 3 g m^{-2} respectively (Lewis et al. 1972). In Alexandra Land (Great Lekhov Is. 73°N) standing crop of a "polar desert" community was about 320 g m^{-2} (Aleksandrova 1970) and in the central Asian Arctic the aboveground standing crop in a mixed grass *Dryas* tundra was *ca*. 60 g m^{-2} with an estimated 40 g m^{-2} of aboveground production (Lavrenko et al. 1955).

In the Canadian Arctic Islands, Arkay (1972) reported 40-50 g m^{-2} of shoot standing crop with *ca*. 8 g m^{-2} of shoot production around Char Lake, Cornwallis Island (75°N).

Data from elsewhere in the Canadian Arctic Archipelago show that standing crop is less but plant production about the same for cushion plant and rosette plant-herb communities on Melville Island (Table 12). On Banks and Victoria islands where *Dryas-Carex* or *Dryas-Carex*-dominated plant communities occupy drier sites, both raised beaches and low elevation uplands, standing crop is again much greater because of the amount of dead aboveground plant material, and net production is somewhat greater than on the northern islands (Table 12). These data (Bliss and Svoboda — unpublished) show that moss-biomass is quite considerable at most sites while lichen biomass is often minor.

Leaf Area Index (L.A.I.)

Leaves of most arctic plants are very small and have a variety of forms. In leaf tufts *(Carex nardina)*, rosettes *(Saxifraga oppositifolia)*, and cones *(Cassiope tetragona)*, single leaves overlap to such a degree that only a small part of each leaf is exposed. The reduced leaf area results in a smaller transpiration surface and this is an important factor in an environment where soil water potential remains below −30 bars for many days. Also plants have low water potentials and high leaf resistances (Addison this volume).

In high arctic cushion plants, green shoots represent a minimal portion (4% to 7%) of the total plant biomass with only a little seasonal change. The L.A.I. is unusually low, *ca*. 0.1, *Dryas integrifolia* being its main contributor: 50.7, *S. oppositifolia* 14, *C. tetragona* 11.5, sedges 16.1 and *S. arctica* 3.1%. *Salix arctica* is the only woody species with prostrate shoots and annual leaves. Table 13 compares three leaf parameters of *S. arctica* leaves from the exposed crest and more favorable habitats in rock outcrops. The differences suggest that leaf size and leaf area are significantly lower on exposed sites.

Table 12. Standing crop and production estimates of some Polar Semi-desert sites of Melville, Banks, and Victoria Islands, N.W.T.

Locality & Stand no.		*Standing crop (g m^{-2})*							*Total* standing crop*	*Production (g m^{-2} y^{-1})*			*Community Type*
		Fine roots & org mat.	*Below ground*	*Above-ground Live*	*Dead*	*Total**	*Bryophytes*	*Lichens*		*Below*	*Above*	*Total*	
				Brown Green									
Melville Is. 75°N													
Rea. Pt.	no. 3	947.2	75.4	48.3 24.8	36.0	184.4	656.4	7.3	848.1	14.4	24.8	39.2	Cushion plant-low shrub
	no. 6		21.0	1.3 11.9	58.4	92.6	776.2	5.8	874.6	5.9	11.9	17.8	Cushion plant-herb
	Mean		48.2	24.8 18.3	47.2	138.5	716.3	6.5	861.3	10.2	18.3	28.5	
Banks Is. 73°N													
Big River	no. 1		202.6	216.8 22.4	602.8	1044.6	0	34.7	1079.4	5.7	15.2	20.9	*Dryas-C. rupestris.*
Thompson River	no. 9	2429.1	94.7	212.6 87.0	522.7	917.0	807.0	1.0	1725.0	12.9	54.8	67.7	Cushion plant-moss
	no. 10	1320.9	68.4	91.4	772.8	940.5	42.1	7.6	990.2	6.0	5.8	11.8	Cushion plant-herb
Johnson Pt.	no. 22	2570.2	335.5	275.3 33.4	693.5	1337.8	1419.5	8.4	2765.6	4.7	21.4	26.1	*Dryas-Salix* spot stripped semi-desert
	Mean	1580.0	175.3	199.0 37.7	647.9	1059.9	567.1	12.9	1639.9	7.3	24.3	31.6	
Victoria Is. 71°N													
	no. 32	2110.2	165.2	82.3 25.7	614.8	888.0	7.3	2.7	898.0	22.6	21.2	43.8	Cushion plant-herb fellfield
	no. 37	4892.0	299.9	165.8 58.1	853.9	1377.7	270.2	57.3	1705.2	30.1	41.4	71.5	*Dryas-C. rupestris.*
	Mean	3501.1	232.5	124.0 41.9	734.3	1132.8	138.7	30.0	1301.6	26.3	31.3	57.6	

*Fine root and soil org. matter are not included.

Values of Leaf Area Ratio, L.A.R.=leaf area (cm^2)÷leaf d wt^1 (g) are very low in leaves of cushion plants (*Dryas* 37:1, *S. oppositifolia* 33:1, *C. tetragona* 33:1). *Salix arctica*, which is not a cushion plant, has a much higher L.A.R. (142:1) (Table 13). Low L.A.R.'s testify to the massive structure and solid architecture of these shoot-leaf systems, designed to resist drought, frost and snow abrasion, and for leaves which function two or more seasons.

There have been no comparative data available to the author concerning the L.A.I. of high arctic cushion plants. Measurements on high arctic plants by Warren Wilson (1966) showed increasing values of L.A.R. with increasing nutrient richness of the soil and with decreasing latitude. Tieszen (1970) found that there was a small difference between low arctic and alpine populations in L.A.R. The leaf area ratio averaged 2.5 g dm^{-2} on the Devon cushion plants compared with 0.5 g dm^{-2} for alpine and low arctic plants.

Table 13. Mean leaf length, leaf area, and leaf area ratio (±SE) for *Salix arctica* plants at the Intensive Raised Beach (Crest Zone) and Rock Outcrop east of the Intensive Study Site. Leaf material was collected on 24 July and 3 August, 1972, respectively.

Site	*Leaf length (mm)*	*Leaf area (mm^2)*	*Leaf area ratio (cm^2 g^{-1})*	*Leaf area ratio (mg cm^{-2})*
Crest	17.1±0.80	129.8±16.3	142.2	7.0
Rock Outcrop	24.6±1.42	129.3± 2.1	132.3	7.5

Chlorophyll

Analyses of chlorophyll *a*+*b* show a rapid increase in these pigments after snowmelt, culminating in early July and then a continuous decrease. In both years (1970 and 1971) the seasonal trend of pigment content was very similar, however in 1971 there was a higher pigment standing crop. The highest values were found in *Carex nardina* (3.8 mg g^{-1} dry green tissue), the lowest in *S. oppositifolia* (1.9 mg g^{-1}). *Dryas integrifolia* was intermediate in chlorophyll content (1.5 to 2.5 mg g^{-1}). This is less than Muc (this volume) determined for *C. stans* (8.5 mg g^{-1}) in the sedge-moss meadows, and also less than Dennis and Tieszen (1972) reported for Barrow, Alaska plants (*ca*. 5.0 to 7.5 mg g^{-1}).

There is a low correlation between chlorophyll content and net production at least on a community basis. Individual species (*C. nardina*) show a higher correlation.

Carbohydrates

Cushion plants have very low levels of total carbohydrate reserves. Roots store less reserves than shoots and they also show less fluctuation during the season (Fig. 9). Carbohydrates were mostly stored in the form of oligosaccharides with extremely low amounts of monosaccharides and low levels of starch. The overall levels of shoot carbohydrate ranged from 0.5% to 3% in *D. integrifolia*, 0.5% to 5% in *S. oppositifolia* and 3% to 6% in *C. nardina*. Green, photosynthetic parts contain 2 to 3 times more carbohydrate than living but non-photosynthetic organs.

Pedicularis lanata renews the entire shoot every year. Total sugars rise to

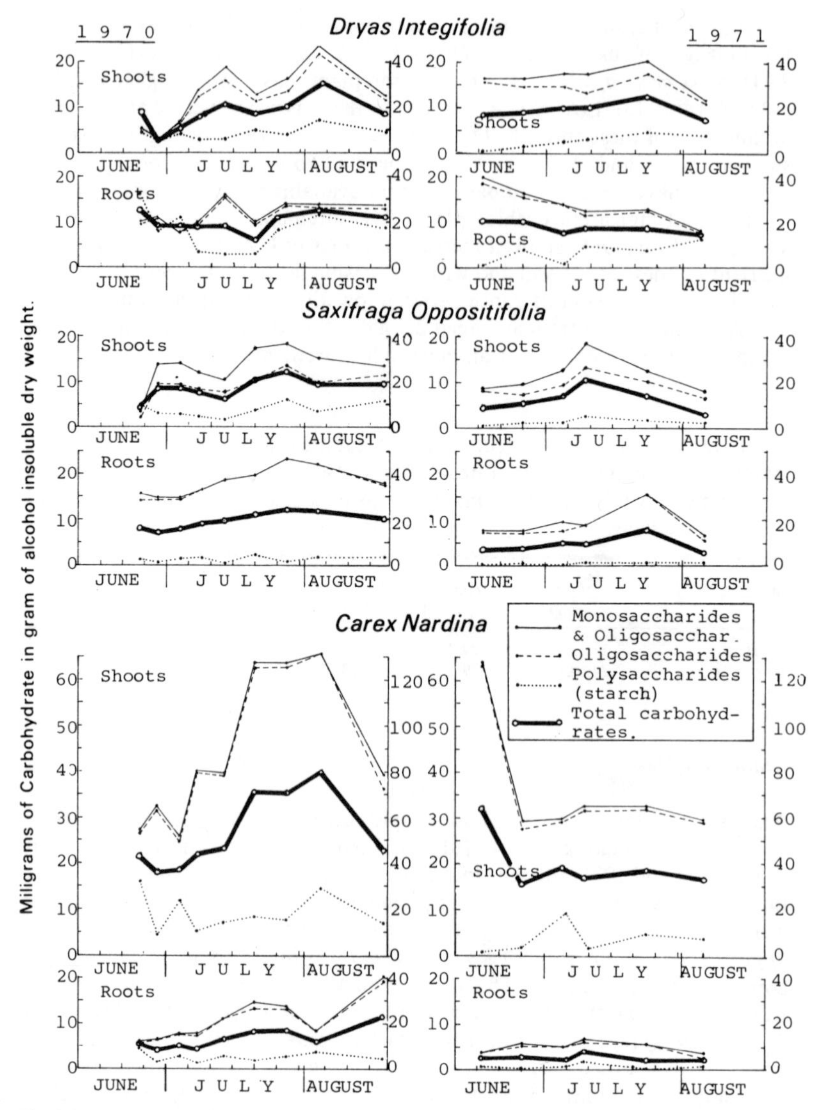

Fig. 9 Seasonal changes in carbohydrate content of three species in 1970 and 1971. Right scale total carbohydrates, mg g^{-1}.

17% in shoots and 47% in roots, expressed on an alcohol insoluble residue basis and are only 30% to 35% lower when expressed per original dry weight. Roots are large and fleshy and thus serve as a carbohydrate storage organ in marked contrast to the wirelike roots of *Dryas* whose function is mainly mechanical and in water and nutrient absorption.

Seasonal fluctuation of carbohydrate was low in roots but higher in shoots, which suggests that mobilization of reserves from shoots is easier in spring when shoots are snow-free while roots remain frozen or experience much lower temperatures than shoots. No significant depletion has been detected immediately after snowmelt but an increase in total carbohydrate has been found in the middle of the growing season.

On 24 July, 1970 five sites with *S. oppositifolia* along the gradient transition-slope-crest were sampled for carbohydrate content. Plants in the transition (late snowbed) were dug out from snow while plants on the crest were developing fruits. Total carbohydrate in *Saxifraga* shoots increased upslope (2.5% to 8.5%) but at the crest sharply declined (5.5%). Starch was almost constant in all five sites (*ca.* 0.5%) and so were the total carbohydrates in roots (*ca.* 2.5%). Although two variables were involved, slope position and phenological stage, other information on seasonal carbohydrate change suggests that it is slope position which is much more important in determining carbohydrate content.

Ratios of carbohydrate components vary between years but their overall trend shows that cushion species differ from meadow species and also from *Pedicularis* in the same habitat. Low levels of starch compared to soluble sugars were found in shoots of *Dryas* (0.15 to 0.34:1), *Saxifraga*, (0.14-0.29:1) and *Carex nardina* (0.10 to 0.20:1). Oligosaccharides were the prominent fraction in all three cushion species tested averaging 32:1 in comparison with monosaccharides. In *Pedicularis lanata* the soluble sugars were 10 times higher in shoots and 20 times higher in roots in comparison with cushion species from the crest. There were more monosaccharides over the major part of the season in shoots of *Pedicularis* while roots maintained high levels of oligosaccharides.

It is evident that some arctic and alpine plants maintain high levels of carbohydrate (Mooney and Billings 1960, Fonda and Bliss 1966), compared with plants from temperate zones. This led to the hypothesis discussed by Warren Wilson (1966), that arctic plants maintain high levels. However, these are exceptions, especially in cushion and tufted plants. High carbohydrate levels were found mainly in high arctic herbs *(Oxyria digyna, Polygonum viviparum, Pedicularis lanata)* (Russell 1940, Warren Wilson 1966 and the author), but these species typically represent less than 1% of the community standing crop. However in most of the older studies, carbohydrate amounts were expressed per unit of "alcohol insoluble residue." The actual carbohydrate percentage per unit of original dry weight may be very different, as was demonstrated by McCown and Tieszen (1972), and also the author, who found the values for *Oxyria digyna ca.* 18% for shoots and *ca.* 15% for roots expressed per unit of insoluble residue. Expressed per original dry weight the values were 20% to 30% lower (12.6% shoots and 12.1% roots). Also leaf temperatures and respiration are not always as low in arctic plants (Mayo et al. 1973) as is still often assumed.

Beach ridge species showed low starch (1% to 2%) and almost constant starch:sugar ratios, and low variation between years. On the sedge-moss meadows the starch:sugar ratio was also low and constant in shoots (0.05) but increased in roots from 0.01 to 1.1, and in rhizomes from 0.2 to 2.2 during the season (calculated from Muc 1972 original data). The ratio of soluble carbohydrates to starch is very high and supports Eagles' (1967) finding that it increases with lower temperature. The main stored component is the oligosaccharide.

The slow growth rates and stunted growth form of high arctic plants seem to be primarily the result of low temperature. Nitrogen deficiency of arctic soils is also a temperature dependent phenomenon. Low temperatures maintain low nitrogen fixation rates (Stutz and Bliss 1975) and though nitrogen uptake is not inhibited by cold soils the uptake of phosphorus is (Haag 1974). Carbohydrate synthesis seems to function as an open system where assimilates are produced and consumed almost at the same rate, a system which is inefficient in the utilization of assimilated products due to the complexity of external and internal limiting factors.

Table 14. Lipid content (%, n=2) of some selected species from the Intensive Raised Beach Site (I.R.B.) (7 July, 1970) and Plateau (1 August, 1970).

		Crest			*Transition*	
Species	*Green*	*Brown*	*Roots*	*Green*	*Brown*	*Roots*
Intensive Raised Beach Site:						
Dryas integrifolia	3.4	3.4	2.8	3.1	2.5	2.6
Salix arctica	4.1	5.6	5.9	4.0	4.5	5.5
Carex nardina	6.2	4.1	2.5	—	—	—
Carex misandra	5.7	6.8	4.1	9.7	5.3	5.5
Saxifraga oppositifolia	5.7	6.8	4.1	3.0	3.9	3.3
Dead plants	—	1.0	1.0	—	1.0	1.0
		Total shoots	*Green*	*Brown*	*Roots*	
Plateau:						
Saxifraga oppositifolia		3.0	2.2	4.3	3.3	
Cerastium alpinum		3.5	—	—	—	
Papaver radicatum		4.5	—	—	—	
Dead plants		1.0	—	—	—	

Lipids, protein, fibre, nutrient status

Lipid content was in all cases higher for plants on the crest than on the transition zone (Table 14). Raised beach plants showed higher lipid levels (3% to 6%) than did meadow species (0.3% to 2.0%).

Beach ridge shrubs were higher in lipids than were the evergreen and deciduous alpine shrubs (2.9%) reported by Bliss (1962). Also sedges and herbs showed higher lipid content than alpine herbs (1.4%, Bliss 1962), and were close to the lipid values of Barrow, Alaska tundra graminoids (5% to 7%, McCown and Tieszen 1972). These authors found unusually high lipid levels at the beginning of the growing season (17% to 20%).

Analyzed plants showed very low phosphorus (*ca.* 0.08%) but high calcium contents (*ca.* 2.0%) which gives an abnormally high Ca:P ratio (*ca.* 25:1) (Table 15), which is expected on a calcareous substrate. Wherever there is a source of phosphorus (exposed whale bone) plants occupy the site and their growth is vigorous.

Protein content averaged *ca.* 5% to 7% (Table 15). Both protein and phosphorus were lower in raised beach than in meadow plants (Muc this volume). Relatively high amounts of crude protein were found in humus and litter, probably due to the presence of soil microorganisms and invertebrates.

A high content of crude fibre (*ca.* 50%) suggests the low digestibility of these plants except for their minute green parts. *Saxifraga* roots showed almost double the amount of phosphorus compared to *Dryas* roots and also the highest caloric value of all species tested. *Carex nardina* has lower calcium than other species and its Ca:P ratio was therefore lowest.

Scotter (1972) presented data on the chemical composition of forage plants from the Reindeer Preserve near Inuvik, N.W.T. Green leaves and twigs of these plants were similar in crude protein and fibre content but lower in calcium and slightly higher in phosphorus than in this study. Ratios of Ca:P were similar to those of *Carex nardina*. Haag (1974) found 13.2% protein, 20.7% fibre and 0.3% phosphorus in the total material from a sedge meadow near Tuktoyaktuk, N.W.T.

Vascular species growing on elevated well drained ridges with virtually no nutrient supply from the surrounding area may utilize only those elements which may become available as a result of direct weathering or as a result of

Table 15. Chemical compositions and mineral nutrient content of plants, litter and soil organic matter on the slope of the I.R.B., 1970; average of harvests No. 1, 3 and 5.

		Percentage (air dry basis)							
Species	*Plant part*	*Sample moist.*	*Crude protein (N×6.25)*	*Crude fibre*	*Ca*	*P*	*Ca/P*	*K*	*S*
Dryas integrifolia	Green	5.9	7.2	32.5	1.8	0.10	21.5	0.20	0.06
	Brown	6.4	7.5	41.8	1.7	0.06	29.8	0.93	0.06
	Roots	6.3	6.1	46.1	1.6	0.07	22.8	0.10	T
	Stand. dead	7.0	6.0	54.2	2.2	0.06	35.2	0.07	T
Saxifraga oppositifolia	Green**	N.D.	5.0	N.D.	1.6	0.10	17.7	0.24	0.13
	Brown	7.5	5.6	40.8	2.5	0.08	32.9	0.10	0.12
	Roots	5.8	6.3	43.4	2.0	0.12	16.9	0.13	T
	Stand. dead	7.1	5.8	41.4	2.8	0.09	31.0	0.08	T
Carex nardina	Green	N.D.	5.0	N.D.	0.9	0.14	9.4	0.47	0.96
	Brown	N.D.	3.7	N.D.	0.5	0.05	9.4	0.18	0.06
	Roots	N.D.	5.1	N.D.	0.9	0.09	16.6	0.16	T
	Dead Plants	4.7	3.3	22.8	1.1	0.07	15.9	0.08	T
Detached unidentified roots (0-10cm)		6.3	7.2	51.1*	2.4	0.10	25.1	0.09	T
Fibric organic matter (0-10cm)		5.4	8.3	44.5*	3.2	0.16	23.0	0.12	T
Litter**		5.7	7.8	N.D.	3.1	0.10	32.0	0.11	T

Harvest 1 (22 June) 3 (7 July) 5 (8 August)
*Only 1 harvest analysed
**Only 2 harvests analysed
N.D.=No data because of insufficient amount of sample material.
T traces only

dung decomposition. Meadow species, on the contrary, may be supplied both by nutrients produced elsewhere within a particular watershed and moved laterally during spring runoff and by the more abundant supply of dung. Physiological aridity (Sørensen 1941, Haag 1974) due to low soil temperatures may inhibit water and nutrient uptake by the plants.

Energy content

Calculation of ash-free caloric values is necessary in calculating efficiency in net production and in energy flow studies. Data analyses lead to the following conclusions: arctic shrubs such as *Dryas integrifolia* (5259 cal g^{-1}) and *Cassiope tetragona* (5816 cal) are higher in caloric content than evergreen alpine shrubs (5098 cal), as are arctic herbs (4840 cal) compared to alpine herbs (4601 cal) (Bliss 1962). Roots of some species *(Dryas* and *Cassiope)* show a decline in caloric content at the beginning of the growing season. Brown shoots of all species tested had higher caloric content than green shoots and leaves. Roots of *Saxifraga oppositifolia* showed very high ash-free values (over 6000 cal) though higher values (7000 cal) were found by Zachhuber (1974) in roots of *S. oppositifolia* in the Austrian Alps. Alpine willows *(Salix herbacea* and *S. uva ursi)* from Mt. Washington had a caloric content *ca*. 4920 cal (Bliss 1962) while Devon Island *S. arctica* showed 5106 in meadows (Muc this volume) and 5317 cal g^{-1} on the raised beaches.

Dryas *clump as a micro-ecosystem*

Growth and decaying processes in high arctic cushion plants occur *in situ*. Soil under the clump is more developed and numbers of microorganisms are greater under clumps (Widden this volume). *Dryas* plants appear to be a semi-closed system which absorbs nutrients from the very limited nutrient pool. These clumps hold and most probably recycle some of these nutrients released in decomposition.

Dryas is the most characteristic and dominant species of the southern Polar Semi-deserts. On raised beaches of Truelove Lowland, *Dryas* comprises 48% of the crest, 78% of the slope, and 71% of the transition zone standing crop. Seedling establishment is rare. Two or more individuals can develop as a result of disintegration of an old clump by means of vegetative radial growth.

A *Dryas* clump is green only on the surface. Old leaves remain attached and after many years they slowly decay *in situ*, i.e. on twigs. Nutrients released are leached to lower strata and may be reabsorbed by new fine roots in the decomposed and humified organic matter. This process operates in other cushion species as well, thus demonstrating a peculiar strategy of survival in a nutrient poor and cold-dry environment. In the cross-section of a cushion, one can distinguish differently colored layers of dead leaves according to layer age. Based upon mean annual shoot increment and mean shoot lengths, *Dryas* clumps were estimated to be 80 to 120 years old. All parameters measured on individual layers showed decreasing values with increasing age (Svoboda 1974). Inside the clump, decomposed organic matter accumulates. The clump works like a sponge which absorbs and holds moisture long after the surrounding area dries in summer. It also buffers the fluctuation of temperature mainly by reducing daily maxima. Clump temperatures are equal to or lower (on sunny days) than the surrounding air, but much lower than surface leaf temperatures when fully irradiated.

Near the ground where there is a noticeable difference in wind speed, *Dryas* clumps grow in a leeward direction and expand unevenly. Data from three measurements of clumps from the transition (1) and the crest (2) zone suggest that the north side of a clump is covered on average by 60% green leaves, the east and west side by 80% to 85% and the south side by 89%. Leaf length and weight were higher on the exposures favoring higher temperatures and consistently lowest on the north exposure (Table 16).

Information on leaf angle orientation was gathered to find out whether a *Dryas* cushion operates as a simple hemispherical surface with its small leaves distributed randomly or whether leaf orientation is non-random. By controlling leaf angle, *Dryas* may control the exposure of its foliage to direct radiation.

While leaf angle orientation appears to be random, actual measurements of a plant from the transition zone showed that most of the leaves are oriented 10° to 30° from the horizontal plane. Leaf azimuth orientation was predetermined by the compass orientation of the particular segment. In other words, new shoots grow in the direction of available space. Measurements of two other *Dryas* clumps on the crest showed that leaf angles in flat clumps are more evenly patterned and that polarization to margins is less pronounced compared with hemispherical *Dryas* clumps from the transition zone. The predominant leaf angle orientation was 30° to 50° in all segments of these crest zone plants.

Leaf angle measurements on *Dryas* suggest the following conclusions. Leaves are oriented in an angle close to that of the sun at summer noon (37° 31' on 21 June). Such an orientation substantially lowers the direct exposure of

Table 16. Mean length (mm) and dry weight (mg), n=15 measurements, of green *Dryas* leaves according to their position on a shoot in different compass segments of one clump.

Segment	*Sequence of leaves on a shoot*								*Total*
	1		*2*		*3*		*4-6*		*green shoot*
	mm	*mg*	*mm*	*mg*	*mm*	*mg*	*mm*	*mg*	*mg*
North	4.4	1.8	7.3	5.2	9.1	7.1	10.5	10.5	12.5
East	6.6	2.6	12.7	8.7	12.3	8.6	15.7	13.8	14.2
South	6.6	3.3	13.4	9.6	14.6	11.9	16.7	16.9	17.7
West	5.8	3.3	10.8	7.2	12.0	10.6	15.2	17.4	15.0

sun-facing leaves. Leaf arrangement can therefore be interpreted as a negatively sun oriented and a positively sky oriented arrangement. This enables the plant to minimize leaf exposure to the direct "hot" sun beam and to retain exposure to "cool" diffuse light.

Dryas integrifolia is the most representative species of this polar semi-desert community which is characterized by a specific growth form and strategy. It maintains a large aboveground biomass in which the standing dead is highly functional as a microclimatic agent and supplementary nutrient pool. It has a small but long-lived root system and a small photosynthetic surface with leaves persisting for two seasons. These adaptations enable *Dryas* to grow in nutrient poor substrates that are relatively dry and warm. In Truelove Lowland, *Dryas* is not a nitrogen fixer (Stutz and Bliss 1975). Its role in humification, however, creates a favorable microclimate for nitrogen fixing bacteria.

Synthesis

The extremely low annual growth of cushion species and their heterogenous distribution made the usual methods of quantitative harvesting per unit area of little use without production estimates from individual species. The phenological pattern of *Dryas* sets the overall community production pattern. For these reasons, the dominant species *Dryas integrifolia* (48% community cover and 65% standing crop) was chosen for further intensive study.

These cushion plants do not produce free litter. Dead shoots and leaves remain attached for many years. By keeping these dead parts, cushion plants protect their growing points and produce higher temperature, moisture, and increased microbial activity within the clump.

Because higher surface temperatures, low soil moisture, and the associated low availability of soil nutrients, the xeromorphic character as demonstrated by low protein and high fibre content and thickening of the cuticle is favored as in the Low Arctic (Haag 1974). Carbohydrate reserves (1% to 5%), protein (5% to 6%), and nutrient content are inadequate even when compared with plants from more favorable arctic habitats. Further detailed ecophysiological studies will be essential if we are to understand how these species are adapted to extreme conditions which are temperature, moisture, and nutrient limiting.

Cushion plants live in an environment below their optimum. This follows from my field greenhouse experiment (unpublished) and fertilizer experiments (Babb and Bliss 1974), and from comparison with individuals of the same species growing in more favorable high arctic habitats. This introduces questions about the degree of adaptation of these plants to high arctic conditions. "Marginal" populations of these species, whose individuals are "sparsely distributed and show effects of physiological stress" (Soule 1973), are closest to extinction but may also be considered the most resourceful genetic material for more resistant new strains if they are able to produce viable seed.

Margalef (1963) suggested the productivity:biomass ratio as an "inverse" index of diversity. Simple ecosystems, such as early stages of succession or man-regulated crop fields, are supposed to be most productive. In Truelove Lowland, Margalef's theory can be applied to meadows and rock outcrops. Raised beaches and the polar desert plateau, however, do not fit such a scheme. Communities of these systems are depauperate in both species diversity and biomass and their production is hardly measurable.

Raised beaches in association with the adjacent meadows display distinct stages along a gradient. In a less extreme environment the inexpressive lowland topography would allow the establishment of a substantial vegetation over the entire area in a short time. Here, however, this process is very slow. Since deglaciation only those crests of the oldest beaches with more sand and less coarse gravel have reached a denser plant cover. Small scale climatic changes such as the Little Ice Age resulted in marked shifts of major vegetation boundaries (e.g. tree line) and undoubtedly affected the existing vegetation in the High Arctic although few data are now available (Jankovska and Bliss this volume and Nichols 1975).

In the Truelove River Valley much thicker remnants of willow stems have been found than occur today on living plants. On the plateau larger areas exist with fully developed stands of cushion plants, all of them dead. Unfavorable climatic conditions most probably killed them. During the Little Ice Age present ice caps expanded and most probably some areas were temporarily covered by thin ice crusts which killed the vegetation previously developed. This suggests that high arctic ecosystems have been often arrested in development or subjected to retrogression during unfavorable climatic episodes.

With respect to the Lowland ecology, raised beaches play an important role in partitioning the Lowland area into many smaller units. This results in slowing down water runoff, leaching away of nutrients, and creating additional niches for plants and animals, especially lemming.

Acknowledgments

I am indebted to Dr. L.C. Bliss for sharing his experience and advice as well as providing sources of assistance for the field, which greatly facilitated the gathering of data. I thank my assistants Miss Dawn Dickinson, Miss Lewina Leung, Mr. Ken Orich, and Mr. Doug Peters for their devoted help. Further thanks go to Kveta Svoboda for the laboratory caloric, lipid, and carbohydrate analyses. Financial support for this study was provided by an N.R.C. grant (A-4879) to L.C. Bliss.

References

Acock, A.M. 1940. Vegetation of a calcareous inner fjord region in Spitsbergen. J. Ecol. 28: 81-106.

Aleksandrova, V.D. 1970. Vegetation and primary productivity in the Soviet Subarctic, pp. 93-114. *In:* Productivity and Conservation in Northern Circumpolar Lands. W.A. Fuller, P.G. Kevan, (ed.). Morges, Switzerland, IUCN N. Ser. No. 16. 344 pp.

Andrews, J.T. 1970. A geomorphological study of Postglacial Uplift with particular reference to Arctic Canada. London, Institute of British Geographers. 1. Kensington Gore. 156 pp. Special publ. no. 2.

———. R. McGhee and L. McKenzie-Pollock. 1971. Comparison of elevations of archeological sites and calculated sea levels in Arctic Canada. Arctic 24: 210-228.

Arkay, K.E. 1972. Species Distribution and Biomass Characteristics of the Terrestrial Vascular Flora, Resolute, N.W.T. M.Sc. Thesis, Dept. Geography, McGill Univ., Montreal. p. 146.

Arnon, D.I. 1949. Copper enzymes in isolated chloroplasts. Polyphenol oxidase in *Betula vulgaris*. Plant Physiol. 24: 1-14.

Association of Official Agricultural Chemists. 1955. Official methods of analysis of the Association of Official Agricultural Chemists. Washington, D.C. Eighth edition. 1008 pp.

Babb, T.A. and L.C. Bliss. 1974. Effects of physical disturbance on arctic vegetation in the Queen Elizabeth Islands. J. Appl. Ecol. 11: 549-562.

Barr, W. 1971. Postglacial Isostatic Movement in Northeastern Devon Island: A reappraisal. Arctic 24: 249-268.

Barrett, P. 1972. Phytogeocoenoses of a Coastal Lowland Ecosystem, Devon Island, N.W.T. Ph.D. thesis. Univ. British Columbia, Vancouver, 292 pp.

Bliss, L.C. 1962. Caloric and lipid content in alpine tundra plants. Ecology 43: 753-757.

———. 1975. Tundra grasslands, herblands and shrublands and the role of herbivores. Geoscience and Man 10: 51-79.

Dennis, J.G., and L.L. Tieszen. 1972. Seasonal course of dry matter and chlorophyll by species at Barrow, Alaska. p. 16-21. *In:* Proceedings 1972 Tundra Biome Symposium, Lake Wilderness Centre, Univ. Washington. 211 pp.

Drury, W.H. 1962. Patterned Ground and Vegetation on Southern Bylot Island, N.W.T. Canada. Contributions from the Gray Herbarium of Harvard University. No. CXC. 111 pp.

Eagles, C.F. 1967. Variation in the soluble carbohydrate content of climatic races of *Dactylis glomerata* (cockfoot) at different temperatures. Ann. Bot. 31: 645-651.

Fonda, R.W., and L.C. Bliss. 1966. Annual carbohydrate cycle of alpine plants on Mt. Washington, New Hampshire. Bull. Torrey Bot. Club. 93: 268-277.

Haag, R.W. 1974. Nutrient limitations to plant production in two tundra communities. Can. J. Bot. 103-116.

Hocking, B. 1968. Insect-flower associations in the High Arctic with special references to nectar. Oikos 19: 359-388.

King, R.H. 1969. Periglaciation on Devon Island, N.W.T. Ph.D. Thesis. Dept. of Geography, Univ. Saskatoon, Saskatoon, 470 pp.

Lavrenko, E.M., V.N. Andreev and V.L. Leontev. 1955. Profile of productivity of the aboveground part of natural plant cover of U.S.S.R. from tundras to deserts. Bot. Zhur. 40: 415-419.

Lewis, M.C., T.V. Callaghan and G.E. Jones. 1972. IBP Tundra Biome. Bipolar Botanical Project. Report on Arctic Research Programme, Phase II. York Univ. 35 pp.

Margalef, R. 1963. On certain unifying principles in ecology. Amer. Naturalist 97: 357-374.

Mayo, J.M., D.G. Despain and E.M. van Zinderen Bakker Jr. 1973. CO_2 assimilation by *Dryas integrifolia* on Devon Island, N.W.T. Can. J. Bot. 51: 581-588.

McCown, B.H., and L.L. Tieszen. 1972. A comparative investigation of periodic trends in carbohydrate and lipid levels in arctic and alpine plants. pp. 40-45. *In:* Proceedings 1972 Tundra Biome Symposium, Lake Wilderness Center, Univ. Washington. 211 pp.

Mooney, H.A., and W.D. Billings. 1960. The annual carbohydrate cycle of alpine plants as related to growth. Am. J. Bot. 47: 594-598.

Nelson, N. 1944. A photometric adaptation of the Somogyi method for the determination of glucose. J. Bot. Chem. 153: 375-380.

Nichols, H. 1975. Palynological and Paleoclimatic study of the late Quaternary displacement of the Boreal forest-tundra Ecotone in Keewatin and MacKenzie, N.W.T., Canada. Institute of Arctic and Alpine Research, Univ. Colorado, Occasional paper No. 15, pp. 84.

Rønning, O. 1965. Studies in *Dryadion* Svalbard. Norsk Polarinst Skr. 134: 1-52.

Russell, R.S. 1940. Physiological and ecological studies on an arctic vegetation. III. Observations of carbon assimilation. Carbohydrate storage and stomatal movement in relation to the growth of plants on Jan Mayen Island. J. Ecol. 28: 289-309.

Savile, D.B.O. 1964. General ecology and vascular plants of the Hazen Camp area. Arctic 17: 237-258.

Shamurin, V. 1966. Seasonal rhythm and ecology of flowering of the plants of the tundra communities in the North Jakutia. pp. 5-125 (in Russian). The vegetation of the far North of the U.S.S.R. and its utilization. (B.A. Tikhomirov, ed.), Fasc. 8. Adaptations of Arctic plants to the environment. Nauka, Leningrad. 272 pp.

Scotter, G.W. 1972. Chemical composition of forage plants from the Reindeer Preserve, Northwest Territories. Arctic 25: 21-27.

Sørensen, T. 1941. Temperature relations and Phenology of the Northeast Greenland Flowering Plants. Medd. om Grønland 125: 1-307.

Soule, M. 1973. The epistasis cycle: A theory of marginal populations. Ann. Rev. Ecol. Syst. 4: 165-187.

Stutz, C.R. and L.C. Bliss. 1975. Nitrogen fixation in soils of Truelove Lowland, Devon Island, Northwest Territories. Can. J. Bot. 53: 1387-1399.

Svoboda, J. 1974. Primary Production Processes within Polar Semi-desert Vegetation, Truelove Lowland, Devon Island, N.W.T. Canada. Ph.D. Thesis. Univ. Alberta. pp. 208.

———. and L.C. Bliss. 1974. The use of autoradiography in determining active and inactive roots in plant production studies. Arctic and Alp. Res. 6: 257-260.

Teeri, J.A. 1972. Microenvironmental Adaptations of Local Populations of *Saxifraga oppositifolia* in the High Arctic. Ph.D. Thesis, Duke Univ.

———. 1973. Polar desert adaptations of high arctic plant species. Science, 197: 496-7.

Tieszen, L.L. 1970. Comparison of Chlorophyll content and leaf structure in Arctic and Alpine Grasses. Amer. Midl. Nat. 83: 238-253.

Warren Wilson, J. 1966. An analysis of plant growth and its control in arctic environments. Ann. Bot. N.S. 30: 383-402.

Zachhuber, K. 1974. Personal communication of some results from his proposed thesis: Entwicklungsrhytmus der reproductiven Organe in verschiedenen Hohenstufen bei *Primulaceae* and *Saxifragaceae*. Austria.

Primary production of Dwarf Shrub Heath communities, Truelove Lowland

L.C. Bliss, J. Kerik, and W. Peterson

Introduction

Plant communities dominated or co-dominated by species within the Ericaceae or closely related families are a common feature of the Low Arctic across mainland Canada and Alaska. This is exemplified by dwarf shrub heath in relatively dry sites of raised centre polygons and ridge tops; cottongrass tussock-heath on imperfectly drained slopes; and low shrub communities of *Salix* and *Betula* with a heath shrub understory on medium drained slopes (Bliss 1975).

Within the High Arctic, dwarf shrub heath with two or more heath species are largely restricted to the southern islands. North of 74°N, *Cassiope tetragona* is the only heath species that provides much cover or production although *Vaccinium uliginosum* may be present in small amounts. In general, this community type is highly restricted to snowbank sites, and within a landscape they seldom occupy more than 1% to 3% of the area.

Within Truelove Lowland, because of granitic rock outcrops with snowbanks that do not melt until early July, this community was better represented (*ca.* 70% of the vegetation within the outcrops which occupy 12.4% of the Lowland).

The objectives of this study were: (1) to describe the plant communities present; (2) to determine alive standing crop and net annual production within the heath communities; and (3) to gather data on the plant phenology and plant growth of *Cassiope tetragona.* Field work was conducted in 1972 by Kerik, and in 1973 by Bliss and Peterson.

Methods

Research was confined to an Intensive Study Site in the granitic outcrop between the other intensive study sites (Fig. 1, Introduction), and to extensive study sites on the east side of Gully River and the outcrops on either side of Icebreaker Beach on the north coast of the Lowland.

In 1972 three temperature profiles (−15, −5, 0, and +3 cm) were established in the Intensive Study Site. Thermocouple sensors were also attached to rock surfaces on the four cardinal compass directions. Spot readings were made periodically. Snow depth measurements were taken at 5 m intervals every three days along three transects (2-175 m, 1-205 m in length) from 20 June through snowmelt (mid-July). Active layer depth was then measured at these locations.

Two transect lines were run across the Intensive Study Site and one each across the Gully River and east Icebreaker Beach outcrops to determine percent plant cover and rock. Using the line intercept method, cover was recorded for vegetated and rock areas. Dominant plant species were listed to aid in establishing plant community types based upon relative frequency and relative plant cover.

Ten plants each of 11 species were tagged for phenological data to be used as a guide in the plant production studies. Dates were recorded for initiation of leaves, flower buds, flowering, and plant senescence. Plants characteristic of different habitats were used.

Standing crop and net annual production (aboveground) in 1972 were based upon 10 samples collected one week after snowmelt (4 July), mid-summer (25 July) and early plant senescence (12 August). Five replicates were collected within 30 cm of large rocks and five 1-3 m from rocks. Cover by species was estimated in a 50×50 cm plot and a nested 20×50 cm plot was then clipped at ground level for plant material. Net annual production was estimated from weight differences between sampling dates in 1972. In 1973 plant phenological information was used to determine green and brown tissue produced the current vs. previous years, in order to refine the weight estimates of the single harvest (4-6 August). The 1972 data were also adjusted using phenological information in addition to weight difference. Two soil cores (6.5 cm diameter) were taken within each clipped plot to a depth of 12 cm (1972 only). Most roots were confined to this depth. Vascular plants were separated by species into green (photosynthetic), brown (living, non-photosynthetic), and dead. Roots were not separated into alive and dead. Lichens were separated from the early July samples and mosses from all three sampling dates in 1972. All samples were ovendried at 80°C.

In 1973, six stands 5×8 m were sampled for community composition using thirty 20×50 cm plots in a stratified random design. Cover was estimated for all species including cryptogams. Prominence values (cover $\times \sqrt{\text{frequency}}$) were then calculated for all species (Beals 1960).

Results and Discussion

Environmental conditions

Variable rock size and resultant snow depth, soil depth, and drainage conditions result in a greater diversity of microhabitats and therefore species and plant community types than occurs elsewhere on the Lowland.

Snow depth on 20 June, 1972 averaged 33±3.6 to 39±2.7 cm along the three transects and by 5 July, most snow had melted. This was the coolest and latest snowmelt summer of the four years (Courtin and Labine this volume). As expected, snow melted earliest on the south-facing exposures and latest on the north-facing exposures, resulting in a 1-2 week delay in initiation of plant growth. This is a significant delay in a 45-55 day growing season. Only 5% of the Intensive Study Site was snow-free on 20 June, 45% was snow-free by 29 June and 95% free by 12 July, 1972. The last areas to melt were frequently ice masses adjacent to rocks where "daytime" melt often refroze at "night." Snowbed communities occur in these sites of prolonged snow-ice melt.

Within 1-2 weeks of snowmelt, the active layer was 15-25 cm in depth, especially near large rocks which provided a heat sink. This rate of soil thaw is equal to or greater than that of the raised beach soils (Svoboda this volume).

Data from Courtin and Labine (this volume) show that rock outcrops are the first sites to warm up on the Lowland, though not the first to become snow-free (raised beaches). They are the last to cool off in late August. Mean weekly air temperatures were often higher here than at the nearby raised beach station (Courtin and Labine this volume, Table 5). Spot readings of temperature were often 3 to 5°C lower at −15 cm, 5° to 25°C lower at the surface, and 2 to 10°C lower at +3 cm on the north-facing vs. the south-facing portions of the Intensive Rock Outcrop Site. Rock surfaces of southern exposures, while warm (10° to 25°C), were often no warmer than the lichen-cushion plant surfaces nearby (10° to 28°C). North-facing rock surfaces were generally 4 to 10°C on sunny days. These temperature differences help to explain plant distribution patterns within these rock masses.

Plant communities

General pattern

Based upon the line intercept data from the transects across the three outcrops, 42% is mostly bare rock, 32% vascular plant dominated communities with lichens and mosses, 10% mosses, 10% lichens, and 6% bare soil. Much of the 20% moss and lichen cover occurs on the rock and soil surfaces.

Using the line intercept cover data and frequency of occurrence of each community (based upon cover estimates of the dominant species) it was possible to calculate the relative importance of the various community types (Table 1). A total of 31 species of vascular plants were sampled in 1972.

Table 1. Plant communities within three granitic rock outcrops on Truelove Lowland. Prominence values are based upon cover × $\sqrt{\text{frequency}}$ of each type.

		Location	
Community type	*Intensive Study Site*	*Icebreaker Beach*	*Gully River*
Dwarf shrub heath-moss	506	119	268
Mixed herb-moss (snowbed)	58	318	25
Graminoid-moss	13	9	146
Crustose Lichen (rocks)	2	31	41
Moss	12	40	3
Cushion plant-lichen	12	2	2

Dwarf shrub heath-moss

The most important community, other than at Icebreaker Beach site, was that dominated by *Cassiope tetragona* with lesser amounts of *Dryas integrifolia*, *Saxifraga oppositifolia*, *Salix arctica*, and *Carex misandra*. In the four sites sampled, 9 to 19 species of vascular plants were found for an average cover of 37%. In the most favorable south-facing slopes and some rocky sites near the Truelove River and in Truelove Valley, small patches of *Vaccinium uliginosum* occur, but they are a minor component where they are found.

Bryophytes contribute 14% cover comprising 17 to 22 species (Table 2). Of these, *Rhacomitrium lanuginosum* is the dominant species. A total of 31 bryophyte species were sampled.

Lichens are very important components, contributing an average 21%

Table 2. Plant communities within granitic rock outcrops, Truelove Lowland. Data are presented as prominance values (cover × $\sqrt{\text{frequency}}$).

Species	Community Type: Dwarf shrub heath-moss (n=4)	Graminoid-moss (n=1)	Cushion plant-moss (n=1)
Vascular Plants			
Cassiope tetragona	213.0	2.3	
Dryas integrifolia	39.8	—	31.5
Saxifraga oppositifolia	20.8	—	5.4
Salix arctica	13.8	0.2	1.0
Carex misandra	5.3	—	—
Oxyria digyna	1.7	—	—
Carex rupestris	1.2	—	16.4
Luzula confusa	0.7	36.3	1.3
Hierochloe alpina	T	24.3	—
Total number species	24	12	12
Bryophytes			
Rhacomitrium lanuginosum	76.4	119.1	161.3
Hypnum revolutum	11.0	1.7	—
Ditrichum flexicaule	9.4	—	6.6
Tomenthypnum nitens	7.5	—	—
Polytrichum juniperinum	6.8	21.2	7.1
Hylocomium splendens	6.6	15.4	—
Dicranum elongatum	5.4	6.3	4.8
Total number species	31	12	16
Lichens			
Cetraria laevigata	19.2	—	—
Cetraria cucullata	18.6	2.6	3.2
Thamnolia subuliformis	14.0	6.1	7.7
Rhizocarpon jemtlandicum	11.3	4.5	1.7
Rhizocarpon geographicum	10.8	16.4	44.3
Alectoria ochroleuca	8.4	67.9	7.2
Dactylina arctica	8.0	—	—
Cetraria nivalis	6.5	1.7	2.6
Alectoria nigricans	6.2	4.1	7.1
Cetraria delisei	5.3	4.1	—
Parmelia separata	0.9	5.2	7.7
Cladonia pyxidata	1.4	5.2	3.6
Hypogymnia subobscura	2.4	3.6	9.1
Umbilicaria vellea	4.3	72.1	203.0
Total number species	20	13	17
Litter	92.5	20.2	12.1
Bareground	4.5	—	0.3
Rocks	147.9	479.6	645.0

cover (13 to 17 species). *Cetraria laevigata*, *C. cucullata*, *Thamnolia subuliformis*, *Dactylina arctica*, and *C. nivalis* are the most important species on soil (Table 2).

Cassiope shrubs, lichens, and mosses are generally best developed within 30 to 75 cm of large rocks, probably the result of more soil water from runoff and warmer soils. Nearer the cooler coast this community is not as common (Icebreaker Beach site).

Graminoid-moss

Small areas (often3×5 m) dominated by *Luzula confusa* and *Hierochloe alpina* often occur near the top of rock masses where winter snow cover is rather minimal. Other characteristic species, though minor in importance, include *Potentilla hyparctica*, *Festuca brachyphylla* and *Poa arctica*. In the one site sampled vascular plants averaged only 12%, mosses 24%, and lichens 6%

(Table 2); rock being of greatest cover (55%). Shrub and cushion plant species are a minor component and species richness is less (10 to 13 species).

Rhacomitrium lanuginosum is again the dominant moss with *Polytrichum juniperinum* and *Hylocomium splendens* being of considerable importance. A total of 12 species of moss and 13 species of lichen were sampled. *Alectoria ochroleuca* and *Umbilicaria vellea* were the most important lichens, both occurring on rocks.

Mixed herb-moss (snowbed)

Snowbed communities were not sampled quantitatively, yet they are an important component within and adjacent to rock masses. The communities generally occupy only 3-10 m^2, yet are quite rich floristically. In the Intensive Study Site outcrop, a total of 20 vascular plant species were recorded with *Oxyria digyna*, *Alopecurus alpinus*, *Saxifraga cernua*, *S. nivalis*, *S. tenuis*, *Luzula nivalis*, *Papaver radicatum*, and *Phippsia algida* the most common species.

Bryophytes are important components of these communities including *Bryum cryophilum*, *B. pseudotriquetrum*, *Cinclidium arcticum*, and *Catoscopium nigritum*.

Cushion plant-lichen

Raised beaches occur in some rock outcrops, especially the Intensive Study Site outcrop and others nearby. *Dryas integrifolia* and *Carex rupestris* are the most important flowering plants. Lichens (9.4%) and mosses (10.7%) provided more cover than the vascular plants (9.7%). These communities are discussed in detail by Svoboda (this volume) (Table 2).

Cryptogam communities

In drier sites among the rocks, masses of *Rhacomitrium lanuginosum* are often found. Rocks, both small and large, often support *Rhizocarpon geographicum*, *R. jemtlandicum*, *Umbelicaria vellea* and other crustose and foliose forms. Small areas of gravelly surface may contain *Thamnolia subuliformis, Dactylina arctica, Cetraria nivalis* and *Alectoria ochroleuca* in addition to crustose species on the pebbles.

Plant Phenology

Owing to their dark surface and higher heat conductivity, the outcrops act as a heat sink, providing habitats that are often more favorable for plant growth. A higher percentage of the plants flower in these habitats than in the sedge meadows. Of the 12 species studied (10 tagged plants per species), flowering averaged 30% to 50% in *Cerastium alpinum*, *Luzula confusa*, *Melandrium affine*, *Poa arctica*, *Potentilla hyparctica*, *Saxifraga nivalis*, and *Stellaria longipes*. Flowering was 70% to 100% in *Arenaria rubella*, *Cassiope tetragona*, *Salix arctica*, *Saxifraga oppositifolia*, and *Papaver radicatum*. Muc (this volume) reported flowering in 3% to 10% of the plants in five common species in the hummocky sedge-moss meadows.

Cassiope tetragona leaves turn from reddish-brown to green 3-8 days after snowmelt and commence flowering 21-25 days after melt. As with most other arctic species, flower initiation occurs the previous year and thus 50% flowering in *Cassiope* shoots in 1973 reflected the cool and late summer in 1972.

A given *Cassiope* stem produces 1-3 flowers *ca*. every 2-4 years. These

Table 3. Aboveground and belowground standing crop (g m^{-2}±SE) for vascular plants and cryptogams in the dwarf shrub heath-moss community.

Component	*1972 Vegetated area n=10*	*1972 Total* area*	*Vegetated area n=15*	*Total area*
Vascular plants				
Aboveground				
Green shoots	109±12	35	50±6	16
Non-green shoots (live)	107±14	34	95±13	30
Dead shoots (attached)	320±30	102	203±27	65
Litter	—	—	181±20	58
Belowground (12 cm)				
Live roots**	1140±270	365	671***	215
Dead roots	380±90	122	224***	72
Mosses (green+brown live)	504±228	166	406±116	130
Lichens	105±34	34	24±4	8
Total live vascular plant	1356	434	816	261
Total live standing crop	1965	629	1246	399
Ratios				
Total Live: dead	1:0.36		1:0.34	
Vascular Live above: live below	1:5.3		1:4.6	
Total Live above: live below	1:1.4		1:1.2	
Total Above: below	1:1.3		1:1.2	

*Based upon 32% with vascular plant cover, 68% rock.
**Assumes 75% live, 25% dead.
***Estimated from 1972 root data.

plants produce 2.2±0.04 pairs of leaves per year and there are 5.2±0.07 years of green leaves per shoot. The two pair of green leaves expand slowly during summer. Based upon total shoot length and mean shoot growth per year, many shoot systems are 30-60 years old.

Salix arctica produces more green biomass per shoot per year and most elongation and leaf expansion occurs within 14-21 days, in contrast with *Cassiope*. Both *Dryas* and *Saxifraga oppositifolia* have leaves that remain green for two years. As on the raised beaches (Svoboda this volume), *Dryas* produces two leaves per shoot per year and *Saxifraga* 4-6 leaves per shoot per year.

Plant production

Standing crop

Because these dwarf shrub heath communities are highly variable in plant cover (the species grow larger near rocks), it is more difficult to obtain accurate estimates of standing crop and net production. In 1972 samples were collected where the vegetation was more lush and thus standing crop estimates were higher than in 1973 where sampling was done in larger areas of heath and away from large rocks (Table 3).

Of the live vascular plant tissue, only 6% to 8% was photosynthetically active (green) and of the green tissue, 83% to 94% was contributed by *Cassiope*. The three woody species, *Cassiope*, *Salix*, and *Dryas* contributed 96-98% of the aboveground live standing crop; *Cassiope* provided 71-85% of the total. The sedge meadows, cushion plant, and heath communities have in common the complete domination of 1-3 species in standing crop and net production although 10 to 15 other species of vascular plants may contribute to the total. Shoot:root ratios are similar to those of the sedge meadows (Muc this volume) and in contrast with the cushion plant communities where root systems are small (Svoboda this volume). Estimates of root standing crop are underestimates because some roots extend below 12-15 cm.

Mosses play a much more significant role than lichens in these communities (Table 3). The much lower estimates for lichen standing crop in 1973 reflect the somewhat drier sites that were sampled in comparison to the previous year. It was estimated that the more lush sites sampled in 1972 represent about 30% of the heath communities and the 1973 sampling the remaining 70%.

Net annual production

Annual production can be estimated in several ways. Two were used in this study. Because of the three sampling periods in 1972, weight differences (ΔW) of green and non-green tissue was used. This method assumes that: (1) sampling errors are small for samples taken on the same date (±10-20% SE only for *Cassiope*, errors were much greater in other species); (2) cover and species size and distribution remain comparable on the different sampling dates (this was not true); and (3) plant phenology was reasonably uniform in the different species (this was true). Using plant weight differences between the 4 July and 12 August sampling dates in most cases, total plant production was 225 g m^{-2} in 1972 (Table 4).

Due to the great variation in green tissue standing crop on 25 July and 12 August (see Bliss and Kerik 1973), it was decided to recalculate estimates of production based on plant phenology as well as peak standing crop data. In this method 20% of *Cassiope* shoots were included as new growth and 50% of *Dryas* and *Saxifraga oppositifolia* green shoots were included. It was assumed that 3.5% of lichen standing crop was new growth and 4.5% of moss standing crop was new growth using method 2. Recalculated total plant production was

Table 4. Net annual production (g m^{-2}±SE) within dwarf shrub heath communities, Devon Island.

Component	*1972 Method 1+*	*1972 Method 2*	*1973 Method 2*	*Average (2 years)* Vegetated area (32% of outcrop)*	*Average (2 years)* Total area*
Vascular Plants					
Aboveground					
Graminoids	1.5	1.5±1.2	0.5±.001	0.8	0.3
Forbs	2.1	2.3±0.4	1.4±.02	1.5	0.5
Woody plants	40.5	24.2±3.9	11.9±1.3	15.6	5.0
Belowground					
Roots**	143.0	117.5	77.6	89.6	28.5
Mosses	28.0	22.7±9.8	16.4±7.7	18.3	7.9***
Lichens	10.0	3.7±0.8	0.8±0.2	1.7	0.5
Total Vascular Plants	187.1	145.5	91.4	107.5	34.4
Total Plants	225.1	171.9	108.6	127.5	43.8

+ Method 1 based upon weight differences between sampling dates, Method 2 used plant phenological information to differentiate green and brown tissue produced the current vs. previous years.

* Assuming equal production in 1972 and 1973 and that 1972 data represent 30% of the heath communities and 1973 data represent 70% of the communities.

** Root production was estimated to be 8-9% of live root standing crop.

*** Includes an estimated 2 g m^{-2} production of *Rhacomitrium lanuginosum* upon rocks.

172 g m^{-2}. Using the same methods for the 1973 data, production was considerably less in areas sampled with less plant cover and with plants of smaller size. The 1973 data are representative of about 70% of the heath type and the 1972 data 30% of this community. The data for the two-year average were weighted accordingly (Table 4).

The plant production data show the clear dominance of *Cassiope* in the vascular plant component and the considerable role that cryptogams play in total production. Due to the extensive development of *Cassiope* roots, their contribution is 70% of total vascular plant production. Although these communities have the highest floristic richness of those in the Lowland, a single species dominates both plant cover and net production as in the other communities intensively studied (Muc, Svoboda both this volume).

Considering that dwarf shrub heath communities are characteristic of the Low Arctic and are near their northern limit at this latitude, species richness, plant cover, and plant production remain unusually high. This attests to the specialized habitats in which the community type occurs here and even to the north on Axel Heiberg and Ellesmere islands. The only data for reasonably comparable communities are for *Betula nana* — dwarf shrub heath in the Khibini Mountains, Kola Peninsula. Chepurko (1972) reported aboveground vascular plant production of 235 g m^{-2} for this community and 50 g m^{-2} for a frost-boil dwarf shrub community at 1000 m. Aboveground vascular plant production of Truelove Lowland heath communities is low by comparison (14 to 28 g m^{-2}).

Acknowledgments

Special thanks go to Lisa Casselmen for her help with the field work and Lewina (Leung) Svoboda for her help in sorting plant material. Financial support for the study was provided by an N.R.C. grant (A-4879) to the senior author.

References

Beals, E. 1960. Forest bird communities in the Apostle Islands of Wisconsin. Wilson Bull. 72: 156-181.

Bliss, L.C. 1975. Tundra grasslands, herblands and shrublands, and the role of herbivores. Geoscience and Man 10: 51-79.

———. and J. Kerik. 1973. Primary production of plant communities of the Truelove Lowland, Devon Island, Canada — rock outcrops. pp 27-36. *In:* Primary Production and Production Processes — Tundra Biome. L.C. Bliss and F.E. Wielgolaski (eds.). I.B.P. Tundra Biome Steering Committee. Edmonton. 256 pp.

Chepurko, N.L. 1972. The biological productivity and the cycle of nitrogen and ash elements in the dwarf shrub tundra ecosystems of the Khibini Mountains (Kola Peninsula). pp 236-247. *In:* Proceedings IV International Meeting On The Biological Productivity of Tundra. F.E. Wielgolaski and T. Rosswall (eds.). Tundra Biome Steering Committee. Stockholm. 320 pp.

The bryophyte vegetation, production, and organic components of Truelove Lowland

Dale H. Vitt and Pekka Pakarinen

Introduction

Bryophytes are an important part of tundra ecosystems and in terms of production, phytomass, and species diversity are sometimes the dominant plant group. Bryophyte dominated communities occur in both the Arctic and Antarctic (Longton 1967, Brassard 1971, Collins 1973). On Truelove Lowland, where 41% of the Lowland is composed of *Carex stans* dominated meadow, bryophytes, particularly mosses, form an almost continuous cover. The water budget, nutrient status, soil formation, depth of active layer, and growth of other plants are strongly influenced by this bryophyte layer (see also Crum 1972). In the drier habitats, mosses and hepatics usually play a minor role, with lichens and various vascular plants often forming a large part of the phytomass (Richardson and Finegan, Svoboda both this volume).

In July and August, 1971 and 1972, field studies were undertaken to: (1) accurately describe the bryophyte vegetation; (2) determine the species diversity and composition of the Lowland areas; (3) analyze the amounts of bryophyte phytomass and production in the major vegetation types of the Lowland; and (4) relate production to differences of chlorophyll, inorganic and organic components, caloric content, microhabitat, and growth of bryophyte species.

Methods

Species composition of Truelove, Skogn, and Sparbo-Hardy Lowlands was studied in 1971 and 1972. About 1,200 specimens were collected in these lowland areas and it is likely that most of the species occurring in the area were recorded. A few collections were obtained by J. Inglis from Sverdrup Lowland. Newman-Smith Lowland was not visited. Voucher specimens have been deposited in the herbarium of the University of Alberta (ALTA) and partial sets distributed to the University of Alaska (ALA), the National Museum of Canada (CANM), the Komarov Botanical Institute in Leningrad (LE), Memorial University of Newfoundland (NFLD), and the New York Botanical Garden (NY).

The moss vegetation of the meadows was studied in 1971, using six sampling areas (stands) of 4×8 m placed in representative communities by subjective estimation. Within each of these 32 m^2 sampling areas, the cover and phytomass of all bryophyte species were estimated in ten random quadrats of 10×10 cm.

In 1972, vegetation studies were expanded to 11 meadows in Truelove Lowland as well as three meadows in Sparbo-Hardy Lowland by using random sampling of 20 plots of 40 cm^2 within an area of 10×10 m in each site.

Bryophyte phytomass was determined (as dry weight), from 40 cm^2 or 100 cm^2 plots after drying at 80°C for more than 12 hrs. The moss layer is composed of three parts: (1) the upper green portion which contains abundant chlorophyll and is biologically active; this layer grades into (2) the lower, brown portion which in most cases contains a limited amount of chlorophyll (see below) and is inactive (and probably alive); and (3) the lower, humified brown layer which is dead. The two presumed living layers (green and brown) are detachable from organic soil in the drier habitats, but in the wetter sites, where decomposition is slower, the lower limit of layer 2 is only gradually delimited, and thus defined indirectly by the presence of the adventitious roots of sedge species. Thus living phytomass includes green active and brown inactive portions. In the wetter sites the boundary between the green and brown layers is clear with the exception of *Drepanocladus* species. In the drier habitats, the transition zone between the two layers is often broad.

Under high arctic conditions only a few bryophyte species produce recognizable yearly increments. In the sedge-moss meadows on Truelove Lowland two species, *Meesia triquetra* and *Cinclidium arcticum*, regularly show yearly increments. The yearly increments are readily distinguished at the boundary between well-developed late season leaves and the smaller leaves of early season growth. Both of these species are acrocarpous and are common in streamside, wet sedge-moss, and hummocky sedge-moss meadows, but occur less frequently in frost-boil sedge-moss meadows (see Muc this volume for details on meadow ecology). The growth of *Meesia triquetra* and hence an estimation of production in certain meadow types (see below) was studied by selecting three sampling sites of 5×5 m in the intensive wet sedge-moss meadow, each representing a different moisture regime. These sites were sampled on 13 July, 1971, and 22 August, 1971, and the length and weight of the current increment of from 20-150 stems were measured in the laboratory from pressed material.

During 2-17 August, 1972, increment lengths of *Meesia triquetra* were measured from a spectrum of meadow types across Truelove Lowland. A representative sampling area of 5×5 m was selected in eleven meadows and 10 to 20 random plots of 40 cm^2 were taken in each. The mean increment length of *Meesia triquetra* was estimated and its growth compared in the three meadow types present on the Lowland.

Production of the bryophyte layer was studied using four methods:

1. Harvesting and dry-weighing the current year's growth, as determined by annual increments, in standard area plots at the end of the growing season:

$$\text{Annual Production} = \frac{\text{mean dry weight per plot (g)} \times 1000}{\text{area of plot (cm}^2\text{)}} \text{ g m}^{-2}$$

2. Determining the dry weight and age, as determined by annual increments, of the upper 0.5 to 1.0 cm of the phytomass in standard area plots:

$$\text{Annual Production} = \frac{\text{mean dry weight per plot (g)} \times 1000}{\text{area of plot (cm}^2\text{)} \times \text{age (yrs)}} \text{ g m}^{-2}$$

In streamside communities where growth is vertical and rapid, the current year's increment can be harvested as in method one. However, in wet sedge-moss meadows where growth is slower, yet still vertical, more than one year's

growth must be used as in method two. It should be noted that in high arctic wet meadow conditions, the moss mat is quite dense. Thus almost all growth is vertical, with individual species growing upward as a unit. Using marker species such as *Cinclidium arcticum* and *Meesia triquetra*, which occur scattered throughout the bryophyte layer, the entire mat can be analyzed.
3. Determining the dry weight of the top 0.5-2.5 cm of the moss layer in standard area plots, and relating this to the length of the current year's increment. This technique was used at intervals during the 1972 summer to estimate current season's production:

$$\text{Production during 1972} = \frac{\text{mean dry weight per plot (g)} \times \text{length of current increment (cm)} \times 1000}{\text{area of plot (cm}^2) \times \text{depth of sample (cm)}} \text{ g m}^{-2}$$

4. Relative production was estimated indirectly by using the length of the annual increments of *Meesia triquetra* as an index of the growth rate of mixed stands (Fig. 1). In extensive surveys the following formula gives an approximation of moss production

$$\text{Annual production} = \text{length of current increment (mm)} \times \text{bulk density of living moss (g dm}^{-3}) \times \text{cover of moss-layer (\%)} \times 0.01 \text{ g m}^{-2}$$

In drier meadow sites (particularly frost-boil meadows) this was the only available method. Production in wet sedge-moss meadows and streamsides was measured by methods one and two, while in hummocky sedge-moss meadows and wet sedge-moss meadows, as given in Table 3, by method 3.

The uppermost 0.5-3.0 cm of nine species of mosses were analyzed for chlorophyll content. Collections made on 5 August, 1971, were taken directly from the field and ground with mortar and pestle in 80% acetone, filtered, and the absorbance of the extract determined with a Bausch and Laumb Spectronic 20 spectrophotometer. Chlorophyll contents are based on the formula of Arnon (1949) and expressed on both a dry weight basis from 40 cm^2 plots and on a surface area basis.

Bryophyte collections for nitrogen, phosphorus, alcohol-soluble matter, petroleum ether-soluble matter, and ash determinations were made in Truelove, Skogn, and Sparbo-Hardy Lowlands during the period of 18-26 July, 1972, (1 to 2 weeks after snowmelt). Caloric, potassium, and calcium values were obtained from specimens collected in Truelove Lowland between 1 July and 18 August, 1971. The green portions of 35 species were analyzed for carbon, nitrogen, alcohol-soluble matter, petroleum ether-soluble matter (lipids), and ash. The green portion consisted of the top 1-2 cm which in high arctic conditions represents several years' growth increment in most habitats. In 21 cases, the lower, brown portion (next 2 cm below the green) was studied for comparative purposes (Pakarinen and Vitt 1974). In a few such cases as *Andreaea rupestris* and *Grimmia ovalis* where no distinction could be made, the top 0.5 cm was used.

Moss samples were washed twice with tap water, air-dried, and ground with mortar and pestle, and finally oven-dried at 80°C immediately before the chemical analyses. The sample size for the determination of petroleum ether- and alcohol-soluble fractions, nitrogen, and ash was 0.5 g dry weight. The total ash content samples were heated in a Thermolyne muffle furnace at 550°C for 12-14 hrs. The residue represented ash, although in a few cases contamination by fine sand or silt was unavoidable despite repeated washing. Total nitrogen was determined using the semimicro Kjeldahl method. The catalyst used in digestion is a commercially available Kel-pak which contains $Hg0$, $CuS0_4$ and K_2S0_4. Total calcium, potassium, and phosphorus values were obtained from

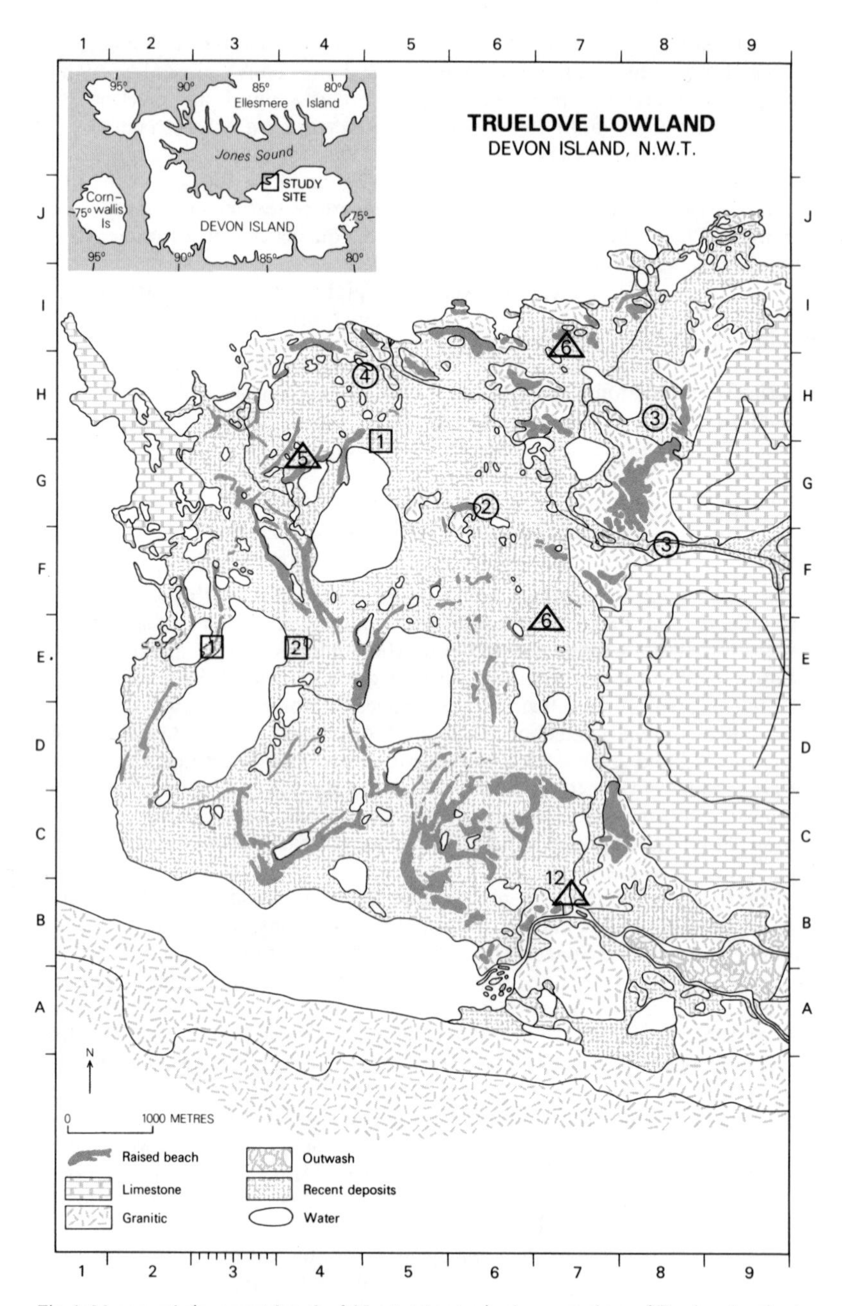

Fig. 1. Mean yearly increment length of *Meesia triquetra* in eleven meadows of Truelove Lowland. Sampling by 40 cm^2 plots taken in August 1972. Figures in mm; circles represent hummocky sedge-moss meadows, squares frost-boil sedge-moss meadows and triangles wet sedge-moss meadows. Mean values are 3.0 for hummocky sedge-moss meadows; 1.3 for frost-boil sedge-moss; and 7.0 for wet sedge-moss meadows.

the Soil and Feed Testing laboratory in Edmonton, Alberta.

Weight loss after six hours extraction with hot petroleum ether in a Soxhlet apparatus represented lipid content. Moss material was oven-dried and weighed before and after the extraction. This residue was further extracted six hours with hot 80% ethanol and the loss in weight determined to calculate the amount of alcohol-soluble material.

Total carbon content was determined by dry combustion using a Leco Carbon Analyzer (sample size 10 mg). Caloric values were obtained with a Parr semimicro oxygen bomb calorimeter (sample size 50-100 mg).

The 35 species of bryophytes are divided into three groups on the basis of habitat. Species collected on granitic or calcareous rock outcrops, growing directly on the rock or in crevices, and those collected on beach ridges were placed in the xeric category (12 species). Species found on hummocks or intermediate levels in hummocky sedge-moss or frost-boil sedge-moss meadows and in transition zones between outcrops and meadows were placed in the mesic group (13 species), while species found either submerged or emergent in wet sedge-moss meadows, pond shores, streams, and streamsides, and those in very wet seepage zones (particularly at the base of late snow areas), were considered as hydric species (10 species).

Results and Discussion

Floristics

In high arctic regions of North America there is a greater species richness of mosses than vascular plants. For example, 151 moss species and 143 vascular plant species have been recorded from northern Ellesmere Island (Brassard and Beschel 1968), while 134 moss species (Holmen 1960) and 106 vascular plants (Fredskild 1966) have been collected from Peary Land, Greenland. The northern lowlands of Devon Island have 132 species (plus two varieties) of mosses (Vitt 1975 and see Appendix 5 this volume) and 93 vascular plants (Barrett and Terri 1973). By comparison, 107 species of mosses and 121 species of vascular plants (Murray and Murray 1973) have been recorded from Pt. Barrow, Alaska, a low arctic tundra site, while in the Mackenzie Delta region of mainland Canada 179 taxa of mosses (Holmen and Scotter 1971) and 420 taxa of vascular plants (Cody 1965) are presently known. It appears that in the High Arctic the number of moss species is greater than the number of vascular plant species, while in the Low Arctic the reverse is true. The reversal in species richness of the two groups is due mainly to a decrease in the number of vascular plant species, while moss species richness remains fairly constant.

In the High Arctic, the species diversity of the genus *Sphagnum* is low. For example, only one species is known from the present study area, while none has been reported from northern Ellesmere Island, Peary Land, Bathurst Island (Brassard and Steere 1968), Cornwallis Island (Steere 1951), or from Prince Patrick Island (Steere 1955). By contrast, Murray and Murray (1973) noted eight species and one variety from Pt. Barrow, while Holmen and Scotter (1971) reported 16 species from the Mackenzie Delta region of the N.W.T. (forest and tundra), and Steere (1939) reported five species from southern Baffin Island. Thus, one of the distinguishing features between high and low arctic sites is the importance of the genus *Sphagnum*.

The number of species per genus is extremely low in the High Arctic. For example, if mosses are considered on a world wide basis there is an average of

Table 1. Bryophyte vegetation in the three major meadow types on Truelove Lowland, Devon Island. Mean cover % (C) and frequency (presence) % (F) in the 1 dm^2 quadrats indicated.

	Hummocky Sedge-moss meadow[1]		Wet Sedge-moss meadow[2]		Frost-boil Sedge-moss meadow[3]	
Species	*C*	*F*	*C*	*F*	*C*	*F*
Cinclidium arcticum	15	100	7	100	9	60
Bryum pseudotriquetrum	3	100	+	30	8	100
Drepanocladus revolvens	12	98	43	100	15	100
Riccardia pinguis	6	93	+	10	1	60
Campylium arcticum	6	90	+	20	10	90
Distichium capillaceum	6	90	+	10	6	90
Orthothecium chryseum	6	88	+	10	10	70
Ditrichum flexicaule	4	85	–	–	10	90
Calliergon giganteum	3	73	9	100	–	–
Drepanocladus brevifolius	7	68	3	50	5	70
Meesia triquetra	6	68	26	100	+	10
Catoscopium nigritum	5	55	–	–	6	70
Philonotis fontana var. *pumila*	+[4]	53	–	–	–	–
Meesia uliginosa	+	43	+	20	3	40
Pogonatum alpinum	+	38	+	40	–	–
Tomenthypnum nitens	2	33	–	–	+	10
Bryoerythrophyllum recurvirostre	+	33	–	–	+	20
Lophozia species	+	33	+	20	–	–
Oncophorus wahlenbergii	2	28	–	–	–	–
Scorpidium turgescens	3	25	+	40	4	50
Total number of species	30		16		19	
Total moss cover (%)	88		93		87	
Number of quadrats studied		40		10		10

1. Additional species (frequency <20, cover +): *Bryum arcticum, Encalypta alpina, Blepharostoma trichophyllum, Myurella julacea, Fissidens adiantoides, Cirriphyllum cirrosum, Timmia norvegica, Cyrtomnium hymenophyllum, Tortella arctica, Plagiomnium ellipticum.*
2. Additional species (C/F): *Bryum neodamense* 2/70, *Cirriphyllum cirrosum* +/10.
3. Additional species (C/F): *Encalypta alpina* 1/10, *Fissidens adiantoides* +/10, *Cirriphyllum cirrosum* +/10, *Grimmia holmenianum* +/10.
4. + = cover less than 1%.

20 species per genus. In North America, north of Mexico there are 4.3 species per genus (Crum 1972), while on the lowland areas of Devon there are only 1.9 species per genus of moss. A somewhat comparable area in size is the Great Lakes Region which was treated by Crum (1974) with 245 species in 106 genera (2.3 species per genus). In the Subantarctic, Campbell Island has a moss flora of 119 species in 69 genera (1.7 species per genus), again an extremely low ratio of species per genus (Vitt 1974). It seems then that the number of species per genus of moss in polar habitats is low and probably represents the extreme situation.

Brassard (1971) reported 42% of the moss flora of northern Ellesmere to contain sporophytes, while in Peary Land, 31% of the moss flora produce capsules (Holmen 1960). On the northern lowlands of Devon Island a comparable figure of 36% of the flora (48 species) are fertile. However, only 29 of these 48 fertile species (or 22% of the total moss flora) produce sporophytes regularly. Very few of the species present on the lowlands have specialized asexual propagules; exceptions to this are such species as *Tortella fragilis* with fragile leaves and *Pohlia annotina* and *Encalypta procera* with axillary brood-bodies. *Aulacomnium palustre*, which commonly produces brood-bodies when found in boreal regions, does so only rarely on these lowlands.

Phytogeographically, the moss flora is composed of four broad components: (1) 21% are circumpolar and generally restricted to the High Arctic, e.g. *Aulacomnium acuminatum*, *Campylium arcticum*, *Drepanocladus brevifolius*, *Funaria polaris* and *Seligeria polaris;* (2) 39% are circumboreal reaching their northern-most point on Devon Island or sometimes extending to a portion of Ellesmere Island, e.g. *Campylium stellatum*, *Encalypta procera*, *Hypnum pratense*, *Tomenthypnum nitens* and *Tortella fragilis;* (3) 30% are arctic-alpine and are more or less restricted to arctic tundra sites with extensions into the alpine and upper mountain zones of more southerly mountain ranges, e.g. *Amphidium lapponicum*, *Aulacomnium turgidum*, *Cirriphyllum cirrosum*, *Dicranoweisia crispula*, *Grimmia torquata* and *Stegonia latifolia;* and (4) 10% are widespread and/or cosmopolitan. In this category can be placed the "weeds of the world" mosses such as *Bryum argenteum*, *Ceratodon purpureus* and *Leptobryum pyriforme* as well as widespread species such as *Dicranum scoparium*, *Fissidens bryoides*, *Grimmia apocarpa* and *Polytrichum juniperinum*.

Vegetation and community structure

Six broad topographic-vegetational units can be recognized in Truelove Lowland based on bryophyte, vascular plant, and lichen components as well as topography (see Muc and Bliss this volume). The dominant vegetation unit is sedge-moss meadow which occupies 41% of the Lowland. *Carex stans* is the most abundant vascular plant in wet sedge-moss and hummocky sedge-moss meadows. *Carex membranacea* along with *Eriophorum triste* dominate the frost-boil sedge-moss meadows (Muc this volume). This meadow vegetation can be divided into three types.

Wet sedge-moss meadows occur in restricted areas of relatively constant water level throughout the season. The topography is slight and the peat-layer often exceeds 30 cm in depth. Hummocky sedge-moss meadows are characterized by well-developed hummocks and depressions, a water level which inundates the depressions directly after snowmelt, but recedes rapidly in mid-season, and a peat layer 10-15 cm in depth. The frost-boil sedge-moss meadows are characterized by dry, well-developed hummocks interrupted by a network of silty frost-boils or non-sorted circles, and have little or no peat layer.

The bryophyte vegetation of these three meadow types is given in Table 1, with 16 species having greater than 1% cover in one or more of the three types. The wet sedge-moss meadows are characterized by *Drepanocladus revolvens*, *Calliergon giganteum*, and *Meesia triquetra*, with a combined cover of 78%. Species richness is low (Fig. 2) and production is high (see below). Hummocky sedge-moss meadows are more common in the Lowland and are characterized by the presence of a range of relatively common bryophyte species rather than a few dominant ones. The three most common species found in this type in Truelove Lowland are *Cinclidium arcticum*, *Drepanocladus revolvens*, and *Drepanocladus brevifolius* together covering only 34% of the area. Ten species have cover values ranging between 2.5% and 6.5% with a remarkably high frequency of occurrence. Thus species diversity is high in hummocky sedge-moss meadows (Fig. 2), while as discussed below production is much lower than in the wet sedge-moss meadows.

Frost-boil sedge-moss meadows are the driest of the three meadow types. *Calliergon giganteum* and *Meesia triquetra* which are common in wet sedge-moss meadows, are rare in frost-boil sedge-moss habitats. Mosses occur here which are found in such drier sites as transition zones and rock outcrops.

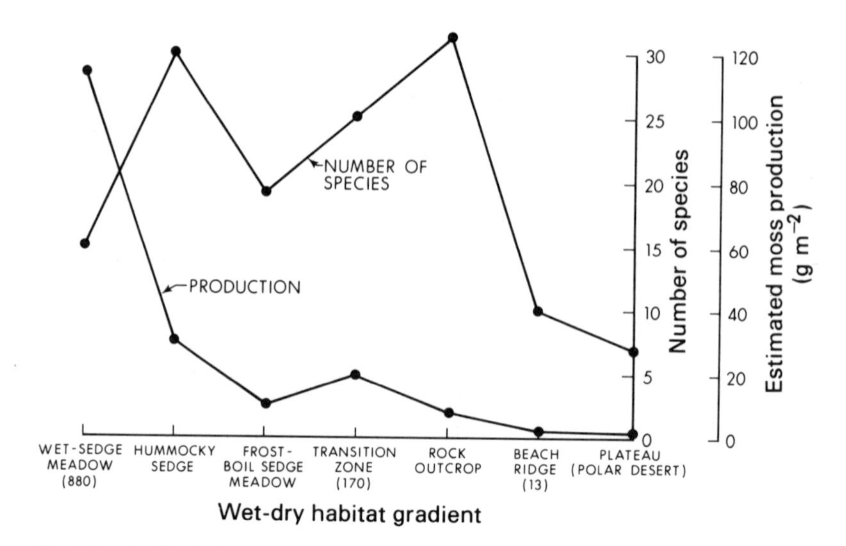

Fig. 2. Species diversity in terms of number of species with greater than 1% cover and production estimated in g m^{-2} in the seven major topographic-vegetational types in and surrounding Truelove Lowland. The vegetational types are arranged in a hydric-xeric sequence with soil moisture given for three of the areas.

Campylium arcticum, *Catoscopium nigritum*, *Meesia uliginosa*, and *Ditrichum flexicaule* reach their maximum cover in frost-boil sedge-moss meadows. In addition, *Seligeria polaris* (along with *Nostoc* and other blue-green algae), occurring exclusively on small stones in the frost-boil circles, is characteristic of this meadow type. Species diversity is intermediate between the other meadow types and production is estimated as the lowest among the meadow types.

Other topographic-vegetational units in which mosses are of less importance have been discussed by Pakarinen and Vitt (1973) i.e. beach ridges, transition zones, and rock outcrops, and by Bliss et al. (this volume) i.e. rock outcrops. Briefly the crest zone of the beach ridges is characterized by *Encalypta rhaptocarpa*, *Distichium capillaceum*, *Mnium lycopodioides* (coll.), and *Hypnum bambergeri*. The transition zones between meadows and beach ridges are dominated by *Tortella arctica*, *Grimmia holmenianum* and *Oncophorus wahlenbergii*, while the transition zones between rock outcrops and meadows include such species as *Tomenthypnum nitens* and *Aulacomnium turgidum*. *Rhacomitrium lanuginosum* is abundant in pure polsters within the rock outcrops. Species diversity (Fig. 2) in terms of number of species is greatest in the mesophytic habitats of the hummocky sedge-moss meadows and in the rock outcrops. It is lowest in hydric and xeric communities and does not correlate with moss production as it occurs over the wet-dry gradient (Fig. 2).

In addition, six moss dominated micro-communities are worth noting.

1. *Scorpidium scorpioides* community (Brassard 1971) occurs submerged in shallow calcareous pools. *Calliergon giganteum* and *Scorpidium turgescens* are common associates.
2. *Grimmia alpicola* var. *rivularis* community occurs submerged in swiftly running streams. *Blindia acuta* is a common associate.
3. *Aplodon wormskjoldii* community (Brassard 1971, Webster and Sharp 1973) occurs on dung of muskox. Additional species usually present are

Table 2. A comparison of bryophyte vegetation in the sedge-moss meadows in Truelove Lowland intensive study site and in Sparbo-Hardy Lowland (A = coastal meadows, B = central lowland, C = at the base of southern cliffs). Frequency % is based on 40 cm^2 plots within an area of 10×10 meters.

Species	*Truelove*		*Sparbo-hardy*	
	(N=20)	*(N=20)* A	*(N=20)* B	*(N=10)* C
Drepanocladus revolvens	100	75	100	80
Bryum pseudotriquetrum	100	100	100	90
Cinclidium arcticum	95	100	95	100
Distichium capillaceum	95	55	95	10
Campylium arcticum	90	65	75	100
Orthothecium chryseum	85	25	70	60
Drepanocladus brevifolius	85	85	100	20
Riccardia pinguis	80	–	45	90
Ditrichum flexicaule	75	–	15	60
Meesia triquetra	70	70	90	70
Philonotis fontana var. *pumila*	50	40	20	100
Tomenthypnum nitens	40	–	–	30
Pogonatum alpinum	35	5	5	100
Meesia uliginosa	35	5	5	40
Oncophorus wahlenbergii	15	–	15	40
Cirriphyllum cirrosum	15	15	5	20
Bryum arcticum	10	5	20	80
Cyrtomnium hymenophyllum	15	–	–	–
Aulacomnium acuminatum	15	–	–	–
Fissidens adiantoides	15	–	–	–
Myurella tenerrima	10	–	–	–
Blepharostoma trichophyllum	5	–	–	–
Catoscopium nigritum	60	30	40	–
Calliergon giganteum	45	55	85	–
Scorpidium turgescens	45	5	40	–
Bryoerythrophyllum recurvirostrum	25	–	15	–
Encalypta alpina	20	–	5	–
Myurella julacea	15	25	10	–
Timmia norvegica	5	–	10	–
Aplodon wormskjoldii	–	–	5	–
Plagiomnium ellipticum	–	–	10	50
Aulacomnium turgidum	–	–	–	100
Drepanocladus uncinatus	–	–	–	100
Calliergon sarmentosum	–	–	–	90
Tritomaria quinquedentata	–	–	–	80
Polytrichum algidum	–	–	–	60
Hypnum plicatule	–	–	–	40
Pohlia nutans	–	–	–	20
Sphagnum orientale	–	–	–	10
Number of species	29	17	25	26
Sampling date	16 Aug. 1972		26 July 1972	

Coef. of similarity between Truelove and A, B, or C of Sparbo-Hardy. A/0.70, B/0.74, C/0.57.

Tetraplodon minoides, *T. pallidus*, *Splachnum vasculosum* var. *heterophyllum* *Voitia hyperborea*, and various *Bryum* species.

4. *Bryum cryophilum* community (Holmen 1955, Brassard 1971) occupies stream sides and snow flush areas. Also common in this community are *Bryum pseudotriquetrum*, *Cinclidium arcticum*, *Philonotis fontana* var. *pumila*, and *Catoscopium nigritum*.

5. and 6. Two commonly encountered disturbed areas have unique moss vegetation. The area surrounding lemming burrows is usually colonized by *Funaria microstoma*,* *F. polaris*, *Desmatodon leucostoma*, and *Desmatodon*

*this is an undescribed variety.

heimii var. *arctica*. These areas are usually gravelly and in all probability high in nutrients. The second successional habitat is formed at the sides of ice wedges and polygonal ice-wedge mounds. In these situations, bare organic soil is exposed and colonized by *Psilopilum cavifolium*, *Dicranella crispa*, *Leptobryum pyriforme*, *Stegonia latifolia*, and *Ceratodon purpureus*, all extremely rare in other habitats.

Truelove Lowland is strongly influenced by the thick calcareous overburden present on the plateau above the Lowland. Farther to the east (20 km), this calcareous overburden ends and Sparbo-Hardy Lowland is decidedly less minerotrophic. The meadow and rock outcrop moss flora of these two lowlands, although similar in the dominant species, are considerably different in their minor components, as indicated by hummocky meadows in Table 2. Such minerotrophic species as *Scorpidium turgescens*, *Distichium capillaceum*, *Drepanocladus brevifolius* and *Catoscopium nigritum* are not as common in Sparbo-Hardy as in Truelove, while such species typically growing in less minerotrophic conditions as *Pogonatum alpinum*, *Calliergon sarmentosum*, *Sphagnum orientale*, and *Drepanocladus uncinatus* are uncommon in Truelove and common in Sparbo-Hardy. Likewise the acidic rock species such as *Andreaea rupestris*, *Dicranoweisia crispula*, *Amphidium lapponicum*, and *Grimmia torquata* are common on outcrops in Sparbo-Hardy Lowland, and extremely rare in Truelove, even though granitic rocks are commonly present. It appears that local differences in substrate play a deciding role in determining the bryophytic species composition of these areas. Coastal meadows in the two lowlands are similar; however, the more inland, protected meadows ("C" in Table 2) are decidedly different in Sparbo-Hardy, perhaps due to a predominance of an acidic substrate.

Phytomass

Green phytomass of bryophytes in hummocky sedge-moss meadows showed a range of 157-210 g m^{-2} (Tables 3 and 4) while total phytomass varied from 843 to 994 g m^{-2}. In wet sedge-moss meadows where production was at least twice that of hummocky sedge-moss meadows, (see below) the green phytomass of 163-385 g m^{-2} was more variable, yet similar to that in hummocky sedge-moss meadows, indicating a faster rate of change from green to brown in the wet sedge-moss sites. Total phytomass in wet sedge-moss meadows was slightly smaller, with a mean figure of 752 g m^{-2} for the intensive site.

In rock outcrops, both in mesic conditions represented by *Hylocomium splendens* (Table 5) and xeric conditions illustrated by *Rhacomitrium lanuginosum* (Table 5), green and total phytomass are high, 1,273 and 2,218 g

Table 3. Production estimates for the intensive study meadows, based on average increment lengths and dry weight/volume of the green moss-layer. Mean values (standard error) are from sampling 20 plots of 40 cm^2 per site.

Component *Sampling date*	*Hummocky sedge-moss meadow* *16 Aug., 1972*	*Wet sedge-moss meadow* *17 Aug., 1972*
Green phytomass (g m^{-2})	175 (14.0)	385 (18.0)
Depth of green phytomass (mm)	10.5 (1.1)	23.9 (1.0)
Weight/volume of green phytomass (g dm^{-3})	16.7	16.1
Average increment length of mosses for 1972 (mm)	2.0	4.8
Estimated production (g m^{-2})	33	77

Table 4. Phytomass and weight/volume of moss stands in different meadows in Truelove Lowland. Means (n=10) with standard error in parentheses are based upon sampling plots placed randomly, each 100 cm^2 in size.

Component *Sampling date*	*Hummocky sedge-moss meadows* *Intensive site* *20 Aug, 1971*	*Coastal meadow* *10 Aug, 1971*	*Plateau base* *11 Aug, 1971*	*Wet sedge-moss meadow* *Intensive site* *13 July, 1971*
Green phytomass (g m^2)	210 (26)	215 (39)	157 (12)	163 (15)
Depth of green phytomass (mm)	8.0 (0.17)	9.2 (1.5)	9.6 (1.4)	11.0 (1.8)
Weight/volume of green phytomass (g dm^3)	26.3	23.3	16.4	14.8
Total phytomass (g m^2)	843 (135)	994 (64)	887 (41)	752 (62)
Depth of total phytomass (mm)	29.4 (3.3)	25.5 (1.6)	21.0 (1.8)	36.5 (2.1)
Weight/volume of total phytomass (g dm^3)	28.7	39.0	42.2	20.6
Ratio of green/total phytomass (%)	25	22	18	22

Table 5. Phytomass and weight/volume for the dominant moss species with 100% cover in the rock outcrops. Data (n=10) were taken from 40 cm^2 samples; means ± (S.E.) are presented.

Component *Sampling date*	*Rhacomitrium lanuginosum* *5 Aug, 1971*	*Hylocomium splendens* *5 Aug, 1971*
Green phytomass (g m^2)	2065 (224)	1273 (148)
Depth of green phytomass (mm)	38.0 (2.4)	47.0 (5.2)
Weight/volume (g dm^3)	54.3	27.1
Total phytomass (g m^2)	4928 (472)	2218 (211)
Depth of total phytomass (mm)	72.5 (4.8)	62.5 (6.8)
Weight/volume (g dm^3)	68.0	35.5
Ratio of green/total phytomass (%)	42	57

m^{-2} respectively for *H. splendens* and 2,065 and 4,928 for *R. lanuginosum*. Subjective evaluations of production for both of these rock outcrop species would seem low.

Thus it would seem that measurement of the green (or so-called living phytomass) is not a useable method of measuring production, as the rate of change between green and brown (or living and dead; active and inactive) is dependent on the microhabitat and not necessarily correlated with production or yearly increments except in a few isolated cases.

Production and growth of bryophytes is dependent not only on the length and weight of individual stems, but also on the density of the stems. In this regard, the actual weight of a particular volume of green phytomass will depend on the robustness of each shoot and density of the shoots. This can be measured as dry weight/volume green phytomass in g dm^{-3}. Results for eight species growing along a water gradient are shown in Table 6. In hummocky sedge-moss meadows the moss stands are more dense (16-26 g dm^{-3}) — Table 4, than the wet sedge-moss meadow (15 g dm^{-3}). Hummock species such as *Tomenthypnum nitens* and *Aulacomnium acuminatum* form loose mats (11-17 g dm^{-3}) while such other species as *Dicranum elongatum* in rock outcrops are extremely dense (105-115 g dm^{-3} — Table 6).

In Abisko, Sweden, a similar range of values has been observed in wet sites with values for *Drepanocladus schulzei* of 11 g dm^{-3} and *Sphagnum balticum* 15-30, whereas on hummocks dense moss stands occur — *Dicranum elongatum* 44-84 g dm^{-3} (Sonesson and Johansson 1974).

Table 6. Dry weight/ volume of the top portion of the living phytomass of bryophyte stands growing in different habitats. Depths are expressed from the top of the moss-layer. Samples of 40 cm^2 surface area were used in the analyses.

Dominant species	*Depth of sample* mm	*Dry weight/volume* g dm^3
Hydric species		
Meesia triquetra	0 - 5	16.8
Meesia triquetra	5 - 10	24.2
Cinclidium arcticum	0 - 5	15.5
Cinclidium arcticum	5 - 10	20.2
Calliergon giganteum	0 - 5	8.2
Calliergon giganteum	5 - 10	10.7
Calliergon giganteum	10 - 20	29.1
Mesic species		
Aulacomnium acuminatum	0 - 5	17.4
Aulacomnium acuminatum	0 - 10	16.4
Aulacomnium acuminatum	10 - 20	29.0
Tomenthypnum nitens	0 - 10	13.0
Tomenthypnum nitens	10 - 20	24.0
Tomenthypnum nitens	0 - 10	11.4
Tomenthypnum nitens	10 - 20	37.4
Xeric species		
Hylocomium splendens	0 - 10	13.5
Hylocomium splendens	10 - 20	30.9
Rhacomitrium lanuginosum	0 - 10	25.1
Rhacomitrium lanuginosum	10 - 20	48.7
Rhacomitrium lanuginosum	0 - 10	41.8
Rhacomitrium langinosum	10 - 20	65.0
Dicranum elongatum	0 - 2.5	114.8
Dicranum elongatum	0 - 2	105.0

Growth and production

The seasonal growth of *Meesia triquetra* in three different microhabitats of the Wet Sedge-moss Meadow Intensive Study Site (see Bliss this volume for site details) are shown in Table 7. In the drier site, *M. triquetra* achieves over 50% of its length increment in the first two weeks of the growing season, while the maximum weight increase is in late season. Since in this dry site, the water level recedes in late season, while in the streamside site it remains constant, it is likely that *M. triquetra* continues to grow predominantly in length as long as there is adequate water, while under the drier conditions of late season, growth takes place largely as a weight increase. Thus, growth of at least this moss, is not a process of uniform increase in length and weight; instead growth takes place first as an increase in length in early season, and predominantly an increase in weight in late season.

It appears that the major factor in controlling growth and production of bryophytes in the High Arctic is the amount of surface water present. This is particularly true in the case of *Meesia triquetra*. As Fig. 1 illustrates, the length increment of this species is related to the type of meadow in which it is found. In wet sedge-moss meadows length increments range from 6-12 mm, in hummocky sedge-moss meadows from 2-4 mm, and in the drier frost-boil sedge-moss meadows from 1-2 mm or sometimes less.

Since the acrocarpous *Meesia triquetra* is common in the wetter meadows, and in most microhabitats the moss mat grows in a vertical direction, it can be used to estimate production in wetter sites. However, it is rare in frost-boil

Table 7. Growth of *Meesia triquetra* in three sites in the wet sedge-moss meadow (intensive study area) and one site of the hummocky sedge-moss meadow (intensive study area) in 1971 (N = number of measured plants).

Site and sampling date	*N*	*Mean increment length (mm)*	*Mean increment weight (mg)*	*Mean weight/length (mg/cm)*
Wet sedge-moss meadow, drier site				
13 July	20	3.28	.164	.500
22 August	150	6.02	.577	.958
Wet sedge-moss meadow, central				
13 July	20	4.43	.227	.513
22 August	150	9.77	.845	.863
Wet sedge-moss meadow, streamside				
12 July	20	3.88	.139	.358
27 July	20	14.80	.884	.598
Hummocky meadow, hollow				
15 July	45	1.20	.100	.820
21 August	45	3.70	.500	1.330

sedge-moss meadows and in this habitat may not give an accurate estimate of total moss production. In the drier habitats, a marking technique using steel rods inserted into the moss mat to a constant level such as used by Sonesson and Johansson (1974) would accurately estimate production; however, this method was not used in our study.

In 1971, using *Meesia triquetra* as a marker species, production of the bryophyte layer was estimated from 33 g m^{-2} to 41 g m^{-2} in the intensive hummocky sedge meadow, while in 1972, production averaged 33 g m^{-2} (Table 4). In 1971, in this same site, production of mosses growing on hummocks was 22 g m^{-2}; intermediate areas 36 g m^{-2}; and depressions 45 g m^{-2} (all with 100% moss cover) indicating a considerably higher production in the wet, depression areas. Likewise production values of 50-60 g m^{-2} were obtained in individual plots in the southern portion of the intensive meadow where there is a considerable flow of water in early and mid-season.

In the wet sedge-moss meadows the annual production of the moss layer varied (in 1971) from 60 g m^{-2} in the drier sites to (in 1972) 77 g m^{-2} (Table 3) in the wetter areas, while along streamsides (included in the concept of wet sedge-moss meadows), bryophyte production increased greatly. On 22 July, 1971, production was 202 g m^{-2} (157 g ashfree) while at the end of the growing season it was 350 g m^{-2}(293 g ash free).

No direct sampling was made of moss production in frost-boil sedge-moss meadows, but occasional observations on the growth increment of *Cinclidium arcticum* as well as *Meesia triquetra* (Fig. 2) combined with bulk density values (Table 6) and cover estimates of the moss-layer suggest a mean range of production of 10 to 20 g m^{-2} for these meadows.

Our results suggest that the highest production of bryophytes in the High Arctic occurs in situations where there is a constant source of water maintained at a constant level throughout the growing season. Furthermore, in these stands, growth takes place in terms of length throughout the growing season; drier sites, with lower production, show a growth maximum in early season with the predominant early season growth being in length and the predominant late season growth a weight increase.

Thus, in a substantial portion of Truelove Lowland, bryophyte annual production is in the 20-45 g m^{-2} range, while in local situations, it reaches as high as 350 g m^{-2}. The latter figure agrees well with other data from localized bryophyte dominated habitats in polar areas. For example, Longton (1970) estimated annual shoot production of *Polytrichum alpestre* ranging from 342 g

m^{-2} on Signy Island, South Orkney Islands, to 507 g m^{-2} on South Georgia. Baker (1972) found a value of 436 g m^{-2} for *Chorisodontium aciphyllum* in similiar situations. Collins (1973) working on Signy Island reported values from 223-893 g m^{-2} in carpets of *Calliergidium austro-stramineum*, while Clark et al. (1971) gave values as high as 1000 g m^{-2} for *Pohlia wahlenbergii* and *Philonotis acicularis* on South Georgia. Rasdorfer et al. (1973) suggested a general estimate for annual bryophyte production in a *Carex aquatilis-Dupontia fisheri-Pogonatum alpinum* var. *septentrionale-Calliergon sarmentosum-Drepanocladus uncinatus* dominated area at Pt. Barrow (Low Arctic) as 250±50 g m^{-2}. At Abisko, Sweden, in a subarctic "elevated bog" dominated by *Betula nana*, *Rubus chamaemorus*, *Dicranum elongatum* and *Sphagnum balticum*, annual production of bryophytes was found to be 64 g m^{-2} (Sonesson and Johansson 1974).

Thus, in restricted localities where conditions are seemingly approaching ideal, bryophyte production is quite high, while in other areas production is much less, even though at times these habitats contain very similar taxa.

Organic and inorganic components

The results of analyses of 35 species of bryophytes for total carbon, total nitrogen, alcohol-soluble fraction, lipids, and ash of the green portion are shown in Table 8. Species growing in hydric habitats have significantly greater total nitrogen and alcohol-soluble fractions than those in mesic or xeric habitats, while the lipid (Pet. ether-soluble fraction) is slightly greater in the wetter habitats. Pakarinen and Vitt (1974) presented data which showed that when the green and brown segments are compared, the nitrogen content and alcohol-soluble portion are greater in the green segment, while the ash content is higher in the brown (based on analyses of 21 species).

Fig. 3 shows a positive relationship between percent phosphorus and percent nitrogen. Likewise Tamm (1953) found a strong correlation between phosphorus and nitrogen contents in *Hylocomium splendens*. Tamm reported phosphorus contents ranging from 0.1 to 0.4%, which are much higher than the values obtained from the Truelove material, where only *Aplodon wormskjoldii* has phosphorus contents above 0.1%. In general, those mosses growing in

Table 8. Contents of total carbon, total nitrogen, alcohol-soluble fraction, pet. ether-soluble fraction (=lipids) and ash in the green portion of hydric, mesic, and xeric groups of mosses on Devon Island 18-26 July 1972. All values (except ash) are on an ashfree dry weight basis, mean ± (SE) is indicated.

Component	*Hydric group*[1]	*Mesic group*[2]	*Xeric group*[3]	*Average for all 35 species*[4]
Total carbon (%)	49.2 (0.9)	49.1 (0.6)	47.6 (0.9)	48.7 (0.5)
Total nitrogen (%)	1.63(0.16)	0.93(0.06)	0.78(0.10)	1.08(0.09)
Alcohol-soluble fraction (%)	17.4 (2.1)	11.1 (0.8)	10.0 (0.7)	12.5 (0.9)
Pet. Ether-soluble fraction (%)	3.7 (0.4)	2.8 (0.2)	2.5 (0.2)	3.0 (0.2)
Ash (%)	8.6 (3.0)	3.4 (0.4)	5.7 (1.3)	5.7 (1.0)

1. Includes 10 collections (species) from wet sedge-moss meadows, streamsides, pond shores, and late snowmelt areas.
2. Includes 13 collections (species) from hummocky and frost-boil sedge-moss meadows and from seepages near rock outcrops.
3. Includes 12 collections (species) from rock outcrops and beach ridges.
4. For a listing of species included in these analyses, see Pakarinen and Vitt (1974).

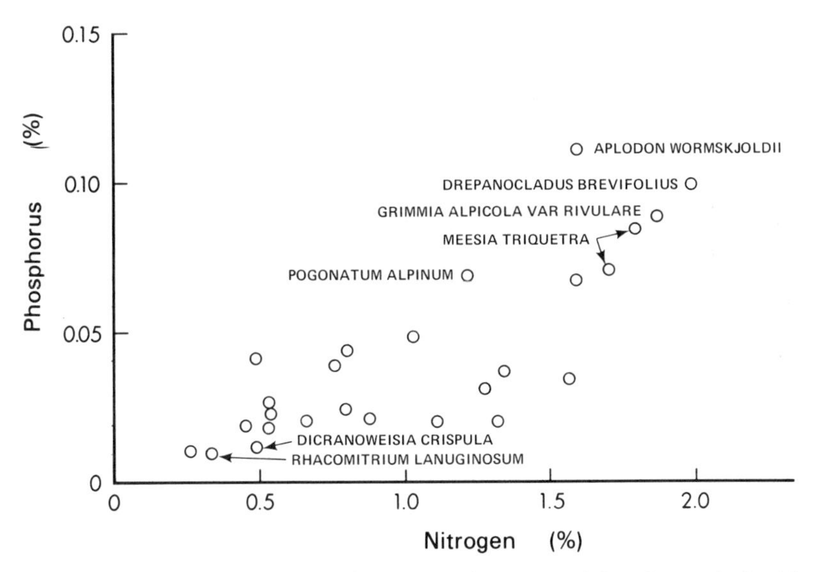

Fig. 3. Total nitrogen content plotted against total phosphorus content in bryophyte species from the Truelove Sparbo-Hardy Lowland area. Data are oven-dry weight and with ash.

hydric habitats have higher phosphorus contents than those growing in xeric habitats (Fig. 3). The low contents of phosphorus in the bryophyte layer might indicate that it could be one of the major limiting factors in plant growth in these arctic meadows.

The vertical distributions of calcium and potassium in the bryophyte layer at the intensive hummocky sedge-moss meadow site show opposite patterns. Potassium is concentrated in the green portion of the bryophyte layer, while calcium becomes more abundant deeper in the brown phytomass (Fig. 4). Similar changes were found in the aging of *Hylocomium splendens* by Tamm (1953).

Caloric contents for eleven species of mosses have been reported by Pakarinen and Vitt, (1974) as ranging between 4.50 and 4.97 Kcal g^{-1} ash-free with a mean of 4.67 Kcal g^{-1}. When caloric contents are considered on a dry weight basis alone, there appear to be distinct differences in contents between species from xeric, mesic, and hydric habitats (Fig. 5 — vertical axis). However, when ash content is taken into consideration the values consistently range between 4.50 and 4.97 Kcal g^{-1} with no significant variation. Thus, even though hydric species apparently have the lowest caloric values on a dry weight basis, they have the highest ash content (Fig. 5) and vice-versa for xeric species.

Chlorophyll contents are highly variable between species and habitats on both dry weight and an area basis (Fig. 6). As with nitrogen, phosphorus, alcohol-soluble matter, growth and production, chlorophyll contents show the same hydric-xeric gradient. Hydric species such as *Calliergon giganteum*, *Bryum pseudotriquetrum*, *Meesia triquetra* and *Cinclidium arcticum* have chlorophyll contents of 2 to 5 mg g^{-1} dry wt, while all mesic and xeric species have less than 1 mg g^{-1}. *Rhacomitrium lanuginosum*, a xeric species, has less than 0.5 mg g^{-1} at the peak season. This species, which forms large mats on rock outcrops, also has the lowest nitrogen, ash, and phosphorus contents found in mosses of the lowlands.

Living material of *Drepanocladus brevifolius* was transported to the University of Alberta in August 1972 and deep-frozen. The absorption

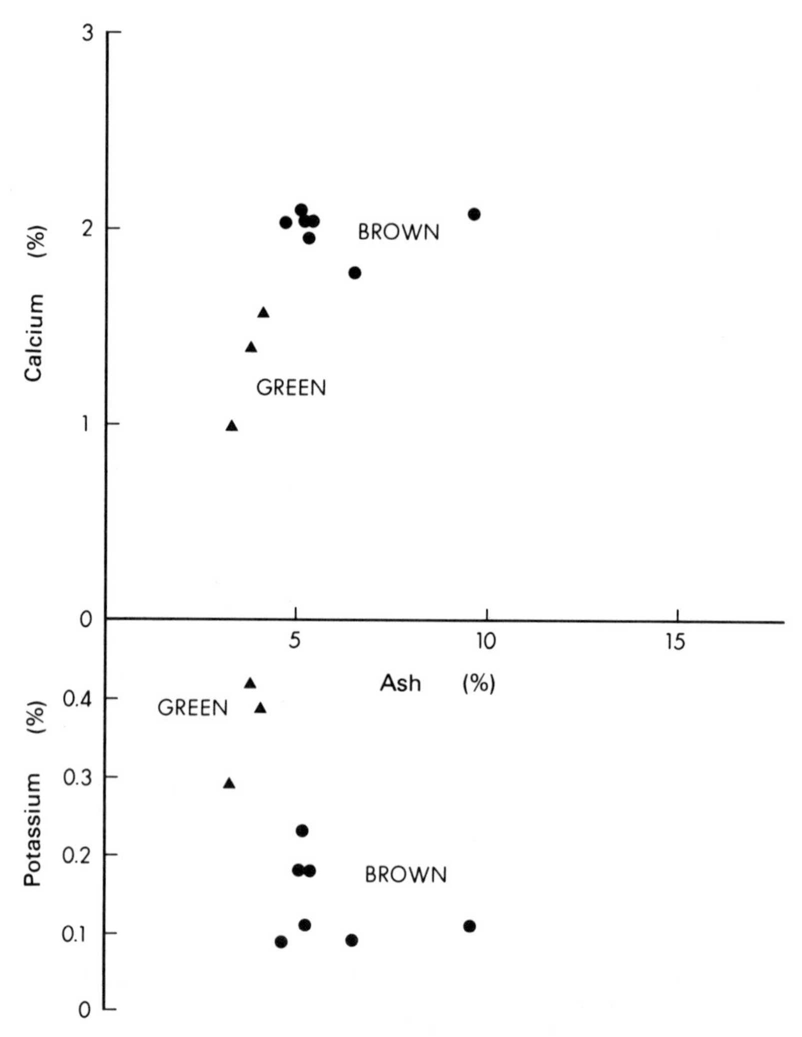

Fig. 4. Calcium and potassium contents in the green (△) and brown (o) portions of the bryophyte layer in the intensive study site of the hummocky sedge-moss meadows 20 August, 1971.

spectrum was determined in alcohol extraction and compared to *Sphagnum warnstorfii* from the Edmonton area. Both spectra were similar with peaks at 412 and 664 nm. When the pigments of both samples were studied more closely with thin-layer chromatography, it was found that lutein and beta-carotene were the major carotenoids in both species as in vascular plants. These results do not suggest any qualitative differences between the pigments of boreal and arctic mosses although higher amounts of carotenoids might be expected in open tundra conditions.

Summary

There are 132 species and two varieties of mosses in 71 genera presently known from the northern lowlands of Devon Island. The number of taxa and species

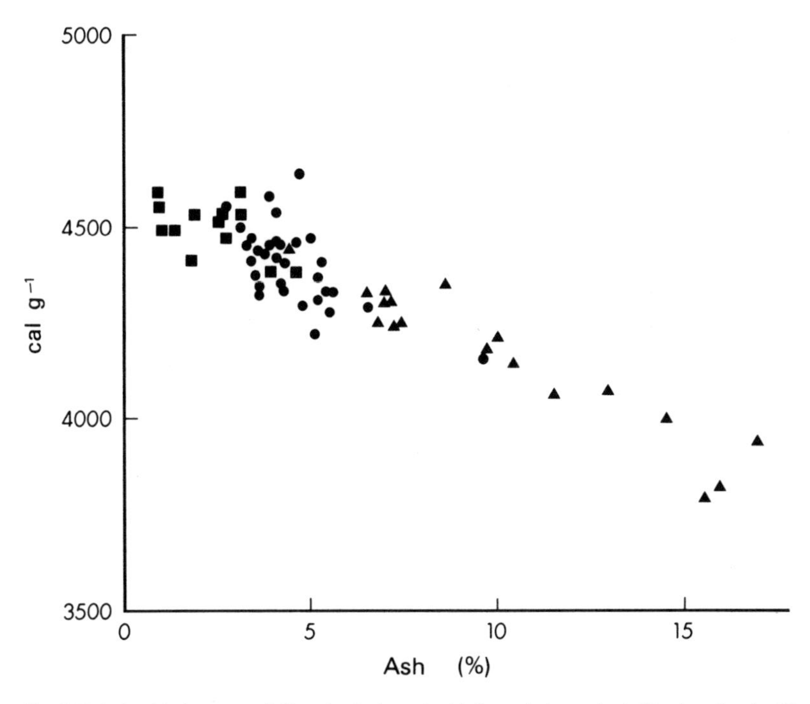

Fig. 5. Relationship between ash % and caloric content in bryophyte species in Truelove Lowland in August 1971 y(Kcal g^{-1}) = 4.60-0.044 × (ash %), Correlation coef.=−0.916. Triangles represent species from wet sedge-moss meadows, circles: species from hummocky sedge-moss meadows, and squares: species from rock outcrops (xeric habitats).

composition is comparable to other High Arctic sites where the number of species per genus is low. In the High Arctic, species richness of mosses is higher than vascular plants due to the marked decrease in the latter. *Sphagnum* is rare in the High Arctic.

The bryophyte vegetation dominates in the wetter areas. Bryophyte composition of sedge meadows is similar to that described by Brassard (1971) on northern Ellesmere and Holmen (1955) in Peary Land. Species richness in meadow habitats ranges from 14-30 species in sampling areas of 10×10 m with total mean cover for the meadow types varying from 87-93%.

Both green and total bryophyte phytomass is extremely variable, depending on the particular habitat. The change from green to brown layers is also variable. In wet habitats the transition from green to brown takes place in less than one year, while in drier habitats two to many years growth may remain green for a considerable period of time. Production cannot be estimated by use of phytomass, as both growth patterns and rate of decomposition vary between habitats.

Production in hummocky sedge-moss meadows is estimated to vary between 22 g m^{-2} to 45 g m^{-2}. In the wet sedge-moss meadows production varies from 60 g m^{-2} in drier sites to 77 g m^{-2} in wetter areas. Along streams production is very high, ranging up to about 350 g m^{-2}.

Analyses of nitrogen, phosphorus, alcohol-soluble matter, and chlorophyll content all show higher values in the wet habitats and correspondingly low values in the more xeric habitats. This correlates well with production estimates for the various communities. Ashfree caloric contents

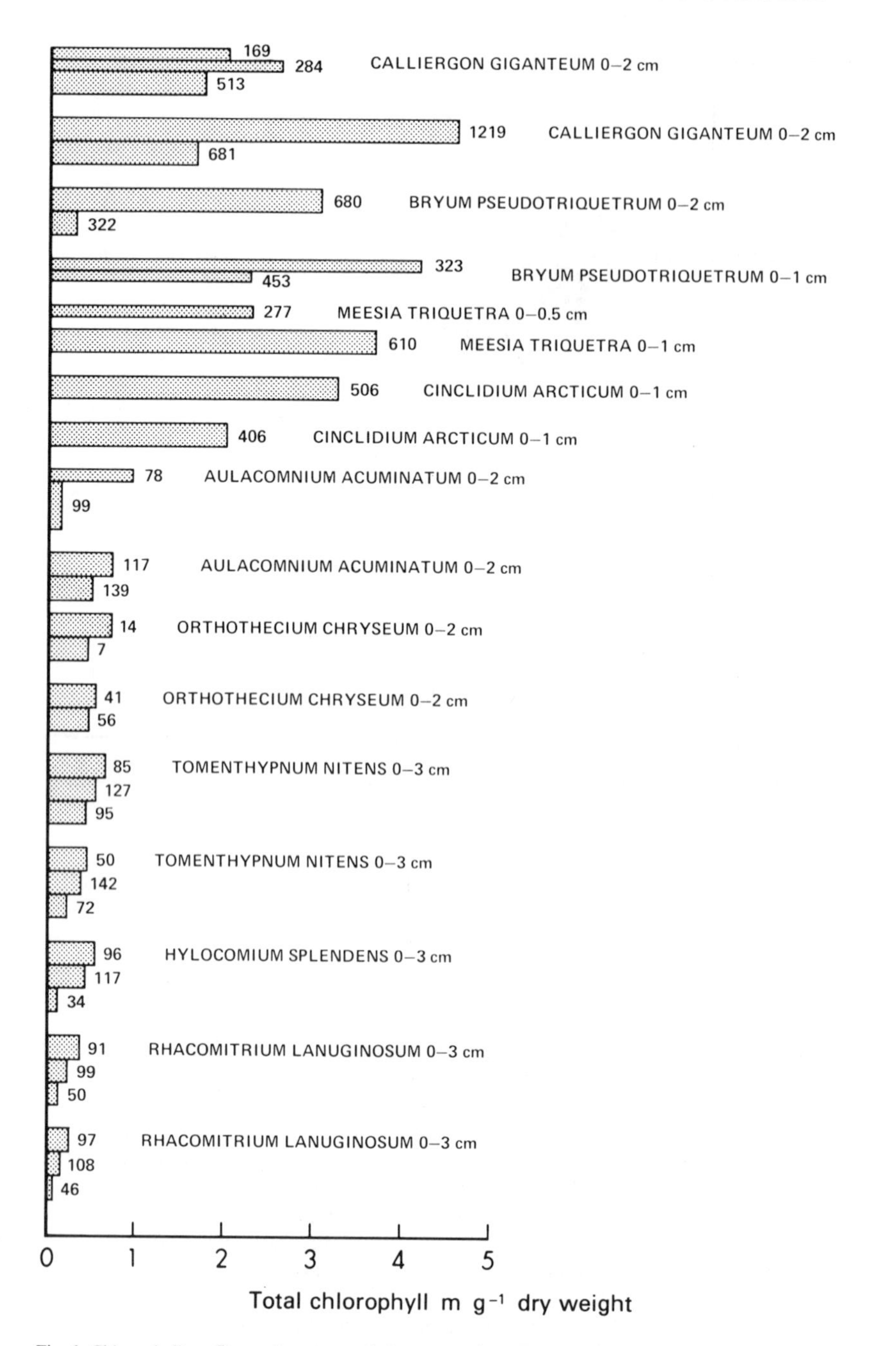

Fig. 6. Chlorophyll profiles and contents of nine mosses from Truelove Lowland. Samples taken 5 August, 1971. Horizontal axis equals chlorophyll content per gram dry weight. The numbers beside the bars indicate amount of chlorophyll (mg m^{-2}). Vertical axis is depth in cm from the top of the moss layer. Species are arranged with those occurring in hydric habitats at the top and xeric at the bottom. Some species are duplicated to show either variability in amount of chlorophyll (e.g. *Orthothecium chryseum*), or to show different vertical components (eg. *Calliergon giganteum*).

consistently range between 4.50 and 4.97 Kcal g^{-1} for all eleven species analyzed.

The High Arctic is a harsh area in which bryophytes often dominate. Bryophytes may equal or sometimes surpass vascular plants in terms of species richness, cover, and, in particular, net production in the wetter habitats, possibly because of lack of competition from vascular plants or a better adaptation to cold soils. In the meadow habitats, it seems to us, that mosses are not forming an inert substrate upon which the flowering plants are scattered (Saville 1972), but rather may be exerting a dominating effect on the depth of the active layer, nutrient dispersal and availability, and water movement and drainage including hummock formation. The interactions and effects of the bryophyte layer in relation to other plants, animals, and soils cannot be ignored as it often has been in the past. The need for intensive studies which include bryophytes as well as all components of the ecosystem is critical.

References

Arnon, D.I. 1949. Copper enzymes in isolated chloroplasts. Plant Physiol. 24: 1-15.

Baker, J.H. 1972. The rate of production and decomposition of *Chorisodontium aciphyllum* (Hook. f. & Wils.) Broth. Brit. Antarctic Surv. Bull. 27: 123-129.

Barrett, P.E. and J.A. Teeri. 1973. Vascular plants of the Truelove Inlet Region, Devon Island. Arctic 26: 58-67.

Brassard, G.R. 1971. The mosses of northern Ellesmere, Arctic Canada. I. Ecology and phytogeography, with an analysis for the Queen Elizabeth Islands. Bryologist 74: 233-281.

———. and R.E. Beschel. 1968. The vascular flora of Tanquary Fiord, northern Ellesmere Island, N.W.T. Canad. Field-Nat. 82: 103-113.

———. and W.C. Steere. 1968. The mosses of Bathurst Island, N.W.T., Canada. Can. J. Bot. 46: 377-383.

Clarke, G.C.S., S.W. Greene, and D.M. Greene. 1971. Productivity of bryophytes in polar regions. Ann. Bot. 36: 99-108.

Cody, W.J. 1965. Plants of the Mackenzie River delta and Reindeer Grazing Preserve — Canada Department of Agriculture, Plant Research Institute. 56 pp. (offset) Ottawa.

Collins, N. 1973. Productivity of selected bryophyte communities in the Antarctic. pp. 177-183. *In:* L.C. Bliss and F.E. Wielgolaski. (eds.) Primary Production and Production Processes, Tundra Biome. Tundra I.B.P. Biome Steering Committee. Edmonton, Alberta. 256 pp.

Crum, H.A. 1972. The geographic origins of the mosses of North America's eastern deciduous forest. J. Hattori Bot. Lab. 35: 269-298.

———. 1973. Mosses of the Great Lakes Forest. Cont. Univ. Michigan Herb. 10: 1-404.

Fredskild, B. 1966. Contributions to the flora of Peary Land, North Greenland. Medd. om Grønland. 178: 23 pp.

Holmen, K. 1955. Notes on the bryophyte vegetation of Peary Land, North Greenland. Mitt. Thuringischen Bot. Ges. 1: 96-106.

———. 1960. The mosses of Peary Land, North Greenland. Medd. om Grønland 163: 1-96.

———. and G.W. Scotter. 1971. Mosses of the Reindeer Preserve, Northwest Territories, Canada. Lindbergia 1: 34-56.

Longton, R.E. 1967. Vegetation in the maritime Antarctic. pp. 213-235. *In:* J.E. Smith (org). A discussion on the Terrestrial Antarctic Ecosystem. Phil. Trans. Roy. Soc. Sec. B, 252. No. 777.

———. 1970. Growth and productivity in the moss *Polytrichum alpestre* Hoppe in the Antarctic regions. 818-837. *In:* M.W. Holdgate (ed.), Antarctic Ecology, Vol. 2. Academic Press, N.Y. 394 pp.

Murray, B.M. and D.F. Murray. 1973. Checklists to the flora of Alaskan U.S. — IBP Tundra Biome Study Sites. U.S. Tundra Biome Data Report 73-30: 1-104.

Pakarinen, P. and D.H. Vitt. 1973. Primary production of plant communities of the Truelove Lowland, Devon Island, Canada — Moss communities. pp. 37-46 (see Collins 1973).

———. and D.H. Vitt. 1974. The major organic components and caloric contents of high arctic bryophytes. Can. J. Bot. 52: 1151-1161.

Rasdorfer, J.R., H.J. Webster, and D.K. Smith. 1973. Physiological ecology of arctic bryophytes: Bryophyte composition, structure, and production of a wet tundra meadow community. U.S. Tundra Biome Data Report 73-15: 1-19.

Savile, D.B.O. 1972. Arctic adaptations in plants. Plant Res. Inst. Monog. No. 6: 1-81. Ottawa.

Sonesson, M. and S. Johansson. 1974. Bryophyte growth, Stordalen 1973. pp. 17-27. *In:* J.G.K. Flower-Ellis (ed.) International Biological Programme — Swedish Tundra Biome Project Technical Report No. 16. Progress Report 1973.

Steere, W.C. 1939. Bryophyta of Arctic America. II. Species collected by J. Dewey Soper, principally in southern Baffin Island. Amer. Midl. Nat. 21: 355-367.

———. 1951. Bryophyta of Arctic America. IV. The mosses of Cornwallis Island. Bryologist 54: 181-202.

———. 1955. Bryophyta of Arctic America. VI. A collection from Prince Patrick Island. Amer. Midl. Nat. 53: 231-241.

Tamm, C.O. 1953. Growth, yield and nutrition in carpets of a forest moss *(Hylocomium splendens)*. Medd. St. Skogsforsk. Inst. 43: 1-140.

Vitt, D.H. 1974. A key and synopsis of the mosses of Campbell Island, New Zealand. N. Zeal. J. Bot. 12: 185-210.

———. 1975. A key and synopsis of the moss flora of the northern lowlands of Devon Island, N.W.T., Canada. Can. J. Bot. 53: 2158-2197.

Webster, H.J. and A.J. Sharp. 1973. Bryophytic succession on Caribou dung in Arctic Alaska. Amer. Bio. Soc. Bulletin 20: 90.

Studies on the Lichens of Truelove Lowland

D.H.S. Richardson and E.J. Finegan

Introduction

On Truelove Lowland lichens are found on the raised beaches and in the rock outcrop areas where they colonize the granite outcrops and form part of the heath vegetation between the boulders. Three areas, a rock outcrop and two raised beaches, were studied in detail. The Intensive Raised Beach site (Fig. 1 Introduction) was typical of the older granite raised beaches while the younger dolomite-based Phalarope raised beach is representative of the younger coastal beaches where weathering is less advanced, as indicated by the much coarser rock material and more poorly developed soils.

Many reports concerning the growth and production of arctic lichens deal with *Cladonia* species which form mats over the ground in low arctic and sub-arctic habitats. Such work has been done by Salazkin (1937), Andreev (1954), Scotter (1963), Igumnova and Shamurin (1965), Barashkova (1967), Karenlampi (1971), and Kershaw and Rouse (1971 a and b). Very little work has so far been reported on lichens from high arctic locations, where *Alectoria* and *Cetraria* species with a variety of crustose species form an important component of the ground flora.

The objectives of the study were to obtain data on the species composition, standing crop and, if possible, production of the lichens of this high arctic ecosystem.

Vegetation and standing crop

Lichen cover

The point quadrat method (see Goodall 1952) using a linear pin frame was chosen as the best suited for vegetation analysis of the low profile, raised beach vegetation, with its heterogeneous micro- and macro-lichen components. Quadrats were chosen randomly on two axes using random number tables (Green 1968). Five 200 cm^2 quadrats were located in each of the zones designated by Svoboda (this volume) as back transition, back-slope, crest, fore-slope, and fore-transition. The plants touched by each of 200 points per quadrat were recorded and the percent cover of the different plant types calculated.

Table 1. Percent plant cover on two raised beaches, Truelove Lowland.

Zone	*Raised beach*	*Bareground*	*Higher Plants*	*Mosses*	*Crustose Lichens*	*Foliose Lichens*	*Fruticose Lichens*	*Total % Lichen cover*
Fore	Intensive	21.4	30.9	22.3	12.0	1.9	11.5	25.4
Transition	Phalarope	2.9	8.7	66.4	7.4	7.2	7.4	22.0
Fore	Intensive	29.2	12.5	8.2	33.3	5.9	10.9	50.1
Slope	Phalarope	32.7	17.8	3.8	22.2	7.9	15.6	44.7
Crest	Intensive	30.5	8.7	2.0	26.8	8.8	23.2	58.8
	Phalarope	31.5	11.5	3.0	36.2	12.7	5.1	54.0
Back	Intensive	32.0	15.7	6.6	28.2	3.2	14.3	45.7
Slope	Phalarope	19.0	18.4	12.4	23.4	6.5	20.3	50.2
Back	Intensive	*—	—	—	—	—	—	—
Transition	Phalarope	14.3	9.7	51.3	14.8	1.9	8.0	24.7

* There is no back transition zone present, as a small pond and marshy area is located directly at the back of the raised beach.

Total percent lichen cover calculated for the two raised beaches studied is shown in Table 1 where it is seen that there is little difference in the percent lichen cover for the two beaches. Three distinct vegetational zones have been described on the raised beaches by Svoboda (this volume), ie., crest, slope, and transition. The percentage of lichen cover was greatest in the crest zones (*ca.* 50%) and least in the transition zones (*ca.* 25%). Cover data were calculated for the three morphological lichen types (crustose, foliose, and fruticose) and for the moss and higher plant components. On the Phalarope beach (age *ca.* 3,300 years, dolomite-based; King 1969) crustose lichens represent the major part of the plant cover for the crest zone, but on the same zone of the Intensive Raised Beach (age *ca.* 7,500 years, granite/sandstone based) fruticose species also form a major component. In the transition zone, mosses predominate on the extensive ridge, but on the intensive ridge there is equal cover by mosses and higher plants. A more detailed analysis of the cover data, indicating the contribution of individual lichen species, is given in Table 2. The higher cover noted for fruticose lichens on the Intensive Raised Beach crest (23%) when compared with a 5% cover on the extensive ridge, reflects the cover by four species of *Alectoria* which only occur with cover values less than 0.5% on the extensive ridge. The consistently high contribution of *T. vermicularis* to the lichen cover in all zones on both beaches should also be noted. Both *Cladonia pocillum* and *Physcia muscigena* show higher cover values for the extensive ridge. In both cases this reflects a substrate preference, *C. pocillum* being a terricolous calcicole and *P. muscigena* a muscicolous species. A considerably higher cover by mosses was noted for the extensive ridge.

The percent cover of lichens in the rock outcrops is very high (Bliss et al. this volume). The foliose lichens *Umbilicaria lyngei* and *U. arctica* together with crustose species cover 85% of the rock surface of granitic boulders. In the heath areas in between, *Dactylina arctica*, *Cetraria* sp. and *Thamnolia vermicularis* are well developed. The percent cover of the various plant groups in this area is shown in Table 3.

Table 2. Lichen cover from the Intensive and Phalarope Beaches, Devon Island.

Species	*Lichen percentage cover*								
	Intensive Beach				*Phalarope Beach*				
	*FT**	*FS*	*C*	*BS*	*FT*	*FS*	*C*	*BS*	*BT*
Lecideaceae									
Bacidia bagliettona							X**		
Rhizocarpon geographicum		X	3.3	1.9		X	X		
Cladoniaceae									
Cladonia pocillum	X	X			1.6	X		0.6	1.7
Umbilicariaceae									
Umbilicaria arctica			4.4						
Umbilicaria lyngei			1.2						
Lecanoraceae									
Lecanora congesta						X			
Lecanora dispersa		1.2	1.0	X					
Lecanora epibryon	1.5	23.2	3.4	13.0		4.3	2.7	5.2	0.6
Pertusariaceae									
Ochrolechia frigida					X				
Pertusaria dactylina				X					
Parmeliaceae									
Parmelia separata			0.6				X		
Hypogymnia sobobscura		5.6	1.7	1.9		4.7	3.5	4.2	X
Cetraria cucullata	3.6	0.5	X	X	1.1	0.6		0.5	X
Cetraria ericetorum					1.2				1.4
Cetraria islandica			X	1.0	X	1.2	X	3.4	0.5
Cetraria nivalis	0.5	1.2	0.5	2.3	0.5	5.6	X	3.4	0.8
Usneaceae									
Cornicularia divergens					X	X	X	X	
Cornicularia muricata							X		
Alectoria chalybeiformis			0.6	X		X	X		
Alectoria nigricans		0.7	2.2	1.7					
Alectoria ochroleuca		X	2.9	0.9			X		
Alectoria pubescens		X	11.4	X					
Thamnolia vermicularis	5.5	7.6	4.4	5.9	3.1	6.8	3.4	8.1	3.0
Dactylina arctica									X
Dactylina ramulosa						X	X	X	X
Physciaceae									
Buellia disciformis							X		
Physcia muscigena	1.9			X	1.2	0.8	3.1	1.0	0.6
Teloschistaceae									
Caloplaca thallincola	X								
Fulgensia bracteata								X	X
Xanthoria elegans							3.8		
Peltigeraceae									
Placynthium aspratile						X	X	X	
Peltigera aphtosa									X
Peltigera rufescens		X	X			X	X		X
Solorina bispora			X	X		X			
Solorina octospora						X	X		
Fungi imperfecti									
Lepraria membranaceae						X			

*FT—Fore transition zone, FS—Fore-slope zone, C—Crest zone, BS—Back-slope zone, BT—Back transition zone.
**X—Species present with cover <0.5%.

Species composition

One hundred and eighty-two lichen species have been recorded from Truelove Lowland (Richardson unpublished, Barrett and Thomson 1975, see Appendix 3). Some idea of the richness of the flora can be gained from the fact that Hale (1951) recorded 260 species from the whole of Baffin Island. While the

Table 3. Percentage plant cover in the Intensive Rock Outcrop area (Bliss et al. this volume).

	Bare rock or ground	*Higher Plants*	*Moss*	*Lichens*
Rocks	15	—	—	85
Heath areas	9	33	31	26

microclimate of the raised beach crests is somewhat similar to the plateau above the Lowland, lichens are abundant on the crests but sparse on the plateau. Seventy-nine species have been recorded from the crests of the raised beaches and 122 from the rock outcrop areas. These plants are also common in the more stable areas around frost-boils but almost totally absent from snow-bed areas (6 species), from meadows (3 species), and from drainage channels where spring run-off occurs (Barrett 1972, Richardson unpublished).

Specimens of the two species used in physiological experiments showed size differences in various locations. Thus *T. vermicularis* thalli tend to be of increasing length and thickness from the raised beach crest zones (length *ca.* 2.5 cm and less), through slope and transition zones, with the largest thalli being found in the rock outcrop heath areas (length *ca.* 5 cm). In New Zealand specimens of this lichen 15 cm in length have been collected. A similar variation in size was noted for the thalli of *Umbilicaria lyngei*, those from the rock outcrop areas being larger than those from the raised beach crest zones. The beach ridge crest zones are the first areas in the Lowland to be snow-free in the spring and often have little snow cover in winter so that they are climatically severe. The transition zone and the rock outcrop areas, apart from the tops of a few outcrop boulders, represent much milder, later lying snow-bed areas.

The lichen *Thamnolia vermicularis* is very abundant on Truelove Lowland. Since 1937 a morphologically identical but chemically different variant has been recognized which is now named *Thamnolia vermicularis* var. *subuliformis*. This chemical strain can be readily distinguished because the thalli fluoresce yellow under ultra violet light. All 300 specimens examined from Truelove Lowland were var. *subuliformis* (U.V. positive).

The two "chemical" varieties of this species have an interesting distribution. Usually, of a number of specimens of *Thamnolia vermicularis* (sensu lato) collected in one locality, some are U.V.+ and others U.V. −. Sato (1965) found that specimens from South America were all *T. vermicularis* (sensu stricto) while a small percentage of var. *subuliformis* was found in Australia. Samples from more northern collection sites in North America proved to contain an increasing percentage of var. *subuliformis;* indeed, specimens from Greenland all proved to be the U.V. positive strain.

Lichen communities

Methods

The percent cover data were further analyzed using two different association extraction techniques in order to investigate the nature of lichen communities on the Intensive and Phalarope raised beach sites. The first technique was subjective while the second (principal component analysis) was objective. In the first analysis, 20 categories were established which represented percent

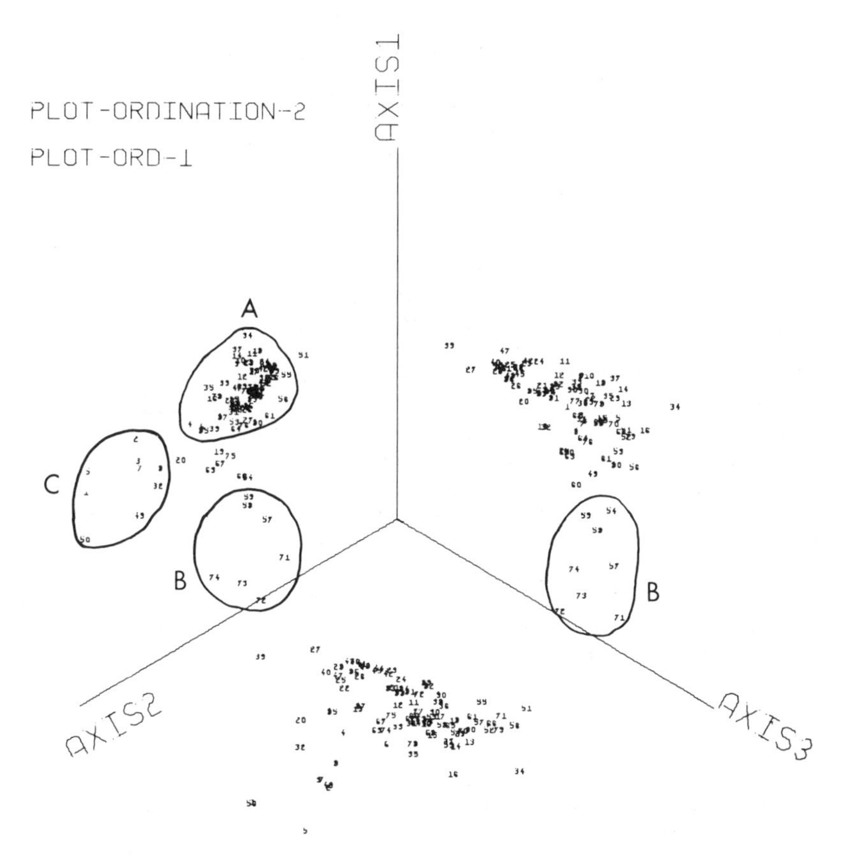

Fig. 1. Ordination of lichen data as a projection on plane surfaces bounded by the first three axes and perspective drawings of the same data together with three associations A.B.C. identified by species overlays.

cover values of 0-4%, 5-9%, 10-14%, 15-19%, etc. These categories were assigned the numerals X, 1, 2, −19 respectively. The percent cover data for individual lichen species, represented in this way, were then sorted, using an association extraction technique of the type described by Poore (1955 a, b, c, and 1956). This sorting of data was carried out by the production of tables in which the quadrat numbers were listed across the top and the lichen species down the left hand side. The cover data representing X−19, were then filled in. Quadrats showing similar species and cover composition were then grouped together, this grouping being repeated to obtain the most satisfactory assemblages of quadrats with a high index of similarity. Poore (1955c) reports that he found it "satisfactory to distinguish most groupings (noda) by constant species, having due regard to other criteria."

For the second analysis, a principal component analysis was carried out. The scale values 1-9 were used to represent percent cover recorded for individual lichen species as well as total moss, total higher plant, and bare ground components. The scale value 1 represented cover values 0-9%, scale value 2 values of 10-19%, etc. No cover values greater than 90% were recorded. The ordination program used, BMD03M (Dixon 1965), carries out a principal component solution and an orthogonal rotation of the factor matrix. The results were graphically ordinated as a projection on three plane surfaces using the program PCAPLOT (Kershaw and Shepard 1972), the factor scores of the ordination being the input. Perspective plane-surface projections produced for the quadrat ordination, and for the species overlays (reflecting the cover of each lichen), are plotted individually for axes 1 and 2, 2 and 3, and 3 and 1. The three plots are then shown together in perspective as the two walls and floor at the corner of a room (see Fig. 1).

Results

Three lichen associations were extracted for the raised beach sites, using the "subjective technique." The *Alectoria pubescens-Rhizocarpon geographicum* association was restricted to the crest zone of the Intensive Raised Beach (Table 4). A markedly different but somewhat less clearly defined association was found to correspond to the crest zone of the Phalarope raised beach. This was termed the *Xanthoria elegans* association. The third lichen association, *Lecanora epibryon-Thamnolia vermicularis* represented the lichen vegetation on the fore- and back-slope zones of the two beaches (Table 5). In the moister, less exposed environment of the slope zones, the increasing presence of fruticose species is noted *(Cetraria nivalis, C. islandica, C. cucullata).*

The *Alectoria pubescens-Rhizocarpon geographicum* association corresponds closely to the *Nardina-Dryado-Alectorictum* association described by Barrett (1972) which is best developed on the crest zones as delimited by Svoboda (this volume). Both he and Barrett (1972) considered

Table 4: Lichen cover for the two lichen communities identified on the crests of two raised beaches on Devon Island.

Alectoria pubescens—Rhizocarpon geographicum Association

Lichen species	*Cover values for each quadrat*							
Alectoria pubescens	3	1	3	1	4	3	2	3
Rhizocarpon geographicum	1	X	1	X	X	X	X	1
Umbilicaria arctica	X	X		X	1	4	1	X
Thamnolia vericularis	X	1	1	1	X	X		
Alectoria ochroleuca	1	X	1		X	X	1	
Hypogymnia subobscura	X	X	1	X			X	X
Lecanora dispersa			X	X	X	X	X	X
Umbilicaria lyngei	X				X	X	X	
Alectoria nigricans	X	X			X			
Lecanora epibryon			2	4				
Alectoria chalybeiformis					X			X
Cetraria nivalis	X							

The Xanthoria elegans Association

Lichen species	*Cover values for each quadrat*								
Xanthoria elegans	1	X	X	X	X	X	1	2	1
Hypogymnia subfusca	1		X	X	X	2	1	X	X
Lecanora epibryon	2	X	X	X	X			X	1
Physcia muscigena	2	X	X	X	X		X		X
Thamnolia vermicularis		1	X	X	1		X	X	1
Alectoria chalybeiformis		1	X		X				1
Physcia sciastra	X		X			X		X	X
Rhizocarpon geographicum			X		X		1		X
Buellia atrata		X	X		X		X		
Parmelia separata						X	1		X
Candelariella arctica								X	X
Cetraria nivalis		X				X			
Alectoria ochroleuca						X			

Note: See text for details.

Table 5. Plant cover for the lichen community identified on the slopes of two raised beaches on Devon Island.

Lichen species	*Lecanora epibryon—Thamnolia vermicularis Association* *Cover values for each quadrat*																																						
Lecanora epibryon	X	2	X	X	1	3	X	1	X	X	2	3	X	X	X	X	1	2	X	6	7	2	1	X	2	2	3	6	3	8	1	3	5	8	2	4	X	1	X
Thamnolia vermicularis	2	2	X	1	2	X	X	1	1	1	X	1	X	X	3	2	1	2	1	1	X	X	2	1	X	X	2	1	X	1	X	X	1	1	2	2	1	1	X
Hypogymnia subobscura	X	X			X	X	X	1	1	2	1		1	3	X	1			X	X	1	X	X	X	X	X	1	1	X	X		X	1	2	2	1			
Cetraria nivalis	X	X	X	X	X	X	1	X	1	2	1	X	1	2	1			X	1	X		1	X	X		X						X	X	X	X	X			X
Alectoria nigricans					1	X	1			X						X		X	X	X	X	X	X		X	X	X								X	X			
Cetraria islandica	1	X	2	X	X	X		1	1	X			X	X	X			X	X						1														
Parmelia fraudans		X		X		1	X	X	X	X			1	X	X								X			1													
Cetraria cucullata	X		X						X				1				X	X					X										X	X					
Physcia muscigena	X	X	X	X	X	X				1							X			X					X	X											1		
Cornicularia divergens	X	X	X								X		X	X	X	X																							
Cladonia pocillum	X	X	X	X																													X					X	1
Alectoria ochroleuca																		X	1	X	X		X			X	X								X				
Dactylina ramulosa	X					X			X					X	X		X																						
Lecanora dispersa																					X						X		X	1	X					X			
Placynthium aspratile					X	X								X		X																							
Alectoria chalybeiformis							1							X					X					X															
Rhizocarpon geographicum																1				X	2															X			
Alectoria pubescens																					X										X	X							

that the environment of this association is the harshest of any plant association on the Lowland, being subjected to considerable snow scouring during late winter. This is probably a factor preventing the development of erect fruticose lichens. The association described above as *Lecanora epibryon-Thamnolia vermicularis* corresponds directly to the *Tetragono-Dryadetum integridfoliae* association reported by Barrett (1972). This association is ecologically equivalent to the type of vegetation which occurs in the transition zone as defined by Svoboda (this volume). This association develops principally on the gently sloping fore-slope of the raised beaches and is a true "snow-patch" (snowmelt, *ca.* 28 June) community as opposed to a "snow-bed" (snowmelt *ca.* 4 July) community, which would be restricted to areas of the very latest lying snow. Scandinavian authors consider *Cetraria nivalis* to be an indicator of chionophobic environments because of its absence from the extreme "snow-bed" habitats, but this species certainly shows greater luxuriance on Devon Island in places where some snow accumulates. This was originally observed by Barrett (1972).

Neal and Kershaw (1973 a and b) and Kershaw and Rouse (1973) describe the plant associations of raised beach systems in north and northwestern Ontario. It is interesting to note the similarity between the Phalarope beach, crest zone association from Truelove Lowland, and the *Dryas-Hedysarum-Xanthoria* association from the highest, most exposed areas of the youngest ridge studied at Cape Henrietta Maria (possibly = 300 years old). In the latter association, *Xanthoria elegans*, *Thamnolia vermicularis*, *Alectoria ochroleuca* and *A. nitidula* were the main lichen species. The lichen associations on the summits of older ridges (up to = 1,000 years old) at Cape Henrietta Maria included many species of *Cetraria* and *Cladonia rangiferina*.

Principal component analysis was also applied to the lichen cover data from Truelove Lowland as subjective methods of association extraction have been criticized. The graphical representation of the quadrat ordination using principal component analysis is shown in Fig. 1 with three associations, A, B, and C identified by means of species overlays. The most compact grouping, A, represents the *Lecanora epibryon-Thamnolia vermicularis* association of the raised beach slopes, but some quadrats from the transition zones of both raised beaches also appear. Most of these additional quadrats are characterized by the presence of *Thamnolia vermicularis* but few have any cover by *Lecanora epibryon*, the other constant species for the association as previously extracted. Groupings B and C respectively correspond closely with the *Alectoria pubescens-Rhizocarpon geographicum* association from the crest of the intensive raised beach and the *Xanthoria elegans* association from the Phalarope crest zone.

The low efficiency of this principal component analysis (percent variation accounted for was 20%) mainly reflected heterogeneity in the sample quadrats, (Greig-Smith 1964), many species being represented in only a small number of the quadrats. Thus, it is seen that of the three associations extracted by the subjective method, two are confirmed by the objective technique. Substantiation of subjectively determined lichen associations using objective methods such as principal component analysis, have recently been published by Kershaw and Rouse (1973), Neal and Kershaw (1973a and b), and Fletcher (1973).

Origin of the lichen flora

While much work remains to be done on the origins of the lichen flora on the Arctic Islands, valuable studies have been completed by Thompson (1972). He recognized that the lichen flora is composed of eight elements. The first element, called the Circumpolar Broad Ranging, is well represented in Truelove Lowland by such species as *Cetraria nivalis*, *Thamnolia vermicularis*, *Cetraria cucullata*, and *Alectoria ochroleuca*. Boreal Forest Outliers, however, seem to be absent from Truelove Lowland, as are the Appalachian Extensions and Arcto-Pacific Elements, though *Cetraria richardsonii* was found on the neighboring Philpotts Island. Typical members of the Amphi-Atlantic Disjuncts which are found in Norway and Greenland are also well represented on Truelove Lowland by *Umbilicaria lyngei*, *Vestergrenopsis isidiata* and *Placopsis gelida*. The hitherto overlooked species *Neuropogon sulphureus* was searched for but not found within the Lowland. None of the American Arctic Endemics or Western states-Eastern Arctic Disjuncts were discovered, but the Great Plains and Arctic Element was present, examples being *Buellia papillata*, *Caloplaca stillicidiorum*, *C. tirolensis* and *Fulgensia bracteata*.

Standing crop

Methods

The standing crop of lichens was estimated by harvesting, sorting, and dry-weighing (24 hr at 80°C) half of each quadrat examined for cover measurements.

Results

The standing crop on the intensive outcrop areas was two to three times that found on the crest of the raised beach sites (Table 6). Bliss et al. (this volume) estimated about 95 g m^{-2} of lichen in the lush vascular plant sites and 24 g m^{-2} on more typical sites within the rock outcrops while Svoboda (this volume) found approximately 70 g m^{-2} on the raised beaches. The differences and similarities between these data and those in Table 6 reflect the differences in lichen biomass in contrasting microsites. Biomass is high between rocks, on back-slopes of raised beaches where there is 20 to 50 cm of snow cover, and in heath communities where there is higher soil moisture in summer, often adjacent to rocks. Lichens comprise about 10% of the total biomass of plant

Table 6. Standing crop values (g d wt m^{-2}) for lichens on raised beaches and rock outcrops, Truelove Lowland (Mean ± SE).

Zone	*Intensive Raised Beach*	*Phalarope Beach*	*Intensive Rock Outcrop area*
Fore-transition	38.20±25.2	16.92± 9.2	—
Fore-slope	33.95±11.6	47.03±26.3	—
Crest	58.13±13.5	31.02± 0.6	—
Back-slope	26.39± 9.5	96.88±18.7	—
Back-transition	—	12.94± 5.5	—
Granite outcrops	—	—	106.07±17.8
Inter-outcrops area	—	—	170.70±34.1

material on the raised beaches. Wein and Bliss (1974) estimated standing crops of 99, 25, and 14 g m^{-2} respectively for cottongrass tundra sites at Dempster, Eagle Creek, and Elliot in the Yukon and Alaska. Although lichens are a very conspicuous feature of the raised beaches and rock outcrops on Devon Island, the standing crop values are much lower than those reported from some low arctic situations where up to 400 g m^{-2} is not uncommon (Bunnell et al. 1973).

Physiology of Arctic lichens

Photosynthetic ^{14}C fixation

Methods

Experiments were carried out in 1971 to examine photosynthetic ^{14}C fixation under conditions of ambient light and temperature. The results showed that the fixation rates (Richardson and Finegan 1972) correlated partially with measured environmental parameters. Two types of study were subsequently undertaken to investigate further the assimilation capacity of lichens. The number of algal cells per unit area of thallus were measured (see following section) and samples of *Thamnolia vermicularis* and *Umbilicaria lyngei* were exposed to standard photosynthetic conditions in the laboratory on Devon Island.

In order to measure ^{14}C fixation, lichen material was washed immediately after collection. Fresh weight samples (100 mg) of *T. vermicularis* and replicates of fifteen 5 mm diameter discs of *U. lyngei* thalli were prepared. Sampling procedures for lichen material are discussed by Richardson (1971). The moist lichen samples were added to glass Petri dishes containing damp filter paper and left in the light overnight. This treatment allows any "resaturation respiration" (Smith and Molesworth 1973) to take place prior to the examination of carbon fixation.

Samples were then immersed in a 3 or 5 ml aqueous solution of carrier free sodium-^{14}C-bicarbonate (10 μ Ci) for 6 hr. This was carried out in 5×2.5 cm glass vials, which were closed with glass coverslips, held in position by silicone grease. For the dark controls, the experimental vials were completely wrapped in aluminium foil. The samples were placed in an illuminated water bath at 15°C with light intensity of 2.9, 0.0, and 1.5 μ w $cm^{-2}nm^{-1}$ in the blue, far red, and red region respectively.

After exposure to the sodium-^{14}C-bicarbonate, the lichen samples were removed from the vials, washed, blotted, and killed by adding to 5 ml hot 80% ethanol for five minutes. Two further extractions with hot 80% ethanol were made and the three extracts combined to form the soluble fraction. Aliquots (0.2 ml) of the soluble fraction were pipetted onto etched planchettes and dried. The amount of radioactivity on each planchette was determined using a Nuclear Chicago, thin end window, gas flow, planchette counter. The insoluble material which remained after extraction formed the insoluble fraction, which was dried, attached to planchettes using double-sided "scotch tape," and counted as above.

Results

The results (Table 7) indicate that the fixation rates of the lichens did not vary significantly from 25 July to 15 August.

Some interesting features are evident from Table 7. *Thamnolia*

Table 7. ^{14}C Fixation (Millions of D p $m^{-1}g^{-1}$d wt) of lichens in two habitats, Truelove Lowland.

Date	Place		*Umbilicaria lyngei* Total fixation	% Insol	*Thamnolia vermicularis* Total fixation	% Insol
26/6/72	Beach Ridge	Light	32.7±6.1	2.5	66.9±4.0	2.6
		Dark	2.7	5.5	11.4	7.1
		% Dark fixation	—	7.9	—	17.1
8/7/72	Beach Ridge	Light	—	—	73.6±5.2	3.5
		Dark	—	—	10.0	6.7
		% Dark fixation	—	—	—	13.6
17/7/72	Beach Ridge	Light	28.3±2.7	4.2	67.9±1.8	5.6
		Dark	—	—	8.3	8.9
		% Dark fixation	—	—	—	12.3
24/7/72	Beach Ridge	Light	—	—	71.2±3.3	5.7
		Dark	—	—	—	—
		% Dark fixation	—	—	—	—
26/6/72	Outcrop	Light	22.4±5.4	4.3	48.8±1.1	2.9
		Dark	3.2	6.6	11.2	7.9
		% Dark fixation	—	8.5	—	22.6
8/7/72	Outcrop	Light	33.7±1.9	3.8	70.7±0.3	2.6
		Dark	2.4	5.7	—	—
		% Dark fixation	—	7.7	—	—
17/7/72	Outcrop	Light	25.6±1.7	1.4	83.3±9.5	2.5
		Dark	2.0	3.3	11.0	6.7
		% Dark fixation	—	7.9	—	13.1
24/7/72	Outcrop	Light	34.6±5.6	5.2	77.9±5.0	3.1
		Dark	3.7	8.0	5.9	12.8
		% Dark fixation	—	10.5	—	7.6

Light fixation values are the mean of three samples. A single dark control was done on each run.

Note: On Devon Island, *Thamnolia vermicularis* when 100% saturated contained 1.41 g of water g^{-1} d wt of thallus while the corresponding figure for *Umbilicaria lyngei* was 1.44 g water g^{-1}d wt.

vermicularis fixed 3 times as much ^{14}C as *Umbilicaria lyngei* g^{-1} d wt under identical photosynthetic conditions. *Thamnolia vermicularis* also showed a generally higher rate of dark fixation averaging about 14% of fixation in the light. The latter feature might be due to death of algae in the lower parts of the podetia even though obviously dead parts of the thalli were removed during sampling. Studies on other lichens have shown dark fixation rates are usually around 5-10% of those in the light, eg., Smith and Drew (1965) found that these rates varied from 4.8-9.4% in *Peltigera polydactyla*.

Algal cell counts Harris (1971), investigating the temperate corticolous lichen *Parmelia caperata*, showed that the number of algae varied in thalli collected in different habitats and at different times of the year. Thalli growing on the lower parts of the tree had a lower net assimilation rate (under laboratory conditions) than specimens from the top of the tree. Correspondingly, thalli at the top contained 34.2×10^5 algal cells whereas those at the base had only 24.0×10^5 cells cm^{-2}. Algal numbers also increased by 1.2-1.7 times between January and September in species collected from the same habitat. Plummer and Gray (1972) also found

Table 8. Number of algae per gram dry weight of two lichen species, Truelove Lowland.

Date	*Thamnolia vermicularis*	*Umbilicaria lyngei*
30 June	97.1 $\times 10^6$*	69.4 $\times 10^6$*
8 July	165.0 $\times 10^6$	43.5 $\times 10^6$*
17 July	113.8 $\times 10^6$	111.8 $\times 10^6$
24 July	126.9 $\times 10^6$	122.2 $\times 10^6$
31 July	105.6 $\times 10^6$	136.8 $\times 10^6$
7 August	114.4 $\times 10^6$	110.4 $\times 10^6$
14 August	110.0 $\times 10^6$	118.1 $\times 10^6$

*Sample collected from beneath snow.

that algal numbers were higher in May than February in three species of *Cladonia*]

In order to see whether such changes occurred in arctic lichens and whether seasonal differences in assimilation rates could be attributed to this cause, lichen samples were collected and algal cell counts made on a haemocytometer after macerating in 2 ml 5% w/v chromic acid or 2 ml 'Decon 75' undiluted concentrate for 24 hr at 37°C. Although it proved difficult to prepare sufficiently separated suspensions of lichen algae, the samples of *T. vermicularis* collected under snow on 30 June contained a mean of 97.0×10^6 algae g^{-1} d wt of thallus, whereas specimens collected later had a mean value of 122.6×10^6 algae g^{-1} d wt. Results from *U. lyngei*, though more variable, showed that this species contained *ca.* 60×10^6 algae under snow and 119.9×10^6 algae g^{-1} d wt later in the season (Table 8). This provides some evidence that the algae in these lichens multiply at the time of snowmelt and then remain numerically constant during the summer. This is reflected by the results of the ^{14}C fixation experiments. However, as the number of algae per g^{-1} d wt of lichen is similar in the two species, the reason for the greater net fixation in *Thamnolia vermicularis* is not clear.

Calorific values and sugar content In 1971 samples of *Thamnolia vermicularis* were collected at weekly intervals during the summer and subjected to calorific analysis. Those from the fore-slope of the raised beach generally had a slightly lower mean calorific value (3616 cal g^{-1}d wt) than samples collected from the crest or back-slope (4011 and 4108 cal g^{-1} d wt respectively), but this difference was not significant at the 5% level. Samples collected on two occasions early in the season from under snow did show a lower calorific value (mean 3648 cal g^{-1} d wt) than samples collected in snow-free areas and this was significant at the 5% level. In the light of these results, a more extensive sampling was undertaken in 1972, the first samples being collected before snowmelt. *Thamnolia vermicularis* and *Umbilicaria lyngei* were collected from the fore-slope of the intensive raised beach and from the rock outcrop area. The samples were dried and later subjected to bomb calorimetry. Table 9 shows the results. Low calorific values were recorded for samples of *Thamnolia vermicularis* collected beneath snow but the difference is not significant. In *Umbilicaria lyngei* neither time of year nor snow cover affect the calorific value of the samples. The collection site seems to make little difference to the calorific value of the lichen in spite of the thalli being, on the average, twice as large in the outcrop area.

Table 9. Calorific values for two lichens collected at different times in 1972 (mean value ± SE).

Collection date	Beach Ridge samples *T. vermicularis*	Beach Ridge samples *U. lyngei*	Rock Outcrop samples *T. vermicularis*	Rock Outcrop samples *U. lyngei*
30 June	*4569	*4158	*4007	*4152
8 July	*3982	4140±221	*4018	*3988± 10
17 July	4254	3969± 26	4231±37	4112±316
24 July	4124	4171± 51	4229±16	3997± 77
31 July	4367± 13	3915±189	4344±52	4272
7 August	4292±212	4040± 98	3980±12	4092± 54
14 August	4161±130	4429±241	4217±62	4081± 61

*Samples collected from beneath snow.

Footnote:
Incombustible residues of lichen samples are given below so that calorific value may be estimated on an ash free basis if desired.

Beach Ridge samples
T. vermicularis 0.0861±0.0082 g g^{-1} oven dry weight of lichen
U. lyngei 0.0693±0.0220 g g^{-1} oven dry weight of lichen

Rock Outcrop samples
T. vermicularis 0.0965±0.0103 g g^{-1} oven dry weight of lichen
U. lyngei 0.0478±0.0040 g g^{-1} oven dry weight of lichen

The calorific values of lichens from Devon Island are slightly lower than those recorded from other tundras. Pegau (1968), working on Alaska lichens, quoted values of 4630-4740 cal g^{-1} d wt for *Cetraria cucullata* and 4077 cal g^{-1} d wt for *C. nivalis*. Bliss (1962) obtained values of 4397 cal g^{-1} d wt for *Cetraria cucullata* from the alpine tundra of Mount Washington, New Hampshire. Corresponding values for these species collected in Truelove Lowland are 3825 cal g^{-1} d wt for *C. cucullata* and 3478 cal g^{-1} d wt for *C. nivalis*.

Samples of the lichen *T. vermicularis*, analysed for soluble sugar using gas liquid chromatography, were found to contain 2.0% sugars g^{-1} d wt. The sugar alcohols, mannitol, ribitol, and arabitol were predominant but small amounts of glucose, fructose, and sucrose were also found. No marked variation in sugar content was found over the snow-free period and this is in contrast to findings in temperate regions where for example *Umbilicaria pustulata* contained 4.2% sugar alcohol in June and only 2.3% in December. In *Peltigera polydactyla* the highest values (10%) were found in late summer (see Lewis and Smith 1967). Since the average calorific value of a range of lichens on Truelove Lowland was 4,100 cal g^{-1} d wt, the total energy trapped in these plants which could be utilized by vertebrate and invertebrate herbivores on the Lowland is estimated as $3{,}600 \times 10^6$ Kcal, from a calculated standing crop of around 880,000 Kg d wt of lichen. Thus, although lichens form only a small part of the total plant biomass and have a low production, they do provide sufficient energy base for populations of mites, Collembola, and perhaps other invertebrates.

Lichen production

Carbon turnover

Single thalli of *U. lyngei*, each attached to a single pebble, were removed from a beach ridge crest zone. These thalli were exposed to 50 or 100 μ Ci of sodium-^{14}C-bicarbonate in aqueous solution for 24 hr in the light. The thalli were then washed and the undersides of the attached pebbles marked before return to the beach ridge. The replacement area was fenced to exclude muskox and lemmings. Immediately after treatment with the sodium-^{14}C-bicarbonate solution, a sample of the thalli was killed with hot 80% ethanol. The remaining thalli were harvested and killed after a period of a year. Ethanol soluble and insoluble fractions were prepared and analysed using the methods given earlier.

The data (Table 10) show that 32% of the radioactivity remained after the end of one year. Twenty-five percent of the ^{14}C fixed originally remained in the ethanol soluble fraction. Thus loss of ^{14}C photosynthate due to respiration by the thallus was remarkably small. In addition much of the fixed ^{14}C remains in soluble storage products (e.g., sugar alcohol) over long periods so that the rate of conversion to insoluble polysaccharides and proteins is extremely slow. This is an indication of the slow metabolic rate of lichens in the arctic environment and is probably reflected in the annual growth of these plants which was not measurable in *Umbilicaria lyngei* over the two summers of field work.

Production

Difficulty was experienced in obtaining sufficiently accurate and reproduceable results when using the IRGA to determine lichen carbon dioxide uptake and release under field conditions using the apparatus described by Mayo et al. (1973). This seems to have been due to the low net fixation rates of lichens which necessitated low flow rates and high amplifier gain, making the apparatus particularly sensitive to changes in pressure and temperature. In a single experimental run, moist thalli of *Thamnolia vermicularis* (80% saturated) showed a maximum net assimilation rate over a ten minute measurement period, of 1.97 mg CO^2 g^{-1} d wt hr^{-1}. The thallus temperature was 7.5°C and the incident light, 0.52 ly min^{-1}.

A measure of production can also be derived from the ^{14}C work by knowing the specific activity of the $NaH^{14}CO_3$ solutions used. Under the admittedly atypical conditions of water saturation and artificial light, the lichens showed the following fixation rates over a 6 hr period. *Thamnolia vermicularis* showed a net fixation rate of 0.137 mg CO_2 g^{-1} d wt hr^{-1} while the figure for *Umbilicaria lyngei* was 0.051 mg CO_2 g^{-1} hr^{-1} over a 6 hr period. These figures were low because: (a) most species of lichen when fully saturated show less than optimal photosynthetic rates; and (b) the light intensity was only 10% of full daylight.

Measurements of lichen production vary considerably but maximum values are around 2.0 mg CO_2 g^{-1} d wt hr^{-1} (see Rundel 1972, Lechowitz et al. 1974). Bliss and Hadley (1964) recorded much lower values of 0.30-0.38 mg CO_2 g^{-1} d wt hr^{-1} in *Cladonia rangiferina*, *Cetraria nivalis*, and *C. cucullata* from alpine habitats. However not all lichens from such habitats show low assimilation rates. For example, *Cetraria sepincola* was able to assimilate

Table 10. *Umbilicaria lyngei* thalli exposed to sodium-^{14}C-bicarbonate and harvested immediately or after one year in the field (counts min^{-1} 100,000's ± SE).

Time of harvest	*Soluble fraction*	*Insoluble fraction*	*Total*	*% Insoluble*
0	447±88	2.8±0.3	450±88	1%
1 Year	113±22	32± 5	145±26	24%

close to 2.0 mg CO_2 g^{-1} d wt hr^{-1} (Wirth and Turk 1973). Thus the little experimental data collected on Devon Island do not allow a reliable estimate of production of the lichens in this area. Production could only therefore be assessed by dividing the standing crop by an estimated growth rate to give an annual production rate accurate within an order of magnitude. This was done and resulted in an estimated net production rate for the rock outcrops, raised beach crests and transition zones of 1.1, 2.5 and 1.5 g m^{-2} $year^{-1}$ respectively.

Conclusions

The diversity of lichens on Truelove Lowland, Devon Island is very rich, 175 species being recorded. This is due both to the range of habitats and to the granitic and dolomitic rocks. Thus, both lichen species with a preference for acidic rock and those with a preference for basic rocks are able to grow in this Lowland. Lichens, particularly crustose, are most conspicuous on the raised beaches and rock outcrops where they develop the greatest luxuriance.

Measurements of various physiological parameters indicate that the foliose and fruticose lichens on Devon Island are not greatly different from those of more temperate regions. However these lichens only have 6-8 weeks during the year when temperatures are equitable and snow cover absent. Even then the thalli are dry and inactive for a considerable part of this time because of the low rainfall and lack of dew. Thus the plants are only physiologically active for periods following snowmelt, rain, or briefly after the water content of the thallus has increased by absorption of water vapour from the humid air from *ca.* 0600 to 1000 hr. These factors in combination result in low annual growth rates and low standing crop.

Muskox have been observed to graze casually on the foliose and fruticose lichens in both of the above areas during the winter and these plants may possibly form a food source for mites and Collembola as has been recorded for other sites (Richardson 1975). However, the mycelium of soil fungi is probably the primary food source for these invertebrates.

The slow growth rate, low biomass, and absence of the large mats of *Cladonia* and *Cetraria* which typify low-arctic and sub-arctic habitats, indicate that lichens, though academically most interesting, do not play an important part in the energy flow through the ecosystem on Devon Island (see Whitfield this volume). However, the role of these plants as an emergency food source for the above animals, in the stabilizing of beach ridge soils, and in the water relations of the raised beaches (Addison this volume) dictates that these plants receive some attention in studies such as those on Devon Island which aim at a greater understanding of high arctic ecosystems.

Acknowledgments

We wish to thank Dr. Pattie for making the calorific value determinations. The help of Dr. K. Kershaw with the computer analysis is gratefully acknowledged and finally we thank Dr. J.W. Thomson for checking the identification of lichens collected on Devon Island.

References

Andreev, V.N. 1954. Growth of forage lichens and methods of improving it. Geobotanika 9: 11-74.

Barashkova, E.A. 1967. Photosynthesis in fruticose lichens *Cladonia alpestris* (L). Rabh. and *C. rangiferina* (L.) Web. in the Taimyr Peninsula. International Tundra Biome Trans. 4. 1971.

Barrett, P.E. 1972. Phytogeocoenoses of a Coastal Lowland Ecosystem, Devon Island, N.W.T. Ph.D. Thesis, Univ. British Columbia, Vancouver. 292 pp.

———. and J.W. Thomson. 1975. Lichens from a high arctic coastal lowland, Devon Island, N.W.T. Bryologist 78:160-167.

Bliss, L.C. 1962. Caloric and lipid content in alpine tundra plants. Ecology. 43: 753-757.

———. and E.B. Hadley. 1964. Photosynthesis and respiration of alpine lichens. Amer. J. Bot. 51: 870-874.

Bunnell, F.L., D.C. Dauphine, R. Hillborn, D.R. Miller, and F.H. Miller. 1973. Preliminary report on computer simulation of barren ground caribou management. Proc. First Int. Reindeer/Caribou Symposium, Fairbanks, Alaska.

Dixon, W.J. (ed.). 1965. B.M.D. biomedical computer programs. *In:* Automatic computation. No. 2. Univ. California Press, Berkeley.

Fletcher, A. 1973. The ecology of maritime lichens on some rocky shores of Anglesey. Lichenologist 5: 401-422.

Green, J.W. 1968. Tables of Random Permutations. Australia Forestry and Timber Bureau Bull. 44: 161 pp.

Goodall, D.W. 1952. Some considerations in the use of point quadrats for the analysis of vegetation. Aust. J. Bot., 5: 1-41.

Greig-Smith, P. 1964. Quantitative Plant Ecology. Second edn., Butterworths London. 256 pp.

Hale, M.E. 1951. Lichens from Baffin Island. Amer. Midl. Nat. 51: 232-263.

Harris, G.P. 1971. The ecology of corticolous lichens 11. The relationship between physiology and the environment. J. Ecol. 59: 441-452.

Igumnova, Z.S. and V.F. Shamurin. 1965. Water regime of lichens and mosses in tundra communities. Bot. Zhur. 50: 702-709.

Karenlampi, L. 1971. On methods for measuring and calculating the energy flow through lichens. Rep. Kevo Subarctic Res. Stat. 7: 40-46.

Kershaw, K.A. and W.R. Rouse. 1971a. Studies on lichen-dominated systems, I. The water relationships of *Cladonia alpestris* in spruce-lichen woodland, in Northern Ontario. Can. J. Bot. 49: 1389-1400.

———. and W.R. Rouse. 1971b. Studies on lichen-dominated systems, II. The growth pattern of *Cladonia alpestris* and *Cladonia rangiferina*. Can. J. Bot. 49: 1401-1410.

———. and W.R. Rouse. 1973. Studies on lichen-dominated systems, V. A primary survey of a raised-beach system in northwestern Ontario. Can. J. Bot. 51: 1285-1307.

———. and R.K. Shepard. 1972. Computer display graphics for principal-component analysis and vegetation ordination studies. Can. J. Bot. 50: 2239-2250.
King, R.H. 1969. Periglaciation on Devon Island, N.W.T., Ph.D. Thesis, Univ. Saskatoon. 470 pp.
Lechowitz, M.J., W.P. Jordan and M.S. Adams. 1974. Ecology of *Cladonia* III. Comparison of *C. caroliniana*, endemic to southeastern North America, with three northern *Cladonia* species. Can. J. Bot. 52: 565-573.
Lewis, D.H. and D.C. Smith. 1967. Sugar alcohols (polyols) in fungi and green plants. I. Distribution, physiology and metabolism. New Phytol. 66: 143-184.
Mayo, J.M., D.G. Despain and E.M. van Zinderen Bakker. 1973. CO_2 assimilation in *Dryas integrifolia* on Devon Island, Northwest Territories. Can. J. Bot. 51: 581-588.
Neal, M.W. and K.A. Kershaw. 1973a. Studies on lichen-dominated systems. III. Phytosociology of a raised-beach system near Cape Henrietta Maria northern Ontario. Can. J. Bot. 51: 1115-1125.
———. and K.A. Kershaw. 1973b. Studies on lichen-dominated systems. IV. The objective analysis of Cape Henrietta Maria raised-beach systems. Can. J. Bot. 51: 1177-1190.
Pegau, R.E. 1968. Reindeer Range Appraisal in Alaska. M.Sc. Thesis. Univ. Alaska. 130 pp.
Plummer, G.L. and B.D. Gray. 1972. Numerical densities of algal cells and growth in the lichen genus *Cladonia*. Amer. Midl. Nat. 87: 355-365.
Poore, M.E.D. 1955a. The use of phytosociological methods in ecological investigations. I. The Braun-Blanquet system. J. Ecol. 43: 226-244.
———. 1955b. The use of phytosociological methods in ecological investigations. II. Practical issues involved in an attempt to apply the Braun-Blanquet system. J. Ecol. 43: 245-269.
———. 1955c. The use of phytosociological methods in ecological investigations. III. Practical applications. J. Ecol. 43: 606-651.
———. 1956. The use of phytosociological methods in ecological investigations. IV. General discussion of phytosociological problems. J. Ecol. 44: 28-50.
Richardson, D.H.S. 1971. Lichens. pp. 267-293 *In:* Methods in Microbiology. Volume 4. C. Booth (ed.). Academic Press, New York. 795 pp.
———. 1975. The Vanishing Lichens. David and Charles. Newton Abbot, London. 231 pp.
———. and E.J. Finegan. 1972. Lichen productivity study, Devon Island. 197-213 pp. *In:* Devon Island, I.B.P. Project, High Arctic Ecosystem Project Report 1970 to 1971. L.C. Bliss (ed.). Dept of Botany, Univ. Alberta, Edmonton. 413 pp.
Rundel, P.W. 1972. CO_2 exchange in ecological races of *Cladonia subtenuis*. Photosynthetica 6: 13-17.
Salazkin, A.S. 1937. Rate of growth of forage lichens. Sov. Olenevod. No. 11.
Sato, M. 1965. Mixture ratio of the lichen genus *Thamnolia*. Bryologist 68: 320-324.
Scotter, G.W. 1963. Growth rates of *Cladonia alpestris*, *C. mitis*, and *C. rangiferina* in the Talston River region, N.W.T. Can. J. Bot. 41: 1199-1202.
———. and S. Molesworth. 1973. Lichen Physiology XIII. Effects of rewetting dry lichens. New Phytol. 72: 525-533.

Smith, D.C. and E.A. Drew. 1965. Studies on the physiology of lichens. V. Translocation from the algal layer to the medulla in *Peltigera polydactyla*. New Phytol. 64: 195-200.
Thomson, J.W. 1972. Distribution patterns of American arctic lichens. Can. J. Bot. 50: 1135-1156.
Wein, R.W. and L.C. Bliss. 1974. Primary production in arctic cotton grass tussock tundra communities. Arctic Alp. Res. 6: 261-274.
Wirth, V. and R. Turk. 1973. Uber Standort, Verbreitung und soziologie der borealen flechten *Cetraria sepincola* und *Parmelia olivacea*. Veroff. Landesst. N.U.L. Bd. Wttb. 41. 88-117.

Primary production processes

Gas exchange studies of *Carex* and *Dryas*, Truelove Lowland

J.M. Mayo, A.P. Hartgerink, Don G. Despain, Robert G. Thompson, Eduard M. van Zinderen Bakker, jr., Sherman D. Nelson

Introduction

Field studies

Field studies of photosynthesis were undertaken during the 1971 and 1972 field seasons with the following objectives in mind: (1) to determine the photosynthetic activity of some of the major plants under the environmental conditions prevailing in the islands of the High Arctic, particularly with respect to continuous light and low temperatures. These studies centered upon *Dryas integrifolia*, a major component of raised beach ridge communities, and *Carex stans* a dominant in meadows; (2) to determine seasonal changes in photosynthetic activity as the plants initiated growth, flowered, and formed seed; (3) to provide CO_2 assimilation data for use in modelling of production processes by other participants in the project (see Whitfield this volume); (4) to compare photosynthetic activity of Canadian high arctic species with information from other polar regions; and (5) to gain sufficient information about the conditions under which the plants are photosynthetically active so that studies could be carried out in the Controlled Environment Facility in Edmonton, Alberta, under near natural conditions.

Controlled environment studies of Dryas

Studies were undertaken in the Controlled Environment Facility at Edmonton for the following reasons: (1) to check field studies under less severe working conditions; (2) to determine the photosynthetic response of single leaves, by age class. Field studies were of the whole plant and it was of interest as well as necessary for the production process model to understand the behavior of single leaves of different ages; and (3) to determine, if possible, leaf and root system resistance to water flux.

Methods

Field studies

Carbon dioxide assimilation was measured with a Mine Safety LIRA infrared gas analyzer (IRGA). Since the study sites were isolated from the main camp, it was necessary to use a portable generator for instrumentation power. A 2.5-KW Onan generator equipped with voltage and frequency meters was used. Frequency fluctuations cause serious problems with infrared gas analyzers (Šesták et al. 1971), thus generator frequency was observed and the whole system trimmed by means of a dummy load. The IRGA amplifier and chopper motor were rewired so that they could be powered separately by an inverter which produced stable 60-Hz power. The combination of an inverter and trimmed system kept the frequency sufficiently stable for good IRGA operation. The IRGA was used as a differential analyzer spanned 100 ppm using standard gases whose CO_2 content had been previously determined by means of gases mixed with Wöstoff pumps as described by Bate et al. (1969). The IRGA and associated equipment were housed in a tent on the study site. Temperature sensitive equipment, including the IRGA, was insulated with 5 cm thick styrofoam to maintain stable temperatures.

An open gas-exchange system was used in which the incoming air stream was divided, with one portion going through the enclosure containing a plant and the other going directly to the IRGA. Enclosures were hemispherial Plexiglas domes, 30 cm in diameter, attached to threaded steel rings driven 12 cm into the ground around the plant at least 24 hr before the measurement period. Enclosure design was based on that of Redmann (1971). This method was used because the compact growth form of *Dryas* made leaf cuvettes impractical with the relatively insensitive M.S.A. analyzer. This method necessitated using a blank dome (i.e., the aboveground portion of the plant removed) to correct for soil CO_2 flux. The amount of soil CO_2 flux was added to the amount of CO_2 removed from the air stream to arrive at the total CO_2 fixed. The magnitude of soil CO_2 flux varied through the 24 hr study period, no doubt because of soil surface temperature variation and lichen and soil algae activity. Smoke tests with simulated *Dryas* plants indicated good mixing of air within the domes. Air flow through the domes was varied so that the CO_2 concentration change was not greater than 15 ppm (*ca.* 10 to 11 l min^{-1}). The domes were air conditioned by means of a Peltier block in the upstream air line. Proportional temperature control was achieved with a controller based on a design by Dobkin (1963).

Use of the domes in the meadow necessitated accounting for moss CO_2 fixation, hence three were used: (1) one dome with all vegetation removed at least 24 hr prior to the measurement period, to evaluate soil CO_2 flux; (2) one with only the *Carex* removed; and (3) one with the complete *Carex-moss* complex. In this way, soil CO_2 flux, moss CO_2 fixation, and *Carex* CO_2 fixation could be measured. Occasionally single leaf cuvettes were used, but these proved impractical because they were flat and at low sun angles reflected too much radiation; however, maximum and minimum rates are reported for times when the cuvettes were at about right angles to the incoming radiation.

The domes were connected to the IRGA by means of polyvinyl chloride tubing. Air stream switching was done with Johnson air valves. Air lines were never longer than 9 m, which resulted in a very responsive system at the air flow rates used.

Copper-constantan thermocouples (.003 in) were used to measure leaf, air, and soil temperatures. Leaf thermocouples were held in place with clips. It was intended that temperatures be taken nearly continuously; however, in 1971 the recorder for this purpose was damaged in shipment, necessitating the use of a voltmeter and cold junction compensator for spot measurements. In 1972 the recorder functioned well, and many more temperature data were obtained.

Radiation was measured periodically (2 to 6 hr intervals) during each 24 hr observation period with an Instrumentation Specialties Company (ISCO) spectroradiometer. ISCO data were taken during 1971 and 1972. In 1971, no other radiation measurements were made, but in 1972 a Kipp and Zonen solarimeter was placed near the domes because it was apparent from the 1971 data, that correlation of net assimilation with the base camp Eppley pyranometer was not always possible, particularly on days with intermittent cloud cover.

Laboratory studies

Characterization of the photosynthetic response of a plant to individual environmental parameters is useful to promote an understanding of how that species responds as it does to the sum of interrelated environmental aspects to which it is subjected in its natural habitat. The Controlled Environment Growth Chamber Facilities in Edmonton were therefore used to carry out such an investigation on single attached leaves of *Dryas integrifolia* (Hartgerink 1975). This information was then used to elucidate the photosynthetic response as measured in the field, and to provide input for the *Dryas* photosynthetic model.

Dryas integrifolia plants were transplanted from raised beaches on Devon Island to pots maintained in a Controlled Environment Chamber in Edmonton. These plants were given 6 to 8 weeks of summer conditions at 100 to 200 $\mu E\ m^{-2}\ sec^{-1}$ PAR (photosynthetically active radiation) and temperatures varying diurnally between −1°C to 30°C. Winter dormancy lasted 3 to 6 months in total darkness at −2°C to −6°C.

Net assimilation of CO_2 was measured with UNOR II infrared gas analyzer in an open system similar to that used in the field, but without a blank. The cuvette consisted of a Plexiglas cube into which a single attached leaf was introduced through a slit and sealed around its petiole with Terostat. A thermocouple was used to monitor leaf temperature. Each experiment consisted of gas exchange measurements at a series of artificially varied leaf temperatures or light intensities (3 to 4 hr at each step) in descending order. Unless otherwise stated, the leaves used were mature and showed no signs of senescence.

Photosynthetically active radiation (PAR, 400-700 nm) was measured with a quantum sensor in $\mu E\ m^{-2}\ sec^{-1}$ and illumination was measured with a photometer (lux), both supplied by Lambda Instruments (Lincoln, Neb.). Conversions to $W\ m^{-2}$, $cal\ cm^{-2}\ min^{-1}$ PAR, and foot-candles were made from these measurements and are therefore approximate.

Results

Field studies — Carex stans

Fig. 1 shows a 24 hr period during 1972 which typifies the response of *Carex* to the environment. Low values for net assimilation were observed during the evening hours and as Fig. 1 indicates, may be negative. Positive values were observed if night time radiation remained above *ca.* 0.01 ly min^{-1}. Midday lag was common, although in the case of Fig. 1 this was delayed because of cloudiness until about 1500 hr when radiation increased but net assimilation decreased.

A light response curve was constructed from 1971 and 1972 data, using hourly means of net assimilation during clear periods and only when net assimilation was ascending with increasing radiation (Fig. 2). While this figure must be considered somewhat subjective; it illustrates the very low light compensation point and quite respectable CO_2 fixation rates by *Carex stans* on Devon Island. The estimate 1/2 maximum net assimilation rate was 7.69 mg g^{-1} hr^{-1} using the Lineweaver-Burk method of plotting the light response data.

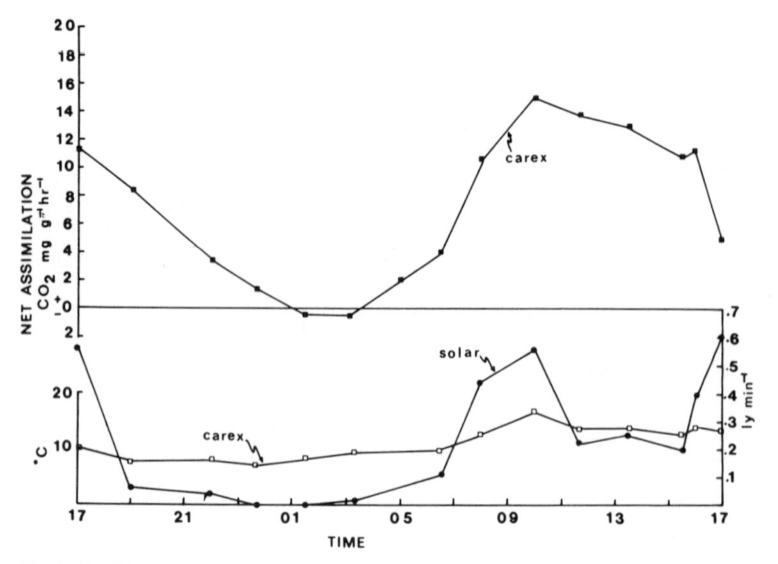

Fig. 1. Net CO_2 assimilation by *Carex stans*; leaf temperature and solar radiation; 8-9 August 1972. A typical 24 hr response illustrating evening depression, midday lag, and high CO_2 fixation.

The data for *Carex stans* are summarized in Table 1. Maximum values are reported for intervals between 15 and 30 min. This accounts for some values being higher than those reported in Fig. 2. Maximum hourly mean values of 15 to 16 mg g^{-1} hr^{-1} were very common whereas the maximum rates reported in Table 1 occurred only occasionally and then for periods of less than one hour. Light compensation was estimated to be approximately 0.01 ly min^{-1} in virtually every instance in which light levels were low. Positive net assimilation was observed at leaf temperatures slightly below 0°C. Leaf temperatures at which maximum net assimilation occurred varied between 10°C and 15°C. The values used in Fig. 1 were from leaves within this temperature range.

Field studies — Dryas integrifolia

The results of the field studies on *Dryas* are shown in Figs. 3 and 4, and Table 2. It can be seen that early in the season, when daytime radiation is

Table 1. Summary of *Carex stans* net assimilation (NA) for 1971 and 1972.

Date	Max NA (mg $g^{-1}hr^{-1}$)	Radiation (ly min^{-1})	Leaf temperature	Time	Min. NA (mg $g^{-1}hr^{-1}$)	Radiation (ly min^{-1})	Leaf temperature	Time
1971								
23-24 July	12.8	0.66	—	1000	3.0	0.09	3°	—
27-28 July	16.4	0.81	—	1600	2.4	0.10	5°	0200
	19.1	0.51	10°	0800				
10-11 Aug.	7.1	0.71	—	1300	−0.7	<0.01	—	2400
17-18 Aug.	22.3	0.67	11°	1200	−1.0	<0.01	—	0200
1972								
2-3 Aug.	22.8	0.32	11°	1900	−2.0	0.00	2°	0030
8-9 Aug.	11.5	0.20	10°	1700				
	15.2	0.56	15°	1000	−0.6	0.00	8°	0130

Date	Estimated light comp. point (ly min^{-1})	Comments
1971		
23-24 July	—	not a full 24 hrs., mixed carex.
27-28 July	—	variable N.A., cloudy, never got near zero.
10-11 Aug.	0.01	leaf temp. often <0°C single leaf cuvette
17-18 Aug.	0.01	single leaf cuvette
1972		
2-3 Aug.	<0.01	positive fixation at 0.01 1
8-9 Aug.	0.01	midday lag offset by clouds

high, the plants tend to fix more CO_2 during the evening and early morning hours and exhibit midday depressions which may be negative. This is especially true for the red-leaved plants just emerged from the snow (Fig. 3). The plants are quite active metabolically, the greatest respiration as well as net assimilation occurring at this stage (Table 2). Later in the season the pattern appears to be evening depression of net assimilation and daytime maxima. This is only partially due to the lower light intensities during the evening later in the season.

Positive net assimilation was observed at 1°C leaf temperature; however, this occurred only at low light intensities (Mayo et al. 1973). The optimum leaf temperature for net assimilation was estimated to be 10°C (Mayo et al. 1973). The leaves tended to become quite warm as radiation increased.

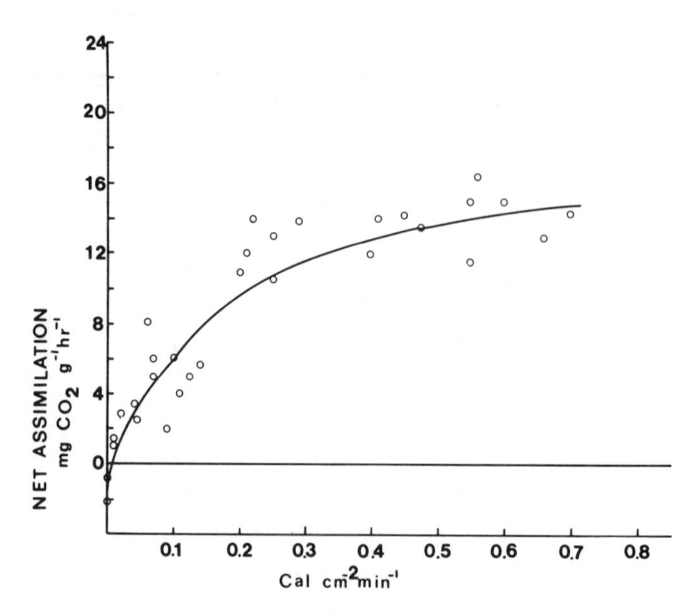

Fig. 2. Response of *Carex stans* net assimilation to radiation. Data from 1971 and 1972. Hourly mean values when light and net assimilation were ascending.

Maximum leaf temperatures of 40°C were observed with numerous readings ranging from 20° to 30°C. Fig. 5 illustrates the response of red-leaved plants to solar and sky radiation. As radiation increases during the day, leaf temperature rises and net assimilation decreases, exhibiting a sharp midday drop which may become negative. Because of the interaction of leaf heating with increasing radiation and the occurrence of low water potentials (see Addison this volume), it was not possible to construct a light response curve from field data. Light compensation was estimated to be less than 0.04 ly min^{-1} (Mayo et al. 1973).

Maximum net assimilation rates tended to be lower in plants that had commenced flowering (Figs. 3 and 4). This is not entirely because flowering occurred later in the season when radiation is less, since the flowering plants studied 9-10 July, 1971 had lower fixation rates than the green-leaved plant

Table 2. Total CO_2 assimilation and maximum net assimilation rate of *Dryas integrifolia* at various development stages during 1971 and 1972.

Date	*Total CO_2 assimilation in 24 hr; mg g^{-1} dry wt. of leaf*	*Maximum net assimilation rate; mg $g^{-1}hr^{-1}$*	*Phenologic stage of development*
30 June - 1 July, 1971	+ 16.54	+ 4.2	green leaves; no sign of flowering
9 - 10 July, 1971	− 3.68	+ 3.6	red leaves; 5 days out of the snow
5 - 6 July, 1972	+ 53.0	+ 7.4	red leaves; 5 days out of the snow
9 - 10 July, 1972	+ 28.12	+ 2.8	green leaves; flowering
22 - 23 July, 1972	+ 19.00	+ 2.2	green leaves; flower buds
27 - 28 July, 1972	+ 22.0	+ 2.1	green leaves; flowering
6 - 7 Aug, 1972	+ 16.40	+ 2.5	leaves reddish; late flower
6 - 7 Aug, 1972	+ 15.76	+ 2.2	leaves reddish; post flower

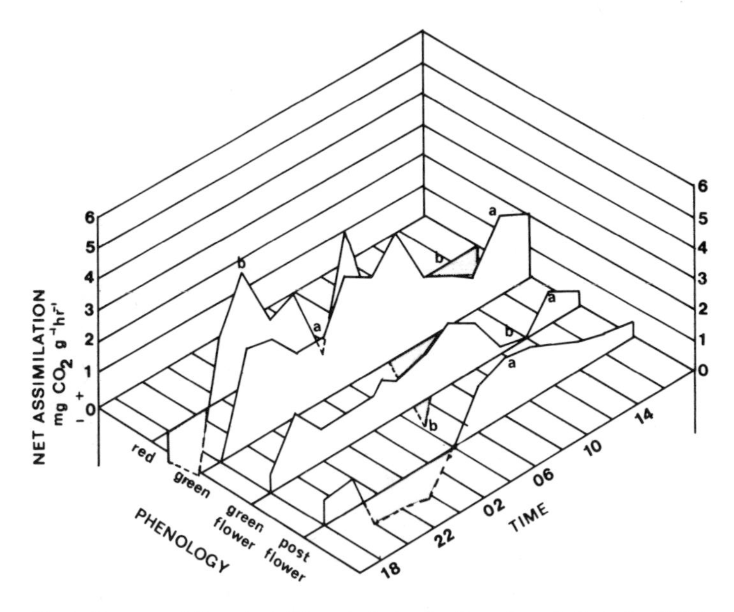

Fig. 3. Summary of diurnal net CO_2 assimilation for *Dryas integrifolia* during 1971 by phenologic stage: red = red-leaved plant just emerged from the snow, 9-10 July; green = green-leaved plant, 30 June-1 July; green flower = green-leaved plant just beginning to flower, 9-10 July; post flower = plants in late-flowering to post-flowering stage, 6-7 August; b = periods of clear weather; a = periods of cloud.

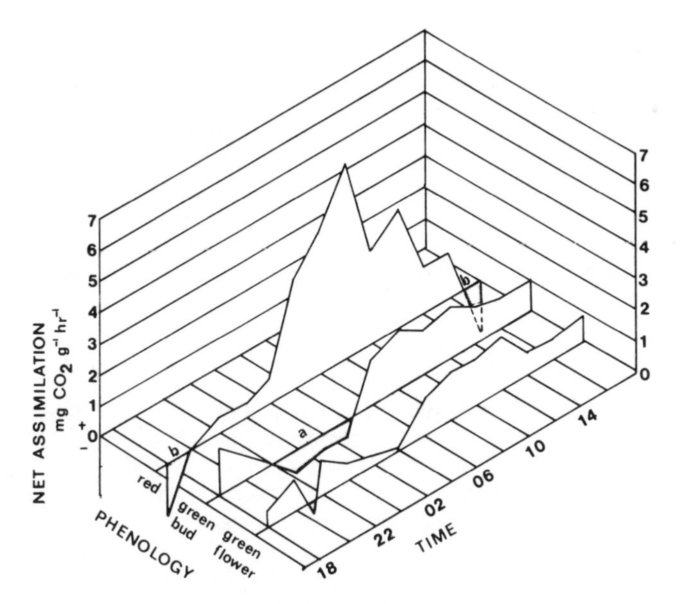

Fig. 4. Summary of diurnal net CO_2 assimilation for *Dryas integrifolia* during 1972 by phenologic stage: red = red-leaved plants 5 days out of the snow, 5-6 July; green bud = green leaves, flower bud stage, 22-23 July; green flower = green leaves, flowering, 27-28 July, b = clear periods; a = cloudy period.

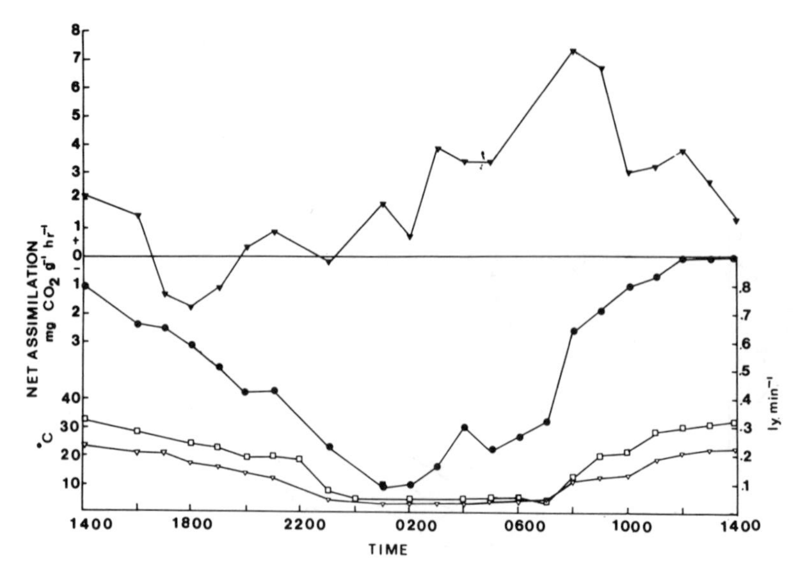

Fig. 5. *Dryas* net assimilation (hourly means) for 5-6 July, 1972; ▼——▼ assimilation average of two plants; •——• solar radiation; □——□ maximum leaf temperature; ▽——▽ average leaf temperature. Plant snow-covered until 30 June.

studied 30 June-1 July, 1971 and the red-leaved plant studied 5-6 July, 1972 (Table 2).

After the 1971 field season it was suggested that *Dryas* might emerge from the snow actively photosynthesizing (i.e., it might commence fixing CO_2 under the snow). During the 1972 season, a snow-covered plant was uncovered and studied. There was no evidence of respiration or net assimilation. The same plant was studied 5 days later and found to be photosynthetically active. The conclusion is that *Dryas* does not commence photosynthetic activity until after it has melted out.

The information gained under field conditions was used to guide controlled environment studies undertaken at The University of Alberta.

Controlled environment studies — Dryas integrifolia

The response of individual *Dryas* leaves to increasing illumination at different temperatures is shown in Fig. 6. Light compensation is very low at 0°C (12 μE m^{-2} sec^{-1} or 70 ft-c) and increases with leaf temperature to 224 μE m^{-2} sec^{-1} (1345 ft-c) at 30°C. The light intensity required for saturation increases with leaf temperature from 550 μE m^{-2} sec^{-1} at 0°C to 2420 μE m^{-2}2 sec^{-1} at 30°C.

The light response curves (Fig. 6) also suggest that 10°C is the optimum temperature for net assimilation. This is more apparent in Fig. 7, which also illustrates the ability of *Dryas* leaves to fix carbon at −5 to −6°C.

Dryas plants grown under controlled environment conditions will go dormant spontaneously without any cold treatment or change in light conditions. Fig. 7 illustrates the reduction of net assimilation associated with the onset of dormancy.

Relatively high rates of dark respiration of mature leaves occur at low

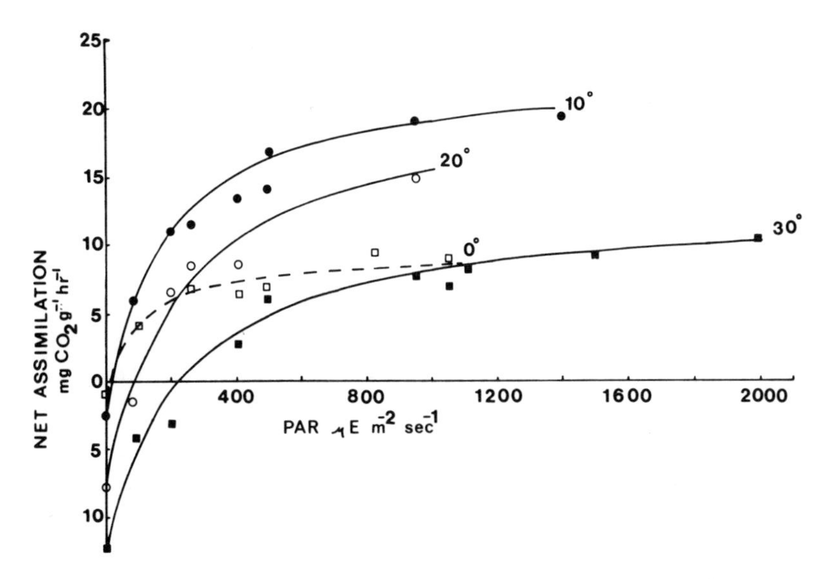

Fig. 6. Individual leaf net CO_2 assimilation of *Dryas* vs photosynthetically active radiation (PAR) at 0, 10, 20, and 30°C.

temperatures (Fig. 7), indicating that the plants are quite active at the temperatures normally found on Devon Island. The dark respiration of plants entering dormancy, is much lower than that of plants during the peak of the growing season, even though the Q_{10} (2.9 to 3.3 respectively) is similar in both cases.

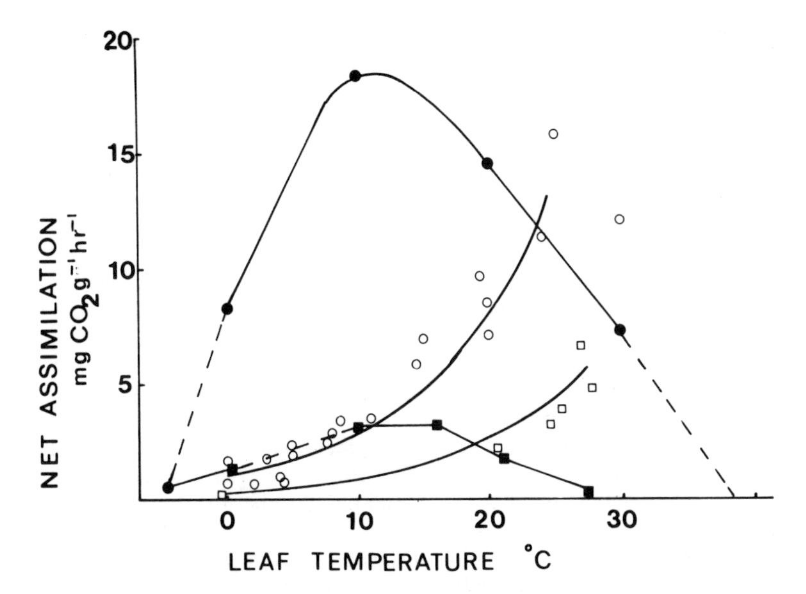

Fig. 7. The effect of temperature on net CO_2 assimilation and respiration of *Dryas* leaves ●——● net assimilation of leaves during "midsummer"; o——o respiration of leaves during "midsummer"; ■——■ net assimilation of leaves on plants going dormant; □——□ respiration of leaves on plants going dormant.

The photosynthetic responses of four leaves of different ages on the same sprig are shown in Fig. 8. Although there are some differences in the temperature responses and in the maximum net assimilation rates; the three oldest leaves are all capable of CO_2 fixation of at least 60% of the average maximum of 18.7 mg g^{-1} hr^{-1}.

Discussion

Carex stans

The light compensation point (0.01 ly min^{-1}) of *Carex stans* is quite low (Fig. 1 and Table 1). This agrees with Tieszen's (1972) results for *Carex aquatilis* at Point Barrow, Alaska. He found the light compensation point to be 0.013±0.001 ly min^{-1}. Tieszen (1973) also reported that the light compensation point for a variety of arctic grasses is between 0.008 and 0.010 ly min^{-1}. This suggests that most arctic grasses would be capable of positive CO_2 assimilation over a 24 hr period if radiation values are above .01 ly min^{-1}. The results with *Carex stans* suggest that this also happens on Devon Island (Fig. 1).

Tieszen (1973) stated that light saturation in arctic grasses was attained at approximately 0.4 ly min^{-1} at Point Barrow. *Carex stans* (Fig. 1) appears to exhibit a similar light response. The light response for *Carex stans* is quite similar to the responses of *Arctophila fulva*, *Alopecurus alpinus*, *Arctagrostis latifolia*, *Calamagrostis holmii*, *Dupontia fischeri*, and *Poa arctica* as reported by Tieszen (1973).

Maximum fixation rates of 22 mg g^{-1} hr^{-1} were observed in *Carex stans*, but hourly mean maxima between 11 and 15 mg g^{-1} hr^{-1} were more common. These rates agree with those reported by Tieszen (1972) for *Carex aquatilis* at Point Barrow. He reported 19 mg dm^{-2} hr^{-1} at 0.6 ly min^{-1} and *Carex stans* on Devon Island fixed 21 mg dm^{-2} hr^{-1} at 0.51 ly min^{-1} in 1971 and 16.8 mg dm^{-2} hr^{-1} at 0.56 ly min^{-1} in 1972. The maximum net assimilation rate for *Carex stans* is similar to the maximum rates for *Alopecurus alpinus*, *Arctagrostis latifolia*, *Arctophila fulva*, *Calamagrostis holmii*, and *Dupontia fischeri;* greater than *Poa arctica*, *Hierochloe alpina*, and *Poa malacantha;* and considerably less than *Elymus arenarius* as reported by Tieszen (1973). The maximum rate for *Carex stans* is higher than those reported by Shvetsova and Voznesenskii (1971) for *Carex ensifolia* and *Eriophorum angustifolium*. They report maximum rates of 9 and 14 mg g^{-1} hr^{-1} respectively. The maximum rate for *Carex stans* is greater than the visible photosynthetic rate for *Alopecurus alpinus* reported by Gerasimenko and Zalensky (1973), but quite comparable to the potential intensities of photosynthesis (18 mg g^{-1} hr^{-1}).

The results with *Carex stans* suggest that it behaves very much like *Carex aquatilis* at Point Barrow and is not greatly different from other arctic graminoid species in that it has a low light compensation point, a respectable maximum net assimilation rate, can fix CO_2 throughout the 24 hour day, and has an evening low, daytime maximum, and may exhibit a midday lag in net assimilation (Fig. 1).

Teeri (1974) has shown that 6 hr of dark or incandescent light (.1 ly min^{-1}) and 18 hr of high intensity fluorescent plus incandescent (.3 ly min^{-1}) light will inhibit flowering of *Saxifraga rivularis*. Since incandescent lamps are rich in the longer wavelengths, he suggests that phytochrome may be involved (i.e., there is a shift in relative amounts of red vs. far-red light).

However, there were no great diurnal shifts in red (650 nm) and far-red (725 nm) radiation throughout the growing season (Tables 3 and 4) (data to be published elsewhere) and incandescent lamps are nearly as rich in 650 nm as 720 nm radiation (ISCO, 1967). This suggests that shifts in red and far-red light are not sufficient to effect flowering on Devon Island.

Table 3. Red (650 nm) and far red (725 nm) radiation at midnight* and noon* throughout the 1971 and 1972 growing seasons***.

	Midnight			*Noon*		
Date	*650 nm***	*725 nm***	*650/725*	*650 nm***	*725 nm***	*650/725*
21-6-71	8.0	5.0	1.60	78	60	1.30
1-7-71	12.0	9.0	1.3	61	34	1.8
1-7-72	8	7	1.4	51	42	1.2
10-7-71	6	5	1.20	71	55	1.3
13-7-72	7.0	6	1.2	49	40	1.2
24-7-71	5	4	1.2	75	59	1.3
29-7-71	5	4	1.2	62	48	1.2
6-8-72	0.49	0.44	1.1	18	15	1.2
12-8-71	1	1	1.0	49	46	1.1
17-8-71	0.44	0.39	1.1	35	28	1.2

* Values taken between 0000 and 0030 and 1205 and 1330, except 1-7-71 (2300 and 1145).
** μ watts cm^{-2} nm^{-1}
*** Data from days with clear sky conditions except for 7-1-71 (alternating sun and cloud) and 8-6-72 (rain with solid overcast).

Table 4. Blue radiation (400-500 nm) at midnight and noon throughout the 1971 and 1972 growing seasons.

	Midnight					*Noon*				
Date	*400*	*425*	*450*	*475*	*500*	*400*	*425*	*450*	*475*	*500*
21-6-71	11	20	13	13	12	82	71	97	103	103
1-7-71	12	11	15	16	16	50	39	62	63	60
2-7-72	8	9	11	12	12	50	48	62	67	66
10-7-71	9	8	10	11	10	64	60	79	87	89
13-7-72	6	6	8	9	9	42	42	12	58	82
24-7-71	6	6	8	8	8	48	52	64	77	81
29-7-71	5	5	7	7	7	57	52	69	77	78
6-8-72	0.57	0.48	0.68	0.78	0.74	14	12	17	20	22*
12-8-71	1.7	1.5	2.0	0.2	0.2	54	49	66	69	73
18-8-72	0.41	0.37	0.51	0.55	0.53	31	31	40	45	45

* Rainy day

Dryas integrifolia

The field studies suggested that *Dryas* has a relatively low light compensation point (<0.04 ly min^{-1}) and could fix CO_2 throughout the 24 hr day (Figs. 3 and 4). The controlled environment studies indicate that light compensation at 0° and 10°C is indeed low (12 $\mu E\ m^{-2}\ sec^{-1}$ or 70 ft-c at 0°C, Fig. 6). However, as the leaf temperature is increased to 30°C the compensation point increases considerably (224 $\mu E\ m^{-2}\ sec^{-1}$ or 1345 ft-c at 30°C). Most alpine and arctic species have light compensation points between 150 and 300 ft-c at 20°C leaf temperature; as measured on whole shoots (Müller 1928, Wager 1941, Mooney and Billings 1961, Mooney and Johnson 1965, Hadley and Bliss 1964), or from 36 to 92 μ E m^{-2} sec^{-1} PAR at 15° and 20°C (Scott and Menalda 1970, Tieszen 1973), converted from cal cm^{-2} min^{-1} and W m^{-2} respectively. The light compensation point of *Dryas integrifolia* at elevated leaf temperatures is therefore high, compared with other arctic and alpine species. *Dryas* leaf temperatures are not often 30°C or higher in the field; except under conditions of high light intensity, thus the high light compensation point at 30°C is not necessarily a disadvantage. This, along with the respiration response (Fig. 7) to temperature, explains the negative net assimilation observed in the red-leaved plants early in the growing season (Figs. 1, 2, and 3).

The net assimilation of some arctic grasses approaches an asymptotic value at 0.4 ly min^{-1} PAR at 15°C (Tieszen 1973). *Dryas* net assimilation as measured in the laboratory approaches an asymptotic value at *ca.* 0.44 ly min^{-1} at 15°C leaf temperature (interpolated value). Many arctic and alpine species exhibit light saturation in the 1000 to 3000 ft-c range (Müller 1928, Wager 1941, Mooney and Billings 1961, Hadley and Bliss 1964, Mark 1975). However, Scott and Billings (1964) reported that several alpine species did not achieve light saturation at 11,000 ft-c. Hadley and Bliss (1964) reported that *Geum peckii* failed to show light saturation at 8,000 ft-c. This compares with saturation by *Dryas* at *ca.* 9000 ft-c (Fig. 6) at 20°C. Gerasimenko and Zalensky (1973) report light saturation for *Dryas punctata* (potential photosynthesis) at 2000 ft-c at 18°C. This is considerably lower than the value for *Dryas integrifolia* or *Geum pickii.*

Mayo et al. (1973) suggested that the optimum temperature for net assimilation was 10°C. The light and temperature response curves (Figs. 6 and 7) tend to bear this out. At this temperature the light compensation point is still very low and net assimilation increases rapidly with increasing radiation. The optimum range (9° to 14°C) as shown in Fig. 7 is lower than the 15° to 20°C range reported for alpine herbs and shrubs (Pisek et al. 1969), arctic grasses (Tieszen 1973), and cold-acclimated subarctic and alpine *Ledum groenlandicum* (Smith and Hadley 1974). Out of nine arctic species examined by Shvetsova (1970), only *Arctophila fulva* exhibited an optimum for potential photosynthesis which was less than 15°C. The potential photosynthesis reported for *Dryas punctata* was greatest in the 20° to 40°C range. *Dryas* is therefore well-adapted to fix CO_2 during the cold evening hours when light levels are relatively low and during the cloudy cold weather which often occurs on Devon Island, particularly late in the growing season.

The effects of temperature on net CO_2 assimilation and respiration shown in Fig. 7 indicate that *Dryas* will assimilate CO_2 down to *ca.* −5° or −6°C. This is lower than the 1°C previously reported by Mayo et al. (1973). However, it is not surprising since light levels in the field would be below the compensation point at these temperatures. The lower temperature limit (Fig. 7) coincides with the freezing point observed in many leaves, particularly north-temperate region conifers (Pisek et al. 1967). This activity at low temperature indicates that *Dryas* maintains considerable frost hardiness even during its period of most active growth, as do other hardy species (Levitt 1972). The upper temperature limit (Fig. 7) is within the range of 40° to 44°C reported for arctic shrubs (Pisek et al. 1968). However, the 40°C limit is not reached by *Dryas* unless the light intensity is 900 to 1000 μE m^{-2} sec^{-1} (60,000 lux). Because *Dryas* leaf temperatures greater than 30°C do occur (Fig. 5; Biebl 1968, Mayo et al. 1973, Addison, pers. comm.) the significance of the relationship between light intensity and the upper temperature limit for assimilation compensation is important as has already been mentioned.

The dark respiration rates of midsummer mature leaves (Fig. 7) compare well with those reported for subalpine *Picea excelsa* (Pisek and Winkler 1959), and are much higher than those reported for *Ledum groenlandicum* (Smith and Hadley 1974). The relatively high respiration at low temperatures indicates that metabolic processes can be carried out at rates sufficient to ensure growth in the relatively cold temperatures experienced on Devon Island. Some plants, adapted to arctic or alpine environments, have been observed to have higher respiration rates than temperate region plants (Pisek and Winkler 1959, Forward 1960, Mooney 1963), thus supporting the concept of metabolic adaptation to the thermal regime of a given habitat (Mooney 1963).

The rate of dark respiration at any given temperature is much lower in plants entering dormancy than during the middle of the growing season (Fig. 7). It is entirely possible that the difference may be even greater when the plants are fully dormant; the field study on a snow-covered dormant plant failed to detect any activity. Pisek and Winkler (1958) observed the same phenomenon with subalpine *Picea excelsa*. It is advantageous that the loss of carbon by respiration be reduced during the autumn before the onset of cold weather. Energy is conserved in this manner.

The average maximum net assimilation rate for eight individual leaves (18.7 mg g^{-1} hr^{-1}) is considerably higher than the maximum 7.4 mg g^{-1} hr^{-1} measured in the field (Table 1). This is not surprising because:

1. The field plants were probably under moisture stress, whereas the laboratory plants were well-watered. Addison (this volume) reported water potentials of −35 bars during July 1971.
2. The field studies were made on entire plants which means that stem, flower, and young leaf (Fig. 8) respiration was included.
3. There was self-shading in the clump during the day.
4. The nutrient status of plants in the controlled environment chamber was probably better since they were fertilized once a year.

The mean maximum rate of net assimilation (18.7 mg g^{-1} hr^{-1}) is lower than the maximum rates (22 to 30 mg g^{-1} hr^{-1}) for alpine herbs (Pisek et al. 1969) and higher than that (2 to 9 mg g^{-1} hr^{-1}) for coniferous trees (Larcher 1969). It is higher than other field measurements for many other arctic herbs and shrubs (Gerasimenko and Zalensky 1973), but is not greater than the potential photosynthesis of *Dryas punctata* reported by Gerasimenko and Zalensky (1973).

There was an indication in the field data (Figs. 3 and 4) that as *Dryas* started to go dormant there was a loss of photosynthetic capacity. Larcher (1969) discussed the decrease of photosynthetic capacity of trees with the onset of winter, but frost injury was not ruled out as a cause of the decrease. In the case of *Dryas integrifolia*, controlled-environment-grown plants spontaneously enter dormancy without a cold temperature stimulus. The decrease in net assimilation (Fig. 7) by these plants is therefore not due to frost injury, but suggests an internal control or degradation of the photosynthetic system.

The assimilation response of four leaves of different ages on the same sprig (Fig. 8) suggests that leaves can recover from the loss of photosynthetic capacity shown as they enter dormancy (Fig. 7). Although there are some differences in the temperature responses and in the maximum rates, the three oldest leaves (1, 2, and 3, Fig. 8) are all capable of high photosynthetic activity relative to the average maximum of 18.7 mg g^{-1} hr^{-1}. The oldest of these leaves (1) was definitely in its second growing season (i.e., it had gone

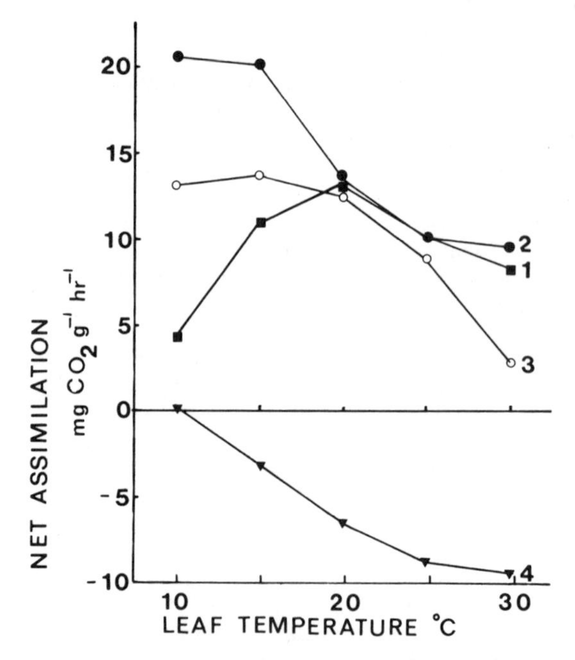

Fig. 8. The effect of leaf age and temperature on *Dryas* net CO_2 assimilation. Four leaves from one branch, (1) being the oldest leaf and (4) the youngest.

through the reduction associated with the onset of dormancy) (Fig. 7). This is in agreement with data on conifers (Freeland 1952). The youngest (4) was the smallest leaf that it was possible to put into a cuvette. These results suggest that the photosynthetic capacity commences early and remains relatively high for a long period of time. Since leaves are initiated at any time throughout the growing season, they can therefore be subjected to winter dormancy at varying ages. And since dormancy itself brings about changes in the photosynthetic activity, it cannot be stated that either first or second year leaves as a group are capable of greater photosynthetic capacity.

The field and growth chamber studies suggest that *Dryas integrifolia* has a low light compensation point and can assimilate CO_2 at low temperatures thus enabling it to fix CO_2 through the 24 hr day or during cloudy periods. It retains a high degree of frost hardiness necessary for survival in a cold environment. It has a relatively high respiratory rate at low temperatures indicating the ability for metabolic functioning in the arctic environment. Green leaves are functional for two years, retaining the ability for positive net assimilation, thus conserving energy that would otherwise be utilized for synthesis of new photosynthetic tissue every year. *Dryas* can fix CO_2 positively within 4 or 5 days after emergence from the snow. The tendency to enter dormancy spontaneously in a controlled environment suggests that *Dryas* will go dormant after a given period of activity regardless of the environment. This may have survival value in that *Dryas* will not be stimulated to put on new growth late in the season even if the weather is unusually favorable.

Acknowledgments

Support for the project by the National Research Council of Canada, Canadian Committee for the International Biological Program, is gratefully acknowledged. The authors are grateful to Dr. L.C. Bliss for his valuable help in making this project possible, Dr. P.R. Gorham for his valuable comments and the use of equipment such as the Wöstoff pumps, and Dr. D.W.A. Whitfield for the computer analysis of data.

References

Bate, G.C., A. D'Aoust and D.T. Canvin. 1969. Calibration of infra-red gas CO_2 analyzers. Plant Physiol. 44: 1122-1126.

Biebl, R. 1968. Heat balance and temperature resistance of arctic plants in West Greenland. Flora B 157: 327-354.

Dobkin, R. 1963. Proportional temperature controller for thermoelectric coolers. Rev. Sci. Instrum. 34: 1277-1278.

Forward, D.F. 1960. Effect of temperature on respiration. W. Ruhland (ed.), *In:* Handbuch der Pflanzenphysiologie XII/2: 234-258.

Freeland, R.O. 1952. Effect of age of leaves on rate of photosynthesis in some conifers. Plant Physiol. 27: 685-690.

Gerasimenko, T.V. and O.V. Zalensky. 1973. Diurnal and seasonal dynamics of photosynthesis in plants of Wrangel Island. Bot. Zh. 58: 1655-1666. IBP Tundra Biome Translation No. 11. 17 pp.

Hadley, E.B. and L.C. Bliss. 1964. Energy relationships of alpine plants on Mount Washington. Ecol. Monogr. 34: 331-357.

Hartgerink, Antoinette P. 1975. Controlled Environment Studies on Net Assimilation and Water Relations of Arctic *Dryas integrifolia*. M.Sc. Thesis, Department of Botany, Univ. Alberta, Edmonton, Alberta. 122 pp.

ISCO. 1967. Iscotables. Instrumentation Specialties Company Inc., Lincoln, Nebraska.

Larcher, W. 1969. The effect of environmental and physiological variables on the gas exchange of trees. Photosynthetica 3: 167-198.

Levitt, J. 1972. Responses of Plants to Environmental Stresses. Academic Press. New York. 697 pp.

Mark, A.F. 1975. Photosynthesis and dark respiration in three alpine snow tussocks (*Chionochloa* supp.) under controlled environment conditions. N. Zeal. J. Bot. 13: 93-122.

Mayo, James M., Don G. Despain, and Eduard M. van Zinderen Bakker, Jr. 1973. CO_2 assimilation by *Dryas integrifolia* on Devon Island, Northwest Territories. Can. J. Bot. 51: 581-588.

Mooney, H.A. 1963. Physiological ecology of coastal, subalpine and alpine populations of *Polygonum bistortoides*. Ecology 44: 812-816.

———. and W.D. Billings. 1961. Comparative physiological ecology of arctic and alpine populations of *Oxyria digyna*. Ecol. Monogr. 31: 1-29.

———. and A.W. Johnson. 1965. Comparative physiological ecology of an arctic and alpine population of *Thalictrum alpinum* L. Ecology 46: 721-727.

Müller, D. 1928. Die Kohlensäureassimilation bei arktischen Pflanzen und die Abhängigkeit der Assimilation von der Temperatur. Planta 6: 22-39.

Pisek, A., W. Larcher, W. Moser and I. Pack. 1969. Kardinale Temperaturbereiche der Photosynthese und Grentztemperaturen des Lebens der Blätter verschiedener Spermatophyten. III. Temperaturabhängigkeit und optimaler Temperaturbereich der Netto-photosynthese. Flora (Jena) B 158: 608-630.

———. W. Larcher, I. Pack and R. Unterholzner. 1968. Kardinale . . . II. Temperaturmaximum der Netto-photosynthese and Hitzeresistenz der Blätter. Flora B 158: 110-128.

———. W. Larcher, and R. Unterholzner. 1967. Kardinale . . . I. Temperaturminimum der Nettoassimilation, Fefrier und Frost-schadensbereiche der Blätter. Flora B 157: 239-264.

———. and E. Winkler. 1958. Assimilationsvermögen und Respiration der Fichte (*Picea excelsa* Link.) in verschiedener Höhenlage und der Zibre (*Pinus cembra* L.) an der alpinen Waldgrenze. Planta 51: 518-543.

———. and E. Winkler. 1959. Licht- und Temperaturabhängigkeit der CO_2 —Assimilation von Fichte (*Picea excelsa* Link.), Zibre (*Pinus cembra* L.) und somnenblume (*Helianthus annuus* L.). Planta 53: 532-550.

Redmann, R.E. Photosynthesis Matador Project. Fourth Annual Report. pp. 35-43.

Scott, D. and W.D. Billings. 1964. Effects of environmental factors on standing crop and productivity of an alpine tundra. Ecol. Monogr. 34: 243-270.

———. and P.H. Menalda. 1970. Carbon dioxide exchange of plants. II. Response of six species to temperature and light intensity. N. Zeal. J. Bot. 8: 361-368.

Šesták, E., J. Čatský, and P.G. Jarvis. (eds.) 1971. Plant Photosynthetic Production. Dr. W. Junk Publisher, The Hague. 818 pp.

Shvetsova, V.M. 1970. Temperature dependence of photosynthesis in certain arctic plants. Bot. Zh 55: 1683-1688.

———. and V.L. Voznesenskii. 1971. Diurnal and seasonal variations in the rate of photosynthesis in some plants of western Taimyr. Bot. Zh. 55: 66-76. 1970. IBP Tundra Biome Translation No. 2.

Smith, E.M. and E.B. Hadley. 1974. Photosynthetic and respiratory acclimation to temperature in *Ledum groenlandicum* populations. Arctic Alp. Res. 6: 13-28.

Teeri, J.A. 1974. Periodic control of flowering of a high arctic plant species by fluctuating light regimes. Arctic Alp. Res. 6: 275-279.

Tieszen, L.L. 1972. CO_2 exchange in the Alaskan Arctic Tundra: measured course of photosynthesis. pp. 29-35. *In:* Proceedings 1972 Tundra Biome Symposium. S. Bowen (ed.). CRREL, Hanover, New Hampshre. 211 pp.

———. 1973. Photosynthesis and respiration in arctic tundra grasses: field light intensity and temperature responses. Arctic Alp. Res. 5: 239-251.

Wager, H.G. 1941. On the respiration and carbon assimilation rates of some arctic plants as related to temperature. New Phytol. 40: 1-9.

Studies on evapotranspiration and energy budgets on Truelove Lowland

P.A. Addison

Introduction

The measurement of evapotranspiration and the determination of energy budgets under various environmental conditions may help to explain how a plant is adapted to maintain itself in a harsh environment. Both arctic and alpine tundra species are exposed to harsh environments and Bliss (1971), Savile (1972), and Courtin and Mayo (1975) have discussed some of the plant strategies that ensure growth and development under such conditions.

In the past only a limited amount of work has been done characterizing the physical and physiological responses of arctic plants to their environment with respect to partitioning of absorbed radiation and retention of water in xeric sites (Ahrnsbrak 1968, Romanova 1971, Miller and Tieszen 1972, Weller and Cubley 1972, Haag and Bliss 1974).

It is important that energy budget studies accompany those of water balance, since plant adaptations for water conservation may have far reaching effects on the overall energy regime of a plant community. Plant adaptations to water deficit, therefore, are limited to some extent, because a balance must be reached between the minimum rate of water loss possible and the maximum leaf temperature that can be tolerated.

The aims of this study were: (1) to determine the energy budgets of the two dominant plant communities in this high arctic ecosystem; (2) to study the plant-soil-water relations of plant communities at the ends of a soil moisture gradient; and (3) to explore the ecological implications of presumed adaptations to environmental conditions.

Two sites, a raised beach and a meadow, were chosen for intensive study as they represented the two dominant vegetation types on Truelove Lowland (Bliss 1972). Soils of the raised beach were coarse textured and ranged from regosols to brunisols (Walker and Peters this volume). The general microtopography and plant community pattern of this site have been described by Svoboda (this volume). The meadow site had a Fibric Organic soil (Walker and Peters this volume), and a high water table maintained by the impervious permafrost below. Plant community pattern and general site characteristics are given by Muc (this volume).

Owing to differences in drainage the sites represented the ends of a soil moisture gradient, and the plant communities reflected this gradient. Cushion plants were the dominant vascular plants on the raised beach, while upright sedges dominated the meadow. Since neither site had a homogeneous plant cover, it was necessary to subdivide each into a number of microsites based upon differences in floristics and water and energy relations. Five microsites comprised the raised beach: (1) vascular plants as represented by *Dryas integrifolia;* (2) crustose lichens; (3) fruticose lichens; (4) fine textured unvegetated soil; and (5) coarsely textured unvegetated soil (20% <2mm). The meadow consisted of three microsites: (1) vascular plants (*Carex stans* with an understory of moss); (2) moss alone (mainly *Drepanocladus brevifolius*); and (3) shallow pools of water.

Energy budget

A description of the energy budget shows the relative importance of each method of dissipating absorbed incoming radiation not used in photosynthesis. Dissipation of energy from the beach ridge and meadow is important as it represents a loss of energy from the ecosystem and relates to the overall study of tracing energy flow through the ecosystem.

Plant photosynthesis utilizes a maximum of 2% of the total energy received (Grable et al. 1966) and the organism must be able to dissipate the remaining energy in order to maintain leaf temperatures below the thermal death point. Equation (1) represents algebraically the partitioning of absorbed or net radiation (R_n) for any surface (Gates 1965).

$$R_n = LE + H + G + M \tag{1}$$

where L is the latent heat of vaporization (580 cal g^{-1}); E, rate of water loss (g cm^{-2} min^{-1}); H, sensible heat flux; G, soil heat flux, and M, plant metabolic energy. The last is only applicable to plant surfaces. The units of H, G, M, and R_n are cal cm^{-2} min^{-1}.

Methods and materials

A Net Radiometer installed at a height of 1 m integrates absorbed radiation over a circular area 10 m in diameter (Reifsnyder and Lull 1965). In areas where there is great heterogeneity in surface characteristics because of a mosaic of microsites, it was important to determine net radiation of each microsite to permit projection of the data to a much larger area. Where microsites were small (often less than 20 cm in diameter), R_n had to be determined indirectly. This was achieved using a Funk Net Radiometer with a black body cup attached to the lower surface. (See Addison 1973.)

Although five microsites comprised the beach ridge, the study of the beach ridge energy relations only required that three be considered. The two lichen microsites were combined as were the two unvegetated soil surfaces to achieve consistency with the work of other researchers who had confined their studies of plant cover at the beach ridge to three components: vascular plants, lichens, and unvegetated soil.

The surface temperatures of both the black body cup and the microsites were measured with copper-constantan thermocouples connected through a cold compensating junction and an automatic stepping switch to a potentiometric millivolt recorder.

Good physical contact between thermocouple and surface is necessary to measure surface temperature accurately. Thermocouples were glued to small pebbles, threaded through lichen thalli and moss stems, and mounted on leaves with thermocouple clips. The leaf thermocouple clip was a modification of that used for Douglas Fir *(Pseudotsuga menziesii)* by Fry (1965). Five thermocouples connected in parallel were used to estimate the mean surface temperature of each microsite.

Soil energy flux (G) was measured with soil heat flux plates (3×5 cm) placed 1 cm below and parallel to the soil surface. These plates measured the rate of energy transfer through the soil beneath *Dryas*, *Carex*, lichen, and water surfaces. It was assumed that the insulative capacity of lichens was negligible because of the sparse cover and, therefore, soil energy flux beneath bare soil and lichens was assumed to be similar. It was assumed also that the soil heat flux beneath a moss canopy was the same as that beneath *Carex* and moss together. At the vascular plant microsites, the metabolic energy flux (M) was assumed to be 2% of global radiation (R_t). Sensible heat flux (H) from all microsites was calculated using equation (1), since R_n, LE, G and M were known.

Results and discussion

Longwave radiation (Ld) comprised over half of the total incoming radiation on 17 and 18 July, mainly as a result of dense clouds during early 18 July (Fig. 1), and the presence of a rock outcrop that emitted radiation to the site. Fig. 1 also shows that longwave radiation compensated for reductions in global radiation (R_t) caused by light cloud. There were sudden drops in shortwave flux at 0800 and 1600 hr on 17 July and at 1200 hr on 18 July. At those times there was a corresponding increase in longwave flux and a resultant smooth curve of total incoming radiation.

Cushion Plant-Lichen community (Raised Beach)

Most of the energy absorbed by the vascular plants on 17 and 18 July was dissipated as sensible heat flux (86%) rather than by either latent heat flux (11%) or soil heat flux (3%). This last term was very small and has been excluded from Fig. 2. There was a reduction in latent heat flux at or shortly after 1200 hr on both days. This drop in latent heat flux was probably a result of "midday stomatal closure" caused by leaf water deficit. This deficit was created by transpiration exceeding the rate of water uptake by the roots. After the plant recovered, the stomata reopened and transpiration increased.

Similar curves for a lichen cover are shown in Fig. 3. Since lichens have no structures controlling water loss, latent heat flux (29%) of lichens was greater than that of vascular plants. Sensible heat flux (64%), however, was the major method of energy dissipation. The cover of lichens on the raised beach (25%) was sparse and so there was little resistance to heat transfer into the ground. The lichen cover therefore had a much larger soil heat flux (7%) than did *Dryas* (3%).

Net radiation of lichens on 17 July was substantially higher than that of *Dryas* during the 6 hr period spanning 1200 hr. The explanation for the difference lies in longwave radiation leaving the two surfaces. The emission of longwave radiation from any object is temperature dependent and, since *Dryas* was 30 C° warmer than the lichens at this time, more radiant energy was emitted by *Dryas*. This resulted in a greater amount of incoming energy remaining at the lichen surface and hence, greater net radiation. When surface temperatures of the two microsites were similar (e.g., 1400 hr on 17 July), net radiation was the same for both surfaces.

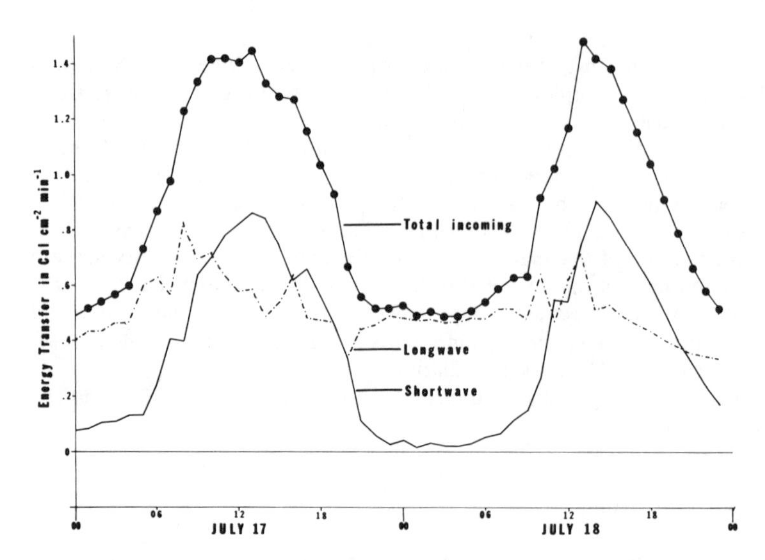

Fig. 1. Total incoming radiation (all wavelengths) and its longwave and shortwave components on 17 and 18 July, 1972.

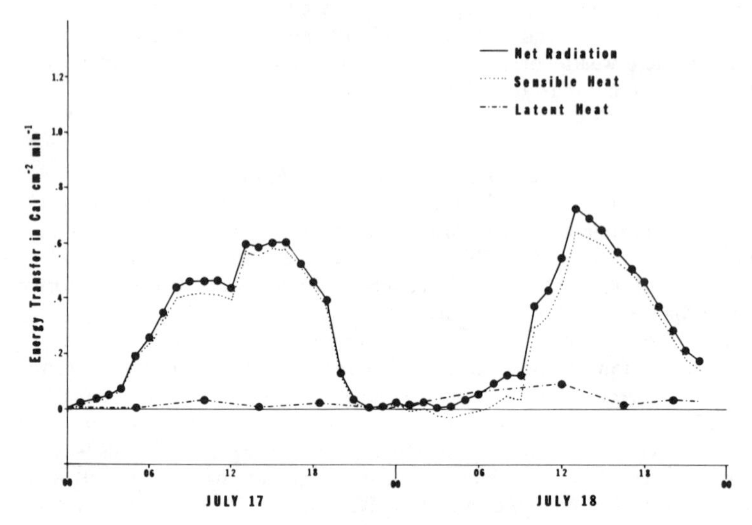

Fig. 2. Net radiation and its components for *Dryas integrifolia* on 17 and 18 July, 1972 (raised beach).

The various fluxes that comprise bare soil R_n lay between those of *Dryas* and lichen microsites. Most absorbed energy was dissipated as sensible heat flux (79%) whereas only 15% was lost as latent heat flux and 6% as soil heat flux. Since the surface temperature of unvegetated soil was similar to that of lichen thalli, emitted longwave radiation was similar for both microsites as was R_n.

Dryas leaf temperature varied much more than did the temperature of either of the other two microsites at the raised beach. On 17 July leaf

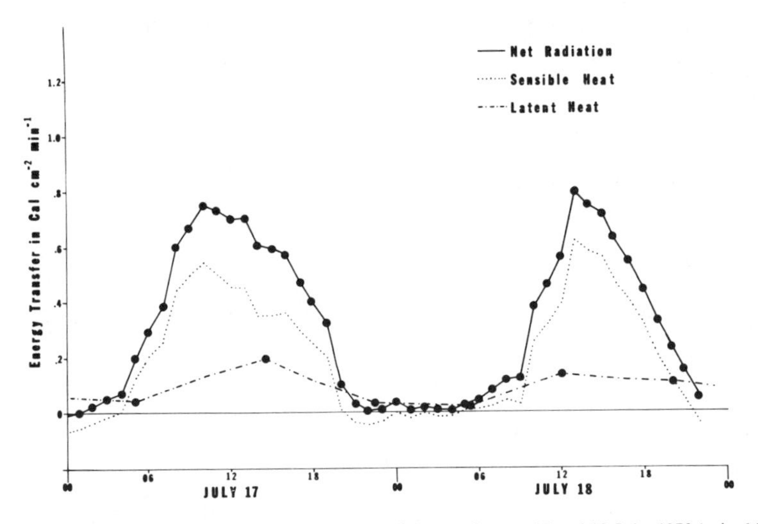

Fig. 3. Net radiation and its components for a lichen surface on 17 and 18 July, 1972 (raised beach).

Table 1. Generalized energy regime of the raised beach microsites under different incoming radiation conditions.

Microsite	*I* *ly min*$^{-1}$	*R*$_n$ *ly min*$^{-1}$	*LE* % *R*$_n$	*H* % *R*$_n$	*G* % *R*$_n$
	17 July (sunny)				
Vascular Plants	0.952	0.304	4.9	92.2	2.9
Lichens	0.952	0.351	29.1	61.1	9.8
Unvegetated Soil	0.952	0.337	13.2	76.6	10.2
	24-25 July (cloudy)				
Vascular Plants	0.609	0.089	24.3	73.9	1.8
Lichens	0.609	0.094	34.2	54.3	11.5
Unvegetated Soil	0.609	0.100	25.0	64.2	10.8

temperature remained above 45°C for three hours. This value is within the range of 45°C to 55°C given by Clum (1926) as the thermal death point of many plant species. The temperature recorded for *Dryas* was a mean of five sensors placed at the centre and four cardinal points of the cushion. Since these were mean temperatures and some of the leaves received more radiation than others (because of their orientation), it is inevitable that some were warmer than the mean. Several days of each year have similar levels of radiation and low wind speed but, nevertheless, the plant had no signs of heat kill.

Each microsite of the raised beach showed a similar diurnal pattern of energy dissipation on both sunny (17 and 18 July) and cloudy (24 and 25 July) days. There were, however, significant differences at all microsites between the two types of days in the ratio of net to incoming radiation. It can be seen from Table 1 that with a drop in incoming radiation (I), there was a greater reduction in R_n. The ratio of R_n to I dropped from 0.34 to 0.15 on cloudy days. Because of the much reduced heat load during cloudy days, it

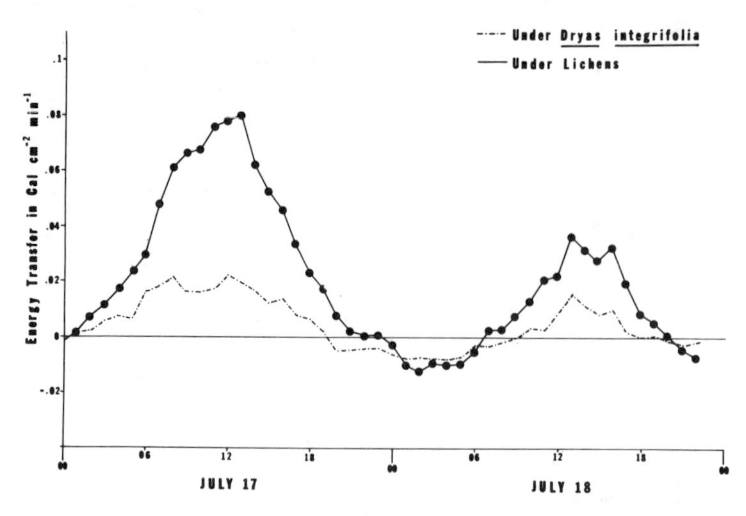

Fig. 4. Soil heat flux beneath *Dryas integrifolia* and lichens during 17 and 18 July, 1972 at the raised beach.

was possible for each microsite to dissipate a greater proportion of absorbed energy by latent heat flux (LE). This resulted in less energy dissipated by the relatively inefficient process of sensible heat flux (H), and hence, much lower surface temperatures on cloudy days than on sunny days at each microsite. There was less variation in R_n among the microsites on dull days than on bright ones, presumably as a result of similar surface temperatures on the dull days.

In all surfaces studied, soil heat flux (G) represented only a small part of energy dissipated. There was, however, a marked difference between soil heat flux beneath vascular plants *(Dryas)* and that beneath a lichen cover (Fig. 4). The very low soil heat flux beneath *Dryas* resulted from a high insulative capacity of the cushion growth form. Other raised beach species such as *Salix arctica* on the other hand, have been shown to have highly fluctuating near-surface soil temperatures as a result of a high soil heat flux (Warren-Wilson 1957). Lichen cover on the raised beach was sparse and offered very little resistance to transfer of energy into the ground. Surface insulation by cushion plants works in both directions and resulted in soil heat flux beneath lichens being greater than that beneath *Dryas* during the day and less during the "night" hours (i.e., 0000 to 0600 hr on 18 July).

Hummocky Sedge-Moss Meadow

Soil heat flux beneath a *Carex* community and that beneath a shallow pool of water are presented in Fig. 5. *Carex* intercepted and dissipated much of the incoming energy before it reached the ground surface and reduced G by 50%. The shallow active layer at this site resulted from insulation of the surface by the plants retarding the melt of frozen soil beneath. The soil energy flux beneath a shallow pool of water was always positive because of the steep gradient between warm surface water and the frost table only 25 cm below.

Water was not limiting at all microsites at the meadow, and all surfaces lost over 40% of absorbed energy through latent heat flux (Table 2). From this it seems that the *Carex* microsite was dominated in water loss by the moss understory. There was, however, 10% more energy dissipated as LE at

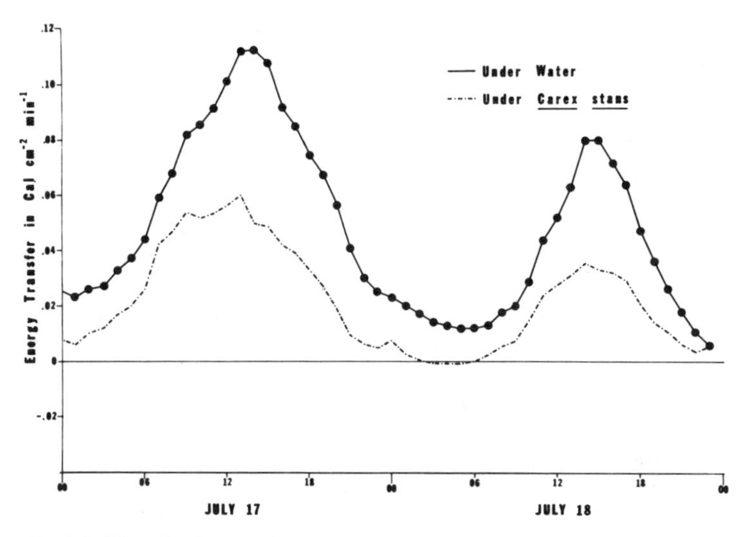

Fig. 5. Soil heat flux beneath *Carex stans* and water during 17 and 18 July, 1972 at the meadow site.

Table 2. Generalized energy regime of the hummocky sedge-moss meadow microsites under different solar radiation conditions.

Microsite	*I* *ly min*$^{-1}$	R_n *ly min*$^{-1}$	*LE* *%* R_n	*H* *%* R_n	*G* *%* R_n
	17 July (sunny)				
Vascular Plants	0.952	0.368	54.6	37.0	8.4
Moss	0.952	0.387	44.0	48.0	8.0
Water	0.952	0.385	43.7	40.0	16.3
	24-25 July (cloudy)				
Vascular Plants	0.609	0.110	49.6	40.6	9.8
Moss	0.609	0.117	54.8	36.1	9.2
Water	0.609	0.119	83.6	-4.0	20.4

the *Carex* microsite as compared with that of moss alone on 17 July. This was a very small difference when related to the substantial increase in surface area (2.5x) but did show that the vascular plants added a component to the energy dissipation at this site. At all three microsites the peak of latent heat flux corresponded approximately to the time when relative humidity was lowest and surface temperatures highest.

Energy dissipation on dull days (24 and 25 July) at all meadow microsites was similar to that on a clear day (17 July). The meadow microsites followed those at the raised beach showing a reduction in the ratio of R_n to I and an increase in the proportion of R_n dissipated as latent heat flux on cloudy vs. sunny days (Table 2). The great differences in the water microsite are owing to a change in the instrument to measure LE.

Both sites general

Figure 6a represents diagramatically the energy regime of the entire raised beach on 17 July, 1972. The beach ridge had a cover of 25% vascular plants

Table 3. Net radiation on sunny days at six arctic sites.

Net Radiation (*cal cm^{-2} min^{-1}*)	*Site*	*Observer*
0.342	Meadow, Devon Island, N.W.T.	Addison
0.310	Beach Ridge, Devon Island, N.W.T.	Addison
0.264	Meadow, Point Barrow, Alaska	Weller & Cubley
0.262	Lichen Mat, Pelly Lake, N.W.T.	Ahrnsbrak
0.224	Lichen Mat, Snowbunting Lake	Ahrnsbrak
0.178	Lichen Mat, Curtis Lake, N.W.T.	Ahrnsbrak

(Svoboda this volume), 25% lichens (Richardson and Finegan this volume) and 50% unvegetated soil. Longwave radiation (L_u) was the most important method of dissipating incoming energy. This observation concurs with that of Mellor et al. (1964) for a wide range of plant species under controlled environments. Net radiation (R_n) of the community was dissipated mainly by sensible heat flux (H). Sensible heat flux was five times greater than latent heat flux (LE) and ten times that of soil heat flux (G).

The meadow was comprised of 70% vascular plants with an understory of moss, 20% moss alone, and 10% standing water. Latent heat flux accounted for over 50% of the net radiation (Fig. 6). The meadow community had a 10% higher R_n than did the raised beach, presumably because of the small amount of reflected shortwave radiation (R_t) and lower emissions of longwave radiation resulting from lower surface temperatures. This compares closely with R_n values presented by Courtin (1972) for these same communities at a similar time in 1971. Soil heat flux at the meadow was slightly larger, comprising 8% on R_n as compared to 5% at the raised beach.

Shortwave incoming radiation on 17 and 18 July was comparable with reported values from Snowbunting Lake, Curtis Lake, and Pelly Lake, N.W.T. (Ahrnsbrak 1968) and from Barrow, Alaska (Weller and Cubley 1972) but net radiation of both sites on Devon Island was higher than any of these other sites (Table 3).

The major cause of the difference between the Devon sites and those studied by Ahrnsbrak (1968) and Weller and Cubley (1972) appears to be in the higher amount of longwave radiation. Large rock outcrops are located close to both sites and, since these stand above the surface, some longwave radiation received was from the rocks. This increased total incoming radiation resulted in higher values of net radiation than would be expected from the incoming shortwave radiation measurements.

In general, the energy regime of the beach ridge reflects the nature of its substrate (a dry gravel) and over 65% of the energy that strikes this surface is either reflected or reradiated immediately upon receipt. The beach ridge is characterized by its cover of lichens and the presence of large areas that are devoid of vegetation. The energy regime of this site also reflects this lack of vegetation because it responds to changes in radiation load in a similar manner to that of the bare soil microsite alone.

The *Carex* microsite typifies the meadow mosaic pattern, a system in which over 50% of R_n is dissipated as LE.

Water relations

Water can become limiting to growth whenever the ratio of precipitation to evaporation is low or where there is rapid drainage or runoff. Both of these conditions exist on raised beaches where *Dryas integrifolia* is one of the

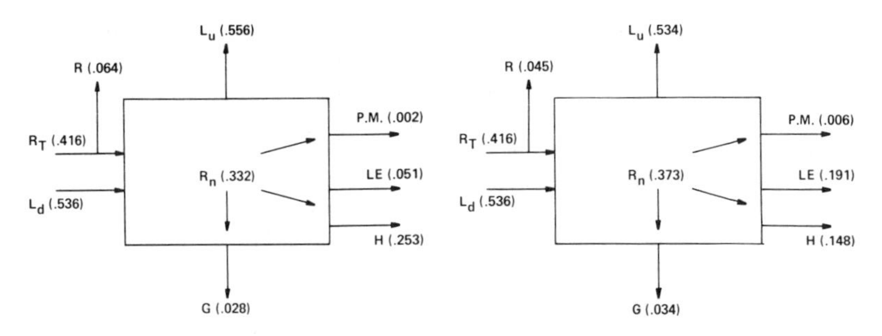

Fig. 6. Energy regime of the raised beach (a) and hummocky sedge-moss meadow (b) intensive study sites under high insolation (17 July, 1972). Units are cal cm^{-2} min^{-1}.

dominant plant species (Svoboda this volume). An understanding of how *Dryas* is adapted to withstand water deficit is only possible by examining the water regime of the soil-plant-atmosphere continuum under various environmental conditions and in the light of the energy regime of the surface.

One of the most important terms in the soil-plant-atmosphere continuum is the rate of water loss from a vegetated surface to the air. This process consists of transpiration from the leaves and evaporation from the soil around the plant. Both of these terms are important when describing water relations of a plant community as both reduce the amount of water that is available to the plant. Various methods of measuring evapotranspiration have been reviewed by Kohnke et al. (1940), Pelton (1961), Van Bavel (1961), Courtin and Bliss (1971), and Mukammal (1971). These methods include the use of potometers, evaporimeters, Bowen Ratio, aero heat method, sod blocks, and lysimeters.

Not only is it important to determine the rate of water loss from a plant community, but it is also necessary to describe the water status of the various parts of the continuum. When water is in the gaseous state, it moves down gradients of vapour density (g H_2O cm^{-3}) but when it is in the liquid state, it moves in response to a gradient in water potential (ψ_W). Water potential is a function of the difference in chemical potential between the water under consideration and pure free water at the same temperature and pressure.

Methods and materials

In this study a lysimeter was developed that could be installed within the active layer of arctic soils (Addison 1973). The active layer in hydric meadow areas can be as shallow as 20 cm and therefore, even the miniature hydrostatic lysimeter described by Courtin and Bliss (1971) was too large to be of use.

Four lysimeters were installed at each site and the soil surface of two was sealed with silicone rubber to prevent evaporation from soil around the plant. The sealed pots gave a measurement of transpiration while the open ones gave evapotranspiration.

Sod blocks were used at each site to supplement lysimetric data and to determine evaporation rates of non-vascular plant surfaces. Sod blocks were the same size as the lysimeter pots (8.25 cm in diameter and 11.5 cm deep) and were positioned in a similar manner with the top flush with the soil surface. Measurements were made by removing sod blocks and weighing them on a triple beam balance (±0.1 g).

The rate of water loss from any surface is directly proportional to its evaporative surface area, so to obtain comparable data from each of the microsites, the surface area of each was measured. Leaf area of vascular plants was determined using the Ballontini glass ball technique of Thompson and Leyton (1971). Since water loss from vascular plants is almost entirely through their stomata, the evaporative surface may be considered as that area where stomata are present. *Carex stans* is amphistomatous and therefore, the evaporative leaf area was the sum of both surfaces. The leaves of *Dryas integrifolia*, on the other hand, have stomata on only the abaxial surface. This surface is pubescent, however, and it was not possible to coat it with glass balls. As a result, the adaxial surface was measured and it was assumed that this area was the same as that of the lower surface. The area of the non-vascular plant surfaces was measured by casting the entire surface of the sod block in silicone rubber and determining the area of the cast with the Ballotini glass ball technique.

Leaf water potential was measured with a Wescor chamber psychrometer and a microvoltmeter that followed the method of Spanner (1951). Soil water potential was measured by a similar method using Wescor soil psychrometers and the same meter. The chamber and soil psychrometers were calibrated with sodium chloride solutions of known water potential.

Plant samples were collected in the field and brought into the laboratory in small sealed bottles. Care was taken to ensure that the bottles were completely full so that water loss from the tissue was minimized. Three readings of the chamber psychrometer were taken for each sample, 10 min apart and after an initial 40 min equilibration time. The soil psychrometers were soaked in water for 24 hr and installed in the raised beach one week in advance of measuring to allow equilibration with the soil.

Results and discussion

It is seen from Fig. 7 that availability of water, as indicated by the water potential of the leaf tissue, limited the rate of transpiration of *Dryas* at the raised beach. Periods of precipitation were accompanied by an increase in both transpiration and leaf water potential. As the soil dried, both parameters decreased until the soil was again replenished by rain. Fig. 8 shows that this relationship did not hold for the *Carex* microsite at the meadow. Precipitation had little or no effect on either of the measured parameters, presumably because of the high water content in the soil (>400% ODW). Variations in transpiration at the meadow were probably a result of the natural variation in aboveground environmental conditions such as radiation load, relative humidity and wind speed.

Cushion Plant-Lichen community (Raised Beach)

Microsites at the raised beach differed greatly from each other in evaporation rate. Table 4 shows that although there was more water lost from crustose lichens than from fruticose lichens, the effect of crustose cover on water loss was small when its rate was compared with that of adjacent bare soil. Fruticose lichens, on the other hand, had an evaporation rate of almost twice that of adjacent bare soil. Water loss from all non-vascular plant microsites was greater than that from *Dryas*.

The main reason why the two lichen covers react differently to the evaporative demand of the air probably lies in their morphology. The vertical

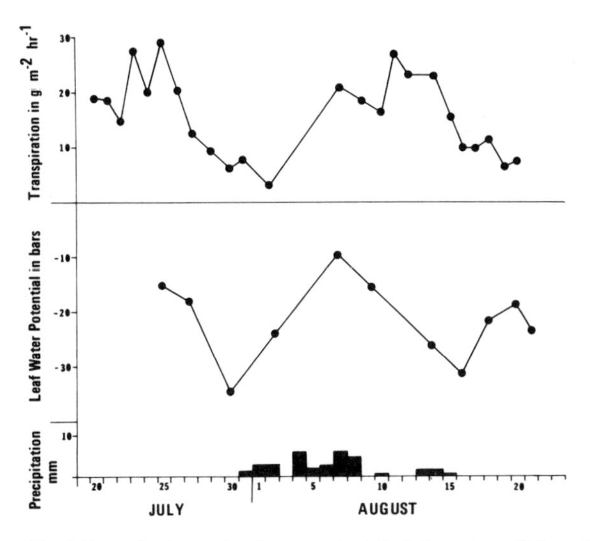

Fig. 7. Transpiration and leaf water potential of *Dryas integrifolia* and precipitation at the raised beach (1971).

Table 4. A comparison of the water status at the raised beach microsites (15 July to 23 August mean, 1971).

Site	Major flora	Soil moisture 0-10 cm (% ODW)	Rate of evaporation ($g\ m^{-2}\ hr^{-1}$) Vegetated surface	Non-vegetated surface
Mesic	*Dryas*	23.81	15.4	44.3
Mesic	Crustose Lichens	23.81	43.6	44.3
Xeric	Fruticose Lichens	1.1	32.4	19.3

and branching nature of fruticose lichens resulted in a 16% larger evaporative surface per unit area of microsite than that of the coarse bare soil. A higher evaporation rate was observed for these lichens during the entire summer. The crustose lichens, on the other hand, formed a comparatively smooth surface closely associated with the ground and had a slightly smaller surface area than did the fine bare soil microsite. This suggests that evaporation from bare soil might always be greater than that from crustose lichens but Fig. 9 shows that this was not the case. When the surface was wet, the larger surface area of the bare soil resulted in a greater evaporation rate but, when the surface was dry, the reverse was true. Under dry conditions, evaporation from bare soil must have been from within the soil rather than from its surface. Since evaporated water would have had to move as vapor through the soil pores to the surface before it could be lost, the water transfer pathway was lengthened. This lengthening of the pathway was probably the cause of the reduction of the evaporation rate at the fine bare soil microsite during dry periods. In the case of crustose lichens evaporation was lower than the bare soil during the wet periods and therefore more water remained in the soil during these periods. When the lichens started to dry, the retained water was available to the plant.

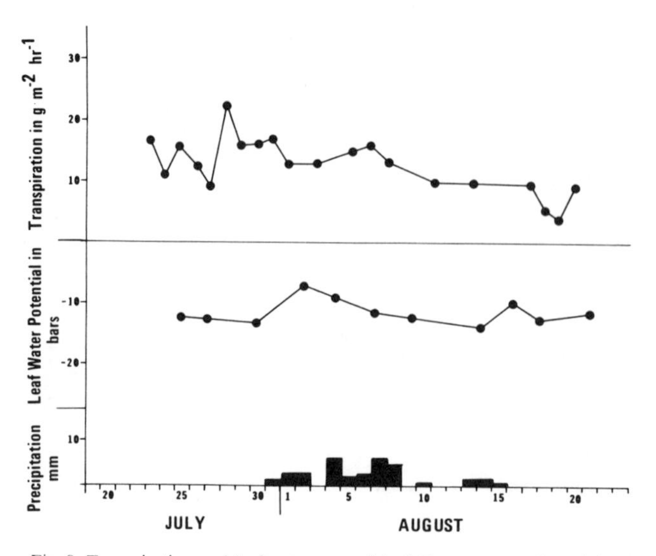

Fig. 8. Transpiration and leaf water potential of *Carex stans* and precipitation at the hummocky sedge-moss meadow (1971).

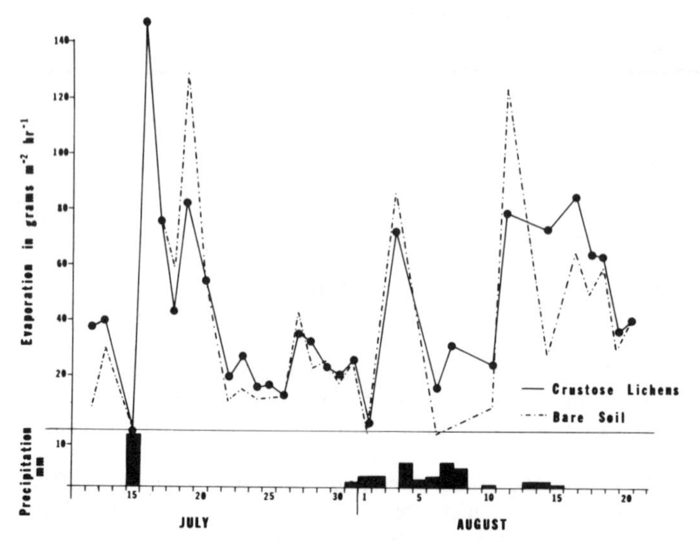

Fig. 9. Seasonal march of evaporation from crustose lichens and bare soil at the raised beach (1971).

The mesic and xeric unvegetated microsites differed greatly with respect to soil moisture, evaporation rate, and net seasonal water flux (Table 4). The probable reason for these differences lies in the soil texture of the two areas. The mesic site had 60% of particles less than 2 mm in diameter whereas the xeric site had only 1.3%. This lack of fine material at the xeric site led to deep percolation of water, rapid surface drying and, ultimately, to a drier soil, less water loss, and a downward net seasonal water flux.

Fig. 10 shows that *Dryas* maintained a water potential gradient between

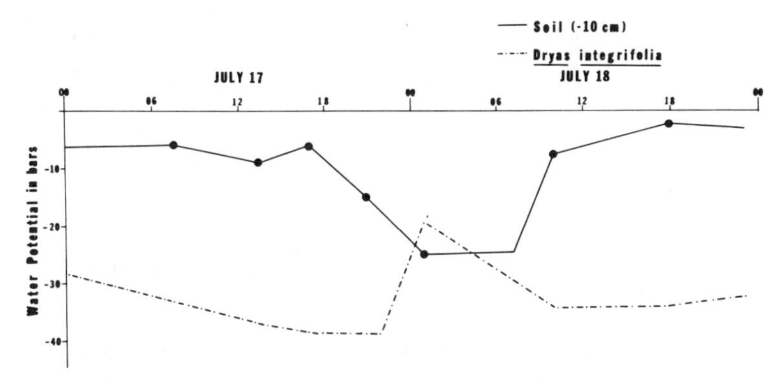

Fig. 10. Soil and leaf water potential during 17 and 18 July, 1972 at the raised beach.

its leaves and the soil of about 25 bars for most of the period presented. Because of stability of soil moisture and low transpiration rates of *Dryas*, the plant had a relatively constant leaf water potential (−35 bars) for most of the period. Such a low water potential was not atypical for *Dryas* and was similar to the seasonal mean (−32 bars). The seasonal mean soil water potential was −20 bars and hence there was usually a substantial gradient through the plant (i.e., 12 bars).

The dramatic increase in leaf water potential (Fig. 10) corresponded to the beginning of precipitation and it is presumed that the measurement of leaf water potential was in error owing to surface moisture on the leaves. Water on the leaf surface evaporated freely as it was not subjected to internal leaf resistances to water vapor transfer that control transpiration. As a result of surface evaporation, there was a much higher latent heat flux and lower leaf temperatures on 18 July than on 17 July.

In general, the four non-vascular plant microsites at the raised beach lost water faster than *Dryas*. These microsites, therefore, are primarily responsible for the dry conditions but temperature gradients in soil are also of importance. Since vapor pressure is temperature dependent, there was a gradient in vapor pressure established from the warm surface to the permafrost resulting in a downward water flux. Extremely dry conditions on the beach ridge are characterized by fruticose lichens that increase evaporation from the soil and coarse bare soil that permits rapid drainage. In areas where the texture of the soil is finer, more moisture is held in the upper layers and crustose lichens predominate.

Hummocky Sedge-Moss Meadow

Carex stans at the meadow had a low transpiration rate even though it was rooted in a soil that was saturated with water. The explanation for this appears to lie in reduction of root permeability to water as a result of cold and poorly aerated soil. This reduction limited the flow of water to the leaves and occasionally caused water potentials as low as −40 bars and stomatal closure.

Evapotranspiration from the *Carex* microsite was 72.2 g m^{-2} hr^{-1} over the summer of 1971 (15 July to 20 August) but of that, only 25% was contributed by the *Carex* even though it represented 75% of the evaporative surface area. In general, *Carex* reduced evaporation from the moss and microsites with moss alone lost 79.2 g m^{-2} hr^{-1}. Evaporation from standing

Table 5. Evapotranspiration from the intensive meadow and intensive raised beach site for 1971-1973. Figures for 1972 and 1973 are estimates.

Site	*Seasonal water loss ($g\ m^{-2}\ hr^{-1}$)*		
	1971	*1972*	*1973*
Meadow	73.1	83.3	89.0
Raised Beach	33.5	38.3	41.0

water averaged 67.4 $g\ m^{-2}\ hr^{-1}$ in 1971 and the difference between this site and the moss resulted from the larger surface area of the moss. Moss surface area was 2.5 times that of standing water relative to a plane of equal perimeter, but when the moss was saturated only a few stems protruded above the water resulting in only a slightly greater surface area. These conditions were maintained for most of the summer. Moss and standing water, therefore, are responsible for most of the water loss from the meadow and there appears to be little biological control on evapotranspiration at this site.

On a seasonal basis, the meadow lost slightly over two times the amount of water as the raised beach ridge (Table 5). Seasonal averages for 1972 and 1973 are estimates based on the energy relations of the two sites during 1971 and 1972 and on radiation data from Courtin and Labine (this volume). It should be noted that there was very little difference in evapotranspiration rate (E) among the years and most of the great difference in E between 1972 and 1973 as shown by Rydén (this volume) was a result of a much longer snow free period (45 days in 1972 vs. 76 days in 1973).

Plant resistances

Resistance to water and heat loss from a plant community is an indication of how well the plants are adapted to withstand water deficit. The role of morphology and canopy structure in the water relations of a plant can be quantified by measuring the resistance offered by the plant to the transfer of water. This is especially important in xeric areas where water is often limiting to plant growth and development (Kramer 1969).

Methods and materials

Resistances to water vapor (R_w) and sensible heat fluxes (R_T) were determined for both vascular species using the hourly temperature readings from a thermister at 15 cm, relative humidity from a hygrothermograph (15 cm), leaf temperature from a thermocouple network, LE from lysimeter measurements, and H from equation 1.

Results and discussion

The resistance measurements, made in an attempt to quantify plant adaptation to water deficits (Table 6), are subject to criticism. The values only give an indication of how the plant responds to various environmental factors such as wind speed, radiation load, and relative humidity. The greatest source of error is probably that leaf temperature was measured on the raised beach and on the meadow roughly 100 m from the intensive study

Table 6. Vascular plant resistances (sec cm^{-1}) to water vapor and sensible heat transfers.

Plant species	Date	R_w		R_T
		$R_a + R_c$	R_l	
Dryas integrifolia	17 July, 1972	.29	20.08	.32
	18 July, 1972	.36	1.80	.40
	24 July, 1972	.47	4.88	.52
Carex stans	17 July, 1972	.26	15.74	.27
	18 July, 1972	.15	7.80	.16
	24 July, 1972	.63	15.71	.69

sites that recorded ambient conditions. The temperature (15 cm) at the stations was comparable to that recorded at the intensive study sites (within 2°C under stable conditions with a 4°C variation under fluctuating thermal regimes). Since H and LE are dependent upon temperature gradients between leaf and air, accurate temperature measurements (i.e. ±0.5°C) are necessary to obtain reliable resistance values and hence the error.

The major resistance to water loss from *Dryas* is that offered by the leaf (R_l) and its tomentose lower epidermis (Table 6). The boundary layer (R_a) and canopy resistances (R_c) represented less than 10% of the total on two of the three days presented. The exception can be accounted for by a reduction in R_l rather than an increase in either R_a or R_c. This reduction in R_l is assumed to be as a result of surface wetting of the leaves.

Internal leaf resistances of *Dryas* varied greatly among the days presented (Table 6). On 17 July, there were fairly high winds and a high radiation load resulting in a water deficit with leaf water potentials of *ca.* −40 bars. Leaf resistance increased until 1300 hrs indicating gradual closure of stomata under moving air conditions (Bange 1953). The maximum resistance recorded on 17 July was 44 sec cm^{-1} and this was comparable with values for White Pine (*Pinus strobus*, 50 sec cm^{-1}) measured by Gates (1966). The resistance values for *Dryas* were also within the range of cuticular resistances for many tree species (37 to 380 sec cm^{-1}) presented by Holmgren et al. (1965). Stomatal closure is not the only factor that results in an increased leaf resistance. *Dryas* leaves curl at the edges as they dry and curling of leaves increases resistance substantially (Slatyer 1967).

The resistances during 24 July were different from those of either of the previous two days. These values represent the minimum resistance possible without surface wetting and were measured under conditions of low light, relative humidity, and moist soil conditions.

Leaf resistance of *Carex* indicates that water deficit at the meadow was relatively constant regardless of environmental conditions (Table 6). No difference between leaf resistances to water vapor transfer on 17 July and 24 July was observed even though the environmental conditions were very different. Since these leaf resistances were both consistent and high, the stomata must have been closed to a similar extent on the two days. This was a result of difficulty in replacing transpired water by absorption from cold and poorly aerated soil and led to low leaf water potentials. The maximum leaf resistance of *Carex* was 28 sec cm^{-1} and this value is much higher than the cuticular resistance of many crop plants (Kuiper 1961; Al-Ani and Bierhuizen 1971). Water on the outside of the leaves on 18 July reduced the measured leaf resistances as it did for *Dryas* but the reduction was less since *Carex* does not have a layer of hairs that can trap precipitation.

In general, both *Carex* and *Dryas* offer high resistances to water transfer from their tissue to the air. *Dryas* leaf resistances vary much more than those of *Carex* and the maximum observed for *Dryas* was twice that for *Carex*. The greater variation of *Dryas* leaf resistances indicates that this plant may be better adapted to withstand water deficit. *Dryas* has a higher resistance when the stomata are closed but it is able to transpire freely on days with moderate light loads and high soil moisture.

Ecological implications

The cushion growth form is predominant on the raised beach ridges. This growth form, exemplified by *Dryas integrifolia*, has a significant effect on how absorbed energy is dissipated from the plant. The cushion is tightly appressed to the ground surface and may be considered to lie within the earth's boundary layer. This results in an increased resistance to sensible and latent heat fluxes and causes high leaf temperatures and most of the incoming energy to be emitted as longwave radiation.

The crests of raised beaches are exposed to wind at all seasons which results in a shallow snow cover during the winter months (Courtin and Addison 1973). Courtin (1968) stated that abrasion by wind-borne ice particles in winter was a major cause of damage to krummholz in the alpine tundra and that the cushion growth form was well adapted to minimize this damage because of reduction of wind speed near the surface. This enables *Dryas* to colonize even the most exposed sites. In many such areas, the centre (higher) portion of the plant is dead, probably either as a result of abrasion of the most exposed parts during the winter or senescence of the oldest parts.

Dryas leaves decompose *in situ* producing an organic layer between live leaves and the surface of the soil. A similar observation was made by Heilborn (1925) for a number of cushion plants of the Ecuadorian Paramos and Courtin (1968) for *Diapensia lapponica* on Mt. Washington, New Hampshire. It appears, therefore, that this may be a relatively common characteristic of the cushion growth form. Although this layer of organic material is probably more important for plant growth in the nutrient budget than in the energy budget of the plant, it significantly reduces the soil heat flux beneath the plant. The insulative properties are sufficiently good to isolate the plant leaves from the soil with respect to temperature.

Dryas leaves are adapted to the extreme environmental conditions to which they are exposed. The cuticle reduces latent heat flux by minimizing cuticular transpiration from the upper surface. The densely tomentose lower epidermis reduces sensible heat flux from the abaxial surface. Thus two major methods of energy dissipation are largely restricted to only one leaf surface and the efficiency of energy transfer is reduced substantially. High *Dryas* leaf temperatures (i.e. 45°C) seen on 17 July as a result of the inability of the plant to dissipate energy. High leaf temperatures increase respiration much more than they do photosynthesis and result in a decrease in net photosynthesis (Mayo et al. this volume). Frequent incidence of cloud in the Arctic reduces radiation load on the surface (Thompson 1967) and hence high *Dryas* leaf temperatures are rare. Leaf temperatures reflected the radiation load and were normally 5° to 15°C above ambient air temperature during the high sun hours (0600 to 1800 hr). Since tissue temperatures were usually above those of ambient air, plant physiological processes were more active when air temperatures indicated that the process should be temperature limited. The probable advantage of this is to give the plant a longer period of adequate growing conditions, an important consideration in a site where plants must withstand severe drought over much of the snow-free period. An increase in latent heat flux has been shown from rain-wetted *Dryas* leaves. Evaporation of this water tends to keep leaf temperature low and it is especially important if a day of high incoming radiation follows precipitation.

The presence of lichens on the beach ridge appears to have a significant effect on the water relations of the site. Fruticose lichens had a higher evaporation rate than did adjacent bare soil. Because both coarse bare soil and lichens had a net gain of water over the summer and because there was little or no vascular plant cover in the areas dominated by these lichens, the impact of fruticose lichens on the vascular plants was minimal. The crustose lichens, on the other hand, were closely associated with vascular plants and were effective in conserving water as compared with adjacent bare soil. These lichens, therefore, improved the soil water relations of the site, a definite advantage to both vascular and non-vascular plants in an area where water appears to be limiting.

The meadow was not, in general, a water limited system. Vascular plants must absorb water through their roots, but both cold soil and low oxygen tensions at this site reduce membrane permeability and, therefore, water uptake. As a result *Carex* plants occasionally showed stress conditions similar to those of *Dryas* on the raised beach (*ca.* −40 bars). Soil factors appeared to impede water uptake sufficiently to cause stomatal closure with only a limited evaporative demand. The characteristic "midday stomatal closure" frequently observed in temperate species under water deficit appears to be expanded to a 12 hr period (*ca.* 0700 to 1800 hr). It was also noted that there was little variation in the leaf resistance of *Carex* during the day, indicating that fluctuations in stomatal aperature were minimal. Although the incident energy received by *Carex* was similar to that of *Dryas*, net radiation was higher because of lower reflectance of shortwave radiation and cooler leaf temperatures resulting in a lower reradiation term. Leaf temperature in itself is a reflection of the energy dissipating capabilities of a plant or plant community. *Carex*, because of its upright growth form, sparse canopy and large effective surface area, was able to dissipate energy much more easily than *Dryas* and, therefore, leaf temperature was lower. A comparison of canopy and boundary layer resistance of the two plants showed that when the wind speed was the same (17 July at the beach ridge and 18 July at the meadow), the resistance to energy dissipation offered by *Dryas* was twice that by *Carex*.

Both physical (boundary layer + canopy) and physiological (leaf) resistances to water vapor and sensible heat transfers are higher in *Dryas* than they are in *Carex*. High resistance to water vapor transfer implies a high resistance to CO_2 entry which, in turn, implies a lower photosynthetic rate. The maximum rate of carbon assimilation by *Dryas* was 4.2 mg CO_2 g^{-1} hr^{-1} as compared with 13 mg CO_2 g^{-1} hr^{-1} by *Carex* (Mayo et al. this volume). A similar pattern was observed when *Diapensia lapponica* was compared with *Carex biglowii* (Hadley and Bliss 1964, Courtin 1968) in an alpine environment. Low photosynthetic rates of *Dryas* were overcome to a large extent by positive net assimilation rates throughout the 24 hr arctic day except under high light loads when leaf temperatures were high (20°C) (Mayo et al. this volume).

The canopy of the *Carex* microsite (*Carex stans* with an understory of moss) had a marked effect on the soil energy flux. Because of the two-layered structure on the plant community, most of the incident radiation was intercepted before it reached the ground. Also, with over 50% of R_n dissipated as LE, surface temperatures were low and there was a shallow gradient between the surface and the permafrost below. This resulted in a reduction of over 50% in the soil heat flux compared with an area devoid of vegetation (i.e. a shallow pool). A self-perpetuating system was thus produced, with the plant community preventing the frost table from thawing, the frost table reducing drainage, and the resulting waterlogged conditions encouraging the establishment of a *Carex*-moss community.

Carex stans, although it does not appear to be well adapted to saturated soils, is able to survive and function under these conditions. Since most vascular species are not able to withstand poorly aerated soil, competition for nutrients, light, and space is much reduced with the exclusion of many species from these areas. This results in a distinct ecological advantage for this species in waterlogged soils. Waterlogging is also of advantage to many moss species as they do not have a root system to replenish the water lost through evaporation. Since the bases of the moss stems are in water at the meadow, water movement to the aerial parts is possible by capillary action on the outside of the stem. This ensures that water is plentiful in the growing tissue at the apex of the shoot.

In general, the raised beach presents a much more hostile environment for plant growth than does the meadow. *Dryas integrifolia* at the beach ridge appears to be well adapted to withstand water deficit and to trap incoming radiation, whereas *Carex stans* at the meadow does not seem to have either of these adaptations. *Carex* is, however, able to withstand cold and poorly aerated soil conditions that characterize the highly organic meadow sites on Devon Island.

Acknowledgments

The author wishes to express his appreciation to G.M. Courtin, Department of Biology, Laurentian University, for his guidance throughout this project and the use of his micrometeorological data and to E.I. Mukammal of the Atmospheric Environment Service (Environment Canada) for his advice on the measurement of radiation fluxes.

Field work was made possible by grants from the National Research Council of Canada (A-5071) from I.B.P.-N.R.C., and by the loan of many instruments from the Meteorology Branch of the Atmospheric Environment Service. Thanks are also due to C.L. Labine for his help with fieldwork.

References

Addison, P.A. 1973. Studies on Evapotranspiration and Energy Budgets in the High Arctic: A Comparison of Hydric and Xeric Microenvironments on Devon Island, N.W.T. M.Sc. thesis. Laurentian Univ., Sudbury, Ontario. 119 pp.

Ahrnsbrak, W.F. 1968. Summertime radiation balance and energy budget of the Canadian tundra. Technical Report No. 37. Dept. Meteorol., Univ. Wisconsin, Madison, Wisconsin. 50 pp.

Al-Ani, T.A. and J.R. Bierhuizen. 1971. Stomatal resistance, transpiration and relative water content as influenced by soil moisture stress. Acta. Bot. Neerl. 20: 318-327.

Bange, G.G.J. 1953. On the quantitative explanation of stomatal transpiration. Acta Bot. Neerl. Z: 255-296.
Bliss, L.C. 1971. Arctic and alpine plant life cycles. Ann. Rev. Ecol. System. 2: 405-438.
Clum, H.H. 1926. The effect of transpiration and environmental factors on leaf temperatures II. Light intensity and the relation of transpiration to the thermal depth point. Am. J. Bot. 13: 217-230.
Courtin, G.M. 1968. Evapotranspiration and Energy Budgets of Two Alpine Microenvironments, Mt. Washington, New Hampshire. Ph.D. thesis. Univ. Illinois, Urbana. 172 pp.
———. and P.A. Addison. 1973. Seasonal and diurnal microclimate above and belowground for a raised beach ridge, Truelove Lowland, Devon Island, N.W.T. *In:* Proceedings of the Symposium on Production Processes and Photosynthesis, C.C.I.B.P. K.M. King (ed.). Guelph, Ontario. 383 pp.
———. and L.C. Bliss. 1971. A hydrostatic lysimeter to measure evapotranspiration under remote field conditions. Arctic Alp. Res. 3: 81-89.
———. and J.M. Mayo. 1975. Arctic and alpine plant water relations. pp. 201-228. *In:* Physiological Adaptation to the Environment. F.J. Vernberg (ed.) Intext Educational Publishers, N.Y., N.Y. 576 pp.
Fry, K.E. 1965. A Study of Transpiration and Photosynthesis in Relation to Stomatal Resistance and Internal Water Potential in Douglas Fir. Ph.D. thesis. Univ. Washington, Seattle, 167 pp.
Gates, D.M. 1965. Energy, plants and ecology. Ecology 46: 1-13.
———. 1966. Transpiration and energy exchange. Quart. Rev. Biol. 41: 353-364.
Grable, A.R., R.J. Hanks, F.M. Willhite and H.R. Haise. 1966. Influence of fertilization and altitude on energy budgets for five native meadows. Agronomy J. 58: 234:237.
Haag, R.W. and L.C. Bliss. 1974. Functional effects of vegetation on the radiant energy budget of boreal forest. Can. Geotech. J. 11: 374-379.
Hadley, E.B. and L.C. Bliss. 1964. Energy relationships of alpine plants on Mt. Washington, New Hampshire. Ecol. Monogr. 34: 331-357.
Heilborn, O. 1925. Contributions to the ecology of the Ecuadorian Paramos with special reference to cushion-plants and osmotic pressure. Svensk Botanisk Tidskrift 19: 153-170.
Holmgren, P., P.G. Jarvis and M.S. Jarvis. 1965. Resistance to carbon dioxide and water vapor transfer in leaves of different plant species. Physiol. Plant. 18: 557-573.
Kohnke, H., F.R. Dreibelbis and J.M. Davidson. 1940. A survey and discussion of lysimeters and a bibliography on their construction and performance. U.S.D.A. Misc. Pub. Washington, D.C. 372.
Kramer, P.J. 1969. Plant and Soil Water Relationships: A Modern Synthesis. McGraw-Hill, New York. 482 pp.
Kuiper, P.J.C. 1961. The effects of environmental factors on the transpiration of leaves, with special reference to stomatal light response. Meded. Landb. Hoogesch. Wageningen 61: 1-49.
Mayo, J.M., D.G. Despain and E.M. van Zinderen Bakker, Jr. 1973. CO assimilation by *Dryas integrifolia* on Devon Island, N.W.T. Can. J. Bot. 51: 581-588.
Mellor, R.S., R.B. Salisbury and K. Raschke. 1964. Leaf temperatures in controlled environments. Planta 61: 56-72.
Miller, P.C. and L. Tieszen. 1972. A preliminary model of processes affecting primary production in arctic tundra. Arctic Alp. Res. 4: 1-19.

Mukammal, E.I. 1971. Comparison of pine forest evapotranspiration estimated by energy budget, aerodynamic and Priestly methods. Unpublished Report of Atmospheric Environment Service. Environment Canada, Ottawa.

Pelton, W.L. 1961. The use of lysimetric methods to measure evapotranspiration. *In:* Canada Department of Northern Affairs and Natural Resources, 1961. Proceedings of Hydrology Symposium No. 2. Evaporation. Ottawa. 263 pp.

Reifsnyder, W.E. and H.W. Lull. 1965. Radiant energy in relation to forests. Tech Bull. No. 1344. U.S.D.A. For. Serv. Wash. D.C. 111 pp.

Romanova, E.N. 1971. Microclimate of tundras in the vicinity of the Taimyr Station. pp. 35-44. *In:* Biogeocenoses of Taimyr Tundra and Their Productivity. B.A. Tikhomirov (ed.). Leningrad: Nauka. 239 pp. Transl. P. Kuchar, Univ. Alberta, Edmonton, Alberta.

Savile, D.B.O. 1972. Arctic adaptations in plants. Monogr. 6, Canada Department of Agriculture, Ottawa. 81 pp.

Slatyer, R.O. 1967. Plant-Water Relations. Experimental Botany, An International Series of Monographs. Vol: 2. Academic Press. New York. 366 pp.

Spanner, D.C. 1951. The Peltier Effect and its use in the measurement of suction pressure. J. Exp. Bot. 2: 145-168.

Thompson, F.B. and L. Leyton. 1971. Method for measuring the leaf surface area of a complex shoot. Nature 229: 572.

Thompson, H.A. 1967. The Climate of the Canadian Arctic. Canada Department of Transport, Meteorological Branch. Ottawa. 32 pp.

Van Bavel, C.H.M. 1961. Lysimetric measurements of evapotranspiration in the eastern United States. Soil Sci. Soc. Am. Proc. 25: 138-141.

Warren-Wilson, J. 1957. Observations on the temperatures of arctic plants and their environment. J. Ecol. 45: 499-531.

Weller, G. and S. Cubley. 1972. The microclimates of the arctic tundra. pp. 5-12. *In:* Proceedings of the 1972 Tundra Biome Symposium. S. Bowen (ed.). Lake Wilderness Center, Univ. Washington. 211 pp.

Biological nitrogen fixation in High Arctic soils, Truelove Lowland

R.C. Stutz

Introduction

Studies of nitrogen fixation in remote natural systems were made feasible by the development of acetylene reduction assays (Stewart et al. 1967). These assays are based on the observation that acetylene is a competitive inhibitor of nitrogen on the nitrogen fixing enzyme (Dilworth 1966). The assay involves incubating biological material in the presence of acetylene and measuring the rate of ethylene evolution. Ethylene can be detected in minute quantities by gas chromatography. Ethylene production can be related to nitrogen fixation by comparing acetylene reduction with N^{15} assays or changes in total nitrogen content. Generally the ratio between acetylene reduction and nitrogen fixation is between 1.5:1 and 25:1 (Hardy et al. 1973). For soil systems the ratio is generally between 3 and 6:1.

Material and Methods

Study plots were established at four sites including the Intensive Raised Beach Study Site (Fig. 1. Introduction; co-ordinates G1-58), Intensive Meadow Study Site (co-ordinates G1-64), northwest of Immerk Lake (co-ordinates G6-43), and on Wolf Hill (co-ordinates C4-67). Each sample area was defined by three 30 m transects. Six cylindrical soil cores 35 $cm^2 \times 10$ cm were taken at random points along each transect at each sampling time. Sampling took place weekly in the intensive study sites and bi-weekly elsewhere (Stutz and Bliss 1975).

Soil cores were placed intact in 1 or 0.5 1 air-tight incubation jars (Stutz and Bliss 1973). The integrity of the seal was tested with a manual vacuum pump. Acetylene, generated from calcium carbide, was injected into each chamber to make 0.1 atm C_2H_2. In order that the natural ethylene production could be monitored, control chambers from each transect contained soil but no acetylene.

The incubation chambers were placed in a soil pit 10 cm deep and covered with soil to maintain incubation temperatures near those of undisturbed soil. Periodically, soil cores were incubated under standing water so the effects of light on the incubation system could be studied.

Acetylene reduction rates were determined from changes in ethylene concentration between hours 12 and 36. Air samples were withdrawn from

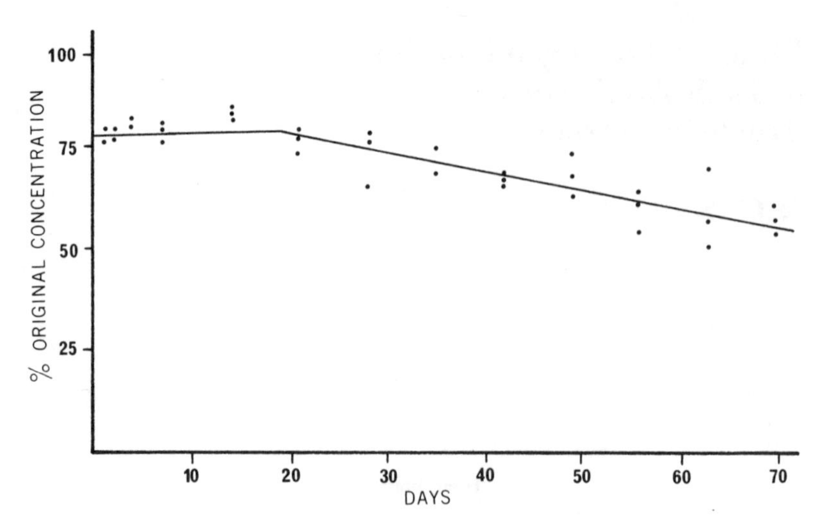

Fig. 1 Percent of original ethylene concentration in serum vials as a function of time.

the incubation chambers at hours 12 and 36, and injected into evacuation serum vials (Stutz and Bliss 1975). Associated with the transfer of gas samples to vials were dilution and leakage factors which were easily accommodated. Fig. 1 demonstrates a dilution factor of 75% and a rate of ethylene leakage of 10% in eight weeks.

To assay the acetylene reducing potential of *Nostoc commune* and certain lichen species, thalli were immersed in pond water for 1 hr at which time they were removed from the water, placed in incubation chambers, and exposed to 0.1 atm C_2H_2 for 12 hr. The chambers were immersed in water during the incubation. Ethylene production was calculated from changes in ethylene concentration between hours 6 and 12. After the incubation period the samples were dried at 105°C and weighed.

Ethylene was detected with a Beckman GC-5 gas chromatograph equipped with a hydrogen flame detector. Dual Poropak R columns were used to separate the hydrocarbons in the sample. Carrier gas (nitrogen) flow rate was 120 ml min^{-1}; column temperature was 60°C, and detector temperature was 95°C. Characteristic retention times were 0.5 min for methane, 1.1 min for ethylene, and 1.3 min for acetylene.

Detector response to methane and ethylene was standardized with commercial analysed gasses. The system was sensitive to concentrations as low as 10^{-12} moles ethylene ml^{-1} and 10^{-13} moles methane ml^{-1}. Acetylene was used as an internal standard to detect major defects in procedures. A six probe Yellowsprings Telethermometer was used alternatively with an Esterline-Angus system to monitor soil temperature.

The Esterline-Angus mv recorder, equipped with a stepping switch, measured output of copper-constantan thermocouples relative to an ice bath. Five probes were read in sequence twice each hour during incubation periods. The probes were placed in the soil at −2 cm and −7 cm and in the incubation jar within the soil pit. Temperature data were reported as daily means derived by dividing degree-hours above 0°C per day by 24 hr.

Duplicate soil samples were collected bi-weekly from the intensive study sites and analysed for KCl-extractable nitrogen using a steam distillation technique (Bremner 1965). KCl-extractable nitrogen was extracted by shaking

20 g soil in 100 ml 6 *N* KCL for 1 hr. The extract was placed in a distillation flask with 5 g Mg, and ammonia was distilled over and collected in borate buffer pH 5.8 which was then titrated with 0.005 *N* H_2SO_4. Nitrate and nitrite were reduced to ammonia and collected in a second distillation.

Populations of algae in the soil were estimated using culture and microscopical observation of soil-water suspensions. Soil was dispersed in nutritive medium, and serial dilutions from 10^{-1} to 10^{-8} g soil ml^{-1} were made in 10 replicates. The cultures were incubated in light at 20°C for 28 days. The number of reproducing units (i.e. cells and colony fragments) was determined from Most Probable Number Tables. The medium was Bold's Basal Medium (Kantz and Bold 1969) which is selective for *Chlorococcum*. Subsequent examination of the algal cultures showed that only *Chlorococcum* was found in the most dilute cultures.

Direct microscopic observations of soil-water suspensions were used to determine algal species composition of fresh soils. A population was characterized by recording frequency of occurrence and colony size of each genus in 30 observations (10 fields of vision on 3 slides). Four frequency classes were used: rare (0-6%), occasional (7-50%), common (51-93%) and abundant (94-100%). Five colony-size classes were used: single cells and clusters (1-5 cells), small colonies and trichomes (6-20 cells), medium colonies (20-100 cells), large colonies (100-500 cells), and macroscopic colonies (more than 500 cells). The product of frequency and colony size for each species divided by the product for all species gives an index of relative abundance of each species.

Results and Discussion

Environmental influences on acetylene reduction

The environmental factors most likely to affect the rate of acetylene reduction are temperature, light, moisture, oxygen, and soluble nitrogen. Moisture was least variable in meadow soils since there was free water in all replicates. Low water potentials may have affected the rates of acetylene reduction on beach ridges. In the closed system of the incubation jar, temperature, oxygen, and light regimes are very much interrelated. Aerobic organisms deplete oxygen levels while photosynthetic organisms, when exposed to light, produce oxygen. The activity of both types of organism is temperature-dependent. The temperature regime of the soil under incubation can be very different from that external to the jar, depending in part on the light regime.

Temperature

Acetylene reduction rates of Truelove Lowland meadow soils (Table 1) and soils of raised beach ridges (Table 2) measured in 1971 were several times higher than rates measured in 1972. This is primarily a result of different incubation methods used for each year. In 1971 incubations were exposed to light (Stutz and Bliss 1973), while in 1972 they were routinely covered by soil. The difference between reduction rates under the two incubation conditions may be considered a function of temperature differences, changes in aeration in the incubation jars, and nitrogen fixation by algae which depend on photosynthesis for energy.

Temperatures of light and dark incubations differed markedly. Temperatures of dark-incubated soil cores closely resembled those of soil at

Table 1. Acetylene reduction rates of meadow soils on Truelove Lowland.

Date	Sample size	Meadow*	μmoles Acetylene m^{-2} hr^{-1} 0-10 cm	0-5 cm	5-10 cm
10 July, 1971	10	Immerk Lake (wet)	1.87		
		ISS (hummocky)	5.18		
		ISS (frost-boil)	7.53		
13 Aug., 1971	7	Wolf (wet)	0.30		
		Immerk Lake (wet)	0.86		
		Wolf (hummocky)	2.47		
		Immerk Lake (hummocky)	2.61		
12 July, 1972	15	Immerk Lake (wet)		0.00	0.00
		Immerk Lake (hummocky)		0.00	0.00
		Wolf (wet)		0.41	0.00
		ISS (hummocky)		0.91	0.12
24 July, 1972	15	Immerk Lake (wet)		0.31	0.00
		Immerk Lake (hummocky)		0.79	0.01
		Wolf (wet)		0.97	0.05
		ISS (hummocky)		2.21	0.23
6 Aug., 1972	15	Immerk Lake (wet)		0.72	0.00
		Wolf (wet)		0.83	0.00
		Immerk Lake (hummocky)		1.26	0.10
		ISS (hummocky)		2.35	0.26

*Areas not scored by the same vertical line are significantly different by Duncans Multiple Range Test, P = 0.95. Soils were incubated in the light in 1971 and in the dark in 1972.

−7 cm. There was relatively small diurnal and seasonal variation and, generally, temperatures remained below 10°C (more so in wet than dry soils). Temperature of soil cores exposed to light showed much diurnal fluctuation, even more than the water under which they were incubated (Fig. 2) and were often much higher than temperatures of dark-incubated cores. Acetylene reduction rates of soil cores were thereby affected.

An experiment in which 18 soil cores selected for uniformity from the Intensive Meadow Study Site were incubated under two temperature regimes, was used to estimate the effect of temperature on acetylene reduction in the Lowland. One group of 9 cores was incubated in a soil pit which had a 24 hr mean temperature of 4.7°C (range 2.6 to 5.2°C); the other group was incubated in a soil pit which had a 24 hr mean temperature of 1.9°C (range 0.0 to 4.8°C). Acetylene reduction was 145±20 $\mu\mu$moles g $soil^{-1}$ hr^{-1} at the higher temperature and 90±8 $\mu\mu$moles g $soil^{-1}$ hr^{-1} at the lower temperature; 1.6 fold increase in rate with 2.8°C rise in mean temperature.

Aeration

While acetylene reduction is an anaerobic process, the anaerobic organisms may depend on high energy substrates associated with aerobic metabolism. Therefore the interface between aerobic and anaerobic soil constitutes an important environment for nitrogen fixation. This interface can be assumed to occur when soil mottling is present. On the Lowland, hummocky sedge-moss meadows have high potential for an aerobic-anaerobic interface. These meadows also show the highest rates of acetylene reduction of any habitat (Table 1). The wet sedge-moss meadows are waterlogged. Methane was produced in large amounts in these meadows (up to 10^3 μmoles m^{-2} hr^{-1}) suggesting that anaerobic conditions prevail. On the other hand, raised beach

Table 2. Acetylene reduction rates of beach ridge soils on Truelove Lowland.

Date	*Sample size*	*Location**	*µmoles Acetylene m^{-2} hr^{-1} (0-10 cm)*
14 June, 1971	17	Wolf Hill	0.23
10 July, 1971	25	Wolf Hill	0.40
		ISS	0.72
13 Aug., 1971	37	Immerk Lake	2.63
		ISS	3.18
20 June, 1972	15	ISS	0.00
		Immerk Lake	0.00
		Wolf Hill	0.13
6 July, 1972	9	ISS	0.02
		Immerk Lake	0.03
		Wolf Hill	0.08
12 July, 1972	15	Wolf Hill	0.08
		Immerk Lake	0.13
		ISS	0.14
24 July, 1972	15	ISS	0.13
		Immerk Lake	0.15
		Wolf Hill	0.16
6 Aug., 1972	15	Wolf Hill	0.10
		ISS	0.41
		Immerk Lake	1.10

*Areas not scored by the same vertical line are significantly different by Duncans Multiple Range Test, P = 0.95. Soils were incubated in the light in 1971 and in the dark in 1972.

ridge soils are so well aeriated that they may lack anaerobic conditions. In these cases acetylene reduction rates were an order of magnitude below those in hummocky sedge-moss meadows. The average (±S.E.) rate of acetylene reduction on raised beach ridges, wet sedge-moss meadows, and hummocky sedge-moss meadows was 0.62 (±0.95), 0.59 (±0.35) and 2.31 (±0.42) respectively.

Soluble nitrogen

The presence of ammonium or nitrate in the soil may repress rates of nitrogen fixation. Analysis of KCl-extractable nitrogen in soils of the Lowland showed up to 10 ppm NH_4^+ −nitrogen on meadow hummocks, but no more than 2 ppm elsewhere (Table 3). No nitrate was detected in freshly extracted soils. The absence of detectable nitrate in the soil afforded the opportunity to assess the effects of soluble nitrogen on acetylene reduction. Replicate soil samples from a hummocky sedge-moss meadow were placed in incubation jars with either 10 ml water or 10 ml solution containing 118 mg $Ca(NO_3)_2$ 1^{-1} (20 µg N ml^{-1}) and incubated in the dark for 36 hr. Subsequent analysis of the soil showed 1 ppm NO= −nitrogen. Acetylene reduction rates were 2.34±0.15 µmoles m^{-2} hr^{-1} in controls and 1.32 0.20 µmoles m^{-2} hr^{-1} in the treated samples; 56% inhibition with 38% increase in a soluble nitrogen.

Biological agents of acetylene reduction

Symbiotic nitrogen fixation by vascular plants is not apparent on the Lowland. Plants commonly associated with nitrogen fixation, e.g.

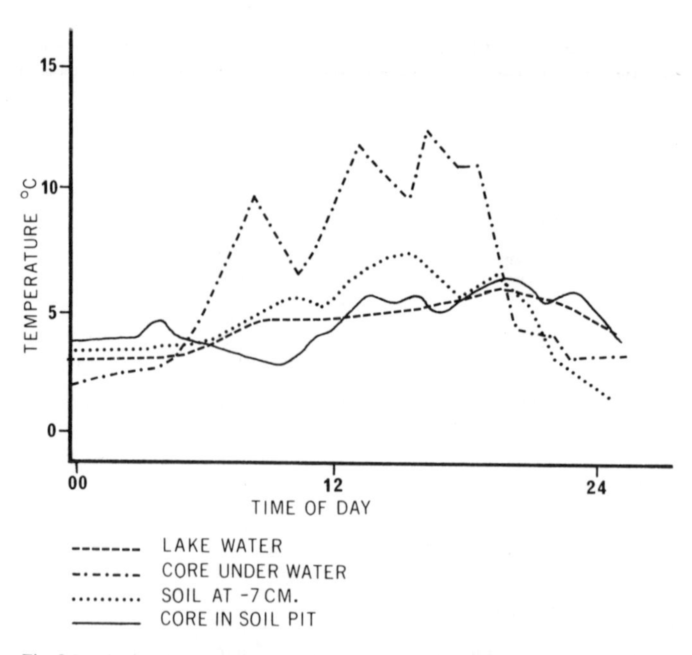

Fig. 2 Incubation temperatures of soil cores in soil pits and under standing water, 8 August, 1971.

Leguminosae and *Alnus* spp. are absent. Conspicuous nitrogen-fixing nodules have been reported on roots of *Dryas integrifolia* in Alaska (Lawrence et al. 1967) and on Disko Island near Greenland (M. Lewis, pers. comm.), but were not seen on the plants growing on Truelove Lowland even though numerous observations of roots were made throughout three growing seasons.

Of lichen species which were tested for acetylene reduction, only *Peltigera aphthosa* reduced acetylene. This species is found only rarely, occupying less than 1% cover on some lakeside meadows.

Blue-green algae are important nitrogen fixers in several types of ecosystems (Cameron and Fuller 1960, Cameron and Devany 1970, Granhall 1970, Paul et al. 1972, Alexander and Schell 1973). On Truelove Lowland, *Nostoc commune* forms a prominent feature of meadows. Its biomass was estimated as high as 39 mg m^{-2} in favorable habitats, averaging 17 mg m^{-2} in hummocky sedge-moss meadows and 9 mg m^{-2} in wet sedge-moss meadows.

Table 3. Soil analysis for ammonium and nitrate nitrogen in beach ridge and mesic meadow soils on Truelove Lowland, 1972.

	ppm Nitrogen					
	Beach Ridge		*Hummocks*		*Interhummock*	
Date	NH_4^+	$NO_3^=$	NH_4^+	$NO_3^=$	NH_4^+	$NO_3^=$
1 July	—	—	4.5	0.0	1.4	0.0
15 July	2.0	0.0	8.4	0.0	1.7	0.0
1 August	0.5	0.0	3.3	0.0	2.3	0.0
15 August	1.6	0.0	9.6	0.0	3.4	0.0

Table 4. Acetylene reduction by *Nostoc commune*.

Date	*Conditions*	*Mean incubation temperature (°C)*	*$\mu\mu$moles C_2H_4 mg^{-1} hr^{-1}*
July 1971	Field	?	6.2
Sept 1971	Growth chamber	0.5	33.4
Sept 1971	Growth chamber	5.5	53.0
July 1972	Field	5.0	44.0
July 1972	Field	12.0	22.8

Table 5. Relative abundance of algae families in soils of Truelove Lowland, Devon Island, 1972.

		% of Total Algae							
Habitat	*Date*	*Chlorococcaeceae*	*Desmidaceae*	*Protococcaceae*	*Naviculariaceae*	*Chroococcaceae*	*Nostocaceae*	*Oscillatoriaceae*	*Scytonemataceae*
Beach Ridge Crest	16 July	34.8	0.3	—	—	—	64.9	—	—
	8 Aug	—	1.5	—	1.6	46.0	4.9	46.0	—
Beach Ridge Slope	11 July	5.5	—	—	0.3	22.1	71.8	0.3	—
	16 July	34.7	—	10.4	26.8	10.4	16.6	1.1	—
	8 Aug	17.8	—	1.2	3.5	29.9	41.4	—	6.2
Meadow Hummock	20 July	13.0	—	—	—	—	87.0	—	—
	26 July	8.3	—	—	—	2.5	8.3	80.9	—
	1 Aug	2.8	—	—	—	33.0	64.2	—	—
	19 Aug	1.3	—	—	—	31.3	67.4	—	—
Meadow Interhummock	20 July	16.0	—	—	1.8	—	81.3	0.2	0.7
	23 July	8.2	—	—	21.2	—	70.6	—	—
	26 July	8.8	—	—	6.8	75.1	6.8	—	2.5
	3 Aug	0.3	—	—	2.2	10.1	87.4	—	—
	8 Aug	0.3	0.8	—	2.9	21.1	73.4	1.5	—

Table 4 shows the results of several assays for acetylene reduction in *Nostoc commune*. The maximum rate observed was 53$\mu\mu$moles C_2H_2 mg^{-1} hr^{-1}. This rate is much lower than the rate reported for the species at Barrow Alaska (1360 $\mu\mu$moles mg^{-1} hr^{-1}) (Alexander et al. 1974), but within the range reported for Abisko, Sweden (Granhall and Selander 1973).

In addition to the large colonies of algae on the surface of meadow soils, numerous microscopic forms were identified in the soil. Blue-green algae (Chrococcaceae, Nostocaceae, Oscillatoriaceae, and Scytonemataceae) dominate the soil microflora, usually providing 75% to 96% of the algal biomass (Table 5). A similar algal composition characterizes arctic soils in the U.S.S.R. (Novichkova-Ivanova 1972).

Dilution culture enumerations of *Chlorococcum* spp. (Chlorococcaceae) are given in Table 6. A maximum population of 1.2×10^4 cells g $soil^{-1}$ was measured in mid-July in a hummocky sedge-moss meadow at which time Chlorococcaceae represented 31% of the soil algae (Table 6). Accordingly, a total algal population of 9.2×10^4 cells g $soil^{-1}$ for 20 July in meadow soil can be deduced. Table 7 gives calculated population (cells g $soil^{-1}$) for each algal family.

Table 6. Population of *Chlorococcum* spp. in soils of Truelove Lowland as determined by dilution culture. Samples were taken in 1972 and were incubated in Bold's Basal Medium at 20°C for 28 days. Number of cells per gram air dry soil was taken from Most Probable Number tables.

		Number g soil^{-1} (air dry wt)		
Habitat	*Depths*	*1 July*	*20 July*	*1 August*
Beach Ridge				
Crest	0-2 cm	0.0	2.5 x 10	0.0
Slope	0-2 cm	9.3 x 10^2	2.0 x 10^3	10
Mesic Meadow				
Hummock	0-2 cm	6.0 x 10	1.2 x 10^4	10
	2-5 cm	ND	1.5 x 10^2	ND
Interhummock	0-2 cm	1.5 x 10^2	5.0 x 10^3	1.0 x 10^3
	2-5 cm	10	ND	ND
Hydric Meadow	0-2cm	ND	3.0 x 10^3	ND
Frost-boil	0-2 cm	10	ND	0.0

ND = Not determined.

The minimum production of algal biomass can be estimated from the difference between the high and low seasonal population levels. Assuming 10^8 algal cells g biomass^{-1}, the minimum production of algae is 1.8 g m^{-2} on meadow hummocks and 6.0 g m^{-2} within the interhummocks (soil density=0.2 c m^{-3}; Brown and Veum 1974). Since hummock and interhummock microsites are equally represented in a hummocky sedge-moss meadow, a minimum production of 3.9 g m^{-2} is estimated for the meadow as a whole.

The importance of blue-green algae as nitrogen fixers in meadow soils was assessed by calculating acetylene reduction rates based on temperature response of soil algae (Table 4). Algae biomass was taken as 10^8 cells g algae^{-1}. An exponential temperature response (Q_{10}) was assumed. Using hourly soil surface temperature measurements (Courtin unpublished data), hourly acetylene reduction rates were calculated. The maximum calculated daily reduction rate was 0.35 μmoles C_2H_4 m^{-2} (0.175 μmoles m^{-2} hr^{-1}), about 10% of the rate measured in the field (Table 1).

The portion of the rate of acetylene reduction in lowland soils which cannot be attributed to soil algae can be assumed to be ascribable to bacteria. Acetylene reduction bacterial populations have not been specifically studied on Truelove Lowland. A census of aerobic bacteria showed no *Azotobacter*. *Bacillus*, a facultative aerobe capable of nitrogen fixation, has been isolated from several soils on the Lowland and is relatively abundant in raised beach ridge soil. Soils from Alaskan tundra contained no aerobic nitrogen fixers, but *Clostridium* and *Bacillus* were isolated anaerobically on nitrogen free medium (Jurgensen and Davey 1971).

Rates of nitrogen fixation

Symbiotic nitrogen fixation appears not to be important on Truelove Lowland. This may be a unique feature of the northern High Arctic since legumes are important in upland sites on Banks and Victoria Islands, Greenland, and northern Baffin Island. Some alder species known to fix nitrogen (Zavitkovski and Newton 1967) are found in low arctic sites, as are legumes in several communities in the Low Arctic of Alaska and the Mackenzie Delta.

Table 7. Algae population in soils of Truelove Lowland, Devon Island, 1972.

Habitat	Date	Number of cells g $soil^{-1}$ (air dry wt)								
		Chlorococaeceae	*Desmidaceae*	*Protococcaceae*	*Naviculariaceae*	*Chroococcaceae*	*Nostocaceae*	*Oscillatoriaceae*	*Scytonemataceae*	*Total*
Beach Ridge	16 July*	2.5×10'	nil	0	0	0	4.7×10	0	0	7.2×10
Crest	8 Aug	0	?	?	?	?	?	?	?	?
Beach Ridge	11 July	?	?	?	?	?	?	?	?	?
Slope	16 July*	2.0×10^{3}	0	6.0×10^{3}	1.5×10^{3}	6.0×10^{2}	9.6×10^{2}	6.3×10	0	6.8×10^{3}
	8 Aug*	1.0×10	0	$1\times10^{\circ}$	$2.0\times10^{\circ}$	1.7×10	2.3×10	0	3.5×10	5.6×10
Meadow	20 July*	1.2×10^{4}	0	0	0	0	8.0×10^{4}	0	0	9.2×10^{4}
hummock	26 July	3.8×10^{3}	0	0	0	1.2×10^{3}	3.8×10^{3}	3.7×10^{4}	0	4.6×10^{4}
	1 Aug*	1.0×10	0	0	0	1.2×10^{2}	2.3×10^{2}	0	0	3.6×10^{2}
	19 Aug	$4.7\times10^{\circ}$	0	0	0	1.1×10^{2}	2.4×10^{2}	0	0	3.6×10^{2}
Meadow	20 July*	5.0×10^{3}	0	0	5.6×10^{2}	0	2.5×10^{4}	6.2×10	2.2×10^{2}	3.1×10^{4}
Interhummock	23 July	1.1×10^{4}	0	0	2.8×10^{4}	0	9.2×10^{4}	0	0	1.3×10^{5}
	26 July	2.0×10^{4}	0	0	1.6×10^{4}	1.7×10^{5}	1.6×10^{4}	0	5.8×10^{3}	2.3×10^{5}
	3 Aug*	1.0×10^{3}	0	0	7.3×10^{3}	3.4×10^{4}	2.9×10^{5}	0	0	3.3×10^{5}
	8 Aug	6.9×10^{2}	1.8×10^{3}	0	6.7×10^{3}	4.8×10^{4}	1.7×10^{5}	3.4×10^{3}	0	2.3×10^{5}

*Chlorococcaceae was enumerated by dilution culture for these time periods. All other data are extrapolations.

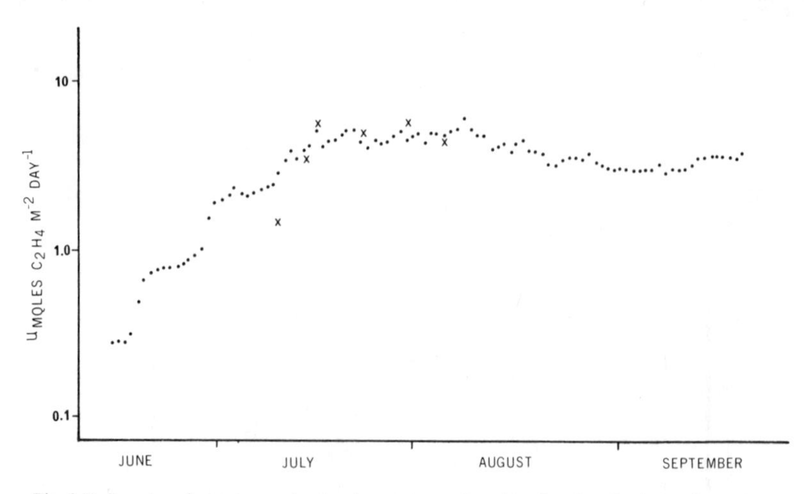

Fig. 3 Daily rates of acetylene reduction in meadow soil on Truelove Lowland as estimated by a temperature-driven model (1972, −7 cm). Measured values are indicated by x.

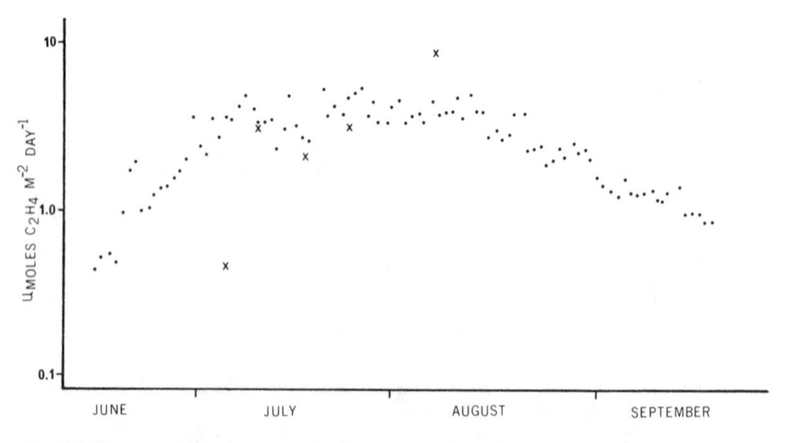

Fig. 4 Daily rates of acetylene reduction in beach ridge soil on Truelove Lowland as estimated by a temperature-driven model (1972, −7 cm). Measured values are indicated by x.

Asymbiotic nitrogen fixation on the Lowland seems to be a function of bacteria as well as algae. This is in contrast to nitrogen fixation in the Low Arctic (Alexander and Schell 1973) and peatlands (Granhall and Selander 1973) where blue-green algae are more important. In Kevo, Finland free-living bacteria and algae are reported to fix about 10% as much nitrogen per year as blue-green algae living epiphytically on sphagnum moss (Kallio 1975). While evidence of epiphytic algae on Truelove Lowland was sought, none was found. For algae on Truelove Lowland to reduce 2.5 μmoles C_2H_2 m^{-2} hr^{-1}, the maximum rate measured in the field (Table 1), a population of 47 g algae m^{-2} performing at maximum rate (53 $\mu\mu$moles mg^{-1} hr^{-1}) would be required. The algal biomass of the Lowland was estimated at 3 g m^{-2} of which potential nitrogen fixers accounted for 10% to 80%.

If, however, the maximum rate of acetylene reduction by algae could be shown to be in the same order of magnitude on Devon Island as was shown in other studies, i.e. 1000 $\mu\mu$moles mg^{-1} hr^{-1} (Alexander et al. 1974, Vlassak

Table 8. Maximum acetylene reduction rates and annual nitrogen fixation rates on Truelove Lowland.

Habitat	*Maximum acetylene reduction rate* μ*moles* $m^{-2}hr^{-1}$	*Annual N Fixation** *(mg* m^{-2} yr^{-1}*)* *1971*	*1972*
Beach ridges	3.18	30	7
Frost-boil sedge meadows	7.53	—	—
Hummocky sedge meadows	5.18	380	120
Wet sedge meadows	1.87	—	—
Soil algae (meadows)	0.35	14×10^{-3}	—
Nostoc commune	(53 $\mu\mu$moles mg^{-1} hr^{-1})	2×10^{-3}	—
Peltigera aphthosa	(5.1 $\mu\mu$moles mg^{-1} hr^{-1})	—	—

*Based on acetylene:nitrogen = 3:1
— Not determined.

et al. 1973), the rates of acetylene reduction measured in the Lowland soils could be attributed to free-living blue-green algae.

Nitrogen fixation of lichens is not important on the Lowland, for only *Peltigera aphthosa* fixes nitrogen and the species is minor in the community composition. In contrast to this high arctic situation, lichens in Kevo, Finland are responsible for most of the biological nitrogen fixation in that sub-arctic ecosystem (Kallio 1975).

The relationship between rates of acetylene reduction and rates of nitrogen fixation is assumed to approach the theoretical value of 3:1 (acetylene:nitrogen). To estimate the annual rate of nitrogen fixation in lowland soils, an empirical equation was used which approximates the response of acetylene reduction to temperature. The equation takes the form of a Q_{10}, although the temperature response of acetylene reduction in soil is more complex than Q_{10}. The value used in the function of Q_{10} was 5.6. When the equation is solved using soil temperatures at −7 cm (the temperature at which the rate of acetylene reduction was measured), calculated rates approximate the measured rates (Fig. 3). It is argued, however, that the greatest nitrogen fixing activity is in the upper 5 cm of soil (Table 1). Using hourly soil temperature at −2 cm, the calculated annual nitrogen input was 380 mg m^{-2} in 1971 and 120 mg N m^{-2} in 1972 (acetylene:nitrogen=3:1).

Attempts to calculate daily and annual fixation rates in beach ridge soils were less reliable (Fig. 4). Calculated fixation rates probably represent maximum potential rates, which are not often obtained in the soil because of low water potentials which develop in these soils. Recognizing this, in beach ridge soils there was an estimated nitrogen input of 30 mg m^{-2} in 1971 and 7 mg m^{-2} in 1972, a cooler summer.

Summary

The acetylene reduction assay for nitrogen fixation suggests the rate at which nitrogen is being injected into the Truelove Lowland ecosystem. Results of this assay are summarized in Table 8 together with our interpretation of what they mean in terms of nitrogen input. The environmental and biological regulations of nitrogen fixation in arctic systems are illusive. Estimated rates of nitrogen fixation range from several grams of nitrogen m^{-2} yr^{-1} in Sweden to less than 1 mg N m^{-2} yr^{-1} in English heaths (Table 9). That the rates of nitrogen fixation on Truelove Lowland are between 7 and 380 mg N m^{-2} yr^{-1} is not a unique feature.

Table 9. Nitrogen fixation by tundra soils.

Location	*mg Nitrogen* m^{-2} yr^{-1}	*Reference*
Abisko, Sweden	200	Rosswall et al. 1975
Hardangervidda, Norway	100-1100	Torsvik 1973
Devon Island, Canada		
(Beachridge)	7-30	This study 1975
(Meadow)	120-380	
Kevo, Finland		
Lichens	150-384	Kallio 1975
Blue-green algae, bacteria	1.8	
Blue-green algae (on moss)	130	
Barrow, U.S.A.	69.3	Bunnell et al. 1975
Moor House, U.K.	0.03-22	Collins 1972

What may be unique is that with respect to nitrogen, Truelove Lowland is a remarkably closed ecosystem. Nitrogen flux resulting from bird migration is insignificant inasmuch as the avian population is so depauperate. With the exception of muskox, the mammalian populations are smaller than bird populations, so nitrogen transport into and out of the ecosystem by animals is virtually nil (see Babb and Whitfield, and Whitfield, both this volume). Nitrogen was not detected in the snow pack or in summer precipitation. Analysis of water from several of the lowland streams shows no trace of ammonium, nitrate, or nitrite, suggesting nitrogen gains and losses from the Lowland via precipitation and erosion are low. Consequently, biological nitrogen fixation represents a major, if not the primary, input of nitrogen into the ecosystem.

Nitrogen losses from the ecosystem have not been studied, but probably include encroachment of permafrost into undecomposed organic matter and biological denitrification of nitrate. However, the absence of nitrate in meadow soils (Table 3) suggests denitrification does not occur.

Taken as a whole the Lowland is a closed ecosystem, but the terrestrial and aquatic phases of the Lowland are very much open to exchange of nitrogen. The major vehicles of nitrogen movement between the two phases may include water- and wind-borne organic debris accumulating in the lakes, insect life-cycles which involve both aquatic and aerial (terrestrial) stages, strong winds which deposit lake sediments over adjacent meadows, and plant succession which may convert lakes to meadows over the years. These factors must be evaluated before an understanding of the nitrogen regime of the Lowland can be complete.

Acknowledgments

This research was developed as partial requirement for a Ph.D. dissertation. Appreciation is expressed to Dr. L.C. Bliss for his supervision and support in all phases of the study. Help with methodology from Drs. Elder Paul and F.D. Cook is gratefully acknowledged. I would also like to thank John Poirier for his field help, and Dr. G.D. Weston for the use of his gas chromatrograph. The study was financially supported by N.R.C. grant, No. A-4879 to L.C. Bliss.

References

Alexander, V. and D. Schell. 1973. Seasonal and spatial variation of nitrogen fixation in Barrow, Alaska tundra. Arctic Alp. Res. 5: 77-88.

————. M. Billington and D. Schell. 1974. The influence of abiotic factors on nitrogen fixation rates in Barrow, Alaska, Arctic tundra. Rep. Kevo Subarctic Res. Stat. 11: 3-11.

Bremner, J.M. 1965. Total nitrogen. *In:* Methods of Soil Analysis. Vol. 2. Am. Soc. of Agron., Madison, Wisconsin. 1965. 1569 pp.

Brown, J. and A.K. Veum. 1974. Soil Properties of the International Tundra Biome Sites. pp. 27-48. *In:* Soil Organisms and Decomposition in Tundra. A.J. Holding, O.W. Heal, S.F. Maclean, P.W. Flanagin (eds.). Tundra Biome Steering Committee. Stockholm. 398 pp.

Bunnell, F.L., S.F. Maclean, Jr. and J. Brown. 1975. Barrow, Alaska, U.S.A. pp. 73-124. *In:* Function of Tundra Ecosystems. T. Rosswall and O.W. Heal (eds.). Ecological Bulletins/N.F.R. No. 20. Stockholm. 450 pp.

Cameron, R.E. and H.J. Fuller. 1960. Nitrogen fixation by some algae in Arizona soils. Soil Sci. Soc. Amer. Proc. 24: 353-356.

————. and J.R. Devany. 1970. Antarctic soil algal crusts: Scanning electron and optical microscope study. Trans. Amer. Microsc. Soc. 89: 264-273.

Collins, V.G. 1972. Nitrogen Fixation. pp. 37-42. 13th Annual Report. Moor House Field Station. Merlewood Res. Stat. Grange-over-Sands, U.K. 46 pp.

Dilworth, J.J. 1966. Acetylene-reduction by nitrogen fixing preparations from *Clostridium pasteurianum*. Biochem. Biophys. Acta. 127: 285-294.

Granhall, U. 1970. Acetylene-reduction by blue-green algae isolated from Swedish soils. Oikos 21: 330-332.

————. and H. Selander. 1973. Nitrogen-fixation in a subarctic mire. Oikos 24: 8-15.

Hardy, F.W.F., R.C. Burns and R.D. Holsten. 1973. Applications of the acetylene reduction assay for nitrogenase. Soil Biology and Biochem. 5: 47-81.

Jurgensen, M.F. and C.B. Davey. 1971. Non-symbiotic nitrogen fixing micro-organisms in forests and tundra soils. Plant and Soil 34: 113-128.

Kallio, P. 1975. Kevo, Finland. 193-223. (see Bunnell et al. 1975).

Kantz, T. and H.C. Bold. 1969. Phycological studies. IX. Morphological and taxonomic investigations of *Nostoc* and *Anabaena* in culture. Univ. Texas Pub. No. 6924, 1969, 67 pp.

Lawrence, D.B., R.E. Schoenike, A. Quispel and G. Bond. 1967. The role of *Dryas drummondii* in vegetation development following ice recession at Glacier Bay, Alaska, with special reference to its nitrogen fixation by root nodules. J. Ecol. 55: 793-813.

Novichkova-Ivanova, L.N. 1972. Soil and aerial algae of polar deserts and arctic tundra. pp. 261-265. *In:* Proc. IV International Meeting on the Biological Productivity of Tundra. F.E. Wielgolaski and T. Rosswall (eds.). Tundra Biome Steering Committee. Stockholm. 320 pp.

Rosswall, T., J.G.K. Flower-Ellis, L.G. Johansson, S. Johansson, B.E. Ryden and M. Sonesson. 1975. Stordalen (Abisko) Sweden. 265-294. (see Bunnell et al. 1975).

Stewart, W.D.P., G.P. Fitzgerald and R.H. Burris. 1967. In site studies on nitrogen fixation using acetylene-reduction technique. Proc. Nat. Acad. Sci. 58: 2071-2078.

Stutz, R.C. and L.C. Bliss. 1973. Acetylene reduction assay for nitrogen fixation under field conditions in remote areas. Plant Soil. 38: 209-213.

———. and L.C. Bliss. 1975. Nitrogen fixation in soils of Truelove Lowland, Devon Island, Northwest Territories. Can. J. Bot. 53: 1387-1399.

Torsvik, V.L. 1973. Norwegian IBP Annual Report. Oslo.

Vlassak, K. E.A. Paul, and R.E. Harris. 1972. Nitrogen fixation. *In:* 5th Annual Rept. pp. 63. Matador Project. I.B.P. Univ. Saskatchewan, Saskatoon. 63 pp.

Zavitovski, J. and M. Newton. 1967. Effect of organic matter and combined nitrogen on nodulation and nitrogen fixation in red alder. pp. 209-233. *In:* Biology of alder. J.M. Trappe, J.F. Franklin, R.F. Tarrant, and G.M. Hansen (eds.). Pacific N.W. For. and Range. Expt. Stat. Portland, Oregon.

Comparison of the estimates of annual vascular plant production on Truelove Lowland made by harvesting and by gas exchange

D.W.A. Whitfield and C.R. Goodwin

Introduction

Annual above and belowground vascular plant production was measured by harvesting methods, for *Dryas integrifolia* and for hummocky sedge-moss meadow gramineae, by Svoboda (this volume) and Muc (this volume), respectively. On the same sites, Mayo et al. (this volume) measured whole plant net CO_2 exchange by *Dryas* and by *Carex stans* using differential IRGA techniques. The latter measurements were made over 24 hr periods at irregular intervals during the growing season. This paper is an effort to compare these two sets of measurements using a simple curve-fit modelling approach. The procedure used is very similar to that of Scott and Billings (1964). This is not a self-consistent model of plant production and was not intended as such. It is a check of consistency between two very different measurement methods which should, within measurement error, give the same answer.

The reader may refer to Mayo et al. for an account of their gas exchange methodology and results. Muc and Svoboda may be consulted for the procedures and results of the harvesting studies. J.M. Mayo (pers. comm.) supplied whole plant net assimilation data (3 to 12 min averages), cuvette air temperature, and, for 1972, incoming shortwave radiation. For 1971, the last-named was obtained from meteorological data supplied by G.M. Courtin (pers. comm.).

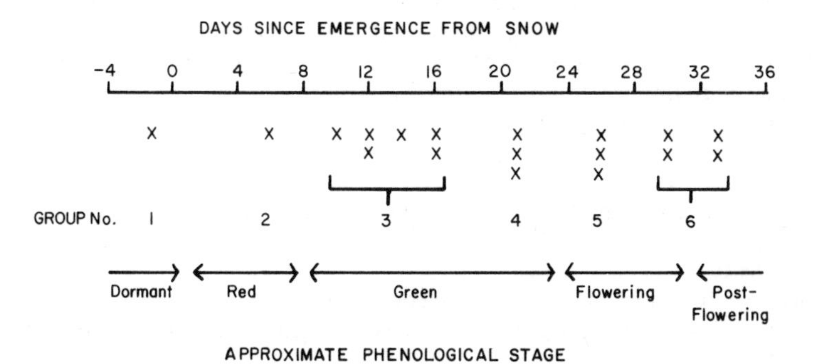

Fig. 1 The location of 1971 and 1972 *Dryas* IRGA runs on temporal and phenological continua, and their aggregation into groups. Each x represents one plant run.

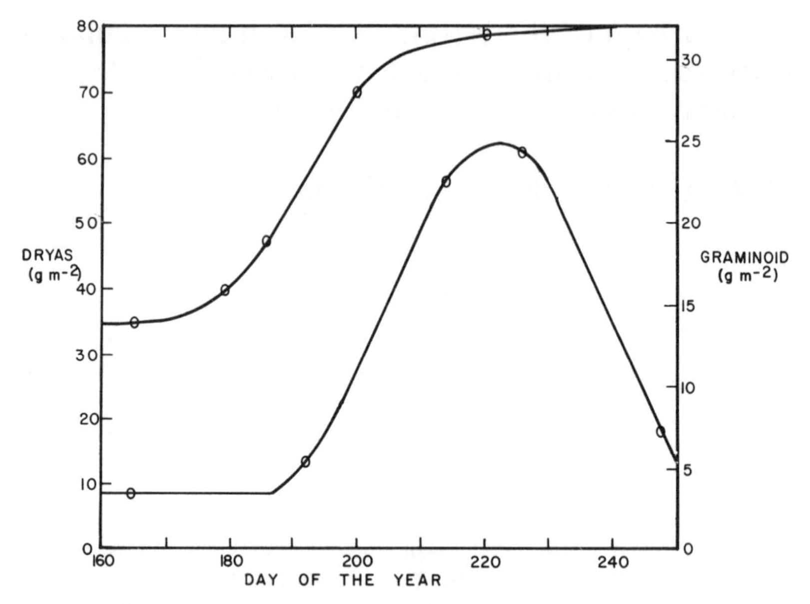

Fig. 2 Seasonal trends in green biomass expressed as g m^{-2}, assuming 100% plant cover, for *Dryas* cushion and graminoids.

Because there were few *Carex* runs, all were pooled under the enforced assumption that net assimilation rates per unit mass of green tissue did not vary over the growing season. The greater volume of data on *Dryas* net assimilation permitted us to take account of phenological changes; the 18 runs were arranged on a phenological continuum according to the estimated number of days since the plant emerged from the snow (Fig. 1). Then the runs were grouped as shown and the data within groups were lumped together.

For the middle four *Dryas* groups and for *Carex*, a stepwise multiple linear regression program was used to fit curves of the form:

$$NA = a+bS+cT+dS^2+eT^2+fST$$

to the measured responses where NA is net similation, S is shortwave radiation, and T is cuvette air temperature. As group 1 of the *Dryas* consisted of plants in dormancy, the NA was zero. The runs of group 6 were made under conditions of very low insolation and over rather narrow temperature ranges and the above procedure led to spurious results. To get around this, we used only S to calculate NA (the S and S^2 terms were those which accounted for the greatest variation in groups 3, 4, and 5).

First the reduced equation:

$$NA=a+bS+dS^2$$

was fitted to groups 5 and 6 data; then a curve was drawn to represent group 6 for $S < 0.1$ and group 5 for $S > .7$ *ly* min^{-1} and to make a smooth transition between the two.

Because the values of T and S were correlated in the field, as can be seen from Figs. 3-5, we could not perform any tests of significance on the results of the curve fitting.

M. Muc (pers. comm.) and J. Svoboda (pers. comm.) supplied data on the seasonal trend of photosynthetic biomass for all graminoids and *Dryas*,

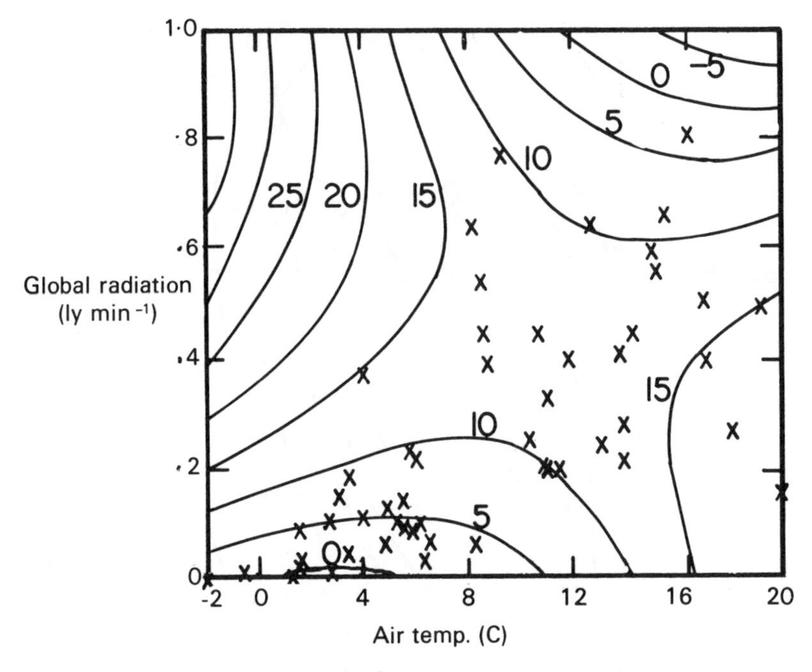

Fig. 3 Isopleths of net assimilation (mg g^{-1} hr^{-1}) by *Carex* as a function of 10 cm air temperature and global radiation. The x's locate the pairs of T and S values at which NA measurements were made, thus indicating the extent to which the isopleths are extrapolations beyond the data.

Table 1. Results of the curve-fittings for *Dryas integrifolia* and *Carex stans* in which a-f are the fitted parameters, R^2 the fraction of the variance removed by the regression, n the number of data points. The equation is
$NA = a+bS+cT+aS^2+eT^2+fST$.

Group	*a*	*b*	*c*	*d*	*e*	*f*	R^2	*n*
Dryas 1			Dormant: net assimilation zero					
2	-.2935	3.5429	.6950	-13.4426	-.1084	1.0516	.583	50
3	-.5675	11.0194	.0880	- 9.0998	-.0085	- .1471	.555	260
4	.4992	10.7727	-.2154	- 6.7056	.0070	- .2511	.752	96
5	.2770	11.8303	-.3620	- 9.1944	.0270	- .4125	.581	99
Carex	0.4018	68.54	-.5190	-37.22	.0834	-3.186	.600	54

respectively. Their data points and the smooth curve drawn through them are shown in Fig. 2. Note that we do not show the very late season partial dieback of *Dryas* leaves, and that in the following we combine data on the NA of one species, *Carex stans*, with green biomass data for all graminoids, under the assumption that *Carex* is representative of them.

Hourly averages of S, and hourly point values of T were extracted from a meteorological data bank supplied by G.M. Courtin (pers. comm.). Temperatures at 10 cm above the soil surface were used.

Daily values for NA on a m^2 basis were determined by calculating an hourly NA per unit photosynthetic biomass, determined from the regressions and 1971 *(Dryas)* and 1972 *(Carex)* meteorological data, and multiplying by the photosynthetic biomass per m^2 for that day. *Dryas* responses were linearly interpolated between phenological groups. Group 6 response was assumed to

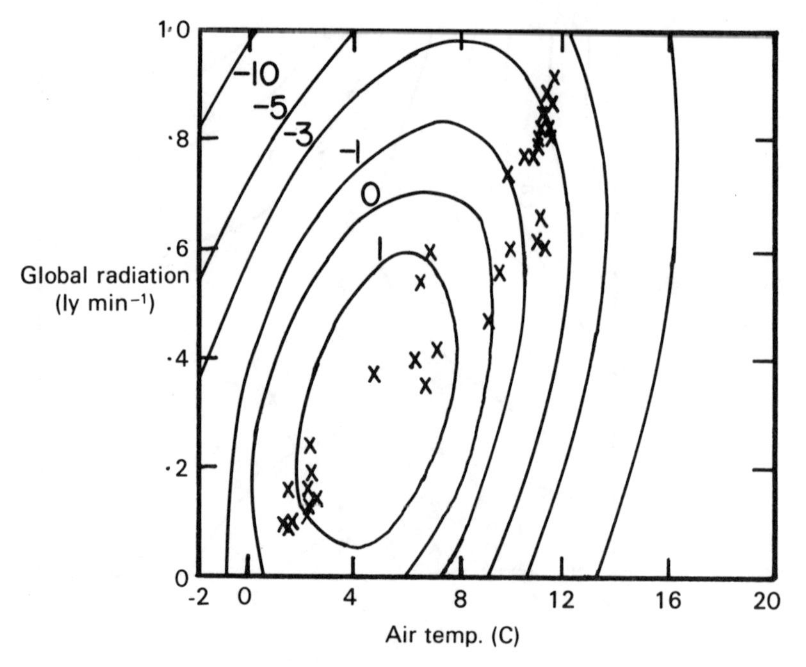

Fig. 4 As Fig. 3, for *Dryas* group 2, with 5 cm temperatures.

continue indefinitely after day 202. CO_2 uptake was converted to biomass accumulated by multiplying by .621, which is appropriate for tissues consisting of 9% protein, 2% lipids, 10% lignin, 77% carbohydrate and 2% minerals. Somewhat surprisingly, both *Dryas* and *Carex* are approximated by this composition. Starting dates for the models were 18 June for *Dryas* and 5 July for *Carex*, corresponding to snowmelt on the beach ridge slope in 1971 and hummocky sedge-moss meadow in 1972, respectively.

Results and Discussion

As representative outputs from the linear regression, Figs. 3, 4, and 5 show isopleths of NA against T and S. Also shown on these figures are the locations of the data points from which the response parameters were derived.

It is evident from Fig. 3 that the *Carex* response for low T and high S is a drastic extrapolation from the data. NA of *Carex* was never observed to reach the high values indicated. Mayo et al. (this volume) report short time peaks of 22 mg g^{-1} hr^{-1}. We therefore imposed an upper limit of 20 mg g^{-1} hr^{-1} on this response. Also, Fig. 3 shows the impossible situation of positive net assimilation at zero light intensity. This is an artifact of the curve fitting procedure brought on by lack of data, under conditions of high temperature and low light intensity. As this region of the S-T plane is very rarely encountered in the meteorological data with which the model was driven, the artifact is unimportant.

Fig. 6 shows the cumulative biomass increment predicted for both species. Most *Dryas* plants were well into dormancy by mid-August. On 15 August the model output was 82% of the harvest estimate (61.3 g m^{-2}, Svoboda, pers. comm.), probably within the compounded errors of harvest sampling and gas

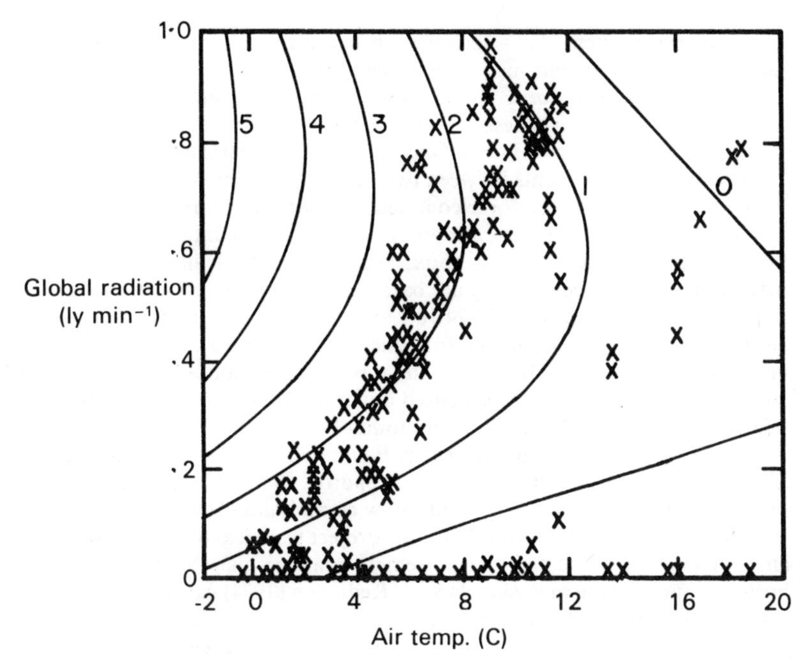

Fig. 5 As Fig. 4, for *Dryas* group 3.

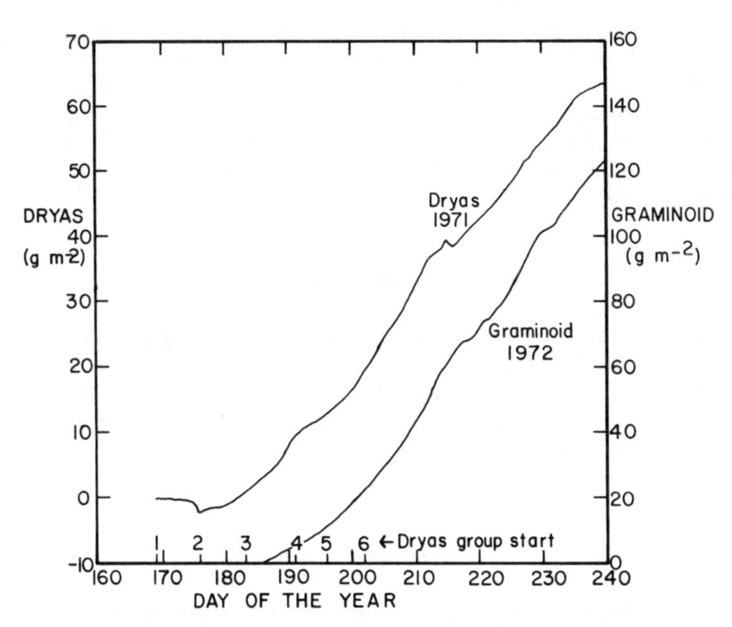

Fig. 6 Cumulative dry biomass increments predicted by the regression model (g m^{-2}). The curve labelled "graminoid" was produced by combining *Carex stans* NA and total graminoid green biomass, as described in the text.

exchange measurement. Thus agreement between the integrated gas exchange and harvest measurements may be considered good.

By 20 August, the *Carex* model showed 103 g m^{-2} biomass accumulation, and by August 30, 126 g. These are 75% and 91% respectively of Muc's estimated 138 g m^{-2} (this volume). Once again we can claim good agreement.

The general agreements between two very different methods of productivity estimation reinforce confidence in both sets of results. However, due caution must be exercised, as both methods could be influenced by systematic errors in the same direction. The graminoid calculations are particularly suspect because they are based on only four 24 hr CO_2 exchange measurements, the earliest on 23-24 July and two in August, and thus they may not represent average seasonal performances. Also, there is a very large statistical uncertainty in Muc's estimate of root production. In fact his present estimate of root production was revised downward from earlier ones (Muc 1973) when very large differences were found between the calculations presented here and the earlier estimates. Examination of the field data led to a different method of calculation, which is reported in his chapter "Ecology and primary production of sedge-moss meadow communities" (this volume).

We can combine data from Matador Project technical reports to make the same whole plant productivity measurement comparison as above, but for a short-grass prairie site in Saskatchewan. Redmann (1974) calculated 1970 net carbohydrate accumulation to be 650 g m^{-2} from his CO_2 exchange measurements. Coupland (1974) found 1048 g m^{-2} dry matter production (corrected for invertebrate consumption) in the same year; thus there is a substantial discrepancy in this case. On the other hand, Botkin et al. (1970) found that gross production in an oak-pine forest measured by CO_2 exchange exceeded harvesting estimates, corrected for forest respiration, by only 22%. Hadley and Bliss (1964) calculated leaf and stem dry matter productions from net assimilation data for several alpine plants and found them to be 50% to 107% of aboveground production determined by harvesting for herbs and 70% to 206% for shrubs. Also, dealing only with aboveground processes, Scott and Billings (1964) found their ratios to be 98% to 172% for *Festuca ovina* and *Deschampsia caespitosa*.

Acknowledgments

At the end of an integrative exercise such as this, it is difficult to give adequate credit to the people who have contributed to it. J. Svoboda, M. Muc, J. Mayo, and L.C. Bliss have all devoted a great deal of time to discussions of their results and ours, and their judgments are woven into the body of the paper in many places. To them we extend our sincere thanks.

References

Botkin, D.B., G.M. Woodwell, and N. Tempel. 1970. Forest productivity estimated from carbon dioxide uptake. Ecology 51: 1057-1060.

Coupland, R.T. 1974. Producers: VI. Summary of studies of primary production by biomass and shoot observation methods. CCIBP Matador Project, Technical Report No. 62. Univ. Saskatchewan, Saskatoon.

Hadley, E.B. and L.C. Bliss. 1964. Energy relationships of alpine plants on Mt. Washington, New Hampshire. Ecol. Monogr. 34: 331-357.

Muc, M. 1973. Primary production of plant communities of the Truelove Lowland, Devon Island, Canada — Sedge meadows. 3-14. *In:* Primary Production and Production Processes, Tundra Biome. Bliss, L.C. and F.E. Wielgolaski (eds.). IBP Tundra Biome Steering Committee. Edmonton. 256 pp.

Redmann, R.E. 1974. Photosynthesis, plant respiration and soil respiration measured with controlled environment chambers in the field: II. Plant CO_2 exchange in relation to environment and productivity. CCIBP Matador Project, Technical Report No. 49. Univ. Saskatchewan, Saskatoon. 97 pp.

Scott, D. and W.D. Billings. 1964. Effects of environmental factors on standing crop and productivity of an alpine tundra. Ecol. Monogr. 34: 243-270.

Invertebrate consumers

Synthesis of energy flows and population dynamics of Truelove Lowland invertebrates

James K. Ryan

Introduction

The quantity of energy released and transformed by invertebrates is presently being investigated in ecosystems throughout the world. Studies of single species provide the most detailed and accurate information on energy flow (Healy 1967, Edgar 1971, Bailey and Riegert 1973, Hodkinson 1973, Hofsvang 1973, MacLean 1973, Richardson 1975). Studies of groups of invertebrate species give a clearer perspective on energy flow through parts of ecosystems (Englemann 1968, Van Hook 1971, and a series of papers in Ekologia Polska 1971). A sophisticated trophic level perspective has been achieved with phytophagous insects (Reichle et al. 1973). There are few published studies evaluating energy flow through all the invertebrates in an ecosystem (Bornebusch 1930, Van Der Drift 1951, *cf.* Odum 1971). The Bornebusch study is deficient, and has been reanalysed several times (Nielsen 1949, Macfadyen 1961, Englemann 1966). The Van Der Drift and Odum results are educated estimates. At each level, such studies are critical to quantitatively understanding how invertebrates function within an ecosystem. While the optimal approach to energy flow is direct measurement of each species, Huhta and Koskenniemi (1975) acknowledge that "to work out detailed energy budgets for all organisms (*sic* soil invertebrates) of an ecosystem would be an overwhelmingly laborious task." Chernov (pers. comm., U.S.S.R. I.B.P.) concurs, and concludes further that an acceptable invertebrate energy flow synthesis appears impossible with our current state of knowledge.

While energy flow through invertebrates may never be accurately measured, the Arctic is the best climatic zone where approximations of it can be made. The main obstacle to assessing invertebrate energy flow is the tremendous diversity of species and habits. The population size and structure must be determined for each species, for a year, to construct a summary annual energy budget. In an arctic ecosystem the invertebrate, and total organismic, species diversities are reduced (Downes 1964, MacLean 1975, *cf.* Appendices). Primary production, the limiting variable for secondary production, is drastically less than that in other ecosystems (Odum 1971). Seasons are short and decisively limited by below 0°C temperature.

This communication investigates which invertebrate taxa are dominant and what trophic levels are most significant in this ecosystem, and estimates the combined invertebrate annual energy flow. The arctic location, data from other energy flow studies, and the integrated research effort, make the

Truelove site ideal for this undertaking. Data are presented in two parts. The first section, "collected data," contains experimentally determined values and direct observations on invertebrate populations. The second section contains energy flow values determined through a computer model. These data are termed "hypothetical" as they are not measured quantities.

Collected data

Materials and methods

Standing crop data are from single 5×5 m plots at two study sites: the slope of the raised beach, and the hummocky sedge-moss meadow. Standing crop data from other years, Collembola data (J. Addison this volume) and emergence data are from similar plots at these same sites. Data given here, unless otherwise noted, are from 1972. When available, data from other seasons (1970-74) are presented for comparative purposes and to contribute to the model's information base.

Standing crops of soil-dwelling invertebrates were determined by O'Connor and modified Tullgren funnel extractions [except Collembola—see J. Addison (this volume)]. O'Connor funnels (O'Connor 1962) were used to extract 5 cm deep × 6.1 cm diameter soil cores split in half to 2.5 cm sections. Sixty-eight full-size soil cores from the raised beach site, and 64 from the meadow, were extracted by this method. Tullgren funnels were used to extract 42 unsplit cores from each site, 7 cores at a time. The Tullgrens were covered with unvented, 31 cm diameter barn lamp shields, and heated by 25 watt light bulbs for 36-48 hr until the soil core was completely dry. After the soil thawed to 5 cm, soil cores were randomly collected on each plot at 5 day intervals throughout the summer. All samples were cut with a tapered corer, and held individually in plastic bags for transport. Animals collected during the extractions were counted and weighed on a taxon per sampling date basis, and subsequently lumped to give mean numbers and biomass m^{-2} for the entire season. The numbers and biomass of each taxon were increased by appropriate factors to estimate the actual mean standing crops. These correction factors were determined by comparative extractions of paired soil cores made with simple Baermann funnels, a Macfadyen Hi-Gradient extractor, Salt Hollick floatation, and the Tullgren and O'Connor funnels; by extracting animals added to sterile cores; extracting cores taken to 15 cm; and from the literature (Edwards and Fletcher 1971, Willard 1972).

Winged insects were collected in 1 m^2 nylon screen pyramid traps similar to those described by Beirne (1955). Each trap was left in place for a full season. The traps were sealed inside with silk strips and glue, and had buried flaps around the edges to stop insects from entering and escaping (Fig. 1B). A funnel mechanism at the top of each trap collected the flying adults into bottles containing 95% ethyl alcohol (Fig. 1A). These bottles were changed every 4 days. In 1971, 4 traps were placed at each study site and in 1972, 5. In 1972, 15 traps (of 35 total) at 5 sites (of 7 total) were cleaned with a battery-powered car vacuum cleaner each time the trap bottles were changed, to determine trapping efficiency.

The standing crop of Protozoa at the meadow site was estimated provisionally by Dr. O.W. Heal. Protozoa washed from 4 cm^3 of soil on 5 August, 1972, were counted in a haemocytomer (volume 400 μl). No correction was made for extraction efficiency. Protozoa weights are from Heal (1971),

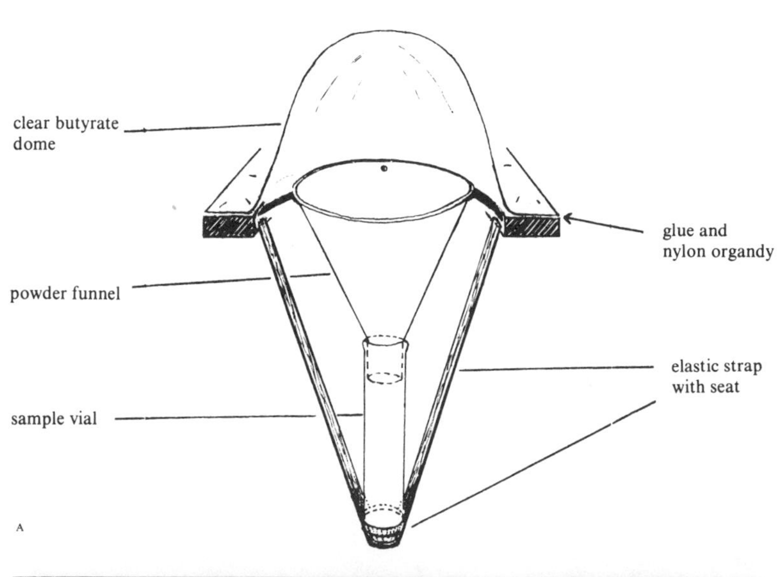

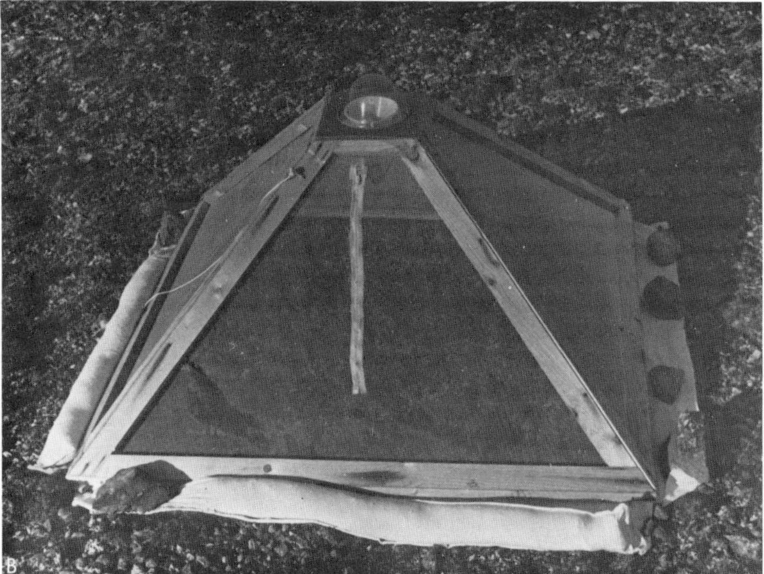

Fig. 1 Emergence trap for winged insects. A. Detail of trap top. Funnel bolted to four tin strips allowed catch vial to be held tightly by elastic strap. B. Trap on raised beach slope, 1971; on the 1972 trap the bottom flaps were buried in place. Note torn screen and cut anchor rope, the result of depredations by an arctic fox.

and the number of generations per year are his estimations. Protozoa increase and decrease rates were estimated in 3 replicates on an experiment at the projcct site. Protozoa were added to a *Carex* infusion and to sterile water, and the individuals in 10 drops were counted every 12 hr for 72 hr.

Field layer standing crops for Araneida, and larval Lepidoptera and Hymenoptera Symphyta were determined from animals found within the 71 m^2 of emergence traps, and from a stratified random sample of 500 m^2 of the

Lowland made with an 0.5 m^2 drop frame. The study site standing crops shown for these groups are means of the lowland raised beach and meadow standing crops.

Weights and caloric determinations were made as follows: animals were oven dried at 80°C for 24-72 hr to a constant weight. They were weighed on a Cahn electrobalance accurate to 4 decimal places, with a maximum sensitivity of ±0.05 μg. All extracted and quantitative animal collections were weighed. All energy values are reported in calories. Caloric determinations were made for 531 samples on a Phillipson microbomb calorimeter. Caloric values for certain taxa are from the literature (principally Cummins and Wuycheck 1971) when Truelove data are insufficient.

The maximum and minimum weights of Enchytraeidae were estimated by sorting animals from 30 extractions into 8 size categories. For the model, the minimum individual mass is the mean of the smallest size category, and the maximum mass is that of the largest individual weighed. The weight range of Collembola was estimated by a similar procedure from 15 Tullgren funnel extractions.

Results

The seasonal standing crops for all invertebrate populations (except for Protozoa and Nematoda) are given in Table 1 as arithmetic means ±S.E., and geometric means (log x+1 transformation) with 95% confidence limits. The values shown are the maximum populations found by either the O'Connor or Tullgren funnels. This table contains estimates of the actual standing crops, and the correction factors used to make these estimates. Nematode populations are treated by Procter [This volume (a)]. There was a Platyhelminth Rhabdocoela population at the meadow site, but these animals preserved poorly and were recognized inconsistently. In 12 samples from hollows in the meadow site, 470 m^{-2} of these turbellarians were found, while none were found from 12 hummock samples in the same plot at the same time. The lumped Copepoda standing crop includes three superfamilies. Thirty meadow O'Connor funnel extractions (30 July - 4 August) were separated to give standing crops of 7695±2125 Harpacticoidea, 3200±2084 Calanoidea, and 116±58 Cyclopoidea. The Calanoidea weighed a mean 6.29 μg per individual. The per sample individual copepod weighed 2.13 μg. As the cyclopoids were approximately the same size as the calanoids (and 3% as abundant), the mean harpacticoid mass was estimated to be 0.34 μg per individual. All standing crop values for endopterygote insects are for larvae only. Parasitic Hymenoptera larvae cannot be counted reliably. Adult counts from the emergence traps are reliable, and are the basis for the larval estimates of 2 individuals m^{-2} at the meadow site, and 0.5 at the raised beach.

The Collembola standing crop value is from Macfadyen hi-gradient funnel extractions. These populations may be compared to those obtained from Tullgren funnels using a log x + 1 transformation. At both sites, the Macfadyen funnels extracted significantly more (P=0.025) Collembola and Acarina than the Tullgren funnels. However, the 1972 Tullgren Collembola populations were not significantly different from the 1973 and 1974 Macfadyen extractions. The efficiency corrected Tullgren populations were not significantly different (P=0.25) from the Macfadyen populations. The corrected arithmetic mean populations from Tullgren funnels are respectively 12% and 3% below the hi-gradient extracted populations at the raised beach and meadow sites.

Table 1. Invertebrate standing crops m^{-2} at the study sites, 1972

Component	*Enchytraeidae*	*Rotifera**	*Tardigrada*	*Copepoda*	*Ostracoda*	*Cladocera*	*Acarina*	*Araneida*	*Collembola*	*Insect larvae: Lepidoptera*	*Coleoptera*	*Diptera: Tipuloidea*	*Other Nematocera*	*Cyclorrapha*	*Hymenoptera*
Raised beach															
Numbers collected,	5791	110	2836	0	0	0	4899	0.3	18221	0.2	0	0	263	8	8
mean ± S.E.	±471		±469				±863						±97		
Numbers collected,	4537	75	1620				2315		11700				193	7	7
geometric mean	(4019	(15	(1147				(1302		(8400				(132	(0	(0
(fiducial limits)	5856)	132)	2244)				3958)		16500)				260)	17)	17)
Dry weight collected, mg	119.8	0.006	2.92				28.4	3.87	96.8	0.774			8.25	7.53	0.51
Correction factor	1.5	2	2				2		—				1.5	1.5	
Estimated actual population	8690	220	5670				9800		18220				395	12	8
Estimated actual biomass, mg	180	0.01	5.8				57		96.8				12.4	11	0.5
Meadow															
Numbers collected,	18632	321	1158	19264	496	5	3231	1.8	8737	0	0	4.9	7076	35	1.8
mean ± S.E.	±1455	±57	±154	±3904	±226		±526						±919	±12	
Numbers collected,	14640	201	754	5838	162	3	1688		4700			3	4447	25	
geometric mean	(11901	(124	(556	(3771	(77	(0	(982		(3400			(0	(3393	(5	
(fiducial limits)	18070)	289)	995)	9010)	263)	10)	2776)		6500)			10)	5800)	43)	
Dry weight collected, mg	395	0.01	1.19	24.6	0.48	0.04	13.2	1.18	42.6			30.7	64.5	2.6	1.33
Correction factor	1.5	20	2	1.2	1.2	1.2	3		—			1.5	1.5	1.5	
Estimated actual population	27500	6400	2300	23100	595	6	9690		8740			7	10600	52	2
Estimated actual biomass, mg	590	0.2	2.4	29.5	0.58	0.05	40		42.6			46	96.8	3.9	

**Rotifer mass is estimated.*

Table 2. Mean numbers and biomass of arthropods collected in 1 m^2 emergence traps at the study sites, 1971 and 1972 seasons*

			Diptera						Hymenoptera		
	Total winged insects	*Lepidoptera*	*Chironomidae*	*Sciaridae*	*Mycetophilidae*	*Empididae*	*Dolichopodidae*	*Muscoidea*	*Tenthredinidae*	*Ichneumonoidea*	*Araneida*
Numbers:											
Raised beach											
1971	92		83	2				5	0.5	1.5	0.5
1972	82	0.2	73	3	0.6	2		3	0.2	0.6	0.2
Meadow											
1971	374		350	7				8.5		5.5	1
1972	909		887.5	5	0.8		0.4	13	0.8	0.8	2
Biomass, mg dry weight:											
Raised beach											
1972	9.36	0.77	(3.26)		1.80	0.42		2.13	0.61	0.37	0.12
Meadow											
1972	54.15		(33.50)		0.20		0.57	18.61	0.74	0.59	1.10

*1971 4 traps per site; 1972 5 traps per site.

The emergence trap data given in Table 2 covers the 1971-72 seasons. The 1971 season traps were set up after insects started emerging, and 5 trap bottles spilled during the first trapping period. The traps were effective before the 1971 season peak, but missed an unknown portion of the emergence. The 1972 season trapping records are complete for the full season. All values in Table 2 refer to adults except for the Araneida. These spiders were vacuumed or aspirated from the traps during, or at the end of, the trapping season. The vacuum procedure, when considered for all 5 trap locations, indicated that there was no significant difference ($P=0.1$; data transformed by $\log x+1$) in the number of insects caught in vacuumed versus non-vacuumed traps. Thus, the collecting mechanism we designed and built was efficient.

Table 3 gives the provisional estimates of meadow site protozoan standing crops, weights, and number of generations yr^{-1}. No estimates were made for the raised beach site.

In the *Carex* infusion experiment, the mean Protozoa doubling time was 9 hr, and in log phase was 4.5 hr. Protozoa put in distilled water initially had a half-life of 16 hr, but this decreased sharply after two days.

The energy flow model of part 2 makes use of several disjointed bits of biological information. These are given here, and treated further in the model discussion. The enchytraeid weight range found in the field was 3.1-66.8 μg, and the mean weight of all standing crop enchytraeids (with dirt in their guts) was 21.1 μg. Tardigrades were weighed on several occasions, from which came the mean 0.3 μg and maximum 1.6 μg weights for the model. Crustacea are found only at the meadow site. Cultures of overwintered calanoid and harpacticoid adults produced eggs almost immediately following spring thaw; these same adults survived a full summer. Two large (compared to meadow dwellers) lake dwelling crustaceans, *Lepidurus arcticus* and *Branchinecta paludosa*, were univoltine; i.e., small individuals were found early, and large egg bearing adults late, in each season. The Acarina of the raised beach and meadow are predominantly Oribatei. Mixed pterygote insects in 128 sweep samples were 5.837 ± 0.534 (2 S.E.) cal mg^{-1}; hence 5.8 cal mg^{-1} were used for all pterygote insects in the model.

Table 3. Protozoa energy flow estimate, meadow site.

Taxon	Est. adults m^{-2}	Adult dry weight (mg)	Est. gen. yr^{-1}	Est. tissue prod. (mg)	Est. cal flow m^{-2}
Flagellates—Naked Amoebae	1.5×10^8	1×10^{-6}	6	900	15000
Ciliates	1.0×10^6	1×10^{-5}	6	60	1000
Testates	5.0×10^7	5×10^{-6}	2	500	8333
				1500	25000

Discussion

Invertebrate standing crops in Table 1 are given as raw, extracted means, and as the corrected estimates of the actual population means. The raw means, arithmetic or geometric, can be compared to population values found at other areas. The corrected mean values should be comparable to those of populations sampled with highly efficient extraction techniques. This was true for the efficiency corrected Tullgren funnel Collembola extractions versus the Macfadyen hi-gradient extractions.

The correction factors used to estimate actual standing crops are necessary evils. Soil inhabiting organisms are rarely (if ever) completely extracted from their substrate. Different extraction techniques recover varying proportions of the animals in the same substrate (Edwards and Fletcher 1971). Injured, small, and immobile stages (e.g. pupae) are likely to be trapped in a soil core and not extracted. Some animals live below the 5 cm extraction depth. Unextracted animals must be included in the energy flow assessment even though their actual populations are not precisely measureable. Willard (1974) extracted animals added to sterile soil cores to determine the absolute efficiencies of his techniques. He then estimated the actual populations of his animals using the correction factors found. Similar procedures were used to estimate the actual mean invertebrate populations at Truelove. These estimated actual population means are the base values from which energy flows were calculated.

Unlike standing crops, winged insect emergence from an area is a direct indication of production. The adults, which leave their larval substrate to mate and disperse, represent some fraction of the energy which flowed through the whole population. Emergence rates depend directly on developmental rates (and population size). The winged insect emergence for the growing season, in Table 2, may be directly compared with the results of emergence studies made elsewhere. Minns (this volume) found the insect emergence from a Truelove lake to be 66.5 mg (372.2 cal) m^{-2} for a partial summer, and suggested that the total emergence was unlikely to exceed twice this figure. This lake value is 1.2- to 2.5-fold the meadow production, and 7- to 14-fold the raised beach production. Winged insect production in Char Lake (Welch 1973), at the same latitude as Truelove, was 2.6-fold the meadow and 15-fold the raised beach production. In two Polish meadow sites (Olechowicz 1971), the emerged insects totalled 1.86 g and 1.22 g dry wt m^{-2} in a 6 month period, which are 34- and 23-fold the meadow site production.

Hypothetical data—energy flow

Energy flow is estimated below by a mathematical model constructed specifically for the Truelove Lowland invertebrates. This model allows production and respiration to be considered in a consistent mathematical framework. It permits the energy releasers to be ranked in a self-consistent hierarchy of energetic importance.

Several concepts must be defined before explaining the model. *Energy flow* in this chapter means total calories assimilated m^{-2} yr^{-1}. *Assimilation* is the sum of two components: calories produced and calories respired (Petruscewicz and Macfadyen 1970). Assimilation is also equal to the consumed calories minus eliminated (feces, urine) calories. *Production* is the total accumulated tissue at the time of *death* of each organism, and hence is that tissue which is available to the next trophic level (Odum 1971). Production depends on growth rates, but is different in the final tally. For example, growth stops at the prepupal stage for an insect, after which it loses weight via metabolism. No tissue (except exuviae) has been produced until the insect dies. *Reproductive tissue* which becomes part of the next generation is not produced tissue: it is not dead. *Respired energy* is energy lost from the ecosystem. This may occur as heat loss, or as other losses, but is interpreted through respiration.

Basic factors and expressions of the model

Since energy flow could not be determined for each species, higher taxonomic units are used in this analysis. These are: Rotifera; Nematoda; Enchytraeidae; Tardigrada; Crustacea: Harpacticoidea, Calanoidea plus Cyclopoidea and Ostracoda; Acarina; Araneida: Erigonidae, Lycosidae; Collembola; Lepidoptera; Diptera: Mycetophilidae, Tipuloidea, Chironomidae and Sciaridae, Cyclorrapha plus; Hymenoptera: Symphyta, and Apocrita (parasitica). The units encompass enough ecological similarity (due to reduced species diversity) that reasonable generalizations can be made in the absence of sufficient actual data for the model. All parameters assigned to each group are meant to approximate those of the most energetically significant species in the group.

Populations within a taxon consist of synthetic cohorts (*sensu* Spinage 1972), subdivisions of the population having mean characteristics of the whole population. A cohort usually contains one fertile female who produces a subsequent identical cohort; it may involve several females contributing to the next identical generation. The populations considered in equilibrium in this analysis are the arithmetic means of the 1972 efficiency corrected extractions of Table 1, and the total season emergence trap catches of Table 2.

Individuals within a synthetic cohort are assumed to behave according to four basic equations. The first of these concerns death rates. When cohort individuals live in the same habitat, use the same resources, and are similar morphologically, the death rate in the cohort is assumed to be exponential with respect to degree day time. The number of individuals, N, present at any time, T, in a synthetic cohort is given by the expression

$$N = N_o e^{-\underline{a}T}$$

where N_o is the initial number of individuals in the cohort and $-\underline{a}$ is the death rate. If an animal no longer behaves according to the original death rate, such as when an insect larva becomes a prepupa, N becomes N_f. In cases where the adult is similar to the immatures, N_f is 0.5. The cohort is extinguished, and has undergone a complete turnover, when less than 0.5 individuals remain alive.

The next basic equation concerns individual respiration rates. At a constant temperature, respiration varies with weight. Among similar animals this relationship is expressed as

$$R = aW^b$$

where R is respiration, "a" is a constant determined by respiration measurements, W is the weight of an individual, and b is an empirical constant. The values of "a" vary with temperature. Normally b is 0.87 for insects (Schroeder and Dunlap 1970).

Developing individuals in a cohort gain weight in increments, punctuated by moults. Weight gain is basically a function of body size, which determines food consumption rates, and hence growth rates. The average rate of weight gain is exponential (or at least can be closely approximated by an exponential relationship) in day degree time. The weight, W, of any individual in a cohort at time T is given by

$$W = W_o e^{\underline{B}T}$$

where W_o is the initial weight of an individual in the cohort, and $\underline{B}$ is the growth rate. When exponential weight gain no longer occurs, such as when an insect larva becomes a prepupa, W becomes W_f. This is the final weight of an animal in the cohort (but *not* the produced tissue weight).

The final basic equation relates respiration and tissue production, P, in developing animals in a cohort. This relationship is expressed as

$$R = cP$$

where "c" is an empirical constant which means that for every calorie respired some caloric quantity of tissue is created.

The basic operation of the model should be apparent, although an example may make it clearer. An insect, between hatching from an egg and becoming a prepupa, gains a measureable amount of weight. Its respiration rate, known for one weight, is calculated for its whole range of weights. The time required to grow to maturity is determined by how much respiration must occur to produce the weight gain observed, given the constant "c" (which is the ratio: Respiration/Production). Some larvae die during development, and this accumulated dead tissue is calculated through the death rate expression.

Energy flow computed by the model

These four equations may be solved simultaneously to determine respiration and production. Only that portion of the synthetic cohort which is actively growing (i.e., immatures) is treated by these equations. Respiration of this portion of the synthetic cohort, R_{sc}, is given by

$$R_{sc} = \frac{N_o W_o^{\,b}}{a(b\underline{B}-\underline{a})}\left[\left(\frac{W_f}{W_o}\right)^b \frac{N_f}{N_o} - 1\right]$$

Production, P_{sc}, for this same group is

$$P_{sc} = \frac{N_o W_o^{\underline{a}/\underline{B}}}{(1-\underline{a}/\underline{B})}\left[W_f^{\,1-\underline{a}/\underline{B}} - W_o^{\,1-\underline{a}/\underline{B}}\right]$$

A correction factor of 0.03 - 0.1, depending on the taxon, is added to the calculated cohort production to estimate tissue loss by moulting. The time required to reach maturity, T_d, is given by

$$T_d = \frac{1}{\underline{B}} \ln \frac{W_f}{W_o}$$

Several other equations are important adjuncts to computing energy flow. The mean weight of an animal in a synthetic cohort, $\overline{W}_{sc}$, is given by

$$\overline{W}_{sc} = \frac{W_o^{\underline{a}}}{(\underline{B}-\underline{a})} \; \frac{\left[\frac{W_f}{W_o}\frac{N_f}{N_o} - 1\right]}{(1-N_f/N_o)}$$

This value predicts the mean weight of immatures collected in the field. The mean number of immatures in a cohort, $\overline{N}_{sc}$, is found by

$$\overline{N}_{sc} = \frac{(N_o - N_f)}{\ln (N_o/N_f)}$$

This value divided into the actual field population of immatures m^{-2} gives the number of synthetic cohorts m^{-2}.

The adult portion of the synthetic cohort is analysed separately. Where adults use the same resources and habitat as the immatures, they follow the same death rate equation used for immatures, although the rate may differ. For these animals, adult respiration, $R_{sc\ ad}$, is

$$R_{sc\ ad} = \overline{N}\,\overline{T}\,aW^b$$

where $\overline{N}$ is the mean number of cohort adults and $\overline{T}$ is the mean length of life of an adult. These adults may gain reproductive tissue weight, which they lose abruptly. They may gain weight allometrically until death, or may become senile and lose weight until starvation death. Because characteristic weight fluctuations are not known, weight in the above expression is W_f, the weight at sexual maturity. Respiration of pupal and adult pterygote insects is calculated indirectly from weight lost during both stages. Adult pterygote insect respiration is increased by assuming that each imbibes 2/3 of a crop load of nectar (many do not, Kevan 1970), which is 58% sugar. This is 29% of the exhausted weight of an adult. They metabolize this at an R.Q. = 1.0 (these values are based on Hocking 1953). Adult production of pterygotes is the tissue remaining after metabolism.

Results from these equations are converted to respiration and production m^{-2} in two steps. The number of synthetic cohorts m^{-2} is found by dividing the mean population m^{-2} by the mean samplable number in a synthetic cohort. This samplable number may include immatures only, e.g. pterygote insect larvae, or adults and immatures, e.g. Collembola. The number of synthetic cohorts developing in one season, Nd_{sc}, is found by dividing the accumulated day °C above 0°C in a season, D°, by the day °C required to complete one synthetic cohort, $D°_{sc}$, thus

$$Nd_{sc} = \frac{D°}{D°_{sc}}$$

Combining these two steps gives the number of cohorts produced m^{-2} in a growing season. This number times the total synthetic cohort production, and respiration, gives the total energy flow for the taxon.

Model time functions

Time and temperature are interdependent in poikilotherms. In the model, time is expressed as accumulated degree days above 0°C. Respiration and marginal activity may occur below this temperature, but the movement of animals, and food availability, will be restricted by ice. Animal activity increases progressively above 0°C. This increase is mirrored by respiration rates.

Meteorological records (Courtin, pers. comm.) show that at −1 cm there is an accumulation of 624°C days above 0°C at the raised beach site, and 313°C days at the meadow. These are the means for the 1971-73 seasons (the 1971 season temperature at −1 cm is extrapolated from adjacent profile temperatures). The yearly mean accumulated day degrees are defined as the growing season at each site for the model calculations.

Respiration rates used in the model are from Scholander et al. (1953) and particularly from Procter [this volume (b)] at this site. The value "a" was calculated from these data from the expression $R = aW^b$. This "a" value for each invertebrate group was used in the model. Procter measured respiration at 2°, 7°, and 12°C. With his data, the developmental time calculated was the mean of three temperature based developmental times. Where no arctic respiration measurements were available for a particular group, estimates were substituted from related groups, or from studies of the same group made at temperate latitudes (Protozoa, Nematoda). Respiration was converted to calories by the factor 4.7 cal ml^{-1} O_2 consumed (Petruscewicz and Macfadyen 1970).

Model parameters by taxon

Each taxon treated by the model requires data inputs beyond those given in the first section. The list of these inputs, and intermediary calculations, for each model taxon are given in Ryan (1977). Manipulations of these inputs require some explanation here. Because the differences between taxa are as great as the taxa themselves, this section is tedious to read. It should be read for information about specific taxa rather than read completely like other sections.

Protozoa were assumed to be 5.8 cal mg^{-1}. It was further assumed that to produce 1 cal of tissue, 2 cal are respired. Therefore the ratio of Respiration/Production, abbreviated R/P ratio, is 2.0. Respiration rates are from Heal (1971). This group was examined with the model, but its energy flow was computed through a simplified analysis.

The treatise by Donner (1966) on Rotifera is helpful, but most of the parameters used to estimate their energy flow are compiled from other invertebrate values given here.

Cummins and Wuycheck (1971) report that enchytraeids are 5.628 cal mg^{-1} ash free and are 9.6% ash, so they are 5.135 cal mg^{-1}; a similar value can be computed from Willard (1974), who reported 6.091 cal mg^{-1} ash free. The R/P ratio was assumed to be 1.0 for immatures. Biological data are from Reynoldson (1939, 1943).

Tardigrade biological information is from Hallas and Yeates (1972), who found 5.4% of the females (no males) in their study contained 1-7 eggs. A synthetic cohort was assumed to consist of 24 hatchlings maturing to 3 adults. Enchytraeid caloric, respiration, and R/P values were used for this group. The weight range for the model, from Truelove specimens, is 0.67 the range reported by Hallas and Yeates.

Crustacea were lumped into two groups. The Calanoidea and Cyclopoidea group cohorts have a 200-fold weight range and a decline from 100 hatchlings to 4 adults. The second group was the smaller Harpacticoidea, which have a 50-fold weight range and decline from 18 hatchlings to 3 adults in a cohort. For both groups the total egg mass was 33% of W_f, the mean adult mass. The adult lifespan was assumed equal to the immature developmental period. The value 5.35 cal mg^{-1} is the mean crustacean value calculated from Cummins and Wuycheck (1971). The R/P value for immatures was assumed to be 0.5.

For Acarina, Block's (1965) study of the life histories of two soil dwelling oribatids, especially *Damaeus clavipes* (Hermann), and Wallwork (1967) are the main information sources for the synthetic cohort data. A cohort consists of 60 hatchlings giving rise to 6 adults; the immature weight range is 30-fold. Cummins and Wuycheck list 5.808 cal mg^{-1} for Acaridae, which are lightly sclerotized; the more heavily sclerotized oribatids are estimated to be 5.5 cal mg^{-1}. The R/P ratio was assumed to be 0.5 for immatures.

Araneida information from Moulder et al. (1970), plus additional information from Riegert et al. (1974), established the overall R/P of 1.7. Edgar (1971) reported 2.5 for *Pardosa (Lycosa) lugubris* Walc., and 5.8 cal

mg^{-1}. The respiration "a" value of Scholander et al (1953) for Lycosidae was used for Erigonidae also. From Leech (1966, pers. comm.), a *Tarentula exasperans* P.-C. (Lycosidae) cohort has 70 hatchlings, while Erigonidae have 10 hatchlings; these give 5 and 3 adults, respectively. Adults die the year they produce eggs. The weight ranges of 100-fold for *T. exasperans* and 75-fold for Erigonidae are estimates based on egg sac size. The lycosid energy flow was increased by a factor of 3.89.

Collembola were investigated by J. Addison (this volume), who found *Hypogastura tullbergi* to be dominant on the raised beach. They survived 2-3 years as adults beyond the 1-2 year maturation period on the raised beach site. Hale (1967) reported that, under sub-arctic conditions, probably fewer than 100 eggs are laid by most species of Collembola during the lifetime of a single female. In the model, a cohort starts with 60 hatchlings. Adults have the same mortality curve as the immatures, and they survive 50% longer than the maturation period. These two facts require that 8.7 adults are produced in the cohort. The weight range observed at Truelove, 28-fold, was used for the population. The value of 5.8 cal mg^{-1} and R/P ratio of 0.5 for immatures are both assumed.

Pterygote insects can be analysed by the model only as larvae. Synthetic cohorts m^{-2} for the model were based on the emergence trap data, and were obtained by 1/4 (number of adults collected in a taxon) except for Hymenoptera parasitica, which were 1/32 (number of adults collected). Adults at death are an estimated 67% the mass of newly-emerged adults. The mass of each adult captured is an estimated 67% of its prepupal mass. Welch (1973) found emerging female chironomids were 40% eggs. This percent of the emergence weight of one of the 4 emerged adults is taken to be the initial mass of hatchlings in a cohort. The R/P ratio was estimated to be 0.5 for all pterygote larvae other than herbivores, whose R/P is 0.8. Caloric values are those measured for Truelove insects.

Diptera were subdivided into 4 units for this analysis. The first is chironomids and sciarids (96% chironomids), which are similar sizes. Oliver (1968) found a mean fecundity of 160 eggs for several arctic chironomids, while Welch (1973) found 160-212 for 4 species. The mean fertility of females in this group is taken as 160 hatchlings for the synthetic cohort. Mycetophilidae from the meadow and Tipuloidea from the raised beach are assumed to have the same reproductive capacities as the first group. The fourth group is primarily Muscidae and Anthomyiidae, but includes the occasional empidid, dolichopodid, and syrphid. Extrapolating from Buei's (1967) laboratory fecundity studies with *Musca domestica* Macq. to field conditions, cohorts in this group have 80 hatchlings.

Lepidoptera and Hymenoptera Symphyta, being herbivorous pterygotes, are given the same reproductive capacity as *Gynaephora* (Ryan and Hergert this volume), 160 offspring/female. Hymenoptera parasitica parameters are also assumed, with 60 hatchlings declining to 40 in a cohort, and R/P=0.5 since they are carnivores.

Results

The energy flow estimates for Protozoa are given in the last two columns of Table 3. The energy flow conclusions for all taxa in this study are shown in Table 4 in cal m^{-2} yr^{-1}. The model predicts the developmental rates, lifespans (Table 4 gives maximum predicted lifespans), and mean weights for each taxon. Some of these data are given below.

Table 4. Summary of energy flow through the study site invertebrates in a mean growing season*

	Raised beach			*Meadow*		
Invertebrate Taxon	*Maximum lifespan, Seasons*	*Respiration*	*Production*	*Maximum lifespan, Seasons*	*Respiration*	*Production*
Protozoa		?	?	0.17	17500	8500
Rotifera	0.3	0.1	0.1	0.5	2.6	3.2
Nematoda		4589	3336		561	408
Enchytraeidae	1.4	1616	1930	2.8	2563	3104
Tardigrada	0.7	151	193	1.3	13.9	18.4
Crustacea		0	0	2.0	1076	410
Acarina	34	106	61	68	57	35
Araneida	8	12	23	4.1	0.8	1.2
Collembola	3.5	552	736	7.0	143	107
Coleoptera		0	0		0	0
Lepidoptera	3.1	16	13		0	0
Diptera	2.0	234	220	4.3	1362	1256
Hymenoptera						
Symphyta	4.0	21	15	7	25	19
Apocrita	2.3	6.0	2.6	5	9.6	4.2
Total		7303	6530		23313	13866

*energy values are cal m^{-2}

Enchytraeid worms were predicted to have a mean weight of 17.5 μg, and a maximum lifespan of 1060 C degree days. Both crustacean groups have slightly more than 1 generation per year. Acarina require 2.6 years to reach maturity on the raised beach, and 5 years on the meadow. The total time until the last individual in a cohort has died is 11-fold this maturation time.

In the Araneida, the Lycosidae require 7 years to mature on the raised beach, and the smaller Erigonidae 3 years in the meadow; both survive one year as adults. The model predicts the mean weight of lycosids collected to be 4.5 mg, while the actual mean weight collected was 17.5 mg; for erigonids the predicted mean was 597 μg, while the actual mean was 658 μg. For all other invertebrates the predicted and actual mean weights were approximately as close as, or closer than, the erigonid value.

For Collembola, the model predicts a 1.4 year maturation period and a 2.1 year adult lifespan on the raised beach. On the meadow a 2.8 year maturation period and a 4.2 year adult lifespan are predicted.

Raised beach chironomid and sciarid larvae require a 2 year maturation period, while Tipuloidea and Cyclorrapha require 3.5 years. At the meadow site, chironomid and sciarid larvae require a 4 year maturation period, Mycetophilidae require 5 years, and Cyclorrapha require 8 years. The Table 4 lifespan values are the weighted means of these lifespans.

One Lepidoptera adult was captured on the raised beach. The model predicts it required 3 years to develop. Hymenoptera Symphyta require 4 years on the raised beach, and 7 years on the meadow, to mature. Parasitic Hymenoptera require 2.5 years on the raised beach, and 5 years on the meadow, to mature.

The overall R/P for invertebrates ranged from 0.6 to 2.3. Enchytraeids were 1.4, Crustacea 2.1, Acarina 1.6, and Collembola 0.8. Among the pterygote insects Diptera were 1.0, Lepidoptera and Hymenoptera Symphyta 1.1-1.2, and Hymenoptera parasitica 2.3. The immature R/P for Araneida was 0.8, predicted from the given overall R/P=1.7.

Discussion

Energy flow through Protozoa was determined by an estimate of the number of generations produced per year rather than by the model. For every tissue calorie produced, two calories are counted as respired [Heal (pers. comm.) suspects this should be 4 cal respired per cal tissue produced]. From the single date Protozoa standing crop count, they are found to have the greatest energy flow in the meadow. This is the most tentative of all the energy flow values reported as it is based on minimal data. The number of generations of Protozoa used for the entire season are 67% of the number observed in the 72 hr *Carex* infusion experiment, and 67% of the number of generations predicted by the model (based on respiration rates in Heal 1971, and R/P=2.0). The great quantity of energy available to Protozoa through bacterial and fungal sources (Whitfield this volume) suggests that their role may be greater than is estimated here. Rapid mortality and natality, great numbers, and minute size makes energy flow through Protozoa almost impossible to estimate in warmer climates; a single rainfall could provide conditions suitable for the development of 1 to several generations. The only other field record of Protozoa production I know is from Antarctic Signy Island (Heal 1965, 1967). At this cold site Protozoa were quite significant energy consumers. Macfadyen (1961) stressed the significance of Protozoa at a temperate site. Energy flow through the raised beach Protozoa was not estimated, but is presumably small in this dry habitat (*cf.* Kuehnelt 1955).

The rotifer energy flow estimate is arbitrary since most of its model parameters are arbitrary. Counts of these animals were highly inefficient, but the true population is believed to be within an order of magnitude of the estimate. Their minute size and low numbers make their impact negligible, especially on the raised beach where they probably undergo cryptobiosis for most of the season.

Enchytraeidae is the only known oligochaete family at Truelove, although a comparison made with other tundra site faunas indicates that additional families are likely to be present (Ryan 1978). The Enchytraeidae are numerically abundant, comparatively large invertebrates. They have the largest standing crop of the meadow invertebrates, and the second largest (exceeded by nematodes) on the raised beach. Three sources of biological data on enchytraeids are contradictory (Reynoldson 1939, 1943; Ivleyva 1953, discussed by O'Connor 1967; and Learner 1972). For two species, Reynoldson had mean values of 10 viable eggs produced per adult worm, with the average adult living 0.35 the timespan required to reach maturity. The mortality curve of his population was exponential, conforming to the model. Ivleyva reported that enchytraeids produced 260 eggs per adult worm, and that these adults lived 12.5 times longer than the time spent as immatures. For this population to be in equilibrium, the survivorship of immatures must *drastically* differ from that of adults. Since immatures and adults are similarly shaped, live in the same habitat, and consume the same resources, their mortality curves should be similar. The data of Learner suffers the same criticism. Both are laboratory studies which indicate the increase potential of enchytraeids, but not their behavior in a field situation. The field study parameters of Reynoldson were used in the model. His study of enchytraeids required approximately 1100 C degree days (above 0° C) for a complete lifespan, while a 1060 C degree days lifespan was predicted by the model for Truelove enchytraeids.

Tardigrades are abundant at both the raised beach and meadow sites. They are presumably inactive (in a cryptobiotic state) during the dry periods on the raised beach, but were rehydrated and extracted in the O'Connor funnels. Their impact on the raised beach is overestimated by using temperature as the only driving variable, and ignoring cryptobiotic inactivity.

The "a" value of the Crustacea at 12°C was 5.0 (mean, calculated from Procter), and is the highest of all the Truelove invertebrates. Computing from Roff (1973), *Limnocalanus macrurus* Sars at Char Lake (same latitude as Truelove) has an "a" value of 4.62 at 12°C; this species matures in 9 months in its lake habitat. Comita (1968) gave data for 5 *Diaptomus* species, including *D. arcticus*, from which an "a" value of 6.6 at 12°C can be computed. Procter's high value predicts a developmental rate consistent with the observed, one season, maturation period of two large crustacean species on the Lowland. For their complete lifespan, Crustacea respire a higher portion of their energy than do the other invertebrates discussed here. Much of this energy is respired by adults. The R/P for the crustacean *Daphnia pulex* Leydig varied from 0.37-32., depending on whether they were reproducing, and secondarily on food concentration (from Richman 1958). The mean R/P for actively reproducing *Daphnia* was 0.48, versus my assumed 0.5 for growing immatures.

The Acarina populations have one astonishing feature. Wallwork (1967) found, on a seasonal basis, that 50% of the oribatids he extracted were adults, and 50% immature. Addison (pers. comm.) found the same age distribution at the raised beach, where 50% of the 1972 and 53% of the 1973 Acarina were heavily sclerotized (=adult). Adults and immatures, which differ in body form, must have different mortality rates. Given that immature and adult mortalities are both exponential, in order to maintain the 50:50 ratio in a cohort with 60 starting individuals, the average adult must live 11-fold as long as the average immature, regardless of the number surviving to adulthood. Thus, some individual mites will survive up to 34 years on the raised beach, and to 68 years at the meadow. This long lifespan may reveal a faulty model assumption. Adults could have drastically reduced mortality rates (compared to immatures) until old age death. Then their total lifespans would be considerably less. Long adult lifespans, or strikingly different adult mortality rates, are the two apparent choices for allowing the 50:50 ratios observed. Long lifespans for mites have been noted in temperate areas (Wallwork 1967). Englemann (1961) reported that 96% of the energy assimilated by his oribatid populations was respired, which necessitates that growing immatures were only a small segment of his populations. Truelove oribatid respiration accounted for 69% of the assimilated energy. The respiration rate of oribatids was the lowest Procter [this volume (b)] measured for all the Truelove invertebrates.

The 7 year maturation period predicted for lycosids is close to the 6 year maturation period postulated for Lake Hazen lycosids (Leech 1966). Two factors should be emphasized. The predicted lycosid mean weight is 0.26 the collected mean weight. This requires that survival to sexual maturity is greater than was assumed, or that the sampling technique favored the collection of mature specimens. Both possibilites are probably correct. This necessitated correcting the model energy flow by 3.89-fold (observed weight/predicted weight).

The spider energy flow is comparatively low, despite this corrective increase. Researchers at other sites have ascribed spiders with a proportionately larger (Heal et al. 1975) to much greater (Mani 1968, Danilov 1972) role than was found here. The Lowland apparently is less favorable to, or less colonized by, spiders. Also, active predators are more frequently collected than their potential prey, and this creates an inflated impression of their importance (Southwood 1966). Pitfall trapping at Truelove yielded 460 spiders/trap in a season, but absolute sampling techniques showed the standing crop to be 0.3 individuals m^{-2} on the raised beach and 1.8 m^{-2} on the meadow. The raised beach spider (=lycosid) energy flow was 34 cal m^{-2}, while the emergence trap insect catch was 53 cal m^{-2}. Allowing that the winged insects metabolize part of their "trap collected" mass before death, and that the spiders do not totally consume their prey, would appear to make the spiders almost 100% efficient in capturing the raised beach Diptera adults. What actually occurred was that Diptera from adjacent meadows and water bodies swarmed over the raised beaches, and were available for consumption by the spiders. This swarming created the illusion that the raised beaches produced more adult Diptera than did meadows. A similar migration may explain Kajak's (1971) report that spiders consumed the entire winged insect emergence of her study area.

The Collembola 1.4 year maturation period on the raised beach is a product of the model; this would become 1.9 years if the immature/adult weight ratio was 50% greater. The maturation time for *Hypogastrura tullbergi* (Schaeffer), the dominant raised beach collembole, is approximately 2 years (J. Addison this volume). Since both juveniles and adults are similarly shaped, moult continuously, live in the same habitat, and consume the same food, the mortality rate should be similar for all stages. Once the initial number in a cohort is determined, and the adult life determined to be 50% longer than the maturation period (J. Addison this volume), the individuals surviving to maturity become a fixed number. The time to decrease from 8.7 individuals to 0.5 is 50% longer than from 60 to 8.7.

Pterygote insects in the emergence traps were not collected immediately after emergence, but remained in the traps up to several days before being collected. They lost weight from the prepupa, which ceased feeding, by metabolism, moulting twice, excreting a meconium, and possibly by depositing eggs before being collected. Leaching into the alcohol preservative is another source of weight loss (Dermott and Patterson 1973). These losses, and the probable death rates, were estimated, and a correction factor for the production computation introduced to make pharate pupae 1.5-fold the weight of the emergence trap adult weights.

Lepidoptera energy flow of 29 cal m^{-2} is comparable to the 4.4 cal estimated for *Gynaephora* alone on raised beaches (Ryan and Hergert this volume). The difference is due to the small samples of 1 Lepidoptera adult in the emergence traps, and 3 *Gynaephora* larvae collected for its standing crop estimate.

The predicted chironomid and sciarid maturation times on the raised beach are equivalent to known arctic lake chironomid maturation times of 2-3 years (Oliver 1968, Welch 1973), while those in the meadow are markedly longer. The Tipuloidea include trichocerids, which mature in 3 years, and tipulids, which require 4. This latter time is equivalent to that found by MacLean (1975) at Barrow, Alaska, for the tipulid *Pedicia hanna* Alexander. The Cyclorrapha energy flows can be computed through the respiration rates of either Procter or Scholander et al. Both give identical maturation times.

R/P ratio

The relationship of calories respired to calories of tissue produced, R/P (=c), is critical as it determines both the time for a cohort to develop and the quantity of energy metabolized. Throughout the model, I have consistently chosen R/P ratios which are the minimum justifiable (*cf.* Waldbauer 1968) and which minimize the development times. These minimized times correspond to observed developmental rates in the field. The minimized ratios are similar, when considered for the complete lifespan, to those theorized by Kozlovsky (1968).

The R/P ratios require careful evaluation if the model energy flow calculations are to be accurate, yet there are major obstacles to doing this. For example, they should account for trophic level differences. Herbivores, which consume discrete quantities of leaf tissue and produce discrete fecal pellets, have been studied most frequently (Waldbauer 1968), and allow these ratios to be measured. Carnivores, omnivores, and saprovores especially, are infrequently examined. Invertebrates, especially microinvertebrates, consuming heterogeneous substrata containing bacteria, fungi, and soil organic matter, are major energy releasers. Their R/P ratios are virtually undetermined; and such determinations would be technological feats. Radioisotopes used in feeding studies appear to be the most logical approach to measuring these. Another problem with feeding tests is that high laboratory survival rates are interpreted as indicating proper techniques, and feeding experiment results are given as means of the survivor's R/P ratios. Consumption and growth vary, while some temperature dependent level of respiration must always occur. This means R/P ratios will vary from some minimum value, when larvae grow rapidly, to greater values for slower growth. Such variation does not affect survival in a laboratory colony, but may determine survival in a field situation. The mean results of laboratory experiments may not be directly applicable to field conditions.

Energy flow pathways

Major pathways for the consumption and release of energy by Truelove invertebrates can be seen in Table 4. The dominant energy releasers at the raised beach site are nematodes, which account for 60% of the energy flow, followed in order by Enchytraeidae, Collembola, Diptera, and Acarina. At the meadow site, Protozoa account for 70% of the energy flow, while among the more accurately assessed groups Enchytraeidae dominate, followed in order by Nematoda, Diptera, and Crustacea. These are all soil dwellers, except for the adult, dispersal stage, of Diptera. The soil is the site of the greatest conversion and release of energy by invertebrates at both sites.

The feeding habits of these invertebrates must be evaluated primarily from the literature (especially Burges and Raw 1967). Lepidoptera and Hymenoptera Symphyta are herbivores, as are some small portion of the Nematoda, Acarina, Collembola, Diptera, and Hymenoptera Apocrita. Herbivory is a minor component of the Truelove energy flow. Carnivory is found in all Araneida, and some portion of the Protozoa, Nematoda, Tardigrada, Crustacea, Acarina, Collembola, Diptera, and Hymenoptera Apocrita. This energy pathway is more important than herbivory, but is still comparatively small. The pattern of energy flow in this ecosystem is dominated by the detritus food chain. Bacteria, fungi, or partially decomposed organic matter dominate the diets of Protozoa (Stout and Heal 1967), Rotifera (Pennak 1953), Nematoda (Nielsen 1967), Enchytraeidae (O'Connor 1967), Tardigrada (Hallas and Yeates 1972), Crustacea (Pennak 1953, Bottrell 1975), Acarina (Wallwork 1967, Luxton 1972), Collembola (Hale 1967) and Diptera (*cf.* Stone et al. 1965, Dowding 1967).

How does invertebrate energy flow compare to other energy flows in this ecosystem? An approximate answer, in Table 5, summarizes production and habitat data in Bliss (1975) and preliminary calculations by Whitfield (this volume). Invertebrates assimilate approximately 5% of the net primary production. This is 4-fold the combined assimilation of all vertebrates.

Table 5. Comparative energy flow yr^{-1} through the Lowland. Averaged from meadows (hummocky sedge moss, 20.6% of lowland) and raised beaches (cushion-plant lichen, 6.8%), 1970-73 period data.

	Kcal m^{-2}
Net primary production	656
Consumer assimilation:	
Muskox	6.9
Lemmings	0.52
Vertebrate carnivores	0.08
Birds (excl. carnivores)	0.005
All invertebrates	30.9

Vertebrates respire a higher proportion of their assimilated energy than invertebrates, so much more invertebrate tissue is produced (from Table 4 components: 1.1 g m^{-2} on the raised beach, and 2.3 g m^{-2} on the meadow). Although these values are tentative, they are sufficient to indicate the relative energetic importance of invertebrates. In this cold dominated ecosystem, where they are inactive almost 10 months yr^{-1}, the invertebrates are the major animal energy releasers by a several fold margin.

Conclusions

The standing crop animals (here excluding Protozoa and Nematoda) of the raised beach were numerically dominated by Collembola. Enchytraeidae dominated in biomass. The total invertebrate standing crop biomass of the raised beach was estimated to be 363 mg dry wt m^{-2}. On the meadow, Enchytraeidae followed by Copepoda were numerically dominant; Enchytraeidae again dominated in biomass. The total invertebrate standing crop on the meadow was estimated to be 852 mg dry weight m^{-2}.

In 1972, insect emergence on the raised beach was dominated numerically by Chironomidae (73 of a total 82 winged insects m^{-2}). The total mass of winged insects was 9.4 mg dry wt m^{-2}, predominantly chironomids (3.3 mg) and Muscoidea (2.1 mg). On the meadow, chironomids accounted for 887 of the 909 total winged insects m^{-2}. The total biomass emerged was 54.1 mg dry wt m^{-2}, which was predominantly chironomids (33.5 mg) and Muscoidea (18.6 mg).

Protozoa, including naked amoebae, ciliates, and testates were estimated to produce a minimum of 1.5 g dry wt m^{-2} yr^{-1} total tissue, and respired 17,500 cal, in the hummocky sedge-moss meadow. Protozoa were the most important invertebrate group in the meadow, and are suspected to be more important than shown here.

Other soil invertebrates, especially Nematoda and Enchytraeidae, were the next most significant energy releasers. Nematoda on the raised beach assimilated 7,925 cal m^{-2}, and at the meadow 969 cal m^{-2}. The enchytraeid energy flow on the raised beach was 3,546 cal m^{-2}, and in the meadow 5,667 cal m^{-2}. Collembola and Diptera were the third and fourth energy releasers of the raised beach invertebrates, and Diptera and Crustacea the fourth and fifth dominant energy releasers on the meadow.

For the raised beach, the yearly sum total of invertebrate tissue produced was 6,500 cal m^{-2}, with a total respired 7,300 cal m^{-2}. At the meadow site these totals were 13,200 cal m^{-2} tissue produced, with 23,300 cal m^{-2} respired. The raised beach total energy flow was 13800 cal m^{-2}, and the meadow total was 36,500 cal m^{-2}.

Herbivory was a minor component of the overall invertebrate energy flow. The major pathway for energy release was the decomposition based food chain.

Acknowledgments

My thanks to Lewina Svoboda and Colin Hergert for their hard work in getting my data together, and to L.C. Bliss and B.S. Heming. I am grateful to Doug Whitfield for his mind-boggling math and to Ross Goodwin. Others who helped were Dennis Procter, Jan S. Addison, Don Pattie, Hal Hamilton, Anne Avramenko, John Willard, George Ball, Josef Svoboda and the late Brian Hocking. I am grateful, too, to my fellow Devon Islanders, I.B.P. tundra researchers, and the Department of Entomology at The University of Alberta.

References

Bailey, C.G. and P.W. Riegert. 1973. Energy dynamics of *Encoptolophus sordidus costalis* (Scudder) (Orthoptera: Acrididae) in a grassland ecosystem. Can. J. Zool. 51: 91-100.

Beirne, B.P. 1955. Collecting, Preparing and Preserving Insects. Can. Dept. Agric. Publ. 932. 133 pp.

Bliss, L.C. 1975. Devon Island, Canada. pp. 17-60. *In:* Structure and function of Tundra Ecosystems. T. Rosswall, O.W. Heal (eds.). Ecol. Bull. 20 (Stockholm). 450 pp.

Block, W. 1965. The life histories of *Platynothrus peltifer* (Koch 1839) and *Damaeus clavipes* (Hermann 1804)(Acarina: Cryptostigmata) in soils of pennine moorland. Acarologia VII: 735-743.

Bornebusch, C.H. 1930. The fauna of forest soil. Forstl. Forsøgsvesen Denmark 11:, 1-224.

Botrell, H.H. 1975. Generation time, length of life, instar duration and frequency of moulting, and their relationship to temperature in eight species of Cladocera from the river Thames, Reading. Oecologia 19: 129-140.

Buei, K. 1967. (The relationship between the body weights of pupae of houseflies and the number of matured eggs in the ovaries.) Jap. J. Sanit. Zool. 18: 18-20.

Burges, A. and F. Raw, eds. 1967. Soil Biology. Academic Press. London. 532 pp.

Chernov, Ju. I. (pers. comm.) A.N. Sewerzoff Institute of Evolutionary Morphology and Animal Ecology, Leninskie prospekt 33, Moscow, U.S.S.R.

Comita, G.W. 1968. Oxygen consumption in *Diaptomus*. Limnol. Oceanogr. 13: 51-57.

Cummins, K.W. and J.C. Wuycheck. 1971. Caloric equivalents for investigations in ecological energetics. Mitt. Int. Ver. Limnol. No. 18. 158 pp.

Danilov, N.N. 1972. Birds and arthropods in the tundra biogeocenosis. pp. 117-121. *In:* Tundra Biome Proceedings IV International Meeting on The Biological Productivity of Tundra. F.E. Wielgolaski and T. Rosswall (eds.). Tundra Biome Steering Committee. Stockholm. 320 pp.

Dermott, R.M. and C.G. Patterson. 1973. Determining dry weight and percentage dry matter of chironomid larvae. Can. J. Zool. 52: 1243-1250.

Donner, J. 1966. Rotifers. (transl. H. Wright). Frederick Warne & Co. Ltd. London. 80 pp.

Dowding, V.M. 1967. The function and ecological significance of the pharyngeal ridges occurring in the larvae of some cyclorraphous Diptera. Parasitology 57: 371-388.

Downes, J.A. 1964. Arctic insects and their environment. Can. Ent. 96: 279-307.

Edgar, W.D. 1971. Aspects of the ecological energetics of the wolf spider *Pardosa* (Lycosa) *lugubris* (Walckenaer). Oecologia 7: 136-154.

Edwards, C.A. and K.E. Fletcher. 1971. A comparison of extraction methods for terrestrial arthropods. pp. 150-185. *In:* Quantitative Soil Ecology, I.B.P. Handbook No. 18. J. Phillipson (ed.). Blackwell. London. 297 pp.

Englemann, M.D. 1961. The role of soil arthropods in the energetics of an old field community. Ecol. Monogr. 31: 221-238.

———. 1966. Energetics, terrestrial field studies, and animal productivity. Adv. Ecol. Res. 3: 73-115.

———. 1968. The role of soil arthropods in community energetics. Amer. Zool. 8: 61-69.

Hale, W.G. 1967. Collembola. pp. 397-412. *In:* Soil Biology. A. Burges and F. Raw (eds.). Academic Press. London. 532 pp.

Hallas, T.E. and G.W. Yeates. 1972. Tardigrada of the soil and litter of a Danish beech forest. Pedobiologia 12: 287-304.

Heal, O.W. 1965. Observations on testate amoebae (Protozoa: Rhizopoda) from Signy Island, South Orkney Islands. Brit. Antarctic Survey Bull. 6: 43-47.

———. 1967. Quantitative feeding studies on soil amoebae. pp. 120-126. *In:* Progress in Soil Biology. O. Graff and J. Satchell (eds.). Braunschweig, Wieweg. 656 pp.

———. 1971. Protozoa. pp. 51-71. *In:* Quantitative Soil Ecology. I.B.P. Handbook No. 18, J. Phillipson (eds.). Blackwell Publ., London. 297 pp.

———. H.E. Jones, and J.B. Whittaker. 1975. pp. 295-320. *In:* Structure and Function of Tundra Ecosystems, T. Rosswall and O.W. Heal (eds.). Ecol. Bull. 20 (Stockholm). 450 pp.

Healy, I.N. 1967. The population metabolism of *Onychiurus procampatus* Gisin (Collembola). pp. 127-137. *In:* Progress in Soil Biology. O. Graff and J. Satchell (eds.). Braunschweig, Wieweg. 656 pp.

Hocking, B. 1953. The intrinsic range and speed of flight of insects. Trans. Roy. Ent. Soc. Lond. Ser. B. 104: 223-345.

Hodkinson, I.D. 1973. The population dynamics and host plant interactions of *Strophingia ericae* (Curt.) (Homoptera: Psylloidea). J. Anim. Ecol. 42: 565-583.

Hofsvang, T. 1973. Energy flow in *Tipula excisa* Schum. (Diptera, Tipulidae) in a high mountain area, Finse, south Norway. Norw. J. Zool. 21: 7-16.

Huhta, V. and A. Koskenniemi. 1975. Numbers, biomass and community respiration of soil invertebrates in spruce forests at two latitudes in Finland. Ann. Zool. Fennici 12: 164-182.

Ivleyva, I.V. 1953. Growth and reproduction of the potworm (*Enchytraeus albidus* Henlé). Zool. Zh. 32: 394-404.

Kajak, A. 1971. Productivity investigation of two types of meadows in the Vistula Valley. IX. Production and consumption of field layer spiders. Ekol. Polska XIX (15): 197-211.

Kevan, P.G. 1970. High Arctic Insect-Flower Relations. Ph.D. thesis. Univ. Alberta, 399 pp.

Kozlovsky, D.G. 1968. A critical evaluation of the trophic level concept. I. Ecological efficiencies. Ecology 49: 48-59.

Kuehnelt, W. 1955. A brief introduction to the major groups of soil animals and their biology. pp. 29-43. *In:* Soil Zoology. D.K. McE. Kevan (ed.), Butterworths, London. 512 pp.

Learner, M.J. 1972. Laboratory studies on the life-histories of four enchytraeid worms (Oligochaeta) which inhabit sewage percolating filters. Ann. Appl. Biol. 70: 251-266.

Leech, R.E. 1966. The spiders (Araneida) of Hazen Camp, 81°49'N, 71°18'W. Quaest. Ent. 2: 153-212.

Luxton, M. 1972. Studies on the oribatid mites of a Danish beech wood soil. I. Nutritional biology. Pedobiologia 12: 434-463.

MacLean, S.F. 1973. The life cycle and growth energetics of *Pedicia hanna antennata* Alex. (Diptera:Tipulidae), an arctic crane fly. Oikos 24: 436-443.

———. 1975. Ecological adaptations of tundra invertebrates. pp. 269-300. *In:* Physiological Adaptation to the Environment. F.J. Vernberg (ed.). Intext Educational Publ., New York. 576 pp.

Macfadyen, A. 1961. Metabolism of soil invertebrates in relation to soil fertility. Ann. Appl. Biol. 49: 215-218.

Mani, M.S. 1968. Ecology and biogeography of high altitude insects. Junk, the Hague. 527 pp.

Moulder, B.C., D.E. Reichle, and S.E. Auerbach. 1970. Significance of spider predation in the energy dynamics of forest floor arthropod communities. Oak Ridge National lab. Tennessee. 170 pp.

Nielsen, C.O. 1949. Studies on the soil microfauna. II. The soil inhabiting nematodes. Nat. Jutland. 2: 1-131.

———. 1967. Nematoda. pp. 197-211. *In:* Soil Biology. A. Burges and F. Raw (eds.). Academic Press, London. 532 pp.

O'Connor, F. G. 1962. The extradition of Enchytraeidae from soil. pp. 279-285. *In:* Progress in Soil Zoology, P.W. Murphy (ed.). Butterworths, London. 398 pp.

———. 1967. The Enchytraeidae. pp. 213-258. *In:* Soil Biology. A. Burges and F. Raw. (eds.). Academic Press, London, 532 pp.

Odum, E.P. 1971. Fundamentals of Ecology. W.B. Saunders & Co., Toronto. 574 pp.

Olechowicz, E. 1971. Productivity investigation of two types of meadows in the Vistula Valley. VIII. The number of emerged Diptera and their elimination. Ekol. Polska 19: 183-195.

Oliver, D.R. 1968. Adaptations of arctic Chironomidae. Ann. Zool. Fenn. 5: 111-118.

Pennak, R.W. 1953. Fresh-water Invertebrates of the United States. Chapt. 17: Copepoda. Ronald Press Co., New York. 769 pp.

Petruscewicz, K. and A. Macfadyen. 1970. Productivity of Terrestrial Animals, Principles and Methods. I.B.P. Handbook No. 13. Blackwells, London. 190 pp.

Reichle, D.E. 1966. Relation of body size to food intake, oxygen consumption, and trace element metabolism in forest floor arthropods. Ecology 49: 538-542.

———. R.A. Goldstein, R.I. Van Hook, Jr., and G.L.J. Dodson. 1973. Analysis of insect consumption in a forest canopy. Ecology 54: 1076-1084.

Reynoldson, T.B. 1939. On the life-history and ecology of *Lumbricillus lineatus* Mull. (Oligochaeta). Ann. Appl. Biol. 26: 782-799.

———. 1943. A comparative account of the life cycles of *Lumbricillus lineatus* Mull. and *Enchytraeus albidus* Henlé in relation to temperature. Ann. Appl. Biol. 30: 60-66.

Richardson, A.M.M. 1975. Energy flux in a natural population of land snail, *Cepaea nemoralis* L. Oecologia 19: 141-164.

Richman, S. 1958. The transformation of energy by *Daphnia pulex*. Ecol. Monogr. 28: 273-291.

Riegert, P.W., J.L. Varley, and B.C. Dunn. 1974. Above-ground invertebrates: IV. Populations and energetics of spiders. C.C.I.B.P. Matador Proj. Tech. Rep. 57. Univ. Saskatchewan. 76 pp.

Roff, J.C. 1973. Oxygen consumption of *Limnocalanus macrurus* Sars (Calanoida, Copepoda) in relation to environmental conditions. Can. J. Zool. 51: 877-885.

Ryan, J.K. 1978. Comparison of invertebrate species lists from I.B.P. tundra sites. *In:* Tundra: Comparative Analysis of Ecosystems. J.B. Cragg and J.J. Moore (ed.). Cambridge Univ. Press, Cambridge. (In press.)

Scholander, P.F., W.F. Flagg, V. Walters and L. Irving. 1953. Climatic adaptation in arctic and tropical poikilotherms. Physiol. Zooel. 26: 67-92.

Schroeder, L.A. and D.G. Dunlap. 1970. Respiration of cecropia moth (*Hyalophora cecropia* L.) larvae. Compr. Biochem. Physiol. 35: 953-957.

Southwood, T.R.E. 1966. Ecological Methods. Chapman and Hall, London. 391 pp.

Spinage, C.A. 1972. African ungulate life tables. Ecology 53: 645-652.

Stone, A., C.W. Sabrosky, W.W. Wirth, R.H. Foote, J.R. Coulson. 1965. A catalogue of the Diptera of America North of Mexico. Government Printing Office, Washington D.C., U.S.A. 1696 pp.

Stout, J.D. and O.W. Heal. 1967. Protozoa. pp. 149-195. *In:* Soil Biology. A. Burges and F. Raw (eds.). Academic Press, London. 532 pp.

Van Der Drift, J. 1951. Analysis of the animal community in a beech forest floor. Tijdschr. Entomol. 94: 1-168.

Van Hook, R.I. Jr. 1971. Energy and nutrient dynamics of spider and orthopteran populations in a grassland ecosystem. Ecol. Monogr. 41: 1-26.

Waldbauer, G.P. 1968. The consumption and utilization of food by insects. Adv. Ins. Physiol. 5: 229-288.

Wallwork, J.A. 1967. Acari. pp. 363-395. *In:* Soil Biology. A. Burges and F. Raw (eds.). Academic Press, London. 532 pp.

Welch, H.E. 1973. Emergence of Chironomidae (Diptera) from Char Lake, Resolute, N.W.T. Can. J. Zool. 51: 1113-1123.

Willard, J. 1972. Soil Invertebrates: I. Methods of sampling and extraction. C.C.I.B.P. Matador Proj. Tech. Rep. 7. Univ. Saskatchewan. 40 pp.

———. 1974. Soil Invertebrates: VIII. A summary of populations and biomass. C.C.I.B.P. Matador Proj. Tech. Rep. 56. Univ. Saskatchewan. 110 pp.

Nematode densities and production on Truelove Lowland

Dennis L.C. Procter

Introduction

The role of free-living soil nematodes in energy flow through high arctic ecosystems is poorly known. This paper is part of a study of the role of nematodes in energy flow in the Truelove Lowland terrestrial ecosystem. The primary purposes of this research were to measure the energy passing through the nematodes, and to determine the relative importance of nematodes in total invertebrate energy flow. Ryan (this volume) has estimated energy flow for the other important invertebrates in this ecosystem.

This paper gives nematode densities for July and August 1972, and for June to September 1973, and presents estimates of standing crop, respiration, turnover time, and production for the 1972 and 1973 growing seasons. Data are given for several habitats in the ecosystem. The dominant genera are also listed, and their trophic relations indicated. Apart from the density estimates, this information is preliminary and more refined estimates are being prepared.

Methods

Sampling

Five 5×5 m quadrats were established in the Intensive Meadow site (Hummocky Sedge-moss Meadow — Muc this volume), and 15 in the Intensive Raised Beach site (I.R.B.). The I.R.B. quadrats were evenly distributed between the Crest, Slope, and Transition zones (Svoboda this volume). The same quadrats were used both years.

Sampling commenced both years when the active layer was approximately 5 cm deep. In 1972 samples were initially collected on alternate days, but during the later part of 1972 and all of 1973 samples were collected every third day. Twenty samples were taken on each sampling date, one from each quadrat. The samples were obtained with a 6 cm diameter bulb planter, and were randomly selected from a grid of 6×6 cm squares in each quadrat. The samples were 5 cm deep, giving a soil volume of approximately 140 cc.

An additional 32 samples to a depth of 10 cm were taken from beneath the 5 cm deep samples, four on each sampling date from 24 July to 14 August 1973. These 5-10 cm deep samples, which were randomly selected from the 5 cm deep samples, were also evenly distributed between the four study areas.

Extraction

Samples were extracted using the 'tray methods' of Whitehead and Hemming (1965). The material was spread over double layers of tissue paper supported on insect screening in 28×22 cm plastic-coated wire baskets standing in 30×27×5 cm plastic dishes. Water was added to the dishes until the soil became wet, and the baskets were left in place for 24 hr. Thirty-two of these samples were extracted for a second 24 hr.

The nematode suspensions (*ca.* 1 *l*) obtained from the collecting dishes were concentrated into 3 dram vials using 40 cm high × 5 cm diameter glass cylinders. The vials were attached to the tapered bottoms of the cylinders with rubber tubing and screw clamps. A tight fitting rubber plunger was run through the cylinders every 2 to 3 hr to prevent nematodes settling on the cylinder sides. The suspensions were allowed to settle for approximately 12 hr. The concentrated nematodes were killed by heat and fixed by addition of formalin-glycerol fixative (Southey 1970).

Extraction efficiency of the 'tray method' was determined by returning counted live nematodes to autoclaved soil (20 min at 15 psi and 245°C). These nematodes were re-extracted for two successive 24 hr periods.

Counting

Counting was done at 25× under a stereo microscope in 120 mm diameter × 20 mm high glass dishes. The dishes were marked into eighths, and a single eighth from each sample was counted. When a subsample contained less than 50 nematodes, an additional subsample was counted, and the mean of the two subsamples was used.

Production

Preliminary estimates of production were obtained using the model developed by Ryan and Whitfield (Ryan this volume). These estimates use the nematode densities presented elsewhere in this paper, and the soil temperature data of Courtin and Labine (this volume). Because research is not complete, the other information used in the model was obtained from the literature.

The Ryan and Whitfield model has the general form:

$$R = cP$$

where R = respiration,
P = production,
and c = constant relating respiration to production.

The 'synthetic cohort' (Ryan this volume) is the basic unit of the model, for which respiration is given by:

$$R = a\frac{NoWo^{b}}{(b\beta-\propto)}\left[\left(\frac{Wf}{Wo}\right)^{b}\frac{Nf}{No}-1\right]$$

where No = number of individuals (eggs) entering the cohort,
Nf = number of individuals reaching maturity,
Wo = mean dry weight in micrograms of the individuals entering the cohort,
Wf = mean dry weight in micrograms of the individuals reaching maturity,

b = rate at which the specific metabolic rate per unit weight decreases relative to weight as weight increases,
β = growth rate,
$\propto$ = death rate,
and a = constant relating respiration to weight, which is

calculated from: $R = aW^b$

where R = microlitres of oxygen per individual per hour at 12°C,
and W = mean individual dry weight in micrograms.

Cohort production is obtained from:

$$P = \frac{NoWo\propto/\beta}{(1-\propto/\beta)}\left[Wf^{1-\propto/\beta} - W^{1-\propto/\beta}\right]$$

with the symbols as in the respiration equation.

The constant c is given by

$$c = F.R|P.1000/4.7$$

where F = calories per milligram dry weight of nematode tissue,
R/P = ratio of respiration to production,
1000 = converts milligrams to micrograms, and millilitres to microlitres,
and 4.7 = number of calories per millilitre of oxygen (Petrusewicz and Macfadyen 1970).

Adult and juvenile production were estimated separately. The following values were used to obtain juvenile production (in lieu of my own data, I used literature information as noted):

No = 500,
Nf = 10,
Wo = 0.005 μg,
Wf = 1.0 μg,
b = 0.67 (Klekowski et al. 1972),
R = 0.78 $\mu l\ 10^{-3}\ ind^{-1}\ hr^{-1}$ (recalculated from Klekowski loc cit., according to Winberg 1971),
W = 0.2186 $\mu g\ ind^{-1}$ (recalculated from Klekowski loc cit., on the assumption that dry weight = 25% of wet weight),
a = 2.161,
F = 5.5 cal mg^{-1},
R/P = 1.17 (recalculated from Marchant and Nicholas 1974, and de Soyza 1970),
c = 1369.1.

Adult production was calculated using the following substitutions (otherwise the juvenile values were used):

No = 10,
Nf = 0.5,

and Wo and Wf were replaced by the mean weight

$\overline{W}$ = 0.92 μg.

The two stages were assumed to be of equal duration.

Production per m^2 was calculated according to Ryan (this volume).

Results

Nematode faunal composition

Preliminary studies produced 18 genera belonging to 11 families and six orders (Table 1). Individuals of the genus *Plectus* occurred in large numbers in 80% of the samples studied, while the genus *Dorylaimus* was important in half of the samples. Significantly less abundant were *Tylenchus* and *Rhabdolaimus*, which were common in 20% of the samples. The remaining genera were common in only 5-10% of the samples.

Table 1. Preliminary list of nematode genera found on Truelove Lowland.

Order	*Family*	*Genus*
Tylenchida	Tylenchidae	*Tylenchus*
		Tylenchorhynchus
	Heteroderidae	*Heterodera*
	Neotylenchidae	*Neotylenchus*
	Aphelenchoididae	*Aphelechoides*
Dorylaimida	Dorylaimidae	*Dorylaimus*
		Carcharolaimus
	Mononchidae	*Mononchus*
		Prionchulus
Rhabditida	Cephalobidae	*Acrobeloides*
		Chiloplacus
		Acrobeles
Teratocephalida	Teratocephalidae	*Teratocephalus*
Araeolaimida	Plectidae	*Plectus*
		Anaplectus
		Rhabdolaimus
	Axonolaimidae	*Cylindrolaimus*
Monhysterida	Monhysteridae	*Monhystera*

Densities at 0-5 cm soil depth

In 1972 the transition zone had the largest seasonal mean number of nematodes per m^2 and the meadow had the smallest (Table 2). Furthermore, the means increased progressively from the crest, through the slope, to the transition. All four habitats had relatively low densities at the start of the season, which increased quickly as the season progressed (Fig. 1).

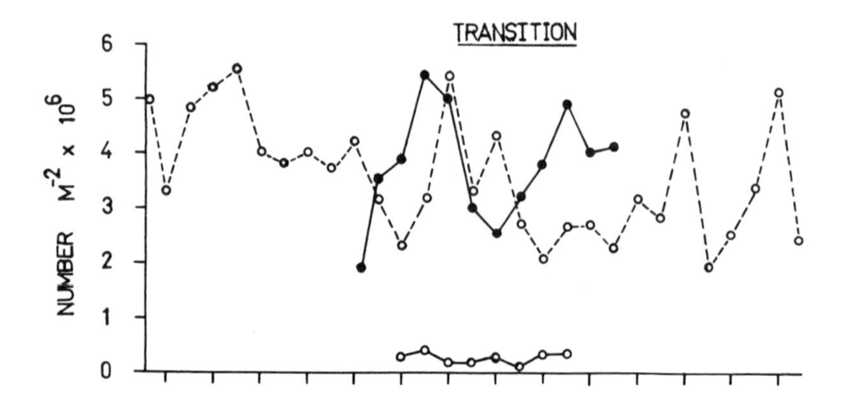

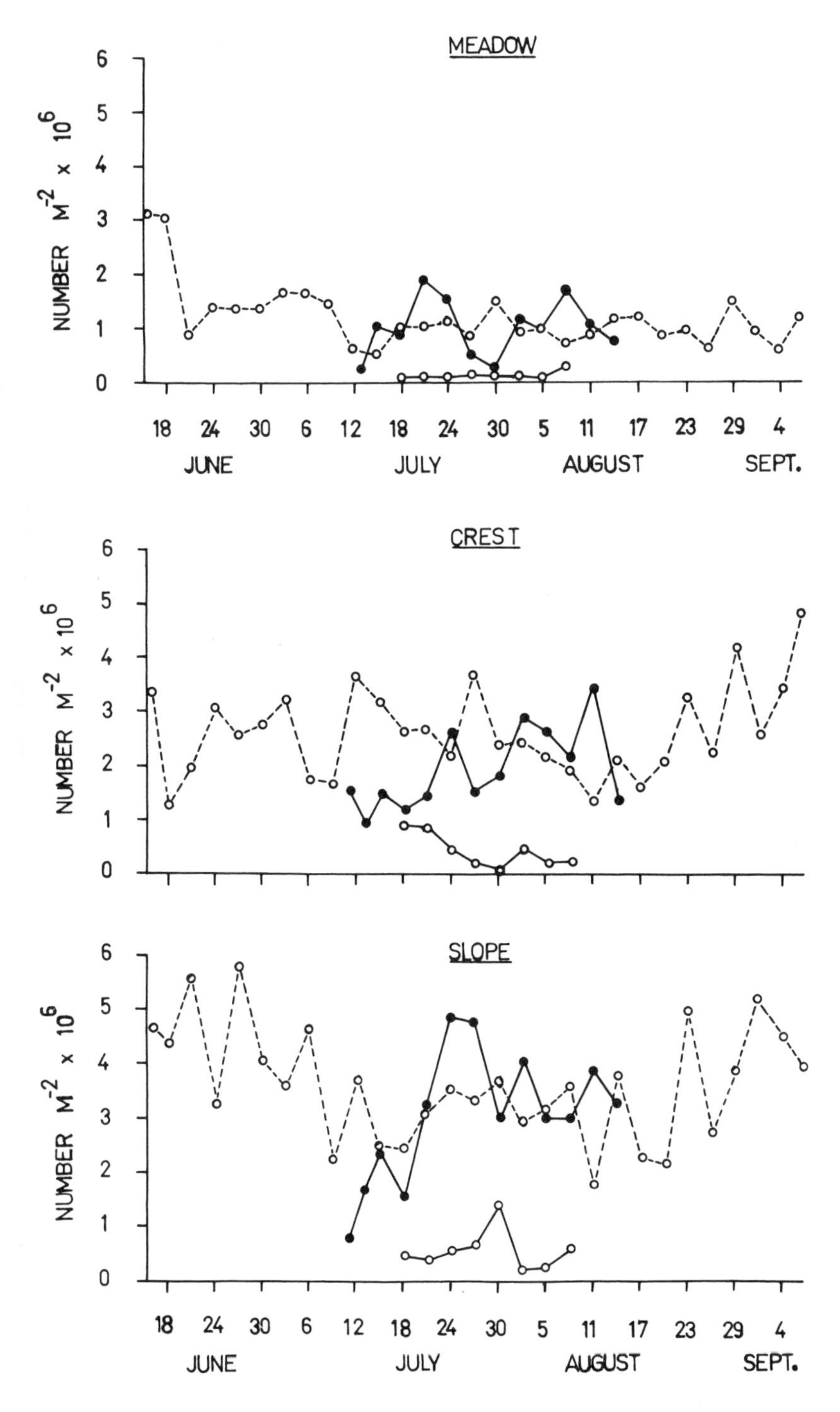

Fig. 1 The 1971 and 1973 seasonal numbers of nematodes per m² at 0-5 cm soil depth, and 5-10 cm depth for Transition, Meadow, Crest, and Slope. Each point is the mean of 5 samples (●—● = 1972 numbers at 0-5 cm depth; o - - o = 1973 numbers at 0-5 cm depth; o—o = 1973 numbers at 5-10 cm depth).

Table 2. The 1972 and 1973 seasonal maximum, minimum and mean numbers of nematodes m^{-2} at 0-5 cm soil depth for the four study areas. n = 1 except for means (n = 64 for crest and slope, n = 59 for transition and meadow in 1972; n = 145 for crest, slope and transition, n = 142 for meadow in 1973).

	Number m^{-2} (X10^{6})			
Range	*Crest*	*Slope*	*Transition*	*Meadow*
1972				
Max	5.520	8.492	9.303	2.868
Min	0.055	0.120	1.151	0.008
Mean±SD	1.983±1.357	3.009±1.814	3.770±1.857	1.008±0.761
1973				
Max	10.992	10.734	9.658	3.557
Min	0.077	0.206	0.947	0.177
Mean±SD	2.736±1.717	3.621±1.835	3.573±1.586	1.168±0.817

The four habitats showed the same relative densities in 1973 that were observed in 1972, except for the fact that the slope mean was slightly larger than the transition (Table 2). The 1973 means were larger than the equivalent 1972 values, except for the transition, which was a little larger in 1972. In contrast to 1972, the 1973 densities were high when sampling began, and either remained high or declined as the season progressed (Fig. 1).

The t-test was used to compare means within years (Li 1964). Except when the slope and transition are compared, all means were different ($p = 0.01$) for both years. In 1972 the transition mean was larger than the slope at $p = 0.05$, while the 1973 means showed no difference at this confidence level.

Plant cover, standing crop, and annual production increased progressively from crest, through slope and transition, to meadow (Table 3). Soil water also increased from crest to meadow, while soil temperature decreased from crest to meadow.

Table 3. Average annual percentage plant cover, standing crop, net production (Svoboda, Muc this volume), mean seasonal soil water content at 0-3 cm depth, and 24 hr mean soil temperature at 5 cm depth (Courtin and Labine this volume) for the four study areas.

Parameter	*Crest*	*Slope*	*Transition*	*Meadow*
Cover	80.5	92.4	100.0	98.0
Standing crop (g m^{-2})	ca. 232	ca. 266	991.0	3207
Production (g m^{-2})	ca. 10.8	ca. 12.3	53.4	230.7
Soil water (% oven dry wt)	11.8	24.3	190.0	746.4
Soil temperature (°C)	8.6	7.4	5.1	3.1

Densities at 5-10 cm soil depth

At 5-10 cm soil depth the slope had the largest seasonal mean, followed by the crest, transition, and meadow with progressively smaller values (Table 4). The crest had the largest proportion of total nematode population in the vertical profile at 5-10 cm depth (29.5%), followed by slope (15.7%), meadow (13.5%) and transition (9.1%). The crest and slope proportions were larger than the transition at $p = 0.05$.

Organic matter declined with increasing soil depth. On the crest organic matter comprised 3.7% of soil dry weight in the top 5 cm of soil, 1.4% at 5-10

Table 4. The 1973 seasonal maximum, minimum and mean numbers of nematodes m^{-2} at 5-10 cm depth for the four study areas (n=1 for maximums and minimums, n=8 for means).

	Number m^{-2} (× 10^6)			
Range	*Crest*	*Slope*	*Transition*	*Meadow*
Max	0.963	0.947	0.391	0.309
Min	0.124	0.239	0.082	0.012
Mean±SD	0.479±0.313	0.533±0.227	0.276±0.107	0.105±0.093

cm depth, and 0.2% at 10-15 cm depth (Svoboda this volume). Temperature also declined with increasing depth. The I.R.B. in 1973 accumulated 833 degree days above 0°C in the top 5 cm, 788 degree days at 5-10 cm depth, and 701 degree days at 10-15 cm depth (Courtin and Labine this volume).

Soil water in the top 3 cm of soil, expressed as percentage of soil dry weight, increased from crest (11.6%) through slope (34.7%) and transition (214.4%) to meadow (745.5%) in 1973 (Courtin and Labine this volume). Soil water at 10-13 cm depth, which was consistently less than in the top 3 cm, also increased from crest (9.0%) to meadow (494.1%), although the slope had a slightly larger value (39.8%) than the transition (37.8%). The slope had the greatest proportion of soil water at 10-13 cm depth (53.4%), followed by the crest (43.7%), meadow (39.9%) and transition (15.0%).

Densities in the top 10 cm of soil

The 1972 seasonal mean numbers of nematodes in the top 10 cm of soil ranged from a high of 4,706,000 (transition) to a low of 1,301,000 (meadow), and in 1973 from 4,658,000 (slope) to 1,564,000 (meadow)(Table 5). The combined seasonal mean was substantially larger in 1973.

Table 5. The 1972 and 1973 seasonal mean numbers of nematodes m^{-2} in the top 10 cm of soil in the four study areas. The combined mean for each year is also given.

	1972	*1973*
Habitat	*Number m^{-2} (× 10^6)*	
Crest	2.557	3.533
Slope	3.902	4.658
Transition	4.706	4.631
Meadow	1.301	1.564
Mean	3.117	3.597

Rate and efficiency of extraction

The four study areas had similar extraction rates, and standard deviations were uniformly high (Table 6). Extraction efficiency was 34.4% after 24 hr, and 42.3% after 48 hr (Table 7). Numbers extracted during the second 24 hr averaged 27.7% of the numbers collected during the first 24 hr.

Table 6. Numbers of nematodes extracted during two 24 hr extraction periods. The 24-48 hr numbers are also expressed as percentages of the 0-24 hr numbers (n=8 for numbers extracted and for percentages).

	Crest		Slope		Transition		Meadow	
	Number extracted							
Time	*Mean±SD*		*Mean±SD*		*Mean±SD*		*Mean±SD*	
0-24hr	2570	1533	2700	939	2840	927	798	312
24-48 hr	604	401	717	552	904	698	246	280
Percentage	29.4		28.5		31.6		28.5	

Table 7. Results of extraction efficiency studies. Given are nematode numbers returned to autoclaved soil, numbers recovered during first and second 24 hr, and total recovery. Numbers recovered are also expressed as percentages of initial numbers, and 24-48 hr number is expressed as percentage of 0-24 hr value (n=8 for means and percentages).

	Number recovered		
Time	*Mean*	± *SD*	%
Initial no.	10040	8539	—
0-24 hr	3917	4394	34.4
24-48 hr	830	779	8.0
0-48 hr	4747	5155	42.4
24-48 hr/0-24 hr	—	—	27.7

Table 8. Production estimates for the theoretical cohort.

Parameter	*Juveniles*	*Adults*	*Combined*
Average no.	70.71	2.24	72.95
Average wt (μg)	0.0432	0.92	0.0701
Production (μg)	18.6647	10.0000	28.6647
Production (cal)	0.1027	0.0550	0.1577
Respiration (cal)	0.1845	0.0324	0.2169
Cohort time (degree-days)	753.5	753.5	1507.0

Production

Juvenile nematodes were numerically the dominant component of the synthetic cohort and consequently, despite their individually smaller biomass, contributed the greater proportion of respiration and production (Table 8). Juveniles and adults contributed equally to cohort duration, as was specified when establishing the model.

Cohort duration was greatest in 1972, ranging from 3.8 seasons on the I.R.B. at 0-5 cm depth, to 36 seasons in the meadow at 5-10 cm depth (Table 9). The equivalent 1973 cohort times were 1.8 and 6.6 seasons.

Standing crops in 1972 varied between 1199 cal on the I.R.B. at 0-5 cm depth, and 52 cal in the meadow at 5-10 cm depth (Table 10). Corresponding 1973 values were a little larger, 1321 cal on the I.R.B., and 63 cal in the meadow.

Production in 1972 ranged from 1769 cal in the I.R.B. at 0-5 cm depth, to 9 cal in the meadow at 5-10 cm depth (Table 10). Production was substantially greater in 1973, with 4089 cal in the I.R.B. and 53 cal in the meadow at the same depths.

Respiration resembled production in variation with microhabitat and season (Table 10). In 1972 nematodes respired 2433 cal on the I.R.B. at 0-5 cm depth, and 12 cal in the meadow at 5-10 cm depth, while the corresponding 1973 values were 5624 and 73 cal.

Nematodes dominated invertebrate standing crop and production on the I.R.B. with 49% and 55% respectively. The Enchytraeidae were the next most important group, accounting for 29% of standing crop, and 35% of production (Ryan this volume). The other groups were relatively unimportant. In the meadow nematodes contributed 9% of standing crop, and 7% of production. The Diptera and Enchytraeidae were the dominant invertebrates in the meadow (Ryan loc cit).

Discussion

Nematode faunal composition

Thirteen of the 18 genera found on Truelove Lowland also occurred at Lake Hazen, N.W.T. (Mulvey 1963). However, Mulvey identified 32 genera from Lake Hazen, including plant- and insect-parasitic nematodes not considered in the present study. The greater number of genera obtained by Mulvey probably reflects his more extensive collecting, and the preliminary nature of the present study.

Of the Truelove Lowland genera not reported from Lake Hazen, the most noteworthy is *Heterodera* (prob. *punctata*). This is the most northerly record of *Heterodera* from Canada. Other genera apparently recorded only from Truelove Lowland in arctic North America are *Carcharolaimus*, *Acrobeloides* and *Anaplectus*. The genus *Rhabdolaimus*, while absent from Lake Hazen, was described from the Canadian Arctic and Alaska by Cobb (See Mulvey 1963).

Cobb (1921) found that 70% of his specimens belonged to the genus *Plectus*. Mulvey (1963) also found plectids abundant and, as in the present study, in nearly all samples taken. *Plectus* was also important in Western Taimyr, U.S.S.R. (Kuzmin 1973). In addition to plectids, genera belonging to the families Dorylaimidae and Cephalobidae were particularly well-represented in all four studies.

Soil-inhabiting nematodes dominate each faunal list. On the other hand, known plant-parasites are poorly represented. While the food of many soil-living nematodes is uncertain, a majority probably feed on micro-flora and fauna. This emphasis may reflect the importance of the micro-flora and fauna compared to higher plants in the productivity of arctic ecosystems.

Densities at 0-5 cm soil depth

The increase in nematode numbers from crest, through slope, to transition (Table 2) is correlated with the amount of vegetation in these sites (Table 3) (although most of the nematodes probably feed on the associated organic matter and micro-organisms). The meadow did not fit this pattern, and had the largest plant biomass and smallest number of nematodes of the four study areas. There are two possible explanations for the meadow's difference. First, the crest, slope, and transition share an essentially common flora dominated by *Dryas integrifolia* (Svoboda this volume), while the meadow is dominated by sedges (Muc this volume). Second, the meadow was the wettest of the study areas (i.e. the soil was saturated, J. Addison this volume).

The extreme wetness of the meadow has two important consequences for nematodes. First, the meadow soil contains relatively little oxygen (Courtin and Labine this volume), which diffuses relatively slowly because the rate of oxygen diffusion is inversely related to soil water content. Nematodes are aerobic, and become quiescent when oxygen is limited or absent (Wallace 1971). Second, small nematodes generally use the surface tension forces of water films to obtain leverage for movement (Croll 1970). In very wet habitats, where surfaces of water films are inaccessible, nematodes are limited to swimming, which is often ineffective. Consequently, despite the meadow's greater plant biomass and production, its extreme wetness makes it the least suitable nematode habitat of the four areas studied.

The low initial nematode densities in 1972 were probably caused by the long preceding winter (the thaw in 1972 was nearly 4 weeks later than in 1973) (Fig. 1). On the other hand, mortality during the short winter following 1972 was negligible, because initial 1973 densities were as high as late 1972 densities.

Initial 1972 densities were probably below habitat carrying capacities, which permitted rapid increase to 1973 levels as the season progressed. With little winter mortality, initial 1973 densities were probably close to carrying capacity, and were therefore relatively unchanged during the season.

Densities at 5-10 cm soil depth

When nematode numbers at 5-10 cm depth (Table 4) are compared with densities at 0-5 cm depth, it is clear that most of the nematodes inhabit the top 5 cm of soil. When this concentration of nematodes is compared with the distribution of organic matter and temperature with soil depth, the reason for the concentration becomes clear. There was approximately twice as much organic matter at 0-5 cm depth compared to 5-10 cm depth, and the shallower soil was warmer. The decline in organic matter and temperature from 5-10 cm depth to 10-15 cm depth was even more marked, and would probably be reflected by an even smaller number of nematodes at 10-15 cm depth.

In contrast to the situation in the top 5 cm, where nematode numbers increased from crest to transition, the proportion of nematodes at 5-10 cm depth decreased from crest to transition. Courtin and Labine's soil water data (this volume) explain this apparent anomally. Both crest and slope were drier than the transition, and both had proportionately more water in the deeper soil (at 10-13 cm depth). The combination of dry habitat with relatively wet deeper soil, accounts for the greater proportion of nematodes at 5-10 cm depth on the crest and slope.

Densities in the top 10 cm of soil

In 1973 the four study areas averaged 480,000 m^{-2} more nematodes than in 1972 (a 12% increase) (Table 5). The 1973 summer was longer and warmer than in 1972 [833 degree-days compared to 397 degree-days at 0.5 cm depth on the I.R.B. (Courtin and Labine this volume)], resulting in increased plant production (*ca.* 10% Bliss this volume), and probably a similar increase in micro-organism production.

My data suggest that few nematodes live below 10 cm depth, and the numbers presented in Table 5 are probably realistic estimates of all nematodes in the soil. These numbers can therefore be compared with estimates from other studies. Yeates (1972) compiled a list of published estimates of nematode numbers, which range from 330,000 m^{-2} in a Danish grass field, to 29,800,000 m^{-2} in a German oak forest. Sohlenius (pers. comm.) has compiled a more extensive list covering 29 temperate, tropical, and alpine locations, which range from 8,100 m^{-2} in a Puerto Rican rain forest, to the German oak forest value previously cited by Yeates. The average rank of my estimates lies above the midpoint of the estimates presented by Sohlenius, and I conclude that high arctic environments can be more suitable for nematodes than many environments at lower latitudes.

Rate and efficiency of extraction

In their original studies Whitehead and Hemming (1965) collected 'few' additional nematodes when extraction was continued beyond 24 hr. Using the same method, I increased initial recovery by 29.5% when I extracted for an additional 24 hr (Table 6). When the maximum number of nematodes is wanted, the extraction period should be extended.

More important is the efficiency of extraction. I obtained a 34.4% recovery after 24 hr, and 42.3% after 48 hr (Table 7). Using a centrifugal-flotation method, Willard (1972) obtained an efficiency of 36.5% for heavy clay soils.

These studies show that if extraction efficiency is not determined, estimates of nematode numbers may be very conservative. Estimates presented in this paper are corrected for extraction efficiency.

Production

Because both the model and the information used in it are being revised, there is little point in extensive comment on the production estimates presented in Table 8. However, two of the model's assumptions warrant comment. First, the model assumes that the number of individuals in the cohort declines exponentially. This assumption is convenient mathematically, but probably unrealistic, in which case the production estimates will be unrealistic. The survivorship curve for one of the common Devon Island species (*Chiloplacus* sp.) is being determined. Second, the model assumes that the juvenile and adult stages are of equal duration. The literature offered no data on either generation time nor length of life of nematodes at low temperatures, and this information is necessary to accurately estimate production. Experiments have been performed which will provide the necessary data.

The differences in cohort duration between 1972 and 1973 (Table 9) reflected the different temperature regimes of the two seasons. For example,

Table 9. The 1972 and 1973 estimates of cohort duration, expressed as seasons per cohort, for the I.R.B. and Meadow at 0-5 cm and 5-10 cm soil depth.

	1972		*1973*	
Depth	*I.R.B.*	*Meadow*	*I.R.B.*	*Meadow*
0-5 cm	3.8	14.6	1.8	4.2
5-10 cm	4.1	36.1	1.9	6.6

the I.R.B. at 0-5 cm depth accumulated more than twice the number of degree-days in 1973 that it had in 1972, and the 1973 meadow value at 5-10 cm depth was five times the corresponding 1972 figure (Courtin and Labine this volume).

The consistently shorter cohort times at the I.R.B. compared to the meadow were caused by the higher temperatures in the I.R.B., particularly in 1972, when the I.R.B. accumulated nearly four times the meadow value at 0-5 cm depth, and nine times at 5-10 cm depth (Courtin and Labine this volume). The greater thermal capacity of the Meadow accounts for its lower temperatures.

Cohort times were consistently longer at 5-10 cm depth than in the top 5 cm, and reflected the temperature gradients in the soil. The gradient effect was most marked in the meadow, especially in 1972, when the number of degree-days at 5-10 cm depth was only 40% of the number at 0-5 cm depth (Courtin and Labine this volume). The temperature gradient in the meadow, which would otherwise conflict with its relatively high thermal capacity, is explained by the shallower active layer in this habitat (Muc this volume).

The 1973 cohort times, and some of the 1972 times, seem reasonable when it is recalled that generation time is approximately half cohort time. However, some 1972 meadow estimates are very long, particularly the 36 seasons at 5-10 cm depth. Nematodes may live this long, but it is unlikely. Two considerations suggest that the very long cohort times do not reflect the true situation. First, 1972 was unusually cold, and the 36 seasons become 11 seasons when the two years are averaged (the two years were the coldest and warmest of the five years of the Devon Island project, and the mean of the two was approximately the mean of the five years). Second, the nematodes may regularly migrate vertically through the soil profile, in which case their distribution shows that they spend most of their lives near the surface where temperatures are highest.

Standing crops are proportional to the numbers of nematodes present, and the 1973 estimates are only slightly larger than the 1972 values (Table 10). Production and respiration are affected by microclimate in addition to numbers and, as a consequence, the estimates for these two parameters are very different for the two years. For example, the 1973 standing crop in the meadow at 5-10 cm depth was 1.2 times the corresponding 1972 value, but the equivalent production and respiration estimates were about six times larger. These results demonstrate the importance of microclimate in determining nematode production.

Wasilewska (1971) studied the nematodes of several habitats and obtained standing crop estimates of 275 to 962 cal m^{-2} (recalculated assuming 5.5 cal mg^{-1} of tissue, and a dry weight equal to 25% of the wet weight), and associated annual respiration of 5,700 to 23,500 cal m^{-2}. Banage (1963) estimated nematode standing crop at 687 to 1,100 cal m^{-2} (recalculated as above) and respiration at 19,800 cal m^{-2} $year^{-1}$. The standing crops of these authors are similar to my standing crop estimates, but their respiration values, when compared with my data on a unit weight basis, average approximately eight times larger per year. Arctic nematodes can be expected to have relatively low annual respiration because they are active for only a small part of the year, and experience relatively low temperatures when they are active.

The dominance of nematodes relative to other invertebrates on the dry I.R.B. (particularly on the crest and slope) can be explained by nematode anhydrobiotic and cryptobiotic capabilities (Cooper and Van Gundy 1971). Dry conditions may induce a resistant quiescent state (anhydrobiosis) which, under extreme conditions, may become cryptobiosis. Ability to survive very dry conditions doubtless gives nematodes an advantage over groups like the Enchytraeidae and Collembola which generally do not have this ability. On the

Table 10. The 1972 and 1973 estimates of nematode seasonal mean standing crop, production and respiration in cal m^{-2} for the I.R.B. and Meadow at 0-5 cm and 5-10 cm soil depth.

Component	*Depth (cm)*	*1972 I.R.B.*	*1972 Meadow*	*1973 I.R.B.*	*1973 Meadow*
Standing Crop	0-5	1198.9	388.9	1320.7	540.5
	5-10	170.7	52.3	198.1	62.7
Production	0-5	1768.9	148.7	4088.5	604.8
	5-10	233.9	8.8	580.8	53.1
Respiration	0-5	2433.2	204.5	5624.1	831.9
	5-10	321.8	12.1	799.0	73.1

other hand, the relative unimportance of nematodes in the meadow is probably caused in part by cold wet soils with low oxygen content.

Cragg (1961) found that nematodes contributed approximately 1% of total invertebrate standing crop and respiration in two Moor House soils, and he concluded that nematodes probably contributed little to energy flow. Bunt (1954), working on sub-arctic Macquarie Island, estimated that nematodes accounted for only 0.9% of total microfaunal activity in the soil. In contrast, Berthet (1964) estimated that nematodes contributed more than 6% of invertebrate biomass, and 30% of respiration in a Belgian grassland, while Wieser and Kanwisher (1961) found that nematodes accounted for 25-33% of oxygen consumed by the mud of Massachusetts salt marsh. By comparison, nematodes contributed 9-49% of invertebrate standing crop, and 7-55% of production, in the Truelove Lowland habitats studied.

The importance of free-living soil nematodes in energy flow clearly varies with the habitat, and they may be very important in some environments and insignificant in others. The present study shows that nematodes are very important in these arctic environments, largely because they can survive conditions which appear to suppress other normally abundant groups.

Summary

Several aspects of nematode ecology were studied on the Truelove Lowland high arctic terrestrial ecosystem. Nematode densities were measured in four habitats and production was estimated for two of these habitats.

The four habitats formed a continuum from the top (crest) of the raised beach to the low-lying meadow, and showed continuous change in topography and microclimate. The crest was exposed, dry and warm, and meadow was sheltered, wet and cold, and the slope and transition were intermediate.

Nematode densities increased progressively from crest, through slope, to transition, and reflected increasing vegetation from crest to transition. The meadow had most vegetation but fewest nematodes, and appeared to be a relatively unfavorable nematode habitat because of its wetness and low temperatures.

The nematodes were concentrated in the top 5 cm of soil, where organic matter was greatest and temperatures were warmest.

Nematode production varied markedly with season, habitat, and soil depth. Differences in climate and microclimate, particularly temperature, appeared to govern production.

Nematodes dominated invertebrate production in the drier habitats, probably because of their anhydrobiotic and cryptobiotic capabilities. They were less important in the wettest areas, because of cold temperatures, limited oxygen, and restricted mobility.

Compared with their role in other ecosystems, nematodes are a relatively important part of this high arctic ecosystem. Above average tolerance of conditions in this ecosystem probably accounts for their importance.

Acknowledgments

I thank my supervisor, Dr. W.G. Evans, for his interest and help in all stages of this study. I am also pleased to thank my wife, Anita Procter, for her assistance in the field and laboratory. Dr. R.H. Mulvey and Dr. R.V. Anderson provided identifications, and Dr. G.E. Ball made helpful criticisms of the manuscript.

References

Banage, W.A. 1963. The ecological importance of free-living soil nematodes with special reference to those of moorland soil. J. Anim. Ecol. 32: 133-140.

Berthet, P. 1964. L'activite des Oribates (Acari: Oribatei). Inst. Royal des Sciences Naturalles de Belgique. Memoire 152. 151 pp.

Bliss, L.C. 1975. Truelove Lowland, a High Arctic ecosystem. pp. 51-58. *In:* Energy Flow—Its Biological Dimensions. T.W.M. Cameron and L.W. Billingsley (eds.). Roy. Soc. Canada, Ottawa, 323 pp.

Bunt, J.S. 1954. The soil-inhabiting nematodes of Macquarie Island. Aust. J. Zool. 2: 264-274.

Cobb, N.A. 1921. Nematodes collected by the Canadian Arctic Expedition under Stefansson. J. Parasitol. 7: 195-196.

Cooper, A.F. and S.D. Van Gundy. Senescence, quiescence, and cryptobiosis. pp. 297-318. *In:* Plant Parasitic Nematodes. Vol. 11. Zuckerman, B.M., W.F. Mai, and R.A. Rohde (eds.). Academic Press, N.Y. 347 pp.

Cragg, J.B. 1961. Some aspects of the ecology of moorland animals. J. Ecol. 49: 477-506.

Croll, N.A. 1970. The Behaviour of Nematodes: their activity, senses and responses. Edward Arnold Ltd., London. 117 pp.

de Soyza, K. 1970. Energy relations in nematodes with particular reference to *Aphelenchus avenae* Bastian, 1965. Ph.D. Thesis. Univ. London. 125 pp.

Klekowski, R.Z., L. Wasilewska, and E. Paplinska. 1972. Oxygen consumption by soil-inhabiting nematodes. Nematologica 18: 391-403.

Kuzmin, L.L. 1973. The fauna of freeliving Nematoda of Western Taimyr. pp. 139-147. *In:* Biogeocenosis of Taimyr Tundra and their Productivity. Soviet National Committee for I.B.P. Nauka, Leningrad. (abstract). 206 pp.

Li, J.C.R. 1964. Statistical Inference. Edwards Brothers Co. Ann Arbor, Michigan. 658 pp.

Marchant, R. and W.L. Nicholas. 1974. An energy budget for the free-living nematode *Pelodera* (Rhabditidae). Oecologia 16: 237-252.

Mulvey, R.H. 1963. Some soil-inhabiting, freshwater, and plant-parasitic nematodes from the Canadian Arctic and Alaska. Arctic 16: 202-204.

Petrusewicz, K. and A. Macfadyen. 1970. Productivity of Terrestrial Animals: Principles and Methods. Blackwell, Oxford. 190 pp.

Southey, J.F. (ed.). 1970. Laboratory Methods for Work with Plant and Soil Nematodes. Ministry of Agriculture, Fisheries and Food. Tech. Bull. 2.

Wallace, H.R. 1971. Abiotic influences in the soil environment. pp. 257-280. *In:* Plant Parasitic Nematodes. Vol. 1. Zuckerman, B.M., W.F. Mai, and R.A. Rohde (eds.). Academic Press, N.Y. 345 pp.

Wasilewska, L. 1971. Nematodes of the dunes in the Kampinos Forest. 11. Community structure based on numbers of individuals, state of biomass and respiratory metabolism. Ekol. Pol. 19: 651-688.

Whitehead, A.G. and J.R. Hemming. 1965. A comparison of some quantitative methods of extracting small vermiform nematodes from soil. Ann. Appl. Biol. 55: 25-38.

Wieser, W. and J.W. Kanwisher. 1961. Ecological and physiological studies on marine nematodes from a small salt marsh near Woods Hole, Massachusetts. Limnol. Oceanogr. 6: 262-270.

Willard, J.R. 1972. Soil invertebrates: 1. Methods of sampling and extraction. Matador Project Tech. Rept. 7. Saskatchewan. 40 pp.

Winberg, G.G. 1971. Methods for the Estimation of Production of Aquatic Animals. Academic Press, N.Y. 172 pp.

Yeates, G.W. 1972. Nematode of a Danish beech forest. I. Methods and general analysis. Oikos 23: 178-189.

Population dynamics and biology of Collembola on Truelove Lowland

J.A. Addison

Introduction

Collembola are amongst the most abundant and widespread of soil arthropods. Species lists for several areas of the Canadian Arctic Islands are now available (Oliver 1963, McAlpine 1965, Danks and Byers 1972), but there has been little attempt to measure the seasonal variation in abundance of different species, and even less to investigate the biological significance of collembolan populations. In the Antarctic, Peterson (1971) carried out detailed population studies on *Gomphiocephalus hodgsoni* Carpenter, and Tilbrook (1970) carried out a similar study on *Cryptopygus antarcticus* Willem, but detailed analyses of the structure and function of arctic populations of Collembola are lacking.

The objectives of this study were: (1) to investigate the population dynamics, life histories, and feeding biology of certain arctic collembolan species, and (2) to attempt to define the impact of the biological activities of these animals on the environment. Field work was carried out during the summer months of 1972, 1973, and 1974, with most of the work taking place during the latter two years. As *Hypogastrura tullbergi* and *Folsomia regularis* were the most widely distributed and abundant Collembola at the Devon Island site, much of the research centred on these two species.

Methods

Distribution and density of Collembola

During the field season of 1972, sampling plots were set up at the Intensive Meadow site, and in the crest and transition zones of the Intensive Raised Beach. A grid was laid out for each sampling plot, allowing the place from which a sample was taken to be located accurately. A table of random grid coordinates was used to ensure that samples were selected randomly from within the plots. At each time of sampling, eight cores of soil were taken to a depth of 7.5 cm, using a corer similar to that described by Macfadyen (1961). Each core was divided into three subsections each 2.5 cm in depth, before being placed in a High Gradient Extractor (modified from Macfadyen 1962). The percentage of plant cover was noted for each sample, and the moisture content of each of the three subsections of every core was determined gravimetrically. Samples from the raised beach (both the crest and transition zones) were left in the extractor for three days, but meadow samples, which had a higher water content, required four days for extraction. After extraction the animals were stored in 95% alcohol. They were subsequently cleared in hot lactic acid, and the species, total body length, sex, and reproductive state of each individual springtail was recorded.

Sampling was carried out at both Intensive Study sites during 1972, but only the transition and crest zones of the raised beach were studied in 1973 and 1974. The frequency of sampling varied from once per week to once every two weeks. In addition to the regular sampling of the Intensive sites, a limited number of samples was also taken from the Plateau, an area of Polar Desert, and from the Intensive Rock Outcrop (Bliss et al. this volume).

Standing crop estimates

For *Hypogastrura tullbergi* and *Folsomia regularis*, the standing crop for each species at each time of sampling was calculated by means of a regression between $\log_{10}$ live weight and total body length. For other species, biomass was estimated by calculating the mean weight/individual for groups of Collembola at the different sites.

Life history and age structure investigations on Hypogastrura tullbergi

A detailed knowledge of the age structure of a population is a prerequisite to the estimation of production. In collembolan populations there are two major problems associated with age structure analysis: (a) Collembola do not have stages of development which are morphologically distinct from one another, making recognition of different instars difficult; (b) recruitment to the population is continuous, or at least prolonged, making the recognition of cohorts difficult.

As instar classes are difficult to define, I have adopted the method suggested by Hale (1965) and Healey (1967) and have classified the animals according to arbitrary size/weight classes. The advantages of using size/weight classes are that there is no overlap between classes, and class widths can be chosen to give the required degree of accuracy. Healey (1967) chose a class interval of 10 μg for *Onychiurus procampatus* Gisin, in a bog site in Wales, but as *H. tullbergi* exhibited a slow growth rate under field conditions, a class interval of 2 μg was chosen for this species.

Pitfall traps

During the 1973 and 1974 field seasons, pitfall traps (25 cm^2) were used to investigate the activity of groups of surface-dwelling arthropods at the Raised Beach Crest and Transition sites. Five traps were placed at each site for 24 hr periods at weekly intervals throughout the field seasons. During May and part of June 1973, and late June to early July 1974, winter pitfall traps, similar to those described by Näsmark (1964), were used.

Feeding studies

Three approaches were used to investigate the feeding habits of Collembola:

Food preference tests

Individuals of *H. tullbergi* and *F. regularis* were introduced into glass petri dishes with plaster of paris and charcoal bases, and were offered a choice of five different food substrates: *Dryas* humus, *Dryas* fragmented leaves (Widden et al. 1972), and three species of fungi. The fungal species used in these experiments were *Phoma herbarum*, *Cladosporium cladosporioides*, *Cylindrocarpon* sp., Sterile dark α116 (Widden, pers. comm.), and *Penicillium notatum*. The number of Collembola at each feeding station was recorded hourly for a period of 14 hr.

Feeding and growth experiments

The hypothesis for this experiment was that animals would survive, grow, and reproduce best on their natural food substrates. Two or three individuals of *H. tullbergi* were placed in small plastic boxes with plaster of paris/charcoal bases, and were cultured on *Dryas* fragmented leaves, *Dryas* humus, *Cylindrocarpon* sp., *Penicillium notatum*, *Phoma herbarum*, or Sterile dark α209. Replicate cultures were set up at laboratory and field temperatures, and the animals were weighed at weekly intervals throughout the experiment.

Gut content analysis

The gut contents of individuals of *H. tullbergi* extracted by the Macfadyen High Gradient extractor were examined using a compound microscope. The percentage of each gut occupied by the following categories of ingested materials was noted to the nearest 10%: particulate organic matter, amorphous organic matter, fungal hyphae, inorganic particles, and animal matter.

Results

Distribution and density of Collembola

Species distribution

Thirty species of Collembola have been collected from various areas on Truelove Lowland, and a list of these species is given in Appendix 7. Table 1 compares the distribution of collembolan species in five different habitats on Devon Island. Using these data, values for Sørensen's Quotient of Similarity (Q.S.) (Sørensen 1948) were calculated to express the degree of faunal similarity between the sites:

O.S. = 2j/a+b
where a = no. of species in habitat "a"
b = no. of species in habitat "b"
j = no. of species both "a" and "b" have in common.

The calculated values of Q.S. were then arranged in a trellis diagram (Table 2) so that the highest values of the index were placed along the diagonal. The resulting arrangement of sites in the trellis diagram was such that those areas with the highest degree of faunal similarity occurred next to one another. This analysis indicated a high degree of affinity between the meadow and rock outcrop sites, and also between the transition zone of the raised beach and

Table 1. Collembola species composition at 5 sites on Devon Island (based on 10 samples from each site).

Species	*Meadow*	*Rock Outcrop*	*Raised Beach Ridge Crest*	*Raised Beach Ridge Transition Zone*	*Plateau*
Poduridae					
Hypogastrura tullbergi	+	+	+	+	+
Anurida granaria	+	+		+	
Micranurida pygmaea		+		+	
Willemia anophthalma	+	+	+	+	
Onychiuridae					
Onychiurus groenlandicus	+	+		+	+
Tullbergia sp. nr. *simplex*			+		
Isotomidae					
Folsomia regularis		+	+	+	+
F. quadrioculata	+	+			
F. bisetosa		+	+		
F. duodecimsetosa		+		+	
Isotoma violacea	+	+		+	
I. sp. nov.	+			+	
I. notabilis v. *pallida*				+	
Isotomurus palustris	+			+	+
Proisotoma sp. nov.			+	+	+
Anurophorus sp.?	+				
Entomobryidae					
Entomobrya comparata			+		
Totals	9	10	7	12	5

Table 2. Trellis diagram indicating affinities between 5 sites on Truelove Lowland, based on Sørensen's Quotient of Similarity, calculated in terms of Collembola species composition. M = meadow, Ro = rock outcrop, RB-TZ = raised beach transition zone, Pl = plateau, RBC = raised beach crest.

	M	*Ro*	*RB-TZ*	*Pl*	*RBC*
M	x	.63	.66	.43	.25
Ro		x	.72	.40	.47
RB-TZ			x	.59	.42
Pl				x	.50
RBC					x

both the meadow and rock outcrop. The fauna of the raised beach crest was more closely allied to that of the plateau than the transition zone, but all values of Q.S. for the raised beach crest were low. According to the collembolan species composition, the plateau was most similar to the transition zone.

The arrangement of the sites according to their degree of collembolan similarity is the same as would occur if the sites were arranged along a moisture-gradient, with the meadow representing the wettest, and the crest zone of the raised beach, the driest site. This would indicate that moisture is probably an important factor influencing the distribution of Collembola species.

H. tullbergi was the most widely distributed and abundant of all the collembolan species, occurring in all habitats studied. *F. regularis* was found in high numbers in all sites except the meadow, from which it was absent. In a discussion of the geographical distribution of this species, Hammer (1955) stated that it did not tolerate the moist climate of South Greenland. In view of this, it is not surprising that *F. regularis* was absent from meadow samples.

Samples taken from the plateau contained only five species of Collembola, all of which also occurred in the raised beach transition site. In the relatively hostile environment of the plateau, Collembola are almost entirely confined to the soil immediately below a vascular plant (Table 3). Therefore

Table 3. Summary of population and standing crop data for Collembola at 3 sites on Devon Island, N.W.T. All figures calculated using a $\log_{10}(n+1)$ transformation of the original data.

Date		$\bar{x}$	*Population Nos m⁻² 95% Confidence limits*	*Standing crop (mg d wt m⁻²)*
Raised Beach Crest—Cushion plant-lichen				
9 July 72	(T1)	12,100	4,600-30,700	101.9
16 July 72	(T2)	11,500	5,200-25,000	82.8
24 July 72	(T3)	9,400	3,200-26,100	103.7
4 Aug. 72	(T4)	13,700	4,600-38,900	91.0
14 Aug. 72	(T5)	12,300	5,400-27,300	104.8
Average		11,700	8,400-16,500	96.8
18 May 73	(T1)	2,400	800- 5,800	17.3
26 May 73	(T2)	2,000	700- 5,200	15.4
2 June 73	(T3)	5,800	1,900-15,800	55.2
10 June 73	(T4)	13,600	6,400-23,300	101.5
17 June 73	(T5)	3,100	1,000- 8,900	26.5
26 June 73	(T6)	7,100	2,200-21,300	79.6
3 July 73	(T7)	3,800	2,300- 6,100	33.7
10 July 73	(T8)	9,200	4,100-19,900	79.6
17 July 73	(T9)	6,300	1,900-21,200	77.3
24 July 73	(T10)	8,100	5,200-12,400	47.6
31 July 73	(T11)	4,100	1,700- 8,900	37.2
7 Aug. 73	(T12)	5,600	2,400-13,400	50.2
14 Aug. 73	(T13)	6,800	4,800- 9,800	66.2
23 Aug. 73	(T14)	6,600	3,200-13,200	65.2
30 Aug. 73	(T15)	3,500	1,100-11,200	51.1
Average		5,000	3,600- 6,300	53.6
Raised Beach Transition Zone—Cushion plant-Moss				
27 July 72	(T1)	16,900	12,400-23,000	70.9
7 Aug. 72	(T2)	9,200	3,800-21,800	50.8
18 Aug. 72	(T3)	20,000	12,100-33,100	86.2
Average		14,700	10,600-20,300	69.3
Meadow—Sedge-Moss				
13 July 72	(T1)	4,800	2,000-11,000	59.9
20 July 72	(T2)	7,800	4,100-14,600	61.9
31 July 72	(T3)	2,400	1,100- 4,600	19.1
10 Aug. 72	(T4)	5,300	3,300- 8,500	36.6
Average		4,700	3,400- 6,500	42.6
Plateau-Polar Desert				
9 July 73	*	29,900	17,100-52,200	203.7
27 July 73	**	84	—	1.2
Average	***	680	—	5.3

* + vascular plant cover
** – vascular plant cover
*** calculated assuming 2% area of the Plateau has a vascular plant cover.

the species list for the plateau contains only species which live in close proximity to a vascular plant species, a situation which resembles that of the transition zone where vascular plant cover reaches 58% (Svoboda this volume). The highest number of species was found in the transition zone, which, as an area of transition between a crest and meadow area, would be expected to include elements from both ends of the ecological gradient.

Density of Collembola

Preliminary statistical analyses of the 1972 Collembola population data using Bartlett's test for homogeneity of variances, showed that a normal distribution of sample means could not be assumed. Several transformations were used to try to normalize the data, and good results were obtained using a $\log_{10}(n+1)$ transformation.

A summary of the population data and biomass estimates for the raised beach crest and transition zones, meadow, and plateau is given in Table 3. Although Collembola were most numerous at the raised beach transition site, because of the predominance of a small species of Collembola *(Folsomia regularis)* at this site, total collembolan biomass was lower than at the raised beach crest. The seasonal average number of Collembola at the crest site was 11,700 m^{-2}, giving an average biomass of 97 m^{-2} in 1972, but in 1973 the mean number of Collembola during those same months dropped to 6,400, with a correspondingly lower biomass (59 mg m^{-2}). The low snow cover during the winter of 1972-73 may have contributed to the significant decrease ($P < .05$) in collembolan population numbers between 1972 and 1973. Analysis of variance was carried out on all data except those from the plateau, in order to determine whether the variations in population numbers at different times in the season were significant. The mean values for each sampling date were calculated on the transformed scale, and the means were compared using Duncan's Multiple Range Test (Table 4). This analysis showed no significant difference ($P=.05$) between the population estimates at the various sampling times at the crest zone of the raised beach in 1972. A significant mid-season decline in the numbers of Collembola at the transition and meadow sites (31 July, 1972 and 7 August, 1972 respectively), could be demonstrated. The 10 June, 1973 sample from the crest contained a significantly higher number of Collembola than the May samples, and corresponded to the time when a large number of juveniles of *H. tullbergi*, the dominant species at this site, appeared in the population. The population estimate for 10 July, 1973 was significantly higher than those

Table 4. Results of analysis of data on seasonal variation. Analyses of variance and multiple range tests (Duncan's Multiple Range Test), were performed on the $\log_{10}(n+1)$ transformed data. The mean number of Collembola m^{-2}, represented by each T (time) value, can be obtained from Table 3. T values not underscored by the same line differ significantly at the 95% level.

Year	*Site*	*Time of sample (T)*
1972	Raised Beach—Crest	T1 T2 T3 T4 T5
1972	Raised Beach—Transition Zone	T2 T1 T3
1972	Meadow	T4 T1 T2 T3
1973	Raised Beach Crest	T2 T1 T15 T5 T7 T11 T12 T3 T9 T14
		T6 T13 T10 T8 T4

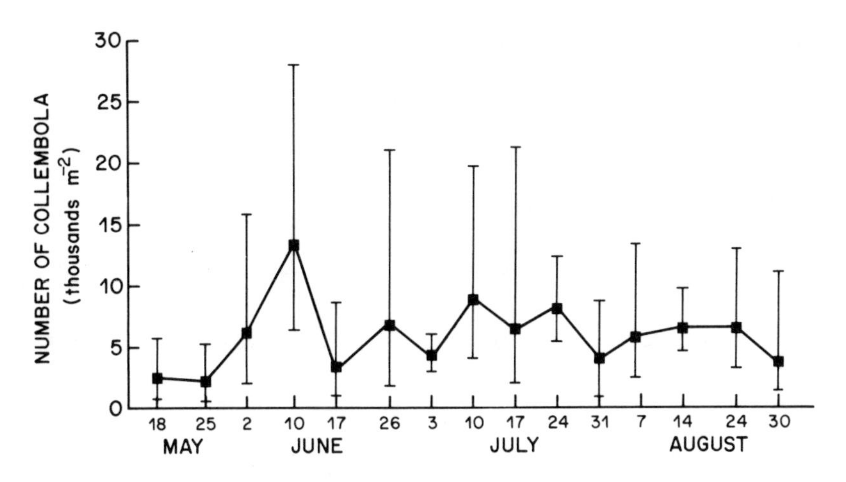

Fig. 1 Population estimates (means ± 95% confidence limits), for Collembola at the Raised Beach Crest during 1973 [based on a $\log_{10}(n+1)$ transformation].

of 17 June and 30 August of the same year, indicating a slight increase in collembolan numbers during the summer months (Fig.1). However, at this site, the maximum population of Collembola in 1973 occurred on 10 June.

Life history and age structure

Collembola effect the transference of sperm from males to females by means of spermatophores. Structures loosely resembling spermatophores described for other species of Collembola (e.g., Sharma and Kevan 1963, Schaller 1971) could be seen clearly within the body of males at certain times of the year. In the present study, the presence of these "spermatophores" together with data on egg production in field-caught females, have been used to indicate the onset and duration of reproductive activity in *H. tullbergi*.

In 1972 "spermatophores" were seen in 58% of the males in the crest samples on 16 July, and in 27% on 24 July. The following year, "spermatophores" were found in 15% of the males in the 10 June samples, and in 69% in the 17 June samples. Field-caught animals laid eggs in laboratory cultures from 22 June to 24 July, 1973, compared with 10-21 July in 1972. The onset of reproductive activity in *H. tullbergi* therefore occurred approximately four weeks earlier in 1973 than in 1972. In both years individuals with "spermatophores" appeared approximately two weeks after snowmelt, and newly hatched juveniles were found 3-4 weeks after the dates when "spermatophores" were observed. The postulated 3-4 weeks taken for eggs to hatch under field conditions is supported by results of laboratory experiments where, under a constant temperature regime, eggs laid by *H. tullbergi* took 21 days to hatch at 10°C, and 47 days at 5°C.

The distribution of size/weight classes for *H. tullbergi* during the 1972 and 1973 field seasons at the raised crest is shown in Fig. 2. As reproductive activity in 1972 did not occur until mid-July, juveniles present in the population in early July must have hatched from eggs produced during the previous year (1971). These juveniles did not become adults (Class 6) until 1973, thus taking two years to reach sexual maturity. Some of the eggs laid in July 1972 hatched by August of the same year, but most of the eggs overwintered, and the

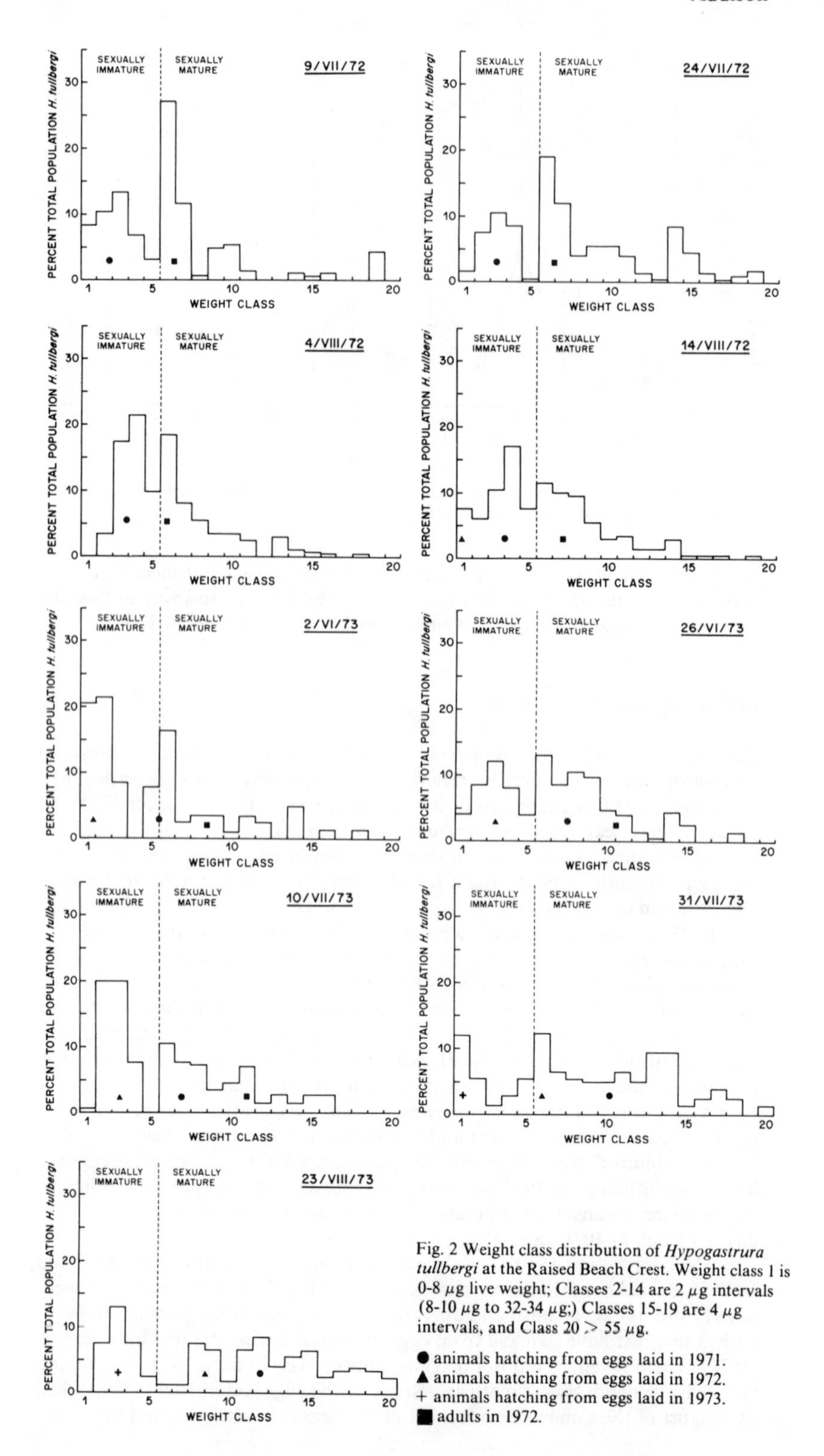

Fig. 2 Weight class distribution of *Hypogastrura tullbergi* at the Raised Beach Crest. Weight class 1 is 0-8 μg live weight; Classes 2-14 are 2 μg intervals (8-10 μg to 32-34 μg;) Classes 15-19 are 4 μg intervals, and Class 20 > 55 μg.

● animals hatching from eggs laid in 1971.
▲ animals hatching from eggs laid in 1972.
+ animals hatching from eggs laid in 1973.
■ adults in 1972.

juveniles appeared in the late May and early June samples of the next year. These juveniles reached Class 6 by July 1973, taking only one year to reach sexual maturity. Peaks in the juvenile size classes are relatively easy to follow, but once the animals reach sexual maturity (Class 6), the picture becomes far more confusing. During the time when the animals develop external genitalia, they apparently do not increase in size. Thus both immature and mature individuals are found in size Class 6, which therefore contains more animals than would be expected. Although individual cohorts of animals are difficult to distinguish in the adult stage, the data presented in Fig. 2 indicate that adult *H. tullbergi* can remain in the population for at least three years after achieving sexual maturity, giving a maximum total life span of approximately five years. The weight class distribution of males with "spermatophores" (Fig. 3), indicates that males can breed at least twice, and possibly three times during their life.

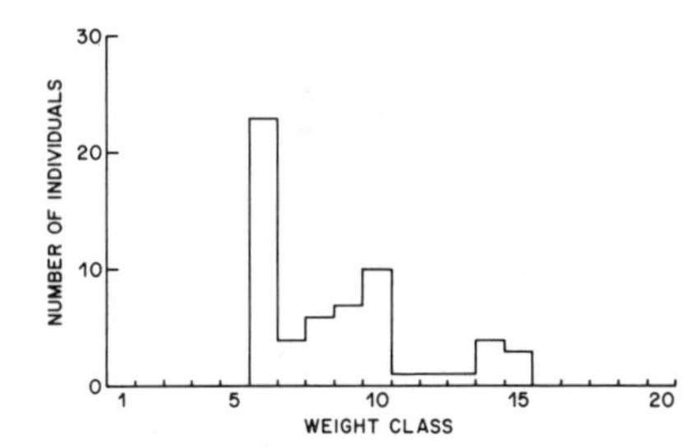

Fig. 3 Weight class distribution of *Hypogastrura tullbergi* males with mature "spermatophores." Weight classes as in Fig. 2

The lack of definition of peaks in adult size/weight classes can be partly explained by the results of the feeding and growth experiments, which showed that under conditions where food was limited, adults were able to survive for a whole season, but did not show any increase in weight.

Vertical distribution

The distribution of collembolan species at the raised beach crest with respect to depth (Fig. 4) shows that *H. tullbergi* was the most abundant species, and in the top 2.5 cm constituted 73% of the total Collembola population. At the second depth (2.5-5.0 cm) *Tullbergia* sp. nr. *simplex*. made up a larger proportion of the total population, and in the 5.0-7.5 cm section, outnumbered *H. tullbergi*. *Proisotoma* sp., *F. regularis* and *F. bisetosa* were present in low numbers at all depths.

At the beginning of the 1973 season, only about 50% of the population of *H. tullbergi* was located in the top 2.5 cm. During the season, this proportion increased to maximum values of over 95% in mid-July to mid-August, after which time proportionally more individuals were extracted from the two lower depths, until the end of August when the vertical distribution of *H. tullbergi* approached that of May (Fig. 5). The relationship between the distribution of *H. tullbergi* and the environmental parameters of soil temperature and moisture obviously requires further consideration. It was expected that as the top few centimeters became drier during July and August, a larger proportion of the population of *H. tullbergi* would be located in lower depths. However, the reverse appears to be true.

At all sites except the crest, Collembola were highly localized in the top 2.5 cm (Table 5). In the crest zone of the raised beach, a larger proportion was

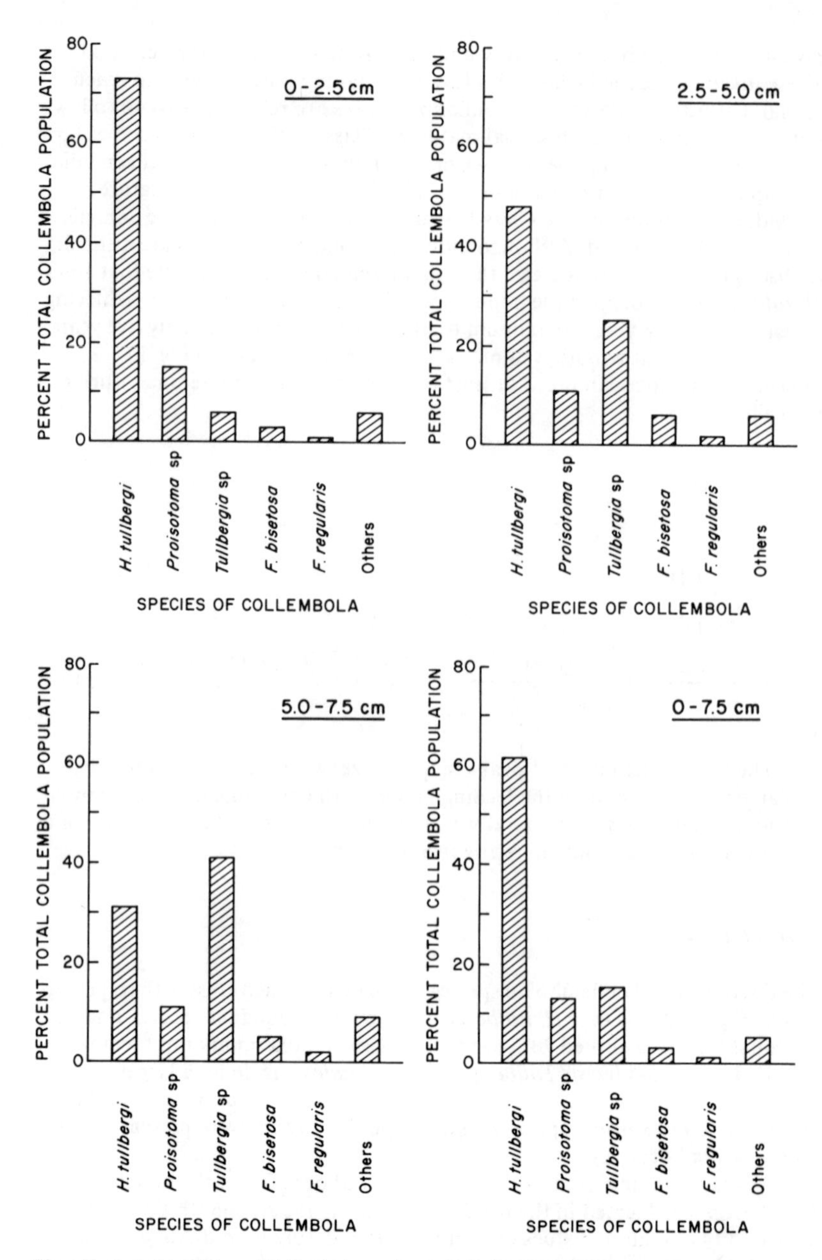

Fig. 4 Vertical distribution of Collembola species at the Raised Beach Crest (based on 1973 data).

found in the 2.5-5.0 cm and 5.0-7.5 cm samples, and it is very likely that the animals were responding to temperature. In the highly organic transition and meadow soils, heat penetration was not as great as in the mineral soil. As the soil at the meadow site was often waterlogged, this could also encourage Collembola to inhabit the surface layer. Chernov et al. (1973) reported that at the Tareya site on the Taimyr Peninsula, large numbers of Collembola were found a few centimetres above the level of the permafrost table. The possible

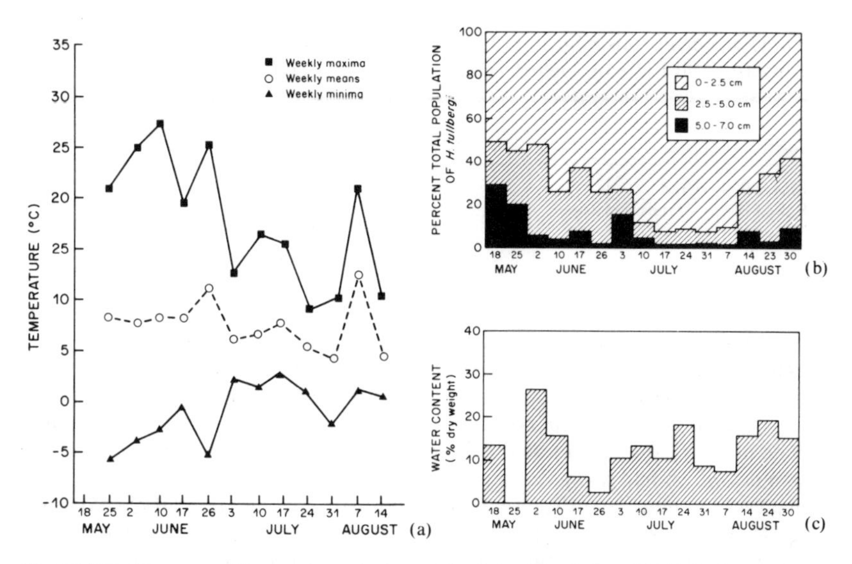

Fig. 5 (a) Weekly mean soil temperatures at −1 cm at the Raised Beach Crest (1973); (b) Seasonal variation in vertical distribution of *Hypogastrura tullbergi* at the Raised Beach Crest (1973); (c) seasonal variation in moisture content of soil samples (0-2.5 cm) at the Raised Beach Crest (1973).

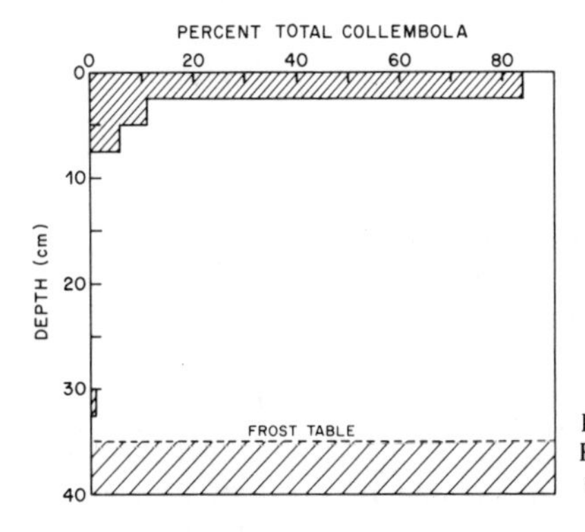

Fig. 6 Vertical distribution of Collembola at the Raised Beach Transition Zone, August 1974 (based on a mean of three samples).

occurrence of this phenomenon at the Devon site was investigated by examining three cores of transition zone soil, taken down to the level of the permafrost table. The results of this study (Fig. 6) show that only a single individual was found below 10 cm, and it is possible that this individual was carried down by the corer. In any case, populations of the size reported by Chernov et al. (1973) did not occur at lower depths on Truelove Lowland.

Horizontal distribution

Blackith (1974) claimed that individual species of Collembola will often show a strong preference for living under a particular plant species.

On the raised beach crest, where plant cushions are usually distinctly separated from each other by areas of bare ground or crustose lichen, it would be expected that such positive associations between collembolan and plant

Table 5. Vertical distribution of Collembola at various sites on Truelove Lowland and Plateau, Devon Island, (based on 1972 data).

Site	*% total at each depth* 0-2.5 cm	2.5-5.0 cm	5.0-7.5 cm
Raised Beach Crest	60	26	14
Raised Beach—Transition Zone	90	8	2
Meadow	85	12	3
Plateau	95	4	<1

species would be easily determined. However, although the animals were extremely clumped in their distribution, no correlation between Collembola numbers or species, and plant species could be demonstrated.

The influence of plant cover on the abundance of Collembola in the crest zone of the raised beach was shown in a very general way by comparing samples which had no vascular plant cover, with those in which a vascular plant was present. Significantly higher numbers of Collembola ($P < 0.05$) were found in those samples with some degree of plant cover. In the plateau samples, there was a dramatic difference between the number of Collembola in samples taken from areas with vascular plant cover, and those taken from areas of bare ground. The density of Collembola within the areas covered by vascular plants was 29,900 m^{-2}, but only one individual was extracted from a total of five samples taken from bare ground, giving a density of 84 individuals m^{-2}.

Pitfall data

In 1973 snowmelt at the raised beach occurred in early June, and soil temperatures at this time were unusually high (Courtin and Labine this volume). In contrast, snow cover remained until early July in 1974, and the season was cooler than that of the previous year. In spite of the relatively large differences in the prevailing weather conditions, the patterns of activity of surface dwelling invertebrates were remarkably similar in both years. On the raised beach crest collembolan activity was greatest during the early part of the season. In 1973 this peak of activity occurred in mid-June (Fig. 7), while the following year, the corresponding peak occurred in early to mid-July (Fig. 9). Similarly, the number of adult flies captured in 1973 was greatest on 24 June, whereas in 1974 this did not occur until 20 July. Similar comparisons can be drawn between the two years of data from the transition site (Figs. 8 and 10), although the difference between the two years is not as pronounced. In general, the peaks of activity of surface dwelling arthropods in 1974 occurred 3-4 weeks later than the corresponding peaks in 1973.

Feeding and growth experiments

Growth experiments carried out during the 1973 field season indicated that *H. tullbergi* could not gain enough nutrition from *Dryas* humus to maintain its body weight, and that in the warmer temperatures in the laboratory (12°C-15°C), a higher metabolic rate caused a faster loss of weight (Table 6). Individuals of this species were able to maintain themselves on a diet of *Dryas* leaves, but only when kept at field temperatures. The animals gained weight

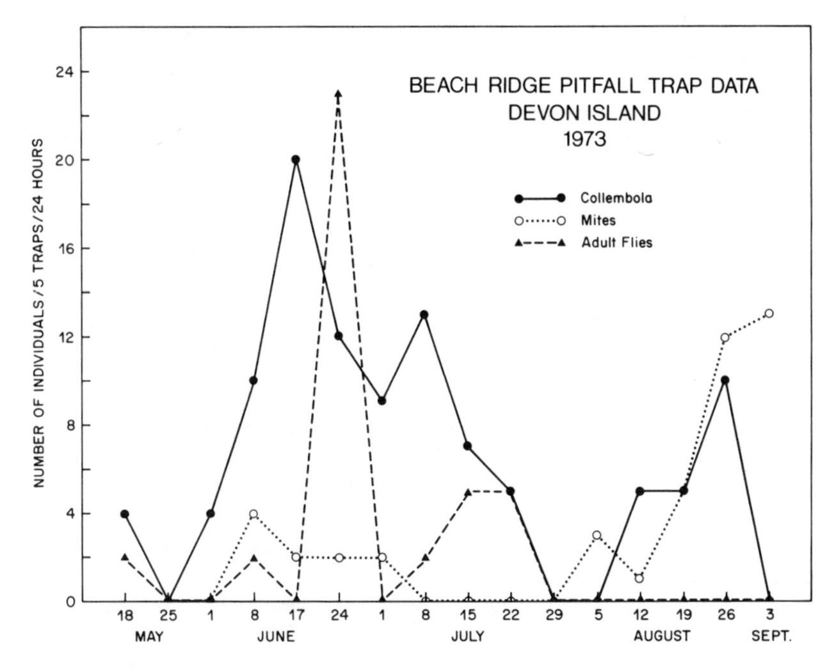

Fig. 7 Raised Beach Crest pitfall trap data, Devon Island, 1973.

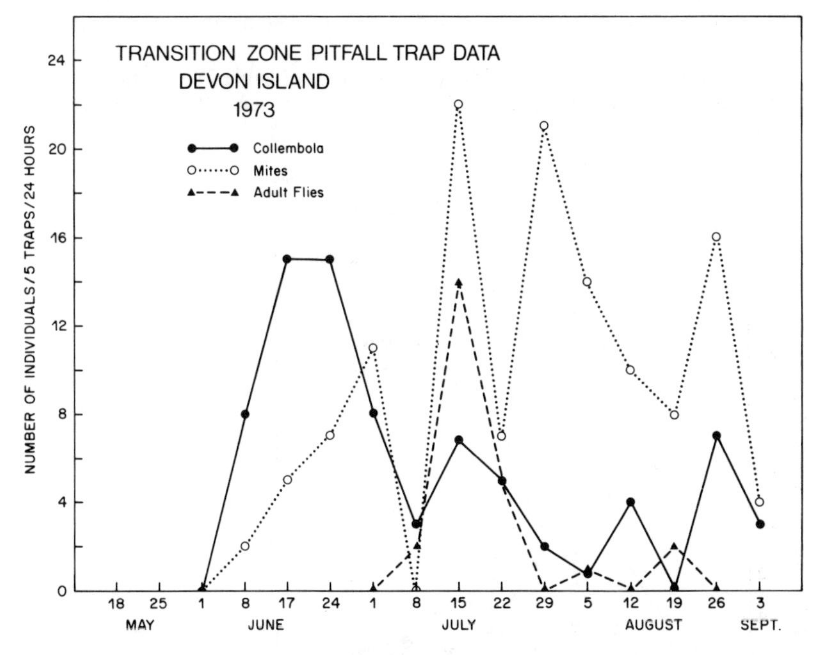

Fig. 8 Raised Beach Transition Zone pitfall trap data, Devon Island, 1973.

Table 6. Summary of feeding/growth experiments using *Hypogastrura tullbergi*, carried out over a 4-week period during the 1973 field season on Devon Island.

Food substrate	*Temperature regime*	*Average change in weight (expressed as % original weight)*
Dryas leaves (fragmented)	Indoors	− 3.9
	Outdoors	+ 5.2
Dryas humus	Indoors	− 7.25
	Outdoors	− 3.4
Cylindrocarpon sp.	Indoors	+14.6
	Outdoors	+68.5
Penicillium notatum	Indoors	− 6.25*
	Outdoors	− 8.4**

*all experimental animals died after three weeks.
**4 out of 5 animals died during experiment.

rapidly when fed on *Cylindrocarpon* sp., but lost weight, and in most cases died when cultured on *Penicillum notatum*.

In similar experiments conducted at field temperatures in 1974 (Fig. 11), *H. tullbergi* gained weight when cultured on either *Phoma herbarum* or a sterile dark species of fungus (α209). The animals were able to survive for at least 10 weeks on a diet of *Dryas* leaves, but did not increase in weight. Individuals kept without food lost weight rapidly, and 50% died within three weeks. The rapid growth rate demonstrated by animals cultured on fungi during early and mid-August, occurred at a time when mean air temperatures were at their maximum values.

The results of the food preference experiments (Table 7) also indicated that *H. tullbergi* is a fungal feeder. In the first experiment, *Phoma herbarum*

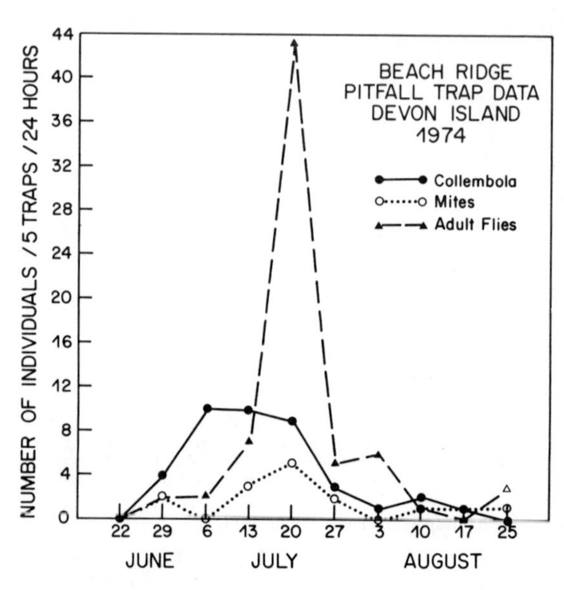

Fig. 9 Raised Beach Crest pitfall trap data, Devon Island, 1974.

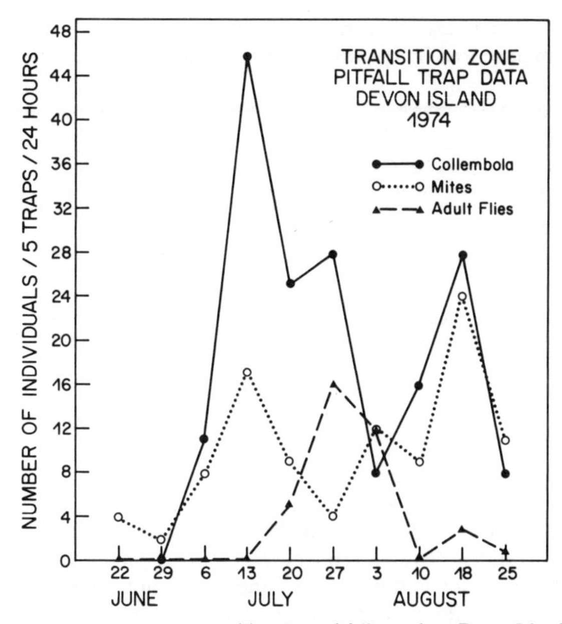

Fig. 10 Raised Beach Transition Zone pitfall trap data, Devon Island, 1974.

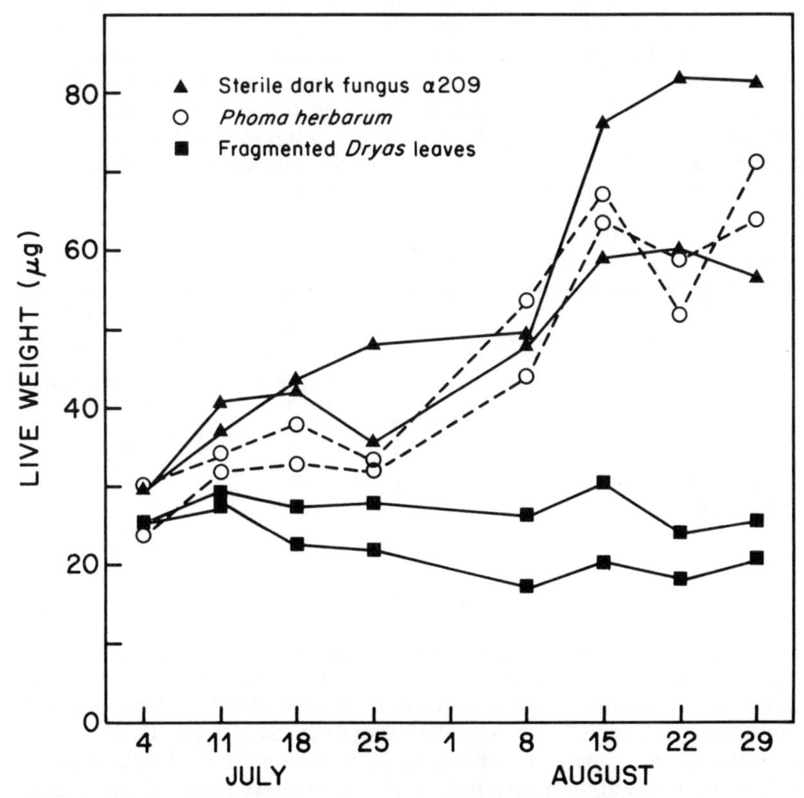

Fig. 11 Growth rates of individuals of *Hypogastrura tullbergi* on three different food substrates, under field conditions.

Table 7. Summary of food preference experiments for two species of Collembola. Means not underscored by the same line differ significantly at the 95% level.

Species of Collembola	Mean no. of animals at each feeding station hr^{-1}				
	Dryas leaves	*Dryas* humus	*Cladosporium cladosporioides*	Sterile dark α116	*Phoma herbarum*
H. tullbergi	0.79	1.43	1.86	2.0	4.14
F. regularis	2.14	1.64	0.14	0.0	0.07

Species of Collembola	*Cladosporium cladosporioides*	*Cylindrocarpon* sp.	*Penicillium notatum*	*Dryas* leaves	*Dryas* humus
H. tullbergi	5.2	1.15	0.0	0.93	0.93
F. regularis	0.21	0.0	0.0	1.42	4.0

was the most frequent choice, although other species of fungi were also eaten. In contrast, under the same conditions, *F. regularis* showed a strong preference for *Dryas* leaves and humus. In the next experiment, *H. tullbergi* preferred *Cladoporium cladosporioides*, which had been its second choice in the first experiment, to all other species of fungi offered. *F. regularis* again showed a preference for *Dryas* leaves and humus. Neither species of Collembola was observed to feed on *Penicillium notatum*, and at the end of the experiment three dead individuals of *H. tullbergi* were seen within 0.5 cm of this fungus.

Although both the growth experiments and food preference studies indicated that *H. tullbergi* is a fungivore, the results of the gut content analysis (Table 8) show that organic matter, both particulate and amorphous, was the most abundant gut content in field populations of this species. Fungal material contributed only 11.7% of the total gut contents at the crest site, and even less (8.2%) at the transition site.

Table 8. Gut contents of *Hypogastrura tullbergi*.

Site	No. of individuals examined	% with gut contents	Gut components (%)				
			particulate organic matter	amorphous organic matter	fungal hyphae	animal remains	inorganic particles
Raised Beach Crest	613	8.8	28.6	42.6	11.7	8.5	8.4
Raised Beach Transition Zone	100	17.0	71.8	15.0	8.2	0	5.0

Discussion

The collembolan fauna of Truelove Lowland consists of 30 species, (see Appendix 7), which is lower than at most other I.B.P. tundra sites. Forty-three species of Collembola have been reported for Barrow, Alaska, 33 species at the Norwegian sites, and 62 species on the Taimyr Peninsula (Ryan 1978). However, when compared to other areas in the Queen Elizabeth Islands and Northern Greenland, the collembolan fauna of Truelove Lowland is species rich. Hammer (1954) collected 15 species of Collembola in Peary Land, N.

Greenland and 19 species from the area around Slidre Fiord, Ellesmere Island (Hammer 1953). Oliver (1963) found 14 collembolan species at Lake Hazen on Ellesmere Island. So far, only six species of Collembola have been collected from Bathurst Island (Danks and Byers 1972), and Ellef Ringnes, one of the Western Arctic Islands, has eight species of Collembola (McAlpine 1964, 1965).

Both *F. regularis* and *H. tullbergi*, the two dominant species on Truelove Lowland, have a northern holarctic distribution. Although *H. tullbergi* is a very common species in many different habitats throughout its distribution, *F. regularis* has been reported from only five sites; Peary Land (Hammer 1954), Ellesmere Island (Hammer 1953), Ellef Ringnes (McAlpine 1964, 1965) the Taimyr Peninsula (Ananjeva 1973), and Spitzbergen (Stach 1962). Two of these areas, the Taimyr and Peary Land, also had large populations of *H. tullbergi*.

The numbers of Collembola found at the different Truelove Lowland sites are of the same order of magnitude as other tundra sites, but in general the abundance of all arthropods, including the Collembola, is lower than figures reported for the other arctic areas.

Matveyeva (1972) reported 4,000 Collembola m^{-2} in a sedge-moss site in the Taimyr, and MacLean (pers. comm.) reported densities of 61,100-171,900^{-2} at various meadow sites at Barrow, Alaska. Those figures can be compared with estimates of 2,400-7,800^{-2} at the Devon Island hummocky sedge-moss meadow. Bengston et al. (1974) found densities of 20,800 and 38,300 Collembola m^{-2} at two plots of lichen tundra on Spitzbergen. The comparable figures for Devon Island are 11,700 and 14,700^{-2} Collembola, representing figures for the raised beach crest (cushion plant-lichen community) and the transition Zone (cushion plant-lichen community) respectively.

As Collembola rely on cutaneous respiration, the humidity of the environment is of utmost importance in governing both their distribution and abundance. Challet and Bohnsack (1968) found that the species composition of populations at Barrow, Alaska could be directly related to the water content of the soil. A similar relationship was demonstrated for the Truelove Lowland populations. Many species of Collembola are unable to survive in habitats which are periodically flooded (Kühnelt 1961), a phenomenon which occurs during the spring run-off at the meadow site on Devon Island. At the other extreme, Wise et al. (1964) considered that a soil moisture level of 2% marked the lower limit of tolerance of an antarctic Collembole *Gomphiocephalus hodgsoni*. Moisture levels on the raised beach crest dropped just below this figure during the summer of 1973. It is therefore not surprising that Collembola were the most abundant, and showed the greatest number of species at the transition zone, where the moisture level was between the two extremes.

A relatively low species diversity, with a few dominant species is a well-known characteristic of the arctic environment. Downes (1965) states that a relationship exists between genetic heterozygosity (physiological versatility to colonize a wide and randomly varying environment) and a restriction of opportunity for more precise adaptation and speciation. The data obtained from the studies on the Collembola of this high arctic area would seem to agree with this statement. The fact that individual species of Collembola at the beach ridge did not appear to be definitely associated with any one particular plant species, may be a reflection of the general lack of specialization of arctic Collembola. Gut content analyses of *H. tullbergi* indicated that this species was able to utilize a wide variety of food resources in the field, although in laboratory studies it showed a clear preference for fungi. Growth rates of this species in the field, when kept on a diet of pure fungus, were much greater than the rates of growth which could be estimated from the size/weight distribution of field populations. This would imply that the low growth rates of field populations were due to a lack of nutrients, rather than a direct result of low temperatures.

Low temperature (and higher latitude) are often quoted as favoring melanin formation in insects (e.g., Vernberg 1962). Comparing complete fauna collections of Collembola, Rapoport (1969) found a clear correlation between percent dark forms and latitude. According to his regression line, the predicted value for percent dark species on Truelove Lowland, based on the latitude of the site, is 80%. The actual figure is 57%, which is the value predicted for a site at 55°N. The Low Arctic affinities of the Truelove Inlet site with respect to vegetation and microclimate have been pointed out by Bliss (1975) and Bliss et al. (1973).

Investigations into the biology of *H. tullbergi* revealed many characteristics which can be considered to have adaptive significance for the species. The animal has a 3-5 year life cycle, compared with 2-12 months in more temperate regions (Christiansen 1964), and each individual can breed in 2 or 3 different years. The long-term breeding success of the species does not depend on the environmental conditions prevailing in any one year. This species is able to overwinter in any stage of development, and can take 1-2 years to reach sexual maturity. During a season in which the thaw occurs earlier than usual (early June), the Collembola are able to take advantage of warmer temperatures and breed earlier in the season. The results of the growth experiments showed that *H. tullbergi* is able to grow rapidly when food is abundant, but under conditions of low food availability can maintain itself for an entire season without growth.

Acknowledgments

I wish to thank my supervisor, Dr. D. Parkinson, for his support and guidance during this study. Special thanks are due to John Bissett for his invaluable assistance in designing and debugging computer programs. I wish to express my appreciation to Neil Colgan, my field assistant during the 1973 field season, and to Jan Marsh for her help in August 1972. Dr. W.R. Richards kindly identified many of the Collembola species mentioned in this paper.

References

Ananjeva, S.I. 1973. Collembola of Western Taimyr. pp. 152-215 *In:* Biogeocenoses of Taimyr Tundra Animals and Their Productivity. Vol. 2. Nauka Pub. Hanse, Leningrad. 207 pp.

Bengston, S.A., A. Fjellberg and T. Solhoy. 1974. Abundance of tundra arthropods in Spitzbergen. Ent. Scand. 5: 2, 137-142.

Blackith, R.E. 1974. The ecology of Collembola in Irish blanket bogs. Proc. Roy. Irish Acad. 74: 203-226.

Bliss, L.C. 1975. Tundra grasslands, herblands and shrublands and the role of herbivores. Geoscience and Man 10: 51-79.

———. G.M. Courtin, D.L. Pattie, R.R. Riewe, D.W.A. Whitfield, and P. Widden. 1973. Arctic tundra ecosystems. Ann. Rev. Ecol. and Syst. 4: 359-399.

Challet, G.L. and K.K. Bohnsack. 1968. The distribution and abundance of Collembola at Pt. Barrow, Alaska. Pedobiologia 8: 214-222.

Chernov, Y.I., S.E. Ananjeva, L.L. Kuzmin, and E.P. Khaiurova. 1973. The pecularities of vertical distribution of invertebrates in soils of tundra zone: pp. 180-186 (see Ananjeva 1973).

Christiansen, K. 1964. Bionomics of Collembola. Ann. Rev. Ent. 9: 147-178.

Danks, H.V. and J.R. Byers. 1972. Insects and Arachnids of Bathurst Island, Canadian Arctic Archipelago. Can. Ent. 104: 81-88.

Downes, J.A. 1965. Adaptations of insects in the Arctic. Ann. Rev. Ent. 10: 257-274.

Hale, W.G. 1965. Observations on the breeding biology of Collembola. Pedobiologia 5: 146-152, 161-177.

Hammer, M. 1953. Collemboles and Oribatids from the Thule District (North West Greenland) and Ellesmere Island (Canada). Medd. om Grønl. 136: 4-16.

———. 1954. Collemboles and Oribatids from Peary Land (Northern Greenland). Medd. om Grønl. 127: 4-28.

———. 1955. Some aspects of the distribution of microfauna in the Arctic. Arctic 8: 115-126.

Healey, I.N. 1967. The energy flow through a population of soil Collembola. *In:* Secondary Productivity of Terrestrial Ecosystems. K. Petrusewicz (ed.). Pol. Acad. Sci., Warsaw. 2: 695-708.

Kũhnelt, W. 1961. Soil Biology. Faber and Faber, London. 395 pp.

Macfadyen, A. 1961. Improved funnel-type extractors for soil arthropods. J. Anim. Ecol. 31: 171-184.

———. 1962. Soil arthropod sampling. Advan. Ecol. Res. 1: 1-33.

Matveyeva, N.W. 1972. The Tareya word model. pp. 156-162. *In:* International Meeting on Biological Productivity of Tundra. F.E. Wielgolaski and T. Rosswall (eds.), Tundra Biome Steering Committee. Stockholm. 320 pp.

McAlpine, J.F. 1964. Arthropods of the bleakest barren lands. Composition and distribution of the arthropod fauna of the Northwestern Queen Elizabeth Islands. Can. Ent. 96: 127-129.

———. 1965. Insects and related terrestrial invertebrates of Ellef Ringnes Island. Arctic 18: 73-103.

Näsmark, O. 1964. Vinteraktivitet under snon has landevande evertebrater. Sartryck ur Zoologisk Revy. 26: 5-15.

Oliver, D.R. 1963. Entomological studies in the Lake Hazen area, Ellesmere Island. Arctic 16: 175-180.

Peterson, A.J. 1971. Population studies on the antarctic collembolan *Gomphiocephalus hodgsoni*. Pacif. Ins. Monogr. 25: 75-98.

Rapoport, E.H. 1969. Gloger's rule and the evolution of the Collembola. Evolution 23: 622-626.

Ryan, J. 1978. Comparison of Invertebrate species lists from I.B.P. tundra sites. *In*: Comparative Analysis of Ecosystems. J.B. Cragg and J.J. Moore (eds.), Cambridge Univ. Press, Cambridge (In press).

Schaller, F. 1971. Indirect sperm transfer by soil arthropods. Ann. Rev. Ent. 16: 407-446.

Sharma, G.D. and D.K. McE. Kevan. 1963. Observations on *Pseudosinella petterseni* and *Pseudosinella alba* (Collembola: Entomobryidae) in eastern Canada. Pedobiologia 3: 62-74.

Sørensen, T. 1948. A method establishing groups of equal amplitude in plant sociology based on similarity of species content and its application to analyses of the vegetation on Danish commons. Biol. Skr. (K. danske vidensk Selsk. N.S.) 5: 1-34.

Stach, J. 1962. On the fauna of Collembola from Spitzbergen. Acta zool. Cracowiensia 7: 1-21.

Tilbrook, P.J. 1970. Biology of *Cryptopygus antarcticus*. pp. 909-918. *In:* Antarctic Ecology. M.W. Holdgate (ed.). Vol. 2. Academic Press, London. 391 pp.

Vernberg, F.J. 1962. Comparative physiology: latitudinal effects on physiological properties of animal populations. Ann. Rev. Physiol. 24: 517-546.

Widden, P., T. Newell, and D. Parkinson. 1972. Decomposition and microbial populations of Truelove Lowland, Devon Island. pp. 341-358. *In:* Devon Island I.B.P. Project, High Arctic Ecosystem. Project Report 1970 and 1971. L.C. Bliss (ed.). Univ. Alberta, 413 pp.

Wise, K.A.J., C.E. Fearon, and O.R. Wilkes. 1964. Entomological investigations in Antarctica, 1962-63 season. Pacif. Ins. 6(3): 541-570.

Invertebrate respiration on Truelove Lowland

Dennis L.C. Procter

Introduction and Methods

This paper presents metabolic rates for representatives of most of the important terrestrial and aquatic invertebrate groups found on Truelove Lowland. Species of Enchytraeidae, Crustacea, Acarina, Collembola, Lepidoptera, Muscidae, and Chironomidae were studied. Field research was carried out during July and August 1972, and from June to September 1973.

These studies are part of a wider investigation to determine the role of invertebrates in energy flow in this high arctic terrestrial ecosystem. Ryan (this volume) obtained the other information required to estimate invertebrate energy flow, and presents a synthesis of this information.

Oxygen consumption was measured at 2°, 7°, and 12°C, using the volumetric microrespirometer described by Gregg and Lints (1967). A Haake Unitherm thermoregulator with cooling coil maintained the temperatures within ±0.25°C of the desired levels. Interchangeable chambers of 100, 250, and 800 μl volume accommodated animals of different sizes. The apparatus was sensitive to change in gas volume of 0.0015% hr^{-1}. When individual oxygen consumption was less than this rate, several specimens were studied together to obtain measurable rates.

Experimental animals were collected immediately before experimentation and were allowed to equilibrate with the system for one hour before readings were taken. The animals were not acclimated to experimental temperatures. Six respirometers were operated simultaneously, with one left empty to serve as control for thermal and barometric changes. A minimum of five readings at 30 or 60 min intervals were taken $respirometer^{-1}$ $temperature^{-1}$. All readings, regardless of level of activity of the animals, were included in the reported results.

Dry weights were obtained with a Cahn G2 Electrobalance accurate to 0.05 μg, after drying for 24 hr at 80°C with calcium chloride.

Results

Mean metabolic rate for all species at 2°C was 0.5366 $\mu l\ O_2$ mg dry wt^{-1} hr^{-1}, with a range of 1.3690 to 0.1322 (Table 1). At 7°C the mean was 1.1531, and

Table 1. Oxygen consumption and dry weight for 16 invertebrate species (n=5).

Species	°C	Dry wt mg mean±SD		Oxygen consumption $\mu l\ O_2\ ind^{-1} hr^{-1}$ mean±SD		$\mu l\ O_2\ mg^{-1} hr^{-1}$ mean±SD	
Enchytraeidae							
Henlea nasuta	2	0.0371	0.0422	0.0308	0.0319	0.7675	0.2594
	7	0.0310	0.0291	0.0344	0.0326	1.1377	0.6097
	12	0.0336	0.0186	0.0689	0.0417	1.9188	0.5791
Crustacea							
Daphnia pulex	2	0.0430	0.0032	0.0587	0.0186	1.3690	0.4186
	7	0.0441	0.0054	0.1090	0.0193	2.4897	0.4933
	12	0.0432	0.0089	0.2058	0.0248	5.0027	1.4051
Prionocypris	2	0.1239	0.0082	0.0416	0.0074	0.3354	0.0535
glacialis (Sars)	7	0.0755	0.0055	0.0569	0.0150	0.7508	0.1891
	12	0.1125	0.0209	0.1196	0.0214	1.0902	0.2148
Cyclops magnus	2	0.1045	0.0062	0.0410	0.0063	0.3982	0.0628
	7	0.1117	0.0127	0.1480	0.0881	1.3080	0.7390
	12	0.0964	0.0106	0.1497	0.0516	1.5615	0.5483
Cyclops magnus	2	0.0194	0.0022	0.0224	0.0079	1.2026	0.4511
(juvenile)	7	0.0182	0.0015	0.0535	0.0182	2.9549	0.9933
	12	0.0189	0.0042	0.1047	0.0239	5.5785	0.9445
Attheyella	2	0.0102	0.0005	0.0119	0.0058	1.1632	0.5493
nordenskioldii	7	0.0083	0.0011	0.0234	0.0143	2.6927	1.4333
	12	0.0089	0.0008	0.0346	0.0076	3.9691	1.0798
Acarina							
Trichoribates	2	0.0147	0.0007	0.0036	0.0014	0.2436	0.0922
polaris Hammer	7	0.0145	0.0008	0.0052	0.0015	0.3589	0.1159
	12	0.0141	0.0008	0.0060	0.0015	0.4254	0.1060
Hermannia	2	0.0402	0.0038	0.0059	0.0015	0.1469	0.0342
subglabra Berlese	7	0.0438	0.0056	0.0108	0.0047	0.2513	0.1123
	12	0.0451	0.0069	0.0144	0.0061	0.3066	0.1059
Lebertia porosa	2	0.2849	0.0314	0.0446	0.0195	0.1596	0.0735
Thor, S. Lat.	7	0.2670	0.0893	0.2063	0.0755	0.8336	0.4114
	12	0.3157	0.0649	0.3612	0.1305	1.1912	0.4978
Collembola							
Hypogastrura sp.	2	0.0188	0.0034	0.0052	0.0028	0.2732	0.2509
nr *trybomi*	7	0.0255	0.0081	0.0172	0.0066	0.6953	0.2436
(Schott)	12	0.0166	0.0052	0.0168	0.0102	1.1960	1.0317
Folsomia	2	0.0025	0.0001	0.0003	0.0011	0.1322	0.4766
agrelli Gisen	7	0.0034	0.0004	0.0021	0.0010	0.5708	0.2869
	12	0.0042	0.0006	0.0050	0.0033	1.2164	0.8730
Lepidoptera							
Gynaephora	2	0.2109	0.0610	0.0714	0.0190	0.3536	0.1079
rossi (Curtis)	7	0.1814	0.0091	0.1506	0.0609	0.8319	0.3373
(larvae)	12	0.1837	0.0051	0.4250	0.1098	2.3117	0.5873
Muscidae							
Spilogona prob	2	2.9590	0.1889	1.3737	0.5012	0.4643	0.1666
tundrae Schnabl.	7	3.6520	0.7211	1.7687	0.7232	0.4817	0.1851
(larvae)	12	3.5752	0.4102	2.0021	1.1566	0.5544	0.3124
Chironomidae							
Psectrocladius sp.	2	0.4829	0.0754	0.2966	0.0847	0.6230	0.1854
(larvae)	7	0.5605	0.0915	0.7966	0.4126	1.4126	0.6748
	12	0.4440	0.0321	1.0759	0.4266	2.4010	0.9053

Table 1. *(Continued)*

Species	*°C*	*Dry wt mg mean±SD*		*Oxygen consumption $\mu l\ O_2\ ind^{-1} hr^{-1}$ mean±SD*		*$\mu l\ O_2\ mg^{-1} hr^{-1}$ mean±SD*	
Procladius	2	0.5831	0.0768	0.2834	0.1065	0.4791	0.1396
culiciformis (Linne)	7	0.3965	0.1636	0.3137	0.2212	0.7975	0.4338
(larvae)	12	0.4378	0.2550	0.8280	0.4131	1.9026	0.7194
Cricotopus sp.	2	0.4597	0.0692	0.2798	0.0832	0.6218	0.2061
(larvae)	12	0.4920	0.0409	1.0675	0.6579	2.1999	1.3752
Orthocladius sp.	2	0.1072	0.0203	0.0412	0.0089	0.3900	0.0866
(larvae)	7	0.0785	0.0129	0.0656	0.0304	0.8687	0.4403
	12	0.0982	0.0292	0.1103	0.0510	1.1278	0.5228

ranged between 2.9549 and 0.2513, and at 12°C the corresponding values were 1.9973, 5.5785 and 0.3066 respectively.

The relation between metabolic rate (R) and body weight (W) for the 16 species at each temperature is represented by the equation:

$$\log R = \log a + b \cdot \log W,$$

where R is in $\mu l\ 0_2\ ind^{-1}\ hr^{-1}$, W is in mg dry wt ind^{-1}, a (0_2 consumption of a 1 mg individual at the specified temperature) is the intercept, and b is the slope of the ln-ln plot of 0_2 consumption in $ind^{-1}\ hr^{-1}$ versus weight ind^{-1} (Table 2, Fig. 1). Analysis of covariance showed that none of the b-values were significantly different ($P=0.05$).

The mean Q_{10} value for the 2-12°C interval was 4.01, with a range of 9.20 to 1.19 (Table 3). The corresponding values at 2-7°C were 6.86, 27.28, and 1.08, and the 7-12°C values were 2.99, 7.72, and 1.32. Eleven species had smaller Q_{10} values at 7-12°C than at 2-7°C, while five species had larger values at 7-12°C.

Table 2. Relation between oxygen consumption (R) and dry weight (W) for 16 invertebrate species at 2, 7, and 12°C: $\log R = \log a + b \cdot \log W$.

Temp (°C)	*a±95% CI*		*b±95% CI*	
2	−0.3290	0.2957	1.0427	0.2232
7	−0.1143	0.2995	0.9343	0.2182
12	0.1160	0.3175	0.9463	0.2404

Discussion

In the present study the Crustacea had the highest metabolic rates on a per mg dry wt basis, followed by the Enchytraeidae, Lepidoptera, and Chironomidae (Table 1). The reproductive condition of two of the crustacean species probably contributed to their high metabolic rates. For example, the *Daphnia pulex* individuals were all females with eggs, as were some of the *Attheyella nordenskioldii* specimens. Developing eggs may have increased metabolic rate while contributing relatively little to biomass. Webb (1969) reported that *Nothrus silvestri* (Acari) adults containing developing eggs had a weight-specific metabolic rate 56% greater than that of non-breeding adults. The inclusion of juvenile *Cyclops magnus*, which metabolised at approximately three times the rate of the adults, also contributed to the high average metabolic rate.

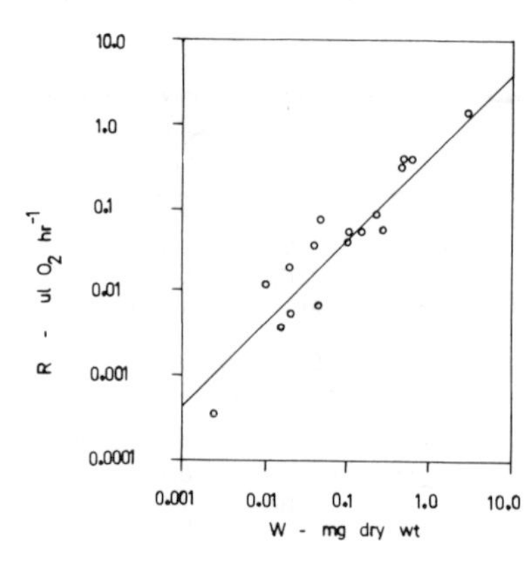

Fig. 1 Relation between respiration rate (R) and body weight (W) for 16 invertebrate species at 2° C. Regression equation: ln R = −0.3290+1.0427 ln W.

The two terrestrial mites, *Trichoribates polaris* and *Hermannia subglabra*, had very low metabolic rates at all temperatures, although both collembolan species and the aquatic mite *Lebertia porosa* had equally low rates at 2° C, and *Spilogona tundrae* metabolised nearly as slowly at 12° C (Table 1). Webb (1970) reported that oribatid mites have lower metabolic rates than most other soil-dwelling invertebrates, including the Enchytraeidae and Collembola. The present study supports this general relationship, but differs in that both collembolan species metabolised more slowly than the one enchytraeid studied. The low metabolic rates of oribatid mites probably reflect their relatively sedentary habits (Webb 1970). Published crustacean metabolic rates resembled my data in being relatively high, particularly at 0° and 2° C (Table 4). The crustacea metabolised more rapidly at these temperatures than did most other species at 5° and 10° C. The exceptions are the smallest individuals of the collembolan *Cryptopygus antarcticus* (Tilbrook and Block 1972), and the very early instar chironomid larvae (Welch 1976), almost all of which had similar rates to the crustacea at the lowest temperatures. *Cryptopygus antarcticus* also metabolised more rapidly than the crustacea at higher temperatures.

Juvenile *C. antarcticus* provide the greatest contrast to my data. For example, *C. antarcticus* weighing 0.0031 mg metabolised at approximately five times the rate of adult *Folsomia agrelli* of the same weight. On the other hand, adult *C. antarcticus* metabolised much more slowly, at a rate similar to adult *Hypogastrura* of the same size. *Cryptopygus antarcticus* provides a striking example of the size-specific differences in metabolic rate frequently observed between developmental stages (e.g., Bierle 1972, Hofsvang 1973, Welch 1976). From my study *Cyclops magnus* show the same phenomenon.

Apart from *C. antarcticus*, Collembola from other studies differed little from the species I studied. Addison (1975) obtained metabolic rates for *Hypogastrura tulbergi* and *Folsomia regularis* at 10° C that are intermediate between the 7° and 12° C rates I obtained for *Hypogastrura* sp. and *Folsomia agrelli*. *Onychiurus groendlandicus* metabolised more slowly than the other species.

Published information for chironomids shows the same size-specific differences in metabolic rate reported for the collembolan *C. antarcticus* (Bierle 1972, Welch 1976). Consequently, the early instar chironomids metabolised much more rapidly than the later instars, including the final instar larvae I studied. However, my data are comparable to the information obtained by Welch for larvae of equivalent development.

In general the metabolic rates I obtained are comparable to data for other arctic, antarctic, and alpine invertebrates at similar temperatures. The few major differences were attributed to studies of different developmental stages, and possibly to different methodology.

With all three weight/oxygen consumption exponents close to unity, I concluded that oxygen consumption is directly proportional to weight when species are compared (Table 2, Fig. 1). Scholander et al. (1953) obtained exponents of 0.80-0.85 and likewise concluded that metabolic rate is directly proportional to weight. In contrast, when different developmental stages of single species were compared, only two cases of direct proportionality were observed — *Lauterbornia* sp. (Welch 1976), and *Henlea nasuta* in the present study. In the other species oxygen consumption increased more slowly than weight (e.g., *C. antarcticus*, *T. excisa*, *C. magnus*, and most chironomid larvae).

The variety of apparently species-specific weight/respiration exponents in the literature has resulted in several hypotheses, including the "direct proportionality" relationship illustrated by my data, and the "surface law" relationship where rate of oxygen consumption is close to the 2/3 power of weight. On the basis of these "metabolic types," von Bertalanffy (1957) determined corresponding "growth types," distinguished by characteristic growth curves. However, Keister and Buck (1973) suggested that because of the number of variables which may affect both metabolic rate and weight, exponents should be considered descriptive only.

The three mean Q_{10} values are similar to the corresponding values estimated from Krogh's (1941) standard metabolic curve (Table 3). However, the means approximate the standard curve primarily because of several very large values, and most of my values are smaller than the standard, including three between 1.19 and 2.09 at 2-12°C, and three between 1.08 and 2.20 at 2-7°C.

Most species had decreasing Q_{10} values with increasing temperature, indicating linear rather than exponential increase in metabolic rate with increasing temperature. Two species, *Henlea nasuta* and *Spilogona tundrae*, had nearly constant values, denoting exponential increase in metabolic rate with temperature, while others, notably *Daphnia pulex*, *Gynaephora rossi* and *Procladius culiciformis*, had increasing Q_{10} values with temperature, and consequently a steeper than exponential rise in metabolic rate.

Table 3. Q_{10} values for 16 invertebrate species at three temperature intervals.

Species	*2°-12°C*	*2°-7°C*	*7°-12°C*
Henlea nasuta	2.50	2.20	2.84
Daphnia pulex	3.65	3.31	4.04
Prionocypris glacialis	3.25	5.01	2.11
Cyclops magnus	3.92	10.79	1.43
Cyclops magnus (Juvenile)	4.64	6.04	3.56
Attheyella nordenskiodlii	3.41	5.36	2.17
Trichoribates polaris	1.75	2.17	1.40
Hermannia subglabra	2.09	2.93	1.50
Lebertia porosa	7.46	27.28	2.04
Hypogastrura sp.	4.38	6.48	2.96
Folsomia agrelli	9.20	18.64	4.54
Gynaephora rossi	6.54	5.54	7.72
Spilogona tundrae	1.19	1.08	1.32
Psectrocladius sp.	3.85	5.24	2.83
Procladius culiciformis	3.97	2.77	5.69
Cricotopus sp.	3.54	—	—
Orthocladius sp	2.89	4.96	1.69
Mean	4.01	6.86	2.99

Table 4. Respiration rates for some arctic, antarctic and alpine invertebrates. For the following species the associated percentages were used to convert live weight to dry weight: *C. antarcticus*, 30%; *H. tullbergi*, 29.9% (Addison 1975); *F. regularis*, 35%; *O. groenlandicus*, 35%; *Z. exulans*, 18.93% (Hågvar and Østbye 1974); *T. excisa*, 25%.

Species	*Temperature (°C)*	*Dry wt (mg)*	*μl O_2 mg^{-1} hr^{-1}*	*Reference*
Crustacea				
Cyclops vernalis	0	0.0025(IV)	1.1204	Taube-Nauwerck
(copepodite IV, adult ♀)	0	0.0059(♀)	0.7705	1972
Limnocalanus macrurus	0	0.012-0.028	1.3490	Roff 1973
	2	0.012-0.028	1.5091	
	4	0.012-0.028	1.8436	
	10	0.012-0.028	2.6545	
	15	0.012-0.028	4.1564	
Mysis relicta	0	0.08-4.50	0.9526	Lasenby and Langford
	2	0.08-4.50	1.0647	1972
	4	0.08-4.50	1.4149	
	6	0.08-4.50	1.6881	
	8	0.08-4.50	1.8562	
Collembola				
Cryptopygus antarcticus	2	0.0009(I)	1.404	Tilbrook and Block.
(sizes: I, II, III, IV, V)		0.0031(II)	0.836	1972
		0.0077(III)	0.564	
		0.0158(IV)	0.416	
		0.0278(V)	0.328	
	6	I	5.520	
		II	2.024	
Cryptopygus antarcticus	6	III	0.956	Tilbrook and Block.
		IV	0.532	1972
		V	0.336	
	10	I	16.108	
		II	4.920	
		III	2.012	
		IV	1.008	
		V	0.588	
Hypogastrura tulbergi	10		0.7448	Addison 1975
Folsomia regularis	10		0.8731	Addison 1975
Onychiurus groenlandicus	10		0.3391	Addison 1975
Lepidoptera				
Zygaena exulans	5	2.5-52.0	0.4437	Hågvar and Østbye
(larvae)	10	2.5-52.0	0.9667	1974
	15	2.5-52.0	1.8542	
	20	2.5-52.0	2.3930	
Tipulidae				
Tipula excisa	5	2-6(II)	0.82	Hofsvang 1972,
(larvae: II, III, IV)	10	2-6(II)	1.14	Hofsvang 1973
	20	2-6(II)	1.75	
	5	6-16(III)	0.44	
	10	6-16(III)	0.80	
	15	6-16(III)	1.30	
	20	6-16(III)	1.44	
	5	16-41(IV)	0.26	
Tipula excisa	10	16-41(IV)	0.38	Hofsvang 1972,
	15	16-41(IV)	0.66	Hofsvang 1973
	20	16-41(IV)	0.82	

Table 4. *(Continued)*

Species	*Temperature (°C)*	*Dry wt (mg)*	*μl 0_2 mg^{-1} hr^{-1}*	*Reference*
Pedicia hannai	0.5	0.1-8.0	0.491	MacLean 1975
	5	0.1-8.0	0.655	
	10	0.1-8.0	1.311	
	15	0.1-8.0	1.886	
	20	0.1-8.0	2.582	
Chiornomidae				
Pseudodiamesa sp.	0	0.002-2.487	1.708-0.136	Welch 1976
Lauterbornia sp.	0	0.0002-0.926	0.222-0.219	Welch 1976
Heterotrissocladius sp.	0	0.001-0.726	0.600-0.115	Welch 1976
Trissocladius sp.	0	0.0003-0.499	1.429-0.234	Welch 1976
Orthocladius sp.	0	0.0004-0.693	1.463-0.264	Welch 1976
Chironomus sp.	5	0.30(III)	0.37	Bierle 1972
(larvae: III, IV)	10	0.30(III)	0.67	
	15	0.30(III)	1.18	
	20	0.30(III)	1.72	
	25	0.30(III)	2.28	
	5	1.90(IV)	0.20	
	10	1.90(IV)	0.34	
	15	1.90(IV)	0.70	
	20	1.90(IV)	0.94	
	25	1.90(IV)	1.30	

Stroganov (1956) presented evidence, based on temperature acclimation in fish *(Gambusia)*, that metabolic rate-temperature curves may be more complex than Krogh's standard curve. An important part of Stroganov's thesis is the presence of a 'zone of relative thermoneutrality' in which Q_{10} is relatively low, and which represents the temperature zone to which the organism is acclimated, or in which it usually lives (Duncan and Klekowski 1975). If Stroganov's curve is generally applicable, the several patterns of Q_{10} values obtained may indicate that measurements were made at different points on the various curves. For example, only the upper part of the 'zone of relative thermoneutrality' was covered by the measurements for *Procladius culiciformis*, only the lower part of the zone was included for *Lebertia porosa*, and all three measurements fell within the zone for Spilogona tundrae (Table 3).

While Stroganov's thesis offers an explanation for the variety of my observations, it must be pointed out that measurement at only three temperatures provides insufficient information to determine the form of a complex metabolic rate-temperature curve. Furthermore, none of the studies cited provide sufficient information for reliable analysis of such a curve, which may require 15-20 measurements taken at 1-2°C intervals.

Published Q_{10} values varied between 30.65 at 2-6°C for immature *Cryptopygus antarcticus*, and 1.06 at 2-6°C for adult *C. antarcticus* (Tilbrook and Block 1972) (Table 5), showing the same very wide range that I obtained (Table 3). Furthermore, the distribution of values was similar, including a large proportion of relatively small values, and a few very large values. The trends in Q_{10} also resembled my results, with most species exhibiting decreasing Q_{10} values with increasing temperature, while a few had constant or increasing values. Of special interest are the low Q_{10} values for *Limnocalanus macrurus*, *Mysis relicta*, *Cryptopygus antarcticus* (sizes IV and V), *Tipula excisa*, and

Pedicia hannai in the 0-10°C range. These values reinforce my observations that some species habitually living in cold climates have low Q_{10} values. Also noteworthy is the extent to which some of these values, both high and low, diverge from Krogh's standard curve. For example, the Q_{10} values for *C. antarcticus* stages I and V at 2-6°C were 30.65 and 1.06 respectively, compared with 8.34 from Krogh's curve. These observations, like those from my experiments, demonstrate that Krogh's curve does not reliably predict invertebrate metabolic rates at low temperatures.

Scholander et al. (1953) suggested invertebrates may have three methods of adaptation to arctic environments: (a) low Q_{10}, whereby temperature sensitivity is low; (b) shift of metabolic curve to maintain high metabolic rate; and (c) selection of favorable microclimates.

Scholander et al. (1953) found no low Q_{10} values in the arctic invertebrates they studied and concluded that this type of temperature adaptation is more likely to occur in temperate animals, which experience greater temperature variation. My studies show that low Q_{10} values occur fairly frequently, and are found in aquatic as well as terrestrial arctic invertebrates. Although arctic environments, and particularly aquatic environments, provide more constant temperatures than temperate or alpine habitats, relatively small temperature changes can be important because temperatures are often only marginally favorable for life cycle processes (Procter this volume). Low Q_{10} values may help maintain adequate metabolic rates in such conditions.

Scholander et al. (1953) found adaptation of metabolic curves only in aquatic invertebrates, which showed relatively high metabolic rates compared to tropical species at low temperatures (they did not measure metabolic rates of tropical species at low temperatures but extrapolated according to Krogh's curve). Scholander et al. (1953) did not suggest why modified metabolic curves should be restricted to aquatic invertebrates. Courtin and Labine (this volume) showed that wet sedge-moss environments are colder than drier beach ridges. The high specific heat of water also causes temperature to vary less. Aquatic animals are therefore confined to colder environments and have access to a smaller range of microclimates than terrestrial animals. Modified metabolic curves may therefore be necessary for aquatic species to attain comparable metabolic rates and life cycles of similar duration.

My preliminary studies produced no unequivocal cases of modified metabolic curves in aquatic species. For example, *Daphnia pulex* had a metabolic rate of 2.4897 $\mu l\ 0_2\ mg^{-1}\ hr^{-1}$ at 7°C, compared with Richman's (1972) 1.8621 μl for similar sized individuals reared at 20° (20° data converted to 7° according to Winberg 1971). These results do not support the five-fold difference Scholander et al. (1953) reported. Lasenby and Langford (1972) compared metabolic rates of the crustacean *Mysis relicta* from an arctic and a temperate lake and obtained no evidence of metabolic adaptation in the arctic population. Holeton (1974) also found little sign of modified metabolic curves in arctic fish, and showed that some reported examples of cold compensation in arctic fish are questionable on methodological and interpretative grounds. Consequently, while modified metabolic curves appear adaptive, particularly in aquatic species, evidence of such adaptation is equivocal at best.

Many of the species studied showed little sign of adaptation of metabolic rate to cold. Lacking adaptation of metabolic rate, these species may have behavioral adaptations which, however, are probably suppressed by the experimental conditions. For example, behavioral thermoregulation of the kind observed in arctic Lepidoptera (Kevan and Shorthouse 1970) would not occur during conventional respirometry. In the present study two species with high Q_{10} values, *Folsomia agrelli* and *Gynaephora rossi*, are active terrestrial

Table 5. Q_{10} values for some arctic, antarctic, and alpine invertebrates.

Species	*Temperature Interval (°C)*	Q_{10}	*Reference*
Limnocalanus macrurus	0-4	2.18	Roff 1973
	4-10	1.84	
	10-15	2.45	
Mysis relicta	0-4	2.69	Lasenby and Langford 1972
	4-8	1.97	
Cryptopygus antarcticus (sizes: I, II, III, IV, V.)	2-6	30.65, 9.12, 3.74, 1.85, 1.06	Tilbrook and Block 1972
	6-10	14.55, 9.21, 6.43, 4.94, 4.05	
Zygaena exulans (larvae)	5-10	4.75	Hågvar and Østbye 1974
	10-15	3.68	
	15-20	1.67	
Tipula excisa (instars: II, III. IV.)	5-10	1.93, 3.31, 2.14	Hofsvang 1973
	10-15	——, 2.64, 3.02	
	15-20	——, 1.23, 1.54	
Pedicia hannai (larvae)	0.5-5	1.90	MacLean 1975
	5-10	4.01	
	10-15	2.07	
	15-20	1.87	
Chironomus sp (instars: III, IV.)	5-10	3.27, 2.89	Bierle 1972
	10-15	3.10, 4.24	
	15-20	2.12, 1.80	
	20-25	1.76, 1.91	

animals which can probably locate favorable microclimates very quickly. Thermoregulatory behavior, including the ability to select favorable microclimates, is presumably shared by all invertebrates and is probably the predominant 'arctic' adaptation to cold (Scholander et al. 1953). In other words, many arctic species probably lack special metabolic adaptations and, using normal capacity for selecting favorable microclimates, are opportunistic in their utilization of arctic environments.

Summary

Metabolic rates and Q_{10} values are given for common terrestrial and aquatic invertebrates from Truelove Lowland, Devon Island. Oxygen consumption was measured at 2°, 7°, and 12°C, and was similar to published information for arctic, antarctic, and alpine invertebrates. When species were compared, metabolic rate was directly proportional to weight, but when different developmental stages within a species were compared, metabolic rate usually increased more slowly than weight. Values of Q_{10} varied between 1.19 and 9.20, with a 4.01 mean, for the 2°-12°C temperature interval. The mean Q_{10} values were close to the equivalent values on Krogh's standard metabolic curve but, because of several very large values, most of the specific values obtained were lower than predicted by Krogh's curve. The smallest Q_{10} values occurred in both terrestrial and aquatic species.

Acknowledgments

I thank my supervisor, Dr. W.G. Evans, for his support and interest. Thanks are also due to my technical assistant, Mr. Peter Meyer, for his work during the study. Dr. L. Wang provided helpful criticism of the manuscript. The following people identified experimental material: Dr. N.C. Dash (Enchytraeidae); Dr. G.R. Daborn, Dr. L.D. Delorme, Dr. H.C. Yeatman and Mr. J.T. Retallack (Crustacea); Dr. E.E. Lindquist and Dr. H. Habeeb (Acarina); Dr. W.R. Richards (Collembola); Mr. J.K. Ryan (Lepidoptera); Dr. H.J. Teskey (Muscidae), and Dr. D.R. Oliver (Chironomidae).

References

Addison, J.A. 1975. Ecology of Collembola at a High Arctic Site, Devon Island, N.W.T. Ph.D. Thesis. Univ. Calgary, Calgary. 212 pp.

Bierle, D.A. 1972. Production and energetics of chironomid larvae in ponds of the arctic coastal tundra. pp. 182-186. *In:* Proceedings 1972 Tundra Biome Symposium. Bowen, S. (ed.). Hanover, New Hampshire. CRREL. 211 pp.

Duncan, A. and R.Z. Klekowski. 1975. Parameters of an energy budget. pp. 97-147. *In:* Methods for Ecological Bioenergetics. Grodzinski, W., R.Z. Klekowski and A. Duncan (eds.). Blackwell, Oxford. 367 pp.

Gregg, J.H. and F.A. Lints. 1967. A constant-volume respirometer for *Drosophila* imagos. Compt. Rend. des Trav. du Lab. Carlsberg 36: 25-34.

Hågvar, S. and E. Østbye. 1974. Oxygen consumption, caloric values, water and ash content of some dominant terrestrial arthropods from alpine habitats at Finse, south Norway. Norsk ent. Tidsskr. 21: 117-126.

Hofsvang, T. 1972. *Tipula excise* Schum. (Diptera: Tipulidae), life cycle and population dynamics. Norsk ent. Tidsskr. 19: 43-48.

———. 1973. Energy flow in *Tipula excisa* Schum. (Diptera: Tipulidae) in a high mountain area, Finse, south Norway. Norw. J. Zool. 21: 7-16.

Holeton, G.F. 1974. Metabolic cold adaptation of polar fish: fact or artifact? Physiol. Zool. 47: 137-152.

Keister, M. and J. Buck. 1973. Respiration: some exogenous and endogenous effects on rate of respiration. pp. 469-509. *In:* The Physiology of Insecta. Vol. VI. Rockstein, M. (ed.). Academic Press, New York. 548 pp.

Kevan, P.G. and J.D. Shorthouse. 1970. Behavioural thermoregulation by high arctic butterflies. Arctic 23: 268-279.

Krogh, A. 1941. The Comparative Physiology of Respiratory Mechanisms. Dover Publ., New York. 172 pp.

Lasenby, D.C. and R.R. Langford. 1972. Growth, life history, and respiration of *Mysis relicta* in an arctic and temperate lake. J. Fish. Res. Bd. Canada 29: 1701-1708.

MacLean, S.F. 1975. Personal communication.

Richman, S. 1972. The transformation of energy by *Daphnia pulex*. Oikos 23: 359-365.

Roff, J.C. 1973. Oxygen consumption of *Limnocalanus macrurus* Sars (Calanoida, Copepoda) in relation to environmental conditions. Can. J. Zool. 51: 877-885.

Scholander, P.F., W. Flagg, V. Walters, and L. Irving. 1953. Climatic adaptation in arctic and tropical poikilotherms. Physiol. Zool. 26: 67-92.

Stroganov, N.S. 1956. Physiological adaptation to environmental temperature in fish. Izd. Akad. Nauk SSSR, Moskva. 151 pp.

Taube-Nauwerck, I. 1972. Cyclopoids. Char Lake Project. Ann. Rep. 48-52.

Tilbrook, P.J. and W. Block. 1972. Oxygen uptake in an Antarctic collembole *Cryptopygus antarcticus*. Oikos 23: 313-317.
von Bertalanffy, L. 1957. Quantitative laws in metabolism and growth. Quart. Rev. Biol. 32: 217-230.
Webb, N.R. 1969. The respiratory metabolism of *Nothrus silvestris* Nicolet (Acari). Oikos 20: 294-299.
———. 1970. Oxygen consumption and population metabolism of some mesostigmated mites (Acari: Mesostigmata). Pedobiologia 10: 447-456.
Welch, H.E. 1976. Ecology of Chironomidae (Diptera) in a Polar Lake. CCIBP Contribution No. 274. J. Fish. Res. Bd. Canada. 33: 227-247.
Winberg, G.G. 1971. Methods for the Estimation of Production of Aquatic Animals. Academic Press, New York. 175 pp.

Energy budget for *Gynaephora groenlandica* (Homeyer) and *G. rossii* (Curtis) (Lepidoptera: Lymantriidae) on Truelove Lowland

James K. Ryan and Colin R. Hergert

Introduction

Two similar moth species, *Gynaephora groenlandica* (Homeyer) and *G. rossii* (Curtis), are the largest invertebrates at Truelove. They are conspicuous as woolly larvae and silken cocoons, but are actually rather uncommon. They are more abundant elsewhere in the Arctic (Curtis 1835, Gibson 1922, Bruggemann 1958). Downes (1962, 1964, 1965) gives the best accounts of *Gynaephora* as arctic and alpine insects, and summarizes their host plant records and early literature. These two species are distinctive in that they are among the approximately two dozen Lepidoptera species found in extreme cold-dominated climates. Their arctic and alpine distributions in North America are evident in Fig. 1, which was compiled from 19 published papers and museum records (Ferguson, pers. comm., McGuffin, pers. comm.).

Byrdia is currently the accepted name for the genus, but Ferguson (pers. comm.), who is reviewing this family in North America, states that the correct name is *Gynaephora*. Kozhanchikov (1950) revised this genus in the U.S.S.R., and considered the Soviet *G. rossii* a new species which he named *G. lugens*. This creates a disjunct distribution, with *G. rossii* in Japan (Ohyama and Asahina 1971) and North America, and *G. lugens* in Eurasia. Chernov (1975) found that specimens of *G. lugens* were frequently parasitized by the tachinid fly *Spoggosia gelida* Coq. and an ichneumonid wasp. These same parasites are common in North American *G. groenlandica* and *G. rossii*. This indicates some taxonomic uncertainties to us, so Fig. 1 was confined to North American records. Both Truelove species seem identical in their food choices and development, and almost identical morphologically. The data for this study have been combined and presented for both as one ecological unit.

The purpose of this study is to determine energy flow through the *Gynaephora* population of Truelove Lowland. Energy flow means calories assimilated m^{-2} yr^{-1}, and is partitioned into respired energy and tissue available for consumption by the next trophic level (= production). Because the larvae are large, consume distinct bites of leaves in measureable amounts, and have discrete, solid excreta, they can be handled and observed with accuracy. Their size and habits make them convenient to study under limiting field conditions. They were intensively studied as model insects to find essential factors governing energy flow within their populations. The energy budget constructed here is the prime example of the methodology used in energy budgets for other invertebrates (Ryan this volume).

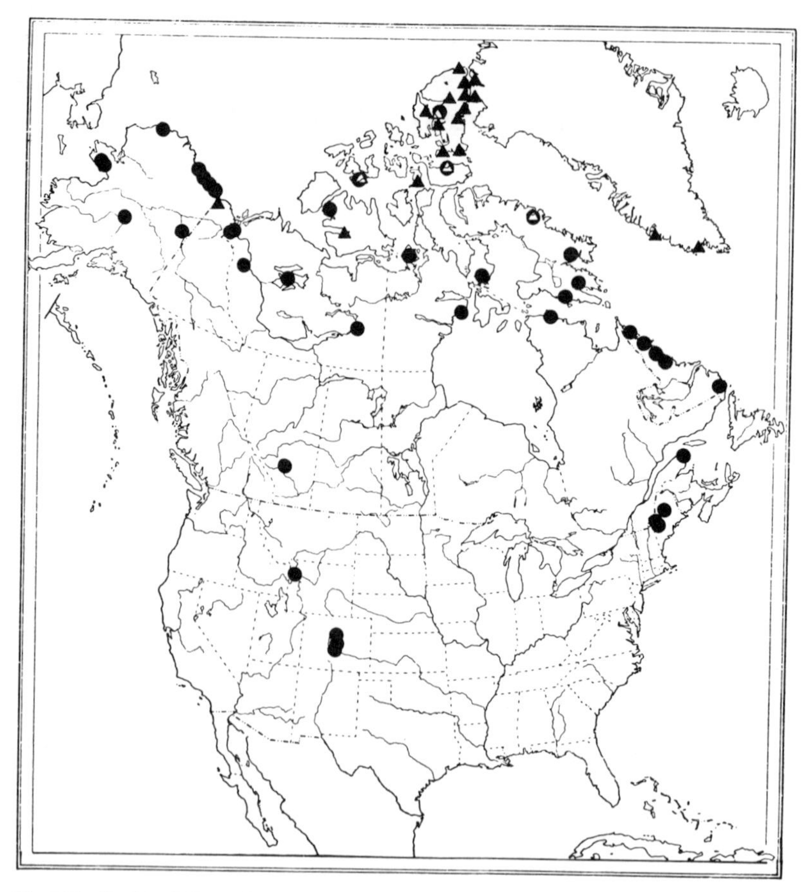

Fig. 1 Distribution of *Gynaephora groenlandica* and *G. rossii* in North America
● *Gynaephora groenlandica*
▲ *G. rossii*
◓ both species

Materials and Methods

We actively sought these animals from 1970 to 1973. For each animal collected we recorded the following factors: location and plant community where found, developmental stage, weight, and whether parasitized or otherwise unhealthy. Some individuals were ovendried at 80°C for 3 days and later burned for caloric content. A Phillipson micro-bomb calorimeter was used for all caloric determinations, including reproductive tissue dissections; dry weights were determined on a Cahn electrobalance. Specimens held for observation and experiments were kept outdoors individually in 40 dram screen covered vials. Emerging adults were sexed, weighed, and determined to species.

Feeding experiments used 4 larvae in 1971, and 20 in 1972. These tests started with 5, and 62, larvae; the difference reflects pupation, parasitization, death, and non-feeding by the experimental larvae. The larvae, maintained in screen capped vials kept outdoors, were offered selected weighed green leaves of *Salix artica*, *Saxifraga oppositifolia*, and *Dryas integrifolia* at 4-5 day intervals; sufficient leaves were added to ensure not all were consumed during the feeding interval. At each food change the uneaten leaves, and feces produced, were collected and ovendried, and the larvae reweighed. Two sets of controls were maintained for leaf wet/dry weight ratios. One set of leaves, for each leaf type, was weighed and immediately ovendried, while the other set was kept in empty larval vials and dried when the food was changed. This procedure compensated for leaf metabolism. The live/dry weight ratio of larvae in this experiment, and others, is from the live/dry weight ratio observed for field collected larvae.

Growth rates were estimated by 6 independent techniques: (1) in 1972 6-10 larvae were placed in each of 3 separate walled exclosures of 4-8 m^2 on a raised beach. The larvae were weighed periodically, and individual weight fluctuations noted; (2) in 1973 10 larvae were weighed and placed on favorable raised beach sites in emergence traps (see Fig. 1, Ryan this volume) between 16 and 31 July. Nine were recovered and reweighed 9 September; (3) growth rates of the feeding experiment larvae are another measure of development. These 3 measures of growth rates were converted to the full season growth rate by multiplying:

$$(C^\circ \text{ total})/(C^\circ \text{ obs})$$

where C° total is the total seasonal °C days above 0°C, and C° obs are the accumulated °C days above 0°C for which larvae were observed; (4) approximately 150 hatchling larvae were reared on *Prunus* leaves in the laboratory, and the weight increase of the 16 largest surviving larvae was measured after 56 days. These were maintained at a constant 23°C. The development rate at this temperature was converted to Truelove rates by the above formula, ignoring Q_{10} factors. (5) collection records from the study site areas indirectly gave a measure of larval development; (6) an estimate of developmental rates was computed from *G. rossii* respiration rates, assimilation rates observed in the feeding experiment, and the mathematical model described in Ryan this volume.

The seasonal temperature used is 645°C days above 0°C, which is equivalent to 53.5×12°C days. This latter value is used when calculating seasonal respiration measurements. Seasonal temperature is the accumulated degrees between −2 and +2 cm at the raised beach site in 1971. This was the warmest of the four monitored seasons; −2 cm was warmer than +2 cm.

Weight declines were measured for larvae, pupae, and adults. Three larvae were weighed, placed without food in an 8°C constant temperature chamber, and reweighed 60 days later. To determine weight loss in the cocoon stage, eight animals were weighed at 4 day intervals as larvae, then throughout the pupal period, and as newly emerged adults. These were kept in screen capped vials, as previously described. Thirteen (6 females, 7 males) newly emerged adults, after excreting their meconia, were weighed and placed in two 20×20×20 cm screen cages and left outside at Truelove. They were offered water but no food. At death they were collected and ovendried; also their lifespans were noted.

Respiration rates are available from two sources. Scholander et al. (1953, our conversions) found that 450 mg live *G. rossii* larvae consumed 0.19 $\mu l\ O_2\ hr^{-1}$ at 12°C. Procter (1975) found first instar larvae, at 0.84 mg live weight, consumed 0.52 $\mu l\ O_2\ hr^{-1}\ mg^{-1}$ at 12°C at Truelove. The data of Procter were used to compute oxygen consumption for any size larva by the equation:

$$R = aW^b$$

where R = O_2 consumed hr^{-1} a is a constant calculated from known values, W is the weight, and $b = 0.85$ (Zeuthen 1953, von Bertelanffy 1957, Reichle 1968). Oxygen consumption for a season was obtained by: 53.5-12° C days multiplied by the O_2 consumed at the mean weight of each larva during the season.

The *Gynaephora* population is considered to be in equilibrium, neither increasing nor decreasing over time. Therefore, the standing crop will replace itself with an identical standing crop in one generation, although in fact all stages of development were observed concurrently in a fraction of the generation time. Such a combined unit is called a synthetic cohort (Spinage 1972). Energy flow through the synthetic cohort is calculated in several steps. First, we calculated the mean fecundity of a female. By assuming an exponential decline in numbers of developing larvae, we then calculated their yearly numerical declines until they reached the prepupal stage. Concommitant weight increases for these larvae were determined from the estimated growth rates. Respiration is the product of the mean number of animals alive in a season times the seasonal respiration of each individual; these values are summed for the whole cohort. Since the field population consists of overlapping cohorts, yearly respiration is expressed as:

$$R_{ys} = Rc/Tc$$

where R_{ys} is the yearly cohort respiration, Rc is the total respiration by the cohort, and Tc is the total seasons required for one cohort to develop. Population respiration m^{-2} yr^{-1}, Rp, is given by:

$$Rp = R_{ys} \cdot N_c$$

where N_c is the number of cohorts m^{-2}.

The larval standing crop was estimated from the number of larvae found in 571 m^2. Most of this count, 500 m^2, was made using a 0.5 m^2 wire frame randomly placed in specific habitats. The 71 m^2 is the population found within emergence traps. The total collectable standing crop of immatures, Scc, in a synthetic cohort is:

$$Scc = \sum_{1}^{t_x} \overline{N_i} \cdot \overline{W_i}$$

where t_x is the time at maturity, and $\overline{N}$ and $\overline{W}$ are respectively the mean numbers, and mean weights, of immature individuals alive in the cohort during each "i" season. The fraction of a synthetic cohort of immatures collectable m^{-2} in one year, Sc, is given by:

$$Sc = Sca/Scc \cdot Tc$$

where Sca is the actual standing crop. Yearly production and respiration similarly are fractions of the total cohort production and respiration. Production and respiration were computed through these combined formulae.

Results

Altogether 374 live animals (not including hatchling larvae) and empty cocoons were collected between 1970-73. Four flying adults were observed in the field, of which 2 were captured.

Life history

The mean live weight of all 156 *Gynaephora* larvae collected was 630 mg; the largest larva weighed 1890 mg. The mean live weight of newly hatched larvae was 1.6 mg. As the larvae hatch late in the summer, they have limited opportunity to feed. One batch of hatchling larvae, offered continuous food during fall, were a mean 1.2 mg live when first weighed after overwintering. This 1.2 mg is used as the larval starting weight in the synthetic cohort discussed later. First stage larvae, carried by wind-blown silk threads, are the dispersal stage.

Seventy-four field collected cocoons from which adults emerged were a mean 751 mg live weight when first weighed. These are assumed to have been randomly collected insofar as development, and, therefore, at the midpoint of weight decline for the pupae. Pupal weight just prior to emergence was 71% of the maximum larval weight (10 obs), and adult weight was 56% of this maximum (n=8). The mean pupal weight was 85.5% of the maximum larval weight. Combining these, the mean maximum live weight of a pharate pupa is estimated to be 879 mg live. This end weight was used in the synthetic cohort.

Adult emergence in vials began in June, peaked in late July, and continued through August. Emergence times of both species overlapped. No pupae (8 examined) were observed to survive a winter at Truelove. Adults emerged from cocoons brought into the laboratory after they would have been killed by frost on Devon. The adult sex ratio is probably 1/1, e.g., in 1972 20 male and 25 female adults were reared or collected.

Reproductive capacity was determined by examining female egg loads, and eggs deposited. Seven females at emergence were 71±2.3 (S.E.)% reproductive tissue, which is equivalent to 240±24 mature eggs. Females at death still held eggs in their ovaries. Thus, they did not deposit the full egg load. Eggs were initially deposited in clumps on the cocoons, and small amounts laid elsewhere later. These clumps on cocoons (n=23) had a mean 139 eggs, while females (n=16), held in vials until death, deposited a mean 159 eggs. Unfertilized eggs from females kept in vials did not hatch. On 19 field collected cocoons, 56% of the eggs failed to hatch, so infertility was high among these insects. The simplest interpretation of these data is that a mean 160 eggs were deposited per reproducing female. Therefore, 160 eggs were assumed to be produced per female in the synthetic cohort. Infertility rates require that two egg laying females are needed to produce one batch of fertile eggs.

The braconid wasp *Rogas* sp. parasitizes larvae of both species with one wasp per larva, but in one instance (n=21). apparently 2 adults emerged from one larva. Of 147 *Gynaephora* larvae collected, 15.6% were parasitized by *Rogas*. The tachinid fly *Spoggosia gelida* emerged from cocoons of both *Gynaephora* species, and in 3 instances from larvae of *G. rossii*. Of 118 examined cocoons, 95 *S. gelida* pupae were found, which yielded 80 adults. The *Gynaephora* cocoons were 17% parasitized, and parasitized cocoons produced a mean of 4 adult flies. No *Apanteles* sp. wasps, nor *Exorista* sp. tachinid flies, were reared from *Gynaephora;* however, 3 *Apanteles* sp. are known from Truelove. Dead larvae were never flaccid, as from bacterial or viral death. Although fungi (*Aspergillus* sp.) occasionally covered the bodies of dead larvae, these apparently did not cause the insect death (Kennedy, pers. comm.). *Gynaephora* and its parasites are shown in Fig. 2.

Feeding studies

In 1971 the mean Efficiency of Assimilation (E.A., which is: food assimilated/food ingested) for the 4 test larvae was 18%, and the Gross Efficiency of growth (G.E., which is: larval growth/food ingested) was 10%. In 1972, the mean E.A. for the 20 test larvae was 24%, and the mean G.E. was 8%.

Apparently healthy larvae intermittently ceased feeding, or ceased

Fig. 2 Stages and parasites of *Gynaephora*, with adults of both species. A. *Gynaephora rossii* (?)♀, wings not spread at emergence. B. Braconid wasp *Rogas* on *G. groenlandica* larvae. C. *Rogas* sp. on *G. rossii* larvae. *D. Rogas* sp. on *G. rossii* larva, larva in hibernaculum, parasite emerged early in spring. F. *G. rossii* larva in hibernaculum, died, no parasite emerged. *G. Rogas* sp. with brachypterous wings (host sp. undetermined). H. Tachinid pupa from *G. groenlandica* larva (only 2 such animals found; tachinid adult did not emerge from either). I. Tachinid flies *Spoggosia gelida* Coquillett. J. Light and dark color cocoons, both forms found with each species. K. Opened cocoon to show pupa and larval exuvium (sp. undetermined). L. Eggs laid on cocoon of *G. rossii* (same for both species).

altogether, and became inactive during these feeding experiments. Three such larvae, inactive for 9 weeks at 8° C, were frozen to −20° C. When thawed 3 months later they pupated, and adults emerged successfully. During the 9 week period these larvae declined 18% from their initial weight. Larvae maintained without food all summer appeared healthy, with little weight loss, at the end of the summer.

Growth rates, weights, and caloric values

The growth rates measured and predicted for *Gynaephora* larvae are summarized in Table 1. *Gynaephora* collection records from the study site vicinity, where there was considerable human activity, are more complete than for other areas. From this area: in 1970, 12 specimens (3 larvae, 9 cocoons) were collected; in 1971 one larva was collected; and in 1972, 10 specimens (4 larvae, 6 cocoons) were collected. If the larvae of the 1972 cocoons developed at maximum field determined rates to 879 mg prepupal larvae, in 1970 they would have weighed between 166-382 mg. If they developed at the average field determined rates, they would have weighed between 391-770 mg in 1970. Since the mean weight of all larvae collected (156) was 630 mg, if larvae grew at the average field determined rates, then more larvae should have been collected during the preceding 2 years. The maximum weight gain rates do not contradict the larval collections, since smaller larvae are most likely to be

Table 1. Growth rates of *Gynaephora (=Byrdia)* larvae adjusted to a 645 °C day (above 0°C) season.

Experiment	*Mean weight increase*	*Maximum weight increase*
Walled exclosures	27%, 29%, 50%	50%
Screen trap exclosures	14% (9)*	226% (2)*
Feeding experiment, 1972	50% (20)	230% (5)
Feeding experiment, lab hatchlings	160% (16)	460% (5)
Production model: prediction		260%

*In parentheses are the numbers of larvae observed.

overlooked. From all the growth rate measurements, we conclude that the average healthy larva grows 2.2-fold in a normal (1285×12°C hr) season.

Nine pupae spent a mean 38 days within the cocoon before adults emerged; 6 of these were pharate pupae during the first 2.2 days in the cocoons. In the adult lifespan experiments, 6 females lived 13±3.3 (S.E.) days, and 7 males lived 16±2.3 days.

The dry weights and caloric values of all stages and tissues of *Gynaephora*, and its parasites, are given in Table 2. All unburned remains were called residues, which are primarily ash but may contain organic material. Thus, the column showing residue free caloric values approximates ash free values. The ovendry weight/live weight ratio of large larvae was 0.225. This is the factor used to convert live weights to dry weights.

The standing crop of *Gynaephora* in the 571 m^2 sampled was 3 individuals, with a mean 0.28 mg dry wt m^{-2}. One *Gynaephora* larva was found in the 500 m^2 sample, and 2 were found under the 71 m^2 emergence trap sample, on one raised beach. The population is contagiously distributed and the sample size is small, so this mean standing crop value is a tenuous approximation of the actual standing crop. The *Gynaephora* records contain 290 precise collection localities which can be partitioned into raised beaches, limestone and rock outcrop areas, and meadows, giving collection frequency ratios of 13:4:1 respectively for these land forms. Applying this to the mean, the tentative standing crop of raised beaches was 0.61 mg m^{-2}, of limestone and rock outcrops 0.5 mg m^{-2}, and of meadows 0.2 mg m^{-2}.

Production

Table 3 presents the extrapolated respiration for all stages of one cohort of *Gynaephora*. Ten years are required to produce one cohort. Table 4 contains the detailed calculation of tissue production for this same synthetic cohort. Table 5 gives the conversion of data in Tables 3 and 4 to calories. The tissue production conversion was made through Table 3 caloric data, while the oxygen consumed is converted to calories expended by the constant 4.7 cal cc^{-1} O_2 (Petruscewicz and Macfadyen 1970).

The total collectable standing crop of non-flying *Gynaephora* in the synthetic cohort is 15768 mg live wt, or 1577 mg yr^{-1} = 355 mg dry wt yr^{-1}. This is computed by summing: (the number of animals alive at mid stage) times (the mean weight/individual in the stage) in Tables 3 and 4. The standing crop found was 0.00079 of this value. Production in one cohort (see Table 5) totals 1672.8 mg (this total excludes the fertile egg mass from which the cohort hatched). The yearly production in one cohort is (0.1)(1672.8) = 167.3 mg. Production m^{-2} yr^{-1} is (0.00079)(167.3 mg), which is 1.13 mg m^{-2} yr^{-1}. In

Table 2. Weights and caloric values for *Gynaephora* and its parasites.

Stage	*mg Oven Dry Wt individual*$^{-1}$*	*No. of individuals*	*No. of measurements*	*cal gm*$^{-1}$ *Oven Dry Wt**	*cal gm*$^{-1}$ *Residue free Oven Dry Wt*
Egg	0.44	1621	2	6208	7127
Larva					
1st Instar					
Starved after hatch, died	0.1	184	1	5632	7000
Newly emerged	0.35	276	1	5902	6576
Large, Healthy	—	12	13	5688±101	6438±137
Larvae preserved in ETOH in field	127.4±50.6	5	6	5695±167	6346±217
Larvae, normal	170.0±37.5	7	7	5682±134	6516±185
Died of Disease?	57.7	3	1	5147	5837
Parasitized					
With 10 tachinid eggs hatched and burrowed in	497.7	1	1	5741	6140
With *Rogas* parasite	33.6	1	1	5310	—
Larval remains after tachinid emergence	39.1	8	1	5421	6234
Larval exuvium	9.96	37	3	5458	5913
From field	11.65	11	1	5206	5612
From cocoon	9.58	26	2	5697	6198
Pupa					
Whole cocoon	196.1±17.7	10	10	5948±102	6776±83
Pupa	178.8±16.2	18	13	5931±74	6787±114
Pupal exuvium	12.3	23	2	4341	5053
Cocoon silk					
Clean	28.8	34	3	4820±67	5343±18
With meconium	56.7±6.9	6	6	4987±82	5944±179
Whole empty cocoon	45.1±4.7	12	12	5260±65	6184±122
Adult					
Male	64.2±5.8	13	10	6555±107	7394±110
Freshly emerged	74.0±4.5	6	5	6772±113	7313±233
Old	49.5±9.0	7	5	6339±124	7459±95
Female	149.0±14.5	25	17	5823±49	6850±75
Freshly emerged	165.8±14.3	12	13	5899±48	6799±56
Old	79.1	13	3	5615	6987
Reproductive tissues of freshly emerged female	106.4±10.7	7	7	6306±41.4	6954±103
Parasites					
Rogas sp. (Braconidae)	5.93	3	1	6152	—
Spoggosia gelida Coq. (Tachinidae)					
Pupa					
Live, new	11.7	10	1	5736	6646
Dead, old	7.8	14	1	3266	4122
Pupal exuvium	1.1	24	1	4805	6155
Adult					
Freshly emerged	12.5	15	1	5842	6780
Starved after Emerged, died	6.3	9	1	4926	6038

*mean ± S. Error

Table 3. Extrapolated oxygen consumption of all stages of *Gynaephora* cohort.

Year	*μl O_2 consumed animal hr^{-1}*	*No. animals live at mid stage*	*ml O_2 consumed/animal/ season (1285-12° C hrs) * × (O_2 animal hr^{-1})*	*Liters O_2 consumed/ season (No. animals) × (O_2 consumed $animal^{-1}$)*
1	0.59	148	0.76	0.112
2	0.84	110	1.08	0.119
3	1.4	77	1.80	0.139
4	3.0	53	3.86	0.205
5	5.5	37	7.07	0.262
6	10.5	25	13.5	0.338
7	19	18	24.4	0.439
8	26	12	33.4	0.401
9	46	8.5	59.1	0.502
10	84	6	107.9	0.645
11	82	4.5 pupae	48.8**	0.220
		3 adults	52.8***	0.158
Total				3.540

* 1971 total positive 12 °C hrs above 0°C on beach ridge, mean +2 + −1 cm

** approximate life of pupa = 640-12°C hrs (=1/2 season)

***386 mg dry wt non-reproductive tissue at emergence minus 191 mg tissue at death = 195 mg tissue metabolized; 0.82 cc O_2 mg^{-1} (Petruscewicz and MacFadyen 1970)

Table 4. Extrapolated *Gynaephora* cohort production, with life table data.

Stage or year	*No. alive start of stage*	*No. died in stage*	*Mean live wt (mg) $indiv.^{-1}$*	*mg dead tiss. produced*	*Exuviae* produced (dry wt)*
Egg→larvae	160	26	1.4	36	—
2	134	42	1.8	77	—
3	92	28	4.0	113	8.8
4	64	20	8.8	177	11.3
5	44	14	19.4	272	—
6	30	9	42.8	385	30.4
7	21	6	94.1	565	—
8	15	5	157.7	779	40.8
9	10	3	276.4	829	—
10	7	2	608.2	1216	30.8
11 pupa	5	1	713	713	181.1
Adults:					
unmated	4	1	372; 162**	534	
reproduced		3		395***	
			Subtotal	6088	
			Dry wt:	1370	303.1
			Total	1673 mg dry; 167 mg dry/yr	

* Larval exuvia dry wt was 5.2% live wt of 1st instars, 1.1% of final instar. Exuviae wt produced = (5.2, 4.4, 3.6, 2.8, 2.0, 1.1%) (larvae starting live wt of those which survive that year)

** Infertile eggs

*** [(mean wt old male + female = 584/2) − (0.1) (584/2)] × (3)

calories, the yearly production is 0.75 cal m^{-2}, and assimilation is 2.02 cal m^{-2}. Taking the observed habitat distribution as the baseline, the yearly dry weight tissue production m^{-2} on raised beaches is 0.22 mg, 0.07 mg in rock outcrops and limestone pavements, and 0.02 mg in meadows. The yearly assimilated cal m^{-2} is 4.4 cal on raised beaches, 1.4 cal on rock outcrops and limestone pavements, and 0.3 cal in the meadows.

Discussion

The results are separable into two parts. The first part includes measured, discrete data. The second part is abstract, and includes calculations of production and respiration. The discussion is arranged this way in sections called "Life history," and "Production."

Life history

The biological information parallels published observations on *Gynaephora*, with two exceptions. Adult emergence overlapped for both species at Truelove. At Lake Hazen *G. groenlandica* adults emerged early in the season and *G. rossii* adults emerged later (Oliver et al. 1964). The pheromones produced by females were hypothesized to be identical for both species, and separate emergence times kept the species isolated. This is not the isolating mechanism at Truelove. We intended to test male-female responses of mixed species pairs, and to examine hybrid viability, but were hampered by shortages of specimens. Two *Gynaephora* parasites found at Ellesmere Island were not found at Truelove. These were *Exorista* sp. (Wood, pers. comm.) and *Apanteles* sp. (Mason, pers. comm.). Their absence is probably due to the low population density of *Gynaephora* at Truelove. This would also explain why no evidence of Truelove birds feeding on *Gynaephora* pupae was ever found. Birds have been observed doing this elsewhere (Kozhanchikov 1950, Bruggemann 1958).

Growth rates and feeding efficiencies of *Gynaephora* larvae are critical to the determination of energy flow. Life cycles of arctic invertebrates are known to be prolonged due to cold MacLean (1975). Downes (1964) estimated the life cycle of *G. groenlandica* required up to 5 years at Lake Hazen, and later (pers. comm.) suggested 8 years. There are 396°C days above 0°C at Lake Hazen (2 year mean from Corbet and Danks, 1974) and 273°C days for the corresponding +2 m. meadow measurement at Truelove. Lake Hazen is approximately 1.4× warmer seasonally than Truelove. By direct conversion of Downes' estimates, the generation time of *Gynaephora* should be 7-11 years at Truelove.

We converted all growth rates to the entire season above 0°C at Truelove. The 0°C temperature is the base for all time span calculations. Larvae are active below this temperature, but we assume the total activity below 0°C is negligible compared to the above 0°C activity. Furthermore, the relationship of the mean temperatures at +2 and −2 cm of the ground surface, to the larval microhabitat temperatures, would be difficult to determine, even if developmental threshold temperatures were known. Thus, 0°C is a point of definable convenience. In spring and fall of each season, no edible leaves are available, so the "eating season" is shorter than the above 0°C season. Conversely, conditions within vials are not optimal. Leaves lying flat on the vial floor rather than held erect on plants, and reduced food choice, are two factors, among many possible, which may reduce growth rates of larvae in vials relative to wild larvae. Rapidly growing larvae are more likely to become adults than slowly developing ones. Slower growth may indicate poor health, and exposes larvae to greater risk of parasitization, predation, and accidents. We assume an increase rate of 220% (becoming 225% when exuvial weights are considered), which is the maximum of the rates observed in the field. One year (year 8) is adjusted to 133% to reach 879 mg at the end of the final season, giving *Gynaephora* larvae a 10 year developmental period. One additional year of development would allow a larva to grow to 1934 mg; the maximum weight larva collected was 1890 mg. The data may support a longer lifespan than 10 years, but not a shorter one given Truelove Lowland conditions.

Larvae are difficult to rear in captivity (Gibson 1922, Bruggemann 1958); we concur. We found some stopped feeding, and became inactive, regardless of the diet offered. This inactivity may be diapause, as with the three pharate pupae, which pupated only after a freezing treatment. In diapause, insects continue to metabolize energy but do not feed, so growth rates and feeding efficiencies are both reduced. Cessation of feeding may be a natural phenomenon, occurring in the field as well as in observation vials. If so, larval growth rates would be lower, and hence generation time longer, than we report. The most rapid growth rate, 460%, occurred with larvae fed on deciduous *Prunus* leaves. Truelove plant leaves were evergreen, except for *Salix arctica*. The Efficiency of Assimilation of Truelove leaves was 24%, while the E.A. for most herbivorous insects is 55-60% (cf. Waldbauer 1968). Larval feeding efficiencies vary with the quality of the food offered (SooHoo and Fraenkel 1966, Schroeder 1972, 1973), and growth rates should vary similarly; so, the more rapid growth on *Prunus* leaves was expected. *Gynaephora* larvae growth rates determine that they slightly more than double in weight each summer. This is quite slow compared to larval *Pieris rapae* L., reared by the same techniques, which doubled every two days (Ryan 1973).

The respiration rates of Scholander et al. and Procter (this volume) are identical when $b = 0.77$ (in $R = aW^b$). The larvae Procter used weighed less than the starting weight of synthetic cohort larvae, but were active throughout the measurement period. Carbon dioxide production measurements at Truelove indicate respiration rates can be 0.1 of the oxygen values cited here.

The standing crop value is based on 3 larvae in 571 m^2 and as such is tentative. The estimated 0.28 mg m^{-2} is high, based on our experiences. Other Lepidoptera were distinctly more abundant, but only one (adult) was collected in an emergence trap. *Gynaephora* were rare in the field, and these larvae would not be expected in emergence traps, whereas other larvae would be. The emergence trap sample contributes heavily to the standing crop. We intended to sample the standing crop extensively with a gasoline driven suction apparatus, but other priorities prevented this.

Production

The survivorship curve starts with 160 hatchling larvae and declines to 5 pharate pupae. The final number in the cohort follows from 2 females being required to produce one batch of fertile eggs, and the equal sex ratio requiring 2 males. Since one pupa in 6 is parasitized by *Spoggosia*, and some die from other causes, a minimum of 5 pharate pupae are required to produce the 4 adults.

Larval mortality in a cohort was considered to be exponential. Morphologically similar animals (larvae), exposed to homogeneous environmental conditions, should have comparable mortality throughout their lives. Exponential mortality, and increase, are the basic assumptions of $e^{r_m t}$ of Andrewartha and Birch (1954), and of other population growth studies (e.g., Manly 1974, Wiegert 1974). Harcourt (1969) reviewed the published data on life tables and cited studies for herbivorous insects. We calculated regression values for each of these studies for exponential survivorship curves, and found (in alphabetical order of the studies he listed) R = 0.96, 0.99, 0.98, 0.99, 0.97, 0.97, 0.99, 0.99, 0.82, and 0.96. The population decreases are highly correlated to exponential declines. The lowest correlation occurred with *Tribolium*

Table 5. Caloric relations in one *Gynaephora* cohort.

Stage	*mg dry wt*	*cal gm^{-1}*	*cal produced*
All larvae	1000.3	5699	5701
All larval exuviae	122.0	5458	666
All cocoons w/o pupae	181.1	5948	1077
Dead pupal remains	160.4	5931	951
All adults	209.0	6032	1261
80 eggs	36.5	6208	226
Total tissue produced:	1709.3		9882

3540 cc O_2 consumed, Total
× 4.7 cal cc^{-1} O_2 (Petruscewicz and MacFadyen 1970)
16540 calories respired, total calories in a cohort, total: 26520 cal

Ratio: tissue produced::total cal 0.37
Ratio: larval growth::larval resp. cal 0.97

confusum, the only laboratory population cited, which was free of predators, parasites, and pathogens. Exponential mortality was assumed in lieu of a life table.

The constructed cohort energy values are summarized in Table 5. These data can be used, in combination with the feeding efficiency data, to calculate total consumption and fecal production in a cohort, the quantity of tissue produced in any stage, and other values. Energy flow per m^2, the end value sought in this study, was determined by comparing the actual standing crop, to the mean predicted standing crop, when one synthetic cohort develops per m^2.

The *Gynaephora* energy flow values may be compared to the net primary production on raised beaches and meadows. Svoboda (this volume), for the raised beach site, reports that the aboveground net vascular plant production, for 1970-72, was 21.6 g m^{-2} yr^{-1}, of which 14.0 grams were woody species. These plants are approximately 4500 cal g^{-1}. The assimilation by *Gynaephora* larvae is 0.007% (4.4/63000) of the net aboveground vascular plant production and 0.02% of the net leaf production. The yearly net aboveground vascular plant production, in the meadows, was approximately 36 g m^{-2} (Muc this volume), which, at 4200 cal g^{-1}, is 150000 cal m^{-2}. *Gynaephora* larvae assimilate 0.0001% of this quantity.

The impact of *Gynaephora* on the aboveground net primary production may be compared to other insect herbivore studies. *Gynaephora* larvae assimilated approximately 0.02% of the net leaf production on raised beaches. At the Kevo I.B.P. tundra site, *Dineura viridiorsata* (Hymenoptera, Tenthredinidae) assimilated 0.3% of the net leaf production of its host *Betula tortuosa* (computed from Haukioja 1973, Haukioja et al 1973). *Strophingia ericae* (Homoptera, Psyllidae) assimilated 0.09% of the net foliage production of its host *Calluna* at the Moor House I.B.P. site (Hodkinson 1973). In a high prairie ecosystem, an acridid population, which produced a mean 0.23 mg dry weight m^{-2} year^{-1} (Riegert and Varley 1973), assimilated approximately 0.5% of the net foliage production (our conversion, assumed NPP of 600 g m^{-2}); these were the dominant insect herbivores of the site. The highest assimilation of net foliage production we located was by *Chrysomela knabi* (Coleoptera, Chrysomelidae), which assimilated 2.6% of *Liriodendron tulipifera* foliage production (Reichle et al. 1973). In the Amazonian tropics, insect herbivory is a comparatively insignificant phenomenon (Fittkau and Klinge 1973).

Insect herbivores in these systems, normally, eat a minor portion of the net primary production. Tropic and arctic plants tend to produce perennial leaves, while temperate plants are mostly deciduous or annual, with thinner leaf walls (MacArthur 1972). We suspect there is a latitudinal pattern of herbivory, with insect herbivores being more significant in temperate regions than in the Tropics and Arctic. Rapidly growing temperate leaves should be more nutritious (abundant enzymes for making the leaves; leaves thinner with thinner cell walls) than in these other areas. This pattern is suggested by the comparisons made, but requires further study.

Conclusions

A generation of *Gynaephora* requires an estimated 10 years to develop on Truelove Lowland. Larvae assimilated 4.4 cal leaf tissue m^{-2} yr^{-1} on raised beaches; which was 0.007% of the net aboveground vascular plant production. This habitat supported the greatest density of larvae. In meadows, *Gynaephora* assimilated 0.0001% of the aboveground vascular plant production yr^{-1}. Vegetation removal by these animals was quite small, and is similar to other reported impact levels of herbivorous insects. *Gynaephora* tissue production was 0.22 mg (=1.15 cal) m^{-2} yr^{-1} on raised beaches, and 0.02 mg (=0.13 cal) m^{-2} yr^{-1} on meadows. Energy respired was 63% of the total energy assimilated.

Determining the steps required to construct the energy budget was a significant part of this study. These were: (1) biological information; (2) weights and caloric values; (3) growth rates; (4) respiration rates; (5) densities; and (6) survivorship rates. Each factor was essential to the energy budget. A survivorship curve was approximated by assuming exponential mortality (log decline) throughout larval life. Following this assumption, the basic values required to estimate production in a synthetic cohort are the initial and final larval weights. Production and growth rates are related to respiration, which is related to temperature.

Acknowledgments

We gratefully acknowledge the interest and help of Dr. Bruce Heming in the preparation of this manuscript. Special appreciation to Dr. Larry Bliss for directing the Devon project, and his personal interest in ensuring this research was funded, executed, and communicated. Dr. Don Pattie supervised the calorimetry. Many have helped, but we shall specify only Marybeth Ryan and D.C. Ferguson. We thank everyone who shared time, energy, or resources with us for this study.

References

Andrewartha, H.G. and L.C. Birch. 1954. The Distribution and Abundance of Animals. Chapman, London. 782 pp.

Bruggemann, P.E. 1958. Insects and Environments of the High Arctic. Proc. X Int. Congr. Entomology. Vol. 1: 695-702. Ottawa.

Chernov, Ju. I. 1975. A review of the trophic groups of invertebrates in typical tundra subzone of western Taimyr. Source: Biogeocenoses of Taimyr tundra and their productivity. Vol II: 167-179. *In:* Intern. Tundra Biome Transl. 12: 17-29. Univ. Alaska, Fairbanks.

Corbet, P.S. and H.V. Danks. 1974. Screen temperatures during the summers 1967 and 1968 at Hazen Camp, Ellesmere Island, N.W.T. Defense Res. Board. Operation Hazen 44. Ottawa. 13 pp.

Curtis, J. 1835. Insects, p. 59-80. *In:* Narrative of a Second Voyage in Search of a North-West Passage, and of a Residence in the Arctic Regions During the Years 1829-1833. Webster, London. 120 pp. plus Appendices.
Downes, J.A. 1962. What is an arctic insect? Can. Ent. 94: 143-162.
———. 1964. Arctic insects and their environment. Can. Ent. 96: 279-307.
———. 1965. Adaptations of arctic insects to their environments. Ann. Rev. Ent. 10: 257-274.
Fittkau, E.J. and H. Klinge. 1973. On biomass and trophic structure of the central Amazonian rain forest ecosystem. Biotropica 5: 1-14.
Gibson, A. 1922. The Lepidoptera collected by the Canadian Arctic Expedition, 1913-18. *In:* Report of the Canadian Arctic Expedition 1913-18. Vol. 3. King's Printer, Ottawa.
Harcourt, D.G. 1969. The development and use of life tables in the study of natural insect populations. Ann. Rev. Ent. 14: 175-196.
Haukioja, E. 1973. Weight development, consumption, and egestion of *Dineura virididorsata* (Hym., Tenthredinidae) larvae. Rep. Kevo Subarctic Res. Stat. 10: 9-13.
———., S. Koponen, and H. Ojala. 1973. Local differences in birch leaf consumption by invertebrates in northern Norway and Finland. Rep. Kevo Subarctic Res. Stat. 10: 29-33.
Hodkinson, I.D. 1973. The population dynamics and host plant interactions of *Strophingia ericae* (Curt.)(Homoptera: Psylloidea). J. Anim. Ecol. 42: 565-583.
Kozhanchikov, I.V. 1950. Orgyids of the U.S.S.R. Zool. Inst. Akad. Nauk SSSR, Moscow (N.S.) 42. Insects—Lepidoptera 12. 582 pp. 296 fig.
MacArthur, R.H. 1972. Geographical ecology. Patterns in the distribution of species. Harper and Row. San Francisco. 269 pp.
MacLean, S.F. 1975. Ecological adaptations of tundra invertebrates. pp. 269-300. *In:* Physiological Adaptation to the Environment. F.J. Vernberg (ed.). Intext Ed. Pub. New York. 576 pp.
Manly, B.J.F. 1974. Estimation of stage-specific survival rates and other parameters for insect populations developing through several stages. Oecologia 15: 277-286.
Ohyama, Y. and E. Asahina. 1971. Frost resistance in an alpine moth, *Byrdia rossi* (Lepidoptera: Lymantriidae) Low Temp. Sci. B. Vol 29: 121-123.
Oliver, D.R., P.S. Corbet, and J.A. Downes. 1964. Studies on arctic insects: the Lake Hazen project. Can. Ent. 96: 138-139.
Petruscewicz, K. and A. MacFadyen. 1970. Productivity of Terrestrial Animals. I.B.P. Handbook No. 13. Blackwells, Oxford. 190 pp.
Reichle, D.E. 1968. Relation of body size to food intake, oxygen consumption, and trace element metabolism in forest floor arthropods. Ecology 49: 538-542.
———., R.A. Goldstein, R.I. Van Hook, Jr., and G.J. Dodson. 1973. Analysis of insect consumption in a forest canopy. Ecology No. 54: 1076-1084.
Riegert, P.W. and J.L. Varley. 1973. Population dynamics and biomass production of grasshoppers. Tech. Rep. No. 16. Above ground invertebrates. I.B.P. Matador grasslands report. Saskatoon, Saskatchewan. 134 pp.
Ryan, J.K. 1973. A quantitative feeding study of *Pieris rapae* L. (Lepidoptera: Pieridae) on *Brassica oleracea* L. Proc. XXI Ent. Soc. Alberta (abst. only).
Scholander, P.F., W.F. Flagg, V. Walters and L. Irving. 1953. Climatic adaptation in arctic and tropical poikilotherms. Physiol. Zooel. 26: 67-92.

Schroeder, L.A. 1972. Energy budget of cecropia moths, *Platysamia cecropia* (Lepidoptera: Saturniidae), fed lilac leaves. Ann. Ent. Soc. Amer. 65: 367-372.

———. 1973. Energy budget of larvae of the moth *Pachysphinx modesta*. Oikos 24: 278-281.

SooHoo, C.F. and G. Fraenkel. 1966. The consumption, digestion, and utilization of food plants by the polyphagous insect *Prodenia eridania* (Cramer). J. Insect Physiol. 12: 711-730.

Spinage, C.A. 1972. African ungulate life tables. Ecology 53: 645-652.

von Bertelanffy, L. 1957. Quantitative laws of metabolism and growth. Quart. Rev. Biol. 32: 217-232.

Waldbauer, G.P. 1968. The consumption and utilization of food by insects. Adv. Ins. Physiol. 5: 229-288.

Wiegert, R.G. 1974. Competition: a theory based on realistic, general equations of population growth. Science 185: 539-542.

Zeuthen, E. 1953. Oxygen uptake as related to body size in organisms. Quart. Rev. Biol. 28: 1-12.

Vertebrate consumers

Population levels and bioenergetics of arctic birds on Truelove Lowland

Donald L. Pattie

Introduction

Information about populations of Canadian high arctic avifauna is largely limited to either intensive studies of one or two species over an extended period (Maher 1970a, Hussell 1972) or seasonal studies of discrete sites or assemblages of birds (Harrington n.d., Sutton and Parmelee 1956, Nicol 1961, Savile 1961, Nettleship 1967, Parmelee 1968). This study succeeded the 1966-69 studies of Hussell and Holroyd (1974) in the same area and was designed primarily to determine densities and population fluctuations of all bird species of a reasonably large area (43 km^2) over four subsequent summer seasons. A secondary goal was the acquisition of bioenergetic and reproductive data from snow buntings *(Plectrophenax nivalis)*. Population levels of all species breeding or summering on this high arctic tundra were needed as a basis for calculating energy flow through the ecosystem (Whitfield this volume). Bioenergetic studies of snow buntings similar to those West (1968) conducted on ptarmigan were carried out to make energy flow estimates more precise. Most of the study was restricted to Truelove Lowland.

Devon Island (74° N, 88° W) in the Canadian Arctic Archipelago is mostly a high plateau Polar Desert surmounted by an icecap that reaches an elevation of over 2000 m. Tundra meadows and Polar Semi-desert areas are restricted to coastal lowlands and fiords or glacial valleys. Truelove Lowland (75° N 84° 40W) is one of five adjacent lowlands along the northeastern coast and is bordered on the north and west by Jones Sound and on the south by fiord-like Truelove Inlet. Cliffs or precipitous slopes 300 m high isolate all five lowlands from the inland plateau, and provide nesting sites for Glaucous gulls *(Larus hyperboreus)*, some snow geese *(Chen caerulescens)*, and peregrine falcons *(Falco peregrinus)*. Plant growth appears more lush and bird densities are higher on Truelove Lowland than on the other four lowlands.

Eleven of the fifteen avian species which breed regularly on Truelove Lowland are in the orders Charadriiformes and Anseriformes. These species as well as the red-throated loon *(Gavia stellata)* are either shorebirds or waterbirds. Abundant surface water is present during summer; three large lakes, numerous smaller lakes and ponds cover 10 km^2 of the 43 km^2 Lowland. Red-throated loons, arctic terns *(Sterna paradisaea)*, oldsquaw *(Clangula hyemalis)*, common eider *(Somateria mollissima)*, and king eider *(Somateria spectabilis)* nest by these water bodies. Several streams with shallow channels dissect the Lowland, providing riparian habitat, and in places spilling out into hummocky sedge-moss meadows. Many of these meadows have a higher moss than Angiosperm cover and are favored feeding grounds for insectivorous birds. Jaegers, both parasitic *(Stercorarius parasiticus)* and long-tailed *(S. longicaudus)*, and white-rumped sandpipers *(Erolia fusicollis)* nest there. Granite outcrops a few meters high form a number of ridges along the northern portion of the Lowland and in the Truelove River Valley. Rock ptarmigan *(Lagopus mutus)*, the only resident herbivorous bird, nests on or near these outcrops.

The two passeriform species, Lapland longspurs *(Calcarius lapponicus)* and snow buntings have distinctly different nesting preferences. The longspurs establish their territories and nest in open cups among frost-boil sedge-moss meadows, which are widely distributed on the Lowland, making up 18.4% of the total area. The buntings, conversely, establish their territories in dryer regions incorporating scree, rocks, or sometimes fractured outcrops. Piles of rocks are frequently associated with raised beach ridges and these, together with the talus along the cliffs and slopes on the east side of the Lowland, provide the cavities favored for bunting nests. The dry gravel tops of the raised beaches serve for nesting sites of black-bellied plovers *(Squatarola squatarola)* and Baird's sandpipers *(Erolia bairdii)*.

Although few bird species breed at these high latitudes (Enemar 1963, Williamson et al. 1966, Savile 1972, Moksnes 1973, Nettleship and Maher 1973, Nettleship 1974, Hussell and Holroyd 1974), no study of terrestrial arctic ecosystems would be complete without their inclusion. Different bird species fill the niche of primary consumers, first level carnivores, top carnivores, and scavengers. This is of particular interest when the limited number of vetebrate species is taken into account. An indication of the sparse number of bird species is derived from Hussell and Holroyd (1974) who recorded 38 species for the north coast of Devon Island; Kaiser et al. (1972) estimate only a combined total of 32 species of birds and mammals in a 322×322 km quadrant including eastern Devon Island. In an equivalent 322×322 km quadrant on Vancouver Island, British Columbia, 267 species are reported. As most insular-nesting arctic bird species are migratory, they provide the sole means of exporting energy fixed by terrestrial arctic autotrophs to lower latitudes. Conversely, they may transport energy and nutrients from more southern regions into the Arctic.

This report covers population levels of Lowland species and the summer ecology of the snow bunting, discusses the niche complexity present in a system supporting relatively few species and, by providing weights and biomass of birds, allows calculation of bird impact on the tundra ecosystem.

Methods

Population levels

A census of all bird species was taken by walking adjacent 100 m wide transects spiraling inward from the outer limits of the Lowland. Two, three, or four observers walking abreast recorded numbers of each species. The person on the innermost transect during the first round recorded landmarks, then took the outermost transect on the subsequent round, with the other observers spacing themselves so that overlapping or duplicate counting was eliminated. Each complete census of the Lowland involved walking 215 km distributed between the observers. Generally the census extended from 0800 to 2400 hr but during the June census the first circle of the Lowland usually took 26 or 27 hours and in those instances was performed around the clock. Because these early counts were performed on snowshoes over a snow-covered landscape even nesting birds could be easily seen. With only 1 or 2 observers and with inclement weather interrupting the count in 1970, the first census took 11 days between 22 July and 6 August. In 1971 only 6 days of counting were needed by 3 observers between 2 and 10 August. Around-the-clock counting by 4 people from 20 to 24 June, 1972 completed the first of 3 complete censuses that year. The other two were taken between 16 and 21 July, and 9 and 15 August. Only 2 observers covered the area between July 21 and 29 in 1973. Weekly census counts were taken along a 5 km route in 1972 to test the reliability of the methods. The results of the counts, together with the data obtained from bioenergetic studies, were used to estimate the proportions of primary production harvested by the birds and to assess their importance in this arctic ecosystem.

All species are not visible to the same extent during census taking, even during comparable times of day or during favorable weather. Colquhoun (1940) described this as the different "conspicuousness" of the species; Enemar et al. (1973) discussed the concept of "species effectivity" which he described in 1959 and which shows, as a percentage, the chance of an isolated bird being observed during a survey. Emlen (1971) developed a "Coefficient of Detectibility" (CD) which made estimates of numbers of each species possible without the repetition of counts required by the first two techniques. Emlen's technique, based upon distances birds flush perpendicular to an observed line of march, a modification of a census, provides an estimate of the percentage of the birds occupying an area which are actually seen by observers. The large size of the Lowland made repetitive counts prohibitive so Emlen's technique was used. The method was generally satisfactory but CD values for many species were different when the land was snow-covered than when snow-free. For this reason and because future observers may wish to compare sightings instead of estimates, actual census counts are given in Table 1. The CD values in Table 2 are for snow-free census periods where sample sizes are generally larger. As each census took several days, wide-ranging birds may occasionally have been counted more than once, especially in 1971. In subsequent seasons increased numbers of observers reduced the number of census days so chances of multiple observations were diminished. To give an idea of the distances involved, 1313 km of census walks in 1972 provided the observations on which population estimates for that year are based.

Banding and color-marking of 64 snow buntings was carried out during 1972. Regular counts and surveys on the Lowland gave some indication of their population dynamics.

Bioenergetics

I attempted to quantify the energy requirements of snow buntings as nestlings and adults. No attempt was made to duplicate Lasiewski and Dawson's (1967) "Standard Metabolic Rate," measurements with postabsorptive birds at rest in thermoneutral surroundings. Instead, food uptake and fecal production were monitored in birds maintained in a small cage where extensive flying was impossible, at temperatures similar to those found in their normal summer environs. The energetic estimates were based upon the mean energy used under these conditions over 3-4 consecutive days with feed available. The figures which resulted are existence requisites rather than "standard metabolic rates," but were more useful for estimating the minimum energy needed by an active bird.

Individually marked snow bunting nestlings were weighed daily in 1970, 1971, and 1972 to determine growth rates. Eggs were collected at various days of incubation for calorimetric analyses as were some nestlings and fledglings. Daily weight records of developing nestlings provided information on growth rates. All material collected for calorimetry was oven dried at 90° C for at least 24 hr before being ground in a Wiley Mill and burned in a Parr 1411A semi-micro oxygen bomb calorimeter. Aliquot samples for calorimetry were prepared from species weighing more than 500 g by grinding the whole carcass several times in a Toledo food chopper powered by a three horsepower motor and equipped with a cutting plate having holes 9.5 mm in diameter. Smaller birds were either dried whole after being opened or dispersed whole in a blender and then dried. Homogeneous samples were then prepared using techniques described by Brisbin and Widdowson (1968). Duplicate, triplicate, or even quadruplicate calorimetric replicates of material from one bird were conducted until two runs gave differences of no more than 50 cal g^{-1} dry weight; this meant less than 2% disparity in the results.

Estimated energy requisites for rearing snow buntings to fledging were developed by rearing 2 two-day old nestlings to fledging on a diet of native insects. Growth rates and daily loss of egg weight during incubation were based upon daily field weighing of eggs and young. By back calculating from hatching date of first young, the date when incubation began was determined. Throughout the incubation and nestling period over 50 eggs and more than 50 young buntings were field weighed daily to the nearest milligram with a Torbal torsion balance. This balance was completely satisfactory except during high winds. Nearly every nest found was watched; since nests were widely scattered it was not unusual to trudge well over 10 km on each round of observations.

Information about energy requisites of adults was obtained from five caged adult buntings maintained on a diet of Hartz Mountain wild bird seed throughout the fall and winter of 1971. All food and feces were weighed to the nearest mg then analyzed using the techniques of Boag and Kiceniuk (1968) and West (1968). Comparable feeding projects in 1972 used two adult geese *(Chen caerulescens)*. In 1973 a recently fledged raven *(Corvus corax)* captured on the Lowland, was maintained for a week to study food consumption.

Estimates of daily energy requisites were checked by determining O_2 consumption of the five adult buntings with a Pauling oxygen analyzer as described by Depocas and Hart (1957) at temperatures comparable to the July mean of 3.7°C on Truelove Lowland and also by using the formula presented by Lasiewski and Dawson (1967).

The amount of energy needed to hatch bunting eggs was determined by placing a 3 mil thermocouple in an egg in an incubated clutch and connecting it to a Rus-trak recorder. Because of low ambient temperatures it was convenient to use an icewater bath in a thermos to maintain the reference thermocouple at 0° C. This apparatus provided a record of incubation rhythms, temperatures, and attentiveness. On occasion a stepping switch was incorporated into the circuit to allow monitoring of brood chamber temperature, ambient temperature, and heat flux through the nest.

Percentages of protein and fat, as well as the relative abundance of magnesium, potassium, calcium, phosphorus, iron, manganese, copper, and zinc in snow buntings and other birds, and the insects, feed, and feces of captive birds were determined by the Provincial Feed Testing Laboratory in Edmonton, Alberta. The results obtained were comparable to those obtained from temperate passerine species by Grimshaw et al. (1958) and Bilby and Widdowson (1971).

Few specimens of the less abundant bird species were collected because of their limited numbers. Instead, when possible, a pair was collected for calorimetric analysis. Data from the collected birds were combined with weight data available from specimens in the National Museum collections from the Canadian Arctic Islands to provide weight estimates of the species. Small populations seem to be the rule for many arctic bird species. Frequently local populations and even species have been brought close to extinction by man's activities (Tikhomirov 1970). Since an annual reduction in the total bird population of Truelove Lowland was recorded from 1970 through 1972 our studies became increasingly non-destructive in nature to prevent depletion of what appeared to be a declining and vulnerable population. It is probable this was unnecessary for in 1973 outstanding breeding success followed an early snowmelt and post breeding census figures reached a high for most species.

Results

Avifauna

Table 1 lists only species seen during the complete lowland census. Besides these birds, lesser scaup *(Aythya affinis)*, gyrfalcon *(Falco rusticolus)*, peregrine

Table 1. Numbers of birds counted on Truelove Lowland in the 1970-73 census.

	1970	*1971*		*1972*			*1973*
Species	*22 July- 6 Aug.*	*21-24 June*	*2-10 Aug.*	*20-24 June*	*16-21 July*	*3-9 Aug.*	*21-29 July*
Regular breeders							
Red-throated loon *Gavia stellata*	55	56	117	0	54	88	72
Snow geese *Chen caerulescens*	25	45	22	4	0	0	26
Oldsquaw *Clangula hyemalis*	172	156	195	0	241	139	298
Common eider *Somateria mollissima*	12	20	64	47	106	2	64
King eider *Somateria spectabilis*	51	44	18	2	23	2	45
Rock ptarmigan *Lagopus mutus*	3	1	10	0	0	0	2
Black-bellied plover *Squatarola squatarola*	18	10	5	6	32	11	25
White-rumped sandpiper *Erolia fusicollis*	1	0	10	0	26	11	14
Baird's sandpiper *Erolia bairdii*	38	89	91	2	9	25	25
Parasitic jaeger *Stercorarius parasiticus*	24	7	8	0	6	2	12
Long-tailed jaeger *Stercorarius longicaudus*	88	33	13	0	21	19	62
Glaucous gull *Larus hyperboreus*	57	34	61	0	29	52	40
Arctic tern *Sterna paradisaea*	57	42	97	0	56	3	105
Lapland longspur *Calcarius lapponicus*	82	153	74	0	37	55	50
Snow bunting *Plectrophenax nivalis*	339	361	268	223	144	266	603
Occasional breeders and visitors							
Brant (V) *Branta bernicla*	0		0		4		0
Ringed plover (OB) *Charadrius hiaticula*	0		4		0		0
Ruddy turnstone (OB) *Arenaria interpres*	0		18		54		14
Knot (OB) *Calidris canutus*	0		1		16		19
Purple sandpiper (V) *Erolia maritima*	5		9		24		7
Sanderling (V) *Crocethia alba*	0		0		0		1
Red phalarope (OB) *Phalaropus fulicarius*	6		7		16		41
Sabines gull (V) *Xema sabini*	0		0		1		0
Snowy owl (V) *Nyctea scandiaca*	0		0		2		0
Common raven (V) *Corvus corax*	0		5		2		0

(OB) — Occasional breeder
(V) — Visitor

falcon *(Falco peregrinus)*, sandhill crane *(Grus canadensis)*, golden plover *(Pluvialis dominica)*, dunlin *(Erolia alpina)*, Thayer's gull *(Larus thayeri)*, ivory gull *(Pagophila eburnea)*, black guillemot *(Cepphus grylle)*, horned larks *(Eremophila alpestris)* and wheatear *(Oenanthe oenanthe)* were seen either on or near the Lowland in 1970-72. Only the peregrine falcons which nested each year on cliffs near the Lowland, and Thayer's and ivory gulls were seen to visit the Lowland annually outside the census periods. Gyrfalcon and black guillemots were seen during two seasons, the other species only once. Thus over the four years 15 species nested regularly on or near the Lowland, another four occasionally, 15 species were considered visitors, and a total of 36 species were seen. Additional species reported for the area by Hussell and Holroyd (1974), include northern fulmar *(Fulmarus glacialis)*, pectoral sandpiper *(Erolia melanotus)*, buff-breasted sandpiper *(Tryngites subruficollis)*, black-legged kittiwake *(Rissa tridactyla)* and thick-billed murre *(Uria lomvia)*. This raises the total number to 41, considerably higher than the 32 species of birds plus mammals Kaiser et al. (1972) predicted.

Birds were observed to occupy quite specific trophic levels. Primary consumers, snow geese and ptarmigan, had a biomass of 1 kg km^{-2}; secondary consumers, the plovers, buntings, longspurs, sandpipers, and ducks, combined to provide 8.2 kg km^{-2}; tertiary consumers, the jaegers, terns, and loons, amounted to 4.9 kg km^{-2}; and finally the quaternary consumers, the snowy owls and peregrine falcons, contributed 0.04 kg km^{-2}. In addition scavenging gulls and ravens amounted to 2.4 kg km^{-2}, giving the entire Lowland 16.5 kg km^{-2}. Fay and Code (1959) studying the avifauna of St. Laurence Island, Alaska found 386 to 3,861 kg km^{-2} along the coast 0.386-386 kg km^{-2} over half the island and 0 to 0.3 kg km^{-2} on 30% of the island. Our results at this more

Table 2. Coefficient of detectibility for Devon Island birds in August, estimated mean post-breeding densities and extremes for 1970-73 plus mean weights of Arctic island birds.

Species	*Summer (C.D.)*	*Mean post-breeding density km^{-2} and range*	*(N)*	*Fresh[3] weight (g)*
Red-throated loon	0.84	2.3 (1.5 - 3.2)	32	1704
Snow goose[1]	—	0.4 (0.0 - 0.6)	20	2443
Oldsquaw	0.95	4.9 (3.4 - 7.3)	31	732
Common eider	0.90	0.9 (0.1 - 1.7)	3	1697
King eider	0.51	1.3 (0.1 - 2.3)	50	1574
Rock ptarmigan[1]	—	0.1 (0.0 - 0.2)	93	526
Black-bellied plover	0.88	0.4 (0.1 - 0.7)	7	206
White-rumped sandpiper	0.40	0.5 (0.1 - 0.8)	19	38
Baird's sandpiper	0.88	1.0 (0.7 - 2.4)	26	42
Parasitic jaeger	0.88	0.3 (0.1 - 0.6)	19	485
Long-tailed jaeger	0.59	1.8 (0.5 - 3.5)	64	319
Glaucous gull	0.80	1.5 (1.2 - 1.8)	50	1617
Arctic tern	0.55	2.8 (0.1 - 4.4)	18	106
Lapland longspur[2]	0.24	8.2 (6.3 - 10.3)	158	28
Snow bunting[2]	0.41	27.2 (19.7 - 44.6)	107	37
Raven	—	0.0 (0.0 - 0.1)	2	1050

Mean post-breeding density of all species 53.6 km^{-2}

1. Since C.D. values could not be calculated for these species, raw numbers counted were used to calculate density.
2. Densities were based upon the 33 km^2 of land area for the passerines. Other densities were calculated on the basis of the entire 43 km^2 lowland with its lakes.
3. All weights given are fresh weight: conversion to dry weight may be achieved since adults of those species analyzed were 65% water.

Table 3. Total bird densities km^{-2} based upon C.D. values for Ellesmere and Devon Island, July 1972 arranged from north to south, plus some comparative 1970 and 1973 estimates from Truelove Lowland.

Site	*Date*	*Habitat*	*Distance km census*	*Density Birds km^{-2}*	*Number of species*
Ellesmere Island					
Otto Fiord 81° 15'N 85°55'W	23 July 1972	glacial outwash plain	8.0	5	5
Marsh area near Sawtooth Mtns. 79° 59'N 82° 32'W	20 July 1972	marshy valley	1.6	49	3
Wolf Valley 79°40'N 82°20'W	21 July 1972	plateau & down river valley	35.0	27	8
Sildre Fiord 79° 1'N 83° 17'W	24 July 1972	moist meadows	19.2	40	12
Hoved Island 77°30'N 85°W	23 July 1972	moist meadow	1.6	38	6
Bauman Fiord 77°23'N 85°37'W	22 July 1972	river bank & gravel beach of sea	19.2	5	8
Devon Island					
Truelove Lowland	22 July-6 Aug. 1971	lowland	215.0	46	17
	20-24 June 1972		215.0	8	6
	3-9 July 1972		215.0	30	13
	16-21 July 1972		215.0	30	21
	2-10 Aug. 1972		215.0	43	21
	21-29 July 1973		215.0	65	20
Truelove River Valley	15 Aug. 1972		33.0	16	6
Plateau above Truelove Lowland	10-11 Aug. 1970	polar desert	54.0	9	1*
	17 Aug. 1972		50.0	0	0

* All were recently fledged snow buntings.

northern Lowland fit into the range they found over most of the inland portion of St. Laurence Island. Mean fresh weights for Lowland species derived from our study plus birds collected in the Queen Elizabeth Islands and housed in the National Museum in Ottawa appear in Table 2.

A combined coefficient of detectibility of 0.84 was calculated for all species of Truelove Lowland birds during census when snow covered most of the ground. When melting was complete many species became less visible and the CD values declined. This is correlated with phenology for snow buntings in Fig. 1. Use of the CD values allowed estimates of species density in several arctic sites and comparison of these figures with other data (Table 3). Inadequate numbers of snow geese and rock ptarmigan were seen to provide a CD value for these species. In calculations two raw counts of these species were used.

Population levels

Based upon the actual counts from the four summers (Table 1) and on population estimates using Emlen's coefficient of detectibility it is apparent Truelove Lowland supports not only few species but also a low density of all species of birds (Table 2). Population levels of passerine birds were low compared with other arctic sites. In Norway, alpine passerines had 34-103 territories km^2 (Lien et al. 1970); each territory was assumed to support a pair of birds. In the Cape Thompson, Alaska region, densities of Lapland longspurs alone ranged from 34 to 160 pair km^2 (Williamson et al. 1966), while at Barrow, Alaska, Lapland longspur territories have numbered 250-400 km^2

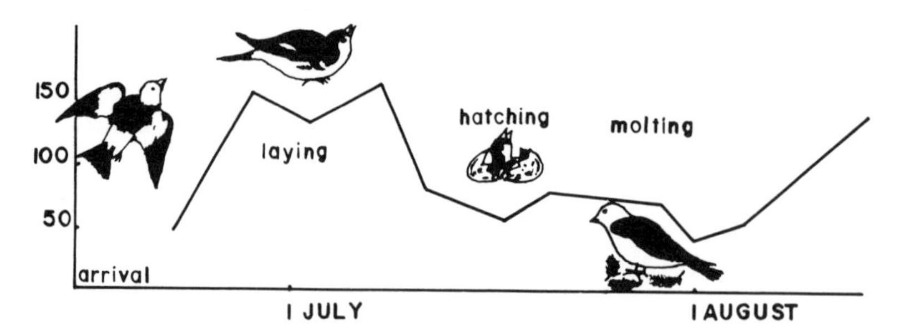

Fig. 1 Seasonal changes in the actual counts of snow buntings *(Plectrophenax nivalis)* on Truelove Lowland related to phenology for the species in 1972. The higher numbers in late June and early July represent migrants accumulating on the north coast of the island. Close attention to the nest during hatching and the secretive habits of adults during molting provide dips in mid-july and early August. Recruitment of fledglings swelled the count into August.

(Pitelka 1971). In the past few years numbers of longspur territories at Barrow have unaccountably declined to 12-25 km^2 (Custer and Pitelka 1972), and Hussell and Holroyd (1974) reported an 80% decline in longspurs on Truelove Lowland between 1967 and 1968. Truelove Lowland populations of snow buntings declined along with most species through 1972. Post-fledgling end of the season estimates of 826, 653, 316 and 1470 buntings in 1970-71-72-73 were derived from a "coefficient of detectibility" of 0.41 (Table 2). Passerine populations were based on the 33 km^2 land area while other species densities were based upon the 43 km^2 area of land plus fresh water. As in other high arctic islands, snow buntings were the most numerous passerine (MacDonald 1954, Bruggemann 1958, Geale 1971) in contrast to Alaska where Lapland longspurs were more numerous (Pitelka 1971). The 1973 post-fledgling snow bunting population estimate soared to 1470. We estimated 46% recruitment of young buntings into the population in 1972, the only year sufficient data were collected to estimate recruitment. The 1972 season was cooler and shorter than the other three. The data suggest much more recruitment in 1973.

Numbers of snow buntings counted per census period during 1972 are indicated in Fig. 1. It is assumed the slopes of the graph are valid for most years. Seasonal changes in CD values may be suggested by the snow bunting graph. Evidence that northward migrating buntings accumulate on the north coast of Devon Island before crossing Jones Sound to Ellesmere came from sightings of snow buntings we had netted and color-marked earlier. Marked birds were seen along the coast and later at Grise Fiord on Ellesmere Island. These build-ups in a region prior to moving on are probably a regular feature of migration, for Nettleship and Maher (1973) reported comparable bunting movements through the Hazen Lake region of Ellesmere Island (81° 49°N, 71° 18°W) on 7 June, 1966. Snow bunting density in early 1972 was estimated at 351 breeding birds or 10.6 km^{-2} of land area based upon a 41% coefficient of detectibility; the estimate increased to 648 or 20 km^{-2} by the end of this short, cold, late season for a recruitment rate of 46%.

Longspur density in early 1972 was estimated at 154 breeding birds or 4.7 km^{-2} of land area based upon a 24% coefficient of detectibility (Table 2); the estimate increased to 229 or 6.9 km^{-2} by the end of the season for 48% recruitment. Longspurs also presented an early season surge in numbers as migrants swept through. Over the entire terrestrial portion of the Lowland then, the total 1972 breeding passerine population was estimated at 15.3 km^{-2}.

Since the passerines derive little of their sustenance from the 10 km^2 of large lakes on the Lowland, it may be desirable for comparing Truelove

Lowland to other sites to provide an ecological density figure for the Lapland longspurs and snow buntings. Mean estimate longspur densities for the 33 km^2 land area of the Lowland at the end of the breeding season were by year: 1970—10.3; 1971—9.3; 1972—6.9 and 1973—6.3 km^{-2}; the cumulative mean of 8.2 for the four years probably approaches the low numbers which followed the high of 1966-67 (Hussell and Holroyd 1974). Mean snow buntings density estimates for the same periods were: 1970—25; 1971—20; 1972—20 and 1973—43 km^{-2}; cumulative mean 27.5 for the four years.

The two species of jaegers provided a problem. Normally considered vertebrate predators, there was seldom sufficient vertebrate prey on the Lowland to support the numbers present. Since they arrived long before any openings appeared in the sea ice, jaegers must have existed partly on stored reserves during much of the season. Maher (pers. comm.) noted they are efficient scavengers and probably get insect larvae and pupae early in the season. They spend a lot of time poking into plant clumps and pulling moss presumably searching for such prey. The large numbers of long-tailed jaegers *(Stercorarius longicaudus)* seen in 1970 and 1973 are not resident. Twice a flock of long-tailed jaegers arrived in the last ten days of July (69 in 1970, 59 in 1971). Hussell and Holroyd (1974) reported large flocks in three of the four years of their study in the area. During their stay the flocks concentrated their activity along the north and west borders of the largest lake (Phalarope). Normally predators of lemmings, they were twice observed feeding there on chironomid flies, but they were not restricted to feeding only along lake borders for they travelled widely as a group during their stay. The stomach of a non-breeding female long-tailed jaeger collected on 22 July, 1972 contained 44 Acarina mites, two adult and one immature (Erigonidae) spiders, 38 adult and four larval chironomids, one adult sciarid, four adult and one larval muscid, five pieces of moss, 18 twigs, and three leaves as large as an adult erigonid spider; feather fragments and a few bones from one lemming completed the count. Danks (1971) reported utilization of arthropods early in the season; Maher (1970b) has also reported consumption of chironomids but feels they are most important to young jaegers. Our data and that of Hussell and Holroyd (1974) indicate considerable summer importance to territorial adults as well. The 62 birds counted in 1973 were widely dispersed over the Lowland. This might have resulted from lemmings *(Dicrostonyx torquatus)* being more abundant in 1973 than in any other year of our study (see Fuller et al. this volume). This was the only year in which at least one pair of long-tailed jaegers was successful in raising a chick to fledging on the Lowland. In 1971 4 pairs established territories and began nesting. Each pair laid 2 eggs, but then lost their nests to foxes. In 1970 and 1972 territories were established but no eggs were laid.

Each season parasitic jaegers *(Stercorarius parasiticus)* arrived in June. Two pairs occupied the same territories 3 successive years. In 1970 2 pairs nested, 3 nested in 1971, 2 pairs in 1972, and at least 1 in 1973. In 1971 at least 1 chick was raised to near fledging and another in 1973. Generally the birds abandoned territories after foxes destroyed the nests and had left the Lowland by 10 August. The density of parasitic jaegers in the Canadian Arctic must be much higher than in Finland, where Hilden (1971) reported only about 225 pairs breed each year. Parasitic jaegers feed mainly on waterfowl eggs and small birds (Hussell and Holroyd 1974) but they too were observed feeding occasionally on chironomid flies. One territorial parasitic jaeger male contained 278 chironomid adults, larvae and pupae, 5 muscids, 2 empedidae, 51 collembolans, one seed, and one plant fragment. Contrary to the situation Maher (1970a) reported for pomarine jaegers where only non-breeding birds had insects as the majority of stomach contents, this parasitic jaeger was the breeding male which had a half-grown chick running about indicating invertebrates were significant in the summer diet.

Rock ptarmigan were never seen in abundance on the Lowland, but in September 1971 a visiting native hunter killed a flock of 13 and we were given the crops. From this sample, qualitative analysis of diet and the energetic value of various components were determined. *Salix*, 67% of the dry wt and 5343 cal g^{-1} dry wt, was the single most important dietary component as it is in other ptarmigan (West 1968). *Saxifraga oppositifolia*, 8% of the diet and 5550 cal gm^{-1} was next, then *Draba* 2% and 5052 cal g^{-1}. Five other species made up a combined 4% and unidentifiable debris and plant material comprised 19% of the crop contents. Combined crop samples had a caloric value of 5334 cal g^{-1} dry wt.

All bird species suffered nest destruction from predation, or adverse weather, or research activity such as the collection of eggs and nestlings. Predators destroyed 25% of all located nests in 1970, 69% of the 1971 nests, and 63% of the 1972 nests. In 1972 bunting nests were guarded by piling rocks around nest entrances. Arctic foxes attempted predation of protected nests but they were unable to reach the eggs or young. Had these guarded nests been taken the combined 1972 predation figures for all species would have been 70%. Arctic foxes accounted for 42%, parasitic jaegers for 35%, inclement weather or flooding accounted for 12%, and research activity was implicated in the destruction or abandoning of 11% of the nests. In 1970 and 1971 long-tailed jaegers and weasels destroyed nests, but in 1972 no evidence of nest predation by these species was recorded. Nest destruction by inclement weather was usually associated with early season melt-water flooding of nest sites, but mid- and late-season mortality was also associated with weather phenomena. In 1971 a violent Föhn wind, blasting waves across a number of low islands in a small lake near the sea by Rocky Point, caused 14 common eiders to abandon their nests. It took the parasitic jaeger population three days to consume the 7000 g of food represented by 78 eggs thus exposed. No gulls or long-tailed jaegers were seen to take the exposed eggs.

Seasonal number profiles of different species show two distinct patterns. Small passerines become less visible during egg laying, incubation, and molting, whereas larger species such as ducks, geese, gulls, and loons are not so easily overlooked. The difference in visibility results in different patterns of population estimates throughout the season, easily compared in Figs. 1 and 2. The seasonal fluctuation of bunting numbers contrasts sharply with the steady oldsquaw population estimates. Oldsquaw numbers were estimated at 164 at the beginning of the 1971 season and 205 at the end, for a recruitment rate of only 25% that year. This amounts to 4.5 and 4.8 km^{-2}. In addition, oldsquaw estimates remained remarkably constant (Fig. 2) from one census period to the next. The increased numbers of oldsquaws in August is caused by the arrival of migrants. Local oldsquaws join the migration by the end of the first week in August and numbers decline (Fig. 2). Since the water birds depend upon the fresh water as well as terrestrial ecosystems for requisites, densities are calculated on number 43 km^2 which includes these aquatic resources.

Phenology of snow buntings

Buntings are present for approximately five months of the year. Flocks of adult males arrive in May. Numbers swell as migrants continue to arrive until late June or early July, then decline slightly when some northbound

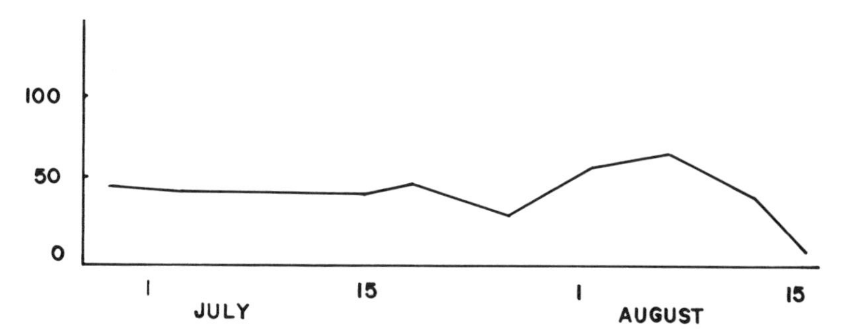

Fig. 2 Seasonal counts of oldsquaw ducks *(Clangula hyemalis)* on Truelove Lowland during 1972. The dip in the graph near the end of July coincided with hatching. Arrival of flocks of migrants swelled the early August counts. Departure of migrants and non-breeding birds left few in the area after mid-August.

populations cross Jones Sound to Ellesmere Island. In early June females begin arriving. Males and females remain in mixed flocks until 10 June when males begin to establish territories. Pair formation continues for 10 days to 2 weeks while nonterritorial birds remain in flocks of 4 to 40. By late June the bunting flocks disperse, territories and pair bonds are established, and nesting begins.

Egg laying ordinarily begins about 20 June, but if snowmelt is accelerated it may begin earlier; conversely, the late melt of 1972 delayed laying until the end of June. Average clutches in 1970 and 1971, like those noted by Hussell (1972), were between 4.9 and 5.6 with a cumulative mean of 5.3. The mean clutch of 23 nests declined to 3.6 in 1972 coincidental with the late melt. If the first clutch is lost to predators or abandoned because of disturbance the female may renest until 7 July. Subsequent clutches never contain more than four eggs, frequently only two or three. There were no indications of double broods nor was the season long enough to accommodate two broods of any species. The incubation period is 12 days for the last egg laid. The nestling period is 12 days but disturbance may cause young to scatter under the rocks at 10 days and a couple of broods stayed up to 14 days in the nests. Males feed females on the nest and both parents feed the young. The male does the bulk of the feeding for the first 2-3 days of the nestling period.

As females begin incubation they become less conspicuous and the population estimate declines until hatching when the females begin to gather food. As the young fledge the adult birds begin to molt and become more secretive, hiding among rocks rather than flying. Adult secretiveness results in a bigger fraction of fledglings in the August count. By mid-August 'Zugunruhe' begins; flocks of young buntings roam widely, even venturing well up onto the plateau near the icecap and over ocean beaches. Departure from the Lowland begins by 15 August, and most buntings are gone by 15 September but a few persist until 7 October.

Bioenergetics of snow buntings

Most feeding is restricted to seeds until mid-June, but with moderation of temperature the birds shift rapidly from a seed diet to two species of spiders, *Hilaria vexatrix* and *Tarentula experasperans*, gleaned principally from the snow. This rapid shift may be related directly to snow bunting bioenergetics for much more biomass of seeds must be gathered than arthropods. There are two

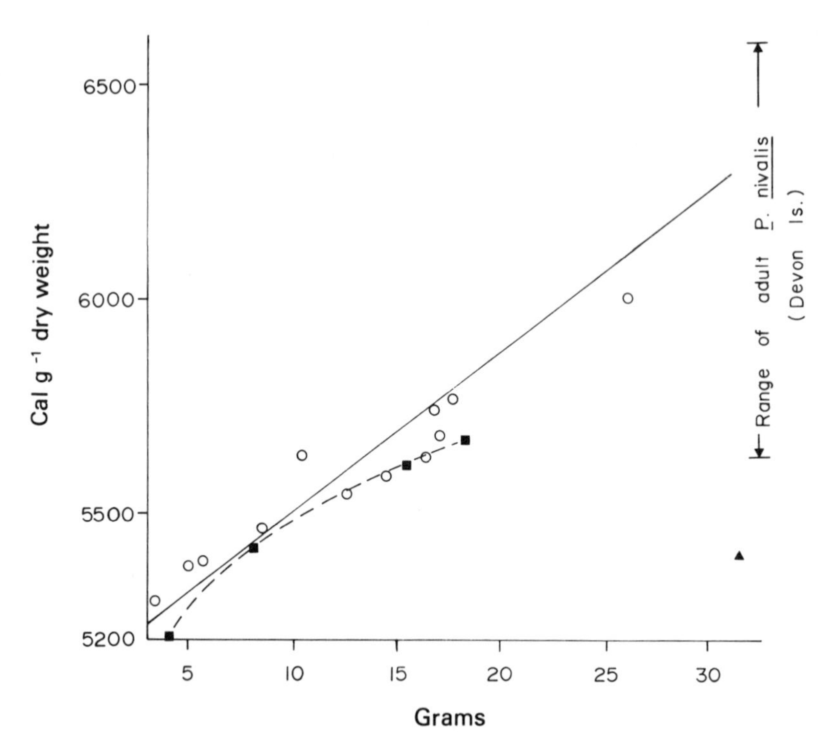

Fig. 3 Caloric content per gram dry wt of *Plectrophenax nivalis* from Truelove Lowland [△ mean cal g^{-1} dry wt of 6 *P. nivalis* adults from Hardangervidda, Norway (Østbye, Skar, Hagen, unpubl.) and □——□ cal g^{-1} dry wt of *Anthus pratensis* from alpine sites in Norway — Hagen, pers. comm.] All of the data are calculated on the basis of ash included.

reasons for this. The first relates to digestive efficiency which is 40% on a seed diet, 73% on an arthropod diet; the second to the actual caloric value of the two dietary components. Dry seeds averaged 5470 cal g^{-1} whereas dry arthropods average 5850 cal g^{-1} dry wt (J. Ryan, pers. comm.). In late June the diet again changes, this time to insects as they become available; nesting begins, and energy stored by the female begins to be directed toward egg production.

Mean caloric value of the contents of eight eggs (without shells) was 6135 cal g^{-1} dry wt on day laid. Energy used by the developing embryo reduces the mean caloric value of eggs to 5377 cal g^{-1} dry wt by one day before hatching. A least squares analysis of 40 eggs collected over the entire incubation period indicated egg contents lost 64 cal g^{-1} dry wt day^{-1} of incubation. This must represent conversion of energy and metabolism by the embryo. During the breeding season the weight and also the energy value per gram of both female and male birds declined (Table 4) suggesting output exceeds input of energy in this critical period.

Nestlings gaining weight are simultaneously dehydrating their tissues and accumulating fat; this is indicated by the slope of the graph in Fig. 3. Insects and spiders comprise nearly all of the diet of both adult and nestling buntings until the end of July or early August when the young fledge. Organic and inorganic residues of nestling bunting food and feces presented in Table 7 reveal the high percentage of protein present in the nestling diet. Stomach and

Table 4. Relative amounts of organic and inorganic residues and caloric content in snow buntings from Devon Island.

Components	Adult males			Adult females			Nestlings and fledglings		
	Dry (%)	Wet (%)	(N)	Dry (%)	Wet (%)	(N)	Dry (%)	Wet (%)	(N)
Water	—	65.0	(11)	—	66.0	(13)	—	73.5	(22)
Protein	59.8	21.0	(11)	60	20.3	(12)	56.6	15.0	(21)
Fat	20.1	7.0	(11)	22.3	7.5	(13)	26.1	6.9	(22)
Calcium	2.35	0.8	(11)	2.36	0.8	(13)	1.7	0.44	(19)
Potassium	0.72	0.26	(11)	0.78	0.26	(13)	0.9	0.24	(18)
Phosphorus	1.7	0.60	(11)	1.12	0.38	(13)	1.5	0.40	(19)
Caloric values for breeding adults cal gm^{-1}	5616	1971	(11)	5704	1925	(13)	5934	1570	(22)
Caloric values for pre-breeding adults cal gm^{-1}	6043	2115	(9)	6631	2321	(8)	—	—	—
Weight	—	35.2	(11)	—	34.7	(13)	—	24.5	(30)

Table 5. Trace element concentrations in snow bunting adults, nestlings and fledglings, insects consumed by nestlings, and nestling feces egested after ingesting insects.

Element	Adults (3 males 2 females)			Nestlings & fledglings			Insect food	Feces
	Wet (ppm)	Dry (ppm)	(N)	Wet (ppm)	Dry (ppm)	(N)	Dry (ppm)	Dry (ppm)
Iron	122.0	335.0	(5)	81.0	304	(5)	191	233
Manganese	3.0	8.8	(5)	16.0	60.2	(5)	36	36
Copper	5.3	15.4	(5)	4.8	18.0	(5)	24	44
Zinc	38.0	110.4	(5)	27.8	105.0	(5)	103	191
Magnesium	421.7	1200	(5)	317.9	1180	(4)	—	—

crop samples show fledglings initially take insects, plant fragments, and stones. By the end of the first week in August they devour an increasing number of ripening fruits and seeds. Adult buntings also begin to select plant material as temperatures decline and the molt begins. There is an abrupt decrease in the percentage of water as buntings mature (Table 4) but there is little change in the concentration of the trace elements (Table 5).

The total energetic impact of snow buntings on the Lowland is small, but not insignificant. Total impact is related to numbers, the time spent on the area and what food is available. Seeds are the only available food from the time the first males arrive in early May until late June. Then spiders are the bulk of the diet for 15 days followed by insects for 45 days, then seeds again. Adult buntings spend 150 days on the Lowland, while young birds spend 12 days as nestlings and another 30 days after fledging. Using a cumulative mean of 15 adults and 13 fledgling snow buntings km^{-2} for the four summers one can give an estimate of energy impact. Adults during 90 days and juveniles for 17 days have a seed diet. Intake averaged 9.7 g dry seed $bird^{-1}$ day^{-1}, having a caloric value of 5470 cal g^{-1} or 53 Kcal $bunting^{-1}$ day^{-1} or 83 cal m^{-2} season. Because of increased digestibility daily intake of animal food was only 36.15 Kcal $bird^{-1}$ day^{-1}. Spiders, taken mostly by adult birds early in the season, accounted for 8.25 cal m^{-2} and insects for 36 cal m^{-2} annually. The snow buntings then gather 127 cal m^{-2} $year^{-1}$ deriving 6% from spiders, 28% from insects, and 66% from seeds. Feces voided on the Lowland account for 62 cal m^{-2}.

Table 6. Major organic and inorganic residues + percent moisture in insect food fed nestling buntings and feces collected. Residues are % dry weight.

Component	*Insect* food (%)*	*Feces (%)*
Moisture	69.0	77.5
Protein**	55.6	100.6
Fibre	—	54.9
Fat	—	1.0
Calcium	0.5	0.09
Phosphorus	0.99	1.62

*Caloric value
**Protein = Nitrogen × 6.25

Seed diet digestive efficiency was 40%; this increased to 73% on an arthropod diet. This agrees closely with the 70% assimilation figure Schartz and Zimmerman (1971) gave for insect assimilation in Dickcissels *(Spiza americana).* Composition of insect food and resulting feces from nestling buntings appears in Table 6. Adult daily metabolic requisites were determined to be 25.7 Kcal $bunting^{-1}$ based upon the results of feeding experiments which revealed 60% of the energy represented by seed intake was eliminated. Testing five buntings with a Pauling oxygen analyzer at 3-5° C and using 4.8 cal cco_2^{-1} as the caloric equivalent gave a value of 22.2 Kcal $bird^{-1}$ day^{-1} for metabolic requisites. Finally, calculation of theoretical standard metabolic requisites for passerines from the formula Lasiewski and Dawson (1967) prepared:

$$\log M = \log 129 + 0.724 \log W \pm 0.113$$

Where M is metabolic rate in Kcal day^{-1} and W is body weight in kilograms.

yield a mean result of 11.75 Kcal $bird^{-1}$ day^{-1}. This was roughly half the two experimental maintenance energy values we found but close to the 11.4 Kcal $bird^{-1}$ day^{-1} Lasiewski and Dawson (1967) listed for snow buntings. The discrepancy undoubtedly arises from our birds being in other than thermoneutral surroundings. Since buntings rarely, if ever, experience thermoneutral temperatures in nature, our results are more applicable for estimating the energy free-living birds consume and their impact on the environment.

During the nestling period young buntings had a mean daily metabolic requisite of 24.5 Kcal. Growth rates of nestlings based upon daily field weighing of more than 50 young birds appear in and are close to those of Maher (1964). The sigmoid growth curve of snow buntings (Fig. 4) is more similar in slope to the alpine horned lark *(Eremophila alpestris)* (Verbeek 1967) or song thrush *(Turdus philemelos)* than it is to the blackbird *(Turdus merula)* (Bilby and Widdowson 1971). Growth curves for Lapland longspurs are given by Maher (1964).

Dry and wet caloric energy values plus organic and inorganic residues from eight species besides the snow bunting are presented in Table 7. Although these represent values from small samples, they are nevertheless the first caloric values published for most of these arctic species. The low sample numbers reflect a reticence to collect many birds during one season in an area with low population densities and minimal recruitment.

Total incubation of an average clutch of 5.25 eggs requires 35.5 Kcal. This was based upon records from two nests in which mean daily ΔT from 9 to 12 total days of incubation was 189.5° C; and assumes that specific heat of the eggs was 1. Eggs lost an average of 0.45% or 0.14 g day^{-1} during incubation,

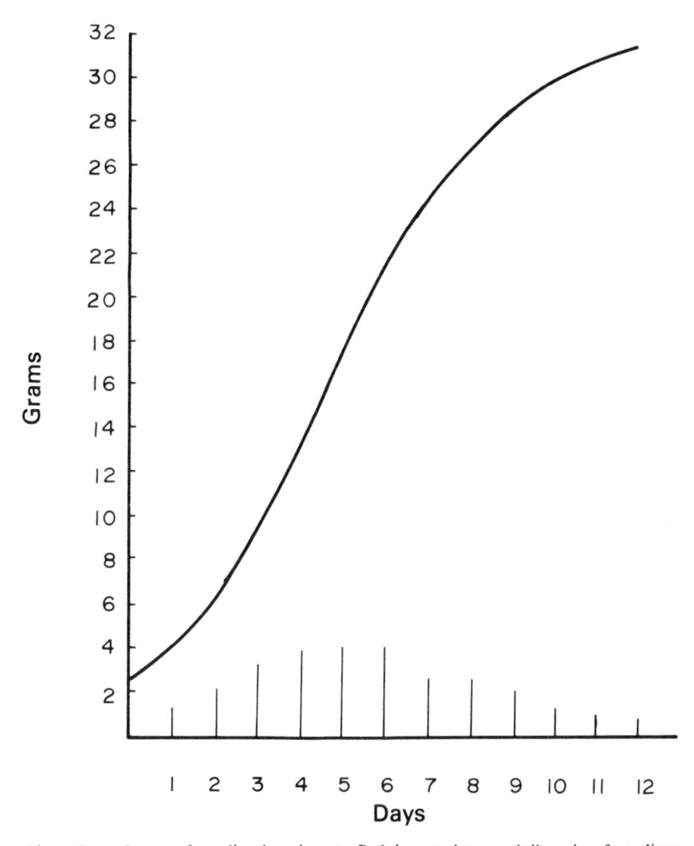

Fig. 4 Growth rate of nestling buntings to fledging, and mean daily gain of nestlings.

part to metabolism, part to dehydration; eggs in a control nest lost 0.002 g day^{-1} without incubation. The significance of this rate of weight loss is not known but the percentage is not constant for all species. Arctic terns with mean egg weights of 18.416 g at laying lose 0.140 g or 0.76% per day during 21 days of incubation. More observation might modify the daily ΔT because records of nests which were lost to predators indicated more frequent absences by some incubating females which could raise ΔT by as much as 50%. Measurements with heat flux plates show no measurable heat lost through nest bottom or sides while the female is brooding. Conduction, radiation, and convection may reduce egg temperatures in excess of a degree a minute for seven consecutive minutes when the female departs, so uncovered eggs do not long retain appreciable amounts of heat.

Goose bioenergetics

About 500 snow geese molted on Sparbo-Hardy Lowland 30 km north of Truelove each summer. Frequently flocks of geese grazed on Truelove Lowland in June or August and September. A few pairs nested successfully on the Lowland tundra or on cliffs beside the peregrine falcon nests in each year, except 1972 when eggs failed to hatch. Two adult captive blue geese *(Chen caerulescens)* weighing 4682 g together were penned for eight days in March to

Table 7. Organic and inorganic residues and caloric values of arctic birds.

Collector's no.	Species	Sex	N*	Date collected	H_2O %	Protein %**
1284	Snow goose	M	1	29 July	—	59.8
1290	Oldsquaw	F	1	11 Aug.	—	54.2
1228	Common eider	M	1	8 July	65	54.5
1208	King eider	F	1	25 June	—	—
1288	Baird's sandpiper	F	1	11 Aug.	—	53.2
1286	Long-tailed jaeger	M	1	6 Aug.	—	64.5
1271	Glaucous gull	M	1	20 July	64	56.5
1261	Lapland longspur	Mixed	12	17 July	—	60.5
1122	Snow bunting	M	1	10 July	66	58.2

Collector's no.	Species	Fibre (%)	Ca %	Phosphorus (%)	Fat (%)	N^2	$K cal\ g^{-1}$ dry wt	$K cal\ g^{-1}$ wet wt
1284	Snow goose	9.1	3.58	2.03	14.3	1	5.4	2.6
1290	Oldsquaw	3.1	3.6	2.03	24.2	2	5.9	2.1
1228	Common eider	11.8	3.5	1.98	26.5	1	5.8	2.0
1208	King eider	—	—	—	—	2	6.5	2.3
1288	Baird's sandpiper	16.2	5.18	3.52	29.1	2	6.1	2.1
1286	Long-tailed jaeger	1.8	4.8	2.5	8.9	1	5.2	1.8
1271	Glaucous gull	16.8	5.35	2.51	18.9	2	6.1	2.1
1261	Lapland longspur	19.6	2.73	1.72	15.2	12	5.6	2.0
1122	Snow bunting	—	2.42	1.74	16.9	24	5.7	1.9

*N = sample number
**Protein = Nitrogen × 6.25

relate food consumption of geese to other herbivores. Temperatures during the study ranged from −2°C to 10°C. During eight days the captive geese consumed 1572 g of dry (4.500 Kcal g^{-1}) feed. They voided 636 g dry (3.855 Kcal g^{-1}) feces. Weight gain of 284g during the feeding experiment represented 728.6 Kcal stored. Daily intake of 960.6 Kcal $bird^{-1}$ was accompanied by 301.7 Kcal egested as fecal matter and 87.2 Kcal stored. On a dry weight basis 69% of all food was assimilated. Requisites during the time spent in reproduction and molting on the tundra are probably closer to the 289 Kcal $bird^{-1}$ day^{-1} that was actually assimilated by the captive birds than to the theoretical, the difference being accounted for by a weight gain in the captives. Again this is double the basic metabolic requirements of 144.6 Kcal $bird^{-1}$ day^{-1} suggested by the formula for non-passerines given by Lasiewski and Dawson (1967).

Raven bioenergetics

The largest passeriform to visit the Lowland each year was the raven *(Corvus corax)*. Although never common and frequently absent from daily observations for weeks at a time, up to as many as 6 would appear within a few days of the death of a muskox whether its remains were scattered by wolves or not. On 7 August, 1972 a fledgling raven which had been in our camp since 5 August was netted. It was maintained outdoors for 10 days in a cage that had two adjacent sides and half the top composed of nylon screen; the remaining sides were plywood. Three days were allowed for the raven to become familiar with its surroundings and routine and for us to judge its food intake. It gained 40g in these three days. Then food intake and fecal production were monitored for seven consecutive days. During the seven days the captive raven consumed 109.5 g of dry (5.800 Kcal g^{-1}) food. It egested 72.4 g dry (3238 cal g^{-1}) feces. Weight gain of 10 g during the feeding experiment represents about 25.6 Kcal stored. Daily intake of 90.7 Kcal was accompanied by 33.48 Kcal egested and 3.7 Kcal stored leaving 53.5 Kcal for maintenance. Digestive efficiency was 63%. Although only a single fledgling was used, these results for maintenance energy are less than the 92 Kcal and 94.9 Kcal listed as Standard Metabolic Rates for Ravens by Lasiewski and Dawson (1967) and also less than the theoretical 140.5 Kcal day^{-1} predicted for a passerine of this weight. This arctic raven was 25% heavier than those reported by Lasiewski and Dawson. Aschoff and Pohl (1970) have estimated an even higher value than Lasiewski and Dawson, based upon the raven being penned in continuous daylight, in their review of metabolic rates. Calder and King (1974) offer little explanation for our low readings, other than a suggestion that differences in displacement from thermoneutrality may make interpretation understandably intricate.

Discussion

The 43 km^2 Truelove Lowland appears lush compared with most of the eastern end of Devon Island. Even in this area population levels of most birds, based upon periodic bird counts made over the entire area, were extremely low compared with more temperate sites. This suggests a low productivity for Truelove Lowland, an inability of the bird species to harvest the producer's energy, or a nutrient rather than a food limitation for the populations.

This lowland is largely ecotonal in nature and the merging of several environments may have considerable effect on avian species. The major transition zones are those between the sea and the land, the lakes and the land, and to a lesser extent the rather sharp demarcation between the dolomitic limestone of the plateau and the basement granite gneisses. Several bird species restrict their nesting to one or another of these transition zones.

Nest initiation is greatly influenced by the time of snowmelt in the High Arctic. Birds do not begin nesting until some exposed ground is present and if the melt is delayed early nests destroyed by predators may not be replaced. Hussell (1972) noted the irregularity of the breeding pattern in several arctic sites and Vinikurov (1971) noted that great changes in numbers and distribution of arctic vertebrates in different years depend upon both biotic and abiotic factors. The early melt in 1973 resulted in an enhanced reproductive effort and success resulting in considerable increases of snow buntings, arctic terns, and red phalaropes. Success was also sometimes modified by predators.

Ecological diversity exists in the role the various species play as well as in nesting sites. The snow geese are primarily herbivores, grazing sedge roots and rhizomes along lake margins. Grazing by geese is frequently so intensive that all vascular plants are removed so that only mosses remain after their feeding. This is especially evident on Sparbo-Hardy Lowland an area used for molt migration by about 500 geese. Rock ptarmigan are also herbivores taking large quantities of willow with lesser amounts of exposed seed heads, flowers, and plant tops from other plant species, but these resident species never seem to denude an area's vegetation the way the migratory geese do. Instead, during the winter the ptarmigan exploit alternative feeding areas such as meadows and sea coasts where they are never seen during the breeding season.

Secondary terrestrial consumers include the largely insectivorous plovers, sandpipers, Lapland longspurs, and snow buntings. The eiders, oldsquaw, red-throated loon and the arctic tern are secondary or tertiary consumers whose primary source of food is the sea but who switch their harvest to lakes for a short time each year. These species also recycle energy via their eggs and sometimes the young into other carnivore portions of the terrestrial system. A young 847 g red-throated loon forced onto the ice by freeze-up on 18 September, 1971 lost 12.5 g day^{-1} until it died on 27 September (R. Riewe, pers. comm.). If we assume 2.3 Kcal g^{-1} wet weight the young loon had a daily energy need of more than 28.7 Kcal, all of which came from lowland lakes or the sea.

Top carnivores include the snowy owl, the peregrine falcon, and the two jaegers. The long-tailed jaeger derives most of its nourishment from lemmings, sitting catlike in wait by burrow entrances for these rodents. They also eat large numbers of insects and a few birds according to Maher (1970b). The parasitic jaeger feeds much more on eggs and young of ducks, loons, and terns plus passerine birds, locating its prey by flying low over the tundra. They no doubt take lemming too when they see them, and thus serve as competitors to the smaller long-tailed jaeger. We noted that both jaeger species fed extensively on insects such as chironomids. Parasitic jaegers frequently regurgitated large masses of chironomids when netted in connection with banding, and a flock of 59 long-tailed jaegers was observed feeding on swarms of chironomids beside Phalarope Lake in late July 1970. Both species take larger insects as well but these tiny flies were frequently the most numerous and available insects. Nesting habitat for the two species of jaegers appears identical; both species defend the nesting site from the other species.

Snowy owls were rarely seen, whether because of the generally low lemming numbers, scarcity of favorable habitat, or the presence of breeding peregrine falcons is not known. When an owl did visit the area its large size and contrasting color during summer insured daily observation. Casts collected from owl perches on large erratic boulders contained remains of lemming, oldsquaw, and jaeger. The few owl casts found were larger than those of the jaegers; casts positively attributable to gulls were not seen, suggesting their diet was largely devoid of large bones, feathers, or fur.

Breeding pairs of peregrine falcons occupied long standing traditional nesting sites, identifiable by feces stained rocks and an orange lichen *(Caloplaca elegans)* growing under eyries, beside at least two of the five contiguous lowlands. Peregrines are chiefly predators of birds. On Devon Island they took snow buntings, Lapland longspurs, rock ptarmigan, sandpipers, and long-tailed jaegers. Adult eiders and snow geese were apparently not molested for both common eiders and two pairs of snow geese nested on narrow ledges high on the cliffs near one eyrie. Since both owls and falcons fed on birds the scarcity of owls may reflect a greater efficiency of falcons in catching birds.

The apparent lack of gull casts, the failure of glaucous gulls to exploit exposed eider eggs or young birds, and an absence of large feeding aggregations near the Lowland were puzzling, since both Tuck (1960) and Wynne-Edwards (1952) reported they take eggs (of murres and fulmars). A very few glaucous gulls (the same ones I suspect) set up summer vigils along shallow streams and occasionally caught small arctic char. The only other feeding by glaucous gulls observed involved meat distributed as bait, viscera, blubber, or scraps left by hunters or dead sea mammals. Baits of 10 kg of frozen meat were devoured by gulls within two days; the viscera and scraps of a butchered ringed seal in one day, and a large ringed seal which died on the ice had most of the meat removed through long rents in the skin within ten days. Camp garbage was not regularly available so gulls seldom stopped at the camp. Despite lengthy periods of time without evident food, at least 30 pairs of glaucous gulls nested on cliffs near Truelove Lowland. Perhaps the Truelove Lowland gulls feed at sea during part of the nesting season as Mathews (1973) noted in the Beaufort Sea. The only evidence of mortality in glaucous gulls were two nestlings dead on rocks below the nests and a two-year-old found dead in an emaciated condition on the sea ice in May (Dr. Earl Godfrey, pers. comm.).

Ravens scavenged muskox remains on the Lowland. Wolf kills occupied the attention of five ravens for more than a week before thaw in 1970. In 1971 ravens fed on maggots around the carcass of a muskox which died near the sea. A docile young raven attempted to scavenge in camp in 1972. Captured, it provided us with estimates of energy requisites before being released. Although a sample derived from a single bird for this short a time is inadequate for a definitive statement, our finding a daily maintenance energy requisite of 53.5 Kcal suggested an intensive study of arctic raven bioenergetics may be desirable to determine whether a resident population of scavenging birds living in this harsh environment could have acquired through selection, a reduced metabolic rate. The advantages of a reduced metabolic rate to large arctic homiotherms by reducing food requisites seems evident. Since scavengers must be opportunists, it seems strange no one reported ravens feeding on dead seals or their remains nor gulls scavenging on muskox carcasses in the four years of the study. The possibility of niche separation of the scavenging gulls and ravens bears further investigation.

Despite the 1973 increase in the estimate of breeding passerines over 1970-72, Devon Island remains the most depauperate of all studied I.B.P. sites in terms of bird fauna, having both fewer breeding bird species and lowest densities. Since species are likely to be "rare" near the margins of their distributions, and outside the distribution they are "extinct" (Andrewartha and Birch 1961), the low population numbers on an island near the limit of the range for many species should be expected. How two small passerine species endure what are marginal conditions for larger species is of interest. Two mechanisms seemed most probable because a third suggestion, an increase in reproductive rate through larger clutch size, had been discounted by Hussell (1972) except possibly for Lapland longspurs. This leaves to be considered: longer life spans for adults, or an ability to capitalize on a particularly good reproductive season. Little information was collected on life spans although band returns indicated some buntings lived at least five years. The enhancement of snow bunting reproductive success in 1973 seems to provide the explanation of the mechanism involved, for it was outstanding and reversed the trend of decline which had been apparent from 1970 through 1972. The 1972 season was especially poor owing to the late melt on the Lowland and across the arctic islands (Alsop and Jones 1973). That year egg production by both buntings and longspurs was significantly lower than that reported by Hussell (1972) for four years or from our 1970-71 season results. We found only one first clutch replaced following early abandonment, and buntings tended to more readily abandon their nests in 1972.

Predators took about 70% of the nests we encountered in both 1971 and 1972. These figures included research activity as predation, limited to 10% of the eggs of buntings in 1971. In 1970 predation was only 25% and, again, the two chief predators were the arctic fox and parasitic jaeger. The arctic fox population on the Lowland had increased during the study even though two were killed in 1970. Five foxes were seen at the camp in late fall 1971. It is noteworthy that both foxes and jaegers were noticeably active in 1973 the year considerable bird population increases were noted.

Numbers of snow buntings reported in the early 1971 and 1972 census figures (Table 1) are high, probably because many birds that were to breed farther north crowded the north side of the island before flying on or because territories on the slopes of the plateau and outside the study area were not yet habitable. The adjoining steep rocky talus slopes supported a small number of buntings that foraged on the Lowland and edge and were counted in the early or late census. Snow bunting territories were estimated at 50-60 on the 33 km^2 of land area. The production of 5.25 eggs per pair of birds amounts to 14.8 Kcal $pair^{-1}$ but since a few nests were destroyed early enough for renesting to take place, the true figure may be slightly higher. The hen has an additional energy load of 35.5 $Kcal^{-1}$ needed for incubation but the extra daily energy demand on the reproducing hen seldom exceeds 10-12% of her average daily metabolic maintenance requisite whether during egg production or incubation. It only exceeds 10-20% during the last days of the nestling period when both parents must collect several times their own needs for the maturing young. Feeding at this time is exclusively on the highly digestible insects. The importance of invertebrates to the buntings cannot be over-emphasized because although only 34% of their summer diet is spiders and insects, nearly 50% of their usable energy is derived from these sources. Having an abundance of insects available at fledging reduces stress on parents. The rapidity with which the arctic birds fledge added to there being but a single clutch per season combine to reduce the time adults must spend in reproduction. The digestive efficiency of buntings may relate to the time of digestion of the various items, for Custer and Pitelka (1975) found invertebrates were digested so completely as to be indistinguishable after 18 to 38 min, whereas small dry seeds could still be distinguished after 180 min.

The male snow bunting feeds the female during laying and at regular intervals during incubation. That such feeding may have an influence on incubation rhythms is indicated by von Haartman (1958) who found that the hen of a pair of pied fly-catchers who was not fed assiduously by her mate, incubated for shorter periods and needed to seek food more often than hens regularly fed on the nest. When von Haartman temporarily removed the male which regularly fed its incubating mate, the hen incubated only 58% instead of 79% of the day, and her weight fell from 16.0 to 14.2 g. The possibility of comparable circumstances in the buntings bears investigation.

The extent of energy export from the Lowland is difficult to ascertain, for predation on buntings by jaegers and falcons continues until late September. In the early August census of 1970, 70% of the buntings enumerated were young; in 1971, 76% were young. In the high production year of 1973, 58% were juveniles. Since this means roughly six young for each pair of adults counted, and as the clutch sizes average less than six, we face the dilemma of explaining what happened to the other adults. Possible explanations include significant predation by jaegers, migration before the August census by adults who lose their nests to predation, or such secretive behavior by those molting during the census period that they either do not fly, or fly when the observers are still so far away that they cannot be recognized as to age or sex. At any rate an average of 170 young buntings each representing 66.2 Kcal were on the Lowland in early August of 1970, and 1971 and these represented 3.4×10^{-1} cal m^{-2} over the 33 km^2 land area of the Lowland.

Another aspect of the high arctic avifauna that bears further comment is the number of trophic levels which are represented. Not only are primary consumers (snow geese and ptarmigan) present along with secondary consumers or insectivores (plovers, buntings, longspurs, sandpipers, and ducks), but the system supports tertiary consumers in the jaegers, terns, and loons and even part-time quaternary consumers in the snowy owl and peregrine falcons which kill and eat jaegers as well as other birds. Even the scavenging raven, although in very low densities, invariably patrols the land. The more numerous scavenger gulls seem more closely tied to marine and fresh water ecosystems. Such complexity, expected in a temperate region, seems surprising in an area which has only 15 regularly breeding species.

Summary

The avifauna of Devon Island in the Canadian high arctic is characterized as having few species and the lowest densities of breeding birds recorded in North America with estimated post breeding densities of all species being only 53.6 birds km^{-2}. Most species appear to be limited in reproductive success by the time of snowmelt with early melting greatly enhancing reproductive success; conversely seasons of delayed snowmelt may enhance predation and greatly diminish or eliminate reproduction for a season. Eleven of the 15 regularly breeding species belong to the Charadriiformes and Anseriformes. Three passeriforms were found on the island and one of these, the snow bunting, was intensively studied during four summers. Snow buntings required 53.1 Kcal $bird^{-1}$ day^{-1} and derived 6% of their food from spiders, 28% from insects, and 66% from seeds. Buntings annually harvested 83 cal m^{-2} in seeds, 8 cal m^{-2} in spiders, and 36 cal m^{-2} in insects on the lowland studied; 62 cal m^{-2} were returned in the form of feces. During reproduction the extra bioenergetic load almost never exceeded 10% of the buntings' daily needs. Limited nutritional studies of a raven and goose suggest further studies of these species may be instructive since the bioenergetic requisite of the arctic raven seems far below the predicted value for a passerine of this weight. Despite the low number of bird species, most of the trophic levels found in temperate situations are represented. Differences in feeding by two jaeger species are described and this may account for their niche separation.

Acknowledgments

As in any project of the magnitude of this effort, many people helped in ways far too numerous to detail. While both deeds and names remain anonymous I am aware of and appreciate all they did. Special recognition is due to Steve Golub, Bruce Grahn, Ron Lett, Lindsay Rackett, Allen Trautman, and Ken Zurfluh who helped collect field data; to Drs. Rick Riewe and S. Wayne Speller who provided early and late seasonal observations and found several nests; and to Dr. Gerard Courtin whose technical assistance, equipment, and ideas greatly improved the study. Thanks are also due Dr. Lawrence Wang who cooperated by donating his equipment and time for determining bunting oxygen consumption; to James Ryan for his assistance in identifying insects and spiders; and to W.J. Maher and R.S. Hoffmann for helpful comments on the manuscript. My special appreciation goes to Emma Bowthorpe and her late husband Bill, for assistance in helping with the bioenergetic study of geese. This research was partially funded from an N.R.C.C. grant No. A6135. Finally, thanks to Dr. L.C. Bliss, who kept the whole project going.

References

Alsop, F.J. and E.T. Jones, 1973. The lesser black-backed gull in the Canadian Arctic. Can. Field. Nat. 87: 61-62.

Andrewartha, H.G. and L.C. Birch. 1961. The Distribution and Abundance of Animals. Univ. Chicago Press. Chicago. 782 pp.

Aschoff, V.J. and H. Pohl, 1970. Der Ruheumsatz von Vögelnals Funktion der Tageszeit und Körpergrösze. J. Orn. 111: 38-47.

Bilby, L.W. and E.M. Widdowson, 1971. Chemical composition of growth in nestling blackbirds and thrushes. Br. J. Nutr. 25: 127-134.

Boag, D.A. and J.W. Kiceniuk, 1968. Protein and caloric content of lodgepole pine needles. Forestry Chronicle 44: 1-4.

Brisbin, L.W. and E.M. Widdowson, 1968. Chemical composition of growth in nestling blackbirds and thrushes. Br. J. Nutr. 25: 127-134.

Bruggemann, P.F., 1958. Insects and environments of the high arctic. pp. 695-702. *In:* Symposium on Distribution of Arctic and Subarctic Insects, International Union of Biological Sciences, Series B, No. 31, Abstract in North of 60: Ecology of the Canadian Arctic Archipelago: Selected References Vol. 1 by N. Merle Peterson. Indian and Northern Affairs Publication No. QS 1572-010-EE-A1, Ottawa.

Calder, W.A. and J.R. King, 1974. Thermal and caloric relations of birds. pp. 268-273. *In:* Avian Biology, Vol. IV, D.S. Farner and J.R. King (eds.). Academic Press, New York.

Colquhon, M.K., 1940. Visual and auditory conspicuousness in a woodland bird community: a quantitative analysis. Proc. Zool. Soc. London 110: 129-148.

Custer, T.W. and F. Pitelka, 1972. Time-activity patterns and energy budget of nesting Lapland longspurs *(Calcarius lapponicus)* near Barrow, Alaska, Tundra Biome Symposium, Washington, D.C., April 3-5.

———. 1975. Correction factors for digestion rates for prey taken by snow buntings. Condor 77: 210-212.

Danks, H.V., 1971. A note on the early season food of arctic migrants. Can. Field Nat. 85: 71-72.

Depocas, F., and J.S. Hart, 1957. Use of the Pauling oxygen analyzer for measurement of oxygen consumption of animals in open-circuit systems and in a short-lag, closed circuit apparatus. J. Appl. Physiol. 10: 388-392.

Emlen, J.T., 1971. Population densities of birds derived from transect counts. Auk 88: 323-342.

Enemar, A., 1963. The density of birds in the subalpine birch forest of the Abisko area Swedish Lapland in 1961. Lunds Univ., Aarskrift Avd 2 Bd 58 N 12., Lund. 21 pp.

———., S.G. Hojman, P. Klaesson, L. Nilsson, B. Sjöstrand, 1973. Bestämning av smafågelbestandets tåthet i fjållbjörkskog genorn boletning och revirkartering i samma provyta. Vär Fågelvarld 32: 252-259.

Fay, F.H. and J.J. Code. 1959. An ecological analysis of the avifauna of St. Laurence Island, Alaska. Univ. Calif. Publ. Zoology 63, No. 2: 73-150.

Geale, J., 1971. Birds of Resolute, Cornwallis Island, N.W.T., Can. Field Nat., 85: 53-59.

Grimshaw, H.M., J.D. Ovington, M.M. Betts and J.A. Gibb, 1958. The mineral content of birds and insects in plantations of *Pinus sylvestris*. Oikos 9: 26-34.

Haartman, L. von, 1958. The incubation rhythm of the female pied flycatcher *(Ficendula hypoleuca)* in the presence and absence of the male. Ornis. Fenn. 35: 71-76.

Harrington, C.R., n.d. Biological observations on northeastern Devon Island, Unpub. Ms. Arctic Institute of North America, Calgary, 58 p.

Hilden, O., 1971. Occurrence, migration and color phases of the Arctic Skua *(Stercorarius parasiticus)* in Finland. Ann. Zool. Fenn. 8: 223-230.

Hussell, D.J.T., 1972. Factors affecting clutch size in arctic passerines, Ecol. Monogr. 42: 317-364.

————. G.L. Holroyd, 1974. Birds of Truelove Lowland and adjacent areas of northeastern Devon Island, N.W.T. Can. Field Nat. 88: 197-212.

Kaiser, G.W., L.P. Lefkovitch and H.F. Howden, 1972. Faunal provinces in Canada as exemplified by mammals and birds: a mathematical consideration. Can. J. Zool. 50: 1087-1104.

King, J.R., and D.S. Farner, 1961. Energy metabolism, thermoregulation and body temperature. pp. 215-288. *In:* Biology and Comparative Physiology of Birds. A.J. Marshall (ed.). Vol. 2.

Lasiewski, R.C. and W.R. Dawson, 1967. A re-examination of the relation between standard metabolic rate and body weight in birds. Condor 69: 13-23.

Lien, L.E., Østbye, A. Hagen, A. Klemetsen, H. Skar, 1970. Quantitative bird surveys in high mountain habitats, Finse, South Norway, 1967-68 Nytt. Mag. Zool. 18: 245-251.

Maher, W.J. 1964. Growth and development of endothermy in the snow bunting *(Plectrophenax nivalis)* and Lapland longspur *(Calcarius lapponicus)* at Barrow, Alaska. Ecology 45: 520-528.

————. 1970 (a). The pomarine jaeger as a brown lemming predator in northern Alaska. Wilson Bull. 82: 130-157.

————. 1970 (b). Ecology of the long-tailed jaeger at Lake Hazen, Ellesmere Island. Arctic 23: 112-129.

Mathews, D., 1973. Reported glaucous gulls feeding at sea. *In:* A Baseline for Beaufort. Exxon U.S.A. 12: 2-7.

Moksnes, A., 1973. Quantitative surveys of the breeding bird population in some subalpine and alpine habitats in the Nedal area in Central Norway (1967-71). Norwegian J. Zool. 21: 113-138.

MacDonald, S., 1954. Report on biological investigations at Mould Bay, Prince Patrick Island, N.W.T., in 1952. Nat. Mus. Canada, Bull. No. 132: 214-238.

Nettleship, D.N., 1967. Breeding biology of Ruddy Turnstones and Knots at Hazen Camp, Ellesmere Island, N.W.T. M.Sc. thesis, Univ. Saskatchewan, Saskatoon, 175 pp.

————. and W.J. Maher, 1973. The avifauna of Hazen Camp, Ellesmere Island, N.W.T. Sonderdruk Polarforschung 43: 66-74.

————. 1974. Seabird colonies and distributions around Devon Island and vicinity. Arctic, 27: 95-103.

Nicol, C.W., 1961. Notes on the avifauna of Devon Island. Unpub. Ms. Arctic Institute of North America, Calgary. 11 p.

Parmelee, D.F., 1968. Snow bunting. pp. 1652-1675. *In:* Life Histories of North American Cardinals, Grosbeaks, Buntings, Towhees, Finches, Sparrows and Allies. A.C. Bent (ed.). Dover Publ. Inc., New York.

Pitelka, F., 1971. Social organization, habitat, utilization and bioenergetics in the Lapland longspur near Barrow, Alaska. pp. 107-110. *In:* The Structure and Function of the Tundra Ecosystem; Vol. 1 Progress report and proposal abstracts. U.S. Tundra Biome Program of the U.S. I.B.P. and U.S. Arctic Research Program.

Savile, D.B.O., 1961. Bird and mammal observations on Ellef Ringnes Island in 1960. Natur. Hist. Pap. Natur. Mus. Can. 9: 1-6.

————. 1972. Arctic adaptations in plants. Can. Dept. Agr. Monogr. 6. Ottawa, 81 pp.

Schartz, R.L. and J.L. Zimmerman, 1971. The time and energy budget of the male dickcissel *(Spiza americana).* Condor 73: 65-76.

Sutton, G.M. and D.F. Parmelee, 1956. On certain charadriiform birds of Baffin Island. Wilson Bull. 68: 210-223.

Tikhomirov, B.A., 1970. Preliminary report of results of biogeocenological investigations on the Taimyr Peninsula pp. 161-170. *In:* Tundra Biome, Working Meeting on Analysis of Ecosystems Kevo, Finland. pp. 297.

Tuck, L.M., 1960. The Murres — their distribution, populations and biology — a study of the genus *Uria*. Can. Wildl. Serv. Ottawa, Can. Wildl. Series No. 1. 260 pp.

Verbeek, N.A.M., 1967. Breeding biology and ecology of the horned lark in alpine tundra. The Wilson Bull. 79: 208-218.

Vinokurov, A.A., 1971. The vertebrate fauna of the region of the Taimyr Station. pp. 212-231. *In:* Biogeocenoses of Taimyr Tundra and their Productivity. Publishing House "Nauka," Leningrad. 239 pp.

West, G.C., 1968. Bioenergetics of captive willow ptarmigan under natural conditions. Ecology 49: 1035-1045.

Williamson, F.S.L., M.C. Thompson and J.Q. Hinds, 1966. pp. 437-480. *In:* Environment of the Cape thompson Region, Alaska. N.J. Wilimovsky and J.N. Wolfe (eds.). U.S.A.E.C. Tech. Inf. Ext. Oak Ridge, Tenn. 1250 pp.

Wynne-Edwards, V.C., 1952. The Fulmars of Cape Searle. Arctic 5: 105-117.

Biology and secondary production of *Dicrostonyx groenlandicus* on Truelove Lowland

W.A. Fuller, A.M. Martell,
R.F.C. Smith, and S.W. Speller

Introduction

Many authors have recognized the pivotal role of lemmings in tundra ecosystems (for reviews see Batzli 1975, and Bliss 1975), not only as primary consumers and a major energy source for higher order consumers, but also as physical agents of change through their burrowing and other activities. Lemming populations are notoriously variable with peaks of abundance apparently occurring on the average every 3 to 4 years. The riddle of lemming cycles provides a second reason for studying the biology of a high arctic population because, in spite of numerous studies conducted in the Low Arctic of both hemispheres, there is no agreement about what factors control or regulate lemming numbers. This paper is based on research on *Dicrostonyx groenlandicus* over parts of four consecutive summers (1970-73). The major objectives were to determine the role they play in energy transfer within this ecosystem and to investigate their population cycles.

Methods

Almost all trapping was carried out by means of snap-traps which killed the animals. Limited live-trapping was conducted to obtain animals for growth and feeding studies but, with the exception of July 1973, not for population studies. We first attempted to sample the population quantitatively by placing traps at the intersections of grid lines, but this method was abandoned because catches were so small. All subsequent trapping was done on trap lines with traps placed at irregular intervals, but always at or near the entrance to a burrow. Most of our analyses are based on an index of catch per unit effort (1,000 trap nights). The trapping method may have produced an upward bias in catches but, since the bias was uniformly present, capture indices should be at least indicative of relative population changes. Absolute densities could only be estimated by calculating the area of each block of similar habitat sampled and assuming that all animals originally present were captured. Since that assumption was probably seldom fulfilled, our density estimates are probably minimum estimates.

Furthermore, trapping techniques and effort varied from observer to observer so that year to year comparisons must be accepted with caution. However, since techniques were constant within any summer, except 1973,

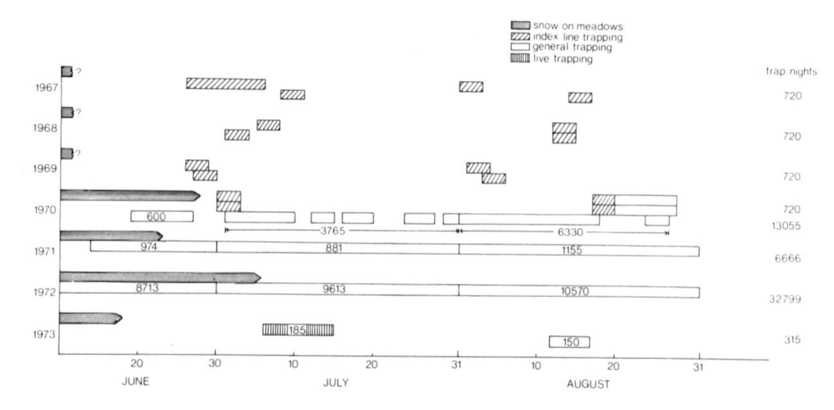

Fig. 1 Distribution of trapping effort during seven summers on Devon Island. Bars indicating snow cover terminate on the date when meadows were 50% snow free.

month to month comparisons probably have a good deal of validity. For 1973, however, it is necessary to compare July live-trapping with August snap-trapping, which may not be valid, and to take account of subjective estimates of observers in the field.

In addition to these data we had access to the results of index trapping on Truelove Lowland in the years 1967-69 through the courtesy of D.J.T. Hussell. Hussell and one of us (A.M.M.) sampled the same area with the same number of traps set at approximately the same locations at two periods each summer. The first sampling began soon after raised beaches were clear of snow and the second took place during the first half of August (Fig. 1). The addition of these data allowed us to follow population trends for seven consecutive summers and the six intervening winters. Altogether, the data are based on some 53,575 trap-nights of effort which yielded 346 *Dicrostonyx groenlandicus* for an average rate of capture of 6.46 per thousand trap-nights. We have data on an additional 46 animals taken by hand, caught in live traps, or found dead.

Standard measurements, weight and sex, were recorded for all animals captured. For the years 1967-69 these are the only data we possess. From 1970-73 we have in addition data on reproductive condition. Males were considered to be in breeding condition when the tubules of the cauda epididymidis were visible to the naked eye. The validity of this criterion was verified in 1970 by means of smears of testis and epididymis which showed that apparently normal sperm were present in abundance when the tubules were visible. Females were classed as anestrous when the vagina was imperforate and dissection revealed no evidence of embryos or placental scars, and as estrous when the vagina was perforate. Embryos and placental scars were counted and evidence of current or recent lactation was recorded.

Flat skins from most autopsied animals were preserved and study of this material suggested that four pelage classes could be recognized. Individuals in class I were small and obviously only recently weaned, whereas those in class IV were in full adult pelage. The other two categories were intermediate in pelage characteristics and, presumably, in age. Class I individuals had the back uniformly brownish-grey except for a pronounced, dark mid-dorsal stripe. Flanks were grey. Class II individuals had a less prominent mid-dorsal stripe, and some reddish hairs mixed with the grey of the flanks. Class III individuals had an inconspicuous mid-dorsal stripe, back slightly mottled with an admixture of brown hairs, and reddish-grey flanks. In class IV individuals the mid-dorsal stripe was absent, the back was distinctly mottled, and flanks were rusty red with the appearance of some white hairs as well.

As an aid to working out habitat preferences and to provide additional data on population density, counts were made of burrows, both active and inactive, in a variety of habitat types. Active burrows were usually easy to recognize by the presence of recently excavated soil at their entrances.

In 1971 samples of lemmings of various sizes were oven-dried to constant weight and the resulting ratio of dry-weight to wet-weight was used to convert live-weight biomass to dry-weight biomass in energy flow calculations. In addition, caloric content of homogenated lemming tissue was determined by means of bomb calorimetry.

Feeding trials were also conducted during 1971. Lemmings were kept in an unheated tent and maintained on a mixture of lab chow and arctic willow *(Salix arctica)*. Energy content of food ingested and feces egested was determined for use in energy flow calculations.

Counts of winter nests were made each spring from 1970 to 1973 just at snowmelt. At the same time, a record was kept of nests probably invaded by predators, predominantly ermines, *(Mustela erminea)*. Nests that had been invaded could usually be recognized because the ermine subsequently modified it to suit its own purposes, and by the presence nearby of lemming remains, such as stomachs and tails, and ermine scats.

Natural history

Habitats

Our approach to summer habitat preference was by recording burrow density in different habitats (See *Introduction* p. 2 for site descriptions). Ice-wedge polygons were obviously preferred summer habitat with raised beaches receiving moderate use and hummocky sedge-moss meadows (mesic meadows) much less (Table 1). Granitic outcrops were rather densely inhabited, but accurate burrow counts were impossible because many burrow entrances were hidden under rocks or in crevices.

A finer subdivision of the location of burrows was attempted in 1970 and 1971. In July 1970 a small (approximately 30×50 m), isolated portion of raised beach and the earth mound areas immediately adjacent to it were mapped for lemming burrows. The total area of the raised beach was 1,700 m^2 and the total area examined was 2,300 m^2. The area examined was bordered by a lake and a hummocky sedge-moss meadow.

The largest concentrations of burrows were in earth mound areas (lower slopes of raised beaches) where, for example, a group of 60 burrows lay in an area of about 18 m^2; however, some earth mound areas contained no burrows. Burrows on the crest and upper slope of raised beaches were strongly

Table 1. Estimates of lemming burrow densities in three habitats on Truelove Lowland; data from all summers combined.

Habitats	*Area samples (ha)*	*Density (burrows ha^{-1})*
Raised Beaches	16.67	115.4
Ice-wedge Polygons	1.44	533.3
Hummocky Sedge-moss Meadows	26.63	26.7

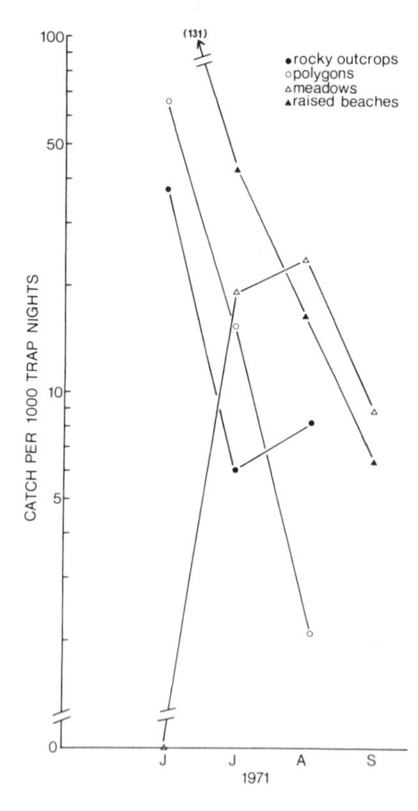

Fig. 2 Changes in snap-trap capture index during summer 1971 in four habitats. Note exponential declines on raised beaches and in polygons, with temporary invasion of meadows in July and August.

associated with frost cracks that served as runways and provided some concealment from avian predators (Pattie this volume).

The second raised beach, approximately 625 m long, was divided into 25 m^2 blocks, and burrow counts were made in each block. Association of burrows with frost cracks on the raised beach and with the cushion plant-moss community was again apparent. As elsewhere in the Arctic, microtopography appears to play an important role in the life of lemmings.

In 1971 the proportion of burrows in a number of microhabitats was tallied (Table 2). Ice-wedges were clearly important on the crests of beach ridges; there was a strong association with rocks on backslopes and mesic meadows; and southern exposures were preferred on polygons.

Some indication of seasonal shifts in habitat was also obtained in 1971 (Fig. 2). Catches declined strongly on raised beaches and polygons; declined, then levelled out in rocky areas, and increased temporarily in hummocky sedge-moss meadows during July and August before declining in September.

Winter habitat preferences were determined from the location of winter nests, which were overwhelmingly located in two areas — transition zones of raised beach foreslopes (cushion plant-moss communities) and those backslopes that terminated abruptly in meadow. These are the places that have the deepest and most persistent winter snow cover.

Winter habitat is within the cushion plant-moss community where *Dryas integrifolia*, *Saxifraga oppositifolia*, *Carex misandra*, and *Salix arctica* predominate (Svoboda this volume). Caloric values for alive aboveground vascular plants were mostly in the range 4.6 to 5.1 Kcal g^{-1} (Svoboda this volume).

Table 2. Proportion of lemming burrows located in different microhabitats in 1971.

Habitats			*Under rocks* %		*Not under rocks* %	*In ice wedges* %
Raised Beaches	Crests		18		29	54
	Foreslope		56		44	
	Backslope		96		4	
Ice-wedge Polygons	Exposure:	North		14		
		South		36		
		East		12		
		West		13		
		Top		25		
Hummocky Sedge-moss	Under rocks			67		
Mesic Meadows	Not under rocks			33		

Food habits

Based upon early summer observations, current feeding and evidence of past feeding was most evident for *Saxifraga oppositifolia* and *Salix arctica* within the transition zone of the raised beaches. Winter nests were constructed from *Carex* shoots but summer feeding trails showed no preference for sedges. *Carex misandra* was found in June at entrances to ventilation shafts but the exact use of this material could not be determined.

Opportunity to study food selection was provided by a light fall of snow (4 cm) on 19 September, 1971. During the next two days lemming trails were frequently encountered and followed. The animals dug feeding craters to reach favored food plants. Of 701 craters examined 95.8% terminated at one of only four species of dicotyledonous plants. The remaining 4.2% terminated at eight additional species. The species most often utilized were *Dryas integrifolia* (42.9%), *Saxifraga oppositifolia* (36.9%), *Salix arctica* (12.3%), and *Pedicularis* sp. (3.7%). Caloric values are high for all four species and the first three species have a high standing crop and plant density.

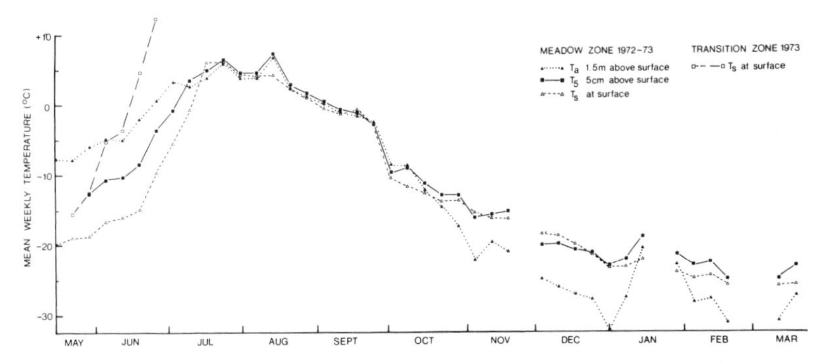

Fig. 3 Weekly mean temperatures in the hummocky sedge-moss meadow (Intensive Study Site) on Truelove Lowland from May 1972 to March 1973 and in a transition zone site during May and June 1973.

Microclimate

Temperature records on the Intensive Meadow (Fig. 3) and snow depths on an adjacent meadow in the winter 1972-73 provided the nearest approach so far obtained to a description of lemming winter microclimate at these latitudes. The record has been analysed in detail (Fuller et al. 1975a). Several striking features merit brief comment here. Spring thermal overturn (Pruitt 1957) occurred before 15 May, but ambient air (Ta) and subnivean (Ts) temperatures did not reach 0°C until 6 weeks later. This is in strong contrast to the taiga (Pruitt *op cit.)* where the two events occur nearly simultaneously. Fall thermal overturn was also delayed well past the time when Ta and Ts fell to 0°C, again in marked contrast with similar events in taiga. Both Ts and Ta fell gradually from the July peak to the winter low (average decrease in weekly mean Ts, 1.3°C; only one week during which mean Ts fell by more than 2°C). This gradual change could be important in relation to physiological acclimatization by the animals. From January to snowmelt, it was slightly warmer 5 cm above

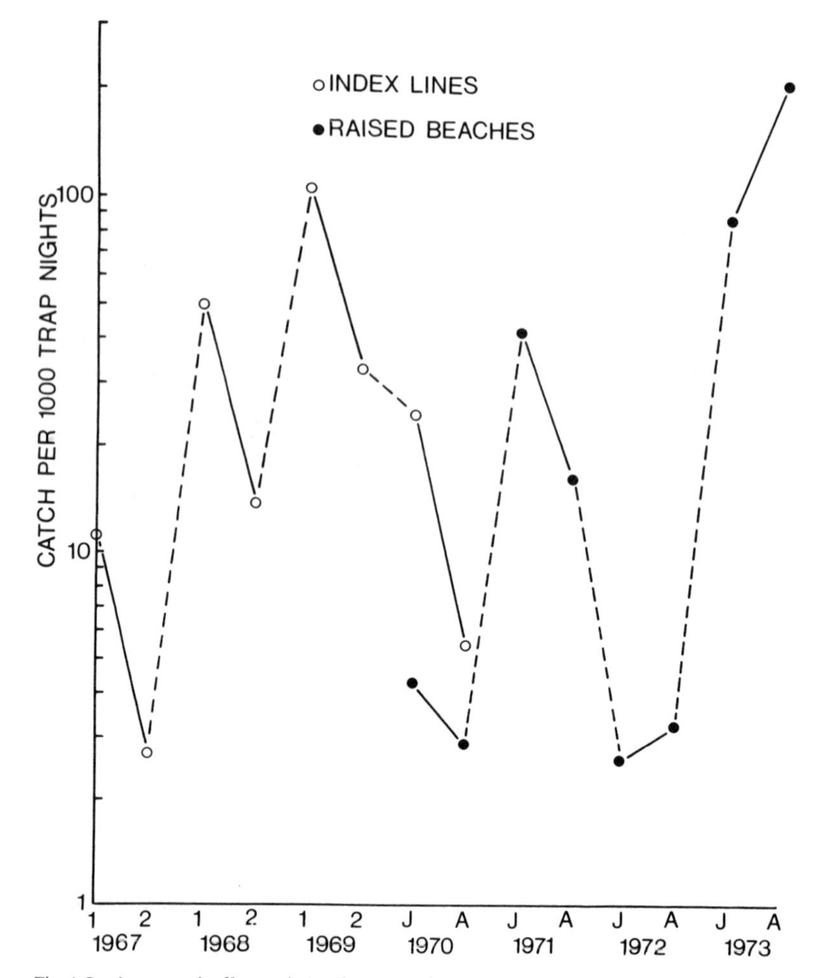

Fig. 4 Catches per unit effort on index lines (o) and all trapping on raised beaches (•) during seven summers. The numbers 1 and 2 in years 1967-69 represent first and second trappings of index lines. The letters J and A represent monthly mean catches on beach ridges for July and August.

Table 3. Predation by ermines *(Mustela erminea)* on winter nests of lemmings over four winters.

Winter	*Nests Examined*	*Attacked by ermines Number*	*%*
1969-70	198	23	11.6
1970-71	80	9	11.3
1971-72	58	9	15.5
1972-73	35	2	5.7
Total	371	43	12.1

the soil surface (T_5) than at the surface, which may explain the location of winter nests — in the api, ("snowpack," Pruitt 1970) not on the soil surface. For a period of at least 11 weeks the most favorable temperature available to lemmings was below −20° C, yet we have evidence that lemmings must have bred vigorously during the winter of 1972-73 (see Fig. 4). This appears to conflict with Quay's (1960) observation that captive female *Dicrostonyx* did not ovulate when held at temperatures below about −8° C. The partial record of Ts from May-June 1973 shows that spring was much earlier in the transition zone that year than in the meadow in 1972.

Predation

No direct observations of either avian or mammalian predation on lemmings were made; however, some idea of the intensity of winter predation by ermine *(Mustela erminea)* was obtained from examination of lemming winter nests in spring (Table 3). Observed rates of ermine predation were not significantly different (chi-square test) from year to year although lemmings were scarce in 1972, declining in 1970, moderately common in 1971, and at peak abundance in 1973. Our average rate of predation is significantly lower than that reported by Maher (1967) on Banks Island (20%) ($p < .05$) and by MacLean et al. (1974) in Alaska (34.7%) ($p < .001$). Only in 1971-72 did our predation rate approach Maher's figure ($p \simeq .5$). If Maher is correct in suggesting that predation accentuates a lemming decline, then ermine predation may have been partly responsible for the extremely low numbers on the Lowland in 1972. Our data, however, give no support to the hypothesis that predation prolongs the depressed phase since our population went from extreme low to peak in a single year.

Demography

Data on numerical changes, and associated changes in other population parameters have been published elsewhere (Fuller et al. 1975b), and we present here the main facts necessary to an understanding of the role of lemmings in energy flow.

Numerical changes

Trapping on index lines (see methods) yielded a regular rise to a peak in 1969 and decline in 1970 (Fig. 4). Unfortunately, the index lines were not run after 1970 and we have been forced to use all trapping on raised beaches to plot

Table 4. Estimates of lemming numbers and mean number of "population units" in four habitats in August of each year of the study.

Year	Raised Beaches (864.3)	Habitat type and area (ha) Polygons (10)	Meadows (1767.3)	Outcrops (533.2)	Total (3174.8)
1970	715	29	0	0	744
1971	605	2	530	267	1404
1972	233	0	18	27	278
1973	5000	100	1600	500	7200
4-year avg.	1638	33	537	199	2406
No. of females (53%)	868	17	285	105	1275
No. of pop. units (40 ♀♀)	21.7	0.4	7.1	2.6	31.9

population trends from 1971 to 1973. The discrepancy between index catches and total catches on raised beaches in 1970 was greater in early summer than in August, and we suggest that all catches during period one index trapping were high for the following reason. During snowmelt, raised beaches were the first areas clear of snow. Lemmings stayed under snow patches on beach ridge slopes and meadows until those areas became too wet, then they were forced out onto raised beaches. Concentration of animals on raised beaches, which probably increases trapability, coincided with period one index trapping.

In spite of these difficulties it is clear that there was a peak in 1969 and 1973, with lows in 1970 and 1972. Numbers in 1971 were anomalously high for the midpoint in a 4-year "cycle."

We draw attention to the fact that sharp declines occurred during at least five of seven summers studied. We consider that no change occurred in 1972 since the difference in capture index between July and August is less than one per 1,000 trap-nights. The summer of 1973 presents a problem. Capture indices conflict with the subjective assessment of several workers that numbers declined throughout the summer. What actually happened makes a large difference in energy flow calculations.

The observed summer declines average about 70%. If early summer indices are biased upward, as we suspect, that would be an overestimate. Making allowance for this bias and for lack of any detectable change in 1972 we estimate 60% decline in what we will call "typical" summers.

The observation that summer declines are typical is contrary to intuition and to experience at lower latitudes where summer is a season of population gain. It follows, then, that summer losses must be made good in most winters. Low summer values in 1970 and 1972 followed decline winters, but the other four winters were marked by strong increases (Fig. 4).

Estimates of absolute August numbers are shown (Table 4) for the main lemming habitats. They are based on moderate to extensive trapping efforts in the years 1970 to 1972, and on very limited trapping in 1973. They are cautious estimates, and the greatest error is likely to be associated with the larger numbers. This is no reflection on the investigators, but merely reflection of the fact that a small percentage error becomes absolutely large when extrapolated to a large population.

Changes internal to the population

Body size

It is now well known that, in a fluctuating population of microtine rodents, a correlation exists between the state of the population (increase, decrease, peak,

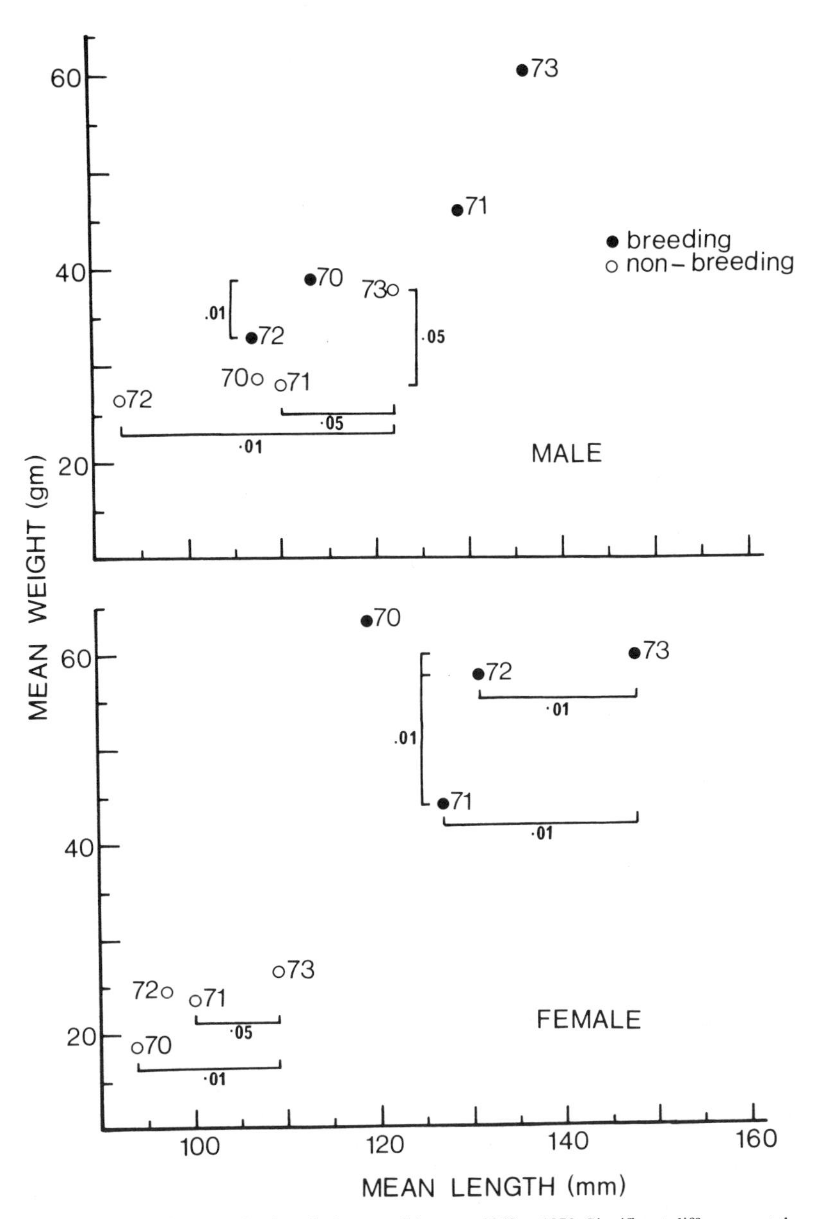

Fig. 5 Body size of *D. groenlandicus* in August of the years 1969 to 1973. Significant differences at the 0.05 and 0.01 level are shown by lines bracketing data points.

depression) and growth rates of the individuals in it (Krebs et al. 1973). Animals born into increasing populations (e.g. 1973) grow rapidly and attain large size whereas those born into decreasing populations (e.g., 1970 and 1972) grow slowly and remain small. Devon Island lemmings fit this general pattern except for one anomaly (Fig. 5). Mature females were not significantly smaller in 1970 and 1972 than in 1973.

To show that growth rates changed, we plotted body size against pelage class (an indicator of age that is independent of length or weight). Clearly,

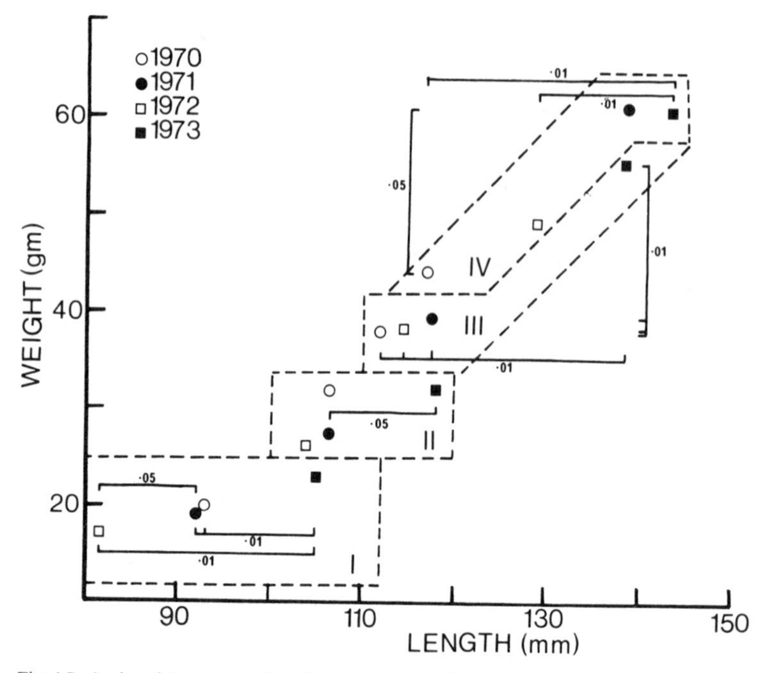

Fig. 6 Body size of *D. groenlandicus* (sexes combined) plotted according to pelage class (Roman numerals). Significant differences at the 0.05 or 0.01 level are shown by lines bracketing data points.

animals of all ages were larger in 1973 than in other years (Fig. 6). For 1971, the youngest animals (born during the summer decline?) were small, whereas the oldest animals (born during the winter increase?) were as large as those in 1973.

Reproduction

Most males less than 105 mm in total length had testes ⩽ 6 mm long and lacked sperm in the cauda epididymidis. For larger males, reproductive condition depended strongly on season. In June, about 90% (n=35) had functional testes, whereas in August the proportion dropped to 34% (n=29). The proportion functional in August varied from year to year, but numbers are too small to demonstrate statistical significance. More, and smaller, males were functional in low density populations (1970 and 1972) than at high density (1973), suggesting inhibition of maturation by some density-related, presumably social or behavioral mechanism. This phenomenon is also well known to students of fluctuating microtine populations (Krebs et al. 1973). When we related sexual maturity and body size to pelage class, it was clear that puberty occurred at a younger age in years of low numbers than in the peak year.

Only 18 visibly pregnant females were taken during the course of this study — too small a sample to yield significant comparisons by years or months. An additional 29 females had at least one set of visible placental scars yielding a sample of 47 breeding females.

Only tentative conclusions can be drawn from the data presented in Table 5. Clearly, there is little chance that females breed until they attain a total

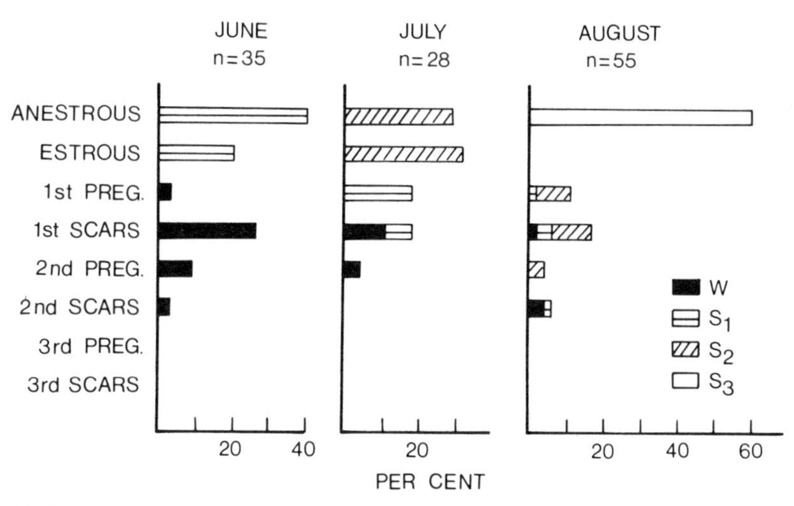

Fig. 7 Reproductive performance of *D. groenlandicus* females by cohorts with data from all years pooled. Cohorts are overwintered animals (W) and three litters of summer young (S_1, S_2, S_3).

Table 5. Proportion of breeding females in different length classes.

Month/Year	*Total length classes (mm)* ≤*109* %	*n*	*110-119* %	*n*	≥*120* %	*n*	*All*≥*110* %	*n*
June								
1971	0.0	13	0.0	6	82.4	17	60.8	23
July								
1970+72	0.0	10	50.0	2	66.7	9	68.8	16
1971	0.0	2	66.7	3	100.0	2		
August								
1970+72	0.0	11	50.0	2	100.0	8	87.5	16
1971	0.0	6	0.0	1	100.0	5		
1973	0.0	9	0.0	4	72.7	11	53.3	15
Total	0.0	51	22.2	18	82.7	52	67.1	70

length of 110 mm. Relatively few breed at lengths of 110 to 119, although some undetected pregnancies among those with perforate vaginae could swell the percentage. Most of the breeding females in this category were taken in July. They were unlikely to breed in June, and may be inhibited in August of years of high numbers. Observed differences in proportion of breeders among the largest animals are interesting even though sample sizes are small. The smallest proportion was found in July of the two years of low numbers, although by August of those years no large females were taken that had not bred. Participation in breeding appeared to be very high in July and August 1971, but may have been reduced in August 1973 when numbers reached a maximum. These trends are difficult to reconcile with a population increase in 1973. Reproductive indices were higher in 1971 than in 1973, yet the 1971 population clearly underwent a summer decline. This reinforces the subjective conclusion that 1973 was also a decline summer.

Analysis of the seasonal progression of reproduction leads to interesting conclusions (Fig. 7). The June population clearly consists of two kinds of females — a cohort (W) of large body size that has recently had at least one litter, and must, therefore, have been present during the preceding winter; and a cohort (S_1) of small body size, showing no visible pregnancies, which must be the late winter/early spring offspring of the large animals.

In July, winter cohort females are reduced in number, but more importantly, their reproductive performance has not improved, i.e., none of them are pregnant a third time, although the most advanced June females could be. Survivors of cohort S_1 are now breeding and account for 62% of total litters born and *in utero* for the month. A second summer cohort of young (S_2) has also appeared.

In August, there are still a few members of the W cohort, but once again they are in the same reproductive categories as in June. Cohort S_1 is also reduced in number and has caught up with cohort W in breeding performance. The bulk of August litters will obviously be produced by S_2 females. The August population is dominated by a third cohort of summer young (S_3), all of which remain in anestrus. We draw two major conclusions from these data: (1) W cohort females did not breed throughout summer at anything like the rate at which they bred in May and June. Had they done so, third (and even fourth) pregnancies would have appeared. Dunaeva (1948) took female *D. torquatus* on Yamal that showed evidence of bearing four summer litters while we have no hard evidence of any female bearing more than two on Devon Island. This difference may be due to the severity of summer conditions on Devon Island and may be one basic reason for the frequency of summer population declines; (2) The bulk of the overwintering population consists of S_3 individuals that do not mature in the summer of their own birth. They, in turn are offspring, in the main, of S_1 and S_2 cohort females that do mature at an early age.

Litter size

To get an estimate of litter size we combined embryo and placental scar counts. The largest litter *in utero* contained nine embryos. We therefore assumed that scar counts of more than 9 and less than 18 represented two litters. We further assumed, unless we could see a difference in the age of the scars, that the two litters represented by scars were equal or subequal (i.e., 12 scars equals two litters of 6; 11 scars equals litters of 6 and 5). This procedure reduces variance and should bias our results toward significant differences, but no significant differences were found between years. Mean litter size was estimated to be 5.7, with a tendency, not statistically significant, for larger size in years of high numbers. Dunaeva (1948) observed larger litters in *D. torquatus* in years of high numbers than during lows, but she did not say whether the difference was statistically significant. Krebs (1964) detected no significant differences in litter size in different phases of a population cycle at Baker Lake. Litter size was about the same as in other studies.

Sex ratio

The overall sex ratio, 50.3% male, was very close to unity (Table 6), but there are significant differences in sex ratio between breeding and non-breeding animals ($X^2 = 11.185$; $p = < 0.005$). Among non-breeders, males were clearly in a minority in all subsamples but one (August 1971, Table 6). On the other hand, breeding males outnumbered breeding females, except, again, in August 1971 and more dramatically in August 1973. Omitting 1973, there was an overall preponderance of males (55%) but this value is not significantly different from a 1:1 ratio.

It is also possible that males are outnumbered by females in peak summers. If such a sex ratio is a causal factor in population increase, however,

Table 6. Summary of sex ratios by year, month, and breeding condition. Figures in the table are percentage of males (sample size).

	Year and breeding condition								
Month	*1970+72*		*1971*		*1973*		*All years*		*All animals**
	Non-br.	*Breeding*	*Non-br.*	*Breeding*	*Non-br.*	*Breeding*	*Non-br.*	*Breeding*	
June	25(4)	100(4)	44(39)	67(42)	–	–	42(43)	70(46)	56 (89)
July	30(20)	76(29)	33(9)	67(12)	31(26)		31(29)	73(41)	49 (96)
August	40(20)	59(22)	58(19)	50(8)	46(28)	10(10)	48(67)	45(40)	47 (107)
All months	34(44)	71(55)	46(67)	65(62)	46(28)	10(10)	42(139)	63(127)	50.3(292)
All animals	55(99)		55(129)		34(64)		50.3(292)		

*Adding May and September animals males constituted 52.5% of a sample of 318.

we would expect the sex ratio in 1971 to be intermediate between that of the low years and that of the peak year, which it was not.

Energy flow

In this section we attempt a dynamic analysis of energy flow, which depends, at several junctures, on some critical assumptions. The first assumption is that lemmings have a 4-year cycle of abundance. This is widely accepted in the literature and our data show highs in 1969 and 1973 with lows in 1970 and 1972 and intermediate numbers in 1971. The 1971 population was higher than expected in mid-cycle, but apart from that the assumption is probably defensible. It is also clear that the population shows cyclic characteristics in parameters such as body size and age at sexual maturity. Given a 4-year cycle of abundance, we then conclude that it is valid to use 4-year mean figures for calculating mean energy flow. It further follows from this that we can assume long term stability in numbers. We need that assumption to make even a crude estimate of energy flow in winter. Given long term stability, population losses in summer must be made good in winter.

A second critical assumption is that rates of gain and loss in numbers are exponential, as are weight gains in post-weaning young. This assumption is probably not always fulfilled, but it is likely more nearly true than an assumption of linear change (see, for example, Fig. 2).

We have accepted certain parameters, with which our limited data tend to be in agreement, from the literature. Mean birth weight is given as 3.8 g by Hansen (1957). We use the round number of 4.0 g. Age at weaning seems to be 14 or 15 days, at which time captive lemmings weigh about 14 g. We use 15 days and 12 g on the assumption that development may be somewhat slower under field conditions in the High Arctic than in a laboratory population. While our figure is below Hansen's mean, it is well above his minimum. The only winter weights we have been able to find for *Dicrostonyx* are for males from Baker Lake on the mainland tundra (Krebs 1964). We have computed a mean weight from his 5-gram weight classes.

Analytical method

The basic method adopted herein was to attempt to follow the fortunes of a "population unit" throughout a year. The unit adopted is a cohort of 50 adult females on 1 May. Once having estimated the numbers of animals in one such unit on different dates, we can estimate the numbers in the corresponding male segment of the population from the observed monthly sex ratios. We then attempt an estimate of the number of "population units" in different habitats, and on the Lowland as a whole, and convert these to a density figure. From the observed body weight and the experimentally determined water and energy content of lemming carcasses, we can determine standing crops m^{-2} in g dry weight and Kcals. From the measured energy intake in feeding trials we estimate total consumption. Production is estimated, for summer only, by the method outlined in Petrusewicz and MacFadyen (1970). Respiration was not measured directly and is therefore estimated from the formula (Morrison et al. 1959) $cm^3\ O_2 g^{-1}\ hr^{-1} = 3.8\ W^{-0.27}$. At an assumed R.Q. of 0.7, one litre O_2 is equivalent to 4.7 Kcal.

Energy content of feces, but not urine, was measured. Animals in feeding trials were fed a mixture of lab chow and arctic willow *(Salix arctica)* which is probably more digestible than native vegetation alone. We can therefore only estimate assimilation and rejecta within wide limits.

Determination of numbers

We began our analysis by observing that the ratio of adult females to young females in late June was 2:3, that is 40% adults, or 40 adults in a population of 100 females. It seems a reasonable assumption that 40 females could survive until late June from a cohort of 50 present on 1 May because for most of that period they are relatively secure under the snow.

We have already indicated that population declines of about 60% occurred in typical summers. Thus, 100 females at the end of June become 40 on 31 August. We further estimated from our trapping data (Fig. 7) that the composition of the late August population (arbitrarily considered to be on 31 August) was as follows:

Cohort	W	S_1	S_2	S_3
% occurrence	5	14	21	59

Where W = winter cohort

S_1, S_2, S_3 = first, second and third summer cohorts.

Assignment to a particular cohort was done on the basis of body size, pelage class, and reproductive history. For example, S_2 individuals could not have produced two litters in August and W individuals were unlikely to have had only one litter. The expected number of females of each cohort on 31 August then becomes:

W		S_1		S_2		S_3		
2	+	6	+	8	+	24	=	40

The number of W females at bi-weekly intervals between 1 July and 31 August, was estimated graphically by plotting numbers on the given dates (40, 2) on semi-log paper and connecting them by a straight line (assumption of exponential decline). In an analogous way, the number of S_1 females was estimated and they declined from 60 to 6.

Given an estimate of females in the population on 15 July, and taking the proportion of those that had produced a litter from our trapping results, it is possible to estimate the number of S_2 females produced and weaned. The number of young in the trappable population is calculated from our trapping data (ratio of young females to all older females). Proceeding in an analogous manner, one can construct a table (Table 7) of probable numbers of females of

Table 7. Estimate of the number of females of different cohorts in a "population unit" at different dates during summer.

Cohort	*May* 1	*May* 20	*June* 5	*June* 30	*July* 15	*July* 31	*August* 15	*August* 31
W	50	47	44	40	19	9	4	2
S_1 born	?	128	—	—	—	—	—	—
weaned	—	—	98	—	—	—	—	—
trappable	—	—	—	60	29	19	11	6
$W + S_1$	50		142	100	48	28	15	8
S_2 born	—	—	—	91	—	—	—	—
weaned	—	—	—	—	70	—	—	—
trappable	—	—	—	—	—	60	23	8
$W + S_1 + S_2$					128	88	38	16
S_3 born	—	—	—	—	—	45	—	—
weaned	—	—	—	—	—	—	35	—
trappable	—	—	—	—	—	—	—	24
Total (excluding nestlings)	50	47	142	100	128	88	73	40

each cohort at fixed dates throughout summer, all of which are related to the original cohort of 50 W females.

Finally, we have projected loss rates through September to get an estimate of numbers remaining. The W cohort goes to extinction, and the S_1 and S_2 cohorts to 1 and 2 respectively. In our small sample of September females (n=5) we have 80% S_3 individuals, which would imply 12 individuals in our model. We have arbitrarily, however, assumed that their survival rate might increase in September because of (a) the departure of avian predators; and (b) appearance of snow cover, and have allowed 17 survivors (85%) for a total end-of-September population of 20 females.

The number of adult males was obtained by multiplying the population of females each month by the sex ratio. The number of young males per "population unit" is assumed to be equal to the number of young females.

The next problem is to estimate the number of "population units." As long as the population loop is closed, i.e., 50 females on 1 May averaged over a 4-year cycle, this estimate can be made in any season. Our best estimate is for August (Table 4), since we have already indicated that estimates made just at snowmelt are biased upwards. In late August, a "population unit" is 40 females. The number of population units in four major habitats, and on the

Table 8. Estimate of absolute numbers per hectare ($m^2 \times 10^4$) by cohort and time.

Month		*W* ♂	*W* ♀	S_1 ♂	S_1 ♀	S_2 ♂	S_2 ♀	S_3 ♂	S_3 ♀	*Total* ♂	*Total* ♀
May	1	0.502	0.502	—	—	—	—	—	—	0.502	0.502
	20	0.472	0.472	1.286	1.286	—			—	1.758	1.758
June	5	0.561	0.441	0.984	0.984	—	—	—	—	1.545	1.425
	30	0.510	0.402	0.602	0.602	0.914	0.914	—	—	2.026	1.918
July	15	0.183	0.191	0.280	0.291	0.702	0.702	—	—	1.164	1.184
	31	0.087	0.090	0.183	0.191	0.602	0.602	0.452	0.452	1.324	1.335
Aug.	15	0.036	0.040	0.098	0.110	0.206	0.231	0.351	0.351	0.691	0.732
	31	0.018	0.020	0.054	0.060	0.072	0.080	0.241	0.241	0.385	0.401

Table 9. Standing crops (Kcal ha^{-1}) of females and both sexes by cohort and time.

Month		Cohort									
		W		S_1		S_2		S_3		*Total*	
		♀	♂♀	♀	♂♀	♀	♂♀	♀	♂♀	♀	♂♀
May	1	59.7	119.4	—	—	—	—	—	—	59.7	119.4
	20	56.0	118.8	8.6	17.2	—	—	—	—	64.6	136.0
June	5	52.3	108.3	19.7	39.4	—	—	—	—	72.0	147.7
	30	48.0	94.2	33.8	62.2	6.2	12.3	—	—	88.0	168.6
July	15	18.6	35.1	17.6	31.3	12.6	25.2	—	—	48.8	91.6
	31	8.8	15.9	10.4	21.4	27.4	47.2	2.7	5.5	49.3	90.0
Aug.	15	3.8	7.7	9.3	15.4	18.6	26.9	6.6	13.2	38.3	63.1
	31	2.2	3.8	4.4	8.2	6.6	10.4	8.8	17.6	22.0	40.0

Lowland as a whole, is given in Table 4, and the number of population units per ha is 1.005×10^{-2}. From estimates of the number of females/population unit (Table 7), sex ratios in different months and age classes (Table 6), and the ratio of 1.005×10^{-2} population units per ha, estimates of absolute numbers per ha for different cohorts were obtained (Table 8).

Determination of standing crop

A small sample of lemmings was used in 1971 to determine caloric content. There was an apparent difference between June values and those from later in the summer, but there were no obvious sex or age differences at a given time. Accordingly, the value 6.143 ± 1.18 Kcal g^{-1} d wt was used to estimate standing crop in May and June, and 5.487 ± 0.81 Kcal g^{-1} d wt was used for the balance of the summer. Water content was determined to average 72.1%. Standing crop (Table 9) was estimated from the product of number of animals (Table 8), mean live weight (from field data), 0.279 (conversion to dry weight) and the appropriate calorific value.

Determination of production

Growth-survivorship curves (Petrusewicz and MacFadyen 1970) were drawn for each cohort (Fig. 8) and areas under the resulting curves were determined graphically to give estimates of production (Table 10). Total summer production is about twice that for *Peromyscus maniculatus* in Michigan (Petrusewicz and MacFadyen 1970; Table 7-1) and about 20% of that of Michigan ground squirrels *(ibid.)*. The production/consumption ratio also falls between those for the species cited, which suggests that we are at about the right order of magnitude, at least.

Table 10. Estimate of production (Kcal ha^{-1}) to 31 August based on growth-survivorship curve (Fig. 8).

Cohort	*Production*	*Elimination*	*"Potential" production*	*Production of newborn*
S_1	87.4	79.3	172.8	17.4
S_2	79.4	68.4	125.8	12.4
S_3	26.7	8.9	34.1	5.4
Total	193.5	156.6	332.7	35.2

Table 11. Results of feeding trials with lemmings maintained at ambient temperatures on lab chow and *Salix arctica* in July 1971.

Class of animal	*Body weight*		*Consumption (C) (Kcal)*	*Feces (F) (Kcal)*	*Assimilated (C - F)*	*Kcal g^{-1} dry wt day^{-1}*	
	Live	*Dry*				*Consumed*	*Assim.*
Males	34.4		62.67	16.56			
	59.5		112.95	30.12			
	34.5		73.46	14.72			
	58.3		104.37	24.65			
	186.7	52.01	353.45	86.05	267.40	6.8	5.1
Pregnant female	62.6	17.46	109.54	21.96	87.58	6.3	5.0
Lactating females	43.3		106.15	22.55			
	44.4		112.80	28.40			
	45.4		131.83	27.04			
	133.1	37.13	350.78	77.99	272.79	9.5	7.3

Determination of consumption

To get total consumption we estimated the mean daily number of adults in the population each month on the assumption of exponential decline. Multiplying by the mean body weight gave us a mean daily biomass for each month. From feeding trials we determined that consumption was 6.8 Kcal g^{-1} dry wt day^{-1} except for lactating females that had a value of 9.5 (Table 11). This enabled us to compute the consumption values in Tables 12 and 13. Calculations of mean daily biomass for animals just entering the population are shown in Table 14, and their consumption values are shown in Tables 12 and 13.

The sum of these values is 1.4135 Kcal m^{-2} from 1 May to 30 September.

Estimate of rejecta

According to feeding trials about 25% of the energy ingested was unassimilated (Table 11). The test food, however, was largely laboratory chow, which is presumably designed for high digestibility; thus, the energy lost in feces in nature is likely to be more than 25% of that taken in. We also have no measurement of the energy content of lemming urine.

Assimilation of monocots by microtines appears to be inefficient (35% for *Lemmus* Melchoir 1972) and 32% to 43% for *M. pennsylvanicus* (Pulkinen 1971) but dicotyledonous material may be easier to handle. Assimilated energy probably lies between 50% and 60% of energy ingested, and we use 55%.

Estimate of respiration

We have no direct measure of Average Daily Metabolic Rate (A.D.M.R.), but have made an estimate from the formula O_2 consumption (cc g^{-1} hr^{-1}) = 2.89 – (0.74°C) (Brown and West 1970, p. 43).

Assuming a summer ambient temperature of 3°C and using mean daily biomass for each month (Tables 13, 14), we calculate A.D.M.R. = 0.2101 Kcal m^{-2}.

A second estimate can be made using the formula of Morrison et al. (1959) $ccO_2\ g^{-1}\ hr^{-1} = 3.8w^{-0.27}$.

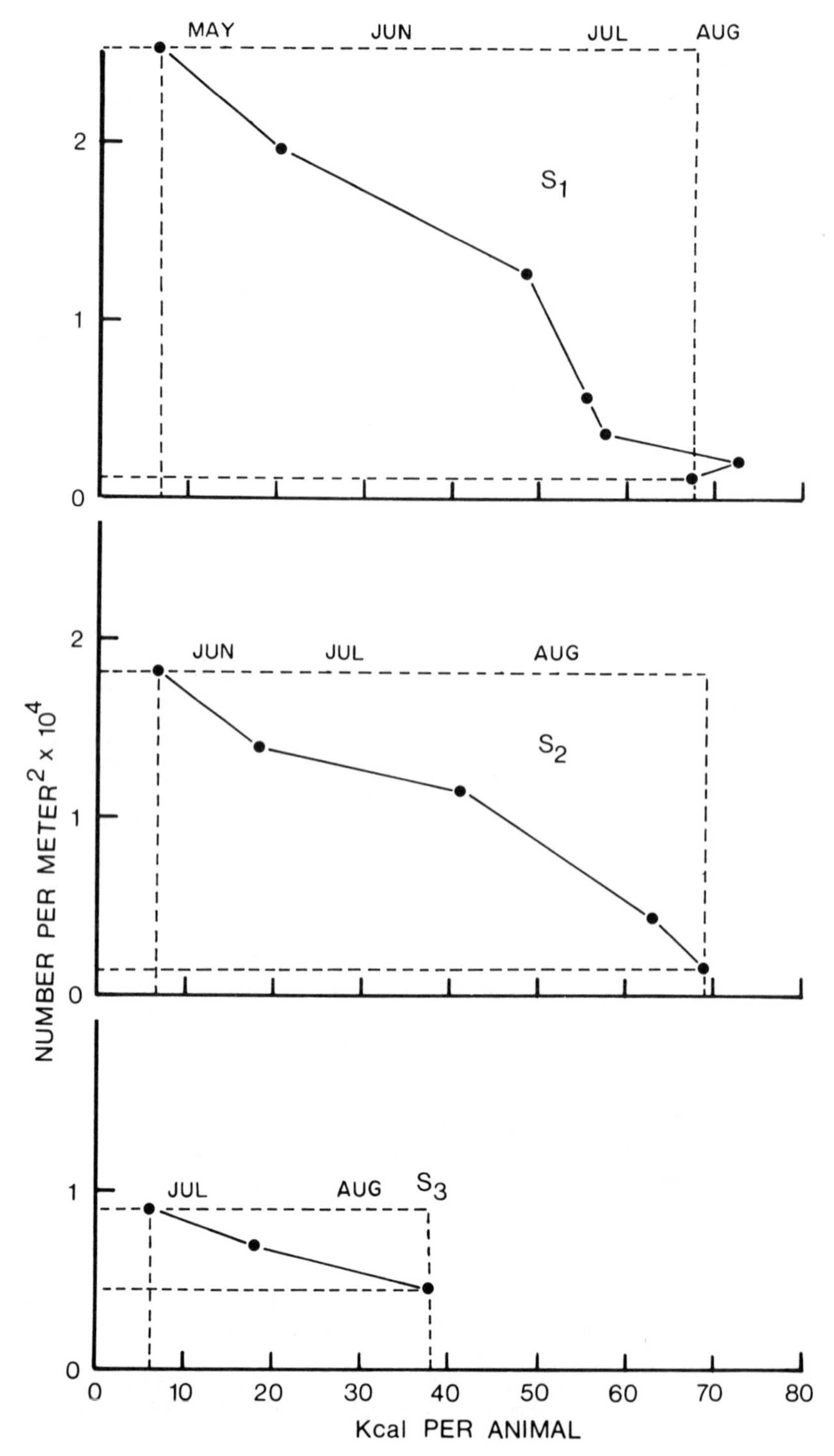

Fig. 8 Growth-survivorship curves for summer generations of lemmings on Devon Island. Estimates of production, based on areas under these curves, are given in Table 14.

Table 12. Estimate of monthly consumption by males from May to September.

Month	Mean daily biomass (g dry wt $m^{-2} \times 10^4$)		Consumption Kcal m^{-2}		
	Adult	*Young*	*Adult*	*Young*	*Total*
May	9.096	0	0.192	0	0.1918
June	5.512	3.665	0.112	0.037	0.1497
July	5.632	2.801	0.123	0.029	0.1515
August	3.213	1.406	0.068	0.014	0.0821
September	2.520	0	0.051	0	0.0514
Total	—	—	0.546	0.080	0.6265

Table 13. Estimate of monthly consumption by females from May to September.

Month	Mean daily biomass (g dry wt $m^{-2} \times 10^4$)	% females		Days during month		Consumption (C) for month Kcal m^{-2}			Monthly total
		Lact.	*Non-lact.*	*Lact.*	*Non-lact.*	*Lact.*	*Non-lact.*	*Young*	
May	9.14	100		10	21	0.087	0.130	0	0.2173
			0						
June	8.24	70		10	20	0.055	0.079	0.058	0.2420
			30				0.050		
July	6.44	75		10	21	0.046	0.069	0.032	0.1811
			25				0.034		
August	3.99	40		10	21	0.015	0.023	0.014	0.1026
			60				0.050		
September	2.15	0		0		0		0	
			100		30		0.044		0.0439
Total	—	—	—	—	—	0.203	0.479	0.104	0.7870

This gives an estimate of Basal Metabolic Rate (B.M.R.), which must be adjusted for ambient temperature. Estimates of B.M.R. at thermoneutrality are given in Table 15. To obtain what might best be called Resting Metabolic Rate (R.M.R.) under summer ambient temperatures on the Lowland, our final figure (0.1562 Kcal m^{-2}) should be adjusted upward by a factor of about 1.5, yielding an estimate of about 0.23 Kcal m^{-2}. This is in fair agreement with the previous estimate, and suggests that the formula in Brown and West is actually an estimate of R.M.R. rather than A.D.M.R.

Total respiration, which includes the energy cost of activities such as searching for food, excavating burrows, reproduction, etc., must be greater than R.M.R. Since we have estimates for Assimilation ($A = 0.55\ C$) = 0.774 and Production ($P = 0.0193$), we can estimate Respiration (R) from the formula $A = P + R$:

$$R = 0.7581 \text{ Kcal m}^{-2}$$

We turn now to consideration of winter energetics. In the complete absence of hard data, we base our calculations on the following assumptions: (1) For long term population stability, a population unit must increase from 20 females in September to the 50 females with which we started the year on 1 May. On the assumption of exponential increase this means a mean daily number of females per population unit of 31.6 or 0.317 females per ha; (2) The sex ratio is 1:1, and the mean weights of males and females are equal. The only winter weights we have found in the literature (58 g) are those of Krebs (1964). The mean daily biomass of lemmings per ha is thus estimated to be

Table 14. Mean daily biomass of males and females from weaning to trappable size.

Cohort	Weanlings ha^{-1} No.	Biomass	Trappable young (per ha) No.	Live weight ♂	♀	Mean daily biomass (g dry wt ha^{-1}) ♂	♀
S_1	0.984	11.81	0.602	24.2	33.1	3.665	4.277
S_2	0.702	7.23	0.602	23.2	29.6	2.801	3.163
S_3	0.351	4.22	0.241	25.0	24.4	1.406	1.395

Table 15. Basal Metabolic Rate of lemmings per ha calculated from the formula $\text{Kcal g}^{-1}\text{ day}^{-1} = \frac{4.7\times24\times3.8\ W^{-0.27}}{1000}$

Month	Adult ♂ day	Mo.	♀ day	Mo.	Young ♂ day	Mo.	♀ day	Mo.	Total
May	5.45	168.95	5.47	169.57	—				338.52
June	3.78	113.40	5.07	152.10	2.81	84.30	3.14	94.20	444.00
July	3.84	119.04	4.24	131.44	2.31	71.61	2.52	78.12	400.21
Aug	2.55	79.05	2.99	92.69	1.40	43.40	1.38	42.78	257.92
Sept	2.14	64.20	1.90	57.00	—	—	—	—	121.20
Summer		544.64		602.80		199.31		215.10	1561.85

$0.317\times2\times58=36.77$ g; (3) Winter activity is more costly than summer activity, since it involves tunneling through snow, but the difference may be offset by a decrease in total daily activity so that the ratio Total respiration/R.M.R. may be about the same in winter as in summer. Using our estimate for summer, that ratio is $0.7581/0.2342 = 3.24$. We can therefore estimate total winter respiration if we can get an estimate of winter R.M.R.; (4) The ratio R/A remains constant from summer to winter. From our summer estimates that ratio is 0.54; (5) Digestibility does not vary from summer to winter, and is estimated to be 55%.

Proceeding from these assumptions, the key is to make an estimate of

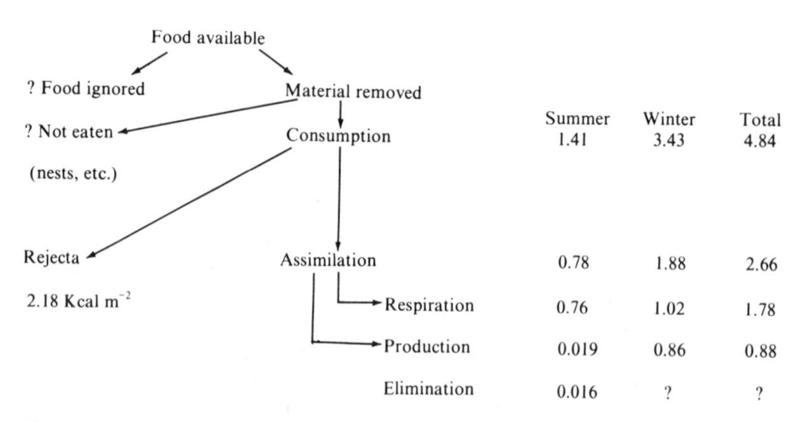

Fig. 9 Energy flow relationships. Aboveground standing crop, vascular plants (alive) Kcal m^{-2}. Raised beaches = 617; sedge moss meadows = 400.

winter B.M.R. Using the formula $ccO_2\ g^{-1}\ hr^{-1} = 3.8\ W^{-0.27}$ and substituting 36.77 for W, we obtain $ccO_2\ g^{-1}\ h^{-1} = 3.8 \times 36.77^{-0.27} = 1.44$.

This is equivalent to $1.44/1000$ litres $O_2 \times 24\ hr \times 211$ days = 7.27 litres $0_2\ g^{-1}\ winter^{-1}$ or $7.27 \times 4.7 = 34.17\ Kcal\ g^{-1}\ winter^{-1}$ or
$34.17 \times 36.77\ g\ ha^{-1} = 1256\ Kcal\ ha^{-1}\ winter^{-1}$
$= 0.1256\ Kcal\ m^{-1}\ winter^{-1}$.
But at prevailing ambient temperatures (about −25C) R.M.R. is about 2.5 times B.M.R. = $0.3141\ Kcal\ m^{-2}$. If total respiration is 3.24 times resting metabolic rate, it ought to be about $1.02\ Kcal\ m^2$ for the winter.

If R:A = 0.54 in winter as in summer, then winter assimilation = $1.02/0.54 = 1.88\ Kcal\ m^{-2}$ and Consumption $A/-0.55 = 3.43\ Kcal\ m^{-2}$
Production can be estimated from the formula P = A-R = 1.88-1.02
$0.86\ Kcal\ m^{-2}$.

Summary and conclusions

1. A population of *Dicrostonyx groenlandicus* was studied for four summers (1970-1973). Snap-trapping (53, 575 trap-nights) yielded 346 specimens for analysis of population phenomena. Feeding trials with live lemmings and calorimetry of lemming tissues gave additional limited information on energy flow.
2. Most lemming burrows were associated with ice-wedge polygons or with ice-cracks on raised beaches. Hummocky sedge-moss meadows were little used in summer although there was evidence of a temporary shift to meadows in July and August. Granitic outcrops were also used, but burrows hidden under rocks were difficult to count.
3. Winter nests were built predominantly along slopes of raised beaches—areas with the deepest and most persistent snow cover. Food plants available during winter are members of the cushion plant-moss community (*Dryas integrifolia*, *Saxifraga oppositifolia*, *Carex misandra* and *Salix arctica* being dominant).
4. Snow tracking and identification of plants in 701 feeding craters showed that *D. integrifolia* (42.9%), *A. oppositifolia* (36.9%), and *S. arctica* (12.3%) are highly selected food plants at the beginning of winter (mid-September).
5. Air temperature seldom, if ever, reaches the lower critical temperature for lemmings which are always, therefore, under some degree of thermal stress during periods of activity outside of insulated burrows or nests. For a period of 11 weeks in the winter of 1972-73 the most favorable subnivean temperature available to them was probably colder than −20° C.
6. Predation on winter nests by *Mustela erminea* averaged 12.1% (n=371). Interannual differences were not significant, but nest predation on Devon is significantly lower than reported for Banks Island (20%; Maher 1967) and for *M. nivalis* at Barrow (34.7%; MacLean et al. 1974). The observed level of nest predation did not seem to prolong the phase of low numbers, but the highest rate (15.5% in 1971-72) preceded the summer of lowest lemming numbers.
7. Lemming numbers on index trap lines increased from 1967 to 1969 and declined in 1970. This was followed by recovery to moderate numbers in 1971, then a deep depression in 1972 and increase to a peak in 1973. Catches per unit effort declined from early to late summer in at least 5 of 7 summers, and increases presumably occurred in the winters as a result of subnivean breeding.
8. Lemmings showed high growth rates in the peak year and attained large body size. This agrees with observations of many other fluctuating populations of microtines (Krebs et al. 1973).

9. Males attained sexual maturity at smaller body size and in more juvenile pelage condition in years of low numbers than in the peak year. This phenomenon also occurs widely in fluctuating microtine populations.
10. Maturation of females was probably delayed in the peak year. No females that had borne more than two litters were taken during the study, although 4 litters have been reported for *D. torquatus* on Yamal Peninsula (Dunaeva 1948). Litter size averaged 5.7 and no significant interannual differences were found.
11. Over-all sex ratio was 50.3% male, but there was a significant shift from a preponderance of females among juveniles to a preponderance of males among the sexually mature. Sex ratio was not well correlated with numerical change, although mature females outnumbered mature males in the peak year.
12. Energy flow relationships can be summarized as in Fig. 9. It must be remembered, however, that calculations are based on rather tenuous assumptions plus a minimum of hard data, and should only be accepted as a very preliminary estimate of the role of lemmings in a high arctic ecosystem.

Acknowledgments

Grateful acknowledgment is made to Sue Martell, Marjorie van Buskirk, and Jim Fuller for their help with the field work. The 1973 data were gathered by Ross Goodwin, Michael Muc, and Doug Whitfield.

References

Batzli, G.O. 1975. The role of small mammals in arctic ecosystems. pp. 243-268 *In:* Small Mammals: Their Productivity and Population Dynamics. F.A. Golley, K. Petresewicz and L. Ryszkowski (eds.). Cambridge Univ. Press. 451 pp.

Bliss, L.C. 1975. Tundra grasslands, herblands, and shrublands, and the role of herbivores. Geoscience and Man 10: 51-79.

Brown, J. and G.C. West. 1970. Tundra biome research in Alaska, U.S. I.B.P. — Tundra Biome Report 70-1. CRRL, Hanover, New Hampshire. 148 pp.

Dunaeva, T.N. 1948. Comparative survey of the ecology of the tundra voles of the Yamal Peninsula. Trudy Inst. Geog. Acad. Sci. U.S.S.R. 41: 78-143 (English Transl. in Boreal Inst., Univ. Alberta).

Fuller, W.A., A.M. Martell, R.F.C. Smith and S.W. Speller. 1975a. High arctic lemmings *Dicrostonyx groenlandicus:* I Natural history observations. Can. Field Nat. 89: 223-233.

———. A.M. Martell, R.F.C. Smith, and S.W. Speller. 1975b. High arctic lemmings *Dicrostonyx grownlandicus:* II Demography, Can. J. Zool. 56: 867-878.

Hansen, R.M. 1957. Development of young varying lemmings *(Dicrostonyx).* Arctic 10: 105-117.

Krebs, C.J. 1964. The lemming cycle at Baker Lake, Northwest Territories, during 1959-62. Arctic Inst. N.A. Tech. Paper 15, 104 pp.

———. M.S. Gaines, B.L. Keller, J.H. Myers and R.H. Tamarin. 1973. Population cycles in small rodents. Science 179: 35-41.

MacLean, S.F. Jr., B.M. Fitzgerald and F.A. Pitelka. 1974. Population cycles in arctic lemmings: winter reproduction and predation by weasels. Arctic Alp. Res. 6: 1-12.

Maher, W.J. 1967. Predation by weasels on a winter population of lemmings, Banks Island, Northwest Territories. Can. Field Nat. 81: 248-50.
Melchior, H.R. 1972. Summer herbivory by the brown lemming at Barrow, Alaska. pp. 136-138. *In:* Tundra Biome Proc. 1972 Symposium. S. Bowen (ed.). CRRL, Hanover, New Hampshire, 211 pp.
Morrison, P.R., F.A. Ryser and A.R. Dawe. 1959. Studies on the physiology of the masked shrew, *Sorex cinereus*. Physiol. Zool. 32: 256-271.
Petrusewicz, K. and A. MacFadyen. 1970. Productivity of Terrestrial Animals. Principles and Methods I.B.P. Handbook No. 13. Blackwell Scientific Pubs. Oxford. 190 pp.
Pruitt, W.O. Jr. 1957. Observations on the bioclimate of some taiga mammals. Arctic 10: 131-138.
————. 1970. Some ecological aspects of snow. Ecology of the subarctic regions. UNESCO. Paris. pp. 83-99.
Pulkinen, D.A. 1971. Food habits of the meadow vole, *(Microtus pennsylvanicus)* and their relationship to nutritional differences in available foods. M.Sc. Thesis, Univ. Saskatchewan.
Quay, W.B. 1960. The reproductive organs of the collared lemming under diverse temperature and light conditions. J. Mammal. 41: 74-89.

Arctic hares on Truelove Lowland

R.F.C. Smith and L.C.H. Wang

Introduction

Arctic hares *(Lepus arcticus)* are confined to the tundra zone of North America (Hall and Kelson 1959). The most northerly subspecies *(L. a. monstrabilis)* occurs throughout the Queen Elizabeth Islands with the exception of Prince Patrick Island.

No scientific studies have been made of population levels of arctic hares in the Canadian arctic. Population sizes are variable spatially and temporally (Cahalane 1961) and anecdotal reports suggest that parts of Ellesmere Island and part of Axel Heiberg Island support relatively high populations in some years while hares are generally rare elsewhere.

This study measured the energy metabolism of *L. a. monstrabilis* under various thermal environments both by conventional feeding experiments and by continuous recording of oxygen consumption in the laboratory. Based on the information obtained, the impact of this species on the production and energy flow on Truelove Lowland is evaluated.

Materials and Methods

No systematic census on a natural population was performed, although occasional investigations were made for signs of hares. Fresh feces from wild hares were collected on the Lowland whenever found.

Two hares, an adult male and a juvenile female, were captured by hand on Fosheim Peninsula, Ellesmere Island (approximately 79° N) during the first week of August 1972 and were flown to the base camp on Truelove Lowland.

Feeding experiments were conducted inside a 3 m by 3 m tent, so that approximate ambient temperatures (Ta) were maintained although the hares were protected from wind and precipitation. Trials were conducted from 6 to 17 August inclusive, although data from the first two days were discarded.

Each hare was placed within a painted wooden box measuring 75 cm × 75 cm × 75 cm with a hinged top and a large mesh floor which fitted over a metal tray. Weighed food, in the form of *Salix arctica*, the only local vegetation acceptable to the animals, was provided every 6 hr. Water was allowed *ad libitum*. Body weight and the weight of feces and uneaten food were determined at the end of each 24 hr period. Urine was measured volumetrically and allowed to air-dry in weighed jars.

Bunches of *Salix* placed inside wire baskets within the boxes were weighed before and after 24 hr trials in order to determine weight loss due to transpiration. Growth tips of willow which could be pinched off with a thumbnail were dried at 70°C for 48 hr. Examination of uneaten food in the feeding trials revealed that growth tips and not woody stems were consumed, so that only those portions of willow that could be removed by pressure from the thumbnail were used in testing transpiration losses and caloric values. An adiabatic calorimeter (Parr Instrument Co.) was used to determine non ash-free caloric equivalents in five samples each of *Salix*, air-dried feces and freeze-dried urine. Urine was initially air-dried to a viscous liquid in the field and subsequently freeze-dried in the laboratory in order to eliminate all water. Fecal and urine samples were selected from daily samples chosen at random.

Following the field tests the animals were transported to Edmonton in late August 1972 for further studies. They were housed individually at the Department of Zoology, University of Alberta in metal cages in a windowless room at a constant Ta of 15±1°C and a photoperiod of 12 hr light/ 12 hr dark. Food consisted of commercial rabbit pellets supplemented with lettuce and carrots, and water was always available. Experiments were conducted after the animals had been in captivity for about one month.

Results

Field experiments

Hares (3 to 4) were observed on Truelove Lowland in both 1971 and 1972. In September 1971 many tracks were observed on the Lowland and a 320 km survey resulted in an estimate of 10 animals (Riewe and Speller 1972). Such evidence suggests that Truelove Lowland supports a low population of arctic hares.

Examination of 480 feeding pits after the 1971 September snow revealed that the animals selected *Salix arctica* 43% and *Pedicularis* spp. 38% of the time. The remaining 19% was shared among 12 plant species (Riewe and Speller 1972).

Transpiration from growth tips of willow in the feeding trials comprised less than 1% loss during a 24 hr period. As feeding was normally completed within 1 hr after the introduction of food, loss of weight due to transpiration was ignored. Oven drying of six samples showed a moisture content of 68.1% (range = 66.8% to 68.9%) in the portions eaten.

Non ash-free caloric equivalents of willow tips showed a mean of 4.747 Kcal g^{-1} (dry wt) (4.734 to 4.758). *Salix* buds averaged 5.115 Kcal g^{-1} (dry wt) (Muc this volume). Randomly-selected feces collected during feeding trials had an energy value of 5.188 Kcal g^{-1} dry wt (5.042 to 5.260), while freeze-dried urine had a mean caloric value of 3.989 Kcal g^{-1} dry wt (3.904 to 4.047).

Values of two fresh samples of hare feces collected near the Lowland (Cape Newman-Smith and Rabbit Hill) in May were lower than those from the feeding trials (4.245 and 4.638 Kcal g^{-1} dry wt, respectively), while a single sample collected near Skogn Camp in July had a value approaching those of the feeding trials (4.716 Kcal g^{-1} dry wt).

The larger of the two hares consumed a mean 1,120.1 g wet wt day^{-1} of plant material (1,016.5 to 1,350.0 g), while the smaller consumed a mean of 706.3 g wet wt day^{-1} (634.5 to 772.0 g). Energy uptake and assimilation rates during the 10-day experiment showed that average amounts of energy used in

Table 1. Daily energy uptake and output (feces and urine combined) in one adult male and one juvenile female arctic hare, under field conditions, early August 1972, Devon Island.

			Adult male					*Juvenile female*				
Day	*Tmax (°C)*	*Tmin (°C)*	*Wt in Kg*	*Total uptake (Kcal)*	*Total output (Kcal)*	*Net uptake (Kcal)*	*% assim-ilation*	*Wt in Kg*	*Total uptake (Kcal)*	*Total output (Kcal)*	*Net uptake (Kcal)*	*% assim-ilation*
1	12.5	5.0	3.74	1761.2	670.3	1090.4	61.9	1.59	1078.0	404.7	673.3	62.5
2	15.0	9.4	3.86	1764.2	755.1	1009.1	57.2	1.70	960.7	434.8	525.9	54.7
3	12.8	0.8	3.86	2044.3	796.1	1248.2	61.1	1.81	1169.5	446.9	722.6	61.8
4	5.1	1.7	3.86	1539.3	930.3	609.0	39.6	1.81	1013.8	508.8	505.0	49.8
5	8.3	2.6	3.86	1782.4	864.0	918.4	51.5	1.81	1041.9	461.0	580.9	55.7
6	5.1	0.6	3.91	1551.4	856.0	695.5	44.8	1.81	1019.9	499.4	520.5	51.0
7	2.6	-1.1	3.86	1688.5	910.1	778.4	46.1	1.87	1132.7	513.4	619.3	54.7
8	1.7	-1.9	3.91	1693.0	946.1	746.9	44.1	1.93	1120.6	535.7	584.9	52.2
9	2.5	-1.1	3.91	1561.3	834.3	727.0	46.6	1.93	1058.5	462.6	595.9	56.3
10	4.4	-2.2	3.91	1576.4	778.6	797.8	50.6	1.98	1099.4	477.2	622.2	56.6
Mean			3.87	1696.2	834.1	862.1	50.8	1.87	1069.5	474.5	595.0	55.6

growth and maintenance were 222.8 Kcal kg^{-1} live body wt day^{-1} in the adult male and 318 Kcal kg^{-1} live body wt day^{-1} in the juvenile female (Table 1).

Laboratory experiments

Results obtained in the laboratory after the field experiments have been published earlier (Wang et al. 1973). These are briefly summarized as follows.

After acclimation in the laboratory, the mean body temperature for the two hares was 38.9°C; the female was consistently about 1°C higher than the male. The thermal neutral zone extended from Ta's of 4°C to 20°C. The BMR was 0.36 cm^3 O_2 g^{-1} hr^{-1}, or 122 Kcal day^{-1} (mean wt = 3004.4 g). Minimum oxygen consumption below the lower critical temperature increased linearly with decreased ambient temperatures. This increase in oxygen consumption was described by the equation:

$$cm^3\ O_2\ g^{-1}\ hr^{-1} = 0.401\text{-}0.010Ta.$$

Loss of weight after 24 hr of food deprivation varied between 1.3% and 16.8%, depending on the Ta to which the animal was exposed and on individual differences. More than 80% of the weight loss was accounted for as water loss, either via the pulmocutaneous route or as urine.

Daily energy consumption was between 133 to 262 Kcal day^{-1} between Ta of 12.5° and −24°C, 6% to 43% higher than the minimum energy consumption observed at the same Ta.

Discussion

Arctic hares must be classed as one of the minor species on Truelove Lowland. The few sightings recorded suggest an estimated population of 5-10 individuals. Fresh snow in the fall permitted easy tracking and observations on feeding habits. An estimate at the upper end of the range is possibly more accurate and is in accord with that by Riewe and Speller (1972) of 10 hares on the Lowland.

Fresh fecal samples obtained from the field in early summer were of lower energetic value than those collected in July, indicating a probable change in diet with time. At neither time were fecal values as high as those obtained with a diet solely of *Salix arctica*. If a diet with a high caloric value is necessary for juveniles to attain a minimal level of growth prior to the onset of winter, the presence of *Salix* may well determine both distribution and abundance of arctic hares. *Salix arctica* is not a major plant species on Truelove Lowland, but may constitute a larger proportion of the vegetation in those areas where populations of hares are much greater. Vegetation composition on Fosheim Peninsula and those parts of Axel Heiberg Island purported to sustain large hare populations is unknown.

Assuming that the adult was fully grown, an allowance can be made for fat deposition during the field experiment. Each gram of fat corresponds to 9.5 Kcal of stored energy (Kleiber 1961) so that an increase of 170 g (Table 1) in the adult is equivalent to storage of 1.615×10^3 Kcal if all the increase was in the form of fat. Energy expenditure in the deposition of neutral fats is difficult to estimate as the composition of neither diet nor stored fats is known. Approximately 80% of the energy of initial components is available in stored fat (White et al. 1973:572), so that $2.4 \times 170 = 0.408 \times 10^3$ Kcals are required for fat deposition. This allows a total energy intake for maintenance of $8.621 \times 10^3 - (1.615 \times 10^3 + 0.408 \times 10^3) = 6.599 \times 10^3$ Kcal, or a mean value of 171 Kcal kg^{-1} live body wt day^{-1}. Such a value is considerably higher than the laboratory-determined figures for maintenance energy metabolism (between 87 and 43 Kcal kg^{-1} live body wt day^{-1}) between ambient temperatures of −24° to 12.5°C (Wang et al. 1973). The high energy expenditure observed in the field is possibly the result of several factors: (1) both animals were extremely active in the boxes; (2) high humidity in the boxes possibly increased thermal conductance and thus energy consumption; and (3) the relatively crude estimations in field measurements. Fluctuating Ta may also have increased energy expenditure as demonstrated in some northern birds which have greater energy requirements with fluctuating temperatures, although the mean temperature remains the same (Steen 1958). Activity cannot account for the increased energy consumption under field conditions. Approximately 2 Kcal kg^{-1} live body wt km^{-1} are used by a 4 kg mammal during running activity (Schmidt-Nielsen 1972). If the adult hare ran 1 km per day within the confines of the box a daily expenditure of approximately 8 Kcal would be required. Increased thermal conductance in association with activity and high humidity would also increase daily energy requirements (Hart and Heroux 1955) but, as no data are available, compensation for that variable cannot be computed.

Rapid growth of the juvenile could explain the difference in energy uptake between the hares in the feeding trials. The net weight gain of the juvenile was more than twice that of the adult and the juvenile also had a higher assimilation rate (Table 1). A rapid growth rate before maturity could be advantageous in an arctic environment for the surface to volume ratio of the animal would continuously decrease, reaching minimum before the advent of cold weather.

The disagreement between field and laboratory results is too large to be resolved. Uncontrolled variables in the field suggest that laboratory results may be more reliable. If one assumes that the mean energy consumption under field conditions is twice that of the maintenance energy measured in the laboratory and that the annual mean Ta of Truelove Lowland is −12.5°C, the energy consumption under field conditions would be 110 Kcal day^{-1} in a 3.5 kg individual (Wang et al. 1973, Table 1). Using an assimilation rate of 50% (Table 1), a 220 K cal day^{-1} uptake is necessary. A population of 10 adult hares would utilize approximately $220 \times 10 \times 60 = 1.32 \times 10^5$ Kcal of plant standing crop during the (maximum) 60-day growing season (Bliss 1975). Of this, approximately one-half (Table 1) would be returned to the environment as feces and urine, allowing a net uptake of approximately 0.66×10^5 Kcal. By reduction of the daily net uptake from 220 Kcal hare^{-1} day^{-1} to, say, 110 Kcal hare^{-1} day^{-1} (i.e. equal to maintenance energy measured in the laboratory) because of reduction of activity for the remaining 305 days of the year, 10 hares would use approximately $110 \times 305 \times 10 = 3.35 \times 10^5$ Kcal. Annual net uptake for a population of 10 hares would then be approximately $1.32 \times 10^5 + 3.35 \times 10^5 = 4.67 \times 10^5$ Kcal. Such an annual energy loss would be negligible in the overall energy flow of the Lowland, but would have serious effects on the standing crop of *Salix arctica*.

If all energy was gained from eating arctic willow [although changes in fecal content and data from feeding pits (Riewe and Speller 1972) suggest that such is not the case], then $4.67 \times 10^5 \div 4.747 = 9.8 \times 10^4$ g dry wt of willow would be needed per year. For an estimated aboveground standing crop of 100 g m^{-2} of willow, 10 hares would consume *ca.* 980 m^2. Since only small shoots are typically harvested, a much larger cover of *Salix* would be grazed each year. Hare, like muskox, graze more marginal areas in summer and they also range over considerable land areas; thus, their actual impact on the Lowland would be less than the above data indicate.

The average weight of 15 animals was 1.316 kg dry wt; thus, 10 animals would average 13.16 kg or *ca.* 3.9 g ha^{-1} on the land area of the Lowland (Riewe and Speller 1972). Assuming the caloric value of the body of an arctic hare to be similar to that of *Lepus europaeus* (5.685 Kcal g^{-1} non ash-free) in central Poland (Myrcha 1968), a biomass of 13.2 kg (dry wt) would provide 7.5×10^4 Kcal for potential predators if all were consumed.

Acknowledgments

We are indebted to Dr. W.A. Fuller for providing both opportunity and support for this study, and to Imperial Oil Limited and Mobil Oil for providing transportation and helicopter assistance in obtaining the hares. Mr. W.J. Fuller and Ms. M. van Buskirk provided invaluable technical assistance.

References

Bliss, L.C. 1975. Devon Island, Canada. pp. 17-60. *In:* Structure and Function of Tundra Ecosystems. T. Rosswall and O.W. Heal (eds.). Ecol. Bull./N.F.R. No. 20. Stockholm. 450 pp.

Cahalane, V.H. 1961. Mammals of North America. Macmillan Co., New York. 682 pp.

Hall, E.R., and K.R. Kelson. 1959. The Mammals of North America. Vol. 1. Ronald Press, New York. 1083 pp.

Hart, J.S., and O. Heroux. 1955. Exercise and temperature regulation in lemmings and rabbits. Can. J. Biochem. Physiol. 33:428-435.

Kleiber, M. 1961. The Fire of Life. John Wiley and Sons, Inc., New York. 454 pp.

Myrcha, A. 1968. Caloric value and chemical composition of the body of the European hare. Acta Theriol. 13: 65-70.

Riewe, R.R., and S.W. Speller. 1972. Arctic hare study. Progress Rept. Devon Island Project. Edmonton. 3 pp.

Schmidt-Nielsen, K. 1972. Locomotion: energy cost of swimming, flying, and running. Science 177: 222-228.

Steen, J. 1958. Climatic variations in some small northern birds. Ecology 39: 625-629.

Wang, L.C.H., D.L. Jones, R.A. MacArthur and W.A. Fuller. 1973. Adaptation to cold: energy metabolism in an atypical lagomorph, the arctic hare *(Lepus arcticus)*. Can. J. Zool. 51: 841-846.

White, A., P. Handler and E.L. Smith. 1973. Principles of Biochemistry. McGraw-Hill, New York. 1296 pp.

Estimated productivity of muskox on Truelove Lowland

Ben A. Hubert

Introduction

The muskox *(Ovibos moschatus)* is native to the tundra of North America and Greenland and is the largest herbivore in those parts. Unlike the caribou, the muskox is not migratory or nomadic, but is sedentary. The muskox is the only ruminant species whose natural distribution is restricted to tundra regions. Many features of muskox anatomy seem to be specialized adaptations for energy conservation in a cold tundra environment. The same object may be served by the behavioral characteristic of muskox first taking a defensive position rather than fleeing when threatened.

Several field studies of muskox biology on their native range had been conducted prior to the onset of this study in 1970. Tener (1965) synthesized the available information and added many details on muskox life history, social behavior, habitat, and clarified muskox taxonomy. Gray (1973) dealt specifically with social organization of muskox on Bathurst Island. Harington (1961) prepared an extensive review of the evolution, distribution, and history of the muskox group.

Recent range of muskox

According to available historical and explorers' records, muskox have never been abundant, but were fairly widely distributed during the current geological era (Hone 1934). Prior to the arrival of explorers and the introduction of firearms, muskox ranged over nearly all the tundra regions of mainland North America and the Arctic Islands, as well as around the northern coastal fringes of Greenland. The notable exception to this distribution is the total lack of evidence suggesting that muskox ever inhabited Baffin Island or the tundra of the Ungava Peninsula.

Archaelogical investigations of pre-Dorset, Dorset, and Thule Inuit campsites at Cape Sparbo on northeastern Devon Island (Lethbridge 1939, Lowther 1960) showed no remains of muskox, but many of caribou. Harington (1964) briefly discussed this phenomenon and suggested that muskox invaded, or possibly reinvaded the area in the past "few hundred years." It is only in the last two decades that caribou have either emigrated or have been extirpated from this part of the island.

Harington (1964) discussed reports of muskox numbers from north

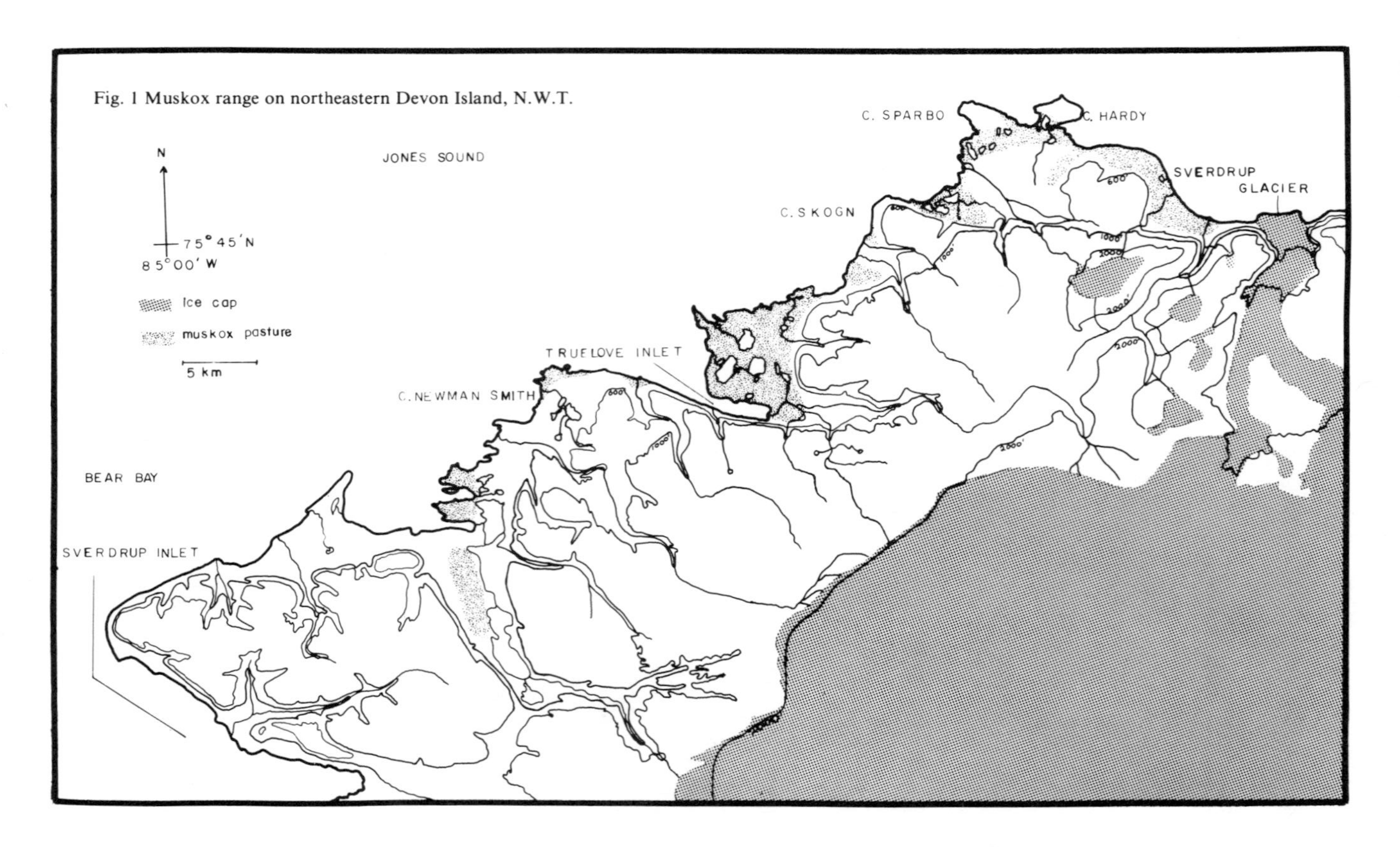

Fig. 1 Muskox range on northeastern Devon Island, N.W.T.

Table 1. Surface characteristics of the Lowland area below 200 m on northeastern Devon Island.

Lowland	*Area* km^2	*Percentage*	*Pond and lake surface (o/o)*
Newman Smith	139.4	46.3	9.4
Truelove	44.4	14.7	22.2
Truelove Valley	13.6	4.5	9.1
Skogn	15.2	5.0	.0
Sparbo-Hardy	66.2	22.0	5.6
Sverdrup	22.6	7.5	5.5
Total	301.4		

Devon supplied by the R.C.M.P. and others from 1908 to 1960. The minimum reported was 20 for 1929, and the maximum was 100 for 1937. The numbers reviewed and discussed by Harington (1964) probably do not reflect the entire population. They probably represent local herds within the total population. Freeman (1971) reported 230 muskox for this area of northeastern Devon.

This small isolated population was studied from May 1970 through August 1973. The objectives of the study were to refine the knowledge of muskox biology with special emphasis on productivity, population characteristics, and winter ecology. This paper elaborates on the productivity aspect of the study.

Study area

Muskox range on northeastern Devon Island consists of 301 km^2 below 200 m between Sverdrup Inlet and Sverdrup Glacier (Hubert 1974). A large portion of the area is rarely utilized by muskox due to its dry, calcareous, and unvegetated surface characteristics. The most westerly Lowland, Newman Smith (139.4 km^2), is predominantly of this type. The Sverdrup, most easterly of the Lowlands group, has few ponds or lakes and is generally well vegetated (Fig. 1, Table 1). It also seems to be free of snow earlier than the other Lowlands, probably because of its easterly exposure and proximity to the polynya ("North Water") near Coburg Island (Nutt 1969). Presumably due to the early melt and well developed vegetation, high muskox concentrations were observed on this Lowland in June and early July 1970 and 1971.

Methods

Muskox distribution

The entire Truelove Lowland and Valley were surveyed at least once a week during the field study to determine number and distribution of muskox. In winter this was done on snowmobile or cross-country skis; in summer on foot. Muskox observed on the surveys, as well as all daily observations, were recorded, with sex and age noted whenever possible (Table 2). Observations made by colleagues were also recorded. From these data maps of animal abundance and distribution were prepared daily; 1,096 in total.

To more accurately map animal movements, individuals were immobilized with succinyl-choline-chloride and ear tags attached (Jonkel et al.

Table 2. Numbers of muskox observed during aerial population surveys on northeastern Devon Island.

Date	*Area**						*Total*
	1	*2*	*3*	*4*	*5*	*6*	
1970							
22 May	14	21	13	32	—	—	70
18 June	—	14	8	33	—	—	55
15 July	23	17	10	17	65	—	132
19 August	11	10	2	40	48	—	111
1971							
27 May	7	28	17	42	62	—	156
2 July	10	22	13	21	50	—	116
1972							
8 May	29	70	1	99	50	2	251
1973							
22 March	16	114	5	49	53	0	237
11 May	25	89	15	53	73	13	268
16 August	57	27	9	167	18	9	287

*1. Newman Smith Lowland
2. Truelove Lowland and Valley
3. Skogn Lowland
4. Sparbo-Hardy Lowland
5. Sverdrup Lowland
6. East of Sverdrup Glacier to Eastern Glacier

1975). Free ranging muskox were marked with a paint pistol either from a helicopter or from the ground *(ibid.)*. By those methods 43 animals were marked. Combining the age and sex structure with presence or absence of marked muskox, movement patterns of individuals and herds seen intermittently could be reconstructed (Table 3).

Age determination

Since horns occur on both sexes but develop at different rates, immature animals can be sexed and aged fairly accurately using criteria already determined (Allen 1913, Tener 1965). Males could be satisfactorily aged to 4.5 years and females to 3.5 years. Sexes could rarely be differentiated before 18 months of age. Guard hairs form a "skirt" on the body of muskox and with increasing age it becomes more prominent. Skirt length was used to differentiate yearlings from 2-year-olds where a comparison of horn development was inconclusive (Table 4).

Live weight determination

Field weights of wild muskox are very rare. Weights reported by exploration expeditions are unreliable since they reported the amount of meat removed from a carcass. Fig. 5 shows the growth reported by Wilkinson (1973) for domestic muskox at Fairbanks.

Using those data and the following assumptions, an estimated growth curve (Fig. 3) was constructed for wild muskox.

1. That calves in the wild are as heavy at birth as calves born in captivity.
2. That the growth rate of calves in the wild is the same as calves in captivity, but that it is not interrupted in the wild by weaning as it is in captivity.

Table 3. Sex and age composition of muskox observed on surveys.

Date	*Adult males*	*Adult females*	*Calves*	*Yearlings*	*2 yr.*	*3 yr. males*	*4 yr. males*
1970							
10 August	36	33	15	5	6	0	0
1971							
27 May	62	52	27	5	10	0	0
12 June	25	37	21	6	(	21	)
2 July	39	47	20	5	5	0	0
1972							
8 May	75	93	48	21	11	3	0
1973							
25 April	40	53	0*	32	26	4	8
11 May	54	84	50	36	34	2	8
16 August	65	95	57	32	28	7	3

* Survey done prior to calving

3. That the period of stability or weight loss in late winter begins in wild muskox at the same time as for captive muskox, but lasts longer due to prolonged snow cover and lack of supplemental feeding.
4. That the weight loss of bulls during the rut in the wild represents the same proportion of body weight as for rutting bulls in captivity and that the effects of rutting are shown in the same cohorts in the wild as in captivity.
5. That the effect of gestation and lactation on body weight is the same for wild and captive females.
6. That in a wild population, in any given year, 50% of the females produce a calf and that the weight loss during lactation is compensated for by the weight gain in early summer by non-lactating females, thus producing a damped-out summer growth rate for an "average" adult female.
7. That the field weight of 177 kg and 259 kg for wild and captive bulls respectively, 33 months old, represents the proportional weight difference between captive and wild muskox, between 12 and 60 months of ages for males, and 12 and 36 months for females.
8. That adult females are two-thirds the size of adult males (Wilkinson 1973).

Food requirements

In February and March 1972, a series of intake and digestibility studies was conducted at the Muskox Farm near College, Alaska. Five pregnant cows, all weighing approximately 230 kg, were used as experimental animals. They were isolated from the herd and fed known amounts of alfalfa-brome-timothy hay once daily. When taking daily live weights on a platform scale, 10 g chromic oxide (Cr_2O_3), an indigestible indicator, were administered to each animal by mixing it with 0.55 kg Omolene (Ralston Purina Co., St. Louis), a commercial ration. The above procedure was followed for 12 days as outlined by Short (1970) and Theurer (1970) to determine fecal output. Beginning day four, fresh feces were collected from each cow twice daily (1000 to 1100 and 1600 to 1700 hrs). Samples of the hay and Omolene were retained for subsequent chemical and energy content analyses. This procedure was repeated with six growing bulls, each 22 months old.

Muskox range relationships

Available forage

Forage available to muskox was defined as that portion of the standing crop above the moss surface. During the snow-free period it would include production of the current season, as well as the standing dead plant biomass. During the period of snow cover the forage above the moss surface would consist mostly of standing dead material. Data reported by Muc (this volume) were used to provide the above baseline for the forage available to muskox (Tables 5 and 6).

Grazing transects

During the winter 1972-73, grazing areas could be readily identified by the disturbed snow cover. All such areas were delineated on a map. Following the melt, areas grazed in winter were sampled by reading paired line intercept transects for frequency of grazing. Winter grazing was identified by the reduced amount of standing dead vegetation. The individual lines of a pair were 1 m apart. Individual transect pairs were spaced at 20 m (See Fig. 9 for transect data and locations). From ten plots in grazed areas, each 20 cm × 50 cm, all remaining graminoid litter was collected in order to estimate forage removed by muskox grazing (Table 7).

Fecal production

If digestibility can be accurately estimated, then daily intake can be determined based on daily fecal output. By observing and noting locations of single adult males during periods of snow cover it is possible to follow their trails in the snow and collect 24 hr fecal samples. This method works if it does not snow, if the feces are not obliterated by windblown snow, if the animal does not fall onto another animal's trail, or if the beast does not double back on itself. All are very probable events. From numerous attempts two 24 hr samples were collected, one in May 1971 and another in October 1971. In summer we had to locate a single animal and observe his every move in such a way so as not to disrupt his activity cycle or alarm him. Out of six such attempts we collected one 12 hr sample. In addition to the above, fresh feces were collected monthly for chemical, energy, and digestibility analyses of composite monthly samples. (See Booth this volume for further details on fecal production and decomposition.)

Results and Discussion

Population size and distribution

From aerial surveys flown in May 1972 and May 1973, the population using the five Lowlands was estimated to be between 250 and 275 animals. Survey results for 1970 and 1971 (Table 2) were incomplete. The results of the August 1973 survey are questionable due to incomplete coverage and poor visibility. Freeman (1971) reported the population of the area to be 230 to 300 animals for 1966-67. His range extended west along the entire shore of Bear Bay, while the range studied here extends along Bear Bay as far west as Sverdrup Inlet. Kiliaan (pers. comm.) reported small isolated herds of muskox near Sverdrup Inlet and Thomas Lee Inlet in April 1973.

Harington (1964) suggest that the average density of muskox for northeastern Devon was 0.28 km^{-3} for the 50 years prior to 1970. The maximum density he reported was 0.53 km^{-3}. If the present population stands at 260 muskox, the density is 0.87 km^{-3}, a 65% increase over Harington's maximum. Density cannot be calculated from the observations of 1966-67 due to imprecise definition of the range (Freeman 1971). Densities would agree with these if his Bear Bay data are not included. This would put the population during 1966-67 at 230 for the area east of Truelove Inlet (1.43 km^{-3}). The density observed in May 1973, for the same area was 1.52 km^{-3}.

The number of muskox occupying Truelove Lowland varied widely from week to week and season to season. However, there appeared to be an annual pattern. In May and June of every year there was a general decrease in muskox on the Lowland. Just prior to, or during, snowmelt and runoff (*ca.* 5 day period), muskox were absent. Muskox left the Lowland via the Truelove Valley, across the frozen Truelove Inlet, and moved along the northeast coast onto Skogn Lowland. During June and July of 1970 and 1971 (see Table 2), high densities were observed on Sverdrup Lowland. Presumably muskox, including those from Truelove Lowland, moved to Sverdrup Lowland due to an earlier snowmelt there.

Muskox returning to the Lowland after the melt entered via the northeast coast and usually either moved along the base of the escarpment south toward the Truelove Valley or west into the outcrops along the north coast. Fluctuations during the snow-free period are due mainly to animals moving into and out of the Truelove Valley from the Lowland. That activity ceased following the first snow fall. Muskox were never observed in Truelove Valley between the end of August and February during the study. Fluctuations observed in September and October of 1971 were due to movements between Skogn and Truelove Lowlands. Animals left the Lowland twice in late 1972 when polar bears appeared there. Although there is no direct evidence from the Canadian Arctic of polar bear preying on muskox, Chr. Vibe and Thor Larsen (pers. comm.) have made observations of such predation on East Greenland. Fig. 2 illustrates the rapid buildup of muskox on Truelove Lowland in December and January. Fluctuations observed in February, March, and April

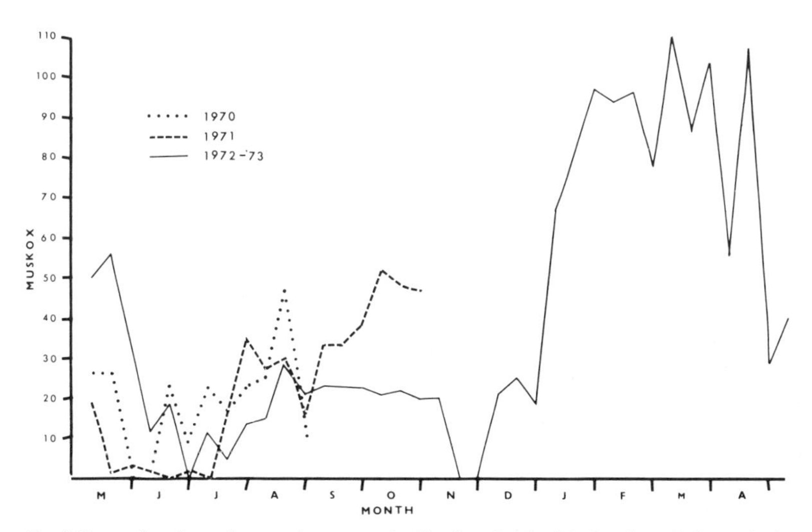

Fig. 2 Fluctuations in muskox numbers occupying Truelove Lowland during the period spent in the field. Points on the curve show the total number of muskox present on day 10, 20, and the last day of each month of the field study.

1973, are due to movement between the Lowland and Truelove Valley. During that period the greatest muskox concentrations were observed in those meadows where the snow profile was free of hard layers of wind slab, despite the fact that these meadows had deeper (25-35 cm) snow than meadows with a harder snow cover.

Reproduction

Tener (1965) reviewed the literature dealing with reproduction and productivity of muskox. Our findings agree with his. Breeding in muskox begins in late July and may last into early October. Wilkinson (pers. comm.) suggested that females are seasonally polyestrous having a 25 to 30 day cycle. This may account for the isolated observation of October rutting on Devon in 1971. The gestation period is eight months (240 to 250 days). On Devon, calving occurred from late April to late May. Tener (1965) reported a calving season from mid-April to early June. From the surveys of May 1972 and 1973, the peak in calving occurred the first week of May. Twinning occurs, but rarely (Tener 1965, Wilkinson 1971). Although direct evidence is lacking apparently under normal conditions in the wild, cows of breeding age bear a calf in alternate years. Cows in captivity at Fairbanks bear a calf annually. In the captive herd the calves are weaned prior to the cow's next breeding season, whereas on Devon, 13- and 15-month-old calves were observed nursing. Nursing may prevent the cows from coming into estrus by increasing energy requirements for maintenance (Pedersen 1958).

The age of sexual maturity in females probably depends on nutrition. Tener (1965) reviewed reports from captive animals in the state of Vermont, U.S.A. and the province of Alberta, Canada. At the Alberta Game Farm near the city of Edmonton a 2-year-old gave birth; she was bred by a yearling bull. On Devon a 3-year-old cow was observed nursing a calf of the year. Tener (1965) suggested that cows in the wild normally give birth to their first calf at 4 years. The same appears true for the Devon Island population, although under favorable nutrient conditions calving may occur at a younger age. Bulls are probably sexually mature long before they are permitted to breed. Most breeding is done by bulls older than 6 years (judging by pelage and horn development).

Population composition and productivity

When flying a census survey with helicopter, we always attempted to classify all animals for sex and age. The same was done on ground surveys. On all surveys except that of 10 August, 1970, more cows than bulls were seen. It is easily possible that some three- and four-year-old bulls were classified as cows. It is also possible that an unequal sex ratio is the normal situation. Gray (1973) reported a near equal adult sex ratio for muskox observed on Bathurst Island during his four-year study. Tener (1965) reported sex ratios in favor of cows in all cases except for Lake Hazen where the ratio was 103.3 bulls to 100 cows. Spencer and Lensink (1970) observed adult males outnumbering adult females on Nunivak Island (209 males: 150 females). Reports on sex ratios in calves are inconsistent. Wilkinson (1971) reported that out of 79 calves born at the Muskox Farm, 45 were females. Spencer and Lensink (1970) reported that of 59 calves captured on Nunivak only 14 were females.

Table 4. Age composition, in percent, of the muskox population on northeastern Devon Island.

Age class (yr.)	*Survey*		
	27 May, 1971	*8 May, 1972*	*11 May, 1973*
≥4	73.1	68.2	54.5
≥2<4	6.4	4.3	13.4
1	3.2	8.4	13.4
Calves	17.3	19.1	18.7
Total	100	100	100

The three census surveys in May of 1971, 1972, and 1973 (Table 3) showed an increase in the proportion of immature animals in the population. Calf crops were higher than Leslie (Appendix 2 in Tener 1965) predicted. Leslie assumed an equal sex ratio, that cows begin breeding at three years, that cows breed in alternate years, and that 50% of the calves die prior to one year of age. Under these conditions the population should have 8% to 15% calves. According to Leslie, with less than 50% calf mortality, the calves would make up 15% to 23% of the population. A calf crop greater than 23% would indicate that cows were breeding at a younger age or that some cows are breeding annually, or both. Spencer and Lensink (1970) suggested the latter conditions for Nunivak Island, Alaska. The data for Devon from 1971 to 1973 show that most calves survived (Table 3), and that occasionally a 2-year-old cow conceived. That situation, however, has not always existed. In August, 1970, we observed 15 calves (15.8%). The following May we saw only five yearlings (3.2%) (Table 4). The low yearling count suggests considerable winter mortality. It is also very possible that the incomplete August 1970 census resulted in an apparently inflated calf crop. Both conditions are possible. Hussel (1970) reported frequent sightings of calves in 1966. In 1967 calves were seen rarely. Twelve fresh carcasses were found, including possibly two calves. No live calves were reported for 1968. Calves were again present in 1969 *(ibid.)*. It seems clear that severe mortality occurred during the winter of 1966-67. The conditions responsible for the low productivity in 1967 may have affected breeding success in 1967 as well. Reductions in cohort production and/or survival account for the lack of sub-adults seen in 1970-71.

We observed only seven cases of natural mortality between 1970 and 1973. Three (one adult male and two adult females) in May 1971 were at a time of deep snow cover. All three carcasses were in an emaciated condition, as were the animals we immobilized that year. The deep (40 cm) snow cover on meadows apparently did not seriously affect 1971 calf survival (Table 4). We found a recently dead adult male after the melt in 1971. The cause of death could not be determined, but the carcass was emaciated. Two dead calves, both males, were found in May 1973. Autopsies on these carcasses showed that one was probably stillborn. The emaciated condition of the other pointed toward starvation as the probable cause of death. An adult bull was found dead in the Truelove River in August 1973. It probably drowned in attempting to cross the river during a particularly heavy discharge following a rain. In 1974 at least four animals died on the Lowland, a year of high muskox mortality elsewhere (Miller and Russell 1975).

From the composition results (Table 4) one can conclude that from 1968 to 1973 the population on northeastern Devon was in a period of increase, and would probably continue to increase until abiotic density-independent factors resulted in mortality and/or reduced or total lack of breeding success similar to that observed by Gray (1973).

The Nunivak Island muskox population in Alaska had both a higher proportion of adult males and immature animals than the northeastern Devon population. The age composition of that population was adults, 53.3%; sub-adults, 30.3%; yearlings, 16.4%. Since 1947 the average calf crop on Nunivak has been 19%. Spencer and Lensink (1970) assumed a 7% annual loss due to mortality which was not sex and age specific. Under those growth conditions the Nunivak population doubled every 4.5 years. On Devon it seems that under favorable conditions the population growth would equal that of Nunivak for one or two years, but that environmentally-induced mortality can be expected to stop population growth and/or cause a decline. Such conditions occur at irregular intervals but could possibly average once every 5 years. Vibe (1967) described periodic muskox population declines produced by severe winter conditions in east Greenland. Miller and Russel (1975) describe the drastic decline of muskox on Bathurst Island, N.W.T. following severe icing conditions there in the winter of 1973-74.

Assuming no mortality and an 18% annual natality, the population of 230 muskox reported by Freeman for 1967 (Freeman 1971) should have increased to 739 by June 1973. The observed population was 236 on 11 May, 1973, and 287 on 16 August, 1973, about 65% lower than the predicted. Assuming from Hussell's reports (1970) that there was 10% mortality in April and May 1967, following Freeman's census; that there was no mortality in the winter of 1967-68 and 1968-69, and an 8% natality in 1969 and 1970 (according to our observations of calves and yearlings in 1970 and yearlings and sub-adults in 1971) the 1973 population should have been 380 muskox (30% below the predicted). Therefore, if there was no consistent rate of mortality, one winter of severe conditions in seven held the population's annual increment to only 9.3%. Including a 5% winter mortality independent of sex or age, the 1973 population would be 287, an annual average increment of 3.5% for seven years. Under the favorable conditions between May 1972 and May 1973, the increment (assuming the censuses are correct) was 6.7%. The above history of occasional increased mortality combined with fluctuating natality suggests "boom" and "bust" rates of population change. Vibe (1967) reported that between 1938 and 1960 there were six winters during which muskox populations suffered considerable losses on northeastern Greenland. He described the winters of 1938-39 and 1953-54 as "catastrophic." Following both winters no calves and very few yearlings were observed among the surviving animals.

Muskox are sensitive to snow conditions. From snow depth and hardness measurements taken in 1972-73 it appeared that muskox preferred meadows with soft snow cover regardless of depth encountered to meadows covered by shallow crusted snow. However, during the spring 1971 when meadows were covered by 40+ cm of snow with varying hardness, muskox distribution was restricted to windswept raised beaches and outcrops. Fig. 3 shows muskox winter distribution in relation to maximum snow depth and Rammsonde hardness recorded in 1972-73.

Growth and change in body weight

Weights of male muskox born dead in captivity were 11 kg (n=9) and females were 12 kg (n=12) (Wilkinson 1973). Fig. 4 shows the growth of captive muskox in 36 months. This curve is based on Wilkinson (1973) and unpublished data of the Muskox Farm at College, Alaska. Weights of one to six cohorts were averaged to produce the curve.

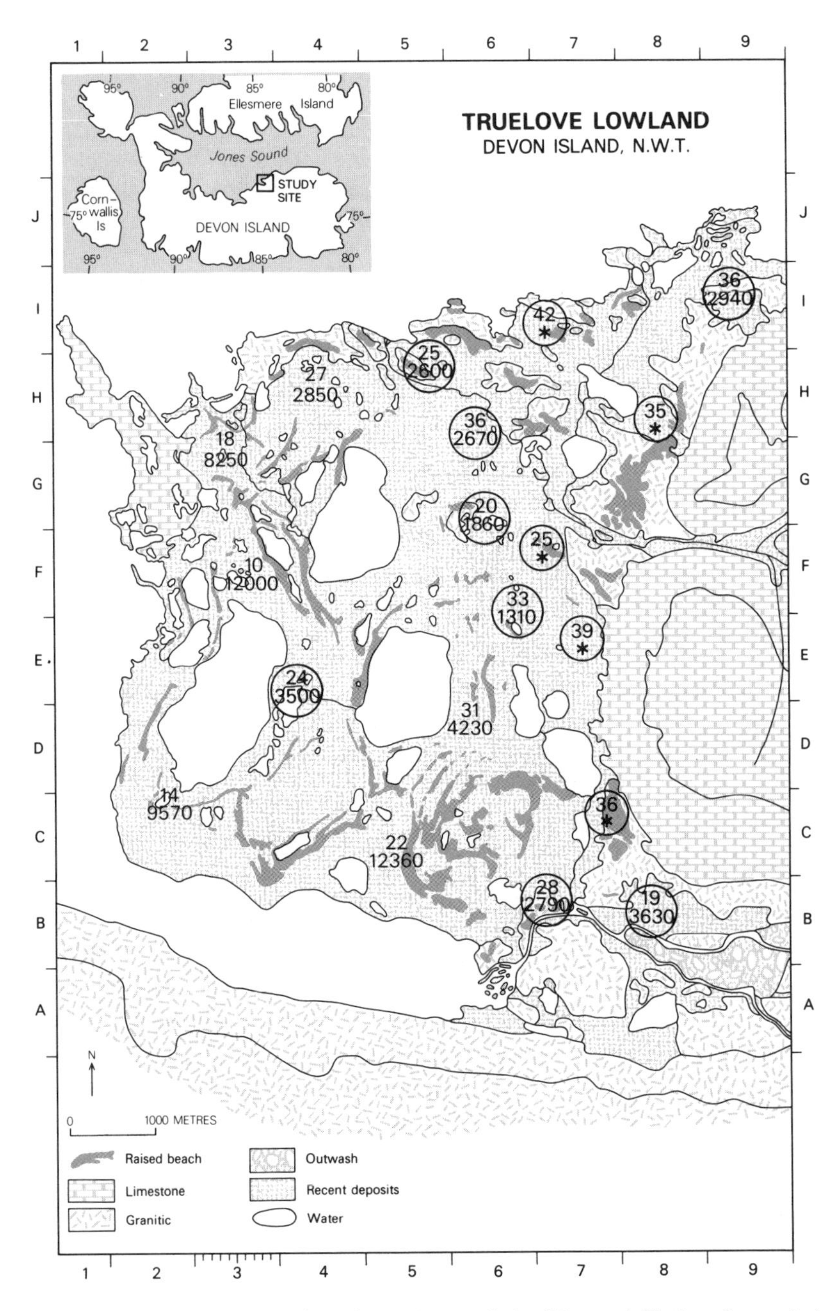

Fig. 3 Representative snow depth in centimetres (upper value) and Ramsonde Hardness (lower value) recorded on meadows in May 1973. An asterisk (*) indicates snow cover was too soft for measurement. Circled values indicate meadows where muskox grazing was observed during the winter of 1972-73.

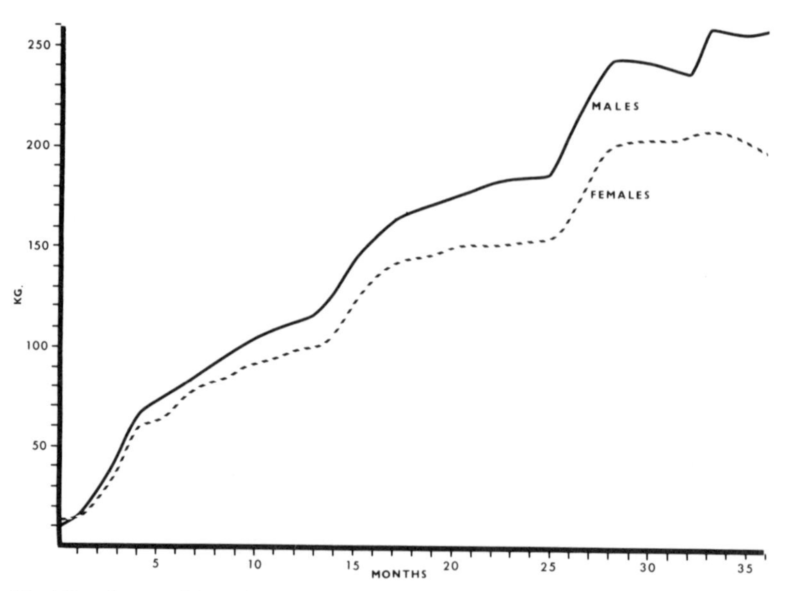

Fig. 4 Growth curve of domestic muskox to 36 months of age, after Wilkinson (1973 and unpublished data).

Sexual dimorphism is observed very early in life. The dimorphism in immature muskox is due to differential growth rates, the rate of males slightly exceeding that of females. Prolonged growth in males probably contributes to adult sex dimorphism. In captivity, muskox females approach adult body weight in their fourth summer, whereas bulls reach adult weights in their fifth or sixth summer. Allen (1913) and Wilkinson (1973) state that adult wild muskox females are 66% the size of adult males. In the wild, females probably attain adult body weight in their fifth or sixth summer while males approach adult weights in their sixth or seventh summer, or perhaps later.

Using Wilkinson's data (1973 and unpublished) for captive muskox and data on wild muskox from available sources (Hubert 1974) a growth curve has been constructed for wild muskox (Fig. 5).

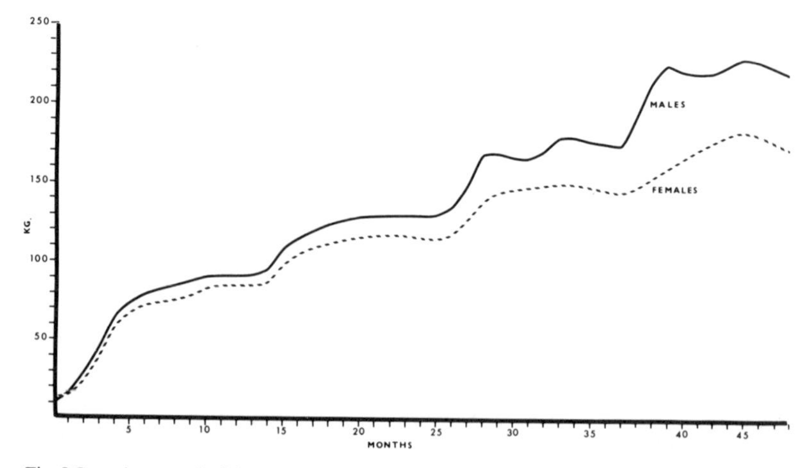

Fig. 5 Growth curve of wild muskox to 48 months of age.

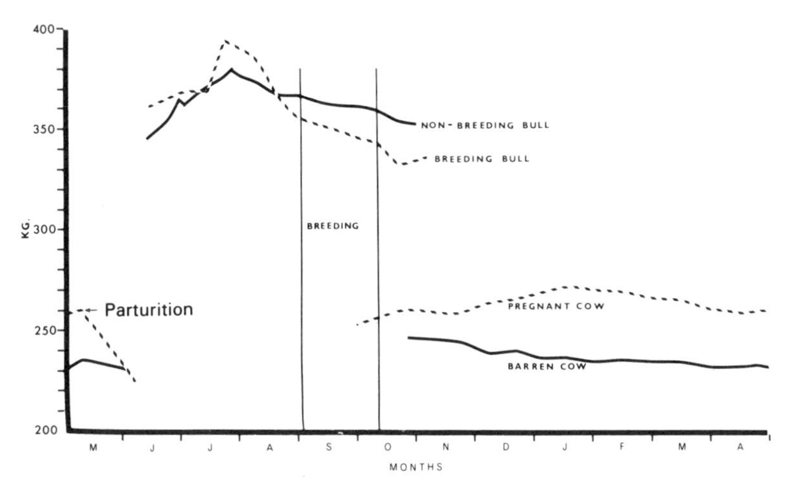

Fig. 6 Effects of reproduction on body weight of adult muskox in captivity, after Wilkinson (1973 and unpublished data).

Seasonal fluctuations in body weight are a combined function of nutrition, environmental stress, reproductive condition, and intrinsic mechanisms regulating "voluntary" food intake (Table 5). Fig. 6 shows the effects of the reproductive condition on captive adult muskox (Wilkinson 1973 and unpublished). Five lactating cows lost an average of 29 kg each in three weeks following parturition (Wilkinson unpublished). During the same period, four non-pregnant, non-lactating cows lost an average of 6.2 kg which could be due to the transition from dry hay and suppliments to fresh pasture.

An intriguing feature of the seasonal fluctuations in body weight of captive muskox is the weight loss during late winter and spring for non-pregnant as well as pregnant females. Fig. 7 shows a weight loss for adult males during the same period. This loss occurs despite the *ad libitum* availability of good quality (10% protein) fodder. White (1975) reported a similar phenomenon for caribou. Segal (1962) suggested that the rate of metabolism in Russian reindeer is 30% lower in winter than in summer. Reducing metabolic rate at a time of low energy and nutrient availability would minimize weight loss. The survival advantage of such a mechanism to arctic ruminants may be very significant. White (1975) also reported that weaned reindeer fawns tend to regulate at a constant dry matter intake regardless of protein content and that the intake may limit winter growth.

The effects of rutting and breeding on body weight of individual adult muskox is shown in Fig. 7. It is clear that there is a greater weight loss associated with rutting, both before and after the breeding period, than with breeding proper. Wilkinson (1973) reported that the immature bulls as young as 28 months showed weight loss during the rut. The loss may have been due to intra-specific aggression of older bulls in the paddock, rather than active rutting on the part of the young bulls. Wilkinson (1973) stated that the weight loss following breeding resulted from aggression in establishing the dominance hierarchy after the breeding bulls were reintroduced to the paddock.

Fig. 6 illustrates the estimated body weight and fluctuations in body weight of adult wild muskox. Due to the lack of age:weight data, all males older than 4 years and all females older than 3 years are considered adult. The estimated weights of pregnant or lactating females and non-reproductive females are averaged, producing a "normal" female.

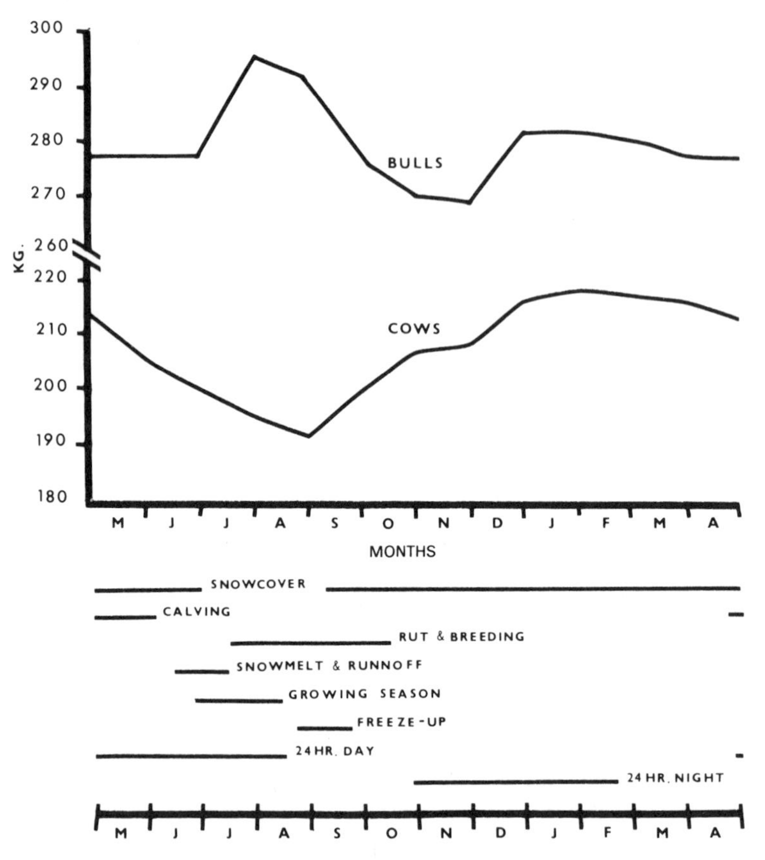

Fig. 7 Seasonal fluctuations in body weight of adult wild muskox in relation to reproductive and environmental conditions.

If the estimated seasonal weights of adult wild muskox are accurate, seasonal energy balance can be determined. Energy balance in this context will be negative during periods of weight loss and positive during periods of weight gain. It appears that bulls are in a positive energy balance only two months of the year, are neutral four months, and negative six months. Cows are positive for five months and negative seven months (Fig. 8). Calculating the monthly percent change in estimated body weight (Fig. 8) shows that annually the net change is +0.5% for bulls and +0.4% for cows. This suggests that adult muskox like most ungulates continue to grow at a slow annual rate.

Forage digestibility and energy intake

Digestibility is an expression of the feed:feces ratio or the efficiency of digestion; and is also a good indication of forage quality (Short 1970). Three independent techniques were used in calculating forage digestibility in the experiments conducted at the Muskox Farm at College, Alaska. All calculations are on a dry weight basis.

1. Chromic oxide (Cr_2O_3) was added to the feed and the daily fecal output determined (Theurer 1970).

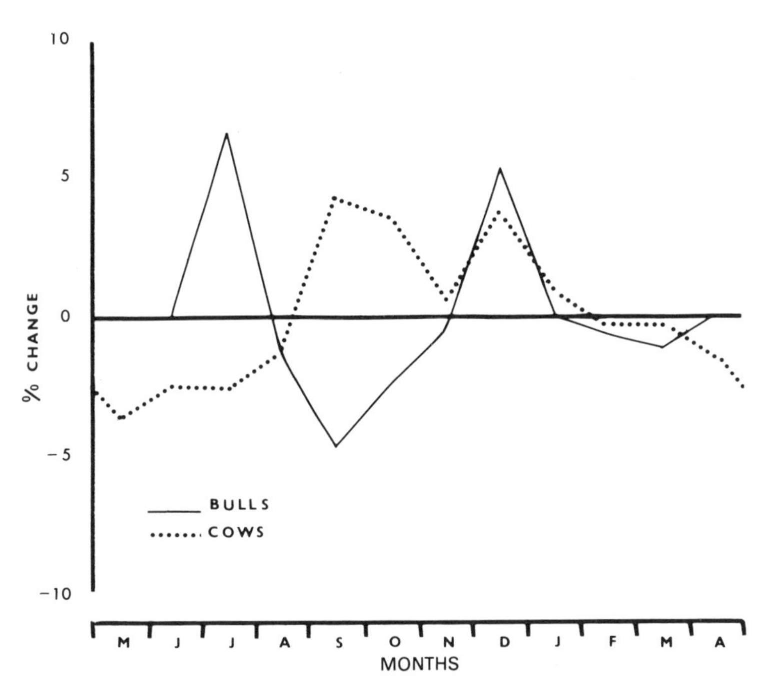

Fig. 8 Percent change in body weight of adult wild muskox. Calculation of percent change was determined as follows: $bw_x(bw_{x,1}) \times 100 - 100$ where bw_x is the body weight of the individual adult muskox in month x.

$$\text{Fecal output (g)} = \frac{\text{g indicator administered}}{\text{g indicator g/fecal dry wt}}$$

2. Nitrogen fecal index (Lancaster 1954) for calculating the feed:feces ratio: $Y = X + 0.9$ where X is the percent nitrogen in the fecal organic matter and Y is the feed:feces ratio. Percent digestibility is calculated by the formula: % digestibility $= Y - 1/Y\,(100)$

3. Lignin ratio. Lignin is used as a naturally occurring indicator since it occurs both in the forage and in the feces.

$$\%\text{OM digestibility} = 100 - 100 \left(\frac{\%\text{ lignin in forage}}{\%\text{ lignin in feces}}\right) \left(\frac{\%\text{ organic matter in forage}}{\%\text{ organic matter in feces}}\right)$$

Using this calculation, lignin need not be indigestible (Theurer 1970).

The results of these experiments showed that percent digestibility ranged from 64.7 to 63.2% for cows and 64.7 to 62.2% for bulls. There was no significant difference in digestibility between cows and bulls. Organic matter intake averaged 13.1 in cows and 15.4 g kg^{-1} day^{-1} in bulls. Gross energy intake and digestible energy intake averaged 62.1 and 40.1 C kg^{-1} day^{-1} respectively in cows and 73.0 and 46.7 C kg^{-1} day^{-1} in bulls (1C = 1,000 calories).

About 64% of the organic matter and 66% of the caloric energy was removed from the feed. The coefficient for metabolizable energy is 0.82 (Blaxter 1967). Therefore, 54% of the energy was metabolized. Metabolizable energy intake (M.E.I.) for the cows was 128.7 C $kg^{.75^{-1}}$ day^{-1}, and for the bulls 146 C $kg^{.75^{-1}}$ day^{-1}. For non-lactating female caribou in summer, White et al. (1974) reported an M.E.I. of 196 C $kg^{.75^{-1}}$ day^{-1}; well above these values. As

Table 5. Estimated daily intake and output of muskox on Devon Island.

Parameters	14 June	Dates (1971) 31 July	22 October
Est. live weight (kg)	227*	281*	268
Daily fecal output			
Total (g)	990	1542	1606
($g\ kg^{-1}$)	4.4	5.5	6.0
Digestibility (%)	75	75**	65
Daily Intake			
Total (g)	3960	6170	4590
($g\ kg^{-1}$)	17.4	22.0	17.1

* On 9 May, 1972, this bull weighed 264 kg
** probably an underestimate

Table 6. Estimated daily organic matter and energy intake for muskox on Devon Island.

Month	Organic matter intake ($g\ kg^{-1}$)	Gross energy intake ($C\ kg^{-1}$)	Digestible energy intake($C\ kg^{-1}$)
May	17.0	81.3	52.8
June	17.5	83.7	62.5
July	22.0	105.2	78.9
August	22.0	105.2	78.9
September	20.0	95.6	62.1
October	20.0	95.6	62.1
November	20.0	95.6	62.1
December	20.0	86.0	55.9
January	18.0	86.0	55.9
February	17.0	81.3	52.8
March	16.0	76.5	49.7
April	16.0	76.5	49.7

suggested earlier, the heat production in winter is possibly much lower for caribou, making a detailed comparison of the above data unreliable.

The average weight of the cows used for the experiment was 235 kg (range 182 to 200). The cows gained an average of 0.14 kg day^{-1} while the bulls gained 0.15 kg day^{-1} during the experiments.

Energy requirements of wild muskox

Using the lignin ratio and the nitrogen fecal index the digestibility of graminoid forage from Devon Island was calculated. Daily fecal output is known from collecting feces produced during a known time interval. As muskox probably select fresh forage when it is available, a digestibility of 75% for the snow-free period is possibly an underestimate. As digestibility and fecal output are known, daily dry matter intake can be estimated. Caloric density of organic matter in the forage was 4.78 C g^{-1} and did not show significant seasonal changes (Muc 1973). Table 6 gives the estimated daily energy intake of the muskox on Devon Island. The summer values are based on June and July data. The winter values are based on the October data and the experimental work from Alaska using captive animals.

The February D.E.I. value estimated for free ranging muskox was 52.8 C

Table 7. Estimated forage consumed on Truelove Lowland by muskox.

*Month**	*Muskox biomass (kg day^{-1})*	*Food requirement (g kg^{-1} day^{-1})*	*Monthly harvest (kg)*
May	6757	17.0	3560
June	2334	17.5	1230
July	2202	22.0	1500
August	3369	22.0	2300
September	3557	20.0	2130
October	3338	20.0	2070
November	1376	20.0	830
December	4842	20.0	3000
January	14935	18.0	8340
February	17343	17.0	8260
March	18764	16.0	9310
April	11388	16.0	5470
Total			48000

* 12 month period ending April, 1973

Table 8. Muskox days for Truelove Lowland, May 1972 - April 1973.

Month	*Bulls*	*Cows*	*Calves*	*Yearlings*	*2 yr.*	*3 yr.*	*Total*
May	165	587	303	224	66	60	1,405
June	62	192	101	64	17	20	456
July	97	125	101	28	34	15	400
August	104	238	130	89	42	17	620
September	90	268	150	114	28	0	450
October	78	261	146	112	25	0	622
November	13	117	65	52	13	0	260
December	197	323	96	58	26	26	726
January	446	1,084	440	234	29	132	2,365
February	483	1,101	478	246	105	103	2,516
March	571	1,208	685	363	153	146	3,126
April	300	805	500	225	8	64	1,902
Total	2,606	6,309	3,195	1,809	546	583	15,048

kg^{-1} day^{-1}, whereas, the same value for pregnant muskox cows in captivity was 40.1 C kg^{-1} day^{-1}. A 30% greater energy requirement for free-living muskox versus non-active captive muskox seems reasonable. Sheep at pasture expend 21% more energy than do their penned counterparts (Blaxter 1967). The D.E.I. for the 22-month-old bulls in captivity was 46.7 C kg^{-1} day^{-1}. Accordingly the above value for wild muskox is probably an underestimate, especially when applied to all age groups.

Having estimated food requirements, growth, and body weight and knowing the number of animals occupying Truelove Lowland (Table 7), one can estimate the amount of forage harvested for the year ending 30 April, 1973.

The results from converting "muskox days" (Table 8) to kilograms per day was multiplied by the daily forage requirement for the particular month (Table 5) which in turn was multiplied by the days per month to predict the total amount harvested during the month (Table 8). The areas most heavily utilized were meadows of the hummocky sedge type. The available standing crop of graminoid forage in this meadow type, averaged over four seasons, was 62 g m^{-3} (Muc this volume and pers. comm.). The other meadow type utilized was the frost-boil sedge-moss type with an available standing crop, averaged over 3 years of 51 g m^{-3}. Muskox grazed over 327 ha between September 1972 and

June 1973. The areas of hummocky sedge meadow and frost boil sedge-moss meadow grazed was 268 and 59 ha respectively. The 327 ha grazed between September 1972 and June 1973 at 60 g m^{-3} (weighted average) would have had 196,250 kg of forage available to muskox. During this period those muskox present required a predicted 44,200 kg (Table 8); 22.5% of the forage available on the 327 ha grazed but only 4.5% of the forage available on all the lowland meadows combined. It is estimated that muskox consumed 4.9% of the forage available in the 12 months ending 30 April, 1973. Of the meadows grazed during the winter of 1972-73 37,386 kg and 6,814 kg of forage was removed from hummocky sedge and frost-boil sedge-moss meadows respectively. The values represent approximately 7% and 1.7% of the aboveground plant biomass in these respective meadow types for the entire Lowland.

Methods used in determining forage removal by muskox were not sensitive enough to determine the extent and amount of summer grazing. This is due to the rapid plant growth in a short growing season. Also, the low muskox densities (Fig. 2) on Truelove Lowland during the snow-free period combined with increased daily movements of muskox in summer make it impossible to locate areas of intensive summer grazing.

The diet of muskox is predominantly graminoid vegetation. In an examination of fecal fragments from Devon Island muskox, Rackette (1974 unpublished) found graminoids made up 98% of the identifiable fecal material in August 1972. Comparing percent graminoid in feces to percent graminoid in rumen contents of three bulls collected from Cape Newman Smith by Inuit hunters in February 1973, Rackette found 84% and 88% graminoids respectively. The non-graminoid material in the feces as well as the rumen contents was almost exclusively *Salix*. *Salix arctica* is found in association with the monocots making up the meadow flora of the study area. Tener (1965) found *Salix* to be a dominant portion of muskox diet on the Fosheim Peninsula as well as in the Thelon Game Sanctuary. *Salix* hummock is a common habitat type on the Fosheim Peninsula. While on the Fosheim Peninsula in July 1972 I found *Salix arctica* made up 32% of the cover where the total plant cover in this habitat type was 36%. Live *Salix* tissue made up 31%, dead *Salix* 66%, and graminoid less than 4% of the total above ground plant biomass. The study area on Devon Island has no such habitat type. Our observations of muskox grazing activities also show almost exclusive grazing

Table 9. Forage removed (mean ± S.E.) from meadows grazed in the winter of 1972-73.

Location	*Forage available** $(g\,m^{-2})$	*Litter remaining+* $(g\,m^{-2})$	*Area grazed (%)*	*Forage++ removed (%)*	*Transect length (m)*
4+ 8	76.0±11.8	11.3±2.0	40±6	30	2400
9	63.6± 2.4	12.7±3.2	26±2	23	3600
10	48.2± 6.7	5.6±1.4	23±2	20	3900
11	63.9**	10.9±1.8	20±2	17	5620
12	67.3± 3.6	6.1±1.0	31±3	28	1380
	$\bar{x} = 63.8$			$\bar{x} = 23.6$	

* Graminoid biomass present September 1972
** Data from Muc (1974)
\+ Litter remaining following grazing and runoff in plot where 100% of area had been grazed prior to runoff
++ Percent forage removed from entire meadow

of meadow sedges. For these reasons the discussion of the effects of forage removal by muskox grazing is restricted to meadows.

The transects sampled large areas where winter grazing had occurred (Fig. 9). The area grazed ranged from 23% to 40% of the total area sampled. The amount of litter remaining in sampling plots that were totally grazed ranged from 9% to 20% of that which was present the previous fall. The amount of vegetation that was removed from each of the meadows sampled ranged from 17% to 30%.

There is close agreement between the observed 23.5% (Table 9) forage removed in those meadows sampled, and the predicted 22.5% based on the estimated energy required by the muskox utilizing the Lowland. This indicates that muskox harvested only a small portion of the total standing crop on Truelove Lowland. Therefore, natural density independent factors regulating population size will probably keep wild muskox from consuming more than the annual production of their food supply.

Significance of Truelove Lowland to the productivity of muskox on northeastern Devon Island

The total area below 200 m between Sverdrup Inlet and Sverdrup Glacier on northeastern Devon Island is 301.4 km^2. Of this area approximately 51 km^2 (18.8%) is meadow habitat. Meadow habitat was determined from aerial photographs. Truelove Lowland represents 14.7% of the land surface of the range, but Truelove Lowland meadows represent 34.5% of the meadow habitat available to the muskox.

To determine the significance of Truelove Lowland to the muskox population occupying northeastern Devon Island, the population is assumed to be 243 animals; based on May 1972 and May 1973 censuses. That means that 243 animals were in the population from 1 May, 1972 to 30 April, 1973. The composition of this "constructed" population is: 73 adult males, 86 adult females, 36 calves, 34 yearlings, 6 two-year-olds, and 8 three-year-old males. The number of calves was set at 36 rather than the total number observed in May 1972 because we know that at least 36 calves were in the population for the entire year. The same reasoning holds for the other cohorts in the above population.

If muskox were distributed randomly over the land surface of their range one would expect approximately 15% occupying the Lowland. On an annual basis 17.0% of the potential muskox days of the population were spent there. If muskox were distributed at random over the meadow area in their range, 34.6% of the population should utilize the Lowland. From Fig. 10 it appears that neither condition seems to hold. The Lowland was under-occupied in relation to the rest of the range as a whole during the entire period of continuous daylight plus the first month of continuous darkness. It was over-occupied most of the period of continuous darkness, maximum snow cover, and the lowest ambient temperature (Table 10). The greatest rate of increase in muskox numbers and biomass on the Lowland coincides with the period when the population approaches and enters the predicted period of zero or negative biomass change. The Lowland appears most important for cows, calves, and yearlings during this period. Since the observed number of 2- and 3-year-old males is so low in this "constructed" population it is difficult to judge the relative importance of the Lowland to these cohorts. The Lowland and valley are also important for calving in May. It would appear that in winter muskox prefer to graze meadows where minimal effort is required to expose forage (Fig. 3).

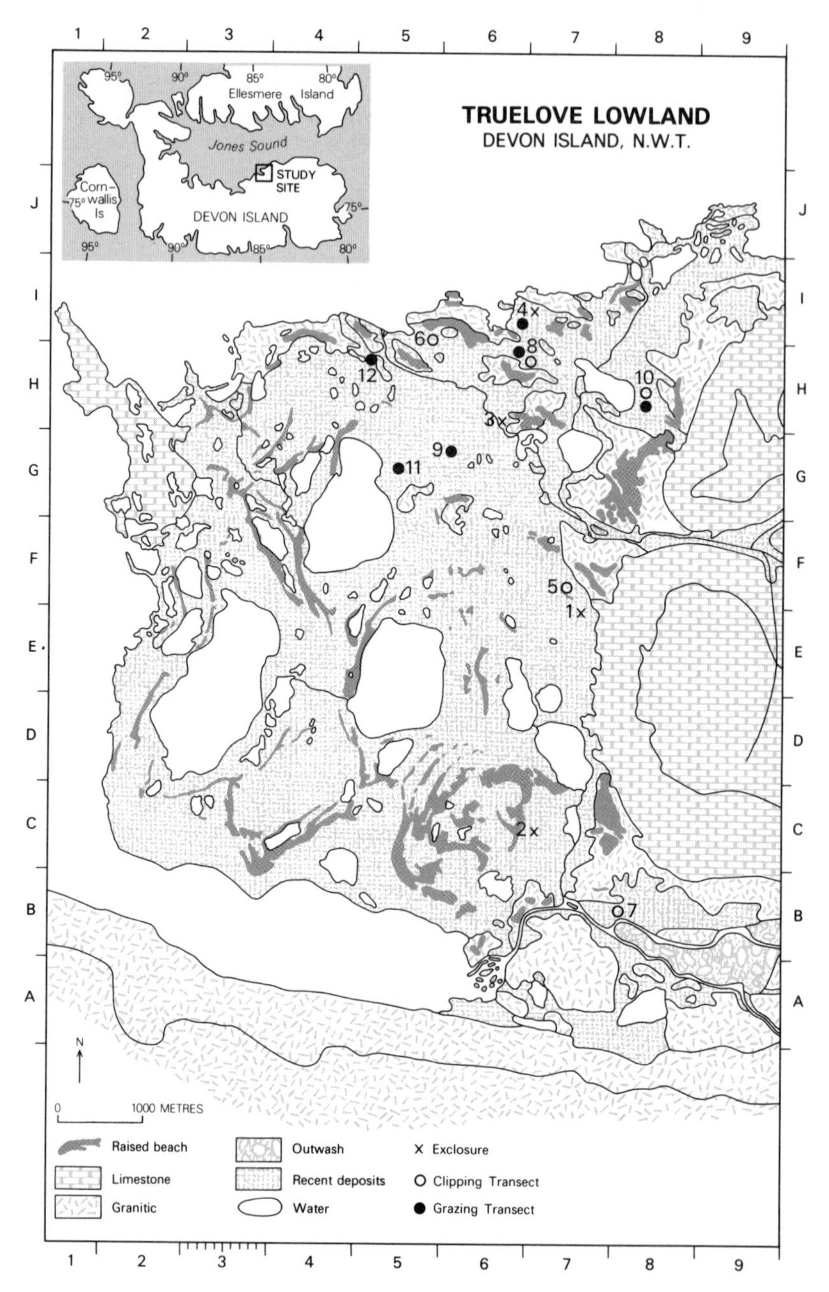

Fig. 9 Locations of grazing investigations on Truelove Lowland.

Productivity of muskox

Productivity in this context is defined as biomass (Table 11) and energy (Table 12) increment rather than increase in population size. Based on the growth and weights of wild muskox the "constructed" population weighed 43,400 kg, the

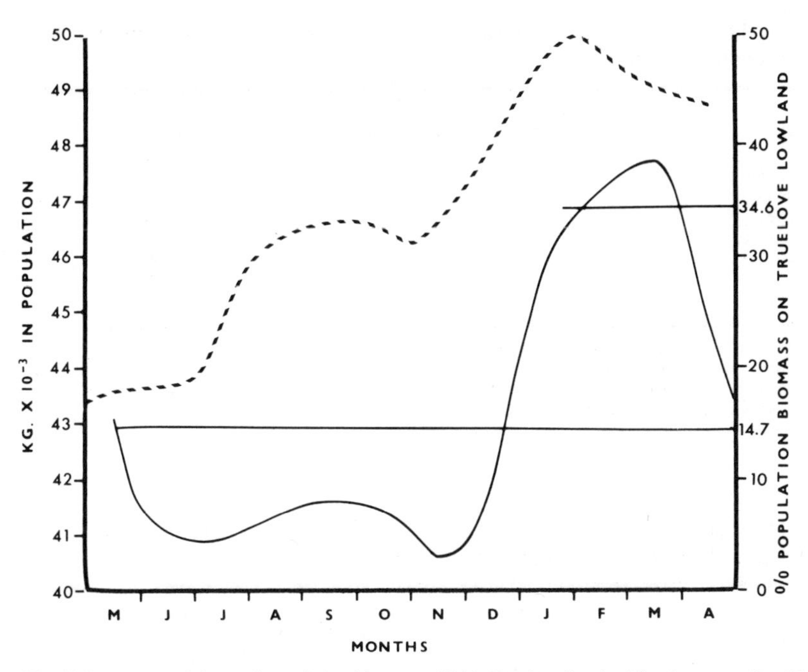

Fig. 10 Percentage of the total population biomass utilizing Truelove Lowland for the year ending 30 April, 1973.

Table 10. Percent of cohort in population occupying Truelove Lowland in the year ending 30 April, 1973.

Month	*Adult males*	*Adult females*	*Calves*	*Yearlings*	*2-yr. olds*	*3-yr. olds*
May	7.3	22.0*	27.2*	21.3*	35.5*	24.2*
June	2.8	7.4	9.4	6.3	9.4	8.3
July	4.3	4.7	9.1	2.7	18.3*	6.1
August	4.6	8.9	11.7	8.4	22.6*	6.8
September	4.1	10.4	13.9	11.2	15.6*	.0
October	3.5	9.8	13.1	10.6	13.4	.0
November	0.6	4.5	6.0	5.1	7.2	.0
December	8.7	12.1	8.6	5.5	14.0	10.9
January	19.7*	40.7**	39.4**	22.2*	15.6*	53.2**
February	23.6*	45.7**	47.4**	25.8*	62.5**	48.5**
March	25.2*	45.3**	61.4**	34.4*	82.3**	58.9**
April	13.7	31.2*	46.3**	22.1*	4.4	26.7*
$\bar{x}$	10.7	20.2*	24.5*	14.6	25.1*	20.3*

* Occupation greater than 14.7%
** Occupation greater than 34.6%

equivalent of 83.5 Mcal* on 1 May, 1972. The population reached its peak weight in January, 1973 at 50,000 kg (96.3 Mcal). Seasonal fluctuations in body weight especially during the rut and late winter for adult males contribute to the net weight gain for the population being less than the January minus May total. This is termed gross production and amounted to 8,930 (17.2 Mcal).

* Caloric determination as follows: Live body weight × dry weight coefficient (.35 after Blaxter 1967) × caloric coefficient of muscle (5.5C g^{-1} after Blaxter 1967).

Table 11. Cohort contribution to estimated production of muskox biomass for year ending 30 April, 1973.

Cohort	*Gross* production (kg)*	*Net* production (kg)*	*Forage intake (kg)*	*Ratio***
Adult bulls	1970	0	137,500	—
Adult cows	2240	685	128,700	0.005
Calves	2760	2760	16,700	0.170
Yearlings	1240	1190	26,500	0.045
2 yr. olds	270	245	6,200	0.040
3 yr. old bulls	450	320	12,100	0.030
Total	8930	5200	327,700	0.016

* Due to seasonal fluctuations in body weight (Fig. 4-7) all cohorts except calves weigh more at some point during the past 12 months than they do on their "birthday" (n). Gross production is the peak cohort weight minus the weight on n−1 "birthday." Net production is the difference in cohort weight between n and (n−1) "birthdays."
** Net cohort biomass production, forage intake by cohort^{-1}

Table 12. Energy equivalent of estimated muskox biomass produced and forage consumed for the year ending 30 April, 1973.

Cohort	*Gross production (C)*	*Net production (C)*	*D.E.I. (C)**
Adult bulls	3,792,250	0	7,562,500
Adult cows	4,312,000	1,318,625	7,078,500
Calves	5,313,000	5,313,000	918,500
Yearlings	2,387,000	2,290,750	1,457,500
2 yr. olds	519,750	471,625	341,000
3 yr. old bulls	866,250	616,000	665,500
Total	17,190,250	10,010,000	18,023,500

* Calculated on an annual mean D.E.I. 55 C.kg^{-1}

The net production over the 12 months ending 30 April, 1973, was 5,200 kg (10 Mcal).

Examining the production of individual cohorts, the expected pattern emerges. The youngest cohort is not only the most productive, but is also the most efficient. The biomass increment/forage consumed ratio for cows is 30 times greater than that of calves. However, in the case of pregnant or lactating females it must be recognized that their net production is largely expressed in the development and growth of their calves. Calves contributed over 50% of the annual (1972-73) biomass increment. The drastic difference between the ratio of calves and other sub-adult cohorts reinforces the role of the cows in reproductive condition. Therefore, in a normal breeding population reproductive cows indirectly assure that population's continued biomass production. It is predicted that the biomass increment in a non-breeding population would be insignificant.

Summary

The object of this study was to determine the productivity of muskox on northeastern Devon Island, N.W.T. The major points emerging from this study are:

1. The muskox population on northeastern Devon Island was in a period of increase and stood at approximately 275 at the close of the field study. Population size is probably regulated by abiotic density independent agents.
2. Female muskox may produce their first calf in their third year but probably most are at least four years old. Calves in the wild are not necessarily weaned before 12 months.
3. Female muskox probably approach adult body weight in their fifth summer while males may grow until their sixth or seventh summer.
4. Cultured hay used in feeding trials with domestic muskox had a digestibility of 64% while sedges on Devon Island had an estimated digestibility of 66% in winter and 84% in summer.
5. Daily organic matter requirements for maintenance of domestic muskox were 13.1 to 15.4 g kg^{-1} body weight in February 1972. For wild muskox the estimated organic matter intake ranged from 16 g kg^{-1} day^{-1} in March and April to 22 g kg^{-1} day^{-1} in July and August.
6. On the basis of body weight and estimated energy requirements of muskox, a predicted 22.5% of the forage was removed from winter grazed meadows on Truelove Lowland for the year ending 30 April, 1973. Examination of these meadows in August 1973 showed that an average 23.6% of the standing crop had been removed.
7. The numbers of muskox utilizing Truelove Lowland ranged from zero during the period of spring melt and runoff to 114 in March 1973. It was most heavily utilized in the season when forage is the most difficult to secure.
8. Net production of muskox over the year was 5200 kg (10 Mcal) or approximately 1 kg ha^{-1} (1925c) of meadow habitat in northeastern Devon Island.

Acknowledgments

The ultimate success or failure of a study such as this depends on more people and organizations than one is permitted to mention. It is a pleasure to acknowledge those who assisted directly in the collection of data. My wife, Linda, assisted and encouraged me during all phases of field work and data preparation. Leo Boukhout, Richard Stardom, Marjorie van Buskirk, Lindsay Rackette, Dennis Magee, and Imosee Nutarajuk assisted at various stages prior to September 1972. Del and Lindsay Rackette, Claude Belcourt, and Mike Hoyer provided assistance and companionship during the winter of 1972-73 on Truelove Lowland. I also thank the many unnamed colleagues while in the field who added to the collection of field data. I thank Larry Bliss for his personal interest, support, encouragement, and criticism throughout the study. W.O. Pruitt gave generously of his time and effort.

Special thanks go to Paul F. Wilkinson and John J. Teal and their staff for permission and assistance in conducting the feeding experiments on domestic muskox at the University of Alaska. Dr. Wilkinson also cheerfully provided many hitherto unpublished data. George Phillips and J. McKirdy of the University of Manitoba advised on the design of feeding experiments and interpretation of the data, and supervised the chemical analyses, respectively. Charles Jonkel of the Canadian Wildlife Service, Ottawa, was generous with helicopter support and personal assistance. Without his help we could not have adequately determined the population size or the adult weight of Devon Island muskox.

The assistance of Kaye Mockford of Yellowknife in the preparation of this manuscript was appreciated.

References

Allen, J.A. 1913. Ontogenetic and other variations in muskox, with a systematic review of the muskox group, recent and extinct. Memoirs of the Amer. Mus. of Nat. Hist. No. 1, Part 4, pp. 103-226.

Blaxter, K.L. 1967. The Energy Metabolism of Ruminants. Hutchinson. London. 332 pp.

Freeman, M.M.R. 1971. Population characteristics of muskoxen in the Jones Sound region of the Northwest Territories. J. Wildf. Managt. 35: 103-108.

Gray, D.R. 1973. Social Organization and Behavior of Muskoxen. PhD. thesis. Univ. Alberta, Edmonton. 212 pp.

Harington, C.R. 1961. History, Distribution, and Ecology of the Muskoxen. M.Sc. thesis. McGill Univ. Montreal. 489 pp.

———. 1964. Remarks on Devon Island muskoxen. Can. J. Zool. 42: 79-86.

Hone, E. 1934. The present status of the muskox. American Committee for International Wildlife Protection. Spec. Publ. No. 5: 1-87.

Hubert, B. 1974. Estimated Productivity of Muskox on Northeastern Devon Island. M.Sc. thesis. Univ. Manitoba, Winnipeg. 118 pp.

Hussell, D.J.T. 1970. Written communication in personal files.

Jonkel, C., D.R. Gray and B.A. Hubert, 1975. Immobilizing and marking wild muskoxen in Arctic Canada. J. Wildf. Managt. 39: 112-117.

Lancaster, R.J. 1954. Measurement of feed intake of grazing cattle and sheep, Pt. V. Estimation of the feed-to-faeces ratio from the nitrogen content of the faeces of pasture fed cattle. New Zeal. J. Sci. and Tech. A. 36: 15-20.

Lethbridge, T.C. 1939. Archaeological data from the Canadian Arctic. J. Royal Anthro. Inst. 69: 203-223.

Lowther, G.R. 1960. An account of an archaeological site on Cape Sparbo, Devon Island. Contributions to Anthropology Part 1. Nat. Museum Can. Bull. No. 180.

Miller, F.L. 1974. Written communication in personal files.

———. and R.H. Russell, 1975. Aerial surveys of Peary caribou and muskoxen on Bathurst Island, N.W.T., 1973 and 1974. Prog. Notes No. 44 Spp. Can. Wildlife Service, Ottawa.

Nutt, D.C. 1960. The North Water of Baffin Bay, Polar Notes 9: 1-25.

Pedersen, A. 1958. Der Moschusochs (*Ovibos moschatus* Zimmerman). A. Ziemsen Verlag. Wittenberg. 54 pp.

Rackette, L.J. 1974. An analysis of muskox *(Ovibos moschatus)* diet by fecal examination. Unpublished report in personal file.

Segal, A.N. 1962. The periodicity of pasture and physiological functions of reindeer. *In:* Reindeer in the Karelian A.S.S.R., Akad. Nauk, Petrozavodsk. pp. 130-150. Dept. Secretary of State Bureau of Translations, Ottawa.

Short, H.L. 1970. Determining forage quality, digestibility trails: in vivo techniques. *In:* Range and Wildlife Habitat Evaluation. U.S. Dept. Agric. Misc. Publ. No. 1147.

Spencer, D.L. and C.L. Lensink, 1970. The muskox of Nunivak Island, Alaska. J. Wildf. Managt. 34: 1-15.

Tener, J.S. 1965. Muskox in Canada. Can. Wildf. Serv. Mongr. Ser. No. 2. Queens Printer, Ottawa. 166 pp.

Theurer, B. 1970. Chemical indicator techniques for determining range forage consumption. pp 111-119. *In:* Range and Wildlife Habitat Evaluation. U.S.D.A. No. 1174.

Vibe, C. 1967. Arctic animals in relation to climatic fluctuation. Medd. om Grønl. 170: 1-227.

White, R.G. 1975. Some aspects of nutritional adaptations of arctic herbivorous animals. pp. 239-268. *In:* Physiological Adaptation to the Environment. F.J. Vernberg (ed.). Intext Educational Pub. New York. 576 pp.

Wilkinson, P. 1971. The domestication of the muskox. Polar Record 15: 683-690.

Mammalian carnivores utilizing Truelove Lowland

R.R. Riewe

Introduction

Mammalian carnivores utilizing Truelove Lowland include short-tailed weasel *(Mustela erminea)*, arctic fox *(Alopex lagopus)*, arctic wolf *(Canis lupus)*, and polar bear *(Ursus maritimus)*. Due to the infrequent use of the Lowland by wolves and bears, negligible data were collected on these carnivores in relation to it. These larger carnivores are discussed briefly on pp. 633-34 of this book. This section deals only with arctic foxes and short-tailed weasels.

Methods

The mammalian carnivores were studied on Truelove Lowland from 1970 to 1972 inclusive. Most of the research was directed toward the arctic fox. In 1970 a total of 144 man-days (9 May - 20 September) were conducted in field work. Over 1,050 fox scats and numerous food scraps were collected on the Lowland. Live trapping was conducted for fox from May through July with box traps and padded leg-hold traps (2,367 trap-nights).

Field investigations on the carnivores were intensified in 1971, 279 man-days being spent between 2 May and 5 October on Truelove Lowland. The main effort was again directed toward foxes, but emphasis was also placed on weasels. Live trapping was conducted from June through September for foxes (649 trap-nights) and weasels (1,123 trap-nights). Whenever good tracking conditions existed, considerable time was spent surveying the Lowland for carnivore sign and following the trails. Sightings of carnivores were also made by many other members of the field party. These sightings were plotted on maps of the Lowland in order to determine areas of carnivore activity, periods of activity, and numbers of individuals present.

Arctic foxes were difficult to capture on Truelove Lowland during the summers, and therefore collecting was done at the National Museum Camp and a Sun Oil Camp on Bathurst Island from 21-28 July and 1-2 September, 1971, respectively. Four foxes were captured and I had the opportunity to become more familiar with arctic fox behavior and ecology.

During August and September 1971, 22 feeding tests were conducted on five captive foxes using bearded seal *(Erignathus barbatus)*, ringed seal *(Phoca hispida)*, arctic hare *(Lepus arcticus)*, collared lemming *(Dicrostonyx groenlandicus)*, oldsquaw *(Clangula hyemalis)*, ptarmigan *(Lagopus mutus)*, and C-ration luncheon meat. The feeding tests were terminated shortly before or just after the foxes had eaten the food presented to them; the tests ran for 13.3 to 145.3 hr, with an average of 47.4 hr. Records were kept on the food and water intake and the feces and urine output. Five feeding experiments were also conducted on one female weasel in September 1971 using collared lemming, snow bunting *(Plectrophenax nivalis)*, longspur *(Calcarius lapponicus)*, ptarmigan, and bearded seal.

In 1972 a total of 132 man-days were spent between 5 May and 18 August on the Lowland searching for fox sign and den sites. All fresh fox and weasel scats and food scraps located were collected. In the hopes of securing additional behavioral information, three arctic foxes captured in 1971 on Bathurst Island were released on Devon Island in May 1972. Prior to release the foxes were marked and fitted with radio-transmitter collars. The area of release was surveyed with both radio and telescope as often as possible after the release.

In addition to the research conducted on Truelove Lowland and Bathurst Island, supplemental data were gathered on carnivores elsewhere in the High Arctic from 1971 through 1973 (see pp. 633-34).

Results

Arctic fox — Jones Sound region

Arctic fox sign may be encountered just about anywhere in the Jones Sound region, but the actual carrying capacity of this region is considerably lower than elsewhere in arctic Canada, such as the Keewatin area (Speller 1972) or Banks Island (Usher 1971). The low carrying capacity is related primarily to the dearth of microtine rodents, and the lack of suitable denning sites may also be a contributing factor to the low fox density. From observations we know of only six fox dens in the entire Jones Sound area with a maximum of two active in any one year.

For most of the year foxes inhabiting this region are directly or indirectly dependent upon the sea for their existence. In winter most foxes scavenge on ringed seal remains left by polar bear or Inuit hunters. When polar bears are well fed, they consume only the seal's skin and blubber (Nelson 1969) and leave the rest of the carcass untouched for the scavengers. When polar bears are having difficulty procuring sufficient food, however, they consume the entire seal carcass, leaving only their own scats for the foxes. During the latter half of April foxes are less dependent upon larger carnivores, because they are able to capture newborn ringed seals by digging them out of their dens (personal observation).

Foxes feed heavily upon nesting birds during the summer months, wherever available (Pattie this volume). On the Bjorne Peninsula and on the plateaus northwest of Vendom Fiord, foxes rely upon collared lemmings as their principal source of nutrition. In some localities foxes concentrate their hunting activities on bands of arctic hare. Where wolves are present, foxes scavenge on their kills.

Weights were recorded from five winter-trapped fox carcasses taken in the Jones Sound area: range 1.58 to 3.39 kg, average 2.35 kg (1.53 kg dry wt).

Arctic fox — Truelove Lowland

Field observations show that foxes utilizing Truelove Lowland lead a subsistence life on the sea ice during the winter months and move onto the land in May when denning activity begins. Initially, foxes find food difficult to obtain, since bird nesting has not begun. When nesting birds are in abundance, between mid-June and late-July, foxes crop and cache more than their immediate needs. Such summer food caching by arctic foxes has been reported from different areas (see McEwen 1951, for review of the literature). During August and September, after most of the young birds are fledged, the foxes again find food increasingly difficult to secure, and in August and September 1971, foxes returned to their egg caches. The lemming study on Truelove Lowland (Fuller et al. this volume) indicates that the Lowland could not support a fox population on lemmings alone, and the foxes leave the land and return to the sea ice in search of food when the sea freezes or when the pack ice runs ashore in autumn. Seasonal movements such as these have been reported elsewhere in the Arctic (Freuchen 1935, Chitty and Chitty 1945, Usher 1971, and others).

By means of field observations and gross scat analyses it has been determined that foxes in the Truelove region prey or scavenge upon snow buntings, Baird's sandpipers *(Erolia bairdii)*, snow geese *(Chen caerulescens)*, glaucous gulls *(Larus hyperboreus)*, ptarmigan, king and common eiders (*Somateria spectabilis* and *S. mollissima*, respectively), oldsquaws, collared lemmings, arctic hares, seals, and muskox *(Ovibos moschatus)*.

D. Hussell (pers. comm.) mentioned that foxes were major predators of nesting birds on Truelove Lowland during the summers of 1966 through 1969. Pattie (this volume) also found that foxes preyed heavily upon nesting birds on the Lowland. Predators destroyed 25% of all nests he located in 1970 and 69% of the nests located in 1971; Pattie attributed the increased predation to an increase in fox utilization of the study area in 1971. Pattie noted that arctic foxes accounted for 42% of nest predation in 1972, a year in which 63% of all observed nests in the Truelove region were predated.

Since 1966 four fox dens, including three active ones, have been discovered on Truelove Lowland. In 1969 D. Hussell (pers. comm.) observed two active dens, one of which produced young. In May 1970 a third active den was discovered, but it was abandoned later in the season and no young were produced. During the summers of 1971 and 1972 no active fox dens were located on the Lowland. In 1973 one active den produced several pups but, unfortunately, research on the carnivores was terminated in 1972 and no information on this site was collected.

On 13 May, 1972 three foxes captured the previous summer on Bathurst Island and kept over the winter for metabolic studies were released on Skogn Lowland, 5 km east of Truelove Lowland. They were released there in hopes of discouraging them from becoming scavengers at Base Camp on Truelove Lowland. Prior to release the foxes were marked with black dye, colored ear tags and radio transmitters. A meat cache was placed on Skogn Lowland in an attempt to keep the foxes in the vicinity for study, but all attempts to locate the foxes during spring and summer of 1972 failed. One Grise Fiord hunter, however, reported trapping one of the marked animals the following winter along the south coast of Ellesmere Island, 90 km northeast of Truelove Lowland.

An analysis of track data collected in the spring of 1972 indicates that foxes frequented various areas of the Lowland with different levels of intensity. The most heavily used areas include the granite outcrops along the north

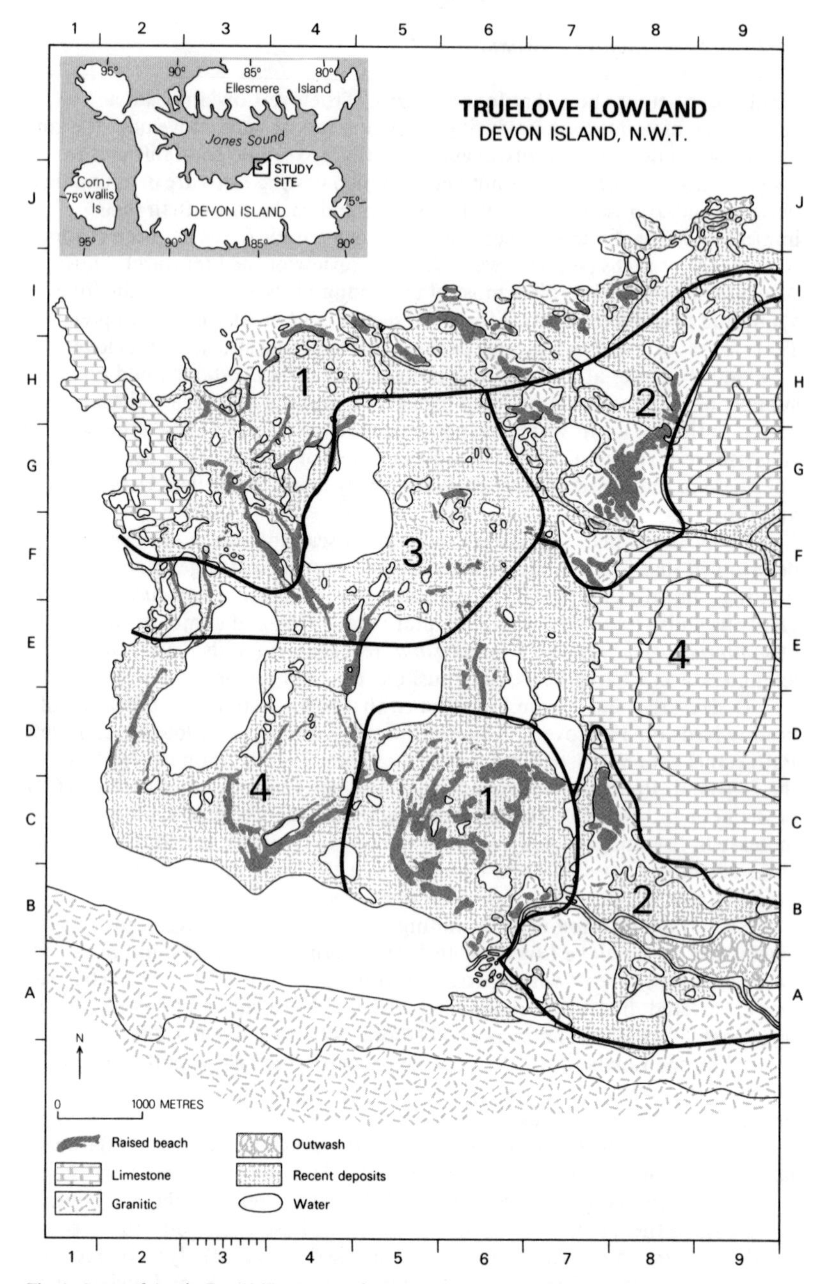

Fig. 1. Areas of Arctic Fox Utilization on Truelove Lowland during Spring 1972. Area 1 was used most intensively by the foxes as hunting grounds. Area 4 was seldom used as hunting grounds and was used only occasionally while they traversed the Lowland. Areas 2 and 3 were areas of intermediate use for both hunting and as travel routes across the Lowland.

shore, Rocky Point, Base Camp[1] and the Wolf Hill areas (Fig. 1). In these areas, observed foxes or their sign indicate that they frequently hunted lemmings, ptarmigan, and birds' eggs and scavenged on seal carcass remains. The areas of intermediate use include the northeast section of the Lowland and Truelove Valley. Here hunting sign was obvious but considerably less frequent than in the areas of heavy utilization. The areas of lowest fox utilization include the southwest portion of the Lowland and the central Lowland at the base of the plateau. The remaining area, in the centre of the Lowland, was often traversed by foxes but rarely was any hunting sign observed.

Since foxes do not possess special adaptations for travel in deep snow, they showed a preference for areas with shallow snow cover and for trampled areas, such as trails left by skis, snowmobiles, and other animals. Most fox trails were noted on beach ridges and rock outcrops, along the base of the plateau or on moraine hills; these were the areas of shallower snow cover. Between these areas, the tracks were straight rather than meandering and back-tracking as in areas where foxes were hunting. Fox movements were influenced to some extent by trails left by other animals or machines since these trails were frequently followed by foxes. When camp was opened in spring 1972 a large herd of muskox was established near Fox Camp (B-7, Fig. 1). Fox tracks were frequently found crisscrossing the large, well-trampled yards left by muskox. The foxes may also have shown an interest in this area because muskox were calving and the foxes may have been scavenging for after-birth or dead calves.

Members of the Devon Island Camp observed foxes on Truelove Lowland 29 times during the spring and summer of 1972; 76% were observed between 2000 and 0800 hr. Most researchers were in the field during both the day and evening hours and it is, therefore, logical to assume that the foxes were more active in the night between 2000 and 0800 than in the other half of the day.

During 1971 there were four foxes utilizing Truelove Lowland during May through July, and six foxes during August and September. From this a rough calculation of biomass can be made. Assuming an average of five foxes $month^{-1}$, an average weight of 2.72 kg (average of five summer-trapped foxes), a 65% water loss, and that the foxes spent all their time on Truelove Lowland, then the monthly average fox biomass (dry wt) on the Lowland between 1 May and 30 September, 1971 would have been 1.46 g ha^{-1}, a considerable increase over 1970, when there was a calculated average of 2.6 foxes $month^{-1}$ (0.76 g ha^{-1}) on the Lowland between 9 May and 27 July.

Feeding tests conducted on five captive foxes in 1971 produced the following metabolic data. Daily meat consumption varied considerably, depending on the type of meat, the amount of roughage, and the individual's appetite. For all food types, daily meat consumption ranged from 82 to 613 g dry wt, with a mean of 206 g. Daily production of feces likewise varied with the type of meat and the amount of roughage. Number of scats day^{-1} ranged from 2 to 64, with a mean of 22. The weight of these feces ranged from 2.7 to 68.4 g dry wt, with a mean of 28.3 g. The digestive efficiency of the foxes was 86.3%. Daily urine production ranged from 82.4 to 492.6 cc, with a mean of 217.4 cc. Daily water consumption ranged from 78.6 to 591.7 cc, with a mean of 298 cc.

Using the above mean daily food consumption, five foxes would have consumed 157.5 kg of biomass (dry wt) between 1 May and 30 September, 1971. For the sake of illustration, this is equivalent to 2,000 28 g (fresh wt) lemmings, 2,000 30 g (fresh wt) snow buntings, and 92 3.76 kg (fresh wt) arctic

1. Foxes often display curiosity toward strange objects on the Lowland. This, plus the fact that they preferred to travel in trampled areas with harder snow (see below), probably increased the number of fox tracks in the Base Camp area.

hare. These estimates indicate that the foxes had a devastating effect upon the Lowland and/or adjacent areas during this time.

In October 1971 one female and three male foxes were transported to Winnipeg, where they were housed in pairs in two outdoor breeding cages, approximately 2.5m × 5m, at the Manitoba Experimental Fur Farm. Between October 1971 and January 1972 the foxes maintained their body weights on a ranch diet of 125 Kcal kg^{-1} body wt day^{-1}. The U.S. National Research Council's Subcommittee on Furbearer Nutrition (1968) estimated that ranch-raised red and arctic foxes could be maintained on 121 Kcal kg^{-1} body weight day^{-1}.

Underwood (1971) found that the rate of energy intake of captive arctic foxes at Barrow, Alaska was significantly higher in summer than in winter. Extrapolation of data from Underwood's Tables 7 and 10 reveals that his foxes ingested a maximum of 370 Kcal kg^{-1} body wt day^{-1} in July and a minimum of 63 Kcal kg^{-1} body wt day^{-1} in January. Speller (1972) found that two captive arctic fox whelps at Aberdeen Lake, N.W.T. consumed an average of 363 g day^{-1} of fresh lemming and caribou meat between 30 June and 1 August. Assuming that the fox whelps weighed an average of 1.3 kg (extrapolated from Speller's Fig. 21), a water loss of 65% and a caloric value of 5 Kcal g^{-1} for the lemming and caribou meat, then Speller's growing fox pups ingested an average of 490 Kcal kg^{-1} body wt day^{-1}.

Short-tailed weasel

Short-tailed weasels are uncommon throughout the Jones Sound region (see p. 633), and the same holds true for Truelove Lowland.

Hussell (1969) mentioned that weasels were not seen on Truelove Lowland during the summers of 1966 through 1968, but that weasels were observed several times in the summer of 1969, a year of a relatively high lemming population. In June 1969 Simpson (1970) watched a pair of weasels with six young prey on snow bunting nestlings on the Lowland.

During most of the 1970 field season weasels or their sign were noted only infrequently. Fuller et al. (this volume) reported that 23 of 198 subnivean lemming nests located in the spring showed evidence of weasel predation. Fresh weasel tracks were seen on 12 May and during the summer and fall two weasels were seen on the Lowland. In September fresh weasel tracks were commonly seen around the Lowland.

In 1971 several tracks were noted on the Lowland when I visited it in February and March. That winter 9 of 80 winter lemming nests had been preyed upon by weasels (Fuller et al. this volume). During May only two tracks were seen. In June only three weasel tracks and one weasel were observed. Weasels preyed on at least two snow bunting nests during the 1971 summer.

On 18 July, 1971 a 160 g adult male was captured northeast of Muskox Lake in the rock outcrops, but he went into shock and died the following day. Two females were captured in the rock outcrop at the Intensive Study Area in September. Both females weighed approximately 100 g when captured. Only one female survived in captivity; the other died after three days.

A weasel survey was conducted on Truelove Lowland between 5 September and 3 October, 1971, when light snow provided ideal tracking conditions. All likely weasel habitats, including rock outcrops, talus slopes, beach ridges, and a considerable portion of the meadows, were searched, approximately 258 km on foot and 61 km on snowmobile. This survey indicated that at least one male and three female weasels occupied the rock outcrops near the Intensive Study Area, three utilized Rocky Point, one hunted in the outcrops near Skogn Camp, and two others were active among the talus slopes between the Gully River and Muskox Hill.

During the 1972 field season weasel tracks were observed only once, on 8 May, when fresh tracks were noted at the base of the plateau north of the Gully River. These tracks were in three distinct locations, but, as each set of tracks was only 300 m or less from the next, it is possible that only one animal was responsible. The remains of a weasel were found at the beginning of August. This animal appeared to have died of natural causes, as the body was dessicated but still intact, and no injuries were noted.

During the entire study the largest weasel population was present on the Lowland in 1971. From field observations and trapping results, it appears that there were 10 weasels occupying the Lowland from 1 May to 3 October, 1971. Assuming a water loss of 65%, the average monthly biomass of weasels on the Lowland between 1 May and 30 September, 1971 would have been 0.14 g ha^{-1} (dry wt).

Three feeding tests conducted in 1971 on a captive female weasel showed consistent results when lemmings and passerines were fed to the weasel. Daily food consumption ranged from 14.3 to 17.4 g dry wt, with a mean of 15.4 g. The number of feces produced per day ranged from 17 to 18, with a mean of 17.6. Digestive efficiency of this weasel was 73.9%. The daily urine production ranged from 5.29 to 13.76 cc, with a mean of 9.45 cc.

Using the above mean daily food consumption, a population of 10 weasels would have consumed 23.56 kg of biomass (dry wt) between 1 May and 30 September, 1971. For the sake of illustration, this is equivalent to 500 30 g (fresh wt) snow buntings and 2,046 28 g (fresh wt) lemmings. This is a conservative estimate, because it is based on the daily consumption rate of a female; an estimate which includes consumption rates of the larger male weasels would probably be considerably more. In turn, consumption for the entire year would be far greater.

Summary

The ecology of arctic foxes and short-tailed weasels was investigated on Truelove Lowland from 1970 to 1972 inclusive. Both carnivores were relatively scarce in the area throughout the study.

The Lowland was most intensively utilized by carnivores during 1971. That year there was an average of five foxes $month^{-1}$ with a biomass of 1.46 g ha^{-1} (dry wt) on the Lowland from May through September. It was estimated that 10 weasels with a biomass of 0.14 g ha^{-1} (dry wt) also occupied the Lowland from May through September of that year.

The arctic foxes preyed or scavenged on snow buntings, Baird's sandpipers, snow geese, glaucous gulls, ptarmigan, king and common eiders, oldsquaws, collared lemmings, arctic hare, seals, and muskox. Short-tailed weasels preyed primarily upon collared lemmings and passerine birds. Despite their small numbers, the carnivores had a great impact on the ground-nesting birds.

Feeding tests conducted on five captive foxes in 1971 gave highly variable results, depending upon the type of meat and the individual's appetite. Daily meat consumption ranged from 81.5 to 613.1 g dry wt, with a mean of 205.9 g. The digestive efficiency of the foxes was 86.3%.

Three feeding tests conducted on one captive weasel showed consistent results when lemmings and passerine birds were fed to the weasel. The daily food consumption ranged from 14.27 to 17.38 g dry wt, with a mean of 15.36 g. The digestive efficiency of this weasel was 73.9%.

An analysis of fox tracks in the spring of 1972 showed that foxes frequented various areas of the Lowland with different levels of intensity. The most heavily used areas included the granite outcrops along the north shore, Rocky Point, Base Camp, and Wolf Hill. Field observations indicated that foxes were nocturnal in their habits during spring and summer.

Acknowledgments

I owe thanks to Dr. W.O. Pruitt, Jr., who acted as my advisor throughout this project. I conducted these investigations while on a post-doctoral fellowship from the University of Manitoba, Zoology Department.

Richard Stardom, Marjorie van Buskirk, Imooshee Nutaraqjuk, and Lindsey Rackette acted as my field assistants. Their contributions to this study have been invaluable. John Owen kindly assisted me with trapping on Bathurst Island in September 1971. Thanks are due to the other members of the I.B.P. Devon Island field party for their stimulation and for reporting carnivore observations to me. Dr. David J.T. Hussell kindly provided me with his carnivore data from Truelove Lowland for the years 1966 through 1969. I wish to thank Dr. Donald L. Pattie for undertaking the exhaustive job of determining the caloric values of my specimens. The Polar Continental Shelf Project and the Sun Oil Company graciously provided me with inter-island air transportation. Sun Oil also extended the facilities of their base camp at Resolute Bay and their field camp at Freeman's Cove, Bathurst Island. The hospitality of Dr. Harold E. Welch and his family at the I.B.P. Char Lake Project always made my visits to Resolute Bay enjoyable. I am most grateful to Dr. Stewart D. MacDonald's field staff for cordially welcoming me to their camp on Bathurst Island and for assisting me in capturing foxes. The Canadian Wildlife Service generously incurred the expense of shipping my specimens south to Winnipeg. Mr. P.A. Kwaterowsky, N.W.T. Superintendent of Game, gave me permission to transport arctic foxes from Devon Island to Winnipeg.

I especially wish to thank my wife, Jane, for her enthusiasm, stimulation, and assistance throughout the course of this study.

References

Chitty, H. and D. Chitty. 1945. Canadian arctic wildlife enquiry, 1942-43. Animal Ecology, 14: 37-41.

Freuchen, P. 1935. Mammals. Part II — field notes and biological observations. Report of the Fifth Thule Expedition (1921-24), II (4-5): 68-278.

Hussell, D.J.T. 1969. Ornithological fieldwork on Devon Island in 1969. Unpublished progress report, Arctic Institute of North America. 2 pp.

McEwen, E.H. 1951. A Literature Review of Arctic Foxes. M.A. thesis, Univ. Toronto. 86 pp.

Nelson, R.K. 1969. Hunters of the Northern Ice. Univ. Chicago Press, Chicago. 429 pp.

Simpson, M. 1970. Due North. London, Victor Gollancz Limited. 191 pp.

Speller, S.W. 1972. Food Ecology and Bunting Behaviour of Denning Arctic Foxes at Aberdeen Lake, Northwest Territories. PhD. thesis, Univ. Saskatchewan (Saskatoon). 145 pp.

Underwood, L.S. 1971. The Bioenergetics of the Arctic Fox. Ph.D. thesis, Penn. St. Univ. College Park. 85 pp.
U.S. National Research Council, Agricultural Board, Committee on Animal Nutrition, Subcommittee on Furbearer Nutrition. 1968. Nutrient requirements of mink and foxes. *In:* Nutrient Requirements of Domestic Animals, No. 7, U.S. Nat. Acad. Sci., Washington, D.C. 46 pp.
Usher, P.J. 1971 The Bankslanders: Economy and Ecology of a Frontier Trapping Community. Vol. 2 — Economy and Ecology. Nor. Sci. Res. Group., Dept. Indian Affairs and Northern Development, Ottawa. 169 pp.

Decomposition and microbiology

Microbiology and decomposition on Truelove Lowland

Paul Widden

Introduction

Although there are a number of reports in the literature on the occurrence of microorganisms in tundra soils, there have been few attempts to gain a complete picture of the microbial populations, their variation, and their functioning in the ecosystem. This report represents an attempt to gain a more complete understanding of the nature and functioning of the microbial community within a tundra ecosystem.

In order to understand the role of the microbial community within an ecosystem, it is first of all necessary to describe the populations within that community. This requires data on the types of organisms present, their numbers and biomass, and their temporal and spatial variation. As the major role of microorganisms is the decomposition of organic matter and the solubilization of mineral nutrients, an understanding of the substrates that the organisms can utilize and of the effects of a changing environment on their ability to degrade these substrates is necessary. It is therefore essential to conduct laboratory studies on the growth responses of representatives of the microbial community under controlled conditions on varied substrates. These data can then be integrated with field data on primary production and decomposition rates in order to gain clearer insight into the functioning of the decomposer system.

The present paper deals primarily with the description of microbial populations as they occur in the field, and lays particular stress on the nature of the fungal populations and the response of individual isolates to varied environmental parameters. Some laboratory studies on bacteria are reported on here, but the responses of specific bacterial isolates have been studied in more depth by Nelson (this volume).

During the present study an attempt was also made to obtain field data on the weight loss rates of natural plant remains (*Carex stans* litter and *Dryas integrifolia* litter). In order to gain some "index" of decomposition which could be compared directly with other I.B.P. study sites, weight loss of cellulose in the soil was also studied. Due to technical problems, no data on *Dryas* decomposition are available.

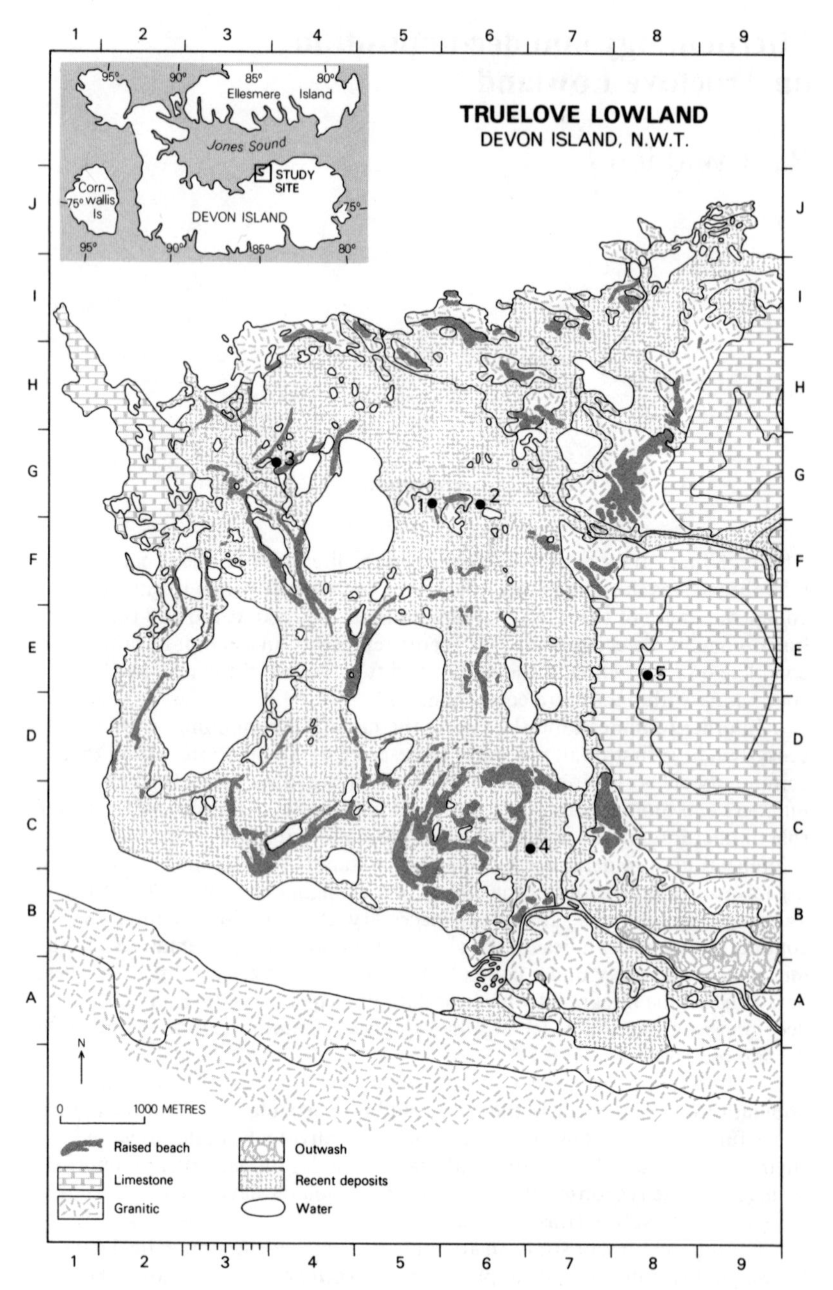

Fig. 1 Map of Truelove Lowland showing the sites where microbiological studies were conducted.

Study areas

Two sites were studied intensively, the Raised Beach and Hummocky Sedge-moss Meadow (Fig. 1, sites 1 and 2 respectively); these sites correspond to the intensive study sites of other workers on the project. For comparative purposes, limited samples were also taken from a wet sedge-moss meadow, a frost-boil sedge-moss meadow and the upland plateau area bordering on the east side of Truelove Lowland (Fig. 1, sites 3, 4, and 5, respectively).

Methods

Sampling

At each of the study sites, a 5m×5m plot was chosen and marked out into 25 $1m^2$ sub-plots from which soil cores could be taken in a random manner. There was no attempt to take random samples over the entire raised beach or the entire hummocky sedge-moss meadow. Thus, all microbial data presented refer to one 25 m^2 area at each site.

Soil sampling

Soil sampling was performed using a 7 cm diameter soil corer. Unless otherwise stated, 5 randomly selected cores were removed at each sampling time and returned to the field laboratory in clean polyethylene bags. Procedures for enumerating microorganisms were initiated the same day, in order to avoid the problem of changes in the microbial populations during storage.

Because of the nature of the cushion plant community on the raised beach, a non-random sampling method was used. Cushions of *Dryas integrifolia* were selected and a transect was established through the centre of each cushion. Soil cores were then taken at a distance of 30 and 10 cm from either side of the centre and a final core was removed from directly under the centre of the cushion. Normally two plants were sampled on each occasion but in one instance four plants were sampled.

Plant leaves

Samples of plant litter were obtained from *Dryas integrifolia* cushions and *Carex stans* plants for microbial studies. The leaves were taken to the laboratory and sorted into the following categories for the study.

Carex stans	Category
Living green leaves	C.1
Standing dead, yellow-brown	C.2
Standing dead, grey	C.3
Dryas integrifolia	
Living green leaves	D.1
Brown leaves	D.2
Brown leaves with a grey cast	D.3
Grey leaves, entire	D.4
Grey leaves, fragmenting	D.5

Moisture and organic matter contents

The moisture content of soil and leaf-litter samples was measured as soon as possible after collection (within 3 days of sampling) by drying to constant weight in an oven at 90°C. Organic matter content of soil samples was measured as ignition loss using a muffle furnace in our laboratory in Calgary. The organic matter determinations were, therefore, delayed for some time and many determinations were not performed due to problems with the furnace. For this reason, numbers of organisms cannot be related directly to soil organic matter content.

Estimation of bacterial populations

Biomass

Bacterial biomass was estimated using the dilution plate method applied as follows. Ten gram sub-samples of soil were taken and transferred to screw-cap bottles containing 90 ml of sterile water and shaken for 10 min on a reciprocating shaker. Dilution series were then prepared from this primary suspension by transfer of a 1 ml aliquot to 9 ml of sterile water in a screw-cap tube. Dilutions were prepared by successive transfer to new 9 ml blanks until the appropriate dilution was achieved. Five 1 ml aliquots of the chosen dilutions were then pipetted separately into Petri dishes and 20 ml of cooled molten peptone yeast extract agar (PYE)* were added. Plates were rotated, cooled, and incubated at room temperature (*ca*. 20°C) for 14 days and then colonies were counted. Colony counts were converted to bacterial biomass using a "standard bacterium" of 2×10^{-13}g dry wt (Satchell 1969).

For plate counts of bacteria from plant leaves, the leaves were suspended in sterile water and then blended in a Waring Blender (previously sterilized by rinsing with alcohol) at high speed for 2 min to prepare the primary suspension.

Taxonomic structure

From each of the two intensive study sites, 250 isolates were picked randomly from the dilution plates prepared in the spring of 1971, (meadow, 28 June, raised beach, 23 June) and transferred to PYE slopes in screw-cap culture tubes. The cultures were flown to our laboratory in Calgary where they were identified to generic level.

Estimation of fungal populations

Biomass

Fungal biomass was estimated using a modified form of the Jones and Mollison (1948) agar film method. Soil samples (1-5 g wet wt) were placed in a mortar and ground with a pestle for 60 sec with 20 ml sterile water. The suspension was allowed to settle for 15 sec and the supernatant poured off and retained. This process was repeated with another 20 ml of water and then with 10 ml of water. The resulting supernatant contained all the soil except for the largest particles which had settled out quickly. To the collected supernatant 50

* PYE agar	"Bacto" peptone	5g
	"Difco" yeast extract	1g
	"Bacto" agar	15g
	distilled water	1,000ml

ml of 3% hot (90°C) water agar was added and mixed thoroughly. Immediately after mixing, samples of the suspension were pipetted onto a haemocytometer slide and a weighted coverglass was placed on top. After the agar had set, the films were removed, placed on slides, dried at room temperature and mounted with glycerine jelly using a number one coverglass. From each soil sample, 5 slides were made. Twenty-five randomly selected fields from each slide were then examined using a Reichart Neopan phase-contrast microscope at a magnification of 400×. The mycelium was drawn using a drawing attachment and measured using a map-measurer. The measurements were converted to metres gram oven-dry^{-1} soil. Biomass was calculated assuming a mean mycelium diameter of 3μ, a specific gravity of 1.123 (Saito 1955) and a water content of 85%.

Leaf material was ground in a Waring Blender for 2 min and 50 ml of the suspension added to 50 ml 3% water agar. The procedure was then the same as for the soil fungi.

Taxonomic structure

To assess the taxonomic structure of fungal populations active in soil and on leaves at the time of sampling, the soil washing method of Parkinson and Williams (1961) was used. After washing, soil or leaf particles were plated on Czapek-Dox agar (Oxoid), acidified to pH 4.5 with lactic acid. Four particles were plated on each plate and 100 particles from each sample were plated. The washing method used was as follows:

Samples of soil (5 g) or leaves (0.5 g) were washed in a washing apparatus similar to that of Bissett and Widden (1972). Preliminary trials indicated that there were few fungal spores in arctic soils and leaf materials, so 5 washings were sufficient. Samples of soil and leaves were then air-dried for 24 hr to reduce growth of bacteria. Leaf materials were then cut into *ca*. 2 mm segments and soil was broken into small particles. Leaves and soil were plated as described earlier. Plates were incubated at room temperature for 14 days and the fungi were then recorded from all particles. Data for percent frequency of occurrence were calculated, using the formula:

$$\% \text{ Frequency} = \frac{\text{No. of particles on which fungus occurred}}{\text{Total no. of particles plated}} \times 100$$

Physiological properties of bacterial population

In order to assess the physiological capabilities of bacteria from the intensive study sites, simple tests were performed on a large number of isolates under varying conditions of temperature, pH, substrate type, and nitrogen source. The results were assessed as either growth (+) or no growth (−). All tests, except the temperature experiments, were conducted at 23°C as this proved to be the highest temperature at which all isolates would grow on PYE.

The 250 isolates from each of the intensive sites randomly selected for identification were used in these tests. During the course of this study, however, some isolates were lost, due to either death or contamination. In no case, however, was any test conducted on less than 100 isolates. The bacteria were tested in the following ways:

Temperature

Bacteria were streaked onto PYE plates and incubated for 7 days at 0°, 5°, 15°, 23°, 30°, or 35°C. After incubation results were recorded as either + or −.

pH

Bacteria were streaked onto PYE plates buffered to pHs of 4.5, 5.0, 6.5, 7.5, or 9.0 and then incubated at 23°C for 7 days and growth recorded as above. This pH range is greater than that found in the field (6.5-8).

Utilization of carbon substrates

Isolates were inoculated into tubes containing phenol red broth + 1% glucose, sucrose, citrate, succinate, lactate or gluconate, and substrate utilization assessed after 7 days at 23°C. Bacteria were streaked on 1% starch or humic acid plates and substrate utilization assessed after 7 days at 23°C.

Utilization of nitrogen sources

Bacteria were grown for 7 days at 23°C on streak plates of solidified Kosers medium + either NO_3 NH_4, Casamino acids or Yeast Extract.

Physiological properties of fungal populations

The effects of temperature and carbon substrate on the growth of fungi was tested on 14 of the common isolates. Four of these isolates were used for more detailed studies on the effects of temperature on growth and respiration.

Effects of temperature and carbon source on growth

Agar Plates. Fourteen of the isolates were grown on a mineral medium* supplemented with 2% sucrose, on agar plates and incubated at 5, 10, 15, and 20°C. The diameter of the colonies was measured every 2 days for three weeks on each of five replicate plates.

Fungi were grown on mineral medium, supplemented with either sucrose, glucose, starch, cellulose, or humic acid at 1%. Five replicate plates were incubated at each of 5, 10, 15, and 20°C and the results recorded as above.

Sand Cultures. Although linear spread of fungi on agar plates is a simple method of assessing growth, it is not completely satisfactory, for it is an artificial system, providing for only 2 dimensional growth on a flat surface, whereas soil is a 3 dimensional matrix. To simulate a more soil-like system the fungi were grown in sand culture, using glucose as a substrate, and their biomass measured by protein measurements using the method of Lowry et al. (1951). The procedure was as follows:

Twenty-five grams of acid-washed quartz sand were placed into "Skrip" ink bottles and the bottles plus sand were sterilized. In the case of fungi which sporulated freely (*Penicillium janthinellum*, *Chrysosporium pannorum*, and *Cylindrocarpon* sp. nov.), cultures were grown in 400 ml medicine bottles until they were sporulating on malt agar. The growing cultures were then flooded with mineral medium plus 1% glucose and shaken with one drop of "Tween 80" to prepare a spore suspension. The ink bottles were then inoculated with 11.5 ml of spore suspension so that the sand, medium, spore mixture formed a stiff paste. The bottles were then incubated at 2.5°, 5°, 10°, or 20°C and sets of 3 bottles were removed at regular intervals, digested with 0.1N NaOH to extract the protein, and the protein measured. *Phoma herbarum* which does not freely sporulate, was prepared for inoculation by growing cultures on dialysis paper

* Mineral medium: g/litre

Ferrous sulphate	0.1	Potassium sulphate	0.35
Sodium nitrate	2.0	Magnesium glycerophosphate	0.5
Potassium chloride	0.5	Agar	12.00

placed on agar. After 10 days the dialysis paper was removed and the colony was blended in a Waring Blender at high speed with the medium. This mycelia suspension was then inoculated into the sand.

To convert protein measurements to biomass, protein was first estimated, for each fungus, as a percentage of dry weight from liquid cultures grown in a glucose-mineral broth at 20°C. This percentage was then used as a conversion factor for the protein measurements.

Decomposition of Carex litter and cellulose

Carex litter

Litter of *Carex stans* was collected in 1970 (August) and 2 g (air dry) samples were weighed and placed in nylon mesh bags with 1 mm mesh. The bags were then placed on the ground in the hummocky sedge-moss meadow and staked down with bamboo stakes. Samples of 3 bags were removed as soon as the surface thawed in spring and just before freeze-up in the fall during subsequent years. In 1971 standing dead (yellow) was removed in August, bagged, placed in the field and sampled in the same manner.

Cellulose

In 1971 Whatman No. 1, 5.5 cm filter papers were oven-dried, weighed, bagged in 1 mm mesh nylon bags, labelled and placed in the field at 0, 5, and 15 cm depth in the sedge-moss meadow and raised beach sites. Samples were placed in the field on 27 June, 27 July, and 27 August on the raised beach and on 5 July, 28 July, and 28 August in the meadow. Samples were removed (5 replicates from each site) 2 July, and 2 September, 1971 and 5 July, 1973. The samples were then air-dried, washed according to the procedure outlined by Fahreus (1947), dried again, weighed, ashed, and re-weighed in order to obtain the actual quantity of cellulose remaining.

Respiration of leaves and litter

Samples of *Dryas integrifolia* leaves and *Carex stans* leaves were removed from the study sites and sorted into the same categories used in the analysis for microbial populations. These samples were taken in July 1973 and flown to our laboratory in Calgary. Sub-samples of *Dryas* weighing 0.1 g (fresh-cut) were placed into 15 ml Warburg flasks and the respiration rates measured at temperatures ranging from 0-25°C using 10% KOH to absorb CO_2. Three replicates were used for each experiment. The same procedure was followed for *Carex stans*, but 0.5 g of materials were used. After the experiments the materials were oven-dried at 95°C for 14 hr and the dry weight recorded.

Results

Microbial biomass

Details of the numbers of bacteria isolated from the two extensive sites are shown in Table 1 and estimates of biomass based on these counts are given in Table 2. From these data it appears that the numbers of bacteria are much higher in the hummocky sedge-moss meadow than the raised beach soils,

Table 1. Numbers $\times 10^6$ g^{-1} dry soil of bacteria isolated from Truelove Lowland soils. Figures in brackets give 95% confidence limits.

Soil depth (cm)	*Distance from Dryas (cm)*	*Sample date*				
Hummocky Sedge-moss Meadow						
		Aug.'70	28 June,'71	14 July,'71	22 July,'71	22 Aug.,'71
0-5	—	63.18	82.08(33-131)	481.02	78.50(0-294)	10.96(0-29)
10-15	—	3.36	—	2.05	2.55(0-9)	3.9 (3-7.5)
Frost-5	—				0.52(0-2)	2.8 —
Raised Beach						
		23 June,'71	21 July,'71	22 Aug.,'71		
0-5	30	8.6 (5.1-12)	12.3 (0-91.7)	12.4 (0-61.8)		
	10	7.3 (4.7-99)	7.7 (6.0-9.3)	11.3 (0-36.6)		
	0	4.7 (4.0-5.3)	12.0 (0-138.4)	36.7 (13.35-60.05)		
10-15	30	—	2.2 (0-12.1)	8.7 (0-42.84)		
	10	—	3.5 (0-32.3)	3.1 (1.31-4.89)		
	0	—	8.8 (0-77.9)	18.3 (0-219)		

Table 2. Biomass ($g\ m^{-2}$) of bacteria isolated from Truelove Lowland soils.

Soil Depth (cm)	*Distance from Dryas (cm)*	*Sample date*				
Hummocky Sedge-moss Meadow						
		Aug. '70	28 June	14 July	22 July	22 Aug.
0-5	—	0.0695	0.0903	0.5291	0.0835	0.0120
10-15	—	0.0037	—	0.0026	0.0028	0.0043
Frost-5	—	—	—	—	0.0006	0.0031
Raised Beach						
		23 June	21 July	22 Aug.		
0-5	30	0.0430	0.0615	0.0620		
	10	0.3650	0.0385	0.0565		
	0	0.0235	0.0600	0.1835		
10-15	30	—	0.0110	0.0435		
	10	—	0.0175	0.0155		
	0	—	0.0400	0.0915		

though errors are large. In both soils there is a decrease in numbers with depth, but this is more marked in the meadow than in the raised beach. The pattern of seasonal variation is very different in the two sites. In the hummocky sedge-moss meadow there is an early spring peak and then the numbers decline, but in the raised beach the numbers increase through the season and presumably decline during winter. In soil away from the plants, numbers of bacteria did not change much during the season. Numbers of bacteria in the lower soil horizons tend to follow numbers in the upper 5 cm in the raised beach soils but the numbers in the lower horizons seem to remain static in the meadow.

Although bacterial numbers as a function of dry weight of soil are fairly high in the meadow, biomass appears to be very low (maximum 0.5291 $g\ m^{-2}$). In the raised beach soils biomass is only 10% of that in the meadow.

The available data on organic matter in the raised beach soils indicate that it varies between 1% and 2% in the deeper layers (10-15 cm) and is approximately 3% in the surface soil. In the meadow, however, organic matter content is between 40% and 70% though occasionally sand lenses occur where organic matter content is very low (3%). The organic matter content in the meadow shows no definite trend with depth. In neither site does the change in

Table 3. Distribution of fungal mycellium in Truelove Lowland soils (metres g^{-1} dry wt^{-1} + 95% confidence limits).

Sampling depth (cm)	*Distance from Dryas (cm)*	*Sampling dates*		
Hummocky Sedge-moss Meadow				
		22 Aug.,1970	28 June,1971	22 July,1971
0-7.5	—	361 (193-584)	902 (517-1308)	1472
7.5-15	—	172 (77-267)		143
Raised Beach (Backslope)				
		22 Aug., 1970		
0-7.5		431 (220-641)		
7.5-15		239 (181-297)		
15-22.5		82 (54-110)		
Raised Beach (Crest)				
		23 June, 1971	6 July, 1971	
0-5	30	39 (28-50)	83 (0-3cm)	
	10	47 (47-61)	78 (0-3cm)	
	0	112 (81-144)	—	
3-6	30		64	
	10		34	
12-15	30		3	
	10		39	
25-28	30		—	
	10		26	
35-38	30		—	
	10		46	

Table 4. Fungal biomass ($g\ m^{-2}$) in soils of Truelove Lowland to a depth of 5 cm.

Site	*Sampling dates*		
	Aug. '70	June '71	July '71
Hummocky Sedge-moss Meadow	4.51	11.39	20.25
Raised Beach	8.72(BS)*	1.32	3.10

*Back slope

bacterial numbers correlate very well with organic matter content but a between-site comparison suggests that on a per gram organic matter basis, bacterial numbers are comparable at both sites.

Mycelium g^{-1} oven dry soil decreases at both sites with depth (Table 3), but this decrease is not as dramatic as that seen in the bacterial numbers. It is clear that there is more mycelium in the meadow soil than the crest of the raised beach. The data from 1970, however, indicate that the amount of mycelium in the soils of the back slope on the raised beach is similar to that found in the meadow. Not enough data are available to see seasonal trends, but it does appear that in the meadow mycelium content increases during spring but drops by August. In the raised beach soils in June there is more mycelium under the plants than at a distance from them. This appears to correlate with organic matter content, which reached 7% under the *Dryas* plants. In July the amount of mycelium increases from June, but no more data are available. Fungal biomass follows the same trend as mycelium content reaching 20 $g\ m^{-2}$ in the hummocky sedge-moss meadow and 3-90 $g\ m^{-2}$ in the raised beach soil (Table 4).

Table 5. Genera of bacteria isolated from Truelove Lowland soils (% isolates) based on a sample of 250 isolates from each site (June 1971).

Taxon	*Hummocky Sedge-moss Meadow*	*Raised Beach*
Alcaligenes/Achromobacter	4	6
Bacillus	2	10
Corynebacterium	29	14
Flavobacterium/Cytophaga	11	19
Pseudomonas	10	7
Streptomyces	<1	18
Coliforms	<1	<1
Unidentified Gram − Rods	6	<1
Unidentified Gram + Cocci	7	9

Distribution of microbial taxa

Bacteria

A range of bacteria was found in the soils of the intensive sites (Table 5). Of these the most important groups found at both sites were the Coryneforms, *Pseudomonas* spp. and the *Flavo-bactorium-Cytophaga* group. *Bacillus* spp. and *Streptomyces* spp. were more frequent in the raised beach soils than the hummocky sedge-moss meadow, but, *Corynebacterium* spp. were more common in the meadow soil. It should be borne in mind that this sample was taken during June, 1971 and, therefore, it is quite possible that the proportions of different genera might change with time.

Fungi

Data on the fungi from the two intensive sites are summarized in Figs. 2 and 3, which show graphically the distribution of microfungi on leaves and litter of *Dryas integrifolia* and *Carex stans* respectively, and the distribution with depth in the two soils. Taxa represented in these figures are those that occurred at any single sampling time with a mean frequency of 5% or more.

The fungal taxa occurring at the two intensive sites were fairly similar. At both sites, sterile mycelia were frequent in the soil and on plant leaves and litter. On the leaves of *Carex stans* an unidentified fungus, E30, was found exclusively on live leaves, *Mortierella* sp. and *Mucor* sp. on standing dead (yellow) and *Trimmatostroma* and a *Penicillium* sp. (E.23) on standing dead (grey); *Acremonium* sp. was found on all classes of *Carex* leaf material. Apart from sterile mycelia, *Phoma herbarum* and *Cladosporium cladosporoides* were the commonest fungi on the *Carex* leaf materials and they were also found, though at reduced frequencies in the soil. Yeasts, *Tolypocladium* spp. *Penicillium janthinellum* and *P. notatum* were found at low frequencies in both *Carex* leaf materials and in the meadow soils. Of the fungi found exclusively in the meadow soils, *Chrysosporium pannorum* and *Cylindrocarpon* sp. nov. were found in the 0-5 cm soil level, whereas *Penicillium* sp., α146 and *Aspergillus ustus* were found only at lower soil depths.

In the raised beach soils and on *Dryas* leaves, there are some similarities in mycoflora. Thus, *Cladosporium cladosporoides* and *Phoma herbarum* were commonest on leaf materials, but were also found in the soil. Unlike the hummocky sedge-moss meadow, no fungi were found exclusively on the *Dryas* leaves and litter except some of the strains included under mycelia sterilia. *Phialophora*, *Pestalotia* and Q.L.2 (an unidentified member of the sphaeropsidales) were more common on leaf materials than in the soil; Q.L.2 was not found on *Carex stans*. Yeasts, *Chrysosporium*, *Penicillium* α146 and

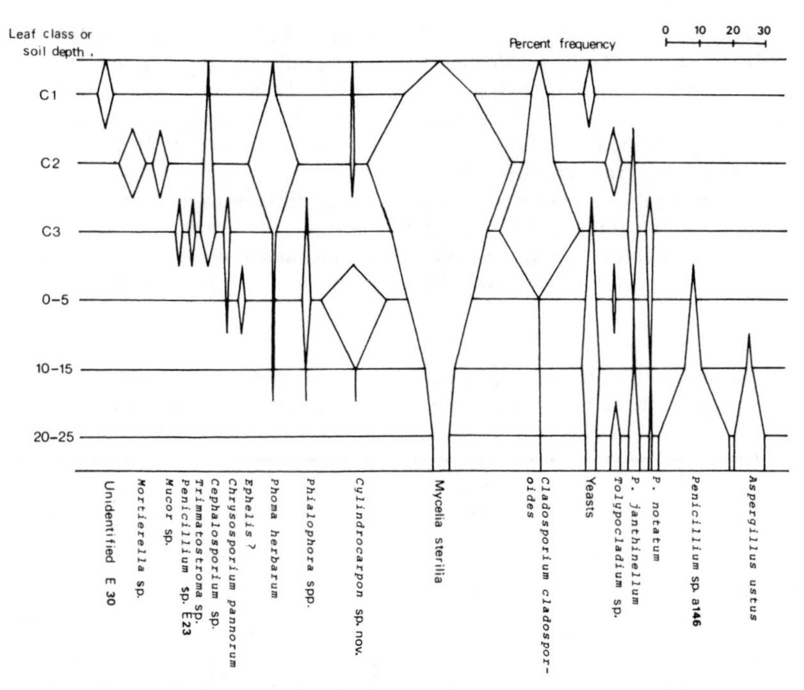

Fig. 2 Distribution of soil, leaf, and litter fungi on the raised beach (crest). Width of kites is proportional to % frequency of fungal taxa.

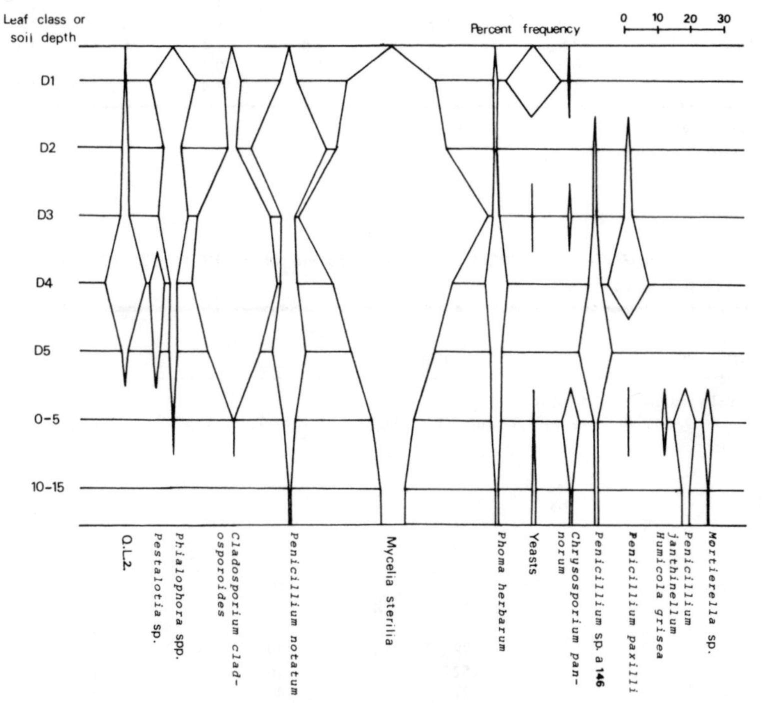

Fig. 3 Distribution of soil, leaf, and litter fungi in the hummocky sedge-moss meadow. Width of kites is proportional to % frequency of fungal taxa.

P. paxilli were found infrequently in both soils and on litter and *Humicola grisea*, *P. janthinellum* and *Mortierella* sp. were only found in the soil. *Penicillium notatum* was common on leaf materials and soil on the raised beach, whereas it was rare in the meadow. *Cylindrocarpon* sp. nov. was very rarely found in the raised beach soils though it was common in the meadow.

When data from the soils of the intensive sites are compared with other sites (Table 6) it can be seen that there are great differences, both in the taxa isolated and the amount of fungal activity as indicated by total percent of colonization (Table 7). The frost-boil sedge-moss meadow soils taken from the centre of the frost-boils show very little fungal activity (1.9% colonization),

Table 6. Comparison of fungal populations from various Truelove Lowland sites. Data are the mean of % occurrence, for 5 replicates on each sampling data.

Taxon	*Hummocky*** sedge-moss meadow*	*Frost-boil* sedge-moss meadow (frost-boil centre)*	*Hydric* sedge-moss meadow*	*Raised*** beach crest*	*Plateau* Middle of** frost-boil	*Plateau* Edge of** frost-boil
Chrysosporium pannorum	0.6	—	0.3	4.7	0.3	0.3
Cylindrocarpon sp. nov	19.9	—	1.3	1.8	1.7	27.0
Ephelis sp. (?)	1.8	—	—	—	—	—
Humicola grisea	—	—	—	1.8	0.7	—
Mortierella 71	—	—	—	2.8	1.7	7.0
Penicillium janthinellum	0.8	—	—	6.3	—	—
Penicillium notatum	1.4	—	—	2.3	3.7	0.3
Phialophora spp.	0.5	—	4.0	0.5	0.7	—
Phoma herbarum	0.3	0.3	0.3	3.8	4.0	24.7
Sterile Dark	9.9	0.3	8.0	5.9	0.3	4.7
Sterile Hyaline	8.8	1.3	2.0	7.4	5.0	7.0
Yeasts	0.2	—	1.0	0.1	9.0	1.7
Total Mean % Colonization	44.2	1.9	16.9	37.4	27.1	72.7

*Based on only one sampling date.
**Based on two sampling dates.
***Based on five sampling dates.

Table 7. Colonization (percent) of soil and leaf-litter particles by fungi from the intensive study areas.

Leaf class or soil depth (cm)	*Sample date* Aug. '70	June '71	July '71	Aug. '71	June '72	Aug. '72	Mean
Hummocky Sedge-moss Meadow							
C.1	—	—	—	31	—	59	45
C.2	—	—	—	88	—	105	97
C.3	—	—	—	92	—	96	94
0-5	60	68	40	46	—	—	54
10-5	8	—	31	32	—	—	24
20-25	—	—	19	75	—	—	47
Raised Beach							
D1.	—	—	55	44	134	—	77
D2.	—	—	14	84	128	—	75
D3.	—	—	78	46	179	—	101
D4	—	—	152	77	126	—	118
D5	—	—	27	77	144	—	83
0-5	33	51	48	34	—	—	41
10-15	12	—	26	24	—	—	20

whereas the soils at the edge of the frost-boils of the plateau show the highest activity (72.7% colonization). Whereas the high mobility and lack of organic matter may explain the lack of activity in the centre of the frost-boils of the frost-boil sedge-meadow, there is no obvious reason why activity on the edges of the plateau frost-boils should be high. In all soils sterile mycelia were very common; *Cylindrocarpon* sp. nov. was common in the hummocky sedge meadow and on the edges of the plateau frost-boils. *Phoma herbarum* was also common (24.7% colonization) in the soils from the edges of frost-boils on the plateau. Of the common fungi, only six were found in five or more of the sites, three of the six being sterile dark, sterile hyaline, and yeasts and thus representing more than one taxon. The others were *Cylindrocarpon* sp. nov., *Chrysosporium pannorum*, and *Phoma herbarum*.

It is clear that the most frequently occurring fungi at all the sites studied, with the possible exception of the plateau, on both leaf materials and in the soil are the mycelia sterilia. At the intensive study sites, where leaves and litter were studied, it can be seen that the percent of particles colonized increased from live leaves to decaying leaves but then sharply decreased in the soil. This suggests that there is a more active mycoflora in decaying leaves than in the soil proper. This is most noticeable on the raised beach in *Dryas integrifolia* cushions. The influence of the *Dryas* cushions on the soil flora can be seen in Table 8. In early spring there is a higher percent colonization in soil close to the cushion but, during the summer, colonization increases in the soil away from the cushions. In August the influence of the cushion is again clear. This may be

Table 8. Percent colonization of washed soil particles by fungi around *Dryas integrifolia* cushions.

Depth of soil (cm)	*Distance from cushion*		*Sampling date*	
		June 1971	*July 1971*	*Aug. 1971*
0-5	0	79	52	61
	10	65	53	33
	30	27	59	40
10-15	0	—	34	30
	10	—	16	27
	30	—	27	26

Table 9. The effects of pH and temperature on the growth of Truelove Lowland soil bacteria. Figures represent the percent of strains that showed growth after 7 days of incubation.

Parameter	*Hummocky sedge-moss meadow*	*Raised beach*
Temperature (°C)		
0	79	55
5	100	100
15	100	100
23	100	100
30	81	80
35	14	30
pH		
4.5	34	25
5.0	64	56
6.5	100	100
7.5	100	100
9.0	89	98

Table 10. Effects of carbohydrates on growth of Truelove Lowland bacteria. Percent of strains 7 days incubation.

Taxon	Substrates							Humic
	Glucose	Sucrose	Citrate	Succinate	Lactate	Gluconate	Starch	acid
Raised Beach								
Actinomycetes	36	0	0	33	11	44	86	0
Alcaligenes	40	30	10	29	29	29	20	0
Bacillus	53	53	0	23	23	15	69	0
Corynebacterium	54	50	0	50	25	33	4	0
Gram +ve Cocci	56	89	4	0	0	0	33	0
Flavobacterium	32	23	17	67	100	67	29	0
Pseudomonas	100	42	75	70	60	60	0	0
Mean	53	41	8	37	35	35	32	0
Hummocky Sedge-moss Meadow								
Alcaligenes	67	67	17	0	20	20	17	0
Bacillus	33	50	0	17	50	33	17	0
Corynebacterium	50	37	4	14	14	17	7	0
Cytophaga	100	20	0	25	0	25	25	0
Enterobacteriales	100	100	25	75	100	75	0	0
Flavobacterium	43	17	0	6	6	6	11	0
Gram +ve Cocci	53	47	6	13	40	0	35	0
Pseudomonas	100	28	98	93	96	89	19	0
Mean	68	45	19	30	41	33	16	0

the result of fungi growing out from the nutrient base of the cushion onto the surrounding soil during summer, but in the fall, mycelium away from the cushion may die more quickly.

Physiological properties of bacteria

Table 9 shows that 55% of the raised beach isolates tested formed visible colonies at 0°C and 30% at 35°C; of the meadow isolates 79% grew at 0°C and 14% at 35°C. Of these isolates, 8% from the raised beach and 20% from the hummocky sedge-moss meadow grew at both temperature extremes tested. All of the isolates tested will grow at pH 6.5 but more grow at pHs above 6.5 than below. From the raised beach and meadow isolates respectively, 11% and 32% of those tested grew at all pHs tried (Table 9).

The bacteria tested are able to use a wide range of simple carbon substrates (Table 10); many could use starch but none could use humic acid. As a group, the Pseudomonads used a wide range of substrates but did not use starch. As for nitrogen sources (Table 11), Pseudomonads could use both mineral and organic nitrogen, whereas most of the bacteria only used organic nitrogen. The *Bacillus* species isolated from the raised beach soils were generally capable of using mineral nitrogen sources, whereas those from the meadow could not. *Cytophaga* species, found only in the meadow, used only organic nitrogen.

Physiological properties of fungi

Effects of temperature and carbon source on growth

Agar Plates. With the single exception of *Cladosporium cladosporoides*, all fungi tested grew faster at 20°C than at 5°C. All fungi showed significant growth at 5°C, though some species (notably *Botrytis cinerea*) showed a much

Table 11. Use of nitrogen source by Truelove Lowland bacteria. Percent of strains growing after 7 days incubation.

Taxon	*Nitrogen source* NO_3	NH_4	*Casamino acids*	*Yeast extract*
Raised Beach				
Alcaligenes	25	50	100	100
Bacillus	60	40	80	100
Corynebacterium	33	42	100	100
Flavobacterium	17	0	67	100
Gram +ve Cocci	0	20	80	100
Pseudomonas	100	78	100	100
Streptomyces	50	25	100	100
Mean	41	36	90	100
Hummocky Sedge-moss Meadow				
Alcaligenes	50	50	100	100
Bacillus	0	0	75	100
Corynebacterium	25	35	90	100
Cytophaga	0	0	67	100
Entecobacteriales	33	67	100	100
Flavobacterium	20	0	80	100
Gram +ve Cocci	25	37	87	100
Pseudomonas	82	88	100	100
Mean	29	35	87	100

Table 12. Effect of temperature on growth of fungi on Czapek-Dox agar from Truelove Lowland soils (colony diameter in mm after 14 days incubation).

Taxon	*Incubation temperature* 5° C	10° C	15° C	20° C
Botrytis cinerea	4.9	18.7	75.0	85.0
Chrysosporium pannorum 1.	11.5	19.2	24.4	26.4
Chrysosporium pannorum 2.	11.7	15.8	20.5	23.5
Cladosporium cladosporoides	6.2	7.8	10.8	8.2
Cylindrocarpon sp. nov.	22.2	39.7	51.8	60.5
Humicola grisea	19.5	31.3	50.0	61.1
Penicillium janthinellum	8.0	21.0	29.4	35.4
P. notatum	6.0	12.6	16.8	40.9
Penicillium sp. α146	8.7	21.5	25.0	31.8
Pestalotia sp.	17.5	40.2	50.5	62.3
Phoma herbarum 1.	16.6	38.0	54.4	69.7
Phoma herbarum 2.	24.9	42.0	52.0	72.8
Sterile dark 116	6.6	10.4	13.8	13.0
Tolypodadium cylindrosporum	4.6	14.4	21.1	32.8

greater difference between growth at 5°C and 20°C than others (*Chrysosporum pannorum* strains 1 and 2 *Cylindrocarpon* sp. nov., *Phoma herbarum* 2 and sterile mycelium α116) (Table 12). All differences between temperatures for each fungus were significant at P=0.05.

Almost all of the fungi tested grew fastest on most substrates at 20°C but *Cladosporum cladosporoides* and sterile dark α116 grew fastest at 15°C on sucrose. On humic acid, however, both *Phoma herbarum* and *Cylindrocarpon* sp. nov. grew faster at 15°C than 20°C. *Cylindrocarpon* showed growth at 20°C that was almost as slow as at 5°C (Table 13).

Sand Cultures. The $\log_{10}$ of the estimated total biomass of fungus in the sand cultures was regressed against time. The resulting linear regressions were all highly significant (P=0.001). Fig. 4 summarizes the results of these experiments by plotting the growth rate [slope factor A of the regressions

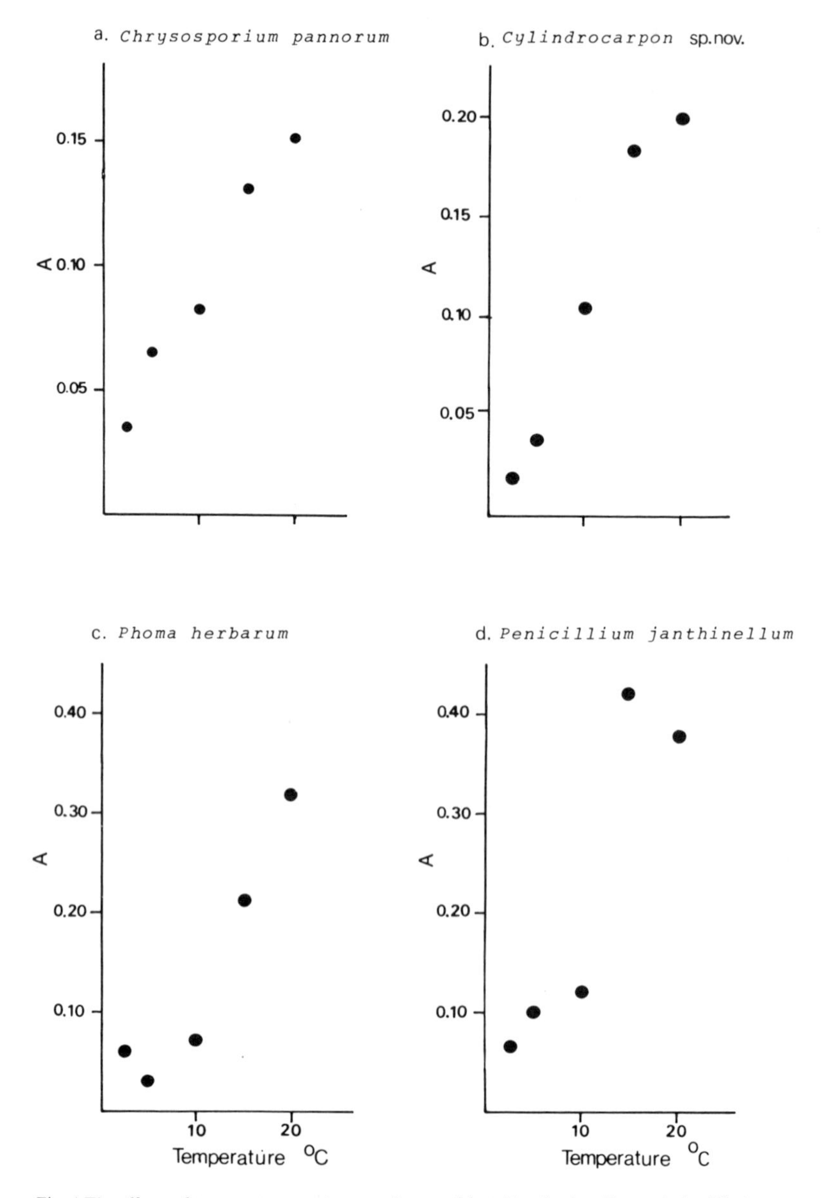

Fig. 4 The effects of temperature on the growth rate of four Tundra fungi in sand plus 1% glucose — mineral medium. A is the slope factor of the calculated regression Log y = Ax + b, where y = fungal biomass, x = temperature and b is the intercept.

calculated from the data (Log Biomass = A) [Time] + B] against temperature. The data indicate that for *Chrysosporium pannorum* the growth rate A decreases in a linear fashion with decreasing temperature *Cylindrocarpon* sp. nov., and *Phoma herbarum* have higher growth rates at 2.5°C than at 5°C and Pencillium janthinellum has a maximum growth rate at 15°C.

Table 13. Effects of temperature and substrate on growth of fungi from Truelove Lowland soils (colony diameter in mm after 14 days incubation).

Taxon	*Temp. (°C)*	*Substrate* *Sucrose*	*Glucose*	*Starch*	*Cellulose*	*Humic acid*
Chrysosporium pannorum	5	9.1	9.1	9.4	9.4	7.0
	10	14.8	16.8	15.6	13.3	18.1
	15	19.8	21.0	19.6	18.9	21.1
	20	25.3	23.3	21.6	22.1	22.3
Cylindrocarpon sp.	5	15.7	22.7	24.2	23.3	12.2
	10	38.1	37.2	42.7	44.1	20.5
	15	51.9	54.8	54.7	57.1	33.8
	20	59.3	66.3	66.0	69.8	12.5
Penicillium janthinellum	5	4.0	4.3	8.5	7.7	—
	10	23.8	13.4	18.2	18.3	19.5
	15	21.0	23.3	25.4	26.4	24.2
	20	27.0	25.5	25.3	26.5	28.1
Phoma herbarum	5	16.3	17.3	18.9	12.6	—
	10	38.7	44.1	41.8	37.0	13.5
	15	51.3	60.5	57.1	49.4	37.5
	20	69.7	69.3	66.6	72.9	24.0

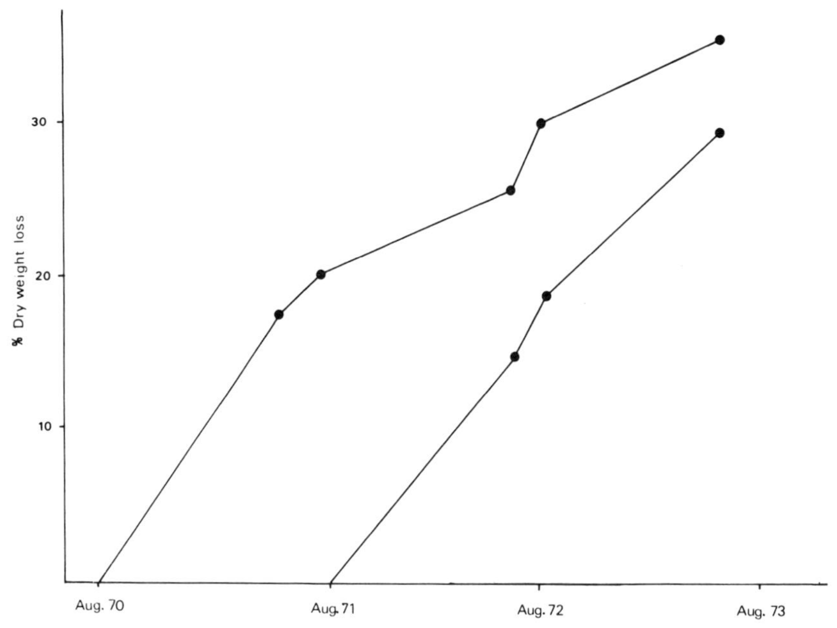

Fig. 5 Dry weight loss of *Carex* litter and standing dead, placed in litter bags on the surface of the hummocky sedge-moss meadow. Data starting August 1970 are for litter and starting August 1971 are for standing dead.

Decomposition rates

Carex litter

From Fig. 5 it can be seen that both the litter and standing dead lost about 20% of their initial weight in the first year and a further 10% during the second year; the decomposition of the standing dead was slightly faster than the litter.

Table 14. Decomposition of Whatman filter papers at two sites, Truelove Lowland, 1971-1973 (% weight loss O.D.W.).

Date placed in field	*Depth (cm)*	*Date removed 1972 2 July*	*2 Sept.*	*1973 5 July*
Raised Beach				
27 June, 1971	0	0	0	0
	5	—	37	71
	15	—	15	44
21 July, 1971	0	0	0	0
	5	—	19	53
	15	0	0	59
27 Aug., 1971	0	0	0	0
	5	0	0	22
	15	—	18	20
Hummocky Sedge-moss Meadow				
5 July, 1971	0	0	0	0
	5	0	0.1	17
	15	0	0.6	32
28 July, 1971	0	0	0	0
	5	0	0	0
	15	0	0	0.3
28 Aug., 1971	0	0	0	0
	5	0	0	0
	15	0	0	13

Cellulose

The data shown in Table 14 indicate that cellulose decomposition is negligible on the surface at both sites. Belowground decomposition is faster in the raised beach soils than the meadow because of higher soil temperatures. In the raised beach decomposition is faster at 5 cm than 15 cm, whereas the reverse is true in the meadow.

Respiration of leaves and litter

In both *Dryas* and *Carex* the live leaves have the highest respiration rates (Figs. 6 and 7). There is a marked tendency in the *Dryas* data for the response with temperature to be bimodal, one peak at 10°-15° and another at, or above, 25°C. the differences in respiration rates between the classes of leaf material are difficult to interpret as field moisture contents varied and this is obviously influencing the response.

Discussion

The data given in Table 1 show that bacteria are more numerous in the meadow soil than in the raised beach soil. In the raised beach soils numbers of bacteria increase during the growing season, whereas in the meadow soils they decline after a brief period of increase in the spring. It is possible that, in the meadow, the prevailing waterlogged conditions allow only a brief period of activity, while the wet soil is aerated after the spring thaw, after which the decline takes place. On the raised beach, however, aeration is always good and

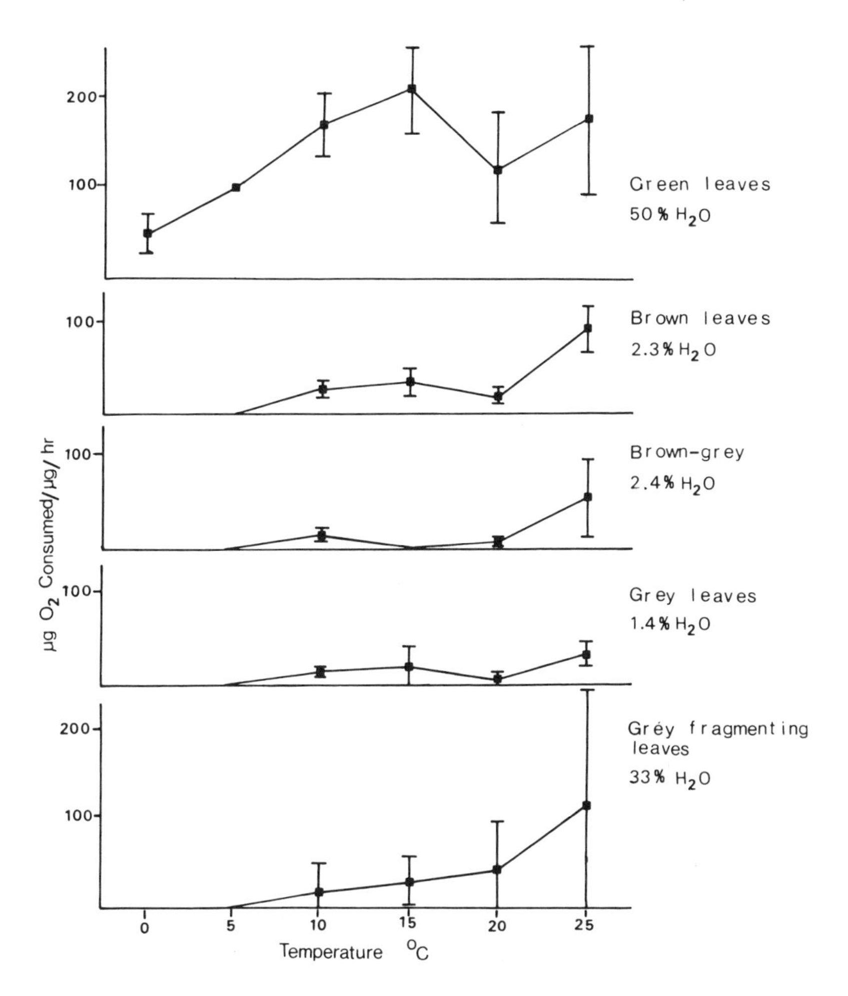

Fig. 6 Respiration of *Carex* leaf materials as a function of temperature. Moisture is given as a percentage of wet weight. Error bars give 95% confidence limits.

soil temperatures are higher than in the meadows, thus bacteria may be active throughout the brief summer. If this hypothesis is correct, the high numbers of bacteria in the meadow soils may reflect a population with a very low turnover rate, whereas the raised beach populations may have a high turnover. This hypothesis is partly supported by the higher decomposition rate of cellulose in the raised beach soils as compared to the meadow (see Table 14).

The data for fungus mycelium show a similar trend to bacterial numbers, presumably for similar reasons, though there are less fungal data (Table 3).

Fungal biomass (Table 2) exceeds bacterial biomass (Table 4) in all the soils examined. Thus to a depth of 5 cm the maximum bacterial biomass figures were 0.53 g m^{-2} and 0.18 g m^{-2} for the meadow and raised beach crest soils respectively, whereas for fungi the figures are 20.25 g m^{-2} and 3.10 g m^{-2} respectively. These figures must, however, be interpreted with care for viable counts (bacteria) are being compared to total counts (fungi). Where viable counts and direct counts for bacteria have been compared, it has been shown

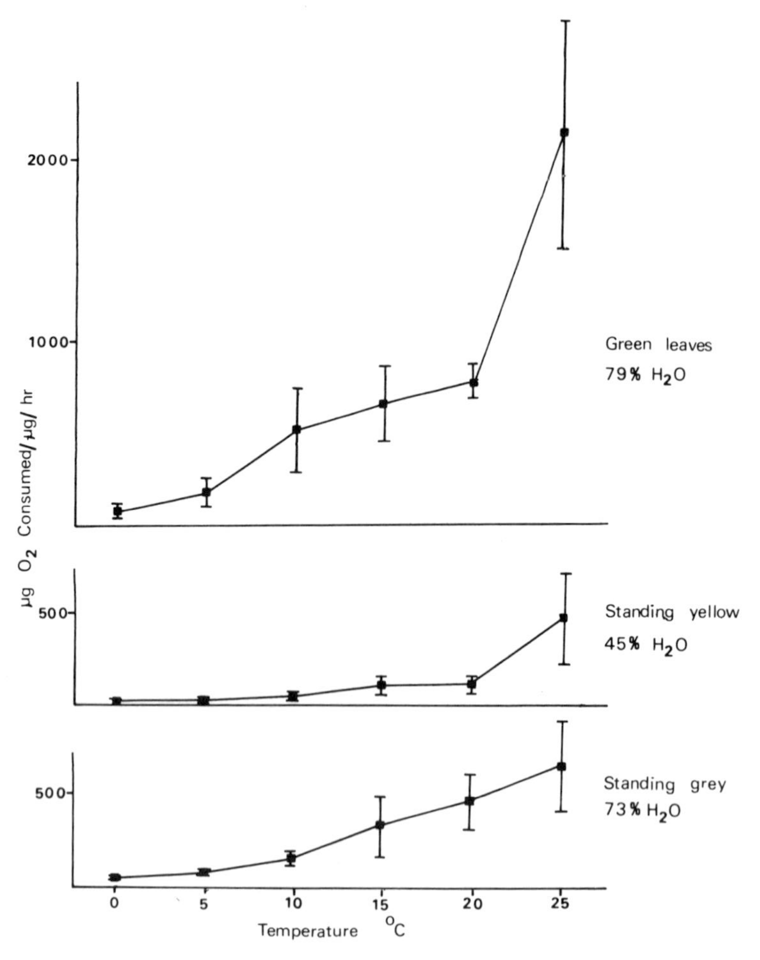

Fig. 7 Respiration of *Dryas* leaf materials as a function of temperature. Moisture is given as a percentage of wet weight. Error bars give 95% confidence limits.

that direct counts of bacteria may range from 20× to 10,000× the viable counts (Aristovskaya and Parinkina 1972).

It appears that in the meadow there is a large microbial population able to grow only for a short period in the spring, possibly due to O_2 limitations and low temperatures. This suggests a slow turnover rate, low decomposition rates, and an accumulation of organic matter due to the higher plant productivity (Muc this volume). Raised beach soils are comparatively warm and well-aerated, thus, micro-organisms may continue to grow throughout the summer as long as moisture is not limiting. Because of this, substrates could be more rapidly decomposed and therefore organic matter would not accumulate. This implies that the raised beach has a low microbial biomass with a rapid turnover rate, whereas the meadows have a high microbial biomass with a low turnover rate. Turnover rates may be especially high under the cushion plants, where, on the raised beach, conditions of moisture and temperature are more favorable than at a distance from them.

It is possible to estimate a minimum microbial production by taking the difference between low and high counts. If this is done, a figure of approximately 16.8 g m^{-2} is obtained for the meadow and 1.8 g m^{-2} for the raised beach. Annual production of aboveground plant materials are, for the meadow, sedges 44 g m^{-2} (Muc this volume) plus 33 g m^{-2} for mosses (Vitt and Pakarinen this volume) giving a total of 77 g m^{-2}, whereas on the raised beaches vascular plant production is 15 g m^{-2} (Svoboda this volume).

In spite of this, the minimum microbial production given above for the meadows is close to 10× the minimum microbial production calculated for the raised beaches. If one assumes that most of the primary production is respired by the microbial populations (thus having very little or no accumulation) it appears that the micro-organisms on the raised beach may be more active than those in the meadows and that their turnover is greater. The large populations of collembolans, which are mycophagous, existing on the raised beaches (see J. Addison this volume and Ryan 1972) may consume much fungal production. Clearly more knowledge of grazing microfauna and better microbial production data are needed to fully understand the dynamics of the decomposer system.

The numbers of bacteria reported from Devon Island are high (12-481$\times 10^6$ g^{-1} oven dry soil) in comparison to those reported from other tundra sites. In Antarctica, Cameron et al. (1970) reported a maximum of 0.3$\times 10^6$ bacteria g^{-1} in dry valley soils, Heal et al. (1967) gave a maximum of 16$\times 10^6$ g^{-1} from Signy Island, and Boyd and Boyd (1971) report 26$\times 10^6$ g^{-1} from a loam soil in Inuvik. Aristovskaya and Parinkina (1972) have, however, reported numbers as high as 108$\times 10^6$ g^{-1} from Taimyr soils. In their analysis of the bacterial counts from I.B.P. tundra sites, Holding et al. (1974) state that, within the high arctic group of sites, Devon Island meadow is a consistent "outlier" because of its high counts. One possible explanation of this is the high moisture and organic matter content and favourable pH (7.6) of the meadow site.

The amounts of fungal mycelium in Truelove Lowland soils compared to other tundra sites are fairly low. Thus the mean for the sedge-moss meadow is 1005 m g^{-1} and for the raised beach is 199 m g^{-1}. These can be compared with other data from polar regions such as Signy Island where 6,328 m g^{-1} soil were found in a snowbank community, and only 4 m g^{-1} in a moraine soil (Bailey unpublished data). Figures for many I.B.P. Tundra Biome sites are between 1,000 and 5,000 m g^{-1} (Dowding and Widden 1974).

Comparison of the taxonomic groups of bacteria isolated from Truelove Lowland soils with reports from other tundra sites indicates that they are similar to those from Antarctica (Boyd et al. 1966) and Alaska (Fournelle 1967), the main difference being the presence of *Corynebacterium* spp. on Truelove Lowland. Bacteria from a heather moor in England (Latter et al. 1967) were 80% *Bacillus* sp., whereas on Truelove Lowland they were only 3% and 12% for the sedge-moss meadow and the raised beach respectively. In a comparison of I.B.P. Tundra sites, Dunican and Rosswall (1974) have shown that Truelove Lowland has a similar range of bacteria to Signy Island, Moor House, and Barrow, but the proportions of genera differ. One major difference is the high frequency of Actinomycetes (*Streptomyces* sp.) in the raised beach soils. Since Actinomycetes are generally considered rare in the Arctic but occur on Truelove Lowland, it was concluded that soil pH was the main factor, rather than temperature. The report of Kriss (1947) that actinomycetes constitute 50% of the microflora of arctic soils which are not of the marshy gley or peat types, agrees with the Truelove Lowland data. This suggests that actinomycetes may be common in the Arctic in neutral to alkaline soils which are well drained.

Fungal species on Truelove Lowland showed, at both sites, a preponderance of sterile fungi. This agrees well with data from Macquarie Island (Bunt 1965), Alaska (Flanagan 1971), the Mackenzie Valley (Ivarsen 1965), and British Moorlands (Latter et al. 1967). Apart from sterile forms, *Penicillium* spp., *Chrysosporium* spp., *Cladosporium* spp. and *Mortierella* spp. appear to be the most common tundra fungi (Dowding and Widden 1974). On Truelove Lowland this appeared to be true, though *Mortierella* spp. were not very abundant and *Cylindrocarpon* sp. nov. was very common in the hummocky sedge-moss meadow. Within the I.B.P., *Cylindrocarpon* spp. were only found in two other tundra sites (Barrow and Prudhoe Bay, Alaska) which, significantly, are both sedge-moss meadow communities. Dowding and Widden (1974) attempted an association analysis of tundra sites based on fungal populations and the Truelove Lowland intensive sites grouped with sites at Fairbanks, Barrow, Moor House, and Mt. Allen (Canadian Rockies). The frost-boil sedge-moss meadow grouped with Macquarie Island soils and the wet sedge-moss meadow grouped with four sites from Alaska. This analysis should not, however, be overstressed as data were generally inadequate.

The growth responses to temperature of fungi and bacteria from Truelove Lowland indicate that the bacteria are generally psychrotrophic as defined by Morita (1975) since the majority of them are capable of growth between 0° and 30°C and all isolates grew at 5°C. Only 14% of meadow isolates and 30% of raised beach isolates would grow at 35°C. Of these isolates 8% of the raised beach isolates and 20% of the meadow isolates grew at both 0° and 35°C suggesting that tundra bacteria are capable of growth over a wide range of temperature.

For the fungi tested (Table 12) a similar pattern showed, all of them growing at temperatures between 5° and 20°C. Only two isolates grew fastest at a temperature below 20°C. These were sterile dark α116 and *Cladosporium cladosporoides*. These two isolates were two of the most commonly found fungi. Of the 14 isolates tested, there were three fungi with a growth rate on agar at 20°C, 2.5× or less the growth at 5°C (*Chrysporium panorum*, *Cladosporium cladosporoides* and sterile dark α116), all of these groups of fungi have been listed as common in tundra soils (Dowding and Widden 1974). It is therefore possible that these organisms are adapted to the cold climate. An opposite case is seen in *Botrytis cinerea* which had a linear growth at 20°C 17× that at 5°C. As this organism was only isolated during July from *Dryas* leaves it may be acting in an opportunistic manner growing rapidly when conditions are suitable during the warmest part of the year.

The pH response of the Truelove Lowland bacteria indicates that they prefer pHs of 6.5 and above, an expected result. Some of these bacteria grew over very wide pH ranges (11% of meadow and 32% of raised beach isolates grow at all tested pHs) but, as pH in the field covered a narrower range (6.5-8), this is not of great ecological significance.

Response of the bacteria to carbohydrates indicated a wide range of substrates could be utilized, but no bacteria used humic acid. This contrasts sharply with the fungi, most of which grew well on humic acid. As a group, the Pseudomonads appeared to grow on most carbon substrates with the exception of starch. The Actinomycetes, however, grew well on starch. Unfortunately the Truelove Lowland bacteria were not tested for cellulose activity.

It is interesting to note that, when the effects of temperature on growth of fungi are examined, *Phoma herbarum* and *Cylindrocarpon* sp. both have an optimum growth at 15°C on humic acid though on other substrates they grow faster at 20°C. This change in response to temperature with a change in substrate has been noted by Flanagan and Scarborough (1974) working with other tundra fungi.

When the carbohydrate metabolism of the Truelove Lowland bacteria was compared with the results obtained by Clarholm and Rosswall (1973) for bacteria from Norwegian, Swedish, Irish, and British I.B.P. tundra sites, no great differences could be found (Rosswall and Clarholm 1974). This is quite remarkable when one considers how different the sites are, especially the moorland soils from Ireland and the U.K. Rosswall and Clarholm (1974) have also compared the aerobic soil bacteria from four "tundra" sites with populations from a pine forest (Goodfellow 1966), a deciduous forest (Hissett and Gray 1973), a forest humus, a grassland, and a cultivated field (Sundman 1970), and found that there are no great physiological differences. It would, therefore, appear that tundra bacterial populations do not differ significantly from those found in other soils, as far as their physiological capabilities are concerned, though cold tolerance was not examined, and that in this respect the Truelove Lowland isolates are probably similar. A word of caution should, however, be added to this conclusion, for many of the initial isolates from Truelove Lowland failed to grow in the laboratory. This apparent delicacy of tundra bacteria appears to be fairly common as it has been noted by other tundra microbiologists (Rosswall, pers. comm.).

The data on the growth rate in sand cultures of four tundra fungi (Figs. 4 a, b, c, d) are of interest in that there are no reports, that I am aware of, showing exponential growth of fungi in soil-like systems. When the growth rates of the various fungi are plotted against temperature, a more or less linear change with temperature is seen with *Chrysosporium pannorum*. If this linear relationship is extrapolated to temperatures below those tested, the data suggest that *Chrysosporium pannorum* may show some growth at temperatures down to −3° C. The data for *Cylindrocarpon* sp. nov., *Phoma herbarum* and *Penicillium janthinellum* do not appear to be so clear. However, the indications are that *P. janthinellum* has a faster growth rate at 15° C than 20° C and that *Cylindrocarpon* sp. nov., and *Phoma herbarum* have decreasing growth rates down to 5° C, but then show an upturn. More data are needed before clearcut patterns of fungal response to temperature in these systems can be confirmed. Some preliminary investigations have been done on the respiration responses of these fungi which indicate that *C. pannorum* and *P. janthinellum* have a linear response between 0° and 20° C, after which rates decline, whereas, *Cylindrocarpon* sp. nov., increases up to 25° C (the highest temperature tested). Data for *Phoma herbarum* are not available.

The data for decomposition of *Carex stans* litter indicate that there is an approximate weight loss of 19% during the first year with a 2nd year loss of 12%. The decomposition of litter from 1970 was slower than that of the 1971 standing dead, possibly reflecting differences in quality of litter vs. standing dead. A considerable weight loss appears to occur during the winter months, but, whether this is due to active decomposition is questionable. Much of this loss may be due to leaching during spring of substances solubilized during the rest of the year. A similar phenomenon has been noted by Wein and Bliss (1974) who studied decomposition of herbaceous litter from *Eriophorum vaginatum* communities in Alaska and the Yukon. First year weight losses of herbaceous materials from these communities ranged from 19% at Dempster to 28% at Eagle Creek. These weight losses are similar to, but somewhat higher than Truelove Lowland data, as are the data on *Dupontia fischeri* litter at Point Barrow reported by Benoit et al. (1972).

Cellulose weight loss data from Truelove Lowland indicate that at neither site was there any decomposition of cellulose lying on the surface, possibly due to lack of moisture. Wein and Bliss (1974) found that, under similar circumstances, weight losses of cellulose in Alaska and the Yukon was approximately 2% or less, and they also attribute this to low moisture. Belowground cellulose decomposition on Truelove Lowland soils was faster in the raised beach than in the hummocky sedge-moss meadow. This may reflect more favorable conditions for microbial activity on the raised beach, as was discussed previously.

The respiration data for selected *Carex stans* and *Dryas integrifolia* leaves shows different patterns for the two species. The rates were higher for the *Carex* litter than the *Dryas* and *Carex* respiration rates increased with temperature. The rates for *Dryas* litter though lower, showed two maxima in all but the grey fragmented leaves. The peaks were generally at 15°C and another at 25°C. This may be a real phenomenon, reflecting the responses of different segments of the microflora, or different respiration rates, or different substrates, or it may be an artifact due to experimental errors. To make any definite conclusions more data are required. Generally, the respiration rates of the plant materials reflect moisture contents, and until more data on the effects of moisture on these rates are obtained, few conclusions should be drawn.

Acknowledgments

The author wishes to thank Dr. D. Parkinson for help, encouragement, and much useful discussion during the course of this work. Thanks are also due to Mr. Ted Prince and Mr. Curtis Johanson for their aid in the field during 1971 and 1972 respectively, Mr. Kent Oliver and Ms. Colleen Hyslop who helped with some of the laboratory work and data analysis during the summer of 1973, to Ms. Teresa Newell whose thankless task it was to perform identifications of bacterial isolates, and to Dr. John Bissett for writing special computer programs for analyzing some of these data.

References

Aristovskaya, A.T.V. and O.M. Parinkina. 1972. Preliminary results of the I.B.P. studies of soil microbiology in tundra. pp. 80-92. *In:* Proceedings IV International Meeting on the Biological Productivity of Tundra. F.E. Wielogolaski and T. Rosswall (eds.). IBP Tundra Biome Steering Committee. Stockholm. 320 pp.

Benoit, R.E., W.M. Campbell and R.W. Harris. 1972. Decomposition of organic matter in the wet meadow tundra. Barrow; a revised word-model. pp. 111-115. *In:* Proc. 1972 Tundra Biome Symposium. S. Bowen (ed.). CRREL. Hanover, New Hampshire, 211 pp.

Bissett, J.D. and P. Widden. 1972. An automic multichamber soil washing apparatus for removing fungal spores from soil. Can. J. Microbiol. 18: 1399-1404.

Boyd, W.L. and J.W. Boyd. 1971. Study of soil micro-organisms Inuvik N.W.T. Arctic 24: 162-176.

————., J.T. Staley and J.W. Boyd. 1966. Ecology of soil micro-organisms of Antarctica. Antarctic Research Series 8: 125-159.

Bunt, J.S. 1965. Observations on the fungi of Macquarie Island. ANARE, Scientific Reports Series B(11) No. 78.

Cameron, R.E., J. King and C.N. David. 1970. Microbiology ecology and microclimatology of soil sites in dry valleys of southern Victoria Land

Antarctica. pp. 702-716. *In:* Antarctic Ecology. M.W. Holdgate (ed.). Vol. 2. Academic Press. 391 pp.

Clarholm, M. and T. Rosswall. 1973. A comparison of bacterial populations from four tundra sites by means of a multipoint technique. I.B.P. Swedish Tundra Biome Project Technical Report No. 10, 32 pp.

Dowding, P. and Paul Widden. 1974. Some relationships between fungi and their environment in tundra regions. pp. 123-150. *In:* Soil Organisms and Decomposition in Tundra. A.J. Holding, O.W. Heal, S. McLean, P.W. Flanagan (eds.). IBP Tundra Biome Steering Committee, Stockholm, 398 pp.

Dunican, K. and T. Rosswall. 1974. Taxonomy and physiology of tundra bacteria in relation to site characteristics. pp. 79-92 (see Dowding and Widden 1974).

Fahreus, G. 1947. Studies in the cellulose decomposition by *Cytophaga*. Symb. Bot. Upsaliensis IX. 2. 128 pp.

Flanagan, P.W. 1971. Decomposition and Fungal populations in tundra regions. pp. 150-155. *In:* The Structure and Function of the Tundra Ecosystem. J. Brown and S. Bowen (eds.). CRREL. Hanover N.H. 282 pp.

————. and A. Scarborough. 1974. Physiological groups of decomposer Fungi on tundra plant remains. pp. 159-183. (see Dowding and Widden 1974).

Fournelle, H.J. 1967. Soil and water bacteria in the Alaskan subarctic tundra. Arctic 20: 104-113.

Goodfellow, M. 1966. The Classification of Bacteria in a Pine-wood Soil. Ph.D. thesis, Univ. Liverpool (quoted from Rosswall 1974).

Heal, O.W., and A.D. Bailey and P.M. Latter. 1967. Bacteria, fungi and protozoa in Signy Island soils compared with those from a temperate moorland. Philosophical Trans. Royal. Soc. London B. 252: 191-197.

Hissett, R. and T.R.G. Gray. 1973. Bacterial populations of litter and soil in a deciduous woodland. 1: Quantitative studies. Revue d'Ecologie et de Biologie du Sol. 10: 495-508.

Holding, A.J., V.G. Collins, D.D. French, B.T. D'Sylva and J.H. Baker. 1974. Relationship between viable bacterial counts and site characteristics in Tundra. pp. 49-64 (see Dowding and Widden 1974).

Ivarsen, K.C. 1965. The microbiology of some permafrost soils in the Mackenzie Valley N.W.T. Arctic 18: 256-260.

Jones, P.C.T. and J.E. Mollison. 1948. A technique for the quantitative estimation of soil microorganisms. J. Gen. Microbiol. 2: 54-69.

Kriss, A.E. 1947. Microorganisms of the tundra and arctic desert soils of the arctic. Mikrobiologija 16: 437-448.

Latter, P.M., J.B. Cragg and O.W. Heal. 1967. Comparative studies on the microbiology of four moorland soils in the northern Pennines. J. Ecol. 55: 445-464.

Lowry, D.H., N.J. Rosebrough, A.L. Farr and R.J. Randall. 1951. Protein measurement with the Folin phenol reagent. J. Biol. Chem. 193: 265-275.

Morita, R.Y. 1975. Psychrophilic bacteria. Bact. Revs. 39: 144-167.

Parkinson, D. and S.T. Williams. 1961. A method for isolating fungi from soil microhabitats. Plant and Soil 4: 347-355.

Rosswall, T. and M. Clarholm. 1974. Characteristics of tundra bacterial populations and a comparison with populations from forest and grassland soils. pp. 93-108 (see Dowding and Widden 1974).

Ryan, J. 1972. Devon Island invertebrate research. pp. 293-314. *In:* Devon Island I.B.P. Project. High Arctic Ecosystem. L.C. Bliss (ed.). Edmonton, Alberta. 413 pp.

Saito, T. 1955. The significance of plate counts of soil fungi and the detection of their mycelia. Ecol. Rev. 14: 69-74.
Satchell, J.E. 1969. Feasibility study of an energy budget for Meathop Wood. Symp. Productivity of the Forest Ecosystems of the World. Brussels, Oct. 1969.
Sundman, V. 1970. Four bacterial soil populations characterized and compared by a Factor analytical method. Can. J. Microbiol. 16: 455-464.
Wein, R.W. and L.C. Bliss. 1974. Primary production in arctic cottongrass tussock tundra communities. Arct. Alp. Res. 6: 261-275.

Muskox dung; its turnover rate and possible role on Truelove Lowland

Tom Booth

Introduction

One feature of the Truelove Lowland ecosystem is the presence of herbivore, particularly muskox, fecal material. Since there have been few studies on abiotic and biotic factor effects on vertebrate dung decomposition processes (White 1960, McCalla et al. 1971, Angel and Wicklow 1974, 1975; Lodha 1974, Wicklow and Moore 1974, Waterhouse 1974), determination of thermal and moisture regimes and turnover times was undertaken. Also, determination of dung contributions of minerals and organic substances to so-called "nutrient poor" tundra ecosystems was considered important.

Several microhabitats occur on the Lowland and, in order to account for variations in decomposition, it was decided to study the Beach Ridge Crest (RC) and Backslope (BS), the Hummocky Sedge-moss Meadow (HSMM), and Wet Sedge-moss Meadow (WSMM) sites. For further details on the microtopography and floristic composition of these microhabitats, the reader may wish to consult Muc and Svoboda (this volume).

Field observations

Methods and materials

Dung accumulation and decomposition

Total accumulated dung was determined by sampling 77 randomly located transects (50m×2m) for all recognizable dung. Collected dung was oven-dried at 80°C to constant weight and subsequently weighed. Sampling areas included hummocky sedge-moss meadows, ridge crests, foreslopes, and backslopes, with at least 10 transects per area. Winter dung, roughly spherical pellets 1-2 cm in diameter, was distinguished from summer dung which is conical, approximately 15 cm long and 19 cm at the base and made up of discernible chips.

Deposition plots, 112 in total, were located in preferred (certain meadows are favored grazing sites) meadow (20), non-preferred meadow (20), random (plots randomly located irrespective of grazing preference) meadow (61) and random ridge (11) sites. Plots, cleared of dung, were 25 m^2 on meadows and varied in area on ridges. To be representative of the location, the plots ran from the foreslope base across the ridge to the backslope base and resulted in areas much greater than on meadows. One year after establishment and clearing, the plots were sampled for accumulated dung. Dung was then oven-dried and weighed to determine total accumulation.

Dung temperature and moisture

Fresh winter dung was spread, one layer of pellets thick, over three separate 1 m^2 plots in each of four microhabitats, i.e., wet sedge-moss meadow, hummocky sedge-moss meadow, ridge crest, and backslope. Epoxy-covered Cu-const. antan thermocouples, 24-gauge, were embedded just below the pellet surface and temperatures were taken bimonthly for a 24 hr period. Dung moisture was gravimetrically determined and expressed as a percent of dry weight.

Hyphal length

Techniques outlined by Jones and Mollison (1948) were applied to 2 g samples from wet sedge-moss meadow, hummocky sedge-moss meadow, ridge crest, and backslope plots on various dates. Unaged random samples from various lowland locations were also collected and similarly treated.

Results and discussion

Dung accumulation and deposition

Accumulated dung is generally greater on meadows than on ridge locations (Table 1). Drier meadow sites have more dung than wet sites. Muskox distribution data (Hubert this volume) and vegetation map information (Muc and Bliss this volume) combined, indicate that muskox presence is more frequent on hummocky sedge-moss meadows than on wet sedge-moss or frost-boil meadows. In fact, Hubert observed that there are wet meadow areas where muskox did not occur in three study seasons. Smaller amounts of dung on beach ridge habitats than on meadow sites is a function of the abundance of suitable forage on meadows as compared with ridges. Foreslope

Table 1: Standing crop of muskox dung in different sites on Truelove Lowland (g dung m^{-2}±SE).

Sites	*Winter dung*	*Summer dung*	*Total*
Hummocky sedge-moss meadows	3.4±0.9	0.6±0.2	4.0
Frost boil meadows	2.5±0.8	0.3±0.1	2.8
Wet sedge-moss meadows	2.3±0.7	0.4±0.2	2.7
Beach ridge crests	1.8±0.7	0.3±0.1	2.1
Foreslopes	0.9±0.4	0.5±0.2	1.4
Backslopes	0.3±0.1	0.1±0.05	0.4

Table 2. Annual deposition of winter and summer muskox dung in various sites (g dung m^{-2}±SE).

Sites	*Winter dung*	*Summer dung*	*Total*
Preferred meadows	2.1±0.4	0.4±0.3	2.5
Random meadows	1.1±0.2	0.5±0.1	1.6
Non-preferred meadows	1.1±0.3	0.1±0.05	1.2
Random ridges	0.05±0.03	0.1±0.05	0.2

and backslope accumulation differences result from the animals' behavior of lying in snowbanks during rumination in warm periods (Tener 1965). Snow remains longer on the north and west-facing foreslopes than on the south and east-facing backslopes.

Deposition data (Table 2) show a greater defecation rate on preferred meadows than on non-preferred or random meadows. This implies that dung is not randomly distributed on the lowland meadows. However, random meadow deposition is around the mean of preferred and non-preferred rates. Random ridge locations received less dung than any of the other plots. This is consistent with accumulation results, where meadows have a greater amount of dung than ridges.

From Table 1 there is a total dung accumulation of 3.2 ± 0.4 g m^{-2} on meadows and 1.3 ± 0.5 g m^{-2} on ridges. Accretion is 1.77 ± 0.4 g m^{-2} $year^{-1}$ on meadows and 0.2 ± 0.04 g m^{-2} $year^{-1}$ on ridges. These latter values are close to a theoretical accretion rate calculated in the following way. Based on 1973 data (Hubert 1976), 243 animals, having a potential of 8.9×10^4 muskox days annually, use the northeastern coast of Devon Island. Occupation of Truelove Lowland from April 1972 through May 1973 1.5×10^4 muskox days, which is approximately 17% of the total population or 42 animals day^{-1}. Adult fecal output was about 1.4×10^3 g animal day^{-1}, and, therefore, 2.1×10^7 g of dung were deposited on the Lowland yearly. Land surface of the Lowland and Truelove Valley is 3.7×10^7 m^2, and, if dung is deposited evenly over the land, there is 0.57 g of dung m^{-2} $year^{-1}$.

Combination of deposition and accumulation results allow postulation of turnover time, $x/y = z$, where x = standing crop (g m^{-2}), y = annual production (g m^{-2} $year^{-1}$) and z = turnover time. Turnover time on meadows is 1.3-2.6 years and 3.3-11.3 years on ridges.

Dung temperature and moisture

Dung temperatures increased from early May to mid-August (Fig. 1) and were constantly above 0°C from early June onward. The regression lines in Fig. 1 are problematical as they are a straight-line representation of a curvilinear relationship and because they fail to account for daily temperature changes since they are based on daily means (Table 3).

Regression data in Table 4 relate daily air temperature changes to dung temperature changes, allowing calculation of dung temperature from air temperature. Air temperature effects on ridge crest dung temperature were greatest on May 14 and leveled off for the remainder of the summer (Fig. 2). Air temperature greatly affected meadow dung temperatures during mid-June. This probably resulted from several factors, including net radiation, dung moisture and substrate age.

Calculation of hours above 0°C for each microhabitat gives 2,054 hr (=86 days) for wet sedge-moss meadow dung, 2,000 hr (=83 days for ridge crest dung, 1,980 hr (=83 days) for backslope dung and 1,926 hr (=80 days) for hummocky sedge-moss meadow dung.

These data are refined in Fig. 3 to percent readings at 5° intervals. As Longton and MacIver (1977) suggest, such percent readings can represent time at particular temperature intervals. To convert percent readings to days above 0°C at each interval requires the product of the fraction of each percent reading over total percent above 0°C times days above 0°C. Fig. 1 indicates that wet sedge-moss meadow dung was warmer than dung in the other microhabitats as time at 10-15°C and 15-20°C intervals was greatest. Ridge crest dung was possibly slightly warmer than hummocky sedge-moss meadow dung and backslope dung as time at 10-15°C and 15-20°C intervals was

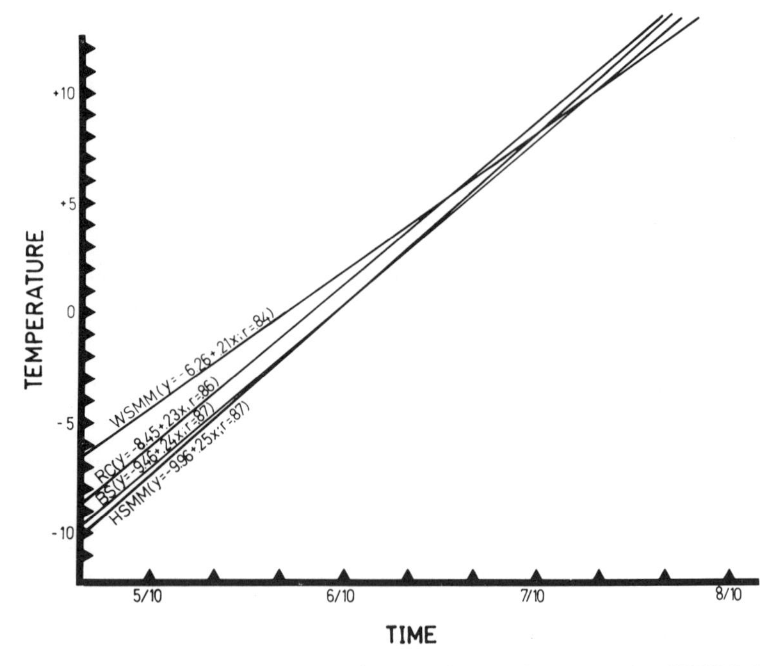

Fig. 1 Regression of dung temperature on time (days) for wet sedge-moss meadow (WSMM), beach ridge crest (RC), backslope (BS) and hummocky sedge-moss meadow (HSMM) dung.

slightly higher. Also as wet sedge-moss meadow dung, 75% of all readings were above 0°C in ridge crest dung, while 71% of all readings were above 0°C in hummocky sedge-moss meadow and backslope dung.

Time above 0°C is very important since temperature can be directly related to decomposition and, hence turnover rates (Heal and French 1975). Decomposition below 0°C was limited. Particular temperature intervals,

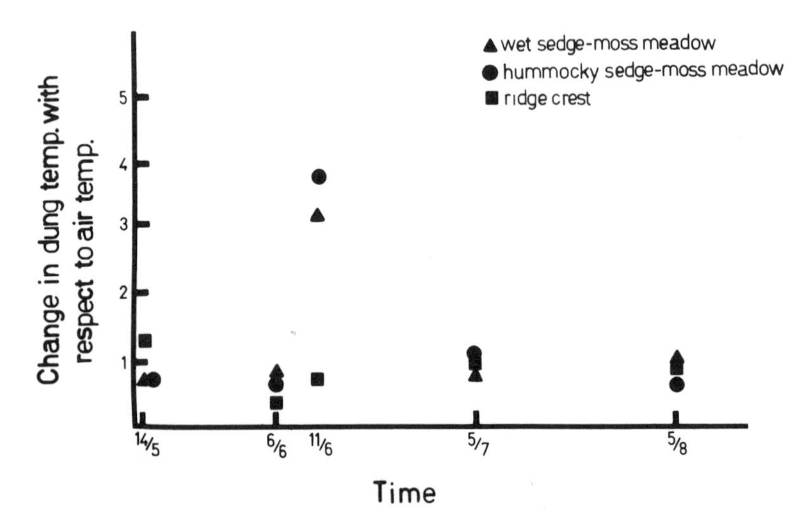

Fig. 2 Effect of change of air temperature on dung temperature during 14 May, 1973 to 5 August, 1973.

especially high, are significant; short periods of high temperature markedly increase turnover rates. However, temperature is not independent of moisture since the two interact to affect decomposition (Flanagan and Veum 1975).

One feature of dung temperature that should be mentioned is the effect of heat release due to microbial activity. Rabbit dung, incubated at 10° C, releases 0.01 cal min^{-1} during the period of greatest caloric loss (Angel and Wicklow 1974). As temperature measurements require no more than a minute and 0.01 cal represents measurement under closed conditions, this amount of heat was negligible in terms of dung temperature in an open system. Also, heat loss from dung pellet surfaces exposed to the natural environment must be significant.

Moisture data (Table 5) indicated that wet sedge-moss meadow dung was wetter (255±39% in 1972, 384±55% in 1973) than dung in any of the other microhabitats. Dung in hummocky sedge-moss meadow sites was next wettest (155±53% in 1972, 226±34% in 1973), followed by backslope (88±61% in 1972,

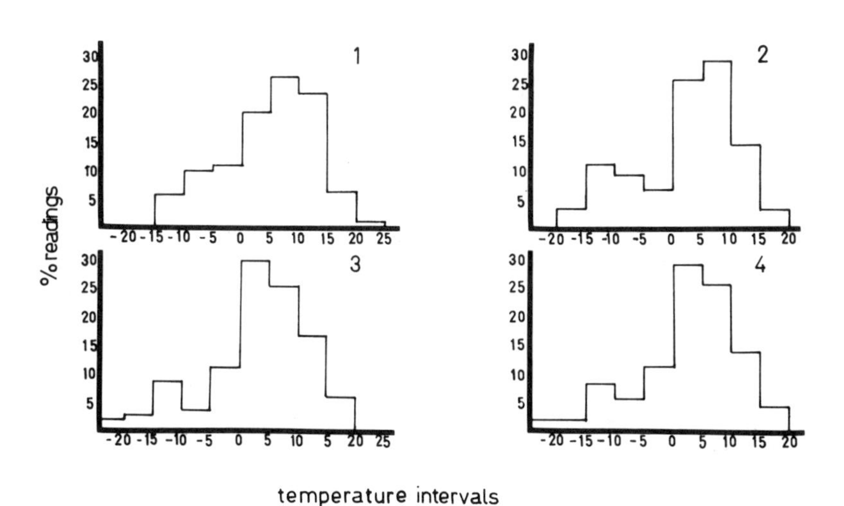

Fig. 3 Percent readings at five degree intervals from 6 May, 1973 to 6 August, 1973. Bars represent days at each interval.

Table 3. Daily mean dung temperatures (° C) in various microhabitats.

Date	*Wet sedge-moss meadow*	*Beach ridge crest*	*Hummocky sedge-moss meadow*	*Beach ridge backslope*
6 May	− 6.7	− 9.0	− 9.9	− 9.5
14 May	− 3.6	− 6.6	− 9.8	− 8.9
19 May	− 8.5	− 11.3	− 10.7	−11.1
22 May	− 3.8	− 3.9	− 7.5	− 4.7
6 June	1.0	1.7	1.0	− 0.5
11 June	7.7	4.0	5.2	4.3
13 June	8.3	6.2	5.9	5.9
21 June	11.9	11.5	8.7	11.5
28 June	3.9	3.4	3.5	3.5
4 July	10.2	8.7	9.2	8.4
12 July	10.2	9.3	9.5	9.1
15 July	11.8	11.2	10.8	10.7
20 July	8.2	8.4	7.7	8.1
27 July	8.2	6.6	6.9	6.7
5 August	10.9	10.1	8.3	9.7

Table 4. Regression of dung temperature on air temperature for wet sedge-moss meadow, hummocky sedge-moss meadow, and beach ridge crest sites at various dates.

Habitats	14 May	11 June	5 July	5 August
Wet sedge-moss meadow	y = 5.5+0.7x	y = 7.7+3.1x	y = 2.8+0.9x	y = 2.5+1.1x
	r = 0.53*	r = 0.68	r = 0.88	r = 0.90
Hummocky sedge-moss meadow	y = -0.4+0.7x	y = 4.8+3.7x	y = 1.7+0.9x	y = 3.0+0.7x
	r = 0.63	r = 0.66	r = 0.94	r = 0.87
Beach ridge crest	y = 6.2+1.3x	y = 1.4+0.8x	y = -5.9+0.9x	y = -5.0+0.9x
	r = 0.76	r = 0.85	r = 0.95	r = 0.98

* all r values significant at P=.05

Table 5. Dung moisture content (% of dry wt) in microhabitats at various dates.

Date	Wet sedge-moss meadow	Beach ridge crest	Hummocky sedge-moss meadow	Beach ridge backslope
30 May 1972	245	257	203	270
10 July 1972	335	47	335	—
1 August 1972	233	18	67	21
22 August 1972	317	92	127	35
5 September 1972	145	24	41	24
5 May 1973	178	203	122	186
15 May 1973	138	138	426	82
22 May 1973	75	104	67	45
7 June 1973	488	37	285	355
12 June 1973	426	14	150	426
22 June 1973	186	9	285	10
30 June 1973	488	194	376	194
5 July 1973	426	43	58	54
14 July 1973	456	88	300	317
22 July 1973	525	300	233	335
29 July 1973	614	85	270	92
7 August 1973	614	33	138	41

178±42% in 1973) and ridge crest (88±44% in 1972, 194±26% in 1973) dung. Moisture levels in 1973 were higher than 1972 levels which is related to the wet 1973 summer versus the cool and dry 1972 summer.

Considering the nature of the microhabitats, it is possible to explain differences in dung thermal-moisture regimes. Direct relationship between temperature and moisture increase of meadow (WSMM and HSMM) dung is a function of constantly provided moisture from melting ice and snow. Backslope and ridge crest sites melt off and dry out much earlier than the other two sites and these environmental conditions combine to result in various associations of temperature and moisture in dung. In a relative sense, wet sedge-moss meadow dung was wet and warm, hummocky sedge-moss meadow dung was wet and cool, ridge crest dung was dry and warm, and backslope dung was dry and cool.

These relative differences affect decomposition and hence, turnover times since (1) temperature is dependent on substrate moisture and quality (Reichle et al. 1973, Heal and French 1975), (2) decomposition and leaching are a function of percent moisture times temperature (Reichle et al. 1973) and (3) high moisture can limit decomposition due to anoxic conditions (Flanagan and Veum 1975).

Table 6. Hyphal length (m±SE) in dung under various conditions.

Habitat	*Dung type*	*15 May**	*12 June**	*30 June**	*Random*
Beach ridge backslope	winter	4367±475	20786±2351	3032± 405	—
Wet sedge-moss meadow	summer				6600±1700
	winter	5151±610	10242±1380	14368±1576	
Beach ridge crest	summer				4200± 500
	winter	5259±414	2316± 270	3747± 401	4300±1000
Hummocky sedge-moss meadow	summer	—	—	—	14100±2000
	winter	—	—	—	9500±5600
	moss-covered				
	summer	—	—	—	5100±2400
	winter	—	—	—	23300±6600

*Dung of known age (collected fresh and studied in the first year).

Hyphal length

A first attempt at measuring suspected differences of dung decomposition in the various microhabitats involved determination of what fungal organisms are present on the dung and how much total fungal mycelium is associated. Types of fungi were reported in Appendix 4 of this volume.

Hyphal lengths are variable according to type and time of the season (Table 6). Lengths, and by inference, biomass, were greatest in moss-covered dung. Of the dung of known age, backslope dung, followed respectively by wet sedge-moss meadow and ridge crest dung, had the greatest fungal biomass in terms of a single date (i.e. 12 June). Also, while reading the slides, it was observed that backslope dung had the greatest length of hyaline hyphae, followed again by wet sedge-moss meadow and ridge crest dung. Darkly pigmented hyphae, nonetheless, were in greater abundance than hyaline hyphae irrespective of the dung habitat. Finally, fruiting bodies of *Sporormia* spp. *Sporormiella* spp., *Coprobia* sp. and *Podospora* sp. appeared in mid-late June on backslope dung and came as late as mid-July on wet sedge-moss meadow and hummocky sedge-moss meadow dung. No fruiting bodies occurred on ridge crest dung in the first year.

From this, it is possible to suggest that turnover time for backslope dung is shorter than wet sedge-moss meadow and hummocky sedge-moss meadow dung. Turnover time is greatest for ridge crest dung. In order to establish this scheme, it is necessary to assume that short periods of optimal conditions are sufficient to account for very significant decomposition of the substrate. During the period of 11-21 June dung temperatures were from 0-5°C for 64 hr, 5-10°C for 116 hr, 10-15°C for 34 hr and 15-25°C for 16 hr. Moisture content during the period was around 70%. Before this period, temperature was below optimal levels and, after the period moisture was not at optimum levels. Thus, there seems to be a relationship between temperature and moisture conditions and hyphal lengths.

A series of laboratory studies of decomposition were undertaken to test the above hypothesis. These included carbon loss, oxygen consumption, caloric loss, and changes in chemical composition.

Laboratory experiments

Methods and materials

Carbon loss

Frozen fresh dung was placed in sterilized cylindrical glass jars with vermiculite in the bottom. Sterile water was used to periodically dampen dung during the course of the experiment. Half of the experimental jars were maintained at 20° C and the other half at 5° C on a 12 hr light and 12 hr dark cycle at 1,000 ft c. Evolution of CO_2 was measured in a closed system Beckman infra-red gas analyser. Production of CO_2 was determined per unit time with five replicate readings at each time.

Oxygen consumption

Oxygen uptake, expressed as μl O_2 g^{-1} hr^{-1} was determined using a Gilson differential respirometer. Random collections of fecal pellets (placed in the 1 m^2 plots) from each of the sites over the various microhabitats were made on five separate dates. Five replicates, each of three to five pellets, constituted collections from each site. Samples were frozen immediately on collection and carried to University of Manitoba laboratories where they were maintained in that condition. Actual respirometry determinations followed methods of Parkinson and Coups (1963). Readings were taken at various temperatures (0° C, 5° C, 10° C, 15° C, and 20° C) for three replications from each site and collection date. Five readings for each replicate were considered adequate. Each sample was weighed for moisture content before and after respirometry. Generally, moisture loss during the experimental run was normally $<2\%$ and rarely as high as 5%.

Nutrient content and caloric determinations

The following nutrient determinations were done in the Alberta Department of Agriculture Soil and Feed Testing Laboratory by standard methods: % protein, phosphorous, fat, fibre, lignin, ash, and ppm Ca^{++}. Nitrogen was derived from protein (% protein ÷ 6.2), and Ca^{++} was converted to milliequivalents. Magnesium determinations involved use of atomic absorption spectrophotometry. Caloric values result from bomb calorimetry determinations.

These chemical parameters were measured in dung from the previously mentioned plots associated with the beach ridge crest and backslope, wet sedge-moss meadow, and hummocky sedge-meadow sites. Samples, randomly collected and immediately frozen, were taken on several dates.

Results and discussion

Carbon loss

Regression of ppm CO_2 on temperature (Fig. 4) indicates an increase in CO_2 loss with increasing temperature. Calculation of ppm CO_2 g $dung^{-1}$ min^{-1} at environmental temperature is derived from the regression formula. As temperatures at or above 0° C seem to produce significant decomposition rates, a correction for time above 0° C is required. To complete expression of g carbon lost, it is necessary to convert ppm CO_2 g^{-1} min^{-1} to g C g $dung^{-1}$ $season^{-1}$.

When all corrections are made carbon loss is 0.206 g yr^{-1} for wet sedge-

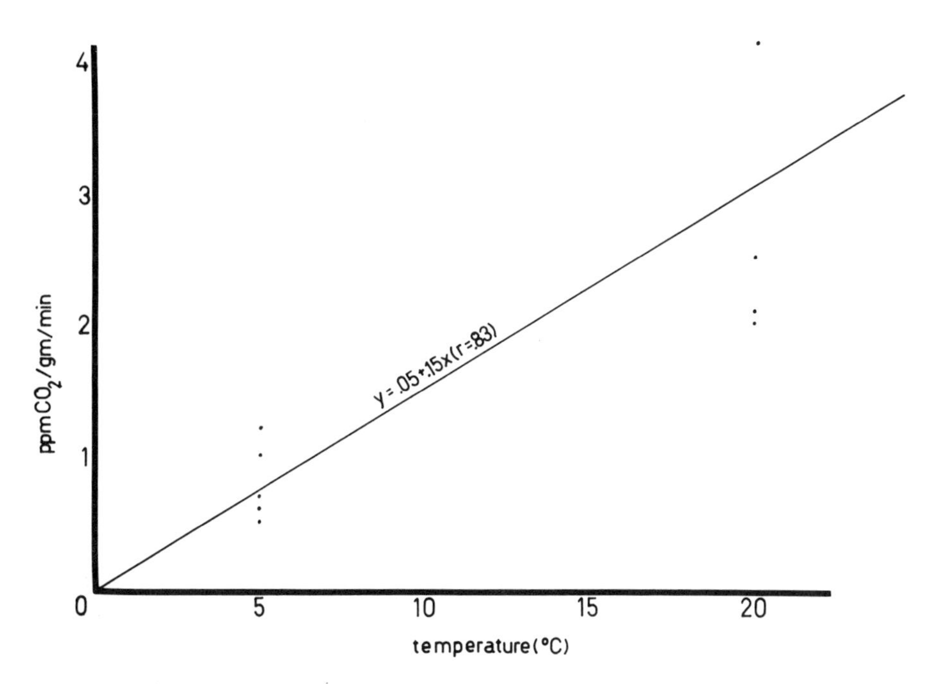

Fig. 4 Regression of ppm CO_2 g^{-1} min^{-1} on temperature.

moss meadow dung, 0.189 g yr^{-1} for ridge crest dung, 0.168 g yr^{-1} for hummocky sedge-moss meadow dung and 0.185 g yr^{-1} for backslope dung. Because the substrate is 40-60% carbon, decomposition will require 1.9-2.9 years for wet sedge-moss meadow dung, 2.1-3.2 years for ridge crest dung, 2.4-3.6 years for hummocky sedge-moss meadow dung and 2.2-3.2 years for backslope dung.

To equate total decomposition and carbon loss, fails to account for the fact that the substrate is made up of more than carbon. Also this presentation assumes that all carbon is as degradable as time progresses as it is in the beginning. It is well-known (Flanagan and Veum 1975) that carbon-containing compounds become less easily decayed as substrate decomposition proceeds. Assumptions of various time factors (i.e. length of decomposition season and time at various temperatures) are also open to question. Finally, there is the underlying supposition that the substrate is exposed to long periods of constant temperature and moisture which is not true for field conditions.

Oxygen consumption

Mean respirometric data at 10°C are presented for each of the microhabitats on various dates (Table 7). By multiplying these values by time at or above 0°C, adding the total, converting to litres, assuming a respiratory quotient of 1, converting 1 O_2 to g C and assuming substrate to be 0.4-0.6 g carbon, turnover times are 6.2-9.2 seasons for wet sedge-moss meadow dung, 11.1-16.7 seasons for ridge crest and backslope dung, and 6.1-9.1 seasons for hummocky sedge-moss meadow dung.

Two important assumptions are made in computing the above turnover rates. These assumptions include an RQ of 1 and that temperature effects alone can account for dung decomposition. Among the reasons for the RQ assumption are: large amounts of carbohydrates which seem to typify herbivore

Table 7. Mean respiration ($\mu l\ O_2\ g^{-1}\ hr^{-1}$) of muskox dung at 10°C in four microhabitats on various dates.

Date	*Wet sedge-moss meadow*	*Beach ridge crest*	*Hummocky sedge-moss meadow*	*Beach ridge backslope*
30 May	121.2	62.4	98.6	38.9
10 July	45.3	17.9	—	—
1 August	45.6	2.6	44.5	13.1
22 August	—	71.3	82.2	25.2
5 September	58.9	19.6	30.7	20.9

Table 8. Weighed respiration means ($\mu l\ O_2\ g^{-1}$ dry wt hr^{-1}) for muskox dung at moisture and temperature grid points.

		Temperature				
Site	*% HOH*	*0°C*	*5°C*	*10°C*	*15°C*	*20°C*
Wet sedge-moss meadow	50	3.7	10.3	43.6	27.6	64.1
	89	56.4	81.3	68.5	61.7	94.4
	158	29.0	41.9	39.6	80.7	55.8
	281	52.7	71.0	60.2	97.4	134.6
	500	43.4	39.3	73.5	74.5	33.7
Beach ridge crest	50	7.5	11.8	13.1	9.1	87.7
	89	34.7	34.7	43.8	78.6	70.0
	158	79.4	49.8	61.9	97.0	62.0
	281	63.4	50.3	46.9	82.9	70.1
Hummocky sedge-moss meadow	50	29.6	30.0	15.5	32.1	33.7
	89	35.6	53.7	65.2	75.4	88.2
	158	46.4	41.5	70.7	77.5	113.9
	281	63.0	40.2	53.4	80.7	83.3
Beach ridge backslope	50	18.2	14.6	28.6	20.7	36.4
	89	28.2	25.3	35.5	77.2	150.0
	158	35.6	32.8	40.0	118.0	235.2
	281	37.5	32.5	44.7	111.1	150.1

dung (Lodha 1974) and low level of fats and organic acids in herbivore dung. Carbohydrates have an RQ of 1, while the RQ of fats is less than 1 and organic acid RQ is greater than unity. However, Nicholson et al. (1966) showed that RQ decreases temporally in response to a diminishing percent of carbohydrate, and the assumption of a constant RQ over the season may not have been entirely representative of the situation in muskox dung. To utilize an expression of temperature as solely controlling respiration is simplistic.

Irrespective of temperature, high moisture levels depress decomposition due to establishment of anoxic conditions (Williams and Gray 1974, Flanagan and Veum 1975, Heal and French 1975), and low moistures also show pronounced suppression of decomposition. As both the oxygen consumption and carbon loss turnover rates were based solely on temperature, more accurate resolution of turnover times should have included moisture effects as well.

Figs. 5-8 are visual representations of data in Table 8. Respirometric data in Table 8 are derived from the GRESP program (Bunnell and Tait 1975) and are, in effect, weighted means.

By interpolating oxygen consumption for mean daily temperature (Table 3) and moisture (Table 5), correcting for time and converting total oxygen

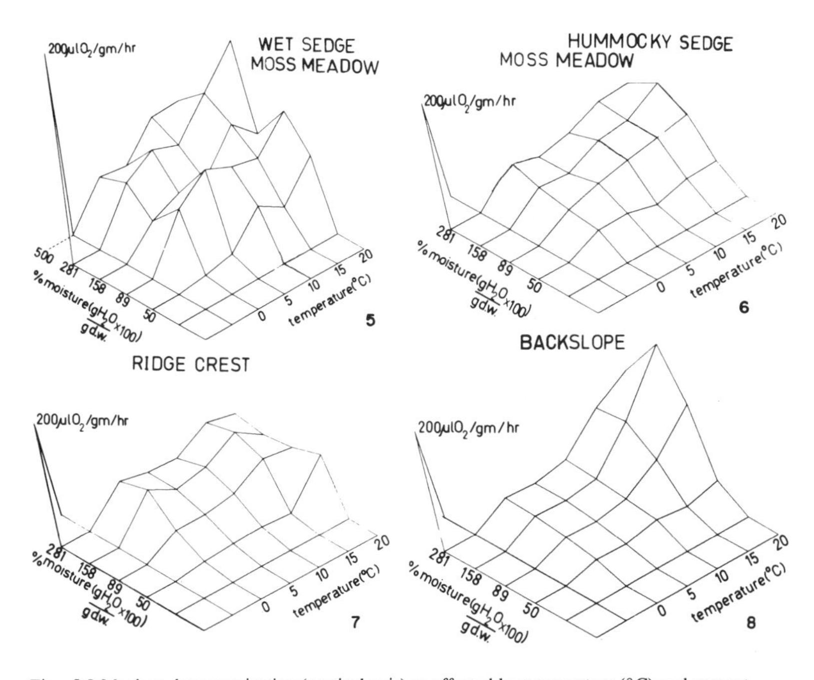

Figs. 5-8 Muskox dung respiration (vertical axis) as affected by temperature (°C) and percent moisture.

Table 9. Summary of turnover times with a mean ± SE calculated for each microhabitat.

Treatment	*Wet sedge-moss meadow*	*Beach ridge crest*	*Hummocky sedge-moss meadow*	*Beach ridge backslope*
Dep.-acc.	1.3- 2.6	3.3-11.3	1.3- 2.6	3.3-11.3
CO_2 (temperature)	1.9- 2.9	2.1- 3.2	2.4- 3.6	2.2- 3.2
O_2 (temperature)	6.2- 9.2	11.1-16.7	6.1- 9.1	11.1-16.7
O_2 (temperature-moisture)	8.3-12.5	18.2-27.3	8.2-12.2	12.0-18.2
Mean:	5.6± 2.0	11.7± 9.0	5.7± 1.9	9.8± 2.9

consumed to carbon loss, 8.3-12.5 seasons are required to decompose dung in wet sedge-moss meadow sites. Dung in hummocky sedge-moss meadow sites requires 8.2-12.2 years to decompose. Backslope dung, with a turnover time of 12.0-18.2 seasons, is followed by ridge crest dung with 18.2-27.3 seasons.

These turnover times suffer many of the limitations of the others. Whether they are more realistic than the others is open to future research, and, for the time being, they are incorporated into a mean turnover time for each microhabitat (Table 9). Turnover times were shortest for wet sedge-moss meadow dung (5.6 years), followed respectively by hummocky sedge-moss meadow dung (5.7 years), backslope dung (9.8 years) and ridge crest dung (11.7 years).

Nutrient content and caloric determinations

When nutrients were corrected for weight loss (Table 10), it became evident that muskox dung is chemically similar to horse dung (Waksman et al. 1939) and cattle dung (Taiganides and Hazen 1966). Increase over one season in ash

Table 10. Nutrient and caloric content of dung corrected for weight loss due to decomposition.

Microhabitat and date or dung condition	*Corrected weight*	*Protein*	*Nitrogen*	*Phosphorus*	*Fat*	*Fibre*	*Lignin*	*Ash*	*Ca*$^{++}$	*Mg*$^{++}$	*Calories*	*Resp. calories*
					% dry wt				*Meq 10 g*		*g dry wt*	
Fresh	1.000	12.9	2.10	0.28	3.2	48.3	25.0	6.0	41.0	23.9	4,900	
Wet sedge-moss meadow:												
16 July 1972	0.898	—	—	—	—	—	—		41.0	16.4	4,849	51
4 August 1972	0.858	—	—	—	—	—	—		36.4	15.2	4,032	867
5 July 1973	0.924	11.6	1.87	0.21	3.0	42.0	25.6	7.2	35.9	—	—	
7 August 1973	0.854	10.8	1.74	0.16	3.2	41.9	28.1	5.9	30.3	—	—	
Beach ridge crest:												
16 July 1972	0.949	—	—	—	—	—	—		43.4	15.0	4,261	638
4 August 1972	0.929	—	—	—	—	—	—		44.3	13.1	4,078	822
5 July 1973	0.961	12.1	1.95	0.20	3.2	43.6	27.8	7.5	38.1	—	—	
7 August 1973	0.926	11.5	1.85	0.20	2.6	43.3	27.1	7.1	31.6	—	—	
Hummocky sedge-moss meadow:												
16 July 1972	0.908	—	—	—	—	—	—		45.5	16.5	—	—
4 August 1972	0.872	—	—	—	—	—	—		38.5	14.6	4,053	847
5 July 1973	0.930	10.0	1.62	0.21	1.7	47.3	28.4	7.3	36.4	—	—	
7 August 1973	0.866	9.7	1.56	0.21	2.6	45.3	26.3	8.7	20.3	—	—	
Backslope:												
16 July 1972	0.939	—	—	—	—	—	—		30.2	13.1	3,894	1,005
4 August 1972	0.915	—	—	—	—	—	—		40.2	15.8	3,939	960
5 July 1973	0.950	11.7	1.89	0.22	2.7	41.3	24.7	7.7	37.7	—	—	
7 August 1973	0.911	11.2	1.81	0.21	2.7	41.9	24.8	7.9	35.2	—	—	

(*ca*. 17%) and lignin (*ca*. 4%) contrasted with decrease in protein (*ca*. 16%), phosphorus (*ca*. 28%), fat (*ca*. 13%), fibre (*ca*. 11%), Ca^{++} (*ca*. 20%) and Mg^{++} (*ca*. 42%). In terms of the "nutrient hypothesis" (Webster 1970), my data differed in that Waksman et al. (1939) observed that soluble organic materials remained constant, and protein levels increased rather than decreased. In horse dung decomposition (Waksman et al. 1939), corrected for activity at 10° C for 67 days, fibre (cellulose + hemicellulose) decreased 13.8% as compared with 11% in muskox dung, and lignin decreased 3.7% as compared with 4% in muskox dung. Muskox dung protein and fat are actively decomposed, and Mg^{++}, phosphorus and Ca^{++} seem to be readily leached away. Although pH is not included in Table 10, it is initially higher in muskox dung (7.8) than is suspected for dung in general (6.5) by Lodha (1974). Muskox dung pH quickly dropped to 6.9-7.1 soon after decomposition began and remained constant throughout the season. Caloric content of fresh muskox dung was 4,900 cal g^{-1} d wt, which is very similar to rabbit dung (Angel and Wicklow 1974). Mean caloric loss of muskox dung over the four microhabitats was 17% $season^{-1}$ and was comparable with a 24% caloric loss determined by correcting rabbit dung data for time and temperature. Difference in loss is perhaps attributable to optimal moisture levels maintained by Angel and Wicklow (1974), which enhanced rabbit dung decomposition, as compared with muskox dung at lower environmental moisture levels.

Table 11. Turnover times for muskox dung in various microhabitats and corrected data of other authors based on percent caloric and fibre loss.

Microhabitat or author	*% calories respired*	*% fibre decomposed*
Wet sedge-moss meadow	5.7	7.5
Beach ridge crest	6.1	9.7
Hummocky sedge-moss meadow	5.8	16.1
Beach ridge backslope	5.1	7.5
Angel and Wicklow (1974)	4.2	—
Waksman et al. (1939)	—	7.3

Discussion

Turnover data derived from field and laboratory studies failed to fit the hypothesis that population levels can be used, even in the relative sense, to predict decomposition rates. Backslope dung, in fact, seemed to take nearly as long to decompose as ridge crest dung. Also, the supposition that short bursts of microbial activity can account for significant decomposition is open to question. Decomposition is most directly a function of the time a substrate is at optimal conditions and of the two principal factors considered here, i.e., temperature and moisture; the latter is most important as turnover times for the wet environments (wet sedge-moss meadow and hummocky sedge-moss meadow) are significantly shorter than for the dry ones (ridge crest and backslope).

By calculating percentage of each of the nutrients lost due to decomposition (Table 10), a total of 9.3 seasons is required for turnover in general (protein=2.0%, nitrogen=2.0%, phosphorus=0.1%, fat=0.4%, fibre=5.3%, and ash=1.0%; total=10.8%). This value is close to the mean of means of all microhabitats (Table 9), 8.2 seasons. Calories respired and fibre decomposed are treated by other authors, i.e., Waksman et al. (1939) and Angel and Wicklow (1974), and can be converted to turnover rates from percentages (Table 11). Caloric loss turnover rates are equivalent with turnover times for wet sites given in the means of Table 9. Dry site values are lower than the calculated means. Angel and Wicklow (1974) data, corrected for time and temperature, result in a turnover time very similar to my values based on caloric loss. With the exception of the turnover time determined for hummocky sedge-moss meadow dung, fibre decomposed turnover rates are similar to those in Table 9. Waksman et al. (1939) corrected data also result in a turnover equalling the other fibre decomposed rates. The turnover times determined from calories respired and fibre decomposed are understatements of reality since the assumption of a constant decompostion rate is incorrect (Angel and Wicklow 1974).

Although there are differences in dung turnover rates determined and calculated by various techniques, it is interesting that, no matter the method, the results are generally similar. The turnover times given herein are possibly realistic, and they are very close to the results of Waksman et al. (1960) and Angel and Wicklow (1974). It can be stated that dung is a substrate that decomposes at various rates depending on environmental conditions. There have been no other studies which establish the effect of environment on microbial populations involved in dung decomposition (Angel and Wicklow 1975).

During decomposition there are nutrient exchanges between dung and the surrounding environment. On Truelove Lowland these nutrients were utilized within a very close physical distance to the dung. Hyphal lengths in soil immediately under dung were an order of magnitude greater than those just 1-3 cm away from the dung (unpublished data). Fresh dung placed on the soil surface of a foreslope base or frost-boil meadow, where there were no bryophytes, stimulated development of a ring of mosses just one year later. During accumulation studies, dung could be found incorporated into thick moss mats *(cf. Drepanocladus).* Finally, during the course of this study, it was observed that *Carex stans* and *C. membranacea* are taller and more dense in the immediate areas where dung plots were established.

Net primary production and heterotrophic respiration data presented by Reichle (1975) can be used to calculate percent production respired. In xeric forest sites 62% of the primary production is utilized by heterotrophic organisms while 65% is used in prairie sites, 76% is used in mesic forest sites and 90% is used in tundra sites. Even though a portion of heterotrophic respiration is due to vertebrate grazers these derived data imply the decomposible substrates are more rapidly colonized in tundra sites than in any of the other biomes. This implication is even more striking when length of decomposition season is considered as tundra sites have the shortest season. Consideration of the importance of moisture in decomposition can also be implied from the above data. All of this points to rapid tie up of nutrients over short physical distances by tundra decomposing populations.

Therefore, dung is a nutrient hot point which, in the case of wet sites, can become more mesic than surrounding microenvironments. Such dual functions of this ubiquitous substrate lead to a final generalization: High Arctic terrestrial ecosystems are not nutrient poor; they are nutrient non-random. Studies of Point Barrow, Alaska owl mounds have also led to a similar conclusion by Scholtz (pers. comm.).

Summary

Muskox dung accumulation was 3.2±0.4 g m^{-2} on meadows and 1.3±0.5 g m^{-2} on ridges, while accretion was 1.77±0.4 g m^{-2} $year^{-1}$ on meadows and 0.2±0.04 g m^{-2} $year^{-1}$ on ridges. Turnover times were calculated from deposition and accumulation data to be: 1.3-2.6 years on meadows and 3.3-11.3 years on ridges. Thermal and moisture regimes were found to differ in dung on beach ridge crest, backslope, hummocky sedge-moss meadow and wet sedge-moss meadow soils. Hyphal lengths varied with time and the habitat in which dung was placed. Carbon loss, oxygen consumption, and caloric loss data, when coupled with accumulation-deposition data, resulted in turnover times of 5.6 years for wet sedge-moss meadow dung, 11.7 years for ridge crest dung, 5.7 years for hummocky sedge-moss meadow dung, and 9.8 years for backslope dung. Dung acts as a type of non-random nutrient hot point.

Acknowledgments

Technical assistance was provided by Ted Prince, Curtis Johanson, Marguerete Roberts, and Jim Richardson. Ben Hubert was an advisor, companion, colleague, and friend throughout the study. Fred Bunnell and David Tait helped in the synthesis stage, and John Whittaker assisted during the writing stage. Productive stimulation was provided at critical times by Paul Widden. Special thanks are due Dennis Parkinson and Larry Bliss for their material support, constant encouragement and helpful advice.

References

Angel, K. and D.T. Wicklow. 1974. Decomposition of rabbit faeces: an indication of the significance of the coprophilous microflora in energy flow schemes. J. Ecology 62: 429-437.

———. and D.T. Wicklow. 1975. Relationships between coprophilous fungi and faecal substrates in a Colorado grassland. Mycologia 67: 63-75.

Bunnell, F. and D. Tait. 1975. Mathematical simulation models of decomposition processes. pp. 207-225. *In:* Soil Organisms and Decomposition in Tundra. A.J. Holding, O.W. Heal, S.F. MacLean, Jr., P.W. Flanagan (eds.). Tundra Biome Steering Committee. Stockholm, Sweden. 398 pp.

Flanagan, P.W. and A.K. Veum. 1975. Relationships between respiration weight loss, temperature and moisture in organic residues in tundra. pp. 249-277. (See Bunnell and Tait 1975.)

Heal, O.W. and D.D. French. 1975. Decomposition of organic matter in tundra. pp. 279-309. (See Bunnell and Tait 1975.)

Lodha, B.C. 1974. Decomposition of digested litter. pp. 213-241. *In:* Biology of Plant Litter Decomposition. C.H. Dickinson and G.J.F. Pugh (eds.). Academic Press, New York. 775 pp.

Longton, R.E. and M. MacIver. 1977. Climatic relationships in Antarctic and northern hemisphere populations of a cosmopolitan moss, *Bryum argentium* Hedw. *In:* Proceedings of the Third S.C.A.R.-I.U.B.S. Symposium on Antarctic Biology. G.A. Llano (ed.). Washington, D.C. In press.

McCalla, T.M., L.R. Frederick and G.L. Palmer. 1971. Manure decomposition and fate of breakdown products in soil. pp. 231-255. *In:* Agricultural Practices and Water Quality. T.L. Willrich and G.E. Smith (eds.). Iowa State Univ. Press. Ames. 415 pp.

Nicholson, P.B., K.L. Bocock and O.W. Heal. 1966. Studies on the decomposition of the faecal pellets of a milliped (*Glomeris marginata* Villers). J. Ecology 54: 755-766.

Parkinson, D. and E. Coups. 1963. Microbial activity in a podsol. pp. 167-175. *In:* Soil Organisms. J. Doeksen and J. van der Drift (eds.). North-Holland Publishing Company, Amsterdam. 453 pp.

Reichle, D.E. 1975. Advances in ecosystem analysis. BioScience 25: 257-264.

———., R.V. O'Neill, S.V. Kaye, P. Sollins and R.S. Booth. 1973. Systems analysis as applied to modelling ecological processes. Oikos 24: 337-343.

Taiganides, E.P. and T.E. Hazen. 1966. Properties of farm animal excreta. Trans. Am. Soc. Agr. Eng. 9: 374-376.

Tener, J.S. 1965. Musk-oxen in Canada. Canadian Wildf. Serv., Monogr. No. 2. 166 pp.

Waksman, S.A., T.C. Cordon and N. Hulpoi. 1939. Influence of temperature upon the microbiological population and decomposition processes in composts of stable manure. Soil Sci. 47: 83-114.

Waterhouse, D.F. 1974. The biological control of dung. Sci. Am. 230: 100-109.

Webster, J. 1970. Coprophilous fungi. Presidential Address, Trans. Br. Mycol. Soc. 54: 161-180.

White, E. 1960. The distribution and subsequent disappearance of sheep dung on pennine moorland. J. Anim. Ecology 29: 243-250.

Wicklow, D.T. and J. Moore. 1974. Effect of incubation temperature on the coprophilous fungal succession. Trans. Br. Mycol. Soc. 62: 411-415.

Williams, S.T. and T.R.G. Gray. 1974. Decomposition of litter on the soil surface. pp. 611-632. *In:* Biology of Plant Litter Decomposition. C.H. Dickinson and C.J.F. Pugh (eds.). Academic Press, New York. 775 pp.

Growth and survival characteristics of three arctic bacteria on Truelove Lowland

Louise Nelson

Introduction

The arctic region is a unique habitat characterized by low temperatures, low precipitation, and the presence of permafrost. As a result of these factors, many tundra soils tend to be poorly drained and retarded in development (Boyd 1967). The microbial populations of these soils play an important role in soil formation, as the primary decomposers. Considerable emphasis in bacteriological studies of tundra soils has been placed on biomass, species lists, and physiological groupings (Boyd 1967, Latter and Heal 1971, Boyd and Boyd 1971). Very little is known of how environmental factors can affect individuals or groups within the microbial populations. Boyd (1967) has stated that such factors as temperature, moisture, pH and low oxygen levels may be limiting microbial growth in polar regions. The concentrations of readily available nurtrients are believed to be low in most soils, thus acting to limit microbial activity (Gray and Williams 1971). Walker and Peters (this volume) have stated that the available nutrient status of Devon Island soils is low when compared with more temperate soils. It was felt that a more intensive study of several arctic bacterial isolates from soil under controlled conditions in the laboratory might provide valuable information for the estimation of potential activity levels in nature as well as some insight into bacterial adaptation to the arctic environment.

The material presented in this chapter represents the preliminary results of laboratory studies performed on three bacterial isolates (*Pseudomonas* sp., *Bacillus* sp., *Arthrobacter* sp.) from the Intensive Hummocky Sedge-moss Meadow Site. The effects of temperature, limiting nutrients, freeze-thawing, and starvation on growth and survival of these three isolates were assessed. Data on the effects of spring thaw on quantitative and qualitative changes in the soil microbial populations in the field are presented.

Methods

Species

The three species of bacteria studied were isolated from the Hummocky Sedge-moss Meadow Intensive Study Site by means of the pour plate technique. The three isolates and their code numbers are *Pseudomonas* sp. (M216), *Bacillus* sp. (M153) and *Arthrobacter* sp. (M51). Pseudomonad and coryneform groups were frequently isolated from the meadow (10% and 29% of the total isolates respectively), whereas the Bacillus group was found rarely in the meadow (2% of the total isolates) and more commonly in the Raised Beach Intensive Study Site (10%) (Widden this volume). Originally M51 was identified as *Corynebacterium* sp. and was included with this group in the study of Widden (this volume). It has more recently been reidentified as *Arthrobacter* sp. by which name it will be referred to in this paper. Stock cultures of M153 and M216 were maintained by freezing to −70° C in 15% glycerol. Stocks of M51 were maintained by lyophilization.

Cultivation media

Peptone-yeast extract medium (PYE) containing 2.5 g peptone, 0.5 g yeast extract and a trace of ferric orthophosphate in 1 *l* distilled water (pH 7.0) was used for growth in liquid culture, as a diluent, and for viable plate counts with the addition of 15 g of agar.

A partially synthetic medium (MK) consisting of 5.0 g NaCl, 0.2 g $MgSO_4$ $7H_2O$, 0.68 g KH_2PO_4, 3.48 g K_2HPO_4, 0.5 g $(NH_4)_2HPO_4$, 0.5 g $NaNO_3$, 0.5 g yeast extract and 0.5 g glucose in 1 *l* of distilled water was used for the growth studies. The pH of MK was about 6.9, which was comparable to the meadow soil pH. Glucose was filter sterilized at 25% concentration and added to the other components which had been previously autoclaved for 15 min at 15 p.s.i. The MK medium was made carbon-limiting by reducing glucose levels to 0.1 g l^{-1} and yeast extract to 0.1 g l^{-1}. It was made nitrogen-limiting by decreasing levels of $(NH_4)_2$ HPO_4, $NaNO_1$ and yeast extract to 0.1 g l^{-1} of each.

Cells in liquid culture were shaken in 250 ml Erlenmeyer flasks containing 50 ml of medium, at 100 rpm on a New Brunswick psychrotherm rotary shaker.

Cell and colony counts

Total cell counts were determined in a Petroff-Hauser counting chamber in which a minimum of 20 fields or 100 bacteria were counted for statistical reliability (Mallette 1969).

Viable counts were determined by the spread plate method using PYE agar and 0.1 ml of inoculum from the appropriate dilution.

Biological analysis

The total protein content of the cells was determined by the method of Lowry et al. (1951). Washed suspensions of whole cells were first heated to 100° C in 1% deoxycholate in 0.5% NaOH for 5 min. Bovine serum albumin was used as a standard.

The total hexose carbohydrate content of the washed cells was determined by the anthrone method as referred to by Herbert et al. (1971). Glucose was used as the standard.

The RNA content of the cells was determined by the orcinol method as modified by Herbert et al. (1971) after extraction with $HC10_4$ by the modified Schneider method (Herbert et al. 1971). Ribose was used as the standard. Cells were harvested by centrifugation and washed with 0.1 M $MgCl_2$ prior to treatment.

Total cell dry weight was determined by filtering the cells onto a Millipore membrane (0.22 μ pore size) which had been pre-dried and weighed. The membrane with washed bacteria was dried overnight at 95° C and stored in a desiccator prior to weighing. A standard curve relating dry weight to optical density was constructed for each species grown in complete MK medium.

Respiration

Oxygen uptake rates were obtained using conventional Warburg flasks attached to a Gilson respirometer. Flasks were shaken at 70 oscillations per minute.

Growth rates

Growth rate studies at varying temperatures were carried out on the New Brunswick rotary shaker (except at −3° C where a reciprocating shaker was utilized) using the complete MK medium. Six to nine replicates of each organism were used at each temperature. Increases in optical density were measured on a Spectronmic 20 at 430 nm and then converted to dry weight using the standard curves.

Growth rates at varying carbohydrate levels were carried out in the carbon-limiting MK medium at glucose levels of 1.2 to 512 mg l^{-1} at 15° C. Four replicates of each isolate were used.

Freezing and thawing experiments

1. Laboratory experiments
Cultures of the three bacteria were grown in MK medium at 15° C, harvested in late exponential phase, washed in phosphate buffer, and resuspended to a density of approximately 10^7 cells per gram of soil when inoculated. Cells were distributed into tubes containing two types of sterile particulate media, 5.0 g of acid-washed sand or 0.5 g of air-dried, sieved soil from the intensive meadow site. The sand was sterilized by autoclaving and the soil by means of ethylene oxide. Both sand and soil tubes were divided into two treatments; one group was saturated with water and the second group was wetted only by the bacterial inoculum, 0.15 ml for the sand tubes and 0.5 ml for the soil tubes. Soil matric potentials were determined using the pressure membrane method (Richards 1965) and were expressed in terms of bars. The inoculated tubes were then equilibrated overnight at 3° C. The temperature was subsequently lowered to 0° C, −5° C, −15° C and finally to −22° C, the tubes being held 24 hours at each temperature. Thawing was accomplished by warming the tubes to 3° C over a 4 hr period or by reversing the freezing procedure (4 days). Microbial viability for each treatment (2-3 replicates per sample) was assessed at 3 times (prior to freezing, immediately after freezing and after 3 months at −22° C) by diluting the particulate media in a 1/4 strength PYE medium and spreading 0.1 ml of an appropriate dilution onto PYE agar plates. Concurrently, unsterilized, air-dried, sieved soil at the two matric potentials was subjected to the same freeze-thaw treatments and the viability of the indigenous bacterial population determined.

2. Field experiments

This work was carried out during late June and early July of 1974 at the intensive hummocky sedge-moss meadow site from which the laboratory isolates (M216, M153 and M51) originally came. The top 5 cm of this site was sampled seven times over a period of 15 days. At the time of the first sample the site was covered with snow 18-30 cm thick. Bare patches occurred 4 days later and 10 days after the initial sample the study site was completely bare of snow. Five replicate soil cores were taken at each sample date; these were kept outside until ready for use and processed within 10 hr of collection. Five gram samples from each core were homogenized with 95 ml of sterile, pre-cooled peptone water (0.13%) in an Osterizer at medium speed for 3 min. Serial dilutions were prepared using pre-cooled 9 ml peptone water blanks and 0.1 ml of an appropriate dilution were spread onto pre-dried half-strength PYE agar plates. Ten replicates were prepared at each dilution, 5 of these were incubated at about 5°C for 4 weeks and 5 at about 20°C for 2 weeks prior to the counting of the bacterial colonies appearing on the plates. A preliminary sample was taken 3 days prior to sample no. 1 in which only 3 replicates were obtained and the plates prepared incubated at 20°C only. A Jena dissection microscope at 7.5 × magnification was used to count the bacterial colonies. Soil moisture was determined for each core and was expressed in terms of percentage of wet weight. Soil temperature data were monitored but are not available and the mean daily minimum and maximum air temperatures at the base camp are presented.

Using a randomized grid method 120 bacteria were selected from each of four sets of dilution plates: (1) before thaw incubated at 5°C; (2) before thaw incubated at 20°C; (3) after thaw incubated at 5°C; and (4) after thaw incubated at 20°C. These 480 bacteria were then restreaked onto half-strength PYE agar plates to obtain pure cultures. Those that grew were inoculated into half-strength PYE broth and shaken at 100 rpm. The isolates were then placed into 11 taxonomic groups on the basis of the following criteria: colony morphology and pigmentation, Gram reaction and morphology of young and old cells, cell motility, oxidase reaction, action on litmus milk and Hugh and Leifson's glucose agar, fluorescein production, and morphology on a rich and poor medium (A.M. Gounot, pers. comm.) for distinguishing *Arthrobacter* sp.

Starvation experiments

Prior to starvation bacterial cells of the three isolates were grown on three media (carbon-limiting MK medium, nitrogen-limiting MK medium, and PYE medium). Cells were harvested in the late exponential phase, washed aseptically 3 times in carbon-free medium and 2 ml aliquots added to 250 ml Erlenmeyer flasks containing 100 ml of carbon-free (CF) medium (2.0 g KNO_3, 0.8 g K_2HPO_4, 0.2 g KH_2PO_4, 0.2 g $MgSO_4$ $7H_2O$ and 0.2 g NaCl in 1 l distilled water). The flasks were shaken at 15°C. Three replicates of each treatment were utilized. Viable cell counts were determined over a period of 30 days. Dry weight, cell protein, carbohydrate, and RNA contents were determined over the first 48 hr for M216 (*Pseudomonas* sp.) only. Respiration rates of growing and starved cells at 15°C were also measured.

Table 1. Maximum specific growth rates and generation times at varying temperatures for three bacterial species grown on MK medium.

Temperature (°C)	*Pseudomonas* sp. $\mu_{max}(hr^{-1})$	*Pseudomonas* sp. g (hr)	*Bacillus* sp. $\mu_{max}(hr^{-1})$	*Bacillus* sp. g (hr)	*Arthrobacter* sp. $\mu_{max}(hr^{-1})$	*Arthrobacter* sp. g (hr)
-3	.02	32.1	no growth		.01	62.7
0	.05	15.3	.02	43.1	.02	34.5
3	.16	4.2	.04	19.2	.06	12.3
10	.26	2.6	.18	3.9	.09	7.6
15	.38	1.8	.25	2.8	.14	5.1
20	.62	1.1	.42	1.7	.15	4.5
25	.65	1.1	.54	1.3	.17	4.1
30	.75	0.9	.55	1.3	.22	3.1
35	.32	2.2	.70	1.0	no growth	
40	no growth		.55	1.3	no growth	

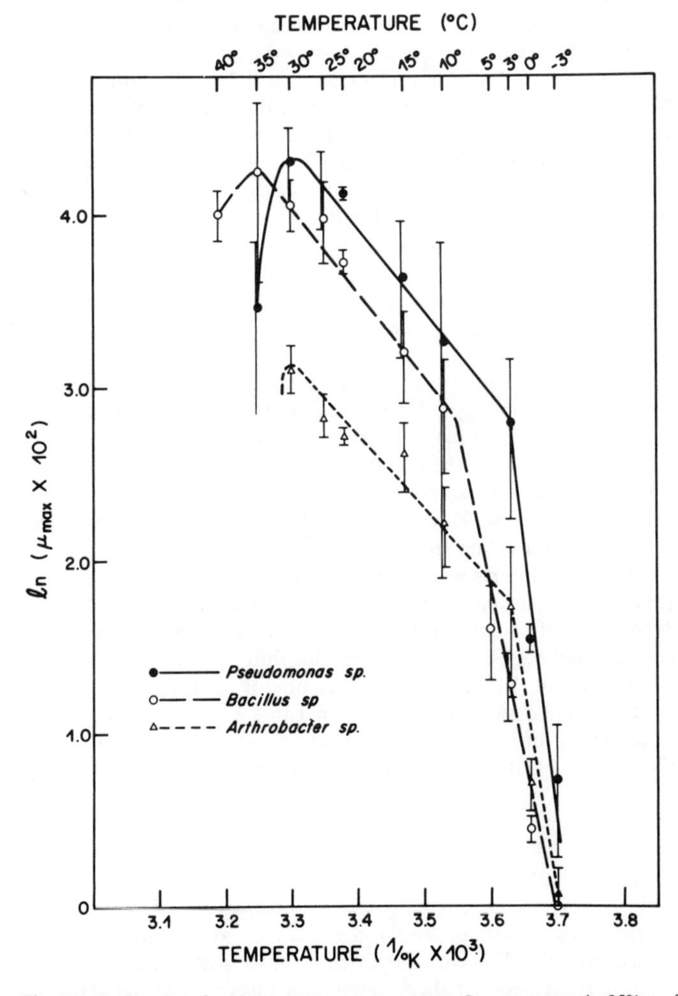

Fig. 1 Arrhenius plot of growth rate versus temperature. Bars represent the 95% confidence limits for each mean maximum specific growth rate determined.

Results

Effect of temperature on growth rate

The maximum specific growth rates and generation times for *Pseudomonas* sp., *Bacillus* sp., and *Arthrobacter* sp. from −3° C to 40° C in MK medium are presented in Table 1. The specific growth rate, u, is the slope of the plot of the natural logarithm of dry weight versus time during the exponential phase of growth. The specific growth rate has been shown to vary with inoculum age, inoculum size, and with different batches of medium. For this reason u was determined 2-3 times over a year and the figures presented are mean values. Only *Pseudomonas* sp. and *Arthrobacter* sp. grew at temperatures below 0° C. *Bacillus* sp., on the other hand, was the only isolate capable of growth at temperatures above 35° C.

In Fig. 1 the natural logarithm of u_{max} was plotted against the inverse of the absolute temperature to obtain an Arrhenius plot. In such a plot a linear portion is obtained which extends from a temperature slightly below the optimum growth temperature to a temperature several degrees above the minimum growth temperature (Harder and Veldkamp 1971). The temperature characteristic of growth is obtained from the slope of the linear portion of the plot by multiplying the slope by the gas constant, R, and corresponds to the energy of activation of a chemical reaction. The temperature characteristics of growth were calculated for the three isolates under study and these are presented in Table 2. *Arthrobacter* sp. had the lowest temperature characteristic and *Bacillus* sp. the greatest.

Effect of carbohydrate level on growth rate

The effect of glucose concentration on growth rate was assessed for each isolate and the results plotted in Fig. 2. On the basis of these data, the K_s values for each organism may be determined. K_s is defined as the concentration of growth-limiting substrate at which the growth rate is equal to half of the maximum rate. It may be viewed as an indicator of the efficiency of growth at low substrate concentrations. The K_s values of the three Devon Island isolates are presented in Table 3 with values from the literature for several other species. The Devon Island isolates were less efficient than a marine isolate, *Spirillum* sp. but were comparable to the other isolates listed.

The plot of the yield of bacteria obtained on a dry weight basis versus the amount of glucose utilized results in a linear relationship, the slope of which is equal to the yield coefficient, Y. The yield coefficients for the three isolates under study were determined and are presented in Table 4. All three isolates exhibited yield coefficients of 35-50%. This is in accord with the figures utilized by Gray and Williams (1971) in their calculation of microbial growth rates in nature.

Freeze-thaw studies

The viabilities of the three isolates in sand and soil at four soil moisture matric potentials, after freezing to −22° C with subsequent thawing after 24 hr or 3 months at −22° C, are presented in Table 5. Two thaw rates were tested at the 3 month sample. The percentage survival of the natural bacterial population in

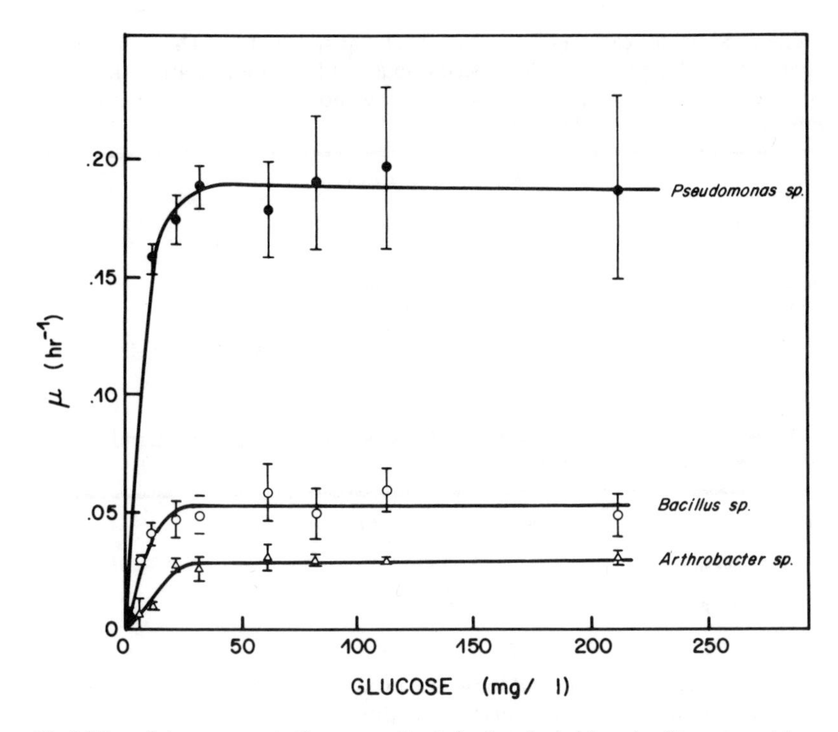

Fig. 2 Effect of glucose concentration on growth rate for three bacterial species. Bars represent the 95% confidence limits for each mean specific growth rate determined.

Table 2. Temperature characteristics of growth for three bacterial species as calculated from an Arrhenius plot.

Species	*Temperature characteristic (Kcal mole^{-1})*
Pseudomonas sp.	9.5
Bacillus sp.	10.0
Arthrobacter sp.	7.7

air-dried Devon meadow soil at two matric potentials also appear in Table 5. The data show that all three isolates had a higher survival rate in drier conditions. Three months at −22° C usually had the effect of lowering the number of survivors by a small percentage. Slow thawing decreased survival rates in most cases, particularly at higher matric potentials when water was more available. Of the three isolates studied *Arthrobacter* sp. appeared least susceptible to freeze-thaw damage.

The fluctuations in the numbers of viable aerobic heterotrophic bacteria during the 1974 spring thaw in the intensive hummocky sedge-moss meadow are presented in Fig. 3. The data shown are for the two incubation temperatures, 20° C and 5° C. The mean soil moisture at each sampling date and maximum and minimum air temperatures during the sampling period are also plotted in Fig. 3. Bacterial numbers reached a maximum on June 26. This coincided with the maximum soil moisture content measured, an indication that snowmelt was at its height. Numbers of bacteria appeared to decline subsequently but variation was high. There were generally slightly lower

Table 3. K_s values calculated for three bacterial species from Devon Island, compared with values for other species obtained from the literature. K_s is equal to the concentration of growth-limiting substrate at which the growth rate is equal to half of the maximum rate.

Species	*K_s (mg l^{-1})*	*CHO Source*	*Reference*
Devon Island *Pseudomonas* sp.	7.4	glucose	
Devon Island *Bacillus* sp.	8.0	glucose	
Devon Island *Arthrobacter* sp.	14.9	glucose	
Aerobacter aerogenes	16.0	sucrose	Canale et al. 1973
Spirillum sp. (marine)	3.0	lactate	Jannasch 1967
Pseudomonas sp. (marine)	8.0	lactate	Jannasch 1967
Aerobacter cloaceae	12.3	glycerol	Herbert et al. 1956

Table 4. Yield coefficients for three bacterial isolates from Devon Island grown on carbon-limiting MK medium.

Species	*Y (mg bacteria produced mg^{-1} CHO consumed)*
Pseudomonas sp.	0.402
Bacillus sp.	0.471
Arthrobacter sp.	0.378

Table 5. Effect of moisture and thaw rate on survival of soil bacteria after freezing and thawing under laboratory conditions.

Treatment		*% Survival (± standard error)*			
		Pseudomonas sp.	*Bacillus* sp.	*Arthrobacter* sp.	*Unsterile soil population*
1. Quick thaw immediately after freezing to −22°C					
	ψ*				
Sand	<−0.01	0.3± 0	N.D.**	19.8±13	—
Sand	−0.03	1.3± 0	52.5±15	49.0± 6	—
Soil	−0.10	59.0±18	65.2±18	109.1±21	22.4± 4
Soil	−1.00	180.2±20	144.2±18	174.5± 7	46.2± 6
2. Quick thaw after 3 months at −22°C					
Sand	<−0.01	N.D.	N.D.	N.D.	—
Sand	−0.03	N.D.	11.5± 4	6.4± 4	—
Soil	−0.10	44.0± 28	8.1± 2	106 ±19	11.4± 2
Soil	−1.00	329.0±140	80.7±10	168 ±64	42.9±22
3. Slow thaw after 3 months at −22°C					
Sand	<−0.01	N.D.	N.D.	N.D.	—
Sand	−0.03	N.D.	0.1± 0	1.8± 0	—
Soil	−0.10	N.D.	0.3± 0	104.4±27	12.5± 3
Soil	−1.00	6.9± 5	33.4±15	104.6±10	32.8±10

*ψ = soil moisture matric potential in bars
**N.D. = non detectable, no colonies present on 10^{-1} dilution.

numbers of bacterial on the plates incubated at 5°C. An analysis of variance was performed on the data presented in Fig. 3. The data were first transformed by taking the natural logarithm of the number plus one to achieve a normal distribution of the data. The results of the analysis are presented in Table 6. There was clearly no significant difference with time. However, the differences with temperature of incubation approached significance with the probability of

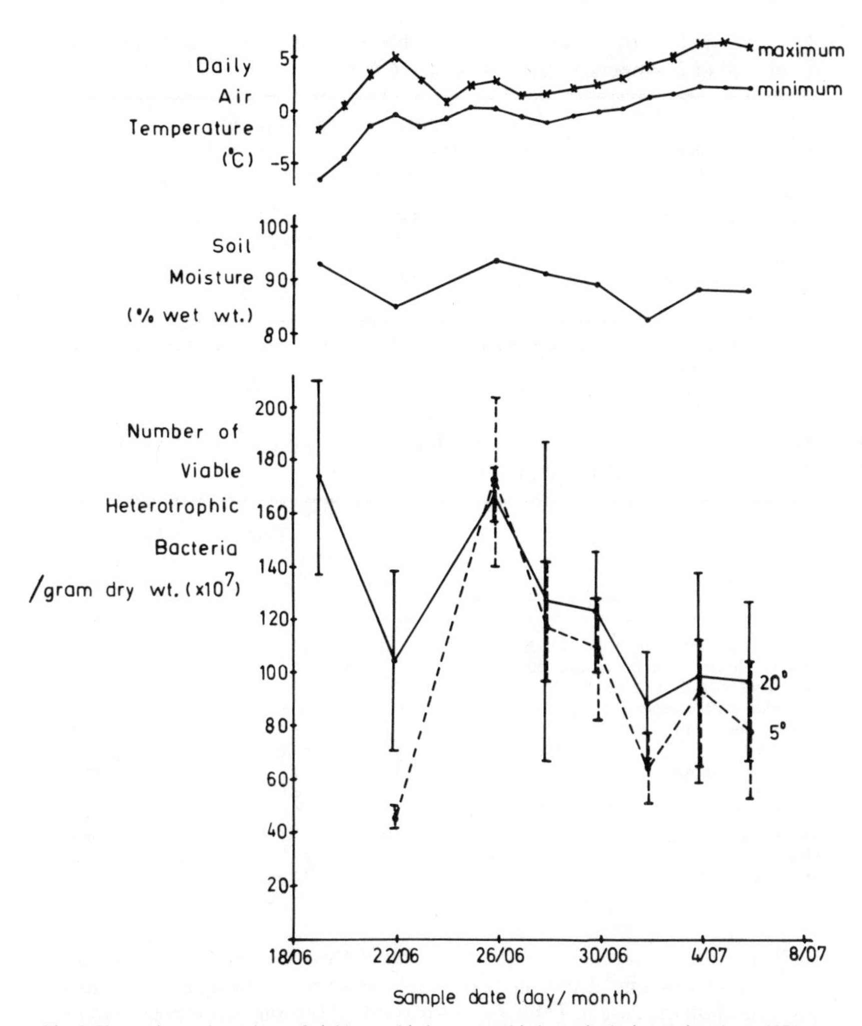

Fig. 3 Change in total numbers of viable aerobic heterotrophic bacteria during spring thaw, 1974 at the intensive hummocky sedge-moss meadow. Plates were incubated at two temperatures, 20° C and 5° C. Bars represent the standard errors for each determination. Soil moisture content (% wet weight) and daily mean maximum and minimum air temperatures during the thaw are also presented.

type I error (α ** = .056) being only slightly greater than the standard level of probability ($\alpha = 0.05$).

Data on the species of bacteria isolated in four treatments (before and after thaw at the two incubation temperatures) are presented in Table 7. The percentage of Gram negative and Gram positive cells are also shown. The majority of the isolates were Gram negative rods but there was a decrease of 7-14% after thaw. The *Pseudomonas* and *Cytophaga/Flexibacter* groups appeared more frequently on plates incubated at 5° C while the nonpigmented Gram negative rods were isolated more often from plates incubated at 20° C. After thaw there was a decrease in the proportions of the *Pseudomonas* and *Cytophaga/Flexibacter* groups and an increase in the proportions of orange pigmented rods, *Arthrobacter* and other coryneform groups from the 5° C-incubated plates. Trends were similar for the isolates from the plates incubated

Table 6. Analysis of variance of total viable bacteria counted with time and incubated at 2 temperatures, 20°C and 5°C.

Source of variation	*D.F.*	*2-Way Analysis of Variance* *Mean square*	*F*	*α*
Time	6	.087	0.90	.537
Temperature	1	.336	3.74	.056
Time × Temperature	6	.168	1.86	.107
Error	48	.389		
Adjusted error		.090		
Total	61			

α = probability of a Type I error (probability of concluding there is a significant difference where none actually exists).

Table 7. Percentage of isolates belonging to different groups before and after thaw from plates incubated at 5°C and 20°C.

	% of Total Viable Isolates			
	Before thaw		*After thaw*	
Genus or group	*5°C plates*	*20°C plates*	*5°C plates*	*20°C plates*
Pseudomonas	28.7	16.3	18.0	6.5
Yellow pigmented Gram negative rods	23.0	27.5	20.0	24.7
Orange pigmented Gram negative rods	5.7	5.0	13.0	12.9
Nonpigmented Gram negative rods	17.2	38.8	22.0	34.4
Cytophaga/Flexibacter	16.1	0.0	5.0	1.1
Chromobacterium	1.1	0.0	0.0	1.1
Arthrobacter	3.4	8.8	9.0	5.4
Other coryneforms	3.4	1.2	10.0	14.0
Streptomyces	1.1	0.0	0.0	0.0
Bacillus	0.0	0.0	1.0	0.0
Gram positive cocci	0.0	2.5	2.0	0.0
Gram negative bacteria	92.0	87.5	78.0	80.6
Gram positive bacteria	8.0	12.5	22.0	19.4

at 20°C with the exception of the *Cytophagal Flexibacter* group which rarely occurred at any time at this incubation temperature. A chi-square test was performed on the data in Table 7 and the result of the analysis is presented in Table 8. For this analysis the *Arthrobacter* and other coryneforms groups were combined into a single group and the *Chromobacterium*, *Streptomyces*, *Bacillus* and Gram positive cocci groups were also combined into a single group. There were significant differences in the proportions of each taxonomic group with temperature and with time of sampling.

Starvation studies

The viabilities of the three laboratory isolates (*Pseudomonas* sp., *Bacillus* sp. and *Arthrobacter* sp.) in carbon-free medium over 30 days were plotted in Fig. 4. The total count of *Pseudomonas* cells previously grown in carbon- and nitrogen-limiting media increased with time in the CF medium. *Bacillus* sp. died off very rapidly whereas *Arthrobacter* sp. cells remained more than 50% viable for a minimum of 10 days. The number of spores formed by *Bacillus* sp. in CF medium was determined by heating samples to 65°C for 15 min, spreading an aliquot of the sample onto PYE agar plates, and counting the

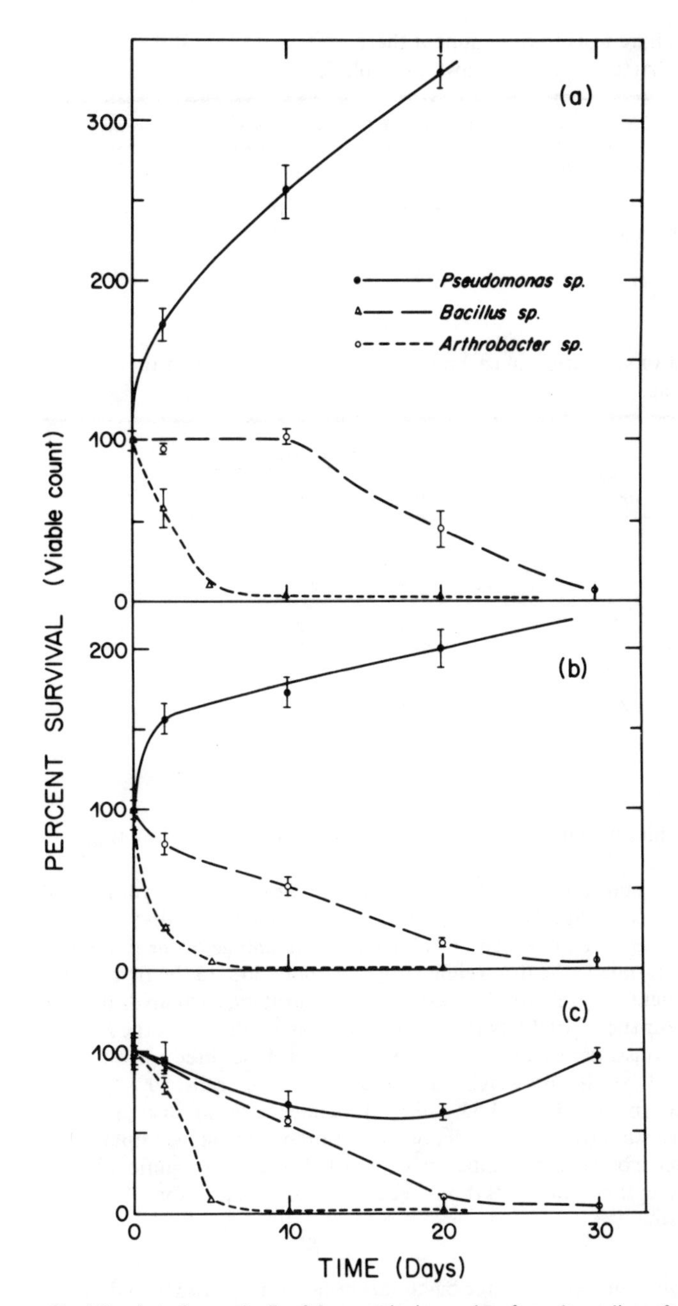

Fig. 4 Survival of starved cells of three species in a carbon-free salts medium after growth on three types of media: (a) carbon-limiting medium; (b) nitrogen-limiting medium; (c) complete medium (PYE). Standard error bars are presented for each point on the graph.

Table 8. Chi-square tests to determine if there were significant differences in distribution of bacterial groups shown in Table 7.

	Before thaw 5° C compared with before thaw 20° C	*After thaw 5° C compared with after thaw 20° C*	*Before thaw 5° C compared with after thaw 5° C*	*Before thaw 20° C compared with after thaw 20° C*
Chi-squared	23.59	11.56	16.42	10.88
Degrees of freedom	6	6	6	6
Level of significance (α)	0.001	0.07	0.01	0.09

Table 9. Effect of starvation in carbon-free medium on cell constituents of *Pseudomonas* sp.

Growth medium before starvation	*Hours after starvation*	*% initial dry weight*			
		Dry Wt.	*Protein*	*CHO*	*RNA*
CL	0	100	24	1.6	9.6
	20	71.5	4.9	0.4	2.2
	45	59.2	4.6	0.1	2.1
NL	0	100	18.8	3.2	7.2
	20	58.4	4.7	0.5	1.1
	45	60.6	4.1	0.2	0.9
PY	0	100	31.1	2.6	11.6
	20	65.9	7.2	0.7	0.5
	45	64.1	8.2	0.8	0.8

number of colonies which appeared. Spores were not detected at any time during the study.

The effect of starvation on dry weight, protein, hexose carbohydrate, and RNA levels in cells of *Pseudomonas* sp. was assessed during the first 48 hr of starvation. The data are presented in Table 9 as a percentage of the initial dry weight. In all treatments the dry weight dropped about 40% in the first 20 hr. Concurrently there was a sharp decrease in cellular protein, carbohydrate, and RNA levels. Over the second day dry weight loss was small as was the further loss of cellular constituents. The loss in dry weight of these three constituents accounted for 80% of the dry weight loss in carbon-limited cells, 50% of the loss in nitrogen-limited cells, and 85% of the loss in cells grown in PYE medium prior to starvation. The effect of starvation on the oxygen uptake rates of the three isolates is presented in Table 10. There were significant decreases in the uptake rates in carbon-free media, particularly for *Pseudomonas* sp.

Table 10. Respiration rates of three bacterial species during growth and during incubation in carbon-free medium at 15°C

Species	*Respiration rate ($\mu l\ O_2\ mg^{-1}\ hr^{-1}$)*		
	Growing state	*Starvation state*	*(% of growing rate)*
Pseudomonas sp.	492	10	(2)
Bacillus sp.	248	69	(28)
Arthrobacter sp.	144	36	(25)

Discussion

Growth rates

All three isolates, *Pseudomonas* sp., *Bacillus* sp. and *Arthrobacter* sp. could be classified as facultative psychrophiles on the basis of the temperature range in which they grew. Their optimum growth temperatures were above 20° C but all were capable of growth below 5° C. In this regard *Bacillus* sp. appeared to be less cold tolerant in that it would not grow at 0° C or below. This may account in part for its smaller representation in the meadow site (Widden this volume). However, as these organisms were originally isolated by the pour plate method with subsequent incubation at room temperature, a population with mesophilic tendencies and a resistance to higher temperatures may have been selected for. The isolates studied here, therefore, may not be truly representative of the average arctic soil microbial populations. The tolerance to low temperature and the relatively fast growth rates at these temperatures, for *Pseudomonas* sp. and *Arthrobacter* sp. in particular, however, do imply some affinity for a cold environment.

Attempts have been made to distinguish mesophiles from psychrophiles on the basis of their respective temperature characteristics of growth, mesophiles reportedly exhibiting higher values than psychrophiles (Ingraham 1958, Baig and Hopton 1969). Others (Shaw 1967, Hanus and Morita 1968) have shown no difference between psychrophiles and mesophiles with respect to temperature characteristic of growth. It is interesting to note, however, that *Bacillus* sp., the least tolerant of low temperatures of the three isolates studied, also had the highest temperature characteristic.

Of the three isolates under study, *Pseudomonas* sp. grew at the greatest rate at all temperatures and possessed the lowest K_s value. *Arthrobacter* sp. had the lowest growth rate of the three isolates, except at temperatures below 5° C, and also the highest K_s value. However, in the study by Widden (this volume), the coryneform group, which would include *Arthrobacter*, made up 29% of the meadow isolates. It would appear that a fast and efficient growth rate does not necessarily in itself provide an advantage in the natural environment. Jannasch (1967) observed the existence of a positive correlation between growth rate and K_s value for several species. In this study there appeared to be a negative correlation. Jannasch (1968) also found that upon repeated transfer onto nutrient rich media, isolates which had formerly possessed low K_s and low growth rate values, lost these characteristics and became adapted to higher substrate concentrations. As a result both K_s and u values increased. As the isolates under study were initially isolated and subsequently cultured on relatively nutrient-rich PYE medium, the K_s and u values found under laboratory conditions may not be representative of those for bacteria in the field.

Freeze-thaw studies

1. Laboratory experiments

The availability of water and the rate of thaw appeared to have a significant effect on the lethality of freezing and thawing for the organisms under study in particulate medium. The mechanism of damage to cells as a result of freezing and thawing is unknown but such factors as dehydration, formation and growth of intracellular ice crystals, and concentration and precipitation of intracellular solutes have been proposed (Mazur 1970). Slow thawing would allow for regrowth of intracellular ice crystals and hence would be more deleterious to the cell (Mazur 1970). This has shown to be of significance only for cells which were initially rapidly cooled (Mazur 1970, Calcott and MacLeod 1974). Although a slow freezing rate was utilized in this study, the slow thaw rate appeared to be significantly different and more deleterious to pure cultures of *Pseudomonas* sp. and *Bacillus* sp. than fast thawing, but had no effect on survival of *Arthrobacter* sp. or the indigenous soil population as a whole. The effect of lower matric potentials and hence more water would be to increase the probability of cell damage due to the transformation of the water to ice both intra- and extracellularly. These factors can be postulated to account to a great extent for the increased mortality of cells at lower matric potentials and slow thaw rates. However, freezing and thawing appeared to have had a more lethal effect on cells in sand as opposed to soil, than can be accounted for by differences in matric potential alone. The cell membrane is believed to be the main target of freezing damage (Mazur 1970). The soil solution may contain a number of macro molecules capable of protecting cells from freeze-thaw damage and adsorption of bacterial cells to soil particles may act to protect the cell surface

The data presented here suggest that different bacterial species respond quite differently to freezing and thawing. Biederbeck and Campbell (1971) found the survival of vegetative bacterial cells and in particular the chromogens in a grassland soil, to be very low following freezing of the soil and subsequent thawing at fluctuating low temperatures. They postulated that freezing may weaken the physiological vigor of the vegetative cell, making it more susceptible than dormant cells to the effects of thawing at alternating low temperatures. Remoistening air-dried soil has been shown to give rise to a burst of microbial activity (Stevenson, 1956). The remoistening of the dried meadow soil prior to freezing may have reactivated the previously dormant microbial cells and increased their susceptibility to freeze-thaw damage.

2. Field studies

Results from the field work during the time of the spring thaw showed some similarities to the laboratory experiments. The chi-square tests (Table 8) showed significant qualitative differences with temperature of incubation and with time, before and after thaw. The *Pseudomonas* and *Cytophaga/Flexibacter* groups appeared less frequently after thaw while the *Arthrobacter*, other coryneform and orange-pigmented Gram negative rod groups occurred more frequently. If the isolates studied in the laboratory can be considered typical of the meadow soil *Pseudomonas* and *Arthrobacter* groups, one would have predicted the percentage of pseudomonads to drop after thaw with little or no mortality for the *Arthrobacter* group. Field studies supported this hypothesis. The reason for the greater tolerance of *Arthrobacter* sp. to freezing and thawing is unknown.

No statistical difference in total numbers of bacteria could be demonstrated from the time immediately preceding the spring thaw until some 12 days after the thaw had occurred. However, a maximum was reached at the time of greatest snowmelt. Variation among replicates was high, due partially to the varying microtopography of the meadow and to the high soil moisture levels to which calculations of bacterial numbers on a gram dry weight basis are particularly sensitive. These factors may have obscured any apparent trends. The total numbers of bacteria which grew on the plates however, were 2-10 fold higher than those obtained by Widden (this volume) sampling the same site and 10-10,000 fold greater than those obtained by workers studying other tundra sites (Holding et al. 1974). The use of the reciprocating shaker and the pour plate method by Widden (this volume) as opposed to the blender and the spread plate method used in this study may account for some of the differences. Clark (1967) and Klein et al. (1974) found that the pour plate technique generally resulted in a total count 2-5 fold less than that obtained using the spread plate method. If the figures for bacterial numbers at the 0-5 cm depth obtained by Widden (this volume) during July and August of 1971 are multiplied 2-5 times, the data are still significantly lower than those obtained during the thaw period of 1974, with the exception of the 14 July, 1971 sample. It is possible, therefore, that the data presented in Fig. 3 may represent the seasonal peak in bacterial numbers, a situation similar to that found on Signy Island (Baker 1970) and at Barrow, Alaska (Boyd 1958). The percentage of Gram negative cells found during the thaw was 40% higher than that found by Widden (this volume) at the same site later in the season. A high percentage of Gram negative cells has also been reported at Barrow (Holding et al. 1974). The use of the pour plate method may have selectively killed many of the Gram negative cells, which are believed to be more sensitive to heat shock than Gram positive cells.

The number of bacteria found in the Devon Island intensive meadow site seem unusually high when compared with other tundra sites (Holding et al. 1974). A positive correlation between high pH, high organic matter content, high phosphorus and calcium, and high bacterial counts in tundra regions has been shown by Holding et al. (1974). Such a correlation is possible for different sites on Devon Island and the data are presented in Table 11. The figures for bacterial numbers reported for the beach ridge and the plateau are the mean of five replicate soil cores and were weighted to account for the average plant cover at each site. In this case the amount of organic matter present seems most highly correlated with bacterial counts. The high pH of the meadow, unusual for a bog, may also allow for high bacterial populations.

The slightly higher counts obtained on plates incubated at 20°C suggest that there is a small proportion of the population which have strictly a mesophilic growth range and grow very slowly if at all at 5°C. Isolation data (Table 7) suggest that this group may be primarily represented by the Gram negative non-pigmented rods. The majority of the taxa, however, grew at both temperatures, indicative of a population which is facultatively psychrophilic. The three isolates under study in the laboratory were also in this category.

Table 11. Comparison of bacterial numbers/gram dry weight* with several soil factors at three sites on Devon Island (0-5 cm depth).

Location	*No. of bacteria/ gram dry weight*	*pH*	*Total organic carbon %*	*Soluble phosphorus*	*Calcium mg m^{-1}*
Intensive hummocky sedge-moss meadow	97.6×10^7	6.2	42.2	0	99.1
Intensive beach ridge	0.85×10^7	7.8	3.7	1	N.A.
Plateau	0.23×10^7	7.8	0.2	N.A.**	N.A.

* Bacterial numbers g dry wt obtained using the spread plate method and an incubation temperature of 5°C.

Data for soil factors from Peters and Walker (this volume).

** Not available.

Starvation studies

The ability of certain soil bacteria to withstand long periods of starvation has been well documented in the literature (Ensign 1970, Robertson and Batt 1973). Data on the survival times of 50% of the population under starvation conditions for several species of bacteria have been obtained from the literature and are presented in Table 12 with the data obtained for the three isolates under study. The presence of various storage substances, glycogen (van Houte and Jansen 1970), or poly-β-hydroxybutyrate (Stokes and Parson 1968), has been shown to increase the survival of some species under starvation stress. The very low rate of oxygen consumption which species such as *Arthrobacter crystallopoietes* (Ensign 1970) exhibit under starvation conditions is also a factor which may contribute to an organism's ability to withstand long periods of starvation.

Media in which different nutrients are limiting, have been shown to influence the quantity and nature of cellular storage products produced by the cell (Neidhardt 1963, van Houte and Jansen 1970). For this reason three types of pre-starvation media were chosen for cell growth. It was expected that a nitrogen-limiting medium might result in the accumulation of some polysaccharide storage product (Neidhardt 1963) which might then be degraded on starvation. The survival curves of each isolate on the three different pre-starvation media are difficult to interpret. Pre-growth in carbon-limiting medium resulted in longer survival for both *Pseudomonas* sp. and *Arthrobacter* sp., while pre-growth in complete medium increased the survival time of *Bacillus* sp. The increase with time in numbers of cells of *Pseudomonas* sp. from carbon-limiting and nitrogen-limiting media was unusual. Mackelvie et al. (1968) have reported a similar phenomenon for starved cells of *Pseudomonas aeruginosa* which were previously grown on complete medium. It is apparent, however, that the three isolates consistently responded differently to starvation stress. *Bacillus* sp. did not form spores under the growth conditions employed and was far more sensitive to starvation than *Arthrobacter* sp. and *Pseudomonas* sp., which were relatively resistant (Table 12).

The loss of cellular constituents from *Pseudomonas* sp. indicated that there was a rapid initial breakdown of cell protein, RNA, and hexose carbohydrate. This loss accounted for much of the total cell dry weight loss measured except in the case of cells previously grown in nitrogen-limiting medium. These cells may have possessed an unidentified storage product which was also being degraded at this time. The additional breakdown of this product had no apparent effect on survival of *Pseudomonas* sp. (Fig. 4). The major loss of dry weight and cellular constituents occurred in the first 20 hr; the rates of decrease stabilized to a low level after this time. This decrease in the rate of loss of cellular material coincided with the establishment of the low basal respiration rate by *Pseudomonas* sp. (Table 10). The ability of an organism to establish a low metabolic rate on starvation may be a key factor in its survival of starvation stress.

The data presented herein represent the results of a rather in depth laboratory study of three bacterial isolates from a tundra meadow soil. The intent of the study was an examination of general growth characteristics of the isolates and of their survival under conditions of starvation and freeze-thawing, factors which might be of importance for their successful survival in the arctic soil habitat. Extrapolation from the laboratory to the field is a dangerous if not impossible task. As mentioned elsewhere in the text, because of the methods of isolation and propagation utilized, it is doubtful that the

Table 12. Survival time of 50% of the cells of several bacterial species under starvation conditions, in carbon-free medium.

Species	*50% survival time (hr)*	*Reference*
Sphaerotilus discophorus	12	Stokes and Parson 1968
Starvation-susceptible strains from a soil	12	Chen and Alexander 1973
Starvation-resistant strains from a soil	12-96	Chen and Alexander 1973
Streptococcus mitis	22	van Houte and Jansen 1970
Escherichia coli	36	Dawes and Ribbons 1965
Sarcina lutea	65	Burleigh and Dawes 1967
Pseudomonas aeruginosa	96	Mackelvie et al. 1968
Nocardia corallina	80	Robertson and Batt 1973
Arthrobacter sp.	1680	Zevenhuizen 1966
Arthrobacter crystallopoietes	2400	Ensign 1970
Pseudomonas sp.	720	Devon Island
Bacillus sp.	78	Devon Island
Arthrobacter sp.	432	Devon Island

isolates under study are identical to or representative of the field populations. Despite these obstacles it appeared that at least several of the natural characteristics of the isolates had been maintained. Field studies during the spring thaw yielded data which correlated well with the results of laboratory freeze-thaw studies in terms of resistance of different species to freeze-thaw damage. All three isolates were tolerant of low temperatures and utilized low concentrations of a carbon source efficiently with respect to their growth rates and yields. The *Pseudomonas* sp. under study was very resistant to starvation stress. All of these factors are distinctive and one might postulate that possession of some of these might indeed be of potential advantage in coping with the arctic environment. While remembering the reservations with regard to extrapolation from the lab to the field, it is interesting, however, to note that the *Bacillus* sp. isolate which was less successful at coping with low temperatures, starvation, and freeze-thaw damage, represented a species which was rarely isolated in the intensive meadow site.

Acknowledgments

I would like to thank Dr. D. Parkinson for supervisory support and the National Research Council who provided support in the form of an N.R.C. post-graduate scholarship and N.R.C. operating grant No. A2257. The field work in the spring of 1974 was supported by the Arctic Institute of North America and the I.B.P. Devon Island Tundra Biome Project. Thanks are due to J.D. Bissett for his assistance with the statistical analyses.

References

Baig, I.A. and J.W. Hopton. 1969. Psychrophilic properties and the temperature characteristic of growth of bacteria. J. Bacteriol. 100: 552-553.

Baker, J.H. 1970. Yeast, moulds and bacteria from an acid peat on Signy Island. pp. 717-722. *In:* Antarctic Ecology, M.W. Holdgate (ed.), Vol. 2, Academic Press, London, 391 pp.

Biederbeck, V.O. and C.A. Campbell. 1971. Influence of simulated fall and spring conditions on the soil system. I. Effect of soil microflora. Soil Sci. Soc. Amer. Proc. 35: 474-479.

Boyd, W.L. 1958. Microbiological studies of arctic soils. Ecology 39: 332-336.
———. 1967. Ecology and physiology of soil microorganisms in polar regions. J.A.R.E. Scientific Reports, Special Issue 1: 265-275.
———. and J.W. Boyd. 1971. Studies of soil microorganisms of North Norway. Astarte 4: 69-81.
Burleigh, I.G. and E.A. Dawes. 1967. Studies on the endogenous metabolism and senescence of starved *Sarcina lutea*. Biochem. J. 101: 236-250.
Calcott, P.H. and R.A. MacLeod. 1974. Survival of *Escherichia coli* from freeze-thaw damage: a theoretical and practical study. Can. J. Microbiol. 20: 671-681.
Canale, R.P., T.D. Lustig, P.M. Kehrburger and J.E. Salo. 1973. Experimental and mathematical modelling studies of protozoan predation on bacteria. Biotech. Bioeng. 15: 707-728.
Chen, M. and M. Alexander. 1972. Resistance of soil microorganisms to starvation. Soil Biol. Biochem. 4: 283-288.
Clark, D.S. 1967. Comparison of pour and surface plate methods for determination of bacterial counts. Can. J. Microbiol. 13: 1409-1412.
Dawes, E.A. and D.W. Ribbons. 1965. Studies on the endogenous metabolism of *Escherichia coli*. Biochem. J. 95: 332-343.
Ensign, J.C. 1970. Long-term starvation survival of rod and spherical cells of *Arthrobacter crystallipoietes*. J. Bacteriol. 103: 569-577.
Gray, T.R.G. and S.T. Williams. 1971. Microbial productivity in soil. Symp. Soc. Gen. Microbiol. 21: 255-286.
Hanus, F.J. and R.Y. Morita. 1968. Significance of the temperature characteristic of growth. J. Bacteriol. 95: 736-737.
Harder, W. and H. Veldkamp. 1971. Competition of marine psychrophilic bacteria at low temperatures. Ant. van Leeuw. J. Microbiol. Serol. 37: 51-63.
Herbert, D., R. Elsworth and R.C. Telling. 1956. The continuous culture of bacteria; a theoretical and experimental study. J. Gen. Microbiol. 14: 601-622.
———. P.J. Phipps and R.E. Strange. 1971. Chemical analysis of microbial cells. pp. 209-344. *In:* Methods in Microbiology, J.R. Norris and D.W. Ribbons (eds.), Vol. 5b, Academic Press, London. 695 pp.
Holding, A.J., V.G. Collins, D.D. French, B.T. D'Sylva and J.H. Baker. 1974. Relationship between viable bacterial counts and site characteristics in tundra. pp. 49-77. *In:* Soil Organisms and Decomposition in Tundra, A.J. Holding, O.W. Heal, S.F. MacLean, Jr. and P.W. Flanagan (eds.), Tundra Biome Steering Committee. Stockholm. 398 pp.
Ingraham, J.L. 1958. Growth of psychrophilic bacteria. J. Bacteriol. 76: 75-80.
Jannasch, H.W. 1967. Growth of marine bacteria at limiting concentrations of organic carbon in seawater. Limnol. Oceanog. 12: 264-271.
———. 1968. Growth characteristics of heterotrophic bacteria in seawater. J. Bacteriol. 95: 722-723.
Klein, D.A. and S. Wu. 1974. Stress: a factor to be considered in heterotrophic microorganism enumeration from aquatic environments. Appl. Microbiol. 27: 429-431.
Latter, P.M. and O.W. Heal. 1971. A preliminary study of the growth of fungi and bacteria from temperate and antarctic soils in relation to temperature. Soil Biol. Biochem. 3: 365-379.
Lowry, O.H., N.J. Rosebrough, Q.L. Farr and R.J. Randall. 1951. Protein measurement with the Folin phenol reagent. J. Biol. Chem. 193: 265-275.

Mackelvie, R.M., J.J.R. Campbell and A.F. Gronlund. 1968. Survival and intracellular changes of *Pseudomonas aeruginosa* during prolonged starvation. Can. J. Microbiol. 14: 639-645.

Mallette, M.F. 1969. Evaluation of growth by physical and chemical means. pp. 521-566. *In:* Methods in Microbiology, J.W. Norris and D.W. Ribbons (eds.), Vol. 1. Academic Press, London. 712 pp.

Mazur, P. 1970. Cryobiology: the freezing of biological systems. Science 168: 939-949.

Neidhardt, F.C. 1963. Effects of environment on the composition of bacterial cells. Ann. Rev. Microbiol. 17: 61-86.

Richards, L.A. 1965. Physical condition of water in soil. pp. 128-152. *In:* Methods of Soil Analysis, Amer. Soc. Agron. Monograph 9. 615 pp.

Robertson, J.G. and R.D. Batt. 1973. Survival of *Nocardia corallina* and degradation of constituents during starvation. J. Gen. Microbiol. 78: 109-117.

Shaw, M.K. 1967. Effect of abrupt temperature shift on the growth of mesophilic and psychrophilic yeasts. J. Bacteriol. 93: 1332-1336.

Stevenson, I.L. 1956. Some observations on the microbial activity in remoistened air-dried soils. Plant and Soil 8: 170-182.

Stokes, J.L. and W.L. Parson. 1968. Role of poly-β-hydroxybutyrate in survival of *Sphaerotilus discophorus* during starvation. Can. J. Microbiol. 14: 785-789.

van Houte, J. and H.M. Jansen. 1970. Role of glycogen in survival of *Streptococcus mitis*. J. Bacteriol. 101: 1083-1085.

Zevenhuizen, L.P.T.M. 1966. Formation and function of the glycogen-like polysaccharide of *Arthrobacter*. Ant. van Leeuw. J. Microbiol. Serol. 32: 356-372.

Limnology

Limnology of some lakes on Truelove Lowland

C.K. Minns

Introduction

Three lakes were investigated, Immerk, Fish, and Loon Lakes, accounting for 22.75% of the surface water in the Lowland. A broad spectrum of measurements was taken from mid-May to mid-August 1973, to obtain at least a qualitative measure of productivity. Most of the work was carried out on Immerk Lake. Comparisons are drawn primarily with data collected in the Canadian I.B.P. project on Char Lake, Cornwallis Island, a moderately deep arctic lake.

Description and morphometry

The three lakes were sounded with a Furuno echo sounder. Depth profiles were taken along eight lines in Immerk and Fish Lakes and along six lines in Loon lake. One-metre contour maps were prepared (Fig. 1a, b, and c). For each lake a set of morphometric parameters was calculated (Hutchinson 1957) (Table 1).

All three lakes are shallow, with maximum depth 7 to 8.5 m. Morphometrically Loon Lake does not appear to be a typical lowland lake resulting from the development of beach ridges during isostatic rebound, but rather from minor faulting of granite adjacent to the granite and limestone cliff on the east side of the Lowland. Similar lakes appear to lie on the fault line running up the Truelove Valley.

The drainage areas of Immerk and Fish Lakes are exclusively within the

Table 1. Morphometric data for Immerk, Fish, and Loon Lakes.

Parameter	*Immerk*	*Fish*	*Loon*
Area (hectares)	95.79	100.52	15.92
Max. depth (m)	8.0	7.0	8.5
Mean depth (m)	3.24	3.07	2.86
Volume (cu m 10^3)	3107.05	3081.80	454.85
Shoreline (km)	4.79	4.36	1.66
Mean/max. depth	0.406	0.438	0.337
Volume development	1.217	1.314	1.009
Shoreline development	1.38	1.23	1.17

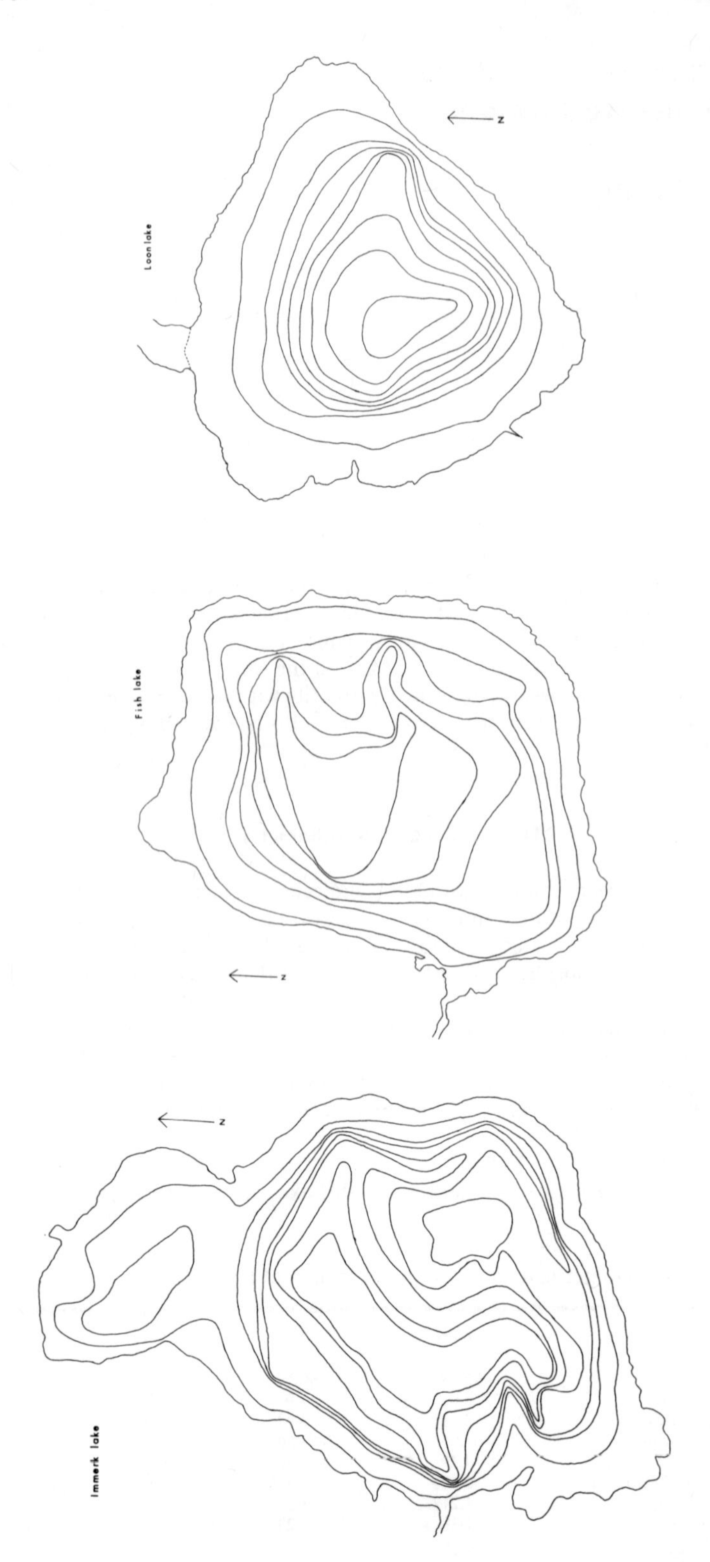

Fig. 1 One-metre contour maps of lakes studied.

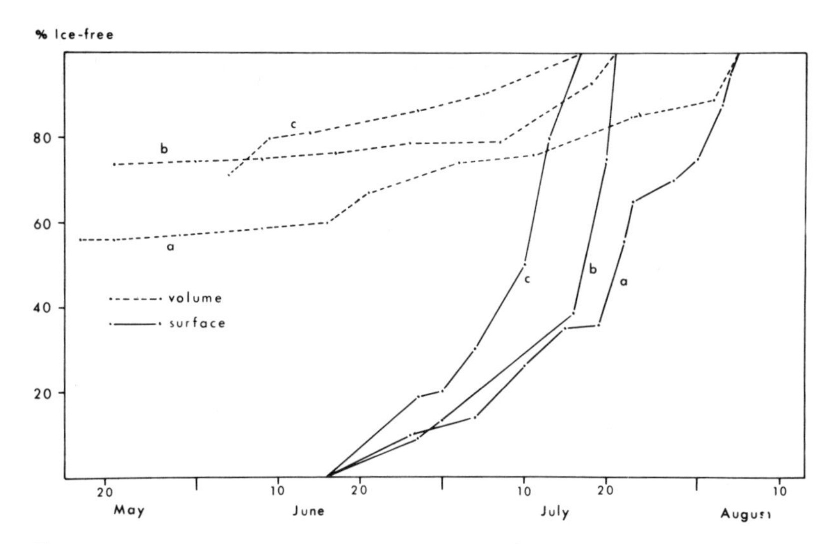

Fig. 2 Progress of ice disappearance by volume and surface area in the 1973 thaw. (a) Immerk Lake; (b) Fish Lake; (c) Loon Lake.

Table 2. Mean snow depth in early summer, prior to snowmelt.

Lake	*Mean snow depth (cm)*	*Number of samples*	*Sampling date*
Immerk	33	59	20 May
Fish	37	66	20 May
Loon	34	70	4 June
Phalarope	32	33	20 May

Lowland. Most of the drainage into Loon Lake, however, comes from the Gully River which drains from the Polar Desert atop the adjacent cliff.

Snow and ice

Snow cover was surveyed on four lakes in late May, just prior to snowmelt. Snow depth was even, with little drifting on the shores apart from some points around Loon Lake where rock outcrops reach the lake edge. The snow depths were similar on the four lakes (Table 2). The snow was well packed, having a melt ratio of 0.247 which represents a melt on the lake surface of 84.8 l m^{-2}.

Ice thickness varied greatly, ranging from 1.0 m on Muskox and Loon Lake to 1.75 m on Immerk and Phalarope Lakes. Within lakes, ice thickness varied up to 0.25 m. In 1972 ice persisted so that old ice made up an important component of the 1973 ice cover. The 1973 summer was warm and windy so that ice went out completely (Fig. 2). Of the three lakes studied, Loon was the first to become ice-free (17 July), followed by Fish Lake (21 July). Immerk Lake was the last lake on the Lowland to become ice-free (5 August).

The area of old ice cover was determined on the three lakes by pacing it out on Immerk and by planimetry of black and white photographs of Fish and Loon Lakes taken from the clifftop.

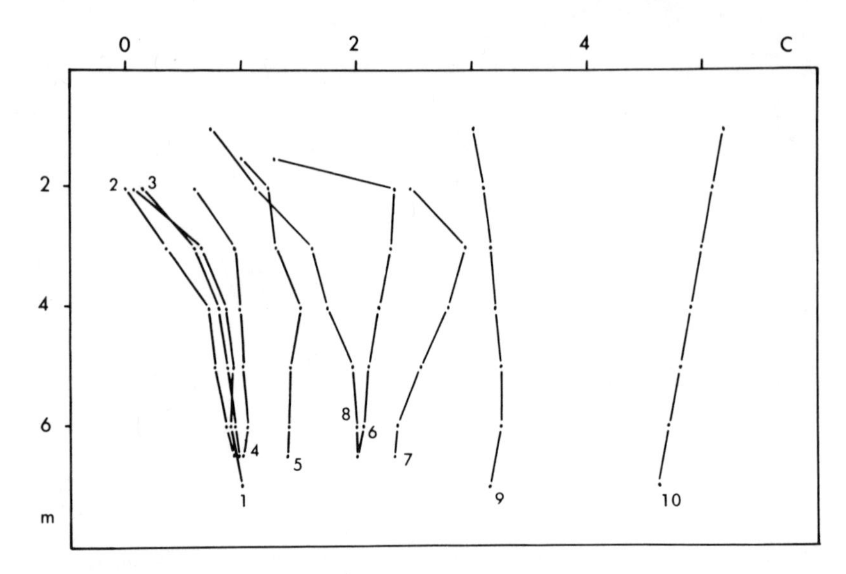

Fig. 3a Thermal regime in Immerk Lake, summer 1973.
1 *17/21 May* 2 *28 May* 3 *9 June* 4 *17 June* 5 *21 June* 6 *28 June* 7 *11 July* 8 *23 July* 9 *2 August* 10 *9 August*

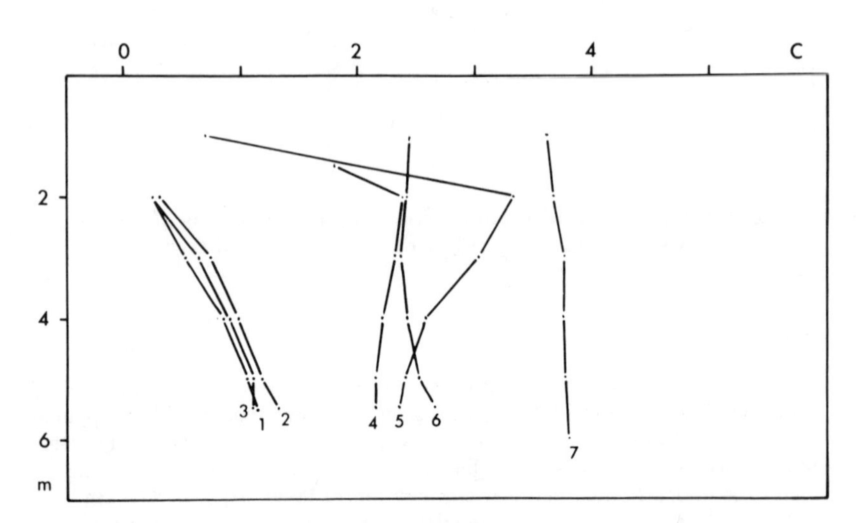

Fig. 3b Thermal regime in Fish Lake, summer 1973.
1 *21 May* 2 *29 May* 3 *8 June* 4 *26 June* 5 *7 July* 6 *18 July* 7 *30 July*

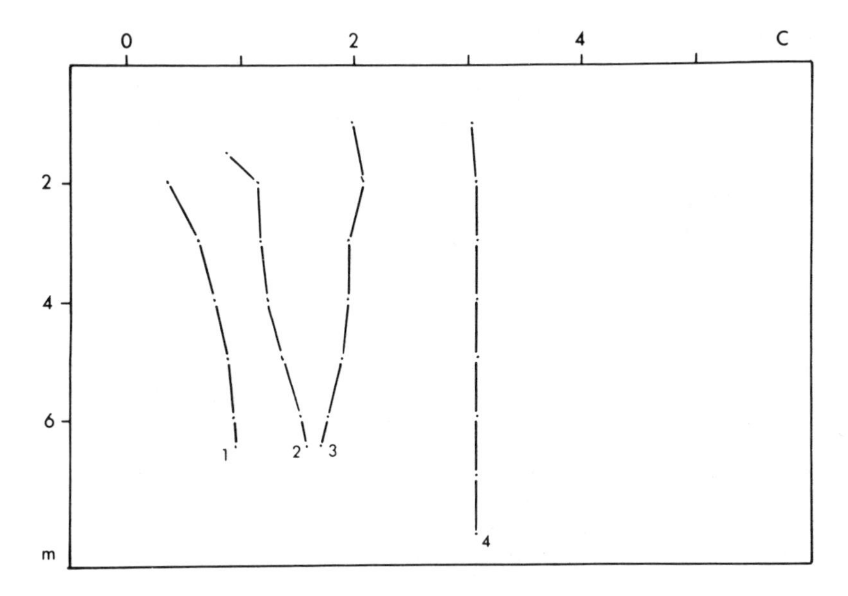

Fig. 3c Thermal regime in Loon Lake, summer 1973.
1 *9 June* 2 *27 June* 3 *5 July* 4 *28 July*

Thermal regime of lakes

Temperature profiles were taken at intervals in all three lakes (Fig. 3a, b, and c) using a thermistor probe. The lakes conform to the general pattern found in other shallow arctic lakes. In the spring the temperature is 0°C under the ice, increasing to approximately 1°C at the lake bottom. As the melt proceeds, the lakes warm with complete mixing taking place once the ice pan becomes a small part of the surface area. All the lakes probably reached 5°+ C before the end of the summer. When the ice pan is still relatively large, mixing is probably incomplete. This is indicated in Immerk and Fish Lakes in the middle of July when there were mid-water temperature maxima. Ponds thawed much faster than the lakes and reached temperatures of 8°+ C.

Hutchinson (1957) presented a review of attempts to classify the 'thermo-circulation' types of lakes. The lakes on Truelove Lowland do not fit into the scheme proposed. 'Cold monomictic' was intended to describe arctic lakes, i.e., one mixing period and the temperature never exceeding 4°C. These Lowland lakes appear to mix continuously in the summer (ice-free period) and exceed 4°C. More recently, Paschalski (1964) presented a definitive classification of circulation. The lakes are 'pleomictic' (continuous summer mixing) under this scheme, the temperature of the lake not being considered.

Conductivity

Conductivity profiles were taken at regular intervals. Samples were taken of inflows and outflows for the three lakes during runoff. The samples were read in Toronto in October and all results were converted to give conductivity at 18°C (Smith 1962). It has generally been found (Hem 1970, Hall 1971), that ln discharge and ln conductivity for runoff are inversely related, so a logarithmic mean of inflow conductivities was calculated in order to obtain an average approximately related to differing discharges. The pattern of time changes in conductivity of inflow, outflow and the lake as a whole are given for each lake (Fig. 4a, b, and c).

Hansen (1967) has reported similar changes in lake conductivity for Greenland lakes. Schindler et al. (1974) observed the phenomenon in Char Lake. Hansen suggested that the annual cycle results from a buildup of solutes in the winter due to freeze-out during ice development which is dissipated by the melting of ice and runoff in the summer. Such changes have also been observed in temperate European lakes but the annual variation is much smaller.

The disparities between lake and outflow conductivity during the main runoff period offer evidence additional to the midwater temperature maxima that mixing of inflow and lake waters is not complete. During the early snowmelt, the main source of outflow water is icemelt (Schindler et al. 1974). This regime is modified by the rate at which the lakes are flushed. In Loon Lake, the outflow and lake conductivities soon match primarily due to the large influx of water from the Gully River which speeds mixing.

The conductivity values are comparable to values for other arctic areas with similar geology (Hansen 1967, Welch 1974).

Lake metabolism

Welch (1974) has developed a method of measuring the total oxygen metabolism of arctic lakes. Since arctic lakes are essentially closed to the atmosphere when the surface freezes, oxygen profiles measured at intervals during the winter months will show the decline in the amount of oxygen in such a lake. When water freezes, there is an almost complete freeze-out of dissolved substances including oxygen. Thus losses from the oxygen pool can be attributed to the metabolism of the lake.

Appolonio (1962) measured a series of oxygen profiles in Immerk Lake in the winter of 1961-62. Using these data, I calculated the total oxygen of the lake at each point in time and then obtained a regression equation describing the total oxygen depletion rate per day for the period 1 October, 1961 to 1 May, 1962. The oxygen content of the lake remains relatively constant for a few weeks before and after that 212-day interval, presumably because of photosynthetic oxygen production. The daily rate of oxygen metabolism was converted to a yearly rate per exposed m^2 of lake bottom, giving a value of 69.4 g O_2 m^{-2} $year^{-1}$.

Tentative estimates of lake metabolism were made by measuring the oxygen content of the three lakes in early May and assuming that the lakes were fully saturated at 0° C at freeze-up the previous fall. I assumed the depletion of oxygen took place over the same interval of 212 days for the winter 1972-73 as was apparent in Appolonio's data for 1961-62. Corrections to the assumed initial oxygen content were applied to allow for lake ice remaining from the winter 1971-72, since that volume of old ice would not provide any freeze-out oxygen. The results are given in Table 3.

Welch (1974) has shown that the metabolic rate of Char Lake was virtually the same for the three years measured. Welch suggests that lake metabolism is a conservative property and will only exhibit a slow response to alteration of the condition of the lake. That the two estimates for Immerk Lake are so similar is probably fortuitous but it does suggest that the lake has

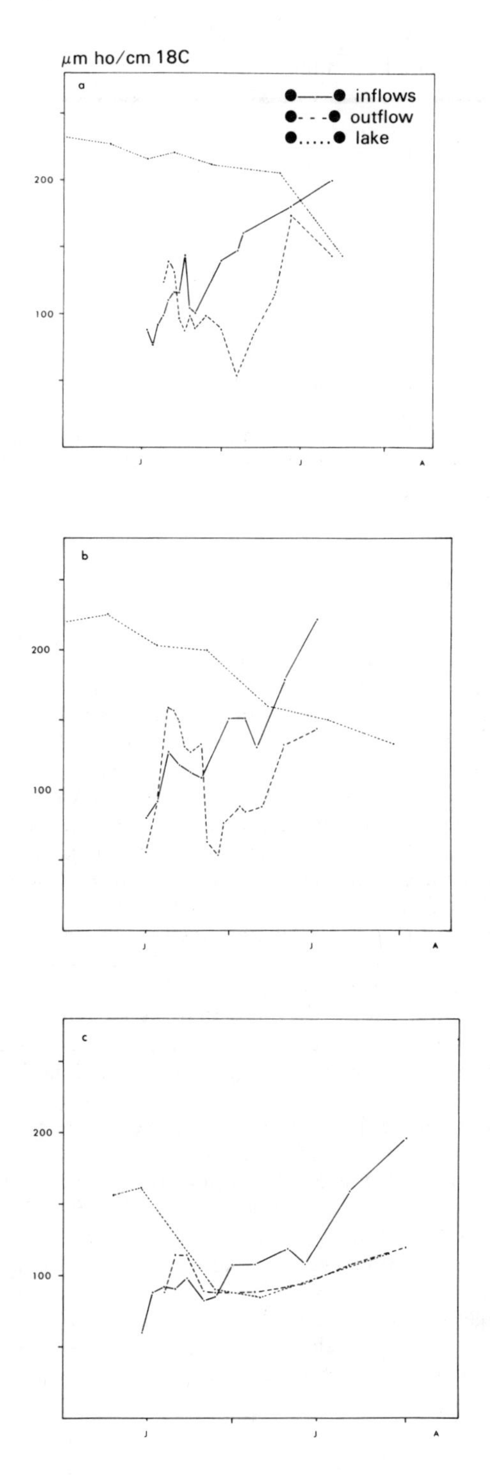

Fig. 4 Changes of inflow, outflow, and in-lake conductivity, summer 1973. (a) Immerk Lake; (b) Fish Lake; (c) Loon Lake.

Table 3. Estimates of lake metabolism.

	Metabolism gm O_2 m^{-2} year^{-1}	
Lake	*1972-1973*	*1961-1962 (Appolonio)*
Immerk	69.2	69.4
Fish	54.2	—
Loon	33.7	—

changed very little in the period from 1962 to 1973, lending support to Welch's hypothesis. Welch (1974) also found a positive correlation between lake metabolic rate and spring (March) conductivity. Such a comparison for the Truelove lakes does not support Welch's relationship though the metabolic rates are similar to that of Char Lake (43.4 gO_2 m^{-2} year^{-1}).

Chlorophyll 'a'

Chlorophyll 'a' concentration was taken as a measure of phytoplankton abundance. Vertical profiles at 1 m intervals were taken regularly in all three lakes. A suitable volume of water was filtered through a 47 mm GF/A glass fibre filter using a millipore filter rig and a hand pump. The filter was rolled and placed in a screwtop 15 ml centrifuge tube and 10 ml of 85% acetone was added. The tube was then screwed tightly shut and shaken vigorously for 15 sec. The tube's contents were then settled by gentle shaking. Chlorophyll was then extracted on ice in a darkened container for at least 24 hr.

The contents of the tube were then filtered into a matched cuvette and the fluorescence read on a Turner 110 Fluorometer. The 85% acetone was used as a blank and a blank was read before every sample to allow for the possible effects of power fluctuations on the fluorometer. The machine used was cross-calibrated with a similar one at the Char Lake Project laboratory in Resolute which had been calibrated against a spectrophotometer (Kalff et al. 1972).

Acid degradation was carried out with 0.1 N HCl to measure the contribution of phaeophytin pigments to the measured fluorescence. The phaeophytin measure was subtracted from the total and the result expressed as mg chlorophyll per m^3.

The resulting vertical profiles were integrated with respect to the volume of each depth contour, allowing for ice where necessary, and a mean concentration computed (Fig. 5). Chlorophyll levels were already high when measurements began in May. They begin to increase in February in Char Lake, and the pattern was most likely repeated in the Truelove lakes (Kalff et al. 1972). Peak levels were obtained in the latter part of May and into June. Subsequently chlorophyll levels declined.

The average summer chlorophyll in Loon Lake is 0.75 of that reported for Char Lake ($\simeq$ 0.4 mg m^{-3}) (Kalff et al. 1972) while those of Immerk and Fish Lakes are respectively 1.78 and 3.2 times the Char Lake value.

Carbon fixation

Carbon fixation was measured on several occasions in Immerk Lake and once in Fish Lake using the ^{14}C radiotracer technique.

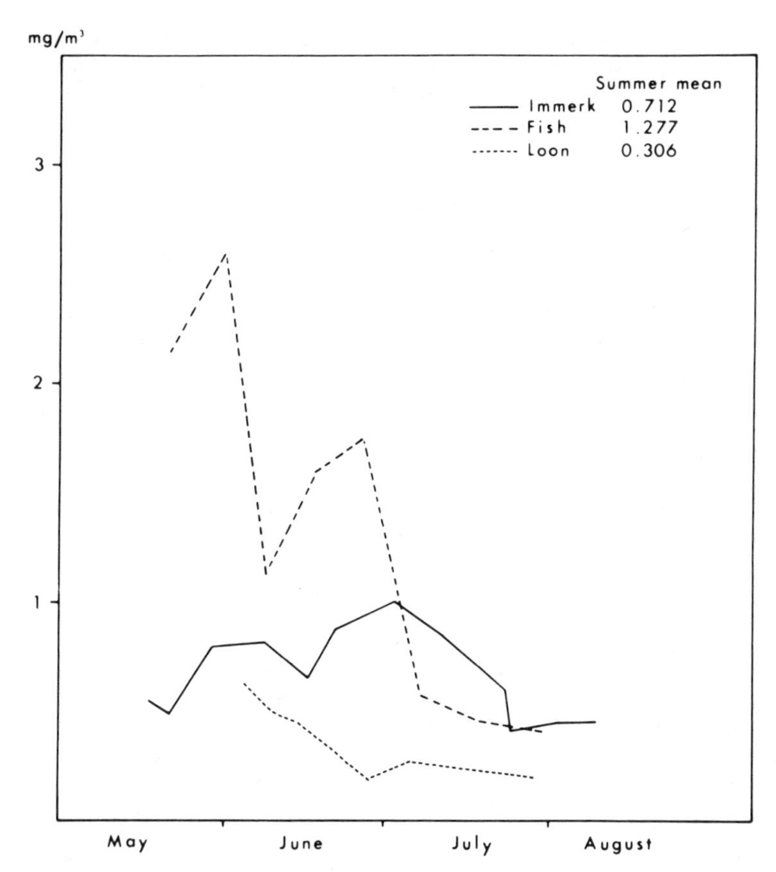

Fig. 5 Chlorophyll levels in three lakes, summer 1973.

Water samples from a series of depths were obtained using a Van Doren bottle. For each depth two 125 ml stoppered bottles, one light and one dark, were filled. A known volume of ^{14}C bicarbonate (activity $-24.83\mu C$ ml^{1}) was added to each bottle and the bottles were shaken to mix the contents. The bottles were then suspended *in situ* at the sample depths with the light bottle above the dark.

After 24 hrs the bottles were removed, taken to a darkened laboratory as quickly as possible and filtered through membrane filters (Millipore HA) using Sartorius filter rigs and a hand vacuum pump. The filters were air dried and affixed to counting planchets with dilute rubber cement.

Experiments generally ran from noon to noon. All field procedures were carried out in semi-darkness under a blanket to avoid shocking the phytoplankton. When the samples were collected, additional water was retained for alkalinity and chlorophyll determinations. Secchi disc measurements were made at the beginning and end of each experiment as a measure of the vertical extinction of light.

Subsequently, the filters were counted on a Geiger-Muller counter. Then the carbon fixation was computed using techniques outlined in the I.B.P. handbook on methods of measuring primary production in aquatic ecosystems (Vollenweider 1969).

Alkalinity was measured by titration with 0.1N HCl to a pH of 4.4 using a pH meter. The pH meter was calibrated using two buffer solutions of known pH. The initial pH was also recorded. Data are given in Table 4.

In Immerk Lake, available carbon declined through the summer and the pH increased. The range of carbon fixation rates is comparable with that found in other arctic lakes (Kalff 1970). An estimate of annual planktonic primary production is obtained by integrating the rates from 1 March to 31 October, assuming initial and final rates of zero. This gives an estimate of 4.3 gC m^{-2} yr^{-1} for Immerk Lake, a slightly higher rate than that reported for Char Lake of 4.1 gC m^{-2} yr^{-1} (Welch and Kalff 1974). The single carbon uptake rate for Fish Lake on 31 May is 40% higher than the 29 May value for Immerk Lake. This information coupled with the higher chlorophyll levels in Fish Lake, allows an estimate of 6.1 gC m^{-2} yr^{-1} planktonic primary production.

Zooplankton

Between 21 July and 3 August, one set of vertical plankton hauls was taken from each of the three lakes. A 20×30 cm net with 140μmesh was used. Samples were preserved in 5% formalin. Duplicate samples were drawn from successive metre depths to the surface and pooled. The zooplankton were identified and total counts made of the crustacea present. For rotifers, presence was recorded and the most abundant species was noted (Table 5). The average number of crustacea per m^3 was estimated. A corresponding zooplankton level in Char Lake is about 2500 animals m^{-3} (Rigler et al. 1974).

As expected (Reed 1963), zooplankton diversity of these high arctic lakes was low. The crustacean species present have been found throughout the Canadian Arctic. The estimates of mid-summer crustacean zooplankton abundance are positively correlated with the mean summer chlorophyll for the three lakes.

Chironomid emergence

Submerged emergence traps, 50×50 cm (Welch 1973) were placed in open water on Immerk Lake as the ice cleared during the summer. Progressively a greater range of water depth was obtained. There were four transect lines, all in the main basin. No attempt was made to sample emergence through candled ice, though it was known that some emergence took place this way.

Table 4. Carbon fixation rates and related data.

Lake	*Date*	*Carbon fixation mg m^{-2} day^{-1}*	*pH*	*C available (by alkalinity) C mg $litre^{-1}$*	*Secchi disc (m)*	*Ice (m)*	*Snow (m)*
Immerk	29 May	27.6	7.1	40	5.5	1.75	0.35
	8 June	13.0	7.0	38	5.8	1.60	0.30
	2 July	55.4	7.5	30	6.1	1.35	—
	11 July	51.1	7.3	25	6.7	1.30	—
	24 July	18.7	7.9	20	6.2	—	—
	2 August	19.8	8.0	20	7.2	—	—
Fish	31 May	38.8	7.3	33	5.3	1.35	0.37

Table 5. The zooplankton of three lakes, 1973.

Sampling date	*Immerk Lake* *3 August*	*Fish Lake* *21 July*	*Loon Lake* *28 July*
Crustacea:			
Limnocalanus macrurus	X	0	0
Cyclops scutifer	0	X	X
Bosmina coregoni	0	X	0
Abundance (number m^{-3})*	1,740	4,784	141
Rotifera:			
Keratella cochlearis	XX	XX	0
Keratella quadrata	X	X	X
Filinia spp.	0	X	XX
Polyarthra spp.	0	X	0

0 — not present; X — present; XX — present and most abundant species
* Net abundance, thus values are only relative.

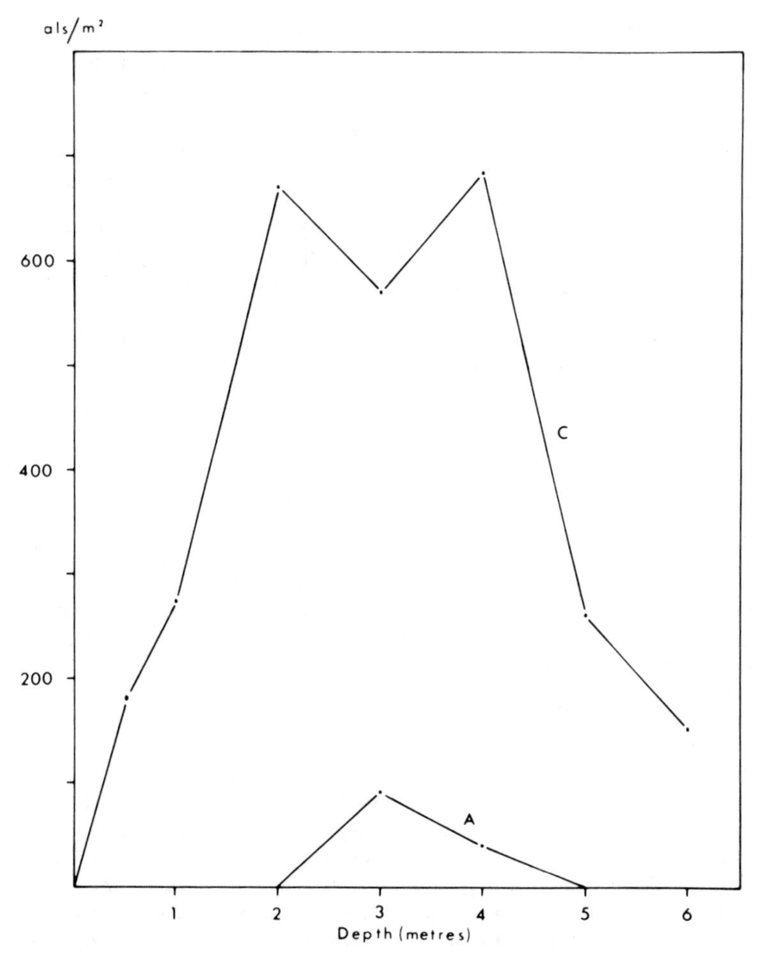

Fig. 6 Depth distribution of chironomid emergence on Immerk Lake, 1973. C — Chironomids; A—Apatania.

Subsequently, the chironomids were counted into a set of arbitrary size categories, distinguishing males and females. Representative individuals from each category were dried and weighed so that the dry weight of emergence could be calculated. Welch (1973) reported a value of 5.6 cal mg^{-1} dry weight for adult chironomids from Char Lake. This value was used to convert the emergence to calories.

Total emergence for the period 19 June to 9 August amounted to 372.24 cal m^{2} (mean for the whole lake surface). The majority of the chironomids are Orthocladiinae of which the genus *Cricotopus* is well represented. The largest chironomid present, *Pseudodiamesa arctica*, accounted for 23% of the recorded emergence calories. The depth distribution of emergence peaks in the 2 to 4 m zone (Fig. 6.)

Two traps in the 'A' transect line at 3 and 4 m caught a few *Apatania zonella*, a caddisfly. Caddis emergence lasted approximately 12 days starting around 25 July. *Apatania* was also observed on Fish and Loon Lakes. The few trap samples were used to estimate a tentative emergence value of 16.52 cal m^{-2} yr^{-1}.

When sampling stopped on 9 August, emergence was not complete. Thus the emergence value of 372.24 cal m^{-2} is only a cumulative value up to the end of sampling. The cumulative plot with respect to time illustrates this point. The plot of average daily emergence indicates two peaks during the summer (Fig. 7).

Welch (1973) gave extensive data on chironomid emergence from Char Lake. He obtained an estimate of 729 cal m^{-2} yr^{-1} for chironomid emergence with the peak occurring around the middle of August. This would suggest that the emergence figure for Immerk is an underestimate. However, emergence in Char Lake in 1971 began a month later than emergence in Immerk in 1973. Thus while the emergence figure for Immerk is incomplete, it is unlikely that total emergence would exceed that of Char Lake.

Welch (1973) found that peak emergence was in the 2 to 4 m zone in Char lake which is similar to the situation found in Immerk Lake.

Fish

The only fish species found was the Arctic char, *Salvelinus alpinus*. The populations are landlocked, none of the lakes having outflows to the sea large enough for a sea run. Fish were collected by ice-fishing at first and later with the aid of a variable mesh gill net. Fork length and wet weight were measured for each fish and the otoliths removed for aging. Before chironomid emergence, some fish guts were removed and preserved for analysis of diet other than emerging chironomids.

During insect emergence, char guts are usually packed with a mixture of pupal and adult chironomids. Before emergence, fish from Immerk Lake contained some *Mysis relicta* and chironomid larvae. Fish from Fish Lake contained cyclops and chironomid larvae. Larger fish generally had empty guts. MacCallum (pers. comm.) in an extensive study has found in Char Lake char that in the insect emergence period, most stomachs are packed with pupal and adult chironomids, while for the rest of the year, empty stomachs are much more frequent and food items are mainly *Mysis* with some larval chironomids and crustacean zooplankton. Thus the few samples from the Truelove lakes indicate a similar seasonal pattern of feeding.

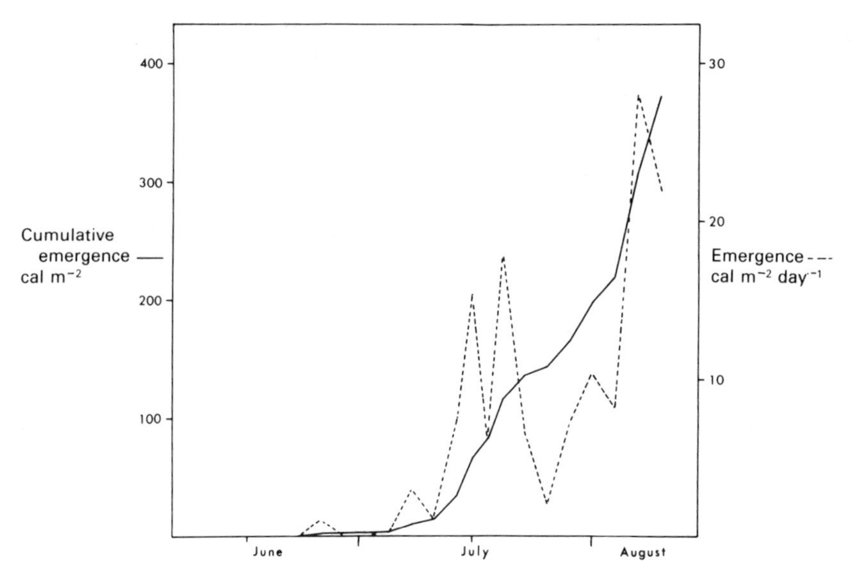

Fig. 7 Cumulative and average daily chironomid emergence on Immerk Lake, 1973.

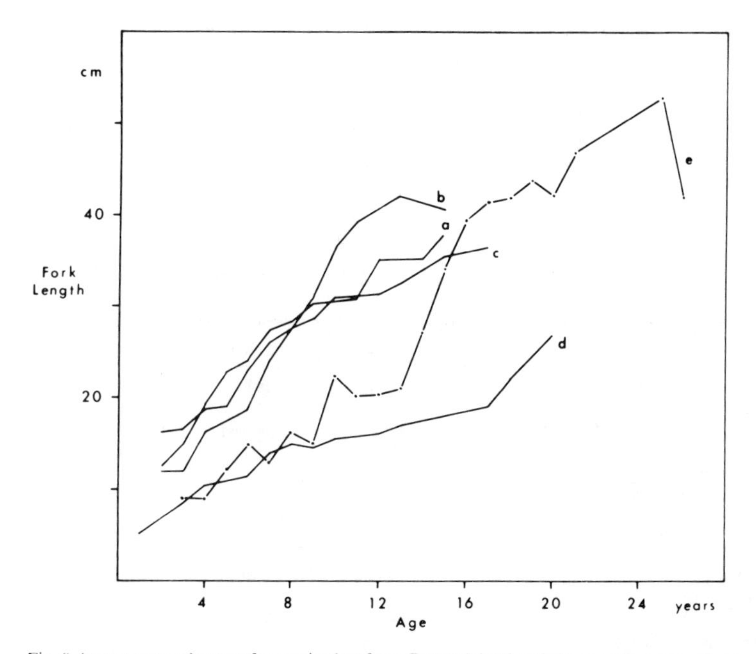

Fig. 8 Average growth curve for arctic char from Devon Island and the growth curves of some other landlocked char populations. (a) Little Fish Lake, N.W.T. — Sprules 1952; (b) Matumek Lake, Quebec — Saunders and Power 1969; (c) Keyhole Lake, Victoria, N.W.T. — Hunter 1970; (d) Char Lake, Cornwallis Island, N.W.T. — MacCallum 1972; (e) Three lakes on Devon Island, N.W.T. — this study.

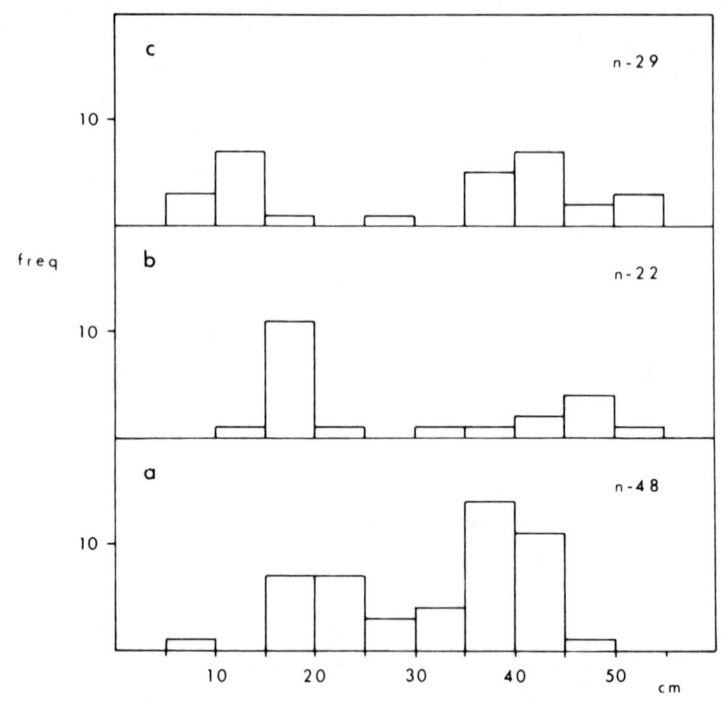

Fig. 9 Size frequencies of arctic char from three lakes, 1973. (a) Immerk; (b) Fish; (c) Loon.

Table 6. Summary of characteristics for the Truelove Lowland lakes and Char Lake.

Characteristic	*Loon*	*Char*	*Immerk*	*Fish*
Area ha.	15.9	52.6	95.8	100.5
Mean depth m.	2.9	10.2	3.2	3.1
Drainage	Polar Desert	Polar Desert	Lowland	Lowland
Vegetation	Sparse	Sparse	Lush	Lush
Mean summer chlorophyll (mg m^{-3})	0.31	0.4	0.71	1.28
Planktonic Primary production (gC m^{-2} yr^{-1})	—	4.1	4.3	6.1
Zooplankton July/August (No m^{-3})	141	2500	1740	4784
Lake Metabolism (gO_2 m^{-2} yr^{-1})	33.7	43.6	69.2	54.2
Estimated age of Arctic char at 20 cm	14+	17+	10+	10+

(Material drawn from text and literature.)

Analysis of the length/weight curves showed no significant difference between fish from the three lakes with the equation for all fish taking the form:

$$\ln \text{Weight (gm)} = 4.75 + 3.00 \ln \text{Length (cm)} \quad (r^2 = 0.997)$$

According to Scott and Crossman (1973), char have a circumpolar distribution in the northern hemisphere and are generally confined to coastal areas. There are sea run and landlocked populations with the latter having much slower growth rates. Comparison of the composite growth curve for Truelove char with those for other landlocked char populations shows that Truelove char grow more slowly up to age 13+ (20 cm) when the curve takes an upward turn (Fig. 8). Fish from Char Lake grow more slowly but their growth curve also turns up at about 20 cm (MacCallum 1972). This could be the size when cannibalism becomes much easier. The growth curves suggest that char are food-limited and that where early growth is very slow, additional growth of larger fish is possible only through cannibalism. Where the food supply is greater, the need for cannibalism appears to be alleviated, giving rise to a different population structure. Both Hunter (1970) and MacCallum (1972) have found evidence of cannibalism. Though the sample sizes are small, the size frequencies suggest there are two size group modes in the Devon Island populations (Fig. 9). This is good supplemental evidence for the shift-to-cannibalism hypothesis.

Discussion

Although the results presented here must be regarded as preliminary, some interpretation is possible. Immerk and Fish Lakes are more productive than Char lake while Loon Lake appears to be less productive (Table 6). This conclusion is consistent with Welch's (1974) suggestion that productivity in arctic lakes depends on the terrestial productivity in their drainage basins. The vegetation around Char Lake is very sparse (Arkay 1972) as is the area draining into Loon Lake, whereas the drainage of Immerk and Fish Lakes supports lush lowland vegetation, as evidenced elsewhere in this volume.

Obviously, other factors influence the productivity of lakes. Brylinsky and Mann (1973) analysed data from 55 I.B.P. freshwater studies. Their main conclusion was that production, particularly at the primary level, depends on latitude, the length of the growing season, and mean annual air temperature. These factors, however, are only approximate indications of productivity to be expected. Other environmental factors as discussed above will influence productivity. The flushing rate of the lake will influence productivity. Loon Lake has a much higher flushing rate than Char Lake and the minimal levels of zooplankton in Loon Lake is evidence of this effect. Lake morphometry can have a considerable impact. Shallow arctic lakes warm up faster than deep ones and temperature plays a primary role in determining production rates. Similarly benthic primary production will inevitably be more important in shallow lakes. Welch and Kalff (1974) report that benthic primary production constituted 80% of the total in Char Lake. Interestingly, in Char Lake a large component of benthic production was due to aquatic mosses; no mosses were found in any of the Truelove Lowland lakes.

Conclusion

These limited data on lakes on Truelove Lowland indicate productivity higher than that of Char Lake. Generally, it is to be expected that arctic lakes with their drainage predominantly in lush lowlands will be more productive than those with drainage predominantly in Polar Desert.

Acknowledgments

I offer sincere thanks to Drs. F.H. Rigler and H.E. Welch for giving me the opportunity to do this study and for their advice and help. I extend my appreciation to Tony Talbot and Joe Daly for their assistance and friendship during the field phase of study. Finally, I wish to thank the director, Dr. L.C. Bliss, for extending the resources of the project to encompass this study.

References

Appolonio, S. 1962. Unpublished report of freshwater research undertaken on Truelove Lowland, 1961-62.

Arkay, K. 1972. Species distribution and biomass characteristics of the terrestial vascular flora, Resolute, N.W.T. Unpublished M.Sc. thesis, McGill Univ., Montreal. 146 pp.

Brylinsky, M. and K.H. Mann. 1973. An analysis of factors governing productivity in lakes and reservoirs. Limnol. Oceanogr. 18: 1-14.

Hall, F.R. 1971. Discharge Solids — Discharge Relationships 2. Applications to field data. Wat. Res. 7: 591-601.

Hansen, K. 1967. The general limnology of arctic lakes as illustrated by examples from Greenland. Medd. om Grønland: 77.

Hem, J.D. 1970. Study and interpretation of the chemical characteristics of natural water. 2nd Edn. U.S. Geol. Surv. Wat. Supp. Pap. 1473.

Hunter, J.G. 1970. Production of Arctic char (*Salvelinus alpinus* L.) in a small arctic lake. F.R.B. Tech Rpt. 231.

Hutchinson, G.E. 1957. A treatise on limnology. Vol. 1. Geography, Physics and Chemistry. John Wiley & Sons Inc., New York.

Kalff, J. 1970. Arctic lake ecosystems. pp. 651-663. *In:* Antarctic Ecology. M.W. Holdgate (ed.). Vol. 2. Academic Press, New York. 391 pp.

———. H.E. Welch, and S.K. Holmgren. 1972. Pigment cycles in two high-arctic Canadian lakes. Int. Ver. Theor. Angew. Limnol. Verh. 18 (1): 250-256.

McCallum, W. 1972. *In:* Annual Report, Char lake Project, 1971-1972.

Paschalski, J. 1964. Circulation types of lakes. Polskie. Archwm. Hydrobiol. 12: 383-408.

Reed, E.B. 1963. Records of freshwater crustacean from Arctic and Subarctic Canada. Nat. Mus. Can. Contr. to Zool., Bull. 199: 29-62.

Rigler, F.H., M.E. MacCallum and J.C. Roff. 1974. Production of zooplankton in Char Lake. J. Fish. Res. Bd. 31: 637-646.

Saunders and Power. 1969. Cited in Crossman, 1973.

Schindler, D.W., H.E. Welch, J. Kalff, G.J. Brunskill and N. Kritsch. 1974. Physical and chemical limnology of Char Lake, Cornwallis Island (75° N Lat.) J. Fish. Res. Bd. 31: 585-607.

Scott, W.B. and E.J. Crossman. 1973. Freshwater fishes of Canada. Fish Res. Bd. Bull. 184.

Smith, S.H. 1962. Temperature correction in conductivity measurements Limnol. Oceanogr. 7: 330-334.
Sprules. 1952. Cited in Scott and Crossman, 1973.
Vollenweider, R.A., ed. 1969. A Manual on Methods for Measuring Primary Production in Aquatic Environments. IBP Handbook No. 12. Blackwell Sci. Publ., Oxford.
Welch, H.E. 1973. Emergence of chironomidae (Diptera) from Char Lake, Resolute, N.W.T. Can. J. Zool. 51: 1113-1123.
———. 1974. Metabolic rates of arctic lakes. Limnol. Oceanogr. 19: 65-73.
———. and J. Kalff, 1974. Benthic photosynthesis and respiration in Char Lake. J. Fish. Res. Bd. 31: 609-620.

Ecosystem models

Mineral nutrient cycling and limitation of plant growth in the Truelove Lowland ecosystem

T.A. Babb, and D.W.A. Whitfield

Introduction

The conceptual emphasis of the Devon Island Study has been on energy cycling, oxidizable carbon compounds being the biological "common denominator" for comparison of taxonomically remote organisms. This focus has been maintained throughout, and research has been directed accordingly. A detailed study of nutrient cycling was therefore not incorporated in the project design. Throughout the course of the project, however, a number of data and biological insights has been gained secondarily which lend themselves to interpretation of nutrient cycling within this high arctic ecosystem. As might be expected, inorganic nutrients, mainly nitrogen and phosphorus, seem to play an important controlling role in the function of the system. Further *a posteriori* analyses therefore appeared warranted.

It has been recognized that plant growth in arctic regions is in part limited by the scarcity of nutrients (Savile 1972, Bliss et al. 1973), but only recently have some of the ecological and evolutionary implications of this scarcity been investigated (Chapin 1974). It is the purpose of this chapter to integrate some of the Devon Island data related to nutrient cycling.

The data upon which the following discussions are based include a range of soil, plant, and water analyses supplemental to other aspects of the project. These and concepts developed through the work of other researchers have been incorporated in a framework which, it is hoped, will elucidate the function of nutrient availability within the Truelove Lowland and other high arctic ecosystems. Included in this chapter are: (1) a discussion of the postulated role of landform and hydrology on mineral nutrition in the Lowland; (2) nitrogen and phosphorus flow schema for the two habitats studied most intensively (the hummocky sedge-moss meadow, "HSM," and the cushion plant-lichen, "CPL," communities); and (3) discussion of possible adaptations and functions of vascular plants and mosses within this system, and general trends in the role of nutrients in the ecosystem's functioning.

Landscape of Truelove Lowland: Postulated effects on nutrient availability

The Lowland is a complex of raised beach ridges, intermediate, poorly drained meadows, granitic and dolomitic rock outcrops, and numerous lakes and

Table 1. Available nutrients and pH in representative Devon Island soil surface horizons.

	Available nutrients (ppm)							
Site characteristics	NH_4^+	NO_3^-	NH_4^+ + NO_3^-	*P*	*K*	*Ca*	*Mg*	*pH*
Rock outcrop organic soil[1]	1.9	0.0	1.9	—	—	—	—	—
Beach ridge upper foreslope[1]	1.0	0.0	1.0	—	—	—	—	—
Beach ridge lower foreslope[1]	3.5	0.0	3.5	—	—	—	—	—
Mesic sedge meadow[1]	1.2	4.0	6.2	—	—	—	—	—
Beach ridge crest, below *Dryas* mat[1]	6.9	0.0	6.9	—	—	—	—	—
Beach ridge crest (control)[1]	3.2	0.0	3.2	—	—	—	—	—
Raised beach (Brunisolic Static Cryosol)[2]	—	1	—	1	12	—	—	7.8
Hummocky Sedge Meadow (Fibric Organo Cryosol)[2]	—	4	—	0	129	110.7*	32.5*	6.5
Muskox trail[3]	—	—	1.0	13.5	163	—	—	6.5
Fox den (Bathurst Island)[3]	—	—	0.0	25.0	65	—	—	7.7
Muskox carcass on beach ridge soil[3]	2.4[1]	4.0[1]	6.4[1]	7.0	56	—	—	8.0
Nostoc peat mounds (Pogonato-Luzulo-Salicetum arcticae)[4]	—	—	—	17	.11*	36.0*	15.4*	6.0

[1]Micro-Kjeldahl field analyses, 1972; [2]After Walker and Peters this volume; [3]After Babb 1972a; [4]After Barrett 1972.
*Exchangeable cations, meq/100 g.

ponds. Unconsolidated parent materials are derived from Pleistocene and near-recent drift and sediment, are sorted to varying degrees, and contain a mixture of calcareous gravels and silts as well as more resistant granitic and other silicaceous rock and fine particles (Krupicka this volume).

The series of beach ridges formed during isostatic uplift of the Lowland creates a regularly repeated pattern of localized site conditions. For the sake of this discussion, the Lowland has been divided into six classes of habitat based on apparent or postulated nutrient regimes. These include: (1) xeric beach ridge crests; (2) mesic fore- and back-slopes of beach ridges; (3) granitic rock outcrops; (4) wet to mesic peaty meadow; (5) mesic, non-peaty meadows; and (6) former lake bottom *Nostoc* peat deposits. Some of these categories have been subdivided further according to plant community variability (Barrett 1972, Muc and Bliss this volume). The correspondence between these classes of habitat and those identified by other workers are indicated below.

Beach ridge crests

Soils in these areas are weakly developed Regosols or Brunisols composed of poorly sorted coarse and fine particles. In small areas significant amounts of humus have accumulated (Walker and Peters this volume), and pH is near-neutral. Organic content is generally low, however, and pH is usually in the range of 7.5-8.2, reflecting the calcareous nature of the largely dolomitic parent material. Aridity and exposure result in sparse vascular plant and moss cover. Lichens, which depend more on surface energy and moisture budgets than on subsurface water (Addison this volume) vary greatly in importance. Generally, the "cushion plant-lichen" community (CPL) is typical of the habitat. Analyses have shown that available macronutrients (N, P, and K) are low (Table 1), and artificial fertilization of vascular plants has verified the assumption that

Table 2. Effects of fertilization on plant growth over a 2 year period following treatment (Summer 1970), Devon Island, 1970-1972. Confidence limits are at the 95% level (after Babb 1972a, b).

						Production (% above controls) Treatment (kg ha^{-1})				
Species	*Year*	*Control*	*K56*	*K336*	*N56*	*N336*	*P56*	*P336*	*NPK56*	*NPK336*
	1971	0±12	-15±10	44±20*	-14±9	53±20**	10±18	15±13	14±23	82±28**
Carex stans	1972	0±13	20±19	63±26*	5±18	-21±12	8±18	2±22	30±30	65±28**
Saxifraga	1971	0±12	34±17*	16±16	131±42**	153±44**	38±17*	201±54**	276±60**	288±42**
oppositifolia	1972	0±12	24±30	60±39*	128±25**	129±24**	29±24	60±23**	110±34**	207±66**
Dryas	1971	0±9	1±9	6±13	44±16*	19±16	12±14	70±25**	50±16**	59±19**
integrifolia	1972	0±8	-16±13	5±10	10±11	38±40	-14±9	57±30*	52±26**	80±40**
Salix arctica	1971	0±16	14±25	0±20	3±18	18±22	-7±15	53±26*	9±18	180±60**
Carex nardina	1971	0±15	-22±10	-17±10	35±25	-25±16	-23±10	35±18*	40±15*	35±15*

*Significantly greater than controls at P=0.95
**Significantly greater than controls at P=0.99

nutrient availability partially limits growth (Table 2). The importance to the nutrient budget of herbivory, downslope leaching, and wind removal of detritus have not been documented, but it is assumed that these net losses are offset by chemical weathering of parent material, precipitation, and deposition of bird and mammal excreta. It is likely that there is a small and local net movement, through wind erosion (Teeri and Barrett 1975) and leaching of nutrients within the soil from the crest to the mesic beach ridge slopes.

Beach ridge slopes

The intermediate slopes of beach ridges appear to enjoy a more optimal balance of moisture availability, winter protection from snow, aeration of soil, and relatively warm temperatures than do either the crests or the meadows. Vascular plant cover is nearly continuous (Svoboda this volume), a considerable amount of well-humified organic material has accumulated (Barrett 1972, Walker and Peters this volume), nitrogen fixation is high (Stutz this volume), and available macronutrients are frequently higher than normally recorded on the beach ridge crest (Table 1). It is thought that these conditions are mutually reinforcing. Stable mesic moisture status is conducive to continuous rather that sporadic plant growth during the warm season. Decomposing organic matter under relatively well aerated but moist conditions tends to depress otherwise high pH to near neutrality, through release of CO_2 and formation of carbonic acid, and probably hastens the release and cycling of phosphorus, which at higher pH in the presence of calcium is normally bound as insoluble phosphates of calcium (Buckman and Brady 1969). Moisture and the availability of phosphorus and cationic nutrients (K^+, Ca^{++}, Mg^{++}, Na^+, etc.) in turn could enhance growth of decomposing bacteria and fungi as well as nitrogen-fixing autotrophs (blue-green algae), and nitrogen-fixing bacteria, thus promoting more rapid nutrient turnover. High plant biomass in the predominant cushion plant-moss (CPM) community, and high levels of soil humic colloids restrict nutrient loss by leaching. If there is a net accumulation of nutrients in dung over those lost to grazing (see Booth this volume), the fore- and back-slope communities are most likely to receive the benefit, as the retention capacity (a combination of living plant uptake capacity and soil humic colloid capacity) of the crest community appears relatively weak.

Granitic rock outcrops

Outcrops on the Lowland are composed of weakly weathered Precambrian granitic rock, fissured occasionally and rounded by erosion, glaciation, and frost action. The fissures and depressions within the outcrops act as entrapments for organic matter and fine mineral particles. Within the outcrops at lower elevations are deposits of the same calcareous materials as form beach ridges, but mineral substrates are otherwise "acidic" in nature in comparison with the rest of those in the Lowland. Soils are high in organic matter (5-10 cm accumulation, Barrett 1972, Walker and Peters this volume), and the communities present (herb-moss snowbeds and dwarf shrub heath) are unique for several reasons: (1) Irregularity within the outcrops results in small warm "islands" of high productivity (Bliss et al. this volume). The lack of water within the underlying granite permits a rapid warming of the rock in the springtime. Depth of thaw is much greater than in porous substrates with a high sensible and latent heat capacity (Brown this volume). The dark color of the rock, variable aspect, localized reflection, and protection from wind enhances localized absorption of radiation and retention of heat. (2) The scarcity of carbonates within the outcrops probably permits a greater mobility of phosphorus within the system once it is present (though as a parent material, granite is very phosphorus poor and weathers very slowly). (3) Winter snow within the outcrops is deep. Pockets that are warm during the summer may remain unfrozen for longer into the autumn following the first permanent snowfalls. The annual period of decomposition and nutrient release could therefore be extended, and the cycling of nutrients hastened. (4) Relative protection from cold and predation make the outcrops a favored habitat for small carnivores, nesting birds, and muskox. It is likely that a small but positive net influx of nutrients is contributed by animals foraging elsewhere and depositing feces and urine in the outcrops.

Mesic to wet, peaty, sedge-moss meadows

Forty percent of the Lowland and 50% of the land area is made up of graminoid-moss meadows. An estimated 15-20% of the meadow area is underlain by peat equalling or exceeding the depth of the active layer. Most of the meadows possessing this feature are wet-mesic hummocky sedge-moss meadows (HSM); not all of the meadows falling into the HSM category are underlain by such deep peat accumulations. A smooth transition to non-peaty, "frost-boil meadows" occurs as the peat layer thins, as wet conditions become more temporal, and frost action becomes more pronounced (Barrett 1972, Muc this volume).

With regard to mineral nutrition, meadows underlain by a deep peat layer (of which the intensive meadow site is an example) are unique in that at least the "mineral" nutrients (P, K, Mg, Ca, etc. but excluding N which may be fixed locally by algae or bacteria) bound or free within the system must initially have been derived from parent materials remote from the sites. A steady state evidently exists in which losses of nutrients occur as the peat aggrades and the base of the layer becomes permanently frozen. Rainwater is a possible source of these nutrients, but the presumed rate of peat accumulation, (see Part II below) and the floristic composition of these meadows indicate that it is more likely that a substantial influx occurs from lateral flow from outside the meadow sites. Minerotrophic moss species such as *Drepanocladus brevifolius*, *Distichium capillaceum*, *Catoscopium nigritum*, and *Scorpidium turgescens* are prominent in the intensive meadow site (Vitt and Pakarinen this volume), and are indicators of a significant lateral flux of cations (mainly Ca^{++} and Mg^{++}), and relative eutrophy in comparison with other boreal and arctic peatlands (Persson and Sjörs 1960).

Non-peaty meadows

In the past much more of the Lowland has been covered by fresh water than at present. The beach ridges form natural dams which, following formation and uplift, were slowly eroded at the "spillways" during summer months. As the lakes have drained, expanses of silty lake bottom have been exposed; much of this area now comprises the "frost-boil meadows" covering 18% of the Lowland, or the transitional "hummocky sedge meadow" communities with some mineral soil in the active layer. The striking difference between these meadows and the former type is that mineral soil predominates, and a net accumulation of organic matter does not appear to occur. Physical properties of the silty soil may be partly responsible, as congeliturbation inhibits plant growth and subsequent peat accumulation. Alternatively or additionally, fluctuating moisture regimes, from moist in the springtime to mesic or xeric by mid-summer, could tip the balance from the continuous net accumulation of peat to an equilibrium situation in which decomposition matches production, once a shallow and discontinuous peat layer has accumulated.

Of the mosses which are prominent, at least *Drepanocladus revolvens*, *Distichium capillaceum* and *Tomenthypnum nitens* (Barrett 1972) indicate mineral richness (Ca^{++} and Mg^{++}) and suggest relative eutrophy (N and P richness) as well (D.H. Vitt, pers. comm.). The structure and location of this community, however, indicate the process of nutrient cycling may be different from that in the physically more stable peaty meadows. The latter appear to depend ultimately on a lateral surface flux of mineral and nutrient rich water from adjacent sites of nutrient release. The "frost-boil" and "hummocky sedge" meadows with shallower peat accumulations appear to be located at lower positions on slope than meadows where peat accumulation is deepest, suggesting that runoff water they receive must in part be depleted of nutrients. It is obvious, furthermore, that the period of lateral percolation is not nearly as great as in the peatier meadows, and that the contribution of flowing surface water to annual nutrient uptake would be correspondingly less. Nutrients must therefore be largely derived locally from the silty, mineral substrate. *Arctagrostis latifolia*, a dominant vascular plant in the frost-boil meadows is thought to be especially adapted to the exploitation of sparse or tightly bound nutrients in cold mineral soil (Younkin 1974). Its importance suggests that it plays a role in enriching these meadows by bringing nutrients to the surface.

Nostoc peat mounds

Within the Lowland is a small area underlain by up to 1 m of peat made up of what appears to be partly humified *Nostoc sp.* (Walker and Peters this volume), apparently deposited on former lakebed and lakeshore.

The community which occupies these areas is unusual, dominated by *Salix arctica*, *Luzula confusa*, *Alopecurus alpinus*, *Carex misandra*, *Dryas integrifolia*, and *Pogonatum alpinum* (Barrett 1972, Muc and Bliss this volume). The importance of *Alopecurus*, a "nitrophilous" grass (Porsild 1964) generally uncommon in the Lowland, and a relatively high level of phosphorus (Table 1) indicate an unusual nutrient regime in these sites. Mosses are much less important in the community than in other meadows, achieving cover values of 20-50% (Barrett 1972) as compared to 95-100% in the vegetated portion of the frost-boil meadows and in the hummocky sedge meadow communities (Muc this volume). The mounds, which are relatively well aerated in comparison with other organic soils, are evidently decomposing as rapidly as low temperature permits, and are providing a slow but steady subsurface release of nutrients. They contrast, therefore, with other peatlands in that: (1) a steady state, in which accumulation of organic material equals decomposition or incorporation into permafrost does not exist; and (2) nutrients are derived primarily from below the surface, being drawn up by vascular plant roots. Foliar extraction of nutrients from surface water by mosses must be a relatively unimportant component of the nutrient cycling process.

Nitrogen and phosphorus cycling patterns with HSM and CPL communities: Phosphorus budget for the Lowland

Methods and Results

With the energy flow patterns in the Truelove Lowland ecosystem as a starting point (Whitfield this volume), community nutrient flow diagrams were developed for the hummocky sedge-moss meadow (HSM) and cushion plant-lichen (CPL) communities in the intensive study sites. Energy standing crops and fluxes were multiplied by the ratio of phosphorus or nitrogen, as appropriate, to energy, using N and P and energy contents of various tissues (Table 3). Wherever possible, we used data from within the project; sources are listed in the table. Because of the large differences in nutrient content among various types of living and dead moss and vascular plant tissues, these are shown separately. Nitrogen fixation rates are from Stutz (this volume). Data on N and P content of precipitation (Table 4) were taken from Schindler et al. (1974). We maintain some reservation about the applicability of these data to Devon Island, mainly because Resolute Bay is the site of a major dirt airstrip handling commercial jets and large transport craft. Dust is probably more important in precipitation. In any case, airborne dust must be the major source of at least phosphorus, as it is non-volatile. We have no estimate of the rate of chemical weathering of phosphorus bearing minerals, and that component is a complete unknown in the schema here. The best we can do is assume that the Resolute data represent a "saturation delivery rate" at which neutral, aqueous precipitation can weather either airborne dust or rock *in situ*, and that the value represents an approximate total for weathering plus precipitation. Precipitation and runoff estimates for Truelove are from Rydén (this volume), where the following average water budget is presented:

143 mm snow + 43 mm rain = 85 mm runoff + 101 mm evapotranspiration

Moss and algal uptake of N and P beyond that available from precipitation and weathering was assumed to be leachate from aboveground

Table 3. Nitrogen and phosphorus content of tissues and their ratios to energy content. Mammals were assumed in the analysis to be the same as birds.

Tissue	*Energy (Kcal g⁻¹)*	*Nitrogen (% of dry mass)*	*Nitrogen mg Kcal⁻¹*	*Phosphorus % of dry mass*	*Phosphorus mg Kcal⁻¹*
Graminoid					
live leaf	4.79[2]	3.12[11]	6.51	0.30[11]	0.63
dead leaf	4.78[2]	1.47[11]	3.08	.098[11]	0.21
stem base	4.79[2]	1.23[11]	2.62	0.12[11]	0.26
rhizome	4.65[2]	1.23[11]	2.62	0.12[11]	0.26
root	4.67[2]	1.23[11]	2.62	0.12[11]	0.26
Shrub					
green leaf	4.82[3]	1.28[12]	2.66	0.12[12]	0.25
brown leaf	4.98[3]	0.96[12]	1.93	0.060[12]	0.12
dead leaf	4.53[3]	0.96[12]	2.12	0.060[12]	0.13
root	4.93[3]	0.98[12]	1.99	0.065[12]	0.13
Moss					
green	4.35[5]	0.95[5]	2.18		
brown	4.35[5]	0.80[5]	1.84		
xeric				.025[9]	.057
mesic				.035[9]	.080
Lichen	4.1[13]	0.22[7]	0.54	.036[7]	.088
Algae	5[1]	1.0[1]	2.0	0.25[1]	0.5
Invertebrates	5.5[1]	10[10]	18.2	0.9[10]	1.64
Protozoa	5[1]	7.5[1]	15	0.75[1]	1.5
Microflora	5[1]	5[1]	10	0.5[1]	1.0
Bird	5.9[8]	9.4[8]	16	1.7[8]	2.88
Soil organic matter					
HSM	4.5[1]	2.43[2]	5.4	0.13[2]	0.29
CPL	4.5[1]		3.9[4]	0.16[3]	0.36
Dung (Muskox)	4.9[6]	2.1[6]	4.29	0.28[7]	0.57

[1]Guess; [2]Muc; [3]Svoboda; [4]Walker and Peters, all this volume; [5]Pakarinen and Vitt 1974; [6]Booth this volume; [7]Scotter 1972; [8]Pattie this volume; [9]Vitt and Pakarinen this volume; [10]Spector 1961; [11]Average from 2 and T. Babb; [12]Average from 3 and T. Babb; [13]Richardson and Finegan this volume.

Table 4. Nitrogen and phosphorus content of 1971 precipitation at Char Lake, N.W.T., from Schindler et al. (1974).

Form of precipitation	*Nitrogen (μg l⁻¹) Dissolved*	*Nitrogen (μg l⁻¹) Suspended*	*Phosphorus (μg l⁻¹) Dissolved*	*Phosphorus (μg l⁻¹) Suspended*
Rain	266	384	10	50
Snow	148	53	6	5

vascular plant tissues. Accumulation rates were estimated by assuming that the observed soil organic matter contents of N and P had been laid down in 7,500 years in the CPL community and 5,600 years in the HSM. The latter figure is an attempt to account for an unmeasured peat accumulation in the permafrost.

Figs. 1 and 2 present the main N and P flux and standing crop estimates. Our estimated uncertainties in the numbers of these figures are indicated as superscripts (a-d) according to the following scheme:

a, .85-1.2 × estimate; b, 0.5-2.0 × estimate;
c, 0.2-5 × estimate; d, 0.1-10 × estimate.

Within the decomposer complex, we attempted to estimate the proportions of N and P mineralized by the various trophic groups discussed by Whitfield (this

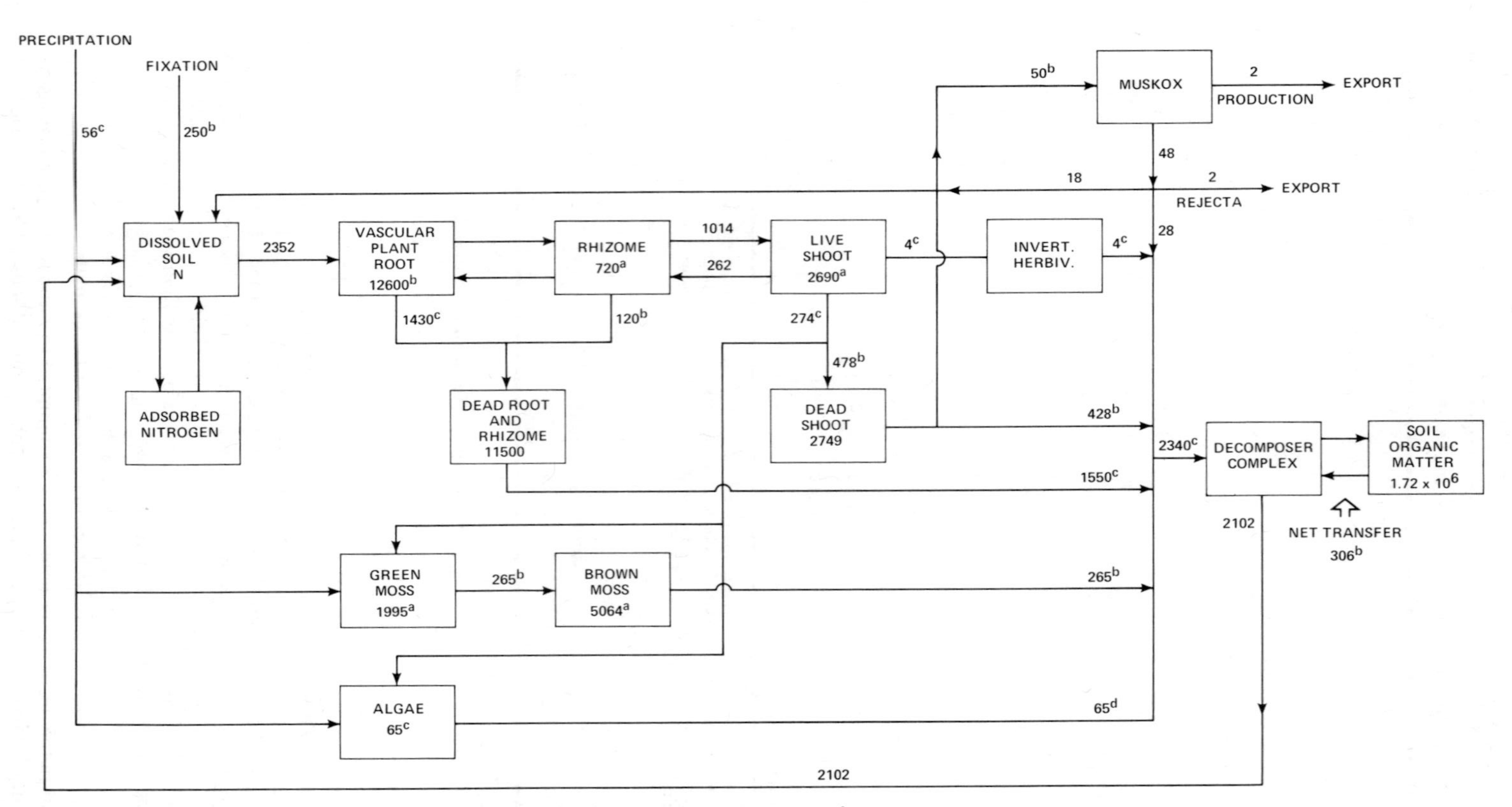

Fig. Ia Nitrogen budget for the hummocky sedge-moss meadow system. Standing crops are in mg m^{-2} and fluxes in mg m^{-2} yr^{-1}.

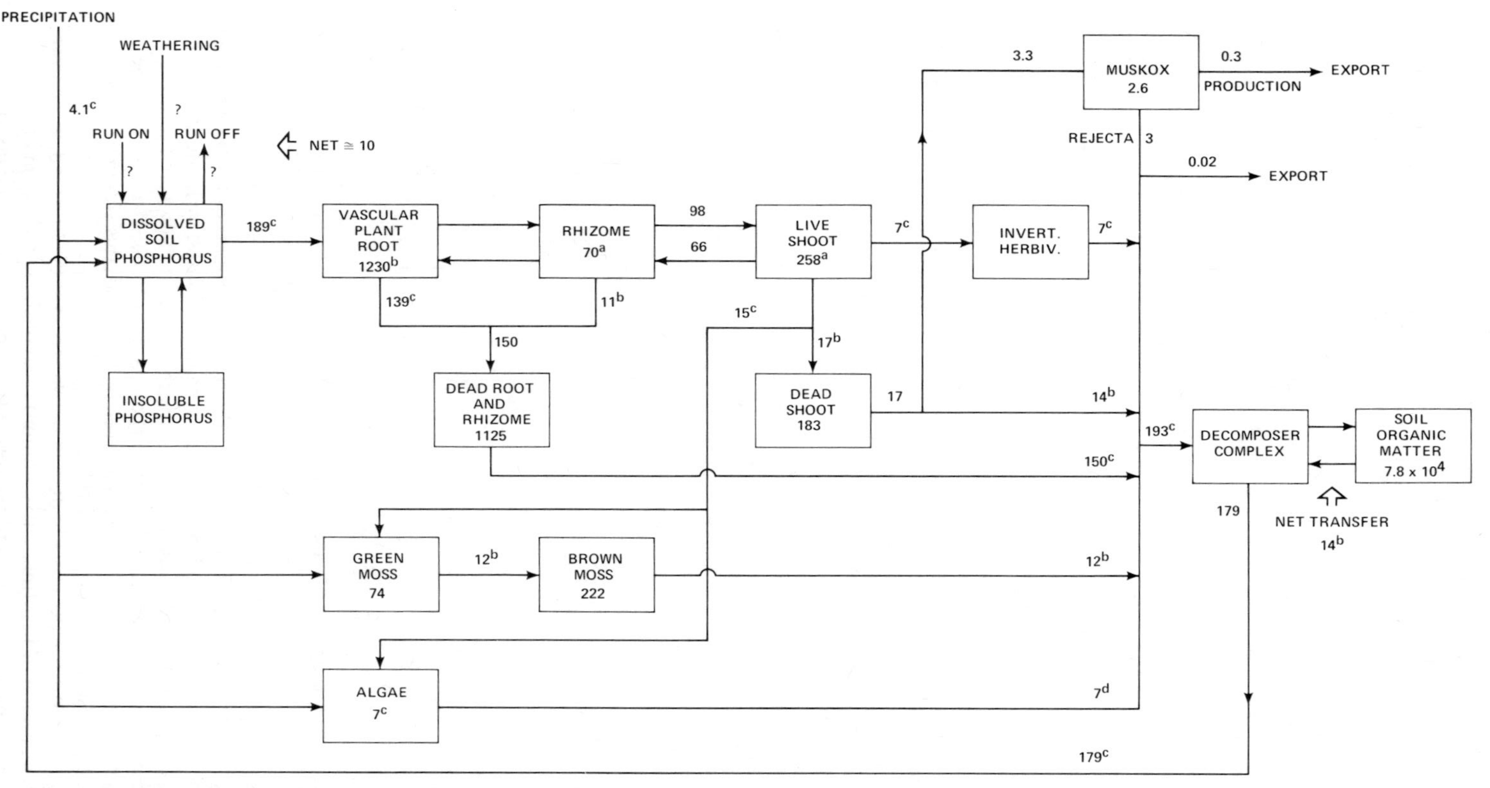

Fig. 1b Phosphorus budget for the hummocky sedge-moss meadow system. Standing crops are in mg m^{-2} and fluxes in mg m^{-2} yr^{-1}.

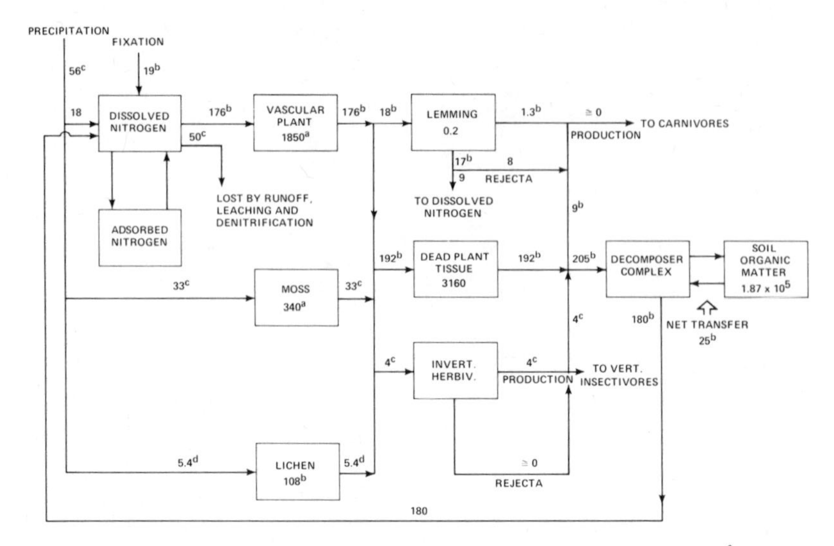

Fig. 2a Nitrogen budget for the cushion plant-lichen system. Standing crops are in mg m^{-2} and fluxes in mg m^{-2} yr^{-1}.

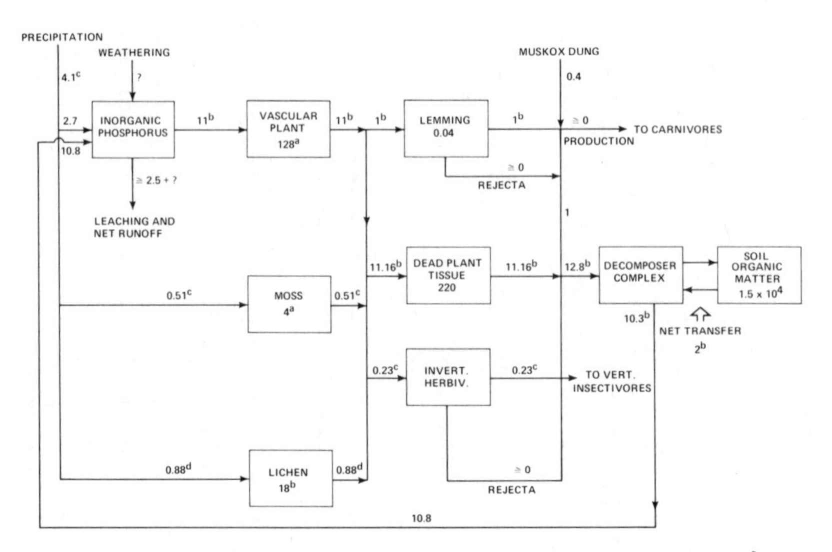

Fig. 2b Phosphorus budget for the cushion plant-lichen system. Standing crops are in mg m^{-2} and fluxes in mg m^{-2} yr^{-1}.

volume). As we have no estimates of the relative amounts of the nutrients mineralized and tied up in fecal material by invertebrates, we made the extreme assumption that either none or all of the rejected materials were tied up in feces. For the HSM the role of the protozoa was estimated at two extremes, a factor of ten apart, for the reasons given by Whitfield (this volume). The results of these estimates appear in Table 5.

To estimate the gross budget of phosphorus for the Lowland, we followed a suggestion of K. Minns (pers. comm.) and considered a system of two compartments: the land and the fresh water bodies (Fig. 3). Inputs via

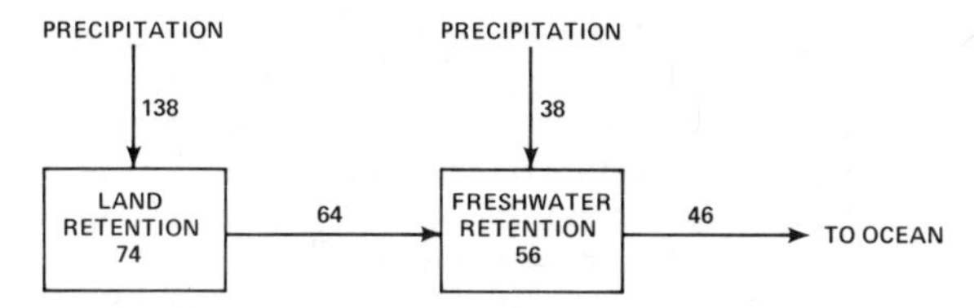

Fig. 3 Average annual phosphorus budget for Truelove Lowland. Units are kg yr^{-1}.

precipitation were estimated from data in Table 4 and the above-quoted water budget of Rydén. Minns (pers. comm.) provided measurements of dissolved phosphorus content of lake inflow (7.5 $\mu g\ l^{-1}$) and outflow (6.4 $\mu g\ l^{-1}$), and pointed out that at Char Lake the ratio of particulate to disolved phosphorus was about 2:1 in the inflow and 1:1 in the outflow. Assuming that these ratios hold on Truelove Lowland, we calculated the fluxes of phosphorus shown in Fig. 3.

Discussion

In interpreting the standing crop and flux estimates of Figs. 1 and 2 it is necessary to keep in mind the very large uncertainties in the estimates.

In neither community does the role of herbivores in nutrient cycling seem very important, unless the removal by lemmings of about 7% of the net primary production of both N and P is significant in the CPL community. As in the case of energy flux, the muskox have little effect on the average. However, where they graze intensively (Hubert this volume), and their consumption is up to five times that shown in Fig. 1, they may actually be quite important.

In generating Fig. 1 and Table 5, we have not dealt with the problem of vertical distribution of root growth or the vertical organization of decomposition. Below 10 cm from the top of the moss the HSM soils are essentially anaerobic (G.M. Courtin, pers. comm.), and we have no estimate of the proportion of decomposition by anaerobic bacteria. The estimate of a bit under 6,000 years accumulation time for organic matter in the intensive meadow site is reasonable in view of the *ca*. 7,500 year time above water of that part of the Lowland (Barr 1971). The accumulation referred to in Fig. 1, furthermore, applies to a depth of 31 cm only (Walker and Peters this volume), whereas we know that there is some organic soil below that depth.

Table 5. Approximate respiration of energy (E) and mineralization of N and P by trophic group expressed as a fraction of net primary production.

	Cushion Plant-Lichen			*Hummocky Sedge Meadow*		
Trophic group	*E*	*N*	*P*	*E*	*N*	*P*
Vertebrate herbivores	2.5	4	4	1	0.8	≃0
Invertebrate herbivores	0.5	≃0	≃0	.02	≃0	≃0
Decomposer complex						
Saprovores	2	≃0	≃0	.2	≃0	≃0
Microbivores	4	23-35	0-54	.5	0-7	0-8
Protozoa	≃0	≃0	≃0	2.1-21	6-61	7-73
Microflora	85	49-61	26-80	72-91	18-80	12-86
Storage in soil						
Organic matter	6	12	16	4.7	13	7

The turnover times of N and P in living plant material may be calculated from Figs. 1 and 2. In the HSM, both come out as 8 years, while in the CPL community they are 10 and 12 years for N and P respectively. Thus, the two communities, very different in many respects, exhibit similar turnover times for these two nutrients. These are longer than the 5.3 years N turnover in live vascular plants and mosses of the Stordalen mire in Sweden (calculated from a diagram in Rosswall et al. 1975) and the 4 year turnover for both N and P in live tissue of the Point Barrow, Alaska ecosystem (calculated from a diagram in Bunnell et al. 1975). In both communities, as at Barrow (Bunnell et al. 1975), most N and P is tied up in soil organic matter, while only about one percent is in living tissue.

The ratios of N to energy content of the soil organic matter in the HSM is about twice that in the living plant tissues, while that of energy to P is about the same in both, indicating that phosphorus is removed more efficiently by physical or biological processes as death and decomposition occurs and suggesting that biological demand for nitrogen is relatively low. We do not have a sufficiently good analysis of organic phosphorus in the CPL soils to draw any parallel conclusions, but we do know that the relationship is similar.

The relative importance of inputs in fixation vs precipitation and chemical weathering is reversed in the two communities: in the HSM, fixation seems to dominate by a factor of five, whereas it is less by 2.5 times in the CPL community.

Possible adaptations to nutrient availability and functions of vascular plants and mosses in nutrient cycling within Truelove Lowland: Experiments and general discussion.

Methods

Field soil analyses on Devon for ammonium and nitrate ions were conducted by a micro-Kjeldahl technique using steam distillation from a KCl soil extract with MgO. Following distillation of NH^+ from an aliquot of extract, powdered Devordas' alloy was used to reduce NO_3^- to NH_4^+, which, in turn, was trapped in dilute boric acid. Titration was conducted using a pH meter and standardized, dilute H_2SO_4. Other analyses were by the Alberta Provincial Feed and Grain Testing Laboratory using standard techniques (Horwitz 1970).

Fertilization trials on native communities were conducted in raised beach ridge and sedge-moss meadow habitats (Babb 1972a, b). Plots 2×3 m were arranged in grid patterns with 2 m wide buffer zones separating plots. Triplicate plots in each habitat were treated with 56 and 336 kg/ ha of N, P, and K and equal mixtures of the three totalling the same intensities. Ammonium nitrate, superphosphate, and potassium oxide fertilizers in granulated form were used. Plant growth in treated plots was compared for two subsequent seasons using, as indices of production, (meadow plots) stem length increments of *Salix arctica*, shoot dry weights of *Carex stans*, (beach ridge plots) green shoot dry weights of *Dryas integrifolia* and *Saxifraga oppositifolia* and leaf dry weight of *Carex nardina*.

Cuttings of *Salix arctica*, *Carex stans*, *Cerasteum alpinum* and *Saxifraga oppositifolia* were rooted and grown in a growth chamber for the equivalent of three growing seasons in sand cultures containing nutrient solutions. Temperatures oscillated from +5° to +15°C on a 24 hr cycle. "Overwintering"

Table 6. Analyses for N, P, and K of plant tissues from plots treated two years previously with fertilizer. Samples taken 31 July - 8 August, 1972.

Treatment		*Carex stans*			*Dryas integrifolia*			*Saxifraga oppositifolia*			*Salix arctica*		
kg ha^{-1}	*Tissue*	*N%*	*P%*	*K%*	*N%*	*P%*	*K%*	*N%*	*P%*	*K%*	*N%*	*P%*	*K%*
Control	green leaves	3.6	.25	1.39	1.6	.19	.63	1.7	.18	.49	2.5	.23	1.07
	stems	—	—	—	—	—	—	—	—	—	0.8	.09	.41
	rhizomes	1.4	.11	.44	—	—	—	—	—	—	—	—	—
	roots	1.0	.06	.33	0.9	.05	.14	0.9	.11	.18	0.8	.08	.30
N56	green leaves	5.3*	.25*	1.33*	2.9	.19	.70	1.7	.17	.58	3.9	.31	1.10
	stems	—	—	—	—	—	—	—	—	—	1.3	.09	.40
	rhizomes	2.4*	.09*	.30*	—	—	—	—	—	—	—	—	—
	roots	2.4*	.10*	.31*	—	—	—	—	—	—	—	—	—
P336	green leaves	3.8	—	—	2.0	.78	.93	1.6	.77	.78	2.8	1.73	1.43
	stems	—	—	—	—	—	—	—	—	—	1.5	.34	.71
	rhizomes	2.1	.36	.84	—	—	—	—	—	—	—	—	—
	roots	1.8	.56	1.23	—	—	—	—	—	—	—	—	—
NPK336	green leaves	4.0	.47	1.55	2.5	.18	.56	2.3	—	—	2.8**	.30**	.60**
	stems	—	—	—	—	—	—	—	—	—	0.7**	.10**	.27**
	rhizomes	1.6	.14	.37	—	—	—	—	—	—	—	—	—
	roots	1.6	—	—	1.2	.09	.23	1.3	—	—	—	—	—

*Treatment = N336 for *C. stans*
**Muskox carcass ± 15 years old, lush *Salix* growth

involved slow dropping of temperature to −1° C followed by 6-8 weeks at −5° C. *Salix* and *Carex* cuttings rooted spontaneously in washed, wet silica sand. *Cerasteum* and *Saxifraga* cuttings 1-2 cm long were treated with a commercial rooting medium (IBA and talcum powder) until rooting was initiated. Following rooting the sand cultures (8×12×5 cm trays containing 6-9 rooted cuttings) were flushed every two weeks with 1/4 strength Hoagland's solution or with solutions modified to provide 1/10 the full complement of N, P, or K found in the complete (1/4 strength) solution. A second series was identical to the first except that nitrogen was provided as ammonium chloride instead of potassium nitrate. Cations were provided as the appropriate chloride salts adjusted to identical molarity to the modified Hoagland's series. Iron was provided as Fe-EDTA. Except for the periodic flushing of cultures with fresh solutions, watering was with distilled water. Sand was kept visibly moist except in the *Carex* cultures, which were kept saturated.

Results and Discussion

Fertilization of natural beach ridge and meadow communities showed that phosphorus and nitrogen, and in some instances potassium, significantly affected plant growth. The effect of nutrients in combination was greater than that of those applied alone (Table 2). This is not surprising as nitrogen and phosphorus especially are frequently, if not usually, present at levels below the physiological optima for species in natural communities. Addition of one macronutrient alone is likely to stimulate growth somewhat until the non-availability of another restricts further production. When N and P are applied in combination, an upper limit is ultimately imposed by insufficiency of other resources.

Although plant tissue analysis data are not complete for all species (Table 6), they show that uptake of either phosphorus or nitrogen, when added alone, is increased relative to total biomass, but that the resulting ratios of N or P to

Table 7. Mean total dry weight increments (N=6–9) of individual cuttings of high arctic plant species grown in sand culture with controlled nutrient regimes for the equivalent of 3 growing seasons.

		Increment Per Plant (g) Modification of Hoagland's Solution (1/4 strength)											
			N as NO_3^-						N as NH_4^+				
Species	*Plant part*	*Complete*	*0.1N*	*N-*	*0.1P*	*P-*	*0.1K*	*K-*	*Complete*	*0.1N*	*0.1P*	*P-*	*0.1K*
Salix arctica	Shoots	.008	.005	.010	.007	.029	.057	.005	.050	.011	.030	.004	.081
	Roots	.051	.037	.062	.041	.051	.081	.051	.045	.041	.049	.045	.123
Cerasteum alpinum	Shoots	.097	.070	.030	.078	.090	.079	.103	.121	.037	.059	.079	.152
	Roots	.156	.232	.060	.060	.064	.193	.099	.194	.055	.050	.093	.106
Carex stans	Shoots	.172	.132	.105	.082	.103	.087	.143	.138	.098	.114	.129	.131
	Roots	.132	.062	.085	.080	.082	.087	.108	.082	.075	.096	.126	.158
Saxifraga	Shoots	.037	.016	.014	.023	.030	.029	.025	.019	.018	.017	.015	.023
oppositifolia	Roots	.030	.012	.009	.024	.017	.029	.014	.031	.010	.016	.005	.035

biomass are lower when the two have been added in combination. This indicates that increased uptake of both nutrients is required before a substantial increase in assimilation of carbon will occur. Increased synthesis of functional proteins, chlorophyll, high-energy phosphates, genetic material, etc., at commensurate levels is necessary to sustain high production.

The growth chamber experiment showed that species of diverse growth habits can maintain slow growth at very low levels of nutrient availability (equivalent to less than 1/40 the level in Hoagland's solution, Table 7). The significant growth noted in even the N– and P– cultures was probably attributable to trace amounts of bound phosphorus not removed from the sand by leaching, and to nitrogen fixation by algae which grew on the culture surfaces. Some nutrients, furthermore, must have been externally or internally recycled from the tissues of the original cuttings.

Although the experiment was inadequate to prove the superiority of either ammonium or nitrate ions as nitrogen sources, results concur with those of other studies (Bunnell et al. 1975) which indicate that arctic species are unusually well adapted to utilize ammonium ions. Although significant amounts of nitrate have been measured in some Devon Island soils (Stutz, and Walker and Peters this volume), it is generally agreed that ammonium is the more abundant source of inorganic N, partly because of the suppression of nitrification at temperatures below +5° C, the likelihood of denitrification under saturated conditions, and perhaps because of the excretion of ammonia by invertebrates, which are an important component of the arctic soil biota (Ryan this volume).

The cultures with the least phosphorus were more vigorous than those deficient in nitrogen, and grew almost as well as those with a complete complement of nutrients. It is worth noting that Hoagland's formula was empirically determined to approximate optimum conditions for a broad range of crop plants, many of which are sub-tropical or temperate region annuals adapted to nutrient rich conditions. It is possible that the optimum ratios of nitrogen to phosphorus for high arctic perennials is different. The results of this experiment suggest that the plants are adapted to a lower P:N ratio than Hoagland's solution provides; this would be a plausible adaptation considering the prevalence of calcareous substrates in the High Arctic and the subsequent unavailability of P.

It has been observed here as well as in other arctic studies (Bunnell et al. 1975) that nitrogen and sometimes phosphorus levels in normal plant tissues are around twice those of more temperate species. The suggestion by Chapin (1974) that this reflects physiological allocation of resources to production of functional (N and P rich) rather than structural (carbon rich) compounds seems most plausible. A greater proportion of biomass would constitute the "machinery" required for such activities as the extracting of nutrients from cold, depauperate soils or photosynthesizing in cold air. Similarly, structural compounds such as cellulose would not be required in as great quantities where tall stature is apparently selected against.

General discussion

An outstanding feature of the lowland system is microsite heterogeneity and the relative scarcity of sites favorable to the release of nutrients, either through the chemical weathering of rock or decomposition of organic material. The presence of permafrost inhibits nutrient release by depressing substrate temperatures and restricting drainage; the latter favors anaerobiosis in soils, denitrification of any nitrates formed, and the lack of oxygen necessary to rapid decomposition. Elevated sites are lithic or are so well drained that lack of moisture can limit nutrient release. The Lowland is thus a mosaic in which slight differences in elevation, moisture status, parent material, etc. can result in important qualitative changes in nutrient regime. It is evident that an understanding of vertical and lateral transfer must be very important to understanding nutrient cycling within the system. Plants and water are active agents in these spatial transfers.

As discussed above, warm mesic areas appear to be the sites of the release of nutrients that are transferred as solids (Muc this volume) or dissolved in moving water. Frozen soil resists percolation at depth, and nutrients are therefore carried mainly near the surface. This condition is favorable to the growth of mosses that depend on the foliar absorption of nutrients from surface water. Such absorption appears extremely efficient, as fairly high rates of production (Vitt and Pakarinen this volume) and hence total nutrient absorption (Pakarinen and Vitt 1974) can be maintained by mosses in contact with soil solutions containing only trace amounts of N and P (less than 1 ppm in many instances). Downslope leaching of substantial amounts of N and P must therefore be a slow process if an intercepting moss carpet occurs.

In the intensive meadow (HSM) site, as new moss overgrows dead organic material, some decomposition and carbon loss occurs, but a resistant fraction, richer in bound nitrogen, and possibly phosphorus, becomes buried. At a depth of 3-5 cm, lack of oxygen, increasingly colder temperatures, and resistance to diffusion effectively tie up these nutrients in a largely irretrievable state. The most likely avenue by which any of the nutrients slowly released at this position could be recycled is through uptake by vascular plant roots. Much of the nutrient material which is eventually reconverted to a usable form must be drawn up by this mechanism. The thick, apparently aerenchymatous roots of *Carex stans* and other graminoids may be adapted not only for respiration in cold, anaerobic soil, but also for the facilitation of local nutrient release in an aerated halo around individual roots.

A seemingly important function of vascular plants in this HSM system is the retrieval of nutrients which would otherwise be lost at depth. It appears that most of the moss production in the intensive meadow site depends on nutrients released in the surface decomposition of vascular plants. The contribution of a nutrient pool outside the area is sufficient to affect a net accumulation of peat, but is not necessarily the source of nutrients required for most production of a given year.

The most likely source of phosphorus in the Lowland, besides rain, is the calcareous rock, mostly dolomite and calcareous sandstone, which is exposed either as bedrock outcrops or as reworked talus, till, or erosional fragments. The chemical weathering of this rock is usually the simple solution of carbonates. As carbonates are etched away, insoluble (and, as such, unavailable) phosphates of calcium, which are likely to be uniformly incorporated in the sediments, are exposed and only slowly rendered soluble. Minerals such as flour-apatite, carbonate apatite, hydroxy-apatite, and tricalcium phosphate do not decompose, releasing available phosphates in soils, unless pH is around 6.5 or lower (Buckman and Brady 1969). The natural condition most conducive to these reactions is the presence of carbonic acid formed from water and respiratory carbon dioxide in the vicinity of decomposing organic matter or living plant roots.

As primary succession on a mineral-dominated landscape proceeds, organic matter accumulates, respiration increases, and the rate of chemical weathering and phosphorus release increases. An equilibrium must eventually be reached at which factors interact to affect a steady rate of production, phosphorus release, nitrogen fixation, and nutrient loss. It appears that low temperatures and topography affect this interplay of phenomena such that much of the High Arctic is stable at a biotically very depauperate level, areas such as Truelove Lowland being very uncommon. It is possible that the topographic heterogeneity, and the function it plays in nutrient cycling and water retention, is largely responsible for the richness of the Lowland.

The correlation between calcium, magnesium, and hydrogen ion concentrations with community structure and floristics in boreal and subarctic peatlands has been well documented, but it is unlikely that the physiological causal relationships will be known for some time (Sjors 1950). It is likely that nutrition (N and P) plays certain important controlling roles, but the perception of these roles is very difficult. Biologically usable N and P are elements required in greatest quantities for most of the better understood physiological functions of plants, but are among the least abundant in nature. In an organically dominated landscape demand for these nutrients makes them the first to be depleted from the abiotic medium (water). The relationship of their availability to community functioning would be determined by their flux through, not their concentration in, the ephemeral "available" state. This is especially true when they occur in concentrations of less than one part per million and difficulties in measurement obscure correlations between availability and actual limitation. Productivity and anatomical and physiological adaptations for uptake, taxon by taxon, would have to be understood before conclusive statements could be made about nutrient availability and community function.

In the high arctic, varied evidence suggests that phosphorus is an "ultimate" (as opposed to "proximal" or physiologically limiting to individual organisms) limiting factor to production. In scarce, fertile upland sites (around carcasses, ancient whalebones, fox dens, etc.) lush plant growth is commonly, if not always associated with unusually high phosphorus availability. Levels of nitrogen and potassium vary, and it is suspected that other nutrients do as well. We believe that a pulse or a sustained "bleeding" of phosphorus from some source can concomitantly stimulate nitrogen fixation, plant growth, decomposition, and subsequently further phosphorus release. These processes could mutually reinforce one another as long as phosphorus does not revert to the unavailable states that prevail in calcarious, mineral-dominated, alkaline substrates. It is possible, therefore, that high arctic plants thought to be "nitrophilous" might appropriately be termed "phosphorophilous."

Acknowledgments

This paper has made use of the data and ideas of nearly everyone who worked in the Devon Island Project, and we wish to extend our thanks to each of these people.

Doug Larson, James Wright, and Lori Babb worked as T.A.B.'s summer assistants.

Dr. L.C. Bliss, the project director, encouraged this integrative viewpoint and provided criticisms of the manuscript.

References

Babb, T.A. 1972a. The Effects of Surface Disturbance on Vegetation in the Northern Canadian Arctic Archipelago M.Sc. Thesis. Univ. Alberta, Edmonton. 71 pp.

———. 1972b. High arctic disturbance studies. pp. 150-162. *In:* Botanical Studies of Natural and Man-modified Habitats in the Mackenzie Valley, Eastern Mackenzie Delta Region, and the Arctic Islands. L.C. Bliss (ed.) ALUR 72-73-N. Information Canada No. R72-8673. Ottawa. 162 pp.

Barr, W. 1971. Postglacial isostatic movement in northeastern Devon Island: a reappraisal. Arctic 24: 249-268.

Barrett, P. 1972. Phytogeocoenoses of a Coastal Lowland Ecosystem, Devon Island, N.W.T. Ph.D. Thesis. Univ. British Columbia, Vancouver. 292 pp.

Bliss, L.C., G.M. Courtin D.L. Pattie, R.R. Riewe, D.W.A. Whitfield, and P. Widden. 1973. Arctic tundra ecosystems. Ann. Rev. Ecol. Syst. 4: 359-399.

Buckman, H.O., and N.C. Brady. 1969. The Nature and Properties of Soils. MacMillan Co. Toronto. 653 pp.

Bunnell, F.L., S.F. MacLean, Jr., and J. Brown. 1975. Barrow, Alaska, U.S.A. pp. 73-124. *In:* Structure and Function of Tundra Ecosystems. T. Rosswall and O.W. Heal (eds.). Ecol. Bull. (Stockholm) No. 20. 450 pp.

Chapin, F.S. 1974. Morphological and physiological mechanisms of temperature compensation in phosphate absorption along a latitudinal gradient. Ecology 55: 1180-1198.

Horwitz, W. (ed.) 1970. Official Methods of Analysis of the Association of Official Analytical Chemists. Assoc. Offic. Anal. Chemists. Washington, D.C. 1015 pp.

Pakarinen, P. and D.H. Vitt. 1974. The major organic components and caloric contents of high arctic bryophytes. Can. J. Bot. 52: 1151-1161.

Persson, H. and H. Sjörs. 1960. Some bryophytes from the Hudson Bay Lowland of Ontario. Svensk. Bot. Tidsk. 54: 247-268.

Porsild, A.E. 1964. Illustrated Flora of the Canadian Arctic archipelago. 2nd ed. Nat. Museum Canada. Bull. 146. Ottawa. 218 pp.

Rosswall, T., J.G.K. Flower-Ellis, L.G. Johansson, S. Jonsson, B.E. Ryden and M. Soresson (eds.). 1975. Stordalem (Abisko), Sweden; pp. 265-294. *In:* Structure and Function of Tundra Ecosystems. T. Rosswell and O.W. Heal (eds.). Ecol. Bull. (Stockholm) No. 20. 450 pp.

Savile, D.B.O. 1972. Arctic adaptations in plants. Can. Dept. Ag. Monog. No. 6. 81 pp.

Schindler, D.W., H.E. Welch, J. Kalff, G.J. Brunskill, and N. Kritsch. 1974. Physical and chemical limnology of Char Lake, Cornwallis Island (75° N Lat.). J. Fish. Res. Board of Canada. 31: 585-607.

Scotter, G.W. 1972. Chemical composition of forage plants from the Reindeer Preserve, Northwest Territories. Arctic 25: 21-27.

Sjörs, H. 1950. Regional studies in North Swedish Mire vegetation. Bot. Notiser 103: 173-222.

Spector, W.S. (ed.). 1961. Handbook of biological data. Saunders, Philadelphia. 584 pp.

Svoboda, J. 1974. Primary Production Processes Within Polar Semi-desert Vegetation, Truelove Lowland, Devon Island, N.W.T., Canada. Ph.D. Thesis, Univ. Alberta, Edmonton. 209 pp.

Teeri, J.A. and P.E. Barrett. 1975. Detritus transport by wind in a high arctic terrestrial ecosystem. Arctic Alp. Res. 7: 387-391.

Younkin, W.E. 1974. Ecological studies of *Arctagrostis latifolia* (R. Br.) Griseb. and *Calamagrostis canadensis* (Michx.) Beauv. in relation to their colonization potential in disturbed areas, Tuktoyaktuk region, N.W.T. Ph.D. Thesis. Univ. Alberta. 148 pp.

Energy budgets and ecological efficiencies on Truelove Lowland

D.W.A. Whitfield

Introduction

The broad approach and detailed integration of the Devon Island Project have permitted an analysis of the energy budget of a high arctic terrestrial ecosystem and comparisons with other terrestrial systems.

The exercise of determining energy budgets for the ecosystem and its subsystems was undertaken initially as an integrating measure within the project, and as such served the valuable function of helping to identify components which needed examination in greater depth. This initial aim was later supplanted by the intention of making as accurate and detailed an energy budget as possible as a contribution toward the understanding of ecosystem structures from diverse biomes.

The level of detail attempted is quite fine, particularly within the decomposer complex. In this sort of study there is an inverse relationship between level of resolution and firmness of results; the more pathways which are studied, the more assumptions and extrapolations must be made. We have gone so far that much of what follows must be regarded as hypothesis. At the same time, the results are more interesting and provocative than they would have been at a higher level of aggregation.

The next section sets out the definitions with which I have worked and this is followed by details of the procedures, data, and assumptions necessary to derive the energy budgets. Then there is a discussion of the energetic pattern, computation of ecological efficiencies, and comparison with other ecosystems.

Definitions

Computation of energy budgets must start with a set of definitions which are to be applied consistently to all trophic levels. The resulting compromises often contradict accepted practice in one or another discipline. For example, I have, as far as possible, worked with yearly average standing crops in a very strict sense: the standing crops present on each day of the year are summed and the result is divided by 365. Thus the migratory bird standing crops used are much different from those for the summer season only.

All fluxes are totals for one year. This applies to consumption, respiration, production, and rejection. Herbivores consume live plant tissue, while those invertebrates which eat and themselves produce the enzymes, or

Table 1. The dry biomass to energy conversion factors applied to various tissue types.

Tissue	*Conversion factor (Kcal g^{-1})*
Vascular plant	
Graminoid	4.7
Shrub	
CPL and CPM	4.9
DSH	5.3
Moss	4.35
Lichen	4.1
Algae	4.0
Invertebrates	5.1-5.8*
Birds	5.86
Mammals	5.0
Dung	4.5
Microflora	4.5

*Depending on taxonomic group. See Ryan (this volume).

have gut symbionts which produce enzymes, to digest dead tissue are called saprovores. Those invertebrates which consume bacteria, fungi, and protozoa are considered together under the name microbivores. The energy unit used is the kilocalorie, and thus standing crops are expressed as Kcal m^{-2} and fluxes as Kcal m^{-2} yr^{-1}.

The biomass to energy conversion factors used are given in Table 1.

Where appropriate, energetics have been determined separately for the major plant community types represented on the Lowland, and as outlined by Bliss (this volume). For the sake of brevity the following shorthand notation is used:

HSM, Hummocky sedge-moss meadow;
WSM, Wet sedge-moss meadow;
FBSM, Frost-boil sedge-moss meadow;
CPL, Cushion plant lichen;
CPM, Cushion plant moss;
DSH, Dwarf shrub heath (= rock outcrop);
LB, Lichen barren.

Energy accounting procedures, data resources, and assumptions

The data bases for the energy budget are mostly available in the various chapters of this book. The intention here is not to repeat the data, except in some few cases, but to describe the procedures and assumptions used, and to indicate the relevant data sources. This section is long and detailed in order to provide the reader with the opportunity of informed criticisms of procedures, and to enhance the value of this paper for inter- and intra-biome comparisons.

Incoming solar radiation

During a full year (19 Aug 1972 to 18 Aug 1973) for which it was measured, the incoming shortwave radiation totalled 8.55×10^5 Kcal m^{-2} (Courtin and Labine this volume). More than half of this arrived outside of the growing season. In order to determine incoming radiation during the growing seasons for the

Table 2. Total incoming shortwave radiation for three lengths of growing season. Units are 10^5 kilocalories m^{-2}.

Lowland unit	*Growing season*	*1971*	*1972*	*1973*	*Average*
CPL and LB	20 June - 18 August	2.94	2.77	2.69	2.80
CPM	25 June - 18 August	2.58	2.46	2.30	2.45
DSH and meadows	30 June - 18 August	2.28	2.16	2.03	2.16

several community types, information on time of spring snowmelt and late summer vascular plant senesence was used to delimit expected average growing seasons. These are 20 June - 18 Aug for CPL and LB, 25 June - 18 Aug for CPM, and 30 June - 15 Aug for meadows and rock outcrops. The incoming radiations for these periods are summarized in Table 2.

Primary production

Bliss, elsewhere in this volume, presents a summary table of plant standing crop and annual net production. Also in that table are the proportions of the Lowland occupied by the various plant community types. I entirely ignore the salt water marshes, which comprise 0.5% of the Lowland.

Gross production is not considered. Insufficient field measurements were made to provide dark respiration estimates for a whole growing season.

Vertebrate herbivores

The vertebrate herbivores of Truelove Lowland are the muskox *(Ovibos moschatus)*, collared lemming *(Dicrostonyx groenlandicus)*, arctic hare *(Lepus arcticus)*, willow ptarmigan *(Lagopus lagopus)*, and snow goose *(Caerulescens)*. Although they may have significant local impact, the latter three contribute so little to energy fluxes on the Lowland compared to muskox and lemming, that we may ignore them in this calculation.

From Hubert (this volume), I calculated yearly average muskox standing crop as 2.73×10^6 kg-days, and reduced it to .0759 g m^{-2} by dividing by 365 days in the year, multiplying by .34 for the conversion to dry mass and dividing by the Lowland land area (3.37×10^7 m^2). He also gave total annual forage removed as 48,000 kg (dry). This is mostly removed from hummocky sedge-moss meadows, with some from frost-boil sedge-moss meadows. The consumption between these two types is assigned in the ratio of their total lowland aboveground live standing crops, which is calculated from Muc (this volume) as 1.7:1, HSM:FBSM.

Hubert also determined that close to 30% of both the matter and the energy ingested appeared in feces. This means 14,400 kg of feces deposited on the Lowland yearly. Booth (this volume) estimated dung deposition as 21,000 kg yr^{-1}. We make use of Hubert's figure to maintain consistency with the consumption data. Hubert quoted Blaxter (1967) in considering 82% of assimilated energy as being respired. Hubert also calculated annual production of the herd which utilizes Truelove Lowland as 12% of the muskox biomass. This production is a very small portion of consumption, so that excreted energy is close to 18% (100-82%) of assimilated.

Fuller et al. (this volume) summarized the Devon Island Project research on lemming populations and energetics. The reader is referred to their paper

Table 3. Fractional assignment of invertebrate groups into functional categories. (J. Ryan, pers. comm.)

	Functional categories				
Taxonomic group	*Bacterivore*	*Herbivore*	*Fungivore*	*Saprovore*	*Carnivore*
Lepidoptera	0	1	0	0	0
Diptera	0	0.04	0.47	0.47	0.02
Nematoda	0	0.1	0.63	0.25	0.02
Acarina	0	0	0.4	0.5	0.1
Enchytraeidae	0	0	0.8	0.2	0
Collembola	0	0	0.5	0.5	0
Hymenoptera:					
Symphyta	0	1	0	0	0
Parasitica	0	0.1	0	0	0.9
Araneida	0	0	0	0	1
Protozoa	0.95	0	0	0	0.05
Rotifera	0.95	0	0	0	0.05
Tardigrada	0	0.1	0.3	0	0.6
Crustacea	0	0.05	0.45	0.45	0.05

for details. Within the beach ridge complex, I assigned the lemming to CPL, LB and CPM according to the Lowland total vascular plant standing crops in these communities.

Invertebrates

Ryan (this volume) used his field data to calculate standing crop and annual production for the invertebrate groups, except the nematoda, which have been considered by Proctor (this volume). Both authors based their results on work in the intensive study sites (HMS and CPL).

Ryan (pers. comm.), working from the literature, assigned the invertebrate groups occurring on the Lowland into functional categories which correspond to the division discussed under definitions above. Table 3 is a presentation of this functional assignment.

The contribution of invertebrate herbivores has been estimated for the whole Lowland by assuming that results, per m^2, for HSM hold in WSM and FBSM, that LB may be represented by CPL and that CPM and DSH contribute as (HSM + CPL)/2.

Vertebrate insectivores

This functional group is essentially made up of two avian species, the snow bunting *(Plectrophenax nivalis)* and lapland longspur *(Calcarius lapponicus)*. The plovers and sandpipers which inhabit the Lowland are also insect eaters but their contribution is quantitatively negligible. (Pattie this volume).

Pattie (this volume) made a detailed study of snow bunting energetics and a less intensive study of the longspurs. He estimated snow bunting standing crop consumption, production, and rejection. I obtained respiration by subtraction. These results were extended to longspurs by the use of two factors: the first, applied to biomass, was the ratio of longspur to bunting standing crop on the Lowland (0.142), the second, applied to all metabolic processes, was their number ratio times their body weight ratio raised to the 3/4 power.

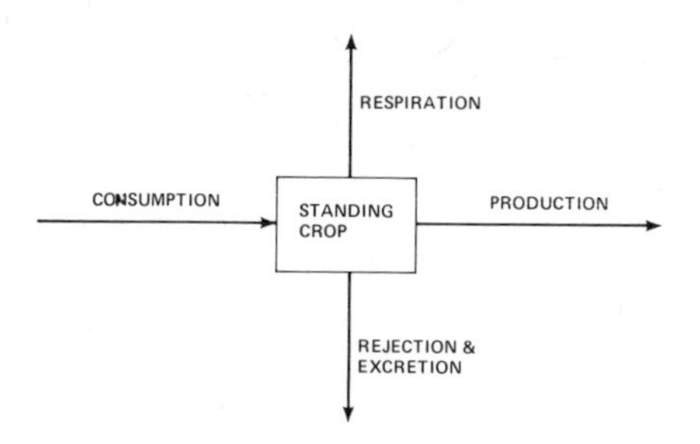

Fig. 1 The general representation of a functional group. Standing crop is a yearly average and the fluxes are yearly totals. For each of these units, the sum of respiration, production and rejection, and excretion equals consumption. Standing crops are Kcal m^{-2} and fluxes are Kcal m^{-2} yr^{-1}.

Vertebrate carnivores

The vertebrate carnivores of the system are the short-tailed weasel *(Mustella erminea)*, arctic fox *(Alopex lagopus)*, arctic wolf *(Canis lupus)*, long-tailed jaeger *(Stercorarius longicaudus)*, and parasitic jaeger *(Stercorarius parasiticus)*.

As well as the estimates of standing crop and rate of food consumption of weasels and foxes in his paper for this volume, Riewe (pers. comm.) calculated the same data for wolves, and the other energy budget components for all three carnivores. He felt that three wolves each spend two weeks on the Lowland, and that average metabolic rate is three, two, and two times the basal rate for weasel, fox, and wolf, respectively. I have further assumed that wolf and fox productivities are negligible compared to weasel and that rejecta and excretia can be obtained by subtraction.

Data on the avian predators are from Pattie (this volume and pers. comm.). He showed that an average of 29 long-tailed and 8 parasitic jaegers were present during the period June-August. Consumption was assumed to be 25% of body mass per day and respiration 60% of consumption (Pattie, pers. comm.). Production, which is nearly all exported from the Lowland, was estimated from nesting records, and rejecta and excretia were determined by subtraction.

Microfloral complex

Widden (this volume) estimated average microfloral standing crop to be 2 g m^{-2} in CPL and 12 g m^{-2} in HSM, to a 5 cm depth. This includes both filamentous fungi and bacteria. His estimates of production are further considered in the discussion.

Budget assembly

The basic structural unit for the energy budget diagrams representing the contribution of a single trophic level is shown in Fig. 1. The arrows represent annual fluxes (Kcal m^{-2} yr^{-1}) and the box is standing crop Kcal m^{-2}). No attempt has been made to represent magnitude of the fluxes or standing crops by size of the arrows and boxes, but rather, in later figures, they are labelled with the quantities they represent.

Included as letters superscript to the fluxes and standing crops of Figs. 2-4 are indications of the estimate accuracies, as follows: a, 0.85-1.2×estimate; b,

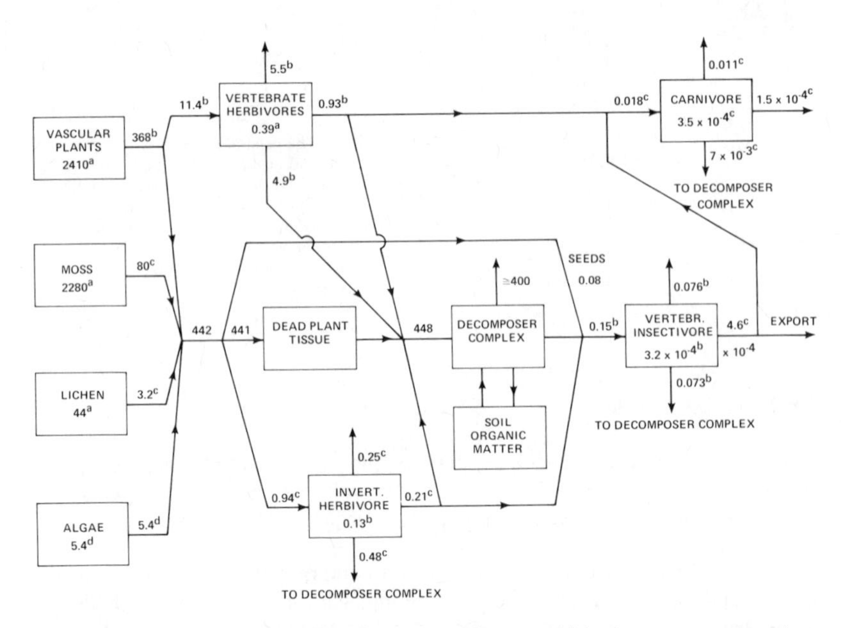

Fig. 2 The Lowland energy budget. Standing crops and fluxes are weighted averages over the various plant communities. Standing crops are Kcal m^{-2} and fluxes are Kcal m^{-2} yr^{-1}.

0.5-2.0×estimate; c, 0.2-5.0×estimate; d, 0.1-10×estimate. These are not statistical uncertainties, but partly subjective judgments on methodology and sampling problems.

At this point it is necessary to take account of suitable ecological units for developing the energy budgets. We have estimates of primary production in all communities, and it is appropriate to treat the aggregated vertebrate herbivores, the carnivores, and the vertebrate insectivores across the whole Lowland (much larger systems are, really appropriate to the wolves, foxes, and muskox). On the other hand, knowledge of the invertebrates involved in decomposition is detailed only for CPL and HSM, and extrapolations to other communities are dangerous, as will be evident in the discussion which follows. Therefore, I have compiled three energy budgets. The first (Fig. 2) is a weighted average for the whole Lowland not considering decomposers and the other two (Figs. 3 and 4) deal only with the CPL and HSM subsystems, respectively.

The Lowland budget

The procedure used to produce this weighted average budget was as follows: the standing crops and productivities of the several plant groups in each community type, obtained from Bliss (this volume), were multiplied by the fractions of the Lowland occupied by these communities and their products were summed for each plant group. The standing crop and productivities of the muskox and lemming were distributed over the whole Lowland land area, while their consumptions from each plant community type were calculated and then averaged, by the above weighing procedure. Invertebrate herbivores were treated as outlined in the above section on Invertebrates. The decomposer complex has been left blank, and no attempt was made to consider soil organic

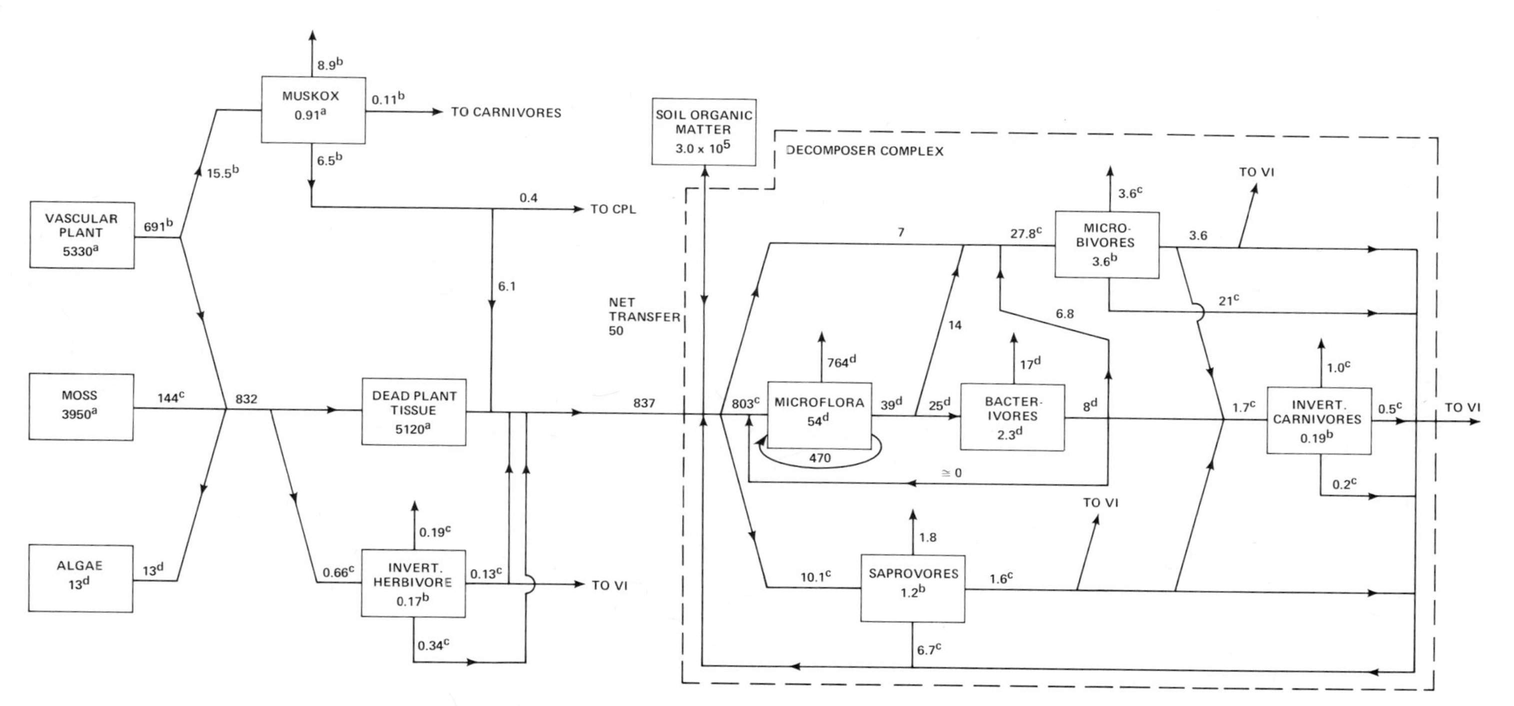

Fig. 3 The hummocky sedge-moss meadow energy budget. The arrows labelled "To VI" are transfers to vertebrate insectivores. Standing crops are Kcal m^{-2} and fluxes are Kcal m^{-2} yr^{-1}.

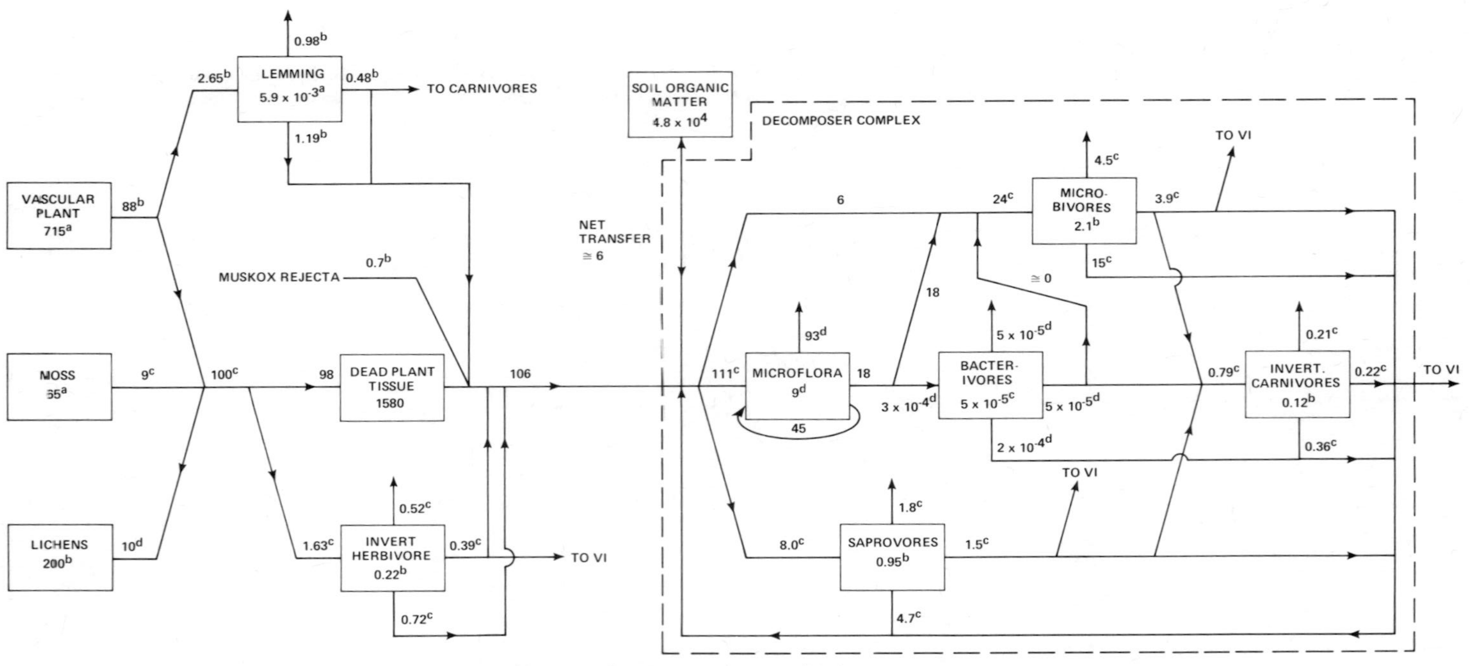

Fig. 4 The cushion plant-lichen energy budget. Standing crops are Kcal m^{-2} and fluxes are Kcal m^{-2} yr^{-1}.

matter. The treatment of carnivores and vertebrate insectivores is straightforward.

Cushion plant-lichen and hummocky sedge meadow budgets

In these budgets the carnivores and vertebrate insectivores are ignored because we are unable to accurately apportion their effects among plant communities.

The mammalian herbivore active in the CPL subsystem is the lemming, and in the HSM it is the muskox. All energy contributions of the former were apportioned among the communities where they feed (CPL, LB, CPM, and DHS) as discussed above. Similarly, the muskox energy components were divided between HSM and FBSM. Muskox dung deposited in the CPL subsystem is the only connection with HSM large enough to consider.

I calculated the annual net flux to the soil organic matter from the measured total accumulation (less integral roots) and assumed accumulation times. Barr (1971) estimated that deglaciation occurred 9,450 years BP and that the Lowland rose rapidly thereafter by isostatic rebound. I took the CPL and HSM accumulations to have occurred over 8,000 and 6,000 years, respectively. The latter figure is an attempt to account for an unmeasured amount of organic matter accumulation in the permafrost. Keeping in mind Heal and MacLean's (1975) assertion that there is probably considerable cycling through decomposer complexes, I show them as loops, and pick off the net transfers to soil organic matter ahead of the decomposers. Any attempt to be more definite about the origins of the accumulated organic matter is unwarranted in these cases.

Microfloral respirations were derived by subtraction of all other decomposer respirations from the input to the decomposer complex less the output to soil organic matter.

I assumed that consumption by microbivores was one-quarter dead plant tissue, to account for ingestion of some substrate along with the sought-after food. In the case of the HSM, I took the ingestion of microflora and bacterivores by microbivores to be in a ratio of 2:1.

Ecological efficiencies

From Figs. 2-4 we can calculate various ecological efficiencies for the Lowland; some are presented in Table 4. Ingestion by primary producers, in other words gross productivity, is not known very accurately, so there are blanks in the table where this would have been used. The vascular plant production is the sum of that above and belowground. This has a particularly strong influence in HSM where the ratio of above to belowground production is about one to three.

Results and analysis

The great danger in looking at Figs. 2-4, is over-interpretation. The estimation accuracy codes must be kept clearly in mind when using any of the numbers.

The ratio of total primary production to total incoming solar radiation during the growing season (Table 2) is .38% for HSM and 0.041% for CPL. These figures may be divided by 0.45 to obtain an estimate of the use of photosynthetically active radiation. The HSM result is at the bottom end of a

Table 4. Various ecological efficiencies, in percent, all formed as 100× ratios of one trophic level to the one below it. WA = lowland weighted average. These ratios are mostly calculated from data in Figs. 2-4.

Item		*Production / Production*	*Ingestion / Production*	*Ingestion / Ingestion*
HSM	Muskox / Vascular plants	0.016	2.3	—
HSM	Invertebrate Herbivores / All plants	0.016	0.08	—
CPL	Lemming / Vascular plants	0.55	3.0	—
CPL	Invertebrate Herbivores / All plants	0.36	1.5	—
WA	Vertebrate Herbivores / Vascular plants	0.25	3.1	—
WA	Vertebrate Carnivore / All other Vertebrates	0.016	1.9	0.16
LOWLAND TOTAL	Weasel* / Lemming	0.016	0.97	0.18
HSM	Invertebrate Carnivore / Other Invertebrates	9.6	33	4.5
CPL	Invertebrate Carnivores / Other Invertebrates	4.0	15	2.5

* Calculated from additional data not shown in the Figures.

range of total solar radiation use covered by many other natural systems which are adequately supplied with water: 0.83% for an English woodland (Satchell 1969), 1.1% in a Michigan old field (Golley 1960), 0.4% (=.45×.88% of visible radiation) in a late successional oak-pine forest (Woodwell and Whittaker 1968), 1.4% in a salt marsh (Teal 1962), 0.73% (=.45×1.62% of visible radiation) in a Polish oak-hornbeam forest (Medwecka-Kornas et al. 1974) and .32-1.1% (=.45×.7−2.4% of visible radiation) in Norwegian alpine tundras (Østbye et al. 1975).

On the other hand, the fact that radiation use efficiency is nearly an order of magnitude lower in CPL follows from the very severe water stress conditions in that community.

The most meaningful comparisons of Truelove Lowland vertebrate herbivores can be made with grassland ecosystems. The values given in Table 4 for vertebrate herbivore consumption as a fraction of net vascular plant production are considerably lower than the range of 28-60% summarized by Golley (1972) for Indian and African grasslands and greater than the 1.4% given by Coleman et al. (1976) for a slightly grazed treatment at the Pawnee grassland site in the United States. At Barrow, Alaska lemmings consume about .02% and 10% of total net primary production in seasons of low and high population, respectively (Bunnell et al. 1975). It seems that in our CPL, lemming consumption of vascular plant production, which fluctuates much less from year to year (Fuller et al. this volume), is on average fully a third of that at the peak at Barrow. If only aboveground plant production is considered, these ratios are 0.1-25% at Barrow and 3.5% in the Devon CPL. Invertebrate herbivory is at much the same level in CPL as that reported for the Polish oak-hornbeam forest study (Medwecka Kornas et al. 1974) (caterpillar consumption/total net primary production = 2.7% and production/production = .95). The results of Coleman et al. (1976), (invert herbivore assimilation/net primary production = 1.5%) are also similar. On

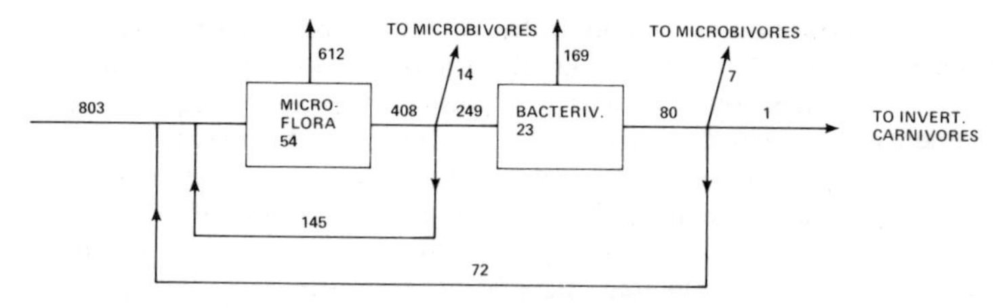

Fig. 5 A subsection of Fig. 4, drawn under the assumption that bacterivore activity is ten times that estimated for Fig. 4. See the text for an explanation. Standing crops are Kcal m^{-2} and fluxes are Kcal m^{-2} yr^{-1}.

the other hand, invertebrate herbivores play a very much smaller role in the HSM community.

In using Hubert's data we conceptually spread muskox effects uniformly over all HSM and FBSM, whereas Hubert determined that some meadows are used more than others, and that their local effect can be as much as five times greater than that shown in Fig. 3. If, in keeping with some other studies, we calculate muskox consumption of aboveground vascular plant production, and moreover consider only these intensively grazed meadows, this consumption is over 20% of production, instead of the 2.3% given in Table 4.

The figures dealing with mammalian carnivores must be treated with particular caution because of lack of detailed knowledge of time spent and food taken on the Lowland by wolves and foxes, and because they do not account for alternative food supplies, such as jaegers eating insects and foxes consuming sea-feeding duck's eggs. In the lemming-weasel subsystem the results are very similar to those for the whole carnivore/herbivore system. A meaningful comparison can be made with Golley (1960), who found that weasel consumption was 30% of the sum of production and immigration by *Microtus* in a Michigan old field; this is 30 times our result.

An examination of Figs. 3 and 4 reveals very interesting differences between their decomposer complex sections. Protozoa are essentially absent in the CPL community, and as these are the dominant bacteria feeders, the bacterivores in CPL are negligible, whereas they play a large role in the HSM. Within the microfloral complexes there is considerable circulation of energy, so that for every unit of production taken by protozoa or microbivores several units are taken by other microflora. If actual microfloral production is assumed (Heal and MacLean 1975) to be 2/3 of respiration, then the ratio of "recycled" to "exported" production is about 2.4 in CPL and 12 in HSM. If the recycling effect is taken into account, the CPL microbe energy turns over about 7 times in a year (ratio of production to standing crop), whereas in HSM the similarly calculated turnover rate is about 9 times per year. If, as is very possible, the contribution of the protoza is underestimated by a factor of 10, the situation could be as in Fig. 5, which shows a tight cycling between microflora and protoza. In this case the microflora turn over 8 times per year, but 64% of their production is consumed by other groups. Most probably, the truth lies somewhere in the middle. If, as he suspects, Widden's (this volume) microfloral biomass estimates are low, then the turnover rates calculated above are high. His microfloral production, 76 Kcal m^{-2} yr^{-1} in HSM and 8.1 Kcal m^{-2} yr^{-1} in CPL, are about seven times lower than those derived from Figs. 3 and 4, while the above speculations about a greater role for protozoa in the meadow is in line with his suggestion of greater activity and more rapid microfloral turnover in the beach ridge soils. In our CPL and HSM decomposer systems the microflora contribute 93% and 78-97% of total respiration. The latter figures depend on the role of the bacterivores. The invertebrate carnivores do not take a greater fraction of their prey production than do vertebrate ones, but presumably because they avoid the high energetic cost of homeothermy, they do produce much more from what they take (Table 4). Reichle (1971) quoted 7% as a common ingestion ratio for predatory invertebrates in forests. Reichle et al. (1975) found that the microflora accounted for 99.7% of decomposer respiration in a mesic deciduous forest, and Coleman et al. (1975) estimated microfloral respiration to be 99.3% of total saprophagic respiration. On the other hand, in a synthetic analysis of secondary productivity, Heal and MacLean concluded that the microbial component of decomposer respiration was often about 83%. Our results are in rough support of the latter. (Note: the Devon Island results quoted by Heal and MacLean have been extensively revised since publication; the present results are to be preferred.)

Conclusions

When the large uncertainties in estimates, indicated on the figures, are kept in mind, it is evident that inter-project, and even intra-project, comparisons are very dangerous, except in the case of very gross differences. This difficulty has been somewhat obscured in the literature by the failure of many authors to provide any estimates of uncertainty. It is very tempting to distinguish one ecosystem from another on the basis of a 2 to 5 times difference in some efficiency, but by the time the large errors in estimates of all the various quantities which go into the comparison are reasonably accounted for, the exercise is likely to be meaningless. Probably in no other scientific field would it be thought desirable (by the authors but not by me) to claim "general agreement" of a quantitative model of trophic structure with observations from a variety of ecosystems, when the predicted and observed quantities differ, frequently, by an order of magnitude or more (Heal and MacLean 1975). Hoping to avoid the temptation toward meaningless comparisons, I will now set forth some thoughts about energy flow in the Devon Island ecosystems.

As observed above, solar energy use efficiency in the HSM is at the low end of those observed in adequately watered systems. This refers to the sum of vascular plant, moss, and algal production. The last, especially, is poorly known and could be badly underestimated. However, the general conclusion seems safe. I feel that an explanation of this low efficiency is to be found in a combination of energetic and nutrient limitation effects: Wide spacing of *Carex* tillers is an evolutionary adaptation to the presence of permafrost. If the tillers were closer together, their shading of the soil surface would retard soil thaw, and hence nutrient release through decomposition. Ng and Miller (1976) modelled the effect of vegetative cover on depth of thaw in the Barrow ecosystem, and concluded that each unit increase in LAI decreased maximum thaw depth by 1.5 to 3 cm.

Moss production, which might otherwise make up for the low vascular plant production, is probably limited by nutrient availability (Babb and Whitfield this volume), while the vascular plants are not so severely limited.

The very low productivities in the CPL community certainly seem to follow from extreme water shortage due to low summer precipitation, very porous, rapidly draining soils, and perhaps vapor phase movement of water downward to the frost table. P. Addison (this volume) measured a seasonal average soil water potential of −20 bars in such an area. This brings into focus the question of why Truelove Lowland is such a productive oasis in the polar desert. It must follow, at least in large part, from trapping of water and nutrients by the many depressions and the basins behind raised beach ridges. Babb and Whitfield (this volume) further comment on this point.

As in apparently all terrestrial ecosystems (Heal and MacLean 1975), the bulk of primary production bypasses herbivores and is reduced by organisms of the decomposer complex.

Within the decomposer complexes, the relative roles of microbes and invertebrates seem clearly to be different in the CPL and HSM systems. Both microbivore and saprovore respirations are nearly the same, despite an eight-fold difference in inputs to the decomposer complexes. I feel that this is explained by the relative inhospitality of the dry CPL soils (and decaying matter in plant cushions) to microfloral and protozoan organisms which exist in water potential equilibrium with their surroundings. During much of the summer they must be inactive in the CPL community, while the more dessication-resistant invertebrates continue to function.

It is important to note that, although the HSM soils are predominantly organic and the CPL are mineral, the ratio of accumulation of organic matter to annual net primary production in the two systems is nearly the same. This is essentially true even if there is a large amount of unaccounted peat in the permafrost in the HSM.

Acknowledgments

My first and greatest debt is to L.C. Bliss, who provided leadership to the Devon Island Project in which cooperation between researchers was encouraged without loss of individual initiative, and who provided much advice and help to me.

This paper is a group effort, not in the writing, but because every member of the Devon Island Project has contributed data and ideas to it, and has given me the benefit of criticism at its various preliminary stages of development. My thanks go particularly to J. Ryan, who, more than any other, has contributed ideas and assistance.

Especially in regard to the speculations presented in the last section, I owe a great deal, impossible to properly acknowledge, to discussions with many people from all Tundra Biome projects.

References

Barr, W. 1971. Post glacial isostatic movement in northeastern Devon Island; a reappraisal. Arctic 24: 249-268.

Blaxter, K.L. 1967. The Energy Metabolism of Ruminants. Hutchinson. London. 332 pp.

Bunnell, F.L., S.F. MacLean, Jr. and J. Brown. 1975. Barrow, Alaska, U.S.A. pp. 73-124. *In:* Structure and Function of Tundra Ecosystems. T. Rosswall, and O.W. Heal, (eds.). Ecol. Bull. (Stockholm) No. 20. 450 pp.

Coleman, D.C., R. Andrews, J.E. Ellis and J.S. Singh. 1976. Energy flow and partitioning in selected man-managed and natural ecosystems. Agro-ecosystems. 3: 45-54.

Golley, F.B. 1960. Energy dynamics of a food chain of an old-field community. Ecol. Monogr. 30: 187-206.

———. 1972. Energy flux in ecosystems. pp. 69-90. *In:* Ecosystem Structure and Function. J.A. Wiens (ed.). Oregon State Univ. Press, Corvallis. 176 pp.

Heal, O.W. and S.F. MacLean Jr. 1975. Comparative productivity in ecosystems-secondary productivity. pp. 89-108. *In:* Unifying Concepts in Ecology. W.H. van Dobben and P.H. Lowe-McConnell (eds.) D.W. Junk, The Hague.

Medwecka-Kornás, A., A. Łomnicki and E. Bandoła-Ciołczyk. 1974. Energy flow in the oak-hornbeam forest (IBP Project "Ispina"). Bull. de l'Académie Polonaise des Sciences. Serie des sciences biologiques Cl. II. 22: 563-567.

Ng, E. and P.C. Miller. 1977. Validation of a model of the effect of tundra vegetation on soil temperature. Arctic Alp. Res. 9.

Østbye, E. (ed.), A. Berg, O. Blehr, M. Espeland, E. Gaare, A. Hagen, O. Hesjedal, S. Hoguar, S. Kjelvik, L. Lien, I. Mysterud, A. Sandhaug, H.J. Skar, A. Skartveit, O. Skre, T. Skogland, T. Solhoy, N.C. Stenseth and F.E. Wielgolaski. 1975. Hardangervidda, Norway. pp. 225-264. *In:* Structure and Function of Tundra Ecosystems. T. Rosswall, and O.W. Heal, (eds.). Ecol. Bull. 20. Swedish Natural Science Research Council. Stockholm. 450 pp.

Reichle, D.E. 1971. Energy and nutrient metabolism of soil and litter invertebrates. Unesco, 1971. Productivity of forest ecosystems. Proc. Brussels Symp., 1969. (Ecology and conservation, 4).

———. J.F. McBrayer and S. Ausmus. 1975. Ecological energetics of decomposer invertebrates in a deciduous forest, and total respiration budget. Progress in Soil Zoology, Proc. 5th Intl. Colloquium on Soil Zoology. Prague. pp. 283-292.

Satchell, J.E. 1969. Feasibility study of an energy budget for Meathop Wood. Unesco, 1971. Productivity of forest ecosystems. Proc. Brussels Symp., 1969. (Ecology and conservation, 4).

Teal, J.M. 1962. Energy flow in the salt marsh ecosystem of Georgia. Ecology 43: 614-624.

Woodwell, G.M. and R.H. Whittaker. 1968. Primary production in terrestrial ecosystems. Am. Zoologist, 8: 19-30.

Inuit

The utilization of wildlife in the Jones Sound region by the Grise Fiord Inuit

R.R. Riewe

Introduction

Since Europeans first began to explore the North American Arctic, the region has held an aura of challenge and mystery which has created an endless demand for both popular and scientific information about it. This demand for arctic literature was filled at first by an ever-expanding number of arctic explorers and later by scientists and journalists. Since man usually possesses a keen curiosity about members of his own species, much arctic literature has centred around indigenous northern inhabitants, the Inuit.

Most anthropological studies dealing with the Inuit and their subsistence strategies have been qualitative and/or theoretical in nature. However, there has been a trend developing among a few northern researchers to quantify the Inuit's dependence upon the land, some of the most-noteworthy examples being Foote and Williamson (1966), Foote (1967), Freeman (1969-70), Kemp (1971), and Usher (1971, 1976).

In following this trend, the objectives of this study were to quantify the dependence of Grise Fiord Inuit upon the land and their utilization of wildlife and to examine their role in arctic ecosystems, both terrestrial and aquatic.

Prehistoric and historic habitation of Devon and Ellesmere Islands

Euro-North American explorers in the Ellesmere-Devon region often remarked on the abundance of archeological ruins left by past inhabitants of the area (particularly Sverdrup 1904 and Humphreys et al. 1936); for a thorough review of the explorations in the region, see Taylor 1955. Due to the remoteness of the region, it is to this day virtually unexplored archeologically. The few sites which have been excavated on Devon (Lethbridge 1939, Lowther 1962, McGhee 1974) and Ellesmere (Speck 1924, Lethbridge 1939, Bentham and Jenness 1941, Maxwell 1960) indicate that peoples with Pre-Dorset, Dorset, and Thule cultures (ancestors of the Inuit) have inhabited the islands. By employing the chronology used by Maxwell (1972) on Baffin Island, it may be assumed that the islands have been occupied by arctic hunters since at least 3,000 B.P., or possibly earlier. For some unknown reason, these early hunters migrated away from the northern islands of the arctic archipelago by at least the 18th century, as indicated by the fact that none were seen by the explorers of the 17th and 18th centuries.

In the mid-19th century there was a temporary occupation of Devon and

Ellesmere islands by a band of Tununirmuit from Baffin Island. In 1856, 40 Inuit led by Qidlak, a shaman, migrated north from Baffin in search of the polar Inuit whom they had heard about from the explorer Inglefield. After spending a few years on Devon, part of the group decided to return to Baffin. The remaining 15 members of the party continued north up across Ellesmere and, in 1860, ended their journey by crossing Smith Sound and reaching Inglefield Land where they settled and mixed with the Greenland polar Inuit (Marie-Rousseliere 1972).

During the next 100 years the only Inuit on Ellesmere and Devon were either seasonal hunters from Greenland or families employed by the Royal Canadian Mounted Police. For reasons of sovereignty the Canadian government began establishing R.C.M.P. posts in the High Arctic in the 1920s; the first post was erected on the southeast coast of Ellesmere at Craig Harbour (Fig. 1) and was in operation for three periods: 1922 to 1927, 1933 to 1940, and 1951 to 1956. The Kane Basin post on Bache Peninsula was the next one built on Ellesmere and was open between 1926 and 1933. This post was replaced by one at Alexandra Fiord in 1953 which remained in operation for at least six years. A post was also set up on the southeast coast of Devon at Dundas Harbour and was occupied by the R.C.M.P. from 1924 to 1933 and from 1945 to 1951; this post was also used by the Hudson's Bay Company from 1934 to 1936 (Dunbar and Greenaway 1956, Bruemmer 1968, Grise Fiord records). Usually one or more Inuit families from either Etah, Greenland, or Pond Inlet, Baffin Island resided with the members of the R.C.M.P. detachments. The Inuit men were hired as guides, general factotums, and often acted as guardians on the trail; the women were hired as seamstresses.

In the 1950s the Department of Northern Affairs and National Resources became concerned about the elimination of game in areas relatively densely populated by Inuit. One such area was Port Harrison on the east shore of Hudson Bay where the caribou herd was at a critically low level. The government offered its assistance to any families willing to settle in the assumed game-rich Far North. In September 1953, the Canadian icebreaker *C.D. Howe* brought six families, principally from Port Harrison, Quebec to southeastern Ellesmere. For these Inuit the move of more than 2,200 km north meant a major adjustment to the unfamiliar long winter nights and mountainous topography. Partly for this reason, families from Pond Inlet, Baffin Island were also induced into migrating to Ellesmere. It was thought that these high arctic Inuit from Baffin could help the more-southern Inuit to adapt to the rigorous conditions of the region (Bruemmer 1968, Freeman 1969).

The families initially took up residence near the R.C.M.P. post at Craig Harbour and lived in tents in the summer and igloos in the winter. Due to the frequent occurrence of violent local winds at Craig Harbour, they soon abandoned the area and built homes 60 km west on the southern shore of Lindstrom Peninsula. In 1956, the R.C.M.P. likewise left wind-swept Craig Harbour and shifted to the present-day site of Grise Fiord. Shortly thereafter the Inuit moved their homes to the R.C.M.P. post, thereby creating the present settlement of Grise Fiord, the northern-most settlement in North America (Bruemmer 1968).

As good news regarding the hunting filtered south, additional families from Pond Inlet and Port Harrison, as well as a few from Pangnirtung, Baffin Island, migrated north to Grise Fiord. Between 1960 and 1972, the number of families fluctuated from 15 to 18, with a total population of 70 to 96. This population has been supported primarily by the activities of the hunters (men 16 years of age or older) who have numbered from 13 to 18.

Methods

The data presented in this paper were gathered during field investigations conducted between 1970 and 1973. During these years, I spent a total of 410 days collecting information on the wildlife of the northern islands of the Canadian Arctic Archipelago, primarily in the Jones Sound region.

Between 1971 and 1973, I resided in Grise Fiord for approximately 106 days and accompanied the men on the trail for an additional 51 days during which we traversed more than 3,500 km by snowmobile and komatik (Inuit sled). While residing in the village, I accompanied the men on their canoe hunting trips for 14 days. An incalculable amount of time was spent with the hunters obtaining information on the wildlife populations, harvesting, and processing techniques and their utilization of the game.

Eight senior hunters and one woman made tape recordings in which they discussed their hunting ranges, their past and present hunting techniques, the wildlife populations, and their utilization of wildlife. They also expressed their frustrations concerning the technological invasion of the Arctic and their anxieties for the future (Riewe 1973a).

In addition to the above data, I was given access to many of the Grise Fiord records, including the R.C.M.P. Game Condition reports from 1951 to 1971; the N.W.T. Traders' fur records from 1964 to 1973; the settlement's vital statistics from 1959 to 1972; the requisitions and inventories for the Craig Harbour Trading Store from 1954 to 1956 and for Thomassie's Trading Store at Grise Fiord from 1957 to 1959; and the requisitions, inventories, and account ledgers for the Grise Fiord Inuit Cooperative from 1960 to 1971. At no time has there been more than one of these trading stores in operation in the Jones Sound region.

I conducted extensive aerial and ground surveys of the wildlife in the northwestern section of Jones Sound during May and July 1973 while under contract to the Department of Indian and Northern Affairs (Riewe 1973b). Three aerial surveys flown during two days in May and four days in July covered 5,814 km of flight lines over southwestern Ellesmere, southeastern Axel Heiberg, Graham, and Buckingham islands. The ground survey traversed more than 1,365 km in the Bjorne Peninsula-Vendom Fiord area between 8 and 28 May.

Results

Inuit travelling and hunting routes

The major routes travelled by the Grise Fiord Inuit during the winter months are depicted in Fig. 1. These routes, which sprawl over an area of approximately 411,800 km^2 (area covered by map in Fig. 1), are differentiated as to the amount of traffic they have borne since 1951. The dates that the various routes were first established by the hunters are also depicted. There are 1,190 km of routes designated as "used regularly"; these have been traversed by at least 16 hunting parties during the last two decades. In fact, ever since these regularly used routes were first established, most of them have been travelled by several hunting parties annually. In addition to these routes, there are 5,170 km of routes which have been used less frequently; since 1951, these have been traversed by 1 to 5 or 6 up to 15 hunting parties. Two of the infrequently

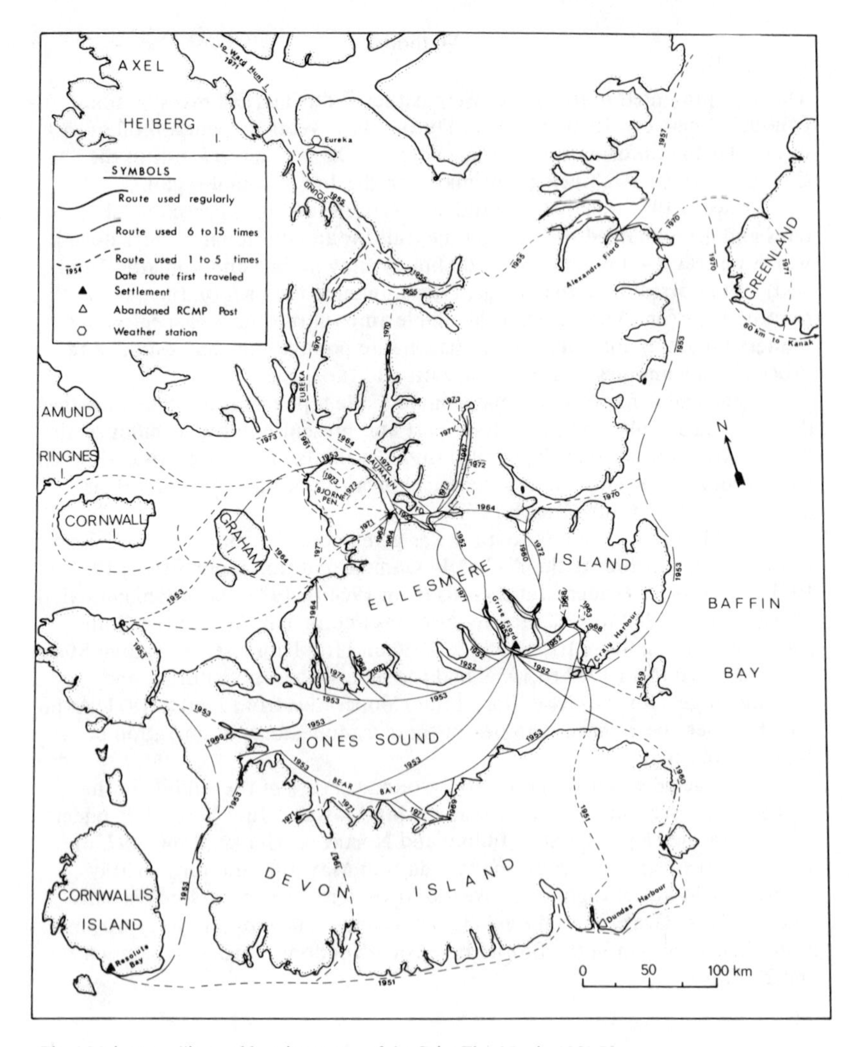

Fig. 1 Major travelling and hunting routes of the Grise Fiord Inuit, 1951-73.

travelled routes are not completely shown, but their destination points are given.

The exact location of routes which traverse bodies of water or lowlands are affected by local ice and snow conditions, as well as the occurrence of game and are therefore highly variable from year to year. Routes crossing overland through mountainous terrain are, by contrast, rigidly fixed from year to year due to the confining topography.

Whenever the Inuit are travelling, they are constantly assimilating information on the terrain, travelling conditions, and game. This information is readily transmitted to their fellow hunters when they return to the settlement. If the areas traversed are productive hunting and fishing territories, then other hunters normally return to these regions at the earliest possible time.

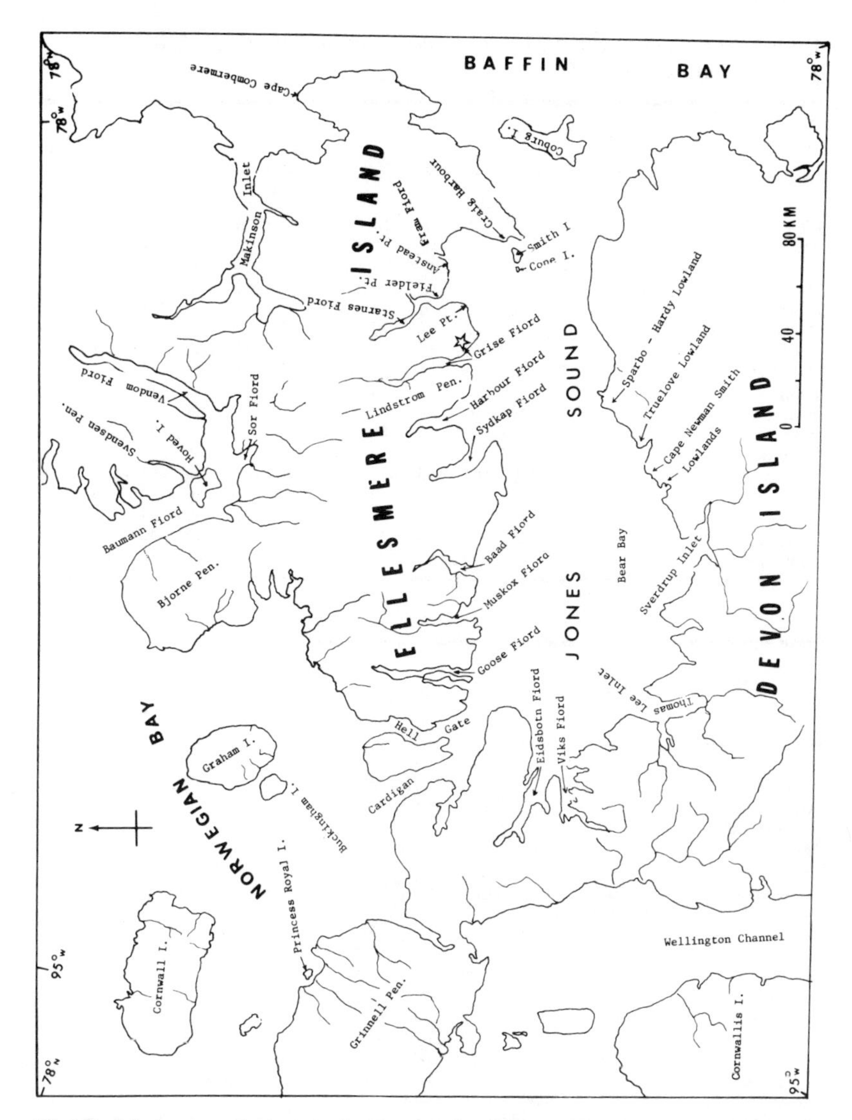

Fig. 2 English place names in the major hunting area of the Grise Fiord Inuit.

Wildlife and its utilization by the Inuit

The majority of the hunting activities of these people are confined to the Jones Sound region proper; Fig. 2 outlines this region and gives many of the pertinent place names. This area encompasses approximately 97,000 km^2 and, except for an occasional Greenlander who hunts along the east coast of Ellesmere or the Resolute Bay Inuit who regularly hunt on Cornwallis Island and Wellington Channel, this vast area is the exclusive hunting domain of the Grise Fiord people. Their main hunting efforts are directed towards harvesting marine mammals inhabiting Jones Sound and the inlets and fiords of southern Ellesmere and northern Devon islands. To a lesser extent they harvest

Table 1. Summary of game taken by Grise Fiord hunters based on R.C.M.P. wildlife records, 1956-1972.[a]

Species	*1956-57*	*1957-58*	*1958-59*	*1959-60*	*1960-61*	*1961-62*	*1962-63*	*1963-64*	*1964-65*	*1965-66*	*1966-67*	*1967-68*
Polar bear	13	8	17	22	16	18	45	16	29	36	45	17
Arctic wolf	1	2	3	2	0	3	1	6	3	3	5	5
Arctic fox	184	664	667	54	534	1,015	38	325	550	63	83	221
Short-tailed weasel	6	5	8	1	3	3	1	0	2	6	5	4
Muskox	0	0	0	0	0	0	0	0	0	0	0	0
Peary's caribou	15	13	0	2	23	20	28	11	12	37	12	24
Arctic hare	90	70	20	50	50	40	100	100	150	62	30	100
Narwhal	0	0	0	0	0	3	0	20	0	27	0	7
Beluga	6	15	5	25	35	10	75	83	20	48	118	0
Walrus	46	39	46	50	38	40	21	19	31	30	35	19
Bearded seal	34	32	27	25	25	40	23	28[c]	20	30	35	25
Harp seal							70		70	25	15	20
Ringed seal	560	485	539	500	600	955	537	620	652	515	400	500
Snow goose	7	15	1	0	1	0	12	17	10	0	9	0
Ducks	23	12	84	41	110	175	35	55	50	57	61	104
Ptarmigan	151	300	50	200	400	125	95	75	40	47	26	79
Arctic char	0	0	0	0	DNA	200	600	300	500	302	250	200

Species	*1968-69*	*1969-70*	*1970-71*	*1971-72*	*Average harvest*
Polar bear	27	29	25	27	24
Arctic wolf	12	9	8	0	4
Arctic fox	274	174	86	324	328
Short-tailed weasel	6	5	2	0	4
Muskox	0	12	9	9	10[d]
Peary's caribou	75	46	60	26	25
Arctic hare	DNA[b]	DNA	DNA	DNA	72
Narwhal	10	49	25	6	9
Beluga	0	0	0	0	28
Walrus	15	7	17	12	28
Bearded seal	20	28[c]	28[c]	28[c]	28
Harp seal	68				45
Ringed seal	1,000	591	901	485	590
Snow goose	10	8	30	DNA	8
Ducks	36	35	25	DNA	60
Ptarmigan	24	92	50	DNA	117
Arctic char	275	1,000	DNA	DNA	279

a. Reporting period = 1 July to 30 June
b. DNA = data not available
c. Since R.C.M.P. wildlife records do not distinguish between ringed, harp and bearded seals, I have assigned this average figure (28) for the bearded seal take and subtracted it from the total seal take to estimate the harp and ringed seal take.
d. Average harvest of muskox calculated only for the years 1969 through 1972.

terrestrial wildlife, most of which occurs on the lowlands and rolling hills of Ellesmere and Devon islands. Much of the mountainous topography in the region is virtually devoid of game. Some fishing is also conducted in the region.

The following is a synopsis of the wildlife in the Jones Sound region and its utilization by the Grise Fiord people.

Table 2. Harvest and distribution of kilocalories by Grise Fiord hunters in the year 1971-72.

Species	*Number of animals harvested*	*Kcal harvested*[a]	*Percent of total Kcal harvested*	*Kcal consumed by Inuit*	*Kcal consumed by dogs*	*Kcal consumed by decomposers and scavengers*	*Kcal shipped south*[b]
Arctic char	279[c]	4.82×10^{5}	0.25	3.61×10^{5}	0	1.21×10^{5}	0
Snow geese	8[c]	5.60×10^{4}	0.03	3.71×10^{4}	0	1.89×10^{4}	0
Ducks	60[c]	1.37×10^{5}	0.01	8.30×10^{4}	0	5.44×10^{4}	0
Ptarmigan	117[c]	1.72×10^{5}	0.09	1.06×10^{5}	0	6.55×10^{4}	0
Polar bear	27	2.27×10^{7}	11.90	1.03×10^{6}	5.92×10^{6}	1.36×10^{7}	2.16×10^{6}
Arctic fox	324	2.50×10^{6}	1.31	9.05×10^{4}	0	2.21×10^{6}	1.95×10^{5}
Walrus	12	1.75×10^{7}	9.17	1.28×10^{6}	1.09×10^{6}	1.51×10^{7}	0
Ringed seal	453	9.02×10^{7}	47.30	6.51×10^{6}	2.12×10^{7}	5.73×10^{7}	5.14×10^{6}
Harp seal	32	1.58×10^{7}	8.30	0	2.05×10^{6}	1.25×10^{7}	1.30×10^{6}
Bearded seal	28[c]	2.49×10^{7}	13.04	6.60×10^{6}	3.40×10^{6}	1.49×10^{7}	0
Arctic hare	72[c]	5.44×10^{5}	0.29	2.58×10^{5}	0	2.86×10^{5}	0
Peary's caribou	26	2.15×10^{6}	1.13	9.35×10^{5}	0	1.21×10^{6}	0
Muskox	9	3.52×10^{6}	1.84	1.71×10^{6}	0	1.71×10^{6}	1.19×10^{6}
Narwhal	6	1.02×10^{7}	5.34	6.78×10^{5}	8.26×10^{5}	8.70×10^{6}	0
Totals		1.91×10^{8}		2.00×10^{7}	3.45×10^{7}	1.28×10^{8}	9.99×10^{6}
Percent utilization of total harvest				10.5	18.2	67.1	5.2

a. The kilocalories harvested have been calculated from calorimetric data collected during this study and from the literature (Foote, 1967; Freeman, 1969-70; Usher, 1971).
b. Kilocalories shipped south are in the form of skins and furs.
c. Data not available on the number of animals harvested for this species. I have used the average number of animals taken between the years 1956 and 1972 for the calculations in this table.

Arctic char (Salvelinus alpinus)

Arctic char are highly valued by the Grise Fiord people not only because of their delicious flesh but also because char were traditional items in the Inuit's diet at their former homes on Baffin Island and Ungava Peninsula. Due to the physiography of the Jones Sound region, the area is very poor in lakes which can support char populations. Not until 1961 were any fishing lakes discovered by the Inuit. Since then, the hunters have located a total of 20 lakes in the region which sustain char; only three of these lakes have been harvested to any extent.

The R.C.M.P. records indicate that, since 1961, an average of 404 char have been taken per year (Table 1). Char from these lakes weigh up to a maximum of about 7 kg; the average is probably about 1.5 kg. Char are extremely important in adding variety to the Inuit's diet, all of the meat being consumed by the people. Their growth rates are discussed by Minns (this volume).

Rock ptarmigan (Lagopus mutus)

Rock ptarmigan are likely to be found in small numbers in any rocky area which supports arctic willow *(Salix arctica)*, the ptarmigan's major food source.

Between 1956 and 1971, an average of 117 ptarmigan have been taken annually (Table 1). Each bird provides about 0.25 kg of meat, all of which is eagerly consumed by the Inuit.

Glaucous gull (Larus hyperboreus), *black guillemot* (Cepphus grylle), *Common eider* (Somateria mollissima), *king eider* (S. spectabilis), *old squaw* (Clangula hyemalis) *and greater snow goose* (Chen caerulescens atlantica)

Sea birds, ducks, and geese play a very minor role in the people's diet. The average annual take of ducks and geese constitutes only 0.04% (Table 2) of the community's total meat harvest; the sea birds are of even less importance. This is due to the relatively limited populations of ducks and geese in the vicinity of the village and to a distaste for gulls and guillemots by most of the people. The few birds taken are shot by the men during the open water season while hunting seals from canoes. Occasionally, a few duck or gull eggs are collected in spring.

Arctic hare (Lepus articus)

Arctic hares occupy most of the lowlands and talus slopes in the Jones Sound region, but usually only in small groups of two to five animals. The largest concentration observed in the region was a band of approximately 70 hares on the northern end of the Bjorne Peninsula. On Raanes and Fosheim peninsulas, northern Ellesmere, hares often gather into tremendous bands numbering into the thousands. These large bands, however, are outside the Inuit's present hunting territory and thus are not harvested.

Hares are usually hunted only when encountered by the Inuit who are searching for caribou and muskox in autumn or late winter. Between 1956 and 1968 an average of 72 hares were shot annually (Table 1). The average weight of 15 specimens shot in the area was 3.76 kg. Each hare yields approximately 2 kg of meat, all of which is consumed by the people. Hare fur possesses excellent insulative properties, but the skin is very weak and rips easily. The fur is, however, sometimes used for ruffs on the parkas of the smaller children.

Peary's caribou (Rangifer tarandus pearyi)

According to archeological evidence, caribou were abundant in the Jones Sound region several hundred years ago (Lethbridge 1939, Bentham and Jenness 1941, Lowther 1962). By the late 19th and early 20th centuries, however, caribou had become scarce in the region (Sverdrup 1904). From aerial surveys conducted in 1973 (Riewe 1973b), I would estimate that there are presently fewer than 250 caribou in the entire Jones Sound region. During the last 16 years, the hunters have harvested an average of only 25 caribou annually (Table 3). Small as this harvest might be, it is still apparently hastening the demise of caribou on Devon Island and southern Ellesmere Island.

The only relatively large concentration of caribou observed during the 1973 aerial surveys was in the vicinity of Blind Fiord on the Raanes Peninsula where an estimated 300 animals existed. This area provides considerably better grazing for the caribou than do the areas further south on Ellesmere, Graham, Buckingham, and Devon islands. Raanes Peninsula is presently just north of Grise Fiord's hunting range, but undoubtedly the men will expand their hunting activities to this region as soon as possible.

Caribou provide bedding skins, sinew, and meat. The average Peary's caribou weighs approximately 45 kg alive and yields about 22.5 kg of meat. All the meat harvested is eagerly consumed by Grise Fiord residents, who hold it in very high esteem.

Table 3. Number of Peary's caribou taken by the Grise Fiord Inuit, 1951-spring 1973.

Area	*Year*	*Number of caribou*
Southeast Ellesmere Island	1953-54	26
	1954-55	29
	1955-56	24
	1957-58	13
	1960-61	1
	1961-62	12
	1962-63	28
	1963-64	6
	1964-65	1
	1967-68	1
		Area total: 141
Sparbo-Hardy Lowland, Devon Island	1959-60	Area total: 1
Coburg Island	1960	Area total: 1
South Grinnell Peninsula, Devon Island	1956	1
	1965	10
	1966	32
	1967	12
	1971	3
		Area total: 58
Southwest Ellesmere Island	1962	8
	1968	6
	1969	5
	1970	15
	spring, 1973	6
		Area total: 40
Graham Island	1961	23
	1968	7
	1969	30
	1970	15
	1971	30
		Area total: 105
Bjorne Peninsula and Vendom Fiord, Ellesmere Island	1964-65	1
	1968-69	40
	1969-70	16
	1970-71	30
	1971-72	26
	fall, 1972	21
	spring, 1973	7
		Area total: 141
Sor Fiord to Makinson Inlet, Ellesmere Island	1965-66	5
	1967-68	10
	fall, 1972	6
	spring, 1973	5
		Area total: 26
		Total caribou harvest: 513

Muskox (ovibos moschatus)

Based on ground sightings, Freeman (1971) made a conservative estimate of 970 muskox for the Jones Sound region, excluding Grinnell Peninsula. By combining my estimates from the July 1973 aerial survey with Freeman's figure, I estimate that the present muskox population in this region, excluding Grinnell Peninsula (no data available), is approximately 1,600 animals. The largest concentrations are located on northeast Devon Island, Bjorne Peninsula, Vendom Fiord, and Sor Fiord.

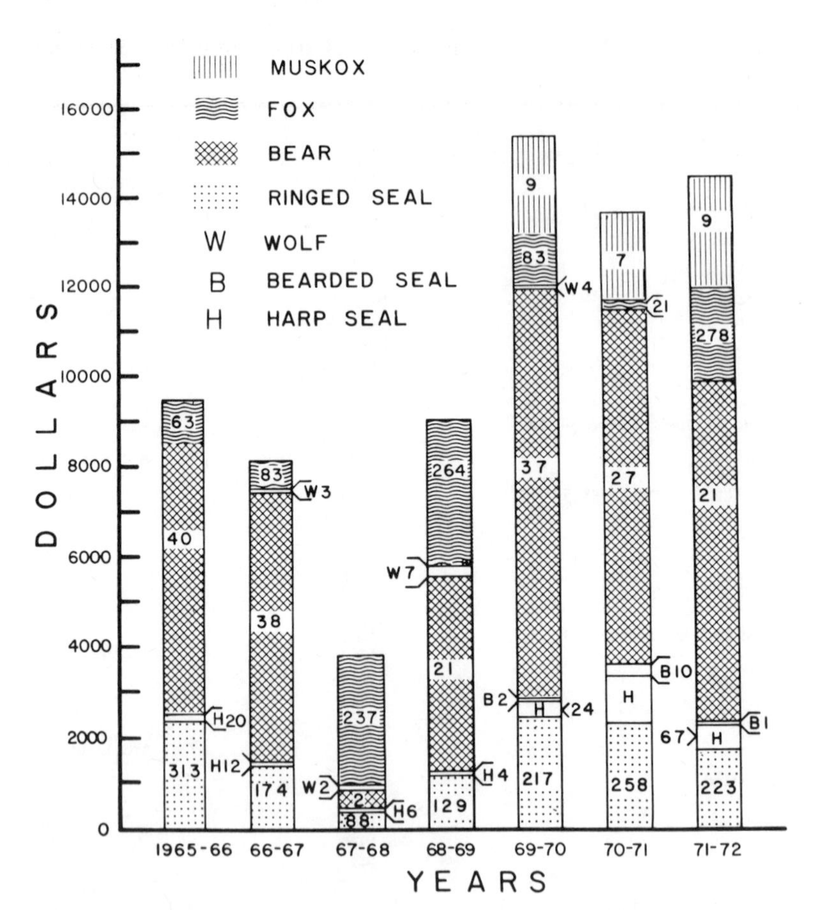

Fig. 3 Total annual income to Grise Fiord hunters from furs and skins sold to the Grise Fiord Co-op. Years run from 1 July to 30 June. Numbers to the sides of the columns indicate the number of furs or skins sold.

In 1917 the Canadian government placed a ban on muskox hunting throughout the country. In 1969 the N.W.T. government opened a muskox hunting season in the Jones Sound region. The Grise Fiord people were allowed to harvest 12 muskox annually from five different zones. The quotas for 1969 through 1972 included three animals to be harvested from two zones on Devon Island and two animals from each of three zones on Ellesmere Island. In 1973 the quota was increased by eight animals.

Grise Fiord muskox hunters selectively shoot adult bulls rather than cows because they provide more meat and possess considerably larger and more-valuable horns. Occasionally, however, cows are shot by mistake. The average adult muskox harvested in this region weighs about 275 kg and yields about 110 kg of meat. All of the meat is avidly consumed by the Inuit, for it adds variety to their diet.

Muskox horns removed from shot animals or scavenged from old carcasses found along the trail, are worth from $10 to $100 a set to the hunters. The muskox pelts have brought the hunters from $100 to $450 apiece. From 1969 to 1972 muskox pelts provided the settlement with 15.6% of its total cash income derived from the sale of furs and skins (Fig. 3).

Short-tailed weasel (Mustela erminea)

Weasels appear to be relatively scarce in the Jones Sound region. During my studies, I have noted weasels or their sign in only six localities. The apparent scarcity of weasels in the region is related to the scarcity of collared lemmings *(Dicrostonyx groenlandicus)*, the weasel's principal prey (see Speller 1972 and Fuller et al. this volume). The weasel's subnivean habits may also account for its apparent scarcity during much of the year.

Weasel pelts have been worth very little during the past couple of decades, only $0.25 to $0.75 apiece, and usually only the children have bothered to trap them. Between 1956 and 1971 an average of only four weasels were captured per year (Table 1); in fact, between 1965 and 1971, only $6.75 worth of weasel pelts were sold to the Co-op.

Arctic fox (Alopex lagopus)

The arctic fox population in the Jones Sound region fluctuates greatly from year to year as it does throughout the rest of the Arctic. Even during years of a low fox population, fox tracks can be encountered just about anywhere on land or on the sea ice.

The hunters have regularly trapped along the south coast of Ellesmere between Craig Harbour and Harbour Fiord since 1953. During years of high fox numbers, the men have also set traps along northeastern Devon, southwestern Ellesmere, and across Ellesmere from the head of Grise Fiord north into Baumann Fiord. Despite a relatively large fox population in Baumann Fiord, the Inuit do not regularly set trap lines there because the trapped foxes are often destroyed by wolves.

Between 1965 and 1972 the sale of fox pelts provided the Inuit with an average 15% of their cash income derived from the sale of furs (Fig. 3). The yearly fox harvest between 1956 and 1972 fluctuated from a low of 38 in 1962-63 to a high of 1,015 (including one red fox, *Vulpes fulva*) in 1961-62, with an average annual harvest of 328 pelts (Table 1).

An average fox in this region weighs approximately 2.7 kg (n = 5) and yields about 1.8 kg of meat and bones. Only a small portion of the fox carcasses are consumed; I estimate that 5% of the foxes harvested are eaten by the former residents of Port Harrison, with the rest of the carcasses being left for the scavengers.

Arctic wolf (Canis lupus arctos)

The arctic wolf is thinly scattered throughout the Jones Sound region and is usually found in those areas occupied by ungulates. On Devon Island wolves or their sign have been recorded on the major muskox grounds around the lowlands between Cape Newman Smith and Sverdrup Glacier during the years 1951-52, 1954-55, 1960-61, 1968-69, 1969-70, and 1970-71. In 1957-58 a single wolf was seen at Viks Fiord, while in 1958-59 tracks of four or five animals were observed near Dragleybeck Inlet. Later, in 1965-66, tracks of three or four wolves were noted in the vicinity of Bear Bay. From my field studies in the Jones Sound region, it appears that, from May 1970 to October 1971, there were only three or possibly four wolves inhabiting the entire north shore of Devon Island. To my knowledge, there were no sightings of wolves or their sign along north Devon from November 1971 to July 1973.

Wolves have been more numerous on Ellesmere Island than Devon Island. The R.C.M.P. records annually reported wolves as being present on southern Ellesmere Island. Most of these reports came from the southeast coast between Craig Harbour and Harbour Fiord, Baumann Fiord and Makinson Inlet (those areas most commonly travelled by the hunters and

R.C.M.P. patrols). My studies indicate that the greatest concentration of wolves on southern Ellesmere Island is in the vicinity of Baumann Fiord.

The hunters have harvested an average of four wolves annually between 1956 and 1972 (Table 1). Since 1964, when the wolf bounty was reimposed in the Northwest Territories, the hunters have turned in 45 carcasses for bounty payment. The reintroduction of the wolf bounty and the acquisition of snowmobiles have contributed to the increase in numbers of wolves taken in the last ten years (see Table 1).

A wolf provides the hunter with a $40 bounty payment and a pelt worth $10 to $100. Some wolf pelts are used locally for parka trim.

The average wolf in the Jones Sound region weighs about 32 kg (n = 4). None of the carcasses is consumed by either the people or their dogs.

Polar bear (Ursus maritimus)

Polar bears are fairly numerous in the Jones Sound region. They may be encountered anywhere on the sea ice where ringed seals, the bear's obligate prey, occur. According to the R.C.M.P. records, one of the largest concentrations of bears in Jones Sound has been Bear Bay along the north coast of Devon Island, where almost 150 bears were shot between 1959 and 1971. There are also concentrations of bears both near Coburg Island and in the vicinity of Hell Gate but, due to treacherous ice conditions, these areas are hunted only infrequently.

The hunters have taken as many as 45 bears in one year; the average annual harvest has been 24 bears (Table 1). In 1967 a quota of 27 bears was imposed upon the hunters. The men have managed to fill their quotas nearly every year. In the spring of 1973, the N.W.T. government allowed the Grise Fiord hunters to harvest an additional six bears from Norwegian Bay.

Fifteen years ago a bear pelt was worth only $10 to $50 to the hunter. Prior to this time, bears were hunted primarily for the status bestowed upon the successful hunter and for the meat. The skins were either sold to whites or used as bedding skins or as tarps to cover their komatiks while travelling. Between 1956 and 1972, bear pelts provided the Inuit with 55% of their cash income derived from the sale of furs to the Co-op (Fig. 3). During the past couple of years, the price of bear skins has been sky-rocketing; in 1973 the hunters received between $500 and $3,000 per skin.

An average polar bear carcass weighs about 270 kg and yields about 100 kg of meat and 100 kg of fat. I estimate that about 20% of all bear meat harvested is consumed by the Inuit, while the dogs consume an additional 30% of all meat, bones, and fat. The remaining meat, bones, fat, and viscera are left for the scavengers on the sea ice.

Beluga (Delphinapterus leucas)

After the ice breaks up in Jones Sound, the beluga move into the Sound from Baffin Bay and the Hell Gate-Cardigan Strait area. This is usually in the month of July. If the shoreline is ice-free, the Inuit can spot the beluga as they migrate past the settlement in vast pods. The R.C.M.P. records reported that a spectacular number of beluga was seen in close concentration as far as the eye could see in mid-September 1959; no other whales were seen that year. In 1963 an estimated 3,000 beluga went past the settlement, and in 1966 a herd was seen that extended nearly 1.6 km in length and 25 to 35 m wide. There were three years, 1968, 1969, and 1972, during which no beluga were observed; this was due to heavy ice in the vicinity of the settlement.

The beluga usually migrate out of Jones Sound during the latter half of September, just ahead of freeze-up. In the winter of 1966-67 a pod of approximately 110 whales was trapped in Jones Sound at the mouth of Starnes Fiord by an early freeze-up (Freeman, 1968).

The hunters harvested an average of 40 beluga annually between 1956 and 1967 (Table 1). Since 1967, beluga have not been hunted, partly because of bad ice conditions for several years and partly because of a diminishing need for dog food.

The average beluga weighs about 450 kg and provides approximately 76 kg of meat, 76 kg of muktuk (skin) and 113 kg of blubber. The Grise Fiord people have used most of the protein and fat as dog food and generally consumed a portion of the muktuk themselves.

Narwhal (Monodon monoceros)

Narwhal migrate from Baffin Bay into Jones Sound as the ice begins to break up in June or July. Because they will penetrate into open leads, they usually arrive somewhat earlier than the beluga, who usually migrate into the Sound only after the ice has broken completely. Narwhal enter the fiords of Jones Sound in pods of 30 to 150 animals and remain there feeding throughout the period of open water. There have been several years in which narwhal were not reported or only in very small numbers.

The first narwhal harvested in Jones Sound were taken in 1961-62 (Table 1). Between 1965 and 1972, an average of nine was harvested annually.

The average narwhal weighs approximately 540 kg. Much of the carcasses have been used for dog food. Narwhal muktuk is considered superior in taste to beluga muktuk and is eaten as a delicacy by the Grise Fiord residents. In 1970 narwhal muktuk was sold to the local Co-op for resale to other Inuit communities where this commodity was not available. This business venture, unfortunately, was not repeated due to the prohibitively high costs of arctic freight transportation.

In recent years, ivory tusks of male narwhal have become a noticeable source of income to the hunters. In 1972 narwhal ivory sold for $44 per kg, which is as much as $300 per tusk. Narwhal sinew is both longer and stronger than caribou sinew and is preferred by the women for sewing skin clothing.

Walrus (Odobenus rosmarus)

Walrus, like narwhal, migrate into Jones Sound from Baffin Bay when the sea ice begins to break up in June or July. Females and young congregate for the summer along the lip of Jakeman Glacier, while the males travel further west towards the Hell Gate area. During open water, walrus have often been seen hauled out on the lowland about 15 km west of South Cape. Just prior to freeze-up, walrus have also been seen in herds numbering into the hundreds in the vicinity of Cape Sparbo as they migrate eastwards out of Jones Sound. Occasionally walrus are stranded in the Sound by a quick freeze-up and are forced to maintain aglus (breathing holes in the sea ice) throughout the winter.

From 1956 to 1972 an average of 28 walrus was harvested. Since 1967, when snowmobiles were adopted by the hunters, there has been a concommitant decline in the demand for dog food. Walrus meat, which is considered by the Inuit to be the best dog food available, has since been harvested in diminishing quantities (Table 1). The average walrus weighs about 675 kg and provides approximately 236 kg of meat (Foote 1967). Back in 1956 the men harvested about seven walrus per family in order to feed their dogs throughout the winter trapping season (R.C.M.P. 1951-71). Some walrus meat, particularly from calves, has also been consumed by the people.

In the past, the Inuit utilized just about the entire carcass for diverse purposes. Now, walrus are killed primarily for their ivory, which brings the hunters from $30 to $50 a set when sold to the Co-op in an unfinished state. If the ivory has been carved into a figurine or cribbage board, the hunter derives a much greater profit. Baccula (penis bones) are in demand by the tourist and are sold by the hunters for $5 to $10 apiece. Usually only 50 to 60 kg of meat is now brought back to the village for human and dog consumption; the remainder of the carcass is left for the scavengers.

Ringed seal (Phoca hispida)

The ringed seal is the most abundant species of seal in the eastern High Arctic. It occurs wherever the ice is sufficiently stable to permit breeding (Mansfield 1967). This seal is found throughout the entire marine area depicted in Fig. 1. In Jones Sound there is a large year-round population of ringed seals which maintain aglus throughout the winters. This population is augmented by large numbers of other ringed seals which migrate from Baffin Bay into Jones Sound for the summer months. Ringed seals, which are solitary animals, are found in highest concentration within 15 km of shore; further from shore their numbers usually decline.

The ringed seal is the primary prey of the Grise Fiord Inuit and has played a major role in their culture. This seal is the preferred food of the people and it is the most stable game population in the region. Approximately 590 ringed seals have been harvested annually between 1956 and 1972 (Table 1). The primary sealing area, where probably 95% of the seals (all species) are harvested, is limited to the area within a 60-km radius of the settlement.

The average ringed seal harvested weighs 48 kg ($n = 33$) and yields approximately 19 kg of meat and bones, 15 kg of blubber, 6 kg of pelt, and 8 kg of viscera.

The people consume ringed seal meat, blubber, liver, intestine, and heart. Young seals are definitely preferred over adult seals; adult male seals in rut or seals drowned in nets are never eaten by the Inuit unless no other seal meat is available. Ringed seal is also a major component of the dogs' diet; they devour prodigious amounts of meat, blubber, and viscera which are unfit for human consumption.

Seal pelts are used in numerous ways by the Inuit. Some skins are cut into ropes used for lashing komatiks and, in the past, for dog traces. Much of the decorative and utilitarian kamik (boot) worn by the Inuit is made from ringed seal skin. In addition to the kamik, innumerable handicraft items, such as rugs, gun cases, and toys are crafted from this species of seal by the women and sold to the Co-op. A large share of the pelts is sold to the Co-op as dried skins; between 1965 and 1972, from 12.9% to 60.8% (average 36.5%) of the annual harvest of pelts was sold to the Co-op. For these years, ringed seal pelts comprized 16.3% of the cash income originating from the sale of furs (Fig. s).

Harp seal (Phoca groenlandica)

During the period of open water, harp seals frequent Jones Sound in large numbers. The herds usually consist of 30 to 50 animals but may be considerably larger. The herds migrate northwards from their breeding and moulting areas on either the Labrador Front or the Gulf of St. Lawrence. By the time the seals enter Jones Sound in July, the herds are composed almost entirely of adult animals three years or older; only rarely are younger animals seen by the Inuit.

The hunters have taken an average of 45 harp seals annually between 1956 and 1972 (Table 1). When harp pelts are reasonably valuable on the fur market, the hunters sell all of their skins to the Co-op. When the pelts have little value, they are usually used locally for tarps, kamiks, and rope. Between 1965 and 1972, harp seal skins provided only 3% of the cash income derived from the sale of furs to the Co-op (Fig. 3).

The average harp seal weighs about 135 kg, with an estimated 40% of all the meat and bones and 10% of the blubber being fed to the dogs. Everything else is left for the scavengers. A few of the hunters actually cut the waste parts of seal and whale carcasses into small chunks so that the sea gulls can consume the waste with greater ease.

Bearded seal (Erignathus barbatus)

Most bearded seals migrate into Jones Sound from Baffin Bay as soon as the leads begin to open up in June. This species is not very social and usually travels as individuals or in pairs. In autumn they usually migrate back to open water in Baffin Bay, but occasionally some remain in Jones Sound and maintain aglus throughout the winter.

There is a great demand for bearded seal skin by the Inuit who prize this strong and durable skin for making harpoon lines, kamik soles, and ropes for lashing their komatiks. Most of the skins harvested by the hunters are used locally, but a few are sold to the Co-op where they are either resold back to the residents or to other Inuit settlements. In 1971 skins sold to the Co-op were worth about $70 apiece.

The average bearded seal weighs about 300 kg and yields approximately 75 kg of meat (Foote 1967), all of which is consumed by the Inuit or their dogs.

Hooded seal (Cystophora cristata)

Hooded seals are extremely rare in Jones Sound; only three have ever been killed by the hunters.

Discussion

Fig. 4 is a highly simplified and purely qualitative schematic representation of the food web in the Jones Sound region. Data for this food web were obtained primarily from my observations and from those of my colleagues working on the Devon Island I.B.P. Project. Despite the fact that this diagram is extremely simplified, it clearly denotes some of the complexities in high arctic ecosystems, as well as some of the interactions between aquatic and terrestrial organisms.

The role of the Inuit in these ecosystems is also clearly illustrated in Fig. 4. From this figure, it is apparent that these people are highly diversified predators, preying upon numerous herbivores, carnivores, and scavengers. The only other predator with as broad a niche is perhaps the arctic fox, which is an opportunistic scavenger as well as a predator. It should also be noted that these people are not directly dependent upon any of the autotrophs in the terrestrial or aquatic systems and are unique in this respect to all other hunter-gatherer peoples of the world, most of whom rely more on plants than on wild game (Service 1966).

In the period prior to contact with Euro-North Americans, Inuit were not prominent importers or exporters of energy or materials; their niche was confined to arctic ecosystems. During this pre-contact era, birds, seals, and whales which annually migrated into and out of the Arctic probably had a greater effect upon the transfer of energy and materials between the arctic ecosystems and the more-southern ecosystems than did man. However, since contact with Euro-North Americans, these people have played a much more dynamic role in the transfer of energy and materials throughout the biosphere. Certainly the Grise Fiord Inuit can now be classified as net importers, since they import vast quantities of fossil fuels, southern foods, and material goods, whereas they export primarily furs, ivory, handicrafts, and a few emigrants to the southern ecosystems.

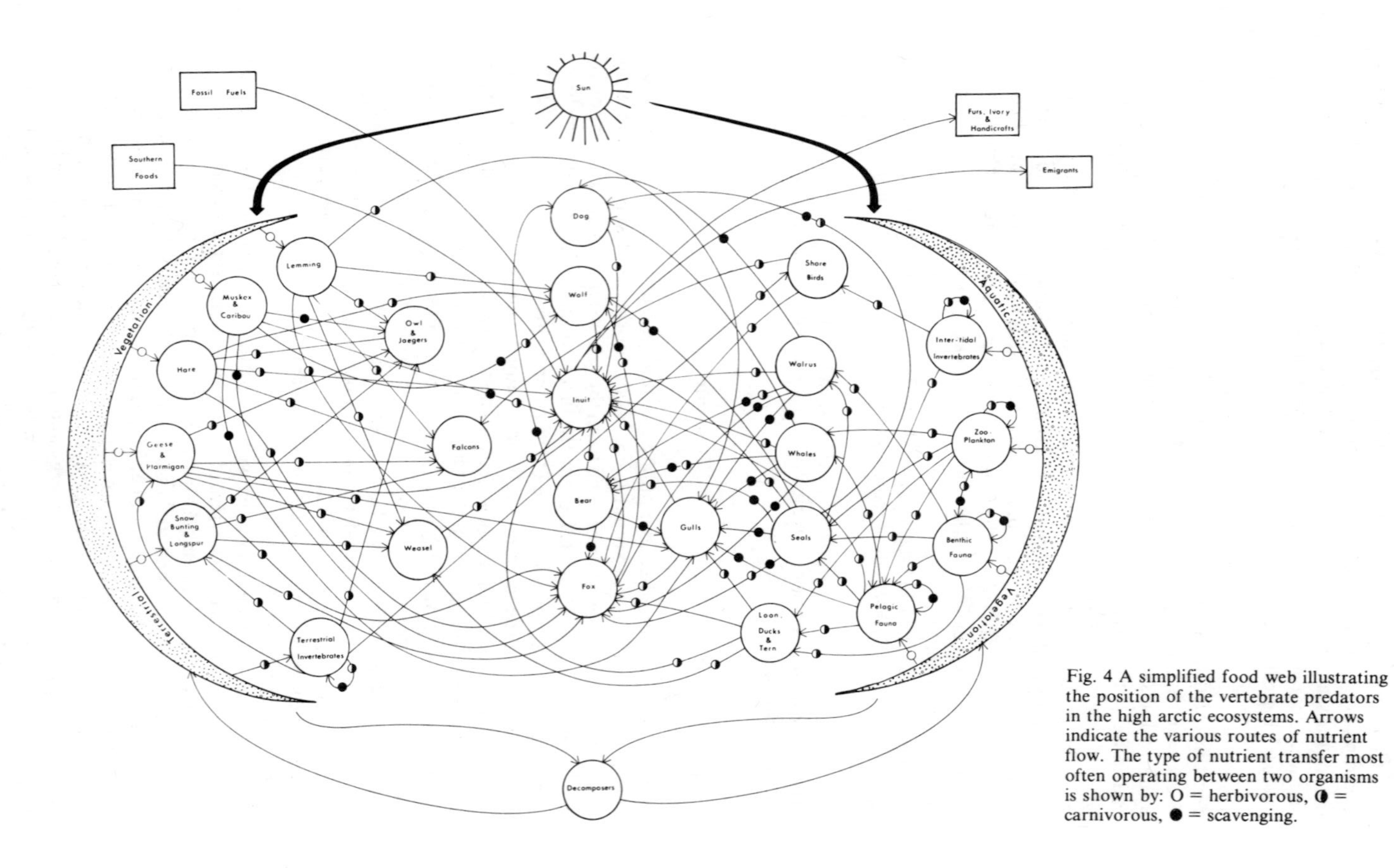

Fig. 4 A simplified food web illustrating the position of the vertebrate predators in the high arctic ecosystems. Arrows indicate the various routes of nutrient flow. The type of nutrient transfer most often operating between two organisms is shown by: O = herbivorous, ◐ = carnivorous, ● = scavenging.

Table 4. Investment in capital goods and operating costs for a Grise Fiord Inuit prior to and after the introduction of the snowmobile.

Capital goods			*Prior to snowmobiles*[a]		*After snowmobiles*[b]	
No.	*Kind*	*Expected life (years)*	*Replacement value*	*Annual depreciation*	*Replacement value*	*Annual depreciation*
1	.22-cal rifle	5	$33.00	$6.60	$50.00	$10.00
1	.222-cal rifle	3	115.00	38.30	130.00	43.40
1	.303-cal rifle	8	25.00	3.12	35.00	4.37
1	.308 or other large cal rifle	5	80.00	16.00	150.00	30.00
2	telescopic sights	4	80.00	20.00	90.00	22.25
1	30x or 40x telescope	6	28.00	5.50	40.00	8.00
35	fox traps	5	55.00	9.15	55.00	9.15
2	butcher knives	1	8.00	8.00	12.00	12.00
1	axe	2	7.00	3.50	8.00	4.00
1	harpoon	2	2.00	1.00	2.00	1.00
1	gas stove	2	18.00	9.00	24.00	12.00
1	gas lantern	2	18.00	9.00	20.00	10.00
1	set of skin clothing	2	15.00	7.50	25.00	12.50
1	sleeping bag	4	45.00	11.25	75.00	18.25
1	tent	5	25.00	5.00	50.00	10.00
1	22-foot canoe	10	537.40	53.74	800.00	80.00
1	20-hp outboard motor	6	340.00	56.70	775.00	129.00
1	14-foot komatik	3	110.75	36.90	75.00	25.00
1	set of dog harnesses, chains, etc.	5	60.00	12.00	0	0
1	snowmobile	2	0	0	900.00	450.00
	miscellaneous gear	5	150.00	30.00	200.00	40.00
		Subtotal:	$1,752.15	$342.26	$3,516.00	$930.92

Operating costs		*Annual expenditures*	*Annual expenditures*
Ammunition		$123.50 (1,350 rounds)	$176.00 (1,440 rounds)
Stove and lantern parts		11.42	16.75
Naphtha and kerosene		28.30 (35 gallons)	22.95 (28 gallons)
Canoe repair supplies		9.55	10.40
Outboard motor parts		10.45	14.00
Snowmobile parts		0	246.00
Gas, oil and methyl hydrate		46.75 (53 gallons)	429.00 (355 gallons)
	Subtotal:	$229.97	$915.10
Total annual costs (depreciation + expenditures):		$572.23	$1,846.02

a. Average for 1965-66 and 1966-67.
b. Average for 1969-70, 1970-71 and 1971-72.

While the total energy budget of this settlement has not been calculated, rough estimates on the energy harvested from wildlife by the people and the distribution of this energy within the biosphere have been made (Table 2). For the year 1971-72, the village (96 people) harvested a total of 191 million Kcal, primarily from ringed seal, bearded seal, polar bear, walrus, harp seal, and narwhal. The Inuit consumed at least 20 million Kcal (approximately 11,400 kg of fresh meat) or about 11% of the total caloric harvest. This is equivalent to approximately 326 gm of fresh meat per person per day. Over 34 million Kcal were consumed by the 50 dogs in the settlement. (Prior to 1967, before the use of snowmobiles, there were about 150 dogs, which would have consumed at least another 70 to 100 million Kcal.) The rest of the energy harvested was either shipped south as furs, skins, and handicraft items (about 10 million Kcal), or was channeled through other cycles in the ecosystems (about 128 million Kcal) by the decomposers and scavengers [glaucous, Thayer's *(Larus argentatus)* and ivory *(Pagophila eburnea)* gulls, raven *(Corvus corax)*, short-tailed weasels, arctic fox, arctic wolf, and polar bear]. There are, in fact, two sizeable gull colonies within 10 km of Grise Fiord which are supported almost entirely by these calories. The hunters are actually taking calories from the relatively productive arctic marine environment and cycling them onto the relatively unproductive terrestrial environment.

Hunting economics

So far, only the hunter's cash income derived from the hunt has been mentioned in this paper. Let us now briefly examine the hunter's balance sheet in greater depth. Table 4 lists the minimum gear required by a Grise Fiord Inuit to conduct his hunting and trapping activities. Several men have considerably more than this minimum list of gear. The replacement values of the various items were ascertained from the Co-op records for the years shown. The expected life of the various pieces of equipment was determined from field observations; these figures tend to be somewhat biased toward the maximum life-span rather than the minimum.

Table 4 indicates that there was more than a threefold increase in the cost of owning and operating hunting gear after the introduction of snowmobiles. The rise in cost was due primarily to depreciation of snowmobiles and the tremendously increased consumption of gasoline and oil.

The low cost of a dog team prior to the snowmobile era was due to the fact that it cost nothing to breed dogs. Dog food could also be considered essentially free, since most hunting efforts were directed toward the harvest of skins and meat for human consumption. This fact is obvious from the 1971-72 harvest data in Table 2, which shows that there was more than ample game (about 128 million Kcal, which was consumed by the scavengers and decomposers) than would have been needed to feed 150 dogs. (Prior to 1967, the 150 dogs would have annually consumed approximately 70 to 100 million Kcal.) Thus, it can be considered that the only cost of a dog team was the depreciation on chains, harnesses, buckles, snaps, and rings.

Between 1965 and 1967 the average hunter's annual cash income derived from the sale of furs was $550. This cash income just about covered the hunter's overhead of $572 in the pre-snowmobile era. From 1969 to 1972 the average hunter's cash income received from furs increased to $890 per year, but this no longer covered his expenses, which had jumped to $1,846 annually due primarily to the use of snowmobiles.

It appears that the Grise Fiord people have adopted so many southern techniques that it has become uneconomical to hunt. The low productivity of the eastern High Arctic can supply them with only so much wildlife, no matter what gadgets are employed. The men were harvesting at as high a rate as possible prior to the snowmobile. This machine did not enable them to increase their take of game (see Table 1), but it did shorten the time required to harvest it.

From this superficial examination of the hunting economics, it looks as if the Inuit would have been better off to give up hunting as an occupation and turn to another form of employment. This actually has been happening in Grise Fiord, as it has in many northern communities (Francis 1969, Hall 1971, Pelto and Muller-Wille 1972, Smith 1972, Pelto 1973). Fig. 5 demonstrates this shift

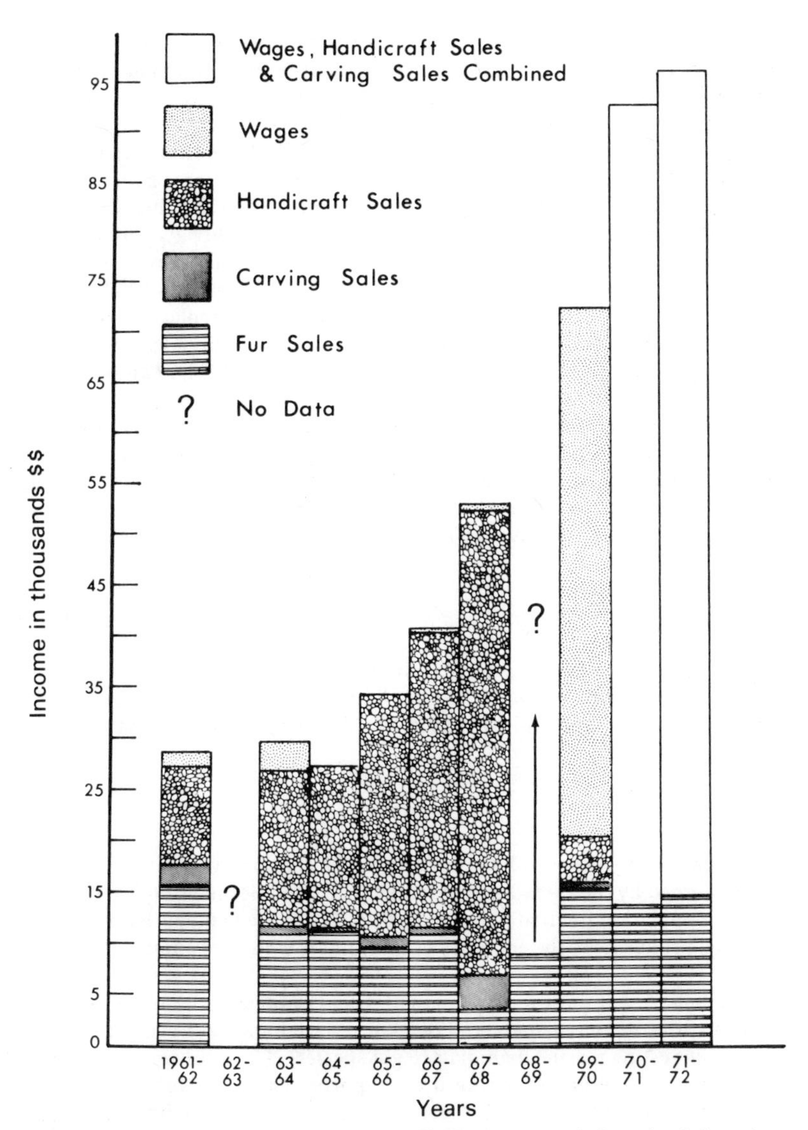

Fig. 5 Grise Fiord's cash income for the years 1961-72. The Co-op records do not break down the income derived from wages and handicraft sales for the years 1970-72; therefore, the data are lumped together for those years.

from a land-based economy to a wage economy in Grise Fiord. From 1961 to 1967 the settlement's cash income was derived primarily from the sale of handicraft items and furs. In 1969-70 the largest portion of the cash income was from wage earnings. During the last two years, 1970 through 1972, the sale of furs provided about the same cash income, while wage income increased dramatically. All men held wage-earning positions by 1972 but continued to hunt after work and on weekends. The mobility of snowmobiles allowed the men to hold wage positions and still harvest a large game bag. In turn, the wages were needed to subsidize the snowmobiles.

Wildlife, however, not only provides the Inuit with a cash income from the sale of pelts, handicrafts, and souvenirs, but it also supplies essentially all of the protein for the maintenance of the human population and the remaining dogs in Grise Fiord. This income in kind is often of greater importance, dollar-wise, than the cash income. For example, in 1971-72, the settlement earned a cash income of $14,493 from the sale of furs, an estimated $4,000 from handicrafts and souvenirs made from wildlife by-products, and $1,125 from payments made by biologists on polar bear bones and muskox stomachs. That same year, assuming that a substitute for wild meat (frozen beef) would have retailed in Grise Fiord for $2.20 per kg, the game consumed by the Inuit was worth $25,080 in kind. Assuming that a substitute for dog food (commercial dry food) would have retailed in the settlement for 40¢ per kg in 1971-72, the replacement value of meat consumed by the dogs would have been worth an additional $4,100 in kind. Thus, the settlement's total income derived from game in 1971-72 amounted to $48,798, 60% from income in kind.

Many Inuit still hold the skills of the hunter and the quality of native game in very high esteem. For this reason alone, they will probably continue to hunt as long as it is feasible. Despite the high cost of operating and maintaining hunting gear, the Grise Fiord people realize that, in order to survive today, they must continue to hunt, even if only on a part-time basis.

Postscript

In recent years, since 1973, the inflated value of skins, most especially polar bear skins, has made it more profitable to hunt than in the past, but the vagaries of the fur market make it impossible to predict whether it will remain so in the future.

Summary

The settlement of Grise Fiord, the northern-most community in North America, was established by the Canadian government in 1956. Between 1960 and 1972 the Inuit population fluctuated from 70 to 96. The settlement has been supported primarily by the activities of the hunters who have numbered from 13 to 18.

The majority of these hunting activities are confined to the Jones Sound region, an area of approximately 97,000 km^2. Main hunting efforts are directed towards harvesting marine mammals (eight species). The most-important of these to the Inuit culture is the ringed seal; approximately 590 have been taken annually between 1956 and 1972. Polar bears have always been important, but recently the inflated value of their skins has made them the most sought-after wildlife in the region. Bear pelts provided the Inuit with 55.4% of their cash income derived from the sale of furs between 1956 and 1972. To a much lesser extent, the people harvest birds (seven species) and terrestrial mammals (six species), Peary's caribou, and muskox being the most important. A few arctic char are also taken from the Jones Sound region.

The settlement harvested an estimated 191 million Kcal in 1971-72. The 96 Inuit consumed at least 20 million Kcal, while their dogs consumed an additional 34 million Kcal. The rest of the energy was either shipped south as furs and handicrafts (about 10 million Kcal), or channeled through other cycles in the ecosystems (about 128 million Kcal) by decomposers and scavengers. In other words, the hunters are actually shunting calories from the relatively productive marine ecosystem into the relatively unproductive terrestrial ecosystem.

The snowmobile revolution, which began in Grise Fiord in 1967, raised the overhead of hunting to such a level that the occupation became uneconomical. All of the men now hold part-time or full-time wage-earning positions; some of this wage income is used to subsidize the cost of maintaining and operating their hunting outfits.

Hunting not only provides the people with a cash income from the sale of furs, handicrafts, and souvenirs but also a significant income in kind from wild meat. Even though it may be uneconomical to hunt on a full-time basis, the Grise Fiord Inuit realize that they must continue to hunt, even if only on a part-time basis, in order to maintain their psychological and physical health.

Acknowledgments

I wish to express my deepest gratitude to Drs. William O. Pruitt, Milton M.R. Freeman, and Lawrence C. Bliss for their support and advice throughout this study.

Without the complete cooperation of the Grise Fiord residents this study would have been impossible. I am especially indebted to Special Constable Pijamini, Mr. and Mrs. Bezal Jesudason, Corporal Allan D. Kirbyson and Corporal Hank Johnston for their warm hospitality, which made my visits there extremely memorable and valuable. I am particularly grateful to the men for allowing me to accompany them on their hunting, trapping, and fishing trips and for providing me with a wealth of ecological information.

The Canadian Wildlife Service generously incurred the expense of shipping many of my specimens south to Winnipeg.

Mr. David Gee and Dr. Robert Morrison granted permission to utilize the data I collected while under contract to the Department of Indian and Northern Affairs during 1973.

I especially wish to thank my family for their enthusiasm and support throughout this study.

References

Bentham, R. and D. Jenness. 1941. Eskimo remains in southeastern Ellesmere Island. Trans. Roy. Soc. Canada, Section II, 3: 41-55.

Bruemmer, F. 1968. Eskimos of Grise Fiord. Can. Geog. J., 77: 65-71.

Dunbar, M. and K.R. Greenaway. 1956. Arctic Canada from the air. Canada Defence Research Board, Ottawa. 541 pp.

Foote, D.C. 1967. The east coast of Baffin Island, N.W.T.: an area economic survey, 1966. Unpub. manuscript, D.I.A.N.D., Industrial Division, Ottawa. 162 pp.

————. and H.A. Williamson. 1966. A human geographical study. pp. 1041-1107. *In:* Environment of the Cape Thompson Region, Alaska. N.J. Wilimovsky, and J.N. Wolfe (eds.). U.S.A.E.C., Div. of Tech. Info., Washington, PNE-481, 1250 pp.

Francis, K.E. 1969. Decline of the dog sled in villages of arctic Alaska: a preliminary discussion. pp. 69-78. *In:* Yearbook of the Association of Pacific Coast Geographers, J.F. Gaines (ed.).

Freeman, M.M.R. 1968. Winter observations on beluga *(Delphinapterus leucas)* in Jones Sound, N.W.T. Can. Field-Nat., 82: 276-286.

————. 1969. Adaptive innovation among recent Eskimo immigrants in the eastern Canadian Arctic. The Polar Record, 14: 769-781.

———. 1969-70. Studies in maritime hunting I. Ecologic and technologic restraints on walrus hunting, Southampton Island, N.W.T. Folk, 11-12: 155-171.
———. 1971. Population characteristics of musk-oxen in the Jones Sound region of the Northwest Territories. J. Wildf. Managt., 35: 103-108.
Hall, E.S., Jr. 1971. The 'iron dog' in northern Alaska. Anthropologica, 13: 237-254.
Humphreys, N., E. Shackleton and A.W. Moore. 1936. Oxford University Ellesmere Land Expedition. Geog. J., 87: 385-443.
Kemp, W.B. 1971. The flow of energy in a hunting society. Sci. Amer., 225: 105-115.
Lethbridge, T.C. 1939. Archaeological data from the Canadian Arctic. Roy. Anthrop. Inst. J., 69: 187-233.
Lowther, G.R. 1962. An account of an archaeological site on Cape Sparbo, Devon Island. Nat. Mus. Can. Contrib. Anthrop. 1960, Part I, Bulletin No. 180, 15 pp.
Mansfield, A.W. 1967. Seals of arctic and eastern Canada. Fish. Res. Bd. Can. Bull. No. 137, 35 pp.
Marie-Rousseliere, G. 1972. On the footsteps of Qidlarjuaq. Eskimo, 29: 3-15.
Maxwell, M. 1960. An archaeological analysis of eastern Grant Land, Ellesmere Island, N.W.T. Nat. Mus. Can. Bull. No. 170, 109 pp.
———. 1972. Archaeology of the Lake Harbour District, Baffin Island. Archaeological Survey of Canada Paper No. 6, Mercury Series, 362 pp.
McGhee, R.J. 1974. The early arctic small tool tradition: a prediction from Arctic Canada. pp. 127-132. *In:* International Conference on the Prehistory and Paleoecology of Western North American Arctic and Subarctic. S. Raymond and P. Schledermann (eds.). Univ. Calgary Archaelogical Assoc. Calgary. 262 pp.
Pelto, P.J. 1973. The Snowmobile Revolution: Technology and Social Change in the Arctic. Cummings Publishing Company, Menlo Park, California. 225 pp.
———. and L. Muller-Wille. 1972. Snowmobiles: technological revolution in the Arctic. Pp. 165-199. *In:* Technology and Social Change. H.B. Bernard, and P.J. Pelto (eds.). Macmillan Company, New York. 354 pp.
Riewe, R.R. 1973a. Grise Fiord hunters. Northern Perspectives, 1: 3-5.
———. 1973b. Final report on a survey of ungulate populations on the Bjorne Peninsula, Ellesmere Island: determination of numbers and distribution and assessment of the effects of seismic activities on the behaviour of these populations. Unpub. Rep., D.I.A.N.D., Ottawa. 59 pp.
Service, E.R. 1966. The hunters. Prentice-Hall, Inc., Englewood Cliffs, New Jersey. 118 pp.
Smith, L. 1972. The mechanical dog team: A study of the skidoo in the Canadian Arctic. Arctic Anthropology, 9: 1-9.
Speck, F. 1924. Eskimo collection from Baffin Land and Ellesmere Land. Heye Foundation Museum of American Indian, Indian Notes, 1: 143-149.
Svendrup, O.N. 1904. New Land: Four Years in the Arctic Regions, vols. I-II. Longmans and Green, New York. 1000 pp.
Taylor, A. 1955. Geographical discovery and exploration in the Queen Elizabeth Islands. Canada Department of Mines and Technical Surveys, Geographical Branch Memoir 3, 172 pp.
Usher, P.J. 1971. The Bankslanders: economy and ecology of a frontier trapping community. Volume 2: economy and ecology. D.I.N.A.D., Northern Sci. Res. Group, NSRG 71-2, Ottawa. 169 pp.
———. 1976. Evaluating country food in the northern native economy. Arctic, 29: 105-120.

Industrial development

High Arctic disturbance studies associated with the Devon Island Project

T.A. Babb

Introduction

Much of the impetus toward a comprehensive study of a high arctic ecosystem was based on the concern that industrial development, particularly gas and oil exploration, was likely to trigger irreversible environmental damage. Low productivity and low species diversity in arctic ecosystems are assumed to result in much more pronounced interdependence of heterotrophic populations than is evident in milder climates. The high arctic landscape with its sparse plant cover, infrequent rainfall, and underlying permafrost is subject to different sorts of physical degradation than are landscapes under low arctic or temperate conditions. The concept of the Arctic as an "extreme" environment, imposing stress at all but the best of times, leads to the notion that very little additional stress is needed to surpass the tolerance limits of organisms. Sensitivity must therefore be anticipated if damage is to be avoided.

Direct examination of some of the effects of human encroachment was thus incorporated as a part of the Devon Island Study. It is the purpose of this chapter to discuss in general terms some of the aspects of industrial activity likely to contribute to habitat degradation.

Methods

Most of this discussion stems from a series of manipulation experiments in combination with the examination of past disturbance at industrial sites throughout the Queen Elizabeth Islands (see Babb and Bliss 1974a, 1974b). Manipulations at the Devon site included removal of surface vegetation from meadow and beach ridge surfaces, fertilization of otherwise undisturbed plant communities with varying combinations of nitrogen, potassium and phosphorus fertilizers, simulated grazing, controlled vehicle passage, and artificial petroleum spills. Regional assessment of surface disturbance was based on visits to Cornwallis, Bathurst, Melville, Prince Patrick, King Christian, Ellef Ringnes, Axel Heiberg, Ellesmere, and Somerset Islands. Emphasis of these studies was on physical processes and on vascular plant succession. The results are applied here to a generalized assessment of some of the problems associated with disturbance.

Fig. 1 Denuded manipulation plots in a *Dryas-Carex*-moss meadow, Devon Island.

Results

Effects of disturbance on permafrost

Much attention has been devoted to the damage caused in low arctic regions, mainly in the Mackenzie Delta, by removal of plant cover and subsequent melting of permafrost. Under conditions of high ice content, exposure of the substrate to insolation, air movement, and running water can result in thermokarst, thermal erosion, or headward subsidence. This is due to a combination of decreased insulation by compressed peat and a downward diversion of energy flow following the removal of the plant canopy. Heat normally dispersed by reflection, convection, and evapotranspiration from the shrub or graminoid canopy instead reaches the soil surface and is conducted towards the permafrost (Haag and Bliss 1974).

In the Queen Elizabeth Islands, even the lushest plant canopy (sedge-moss meadow, *Dryas* or *Cassiope* heath) contributes relatively little to latent and convective heat loss. Removal of plant cover and upper soil layers (Fig. 1) results in a slight decrease in albedo and a corresponding increase in downward heat flux, but the net effect is much less than when a 10-20 cm shrub canopy is removed.

It has been known for some time that there is an inverse correlation between density of plant cover and active layer depth in the High Arctic (Brown 1972, Babb 1972) but the relationship is evidently not direct. It appears that soil moisture content is largely responsible for depth of the active layer, absorbing considerable energy in latent heat of thaw and evaporation. Continuous plant cover, which requires this moisture for growth, is more coincidental than causal.

An apparently widespread but little-studied phenomenon in the arctic islands, especially in areas of low to moderate relief underlain by fine-textured colluvium, is the segregation of a horizontal, ice-rich layer immediately below the active layer (Babb and Bliss 1974a, French 1974). This contrasts with ice lenses or ice wedges which have been studied in detail (Brown 1970).

A number of problems have been associated with such deposits, mainly the

Fig. 2 Erosion of peat initiated by vehicle traffic in a hummocky sedge-moss meadow, Devon Island.

summer softening of lightly bladed surfaces in some types of terrain (Price et al. 1974). A dense, relatively dry active layer abuts abruptly against an ice-rich layer which grades into ice-free parent material 10 to 100 cm below the limit of seasonal thaw. Compaction or removal of part of the active layer results in slightly increased thermal conductivity. Subsequent release of moisture tends to maintain high thermal conductivity past the period following runoff when drying normally occurs. Accelerated thaw and moisture release can persist for several years, until the ice-rich layer has dissipated. During this period, roads, airstrips, etc. are unusable in summer months. It is possible, furthermore, that when the ice layer has dissipated the process could reverse, and the site could revert to a xeric condition, possibly affecting plant recolonization. The energy-water relations involved in the aggrading or dispersal of these layers are not known.

Erosion

Thermal erosion of organic substrates appears to be uncommon in regions where petroleum exploration is most intense. On irregular landscapes where depressions and lower slopes are occupied by eutrophic sedge-moss meadows, peat deposits may contain or be underlain by accumulations of lens or wedge ice. Rutting, initiated by summer vehicle movement, over deep organic soils can result in considerable erosion during runoff periods (Fig. 2). Fortunately, this condition appears most common in mountainous regions on or near the Canadian Shield and is encountered infrequently in petroleum exploration in the western part of the region. Utmost caution, however, is obviously needed in areas of high plant cover.

Most of the barren high arctic landscape is strongly subject to wind, sheet, and gully erosion. Mass wastage in the form of earthflows and solifluction is prominent in many areas. In comparison to these, man's effect on the landscape is inconsequential, but disturbance tends, nevertheless, to be locally conspicuous. Entrapments of blowing snow provide sources of water for subsequent gullying during spring runoff. Natural mass wastage can be accelerated by disruption of surfaces along stream margins or on slopes.

Because of the sparseness of plant cover, biological consequences are generally minor on disturbed Polar Desert or Polar Semi-desert sites. Visible effects, however, are long-lived. Uplands tend to be blown free of winter snow and disturbed areas rarely experience the obliterating effects of erosion during runoff. Rutted or ridged surfaces are either dry or frozen during most of the year and are consequently erased only slowly by wind erosion and frost action. The longevity of surface scars is borne out in the tracks visible today which were left by 19th century explorers on Southern Melville Island.

Superficial damage to vascular plants:
Physical disruption, oil spills

In a simulated grazing experiment it was found that sedges and grasses in a meadow displayed substantial capacity for recovery from clipping. Bi-weekly clipping of small plots over an entire growing season resulted in only minor diminishing of growth the following year. Energy and nutrient reserves maintained in the roots and rhizomes may protect graminoids against periodic damage by intense grazing, unseasonal frosts, or unusually short summers. High root:shoot ratios may in part be an adaptation to such stresses. From this it appears unlikely that brief periods of heavy grazing or other temporary and superficial physical damage to meadow communities could trigger unforseen detrimental effects on plant growth in later years.

Although upland communities were not subjected to the same experiments, several general conclusions can be drawn from other observations. Although meso- or xerophytic species such as *Saxifraga oppositifolia* or *Dryas integrifolia* are physically much tougher than many meadow species, they tend also to be more exposed to damage by vehicle movement, foot traffic, etc. A much greater proportion of a plant lies above the surface, and dry, elevated areas are more heavily travelled. While woody plants in some instances show a "pruning effect" (plants at a site on Ellesmere Island which had been damaged by cleated tracks showed vigorous new growth, *Dryas* leaves of 3 cm and *Salix* leaves of 6 cm being noted), the phenomenon is generally uncommon. Plant growth is usually very slow, and a cushion plant which has been sheared away is unlikely to recover for a considerable period.

The likelihood of land oil spills has been a matter of concern for some time. Controlled applications of diesel fuel at low levels (.05 and .25 l/m^2) to beach ridge and meadow communities showed the expected harmful effects (Baker 1970) to aboveground plant parts. Damage was more intense on the beach ridge sites, while the absorptive properties of moss and litter in the meadow contributed to a considerable degree of recovery (50-80% at the higher intensity) the following year (Babb 1972). Bacterial populations, as in most studies, were found to increase (Widden, pers. comm.), presumably because of the multiplying of oil-decomposers. Damage to plant cover can be of short duration in wet sedge sites.

It is most likely that major surface spills of fuel or crude oil would have much longer lasting effects than in warmer regions. Although breakdown of oil evidently occurs, the typical low availability of inorganic nutrients, and low soil temperatures limit the rate of decomposition. The same factors limit the rate of re-colonization of the area by plants. Sulfur compounds and relatively inert aromatic constituents, furthermore, could leave toxic residues which would further inhibit recovery.

Fig. 3 Surface damage caused by track vehicles, Sherard Bay, Melville Island.

Secondary succession and capacity for rehabilitation

Plant growth in the region is generally very slow (see Muc, Svoboda, Pakarinen and Vitt, this volume). The natural limitations imposed by low temperatures, short growing seasons, and scarcity of nutrients slow natural recovery of disturbed areas.

In peaty meadow areas, if erosion has not been initiated, recovery of disturbed areas is likely to be a process of concurrent in-filling by detritus and recolonization by mosses and vascular plants. On slopes, where vehicle tracks or other disturbances form channels for flow of spring meltwater, recovery is slow. It is possible, however, that once the more erodable, substrate is removed, and the channel is at equilibrium, slow "healing" may occur. Lateral encroachment by the moss mat could gradually restrict flow of water until continuous cover was again established. If the cutting of the channel has affected the local water table, succession may be a complex process involving a considerable area to either side.

In meadow sites where damage has occurred, attempts at rehabilitation are clearly warranted (see Fig. 3). The most promising techniques are simply the "patching" of denuded areas with blocks of native sod and peat. Attention should be paid to microtopography and species composition to assure survival of the sod blocks in as intact a state as possible. Fertilization can in some instances hasten plant growth, but patches of lush plant cover on an otherwise sparse and rapidly eroding landscape serve no particular purpose. In the relatively rare areas where erosion and thermokarst are initiated by removal of plant cover, fertilization in conjunction with physical rehabilitation would probably hasten recovery.

Because plant cover in upland areas is generally sparse, the purely biological consequences of disturbance are minor. The visual effects, however, can be very long-lived. While removal of litter, filling of gullies and sumps, and levelling of mounds of earth is an obvious and simple part of the solution, biological rehabilitation is more difficult to achieve. Reseeding by non-native species may serve some purpose in binding the loosened soil until native species can regain a foothold. Fertilization could hasten the binding process, and is almost certain to speed up growth by some of the potential native reinvaders. It should be noted that successful reseeding by native species appears generally difficult to achieve. Production and establishment of seeds of native high arctic species occurs only under unusual conditions (Bell 1975). Techniques of seed-gathering and cultivation require considerable refinement before artificial rehabilitation to natural conditions will be practical.

It is unlikely that species composition in rehabilitated sites would be similar to that in surrounding areas for a considerable period. The differential response of native high arctic species to fertilization is great; some are actually inhibited by high levels of fertilizer (e.g., *Dryas integrifolia* and most, if not all, lichens) while growth rates and flowering of others are considerably enhanced (*Cerastium alpinum* and *Saxifraga oppositifolia*) (Babb 1972). The merits of establishing a plant cover, regardless of its composition, versus the permitting of slow, natural recolonization, should be weighed against the implications of bare surfaces and the costs of attempted restoration.

It is likely that a good general practice for the rehabilitation of disturbed upland areas would be a combination of: (1) re-sculpting an area to its previous contours and checking erosion; (2) light mulching with organic matter such as pulverized straw or fine sawdust; (3) light seeding with cold-hardy, native grass species; and (4) applying fertilizer at a rate of *ca.* 50 Kg/ha of combined N, P and K.

Pollution of water from terrestrial sources

Although the Devon Island Project concentrated on terrestrial systems, the subject of water pollution should not be bypassed completely. While permanent lakes and rivers are uncommon in the Queen Elizabeth Islands, considerable potential for harm to aquatic and marine systems remains. The most substantial danger is from the release of sewage and other wastes into temporary watercourses which favor algal blooms, higher water temperatures, and decreased oxygen (Rigler 1974). Human wastes are customarily treated with recyclable chemical preparations which liquify and disinfect organic matter, but do not promote biotic decomposition. The "sludge" is buried, and much, if not all of the liquid material is eventually released. The latter, which is rich in nitrogen and phosphorus, promotes algal growth and possibly eutrophication, warming, and anaerobiosis in lakes and ponds, while toxic materials of disinfectants could adversely affect consumer and decomposer populations. Instances of these effects are rare; one documented monitoring of a moderately polluted high arctic lake (Rigler 1974) showed no unusual or extremely harmful changes.

Another preparation, "Biocide D," which at one time was in widespread use, is an antibiotic which was added to drilling mud to prevent the breakdown of organic polymers used to impart proper consistency to the mixtures. This material found its way very easily into surface water, and was known to have considerable toxic effects on aquatic and marine organisms. Fortunately, the use of this material has been discontinued.

Noise pollution: Effects on wildlife

This matter is similarly outside the investigations of the author, but is worthy of mention here. The feeding and movement of mammals, and the nesting of waterfowl is almost certain to be modified by the presence of man, even if damage to habitat is scrupulously avoided. Some effects could be: (1) stress on large mammals caused by aircraft affecting breeding success or winter survival; (2) attraction of carnivores (foxes, wolves, and bears) to garbage dumps, redistributing these populations with great likelihood of repercussions affecting prey populations. It is therefore erroneous to conclude that disruption of habitat is the only potential source of environmental damage.

It should be noted that government and industry have taken note of many of these sources of danger and imposed appropriate restrictions on certain activities. Lower altitudinal limits for flight, for example, have been attached to some land use permits in the region; seasonal bans on activity have been enforced in some areas; incineration and burial of waste material is now required of industrial operations.

Summary

A study of the physical effects of disturbance at an intensive scale on the Devon Island site and at an extensive scale elsewhere in the Queen Elizabeth Islands has permitted a generalized evaluation of potential habitat damage.

Difficulties with thermokarst and thermal erosion are much less important than in the Low Arctic in the Mackenzie Delta region, partly because of differences in physiography, and largely because of differences in surface energy budgets. The general sparseness of vegetation in the High Arctic in combination with cooler summer temperatures and lower permafrost temperatures are prime contributors to the contrast. The occurrence of thin but extensive subsurface ice layers are a permafrost feature which may be unique to the region. These layers can lead to summer softening of lightly disturbed areas and seriously hamper summer use of some airstrips, roads, and campsites until the water evaporates several years later.

Vascular plants have displayed no particular sensitivity to physical disturbance; in fact, adaptations to harsh aspects of the environment appear to impart an unusual degree of resilience. Slow growth rates, however, disfavor rapid recovery of vegetation on disturbed sites. In some instances artificial rehabilitation requiring several successive years' work may prove necessary.

Pollution of water from terrestrial sources and the potential effects of noise and movement on birds and mammals are two topics not studied in the Devon Island Project, but touched upon briefly here.

Acknowledgments

Research was supported financially by grants to the Devon Island Project by the National Research Council of Canada, members of the Arctic Petroleum Operators' Association, and the Department of Indian Affairs and Northern Development. Logistic support was provided by the Polar Continental Shelf Project, the Arctic Institute of North America, King Resources, Sun Oil, Mobil Oil, and Panarctic Oils. Assistance by Doug Larson, James Wright, and Lorita Babb are gratefully acknowledged, as is the supervision and assistance by Dr. L.C. Bliss.

References

Babb, T.A. 1972. The Effects of Surface Disturbance on Vegetation in the Northern Canadian Arctic Archipelago. M.Sc. Thesis, Univ. Alberta, Dept. of Botany. 71 pp.

————. and L.C. Bliss. 1974a. Effects of physical disturbance on arctic vegetation in the Queen Elizabeth Islands. J. Appl. Ecol. 11: 549-562.

————. and ————. 1974b. Susceptibility to environmental impact in the Queen Elizabeth Islands. Arctic 27: 234-237.

Baker, J.M. 1970. The effects of oils on plants. Environ. Pollut. 1: 27-44.

Bell, K.L. 1975. Aspects of seed production and germination in some high arctic plants. pp. 62-72. *In:* Plant and Surface Responses to Environmental Conditions in the Western High Arctic. Bliss L.C. (ed.). A.L.U.R., 74-75-73. Ottawa. 72 pp.

Brown, R.J.E. 1970. Permafrost in Canada. University of Toronto Press, Toronto. 234 pp.

————. 1972. Permafrost in the Canadian Arctic Archipelago. Zeitschr. Geomorphol., Suppl. No. 13: 102-130.

French, H.M. 1974. Active thermokarst processes, eastern Banks Island, Western Canadian Arctic. Can. J. Earth Sci. 11: 785-794.

Haag, R.W. and L.C. Bliss. 1974. Energy budget changes following surface disturbance to upland tundra. J. Appl. Ecol. 11: 355-374.

Price, L.W., L.C. Bliss and J. Svoboda. 1974. Origin and significance of wet spots on scraped surfaces in the High Arctic. Arctic 27: 304-306.

Rigler, F.H. 1974. Char Lake Project PF-2, Final Report. Canadian Committee for the International Biological Programme. 96 pp.

Project Summary

General summary Truelove Lowland ecosystem

L.C. Bliss

Introduction

The chapters of this book have discussed in detail various components of this high arctic ecosystem. Many aspects of the study have been synthesized in the chapters on nutrient cycling (Babb and Whitfield this volume), and on energy budgets and ecological efficiencies (Whitfield this volume). General summaries of the study have been included in the synthesis volume of the Tundra Biome I.B.P. projects (Bliss 1975a) and the synthesis volume of the Canadian Contribution to the I.B.P. (Bliss 1975b).

It is the purpose of this final chapter to highlight some of the major integrative findings of this study, to relate these findings to general ecosystem function in the High Arctic, and to discuss their implications for land use management.

Environmental setting

Areas such as Truelove Lowland are a minor feature within the Queen Elizabeth Islands (1%), yet they are essential to the maintenance of diverse populations of wildlife. With the exception of Melville Island, most of these biologically rich areas are found in the central and eastern islands.

The environmental parameters that appear most influential here and elsewhere are: (1) lowlands usually near the coast; (2) an abundance of surface water resulting from blocked drainage generally via a series of raised beaches, the result of isostatic rebound; (3) higher summer radiation and therefore temperature than in surrounding lands; and (4) higher precipitation in the eastern islands, probably the result of the nearby ice-free North Water area. To these factors is the added influence of föhn winds from the plateau and its ice cap. These warming and drying winds extend late-summer/early-fall into September (Courtin and Labine this volume).

The hydrology of this Lowland is probably representative of similar areas at this latitude. Of the estimated annual precipitation of 185 mm, *ca*. 40 mm occurs in the liquid form. Runoff accounts for 84 mm and evapotranspiration 101 mm, the latter consists of 20 mm of snow evaporation (sublimation), 8 mm open water, and 73 mm plant and soil evaporation. Total lowland evaporation from open water is greater than this, because ponds and lakes occupy 22% of the lowland area while the above figures are based on the Intensive Watershed

Study Site with only 12% water (Rydén this volume). Evapotranspiration is a larger component here than in the Alaskan Arctic, probably the product of the unusually high net radiation (18 kly yr^{-1}) compared with the expected values of 2-3 kly yr^{-1} at this latitude and the measured 11 kly yr^{-1} at Barrow (see Courtin and Labine this volume). Net radiation comprises 80% to 90% of global radiation on sunny days but only 57% to 67% on cloudy days (Addison this volume). High values for net radiation help to explain why these areas are more favorable habitats for plants and animals.

The soils of the lowland and plateau are more weakly developed than their counterparts in the Low Arctic. Those of the plateau are Gleysolic Turbic Cryosols with no surface organic matter owing to very low plant cover. These same soils occupy about 50% of the lowland wetlands. Poorly and very poorly drained areas of sedge-moss meadow with calcareous coarse-textured to noncalcareous fine-textured materials have Gleysolic Static to Fibric Organo Cryosols. Well drained sites on raised beaches and rock outcrops with cushion plant-lichen and dwarf shrub-moss communities have Regosolic Static and Brunisolic Static Cryosols. The turbic phases of these soils associated with frost-boils occur in the cushion plant-moss communities (Walker and Peters this volume).

The processes of humification, melanization, and brunification are highly modified by the severe high arctic climate resulting in shallow active layers, weak translocation of minerals, and weak gleization processes. Cryoturbid soils are common throughout the area (Walker and Peters this volume). Soil nutrient regimes are low, especially phosphorus and nitrogen (Babb and Whitfield this volume).

The pattern of warming of the Lowland and adjacent Truelove Valley in May and June (Courtin and Labine this volume) influence animal activity, especially calving grounds and early grazing areas for muskox (Hubert this volume). Snow melts first along the west-facing cliffs, the granitic rock outcrops, and the crests of the raised beaches. Much of the early activity of migratory birds is concentrated in these areas of available seed, insects, and nesting sites (Pattie this volume). Lemming move from their winter nests at the base of the raised beaches into summer burrows under rocks, etc. as these raised beaches warm and the active layer increases in depth.

Winter snowdepth averaged 15 to 20 cm on crests of raised beaches and lichen barrens on Rocky Point, 45 cm in the meadows and surfaces of lakes and ponds, and 60 to 100 cm on the lower slopes and transition zone of raised beaches and the rock outcrops (Rydén this volume). Most snow typically melted over a 10 to 14 day period, though in 1970 more than 20 cm of snow melted in 8 hr during a period of Föhn winds (Courtin and Labine this volume). During the snowmelt period, water covers most meadows. Runoff accounts for most water loss, although evapotranspiration (little plant growth has occurred and therefore little transpiration) was greatest (3 mm day^{-1}) during this period (Rydén this volume). It is during this period of surface water that masses of *Nostoc commune* and smaller amounts of *Peltigera aphthosa* are effective in fixing nitrogen (Stutz this volume). Waterfowl and shore birds begin nesting at this time (Pattie this volume).

The two major habitats of the Lowland, sedge-moss dominated meadows (41%) and raised beaches (11.4%), were studied most intensively both environmentally and biologically. These are very contrasting habitats, although usually in juxtaposition. As can be seen from Table 1 and Figs. 1 and 2, hummocky sedge-moss meadows have a shorter growing season based upon temperature, a significantly smaller number of accumulated degree days, deeper snow cover, later snowmelt, and because of organic soils that are water

Table 1. Length of growing season and accumulated degree days for the raised beach and hummocky sedge-moss habitats, Truelove Lowland. Data are based upon 5-day running means at the 150 cm level.

	Length of Growing Season (Days)		*Accumulated Degree*
Year	*Above 0° C*	*Phenology**	*Days above 0° C*
Raised Beach			
1971	80	55	384
1972	72	45	208
1973	99	60	705
Hummocky Sedge-moss Meadow			
1971	85	55	394
1972	48	45	108
1973	86	55	409

*Snowmelt and initiation of plant growth to 50% leaf coloration.

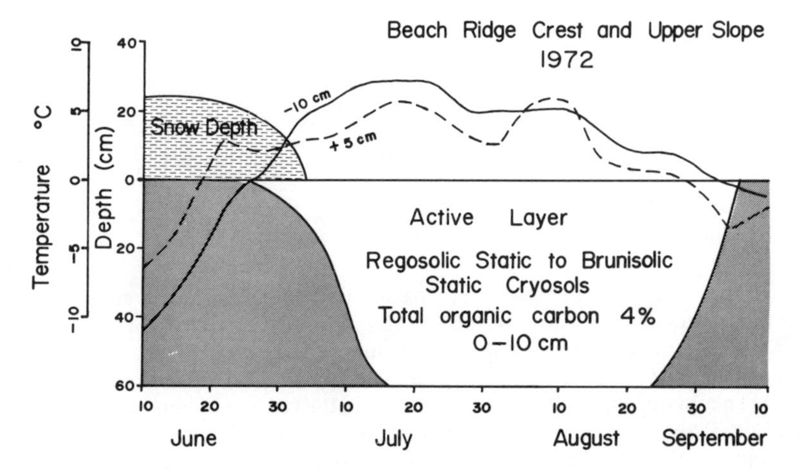

Fig. 1 Pattern of snowmelt, thaw of the active layer, and temperature profile for air and soil in the cushion plant-lichen community of the raised beach crest and slope, 1972.

saturated, only a shallow active layer compared with the raised beaches. Soil water content, aeration, and organic and nutrient content are very different (see Courtin and Labine, Muc, Svoboda, and Walker and Peters — all this volume — for details). Consequently their vascular, cryptogamic, and microfloras as well as vertebrate and invertebrate faunas are very different structurally and functionally. Much of the remainder of this chapter is devoted to details of the similarities and dissimilarities of these habitats and how they are related to other areas within the High Arctic.

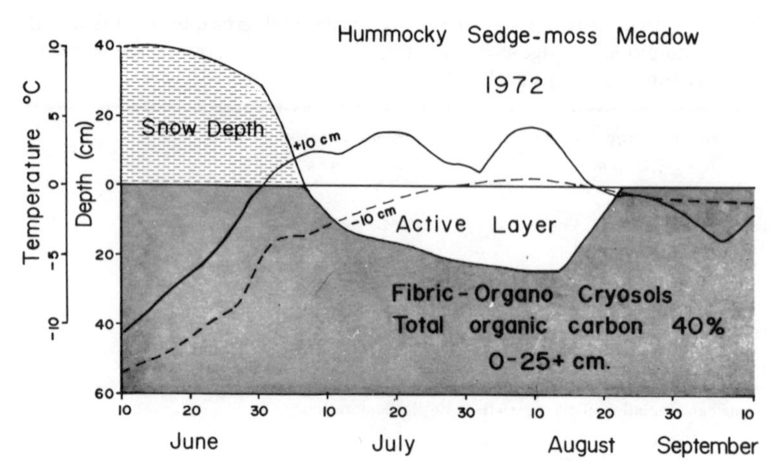

Fig. 2 Pattern of snowmelt, thaw of the active layer, and temperature profile for air and soil in the hummocky sedge-moss meadow, 1972.

Structure and function

Primary producers

Plant cover, standing crop (alive), and net annual production are all highest in the three types of sedge-moss meadow (Tables 2 and 3). Bryophytes are an important component but lichens are very minor. The estimates for algae are minimal and their standing crop and production may be 2 to 5 times the amounts cited. They are probably more abundant within the moss mat than our studies indicate. Plant production in the frost-boil sedge meadow community is about the same for the vegetated areas (49% is frost-boils) as in the other sedge meadows.

Within the literature, frost-boil sedge meadows (spot medallion sedge meadows of the U.S.S.R. literature) are frequently believed to represent an earlier stage of plant succession that leads to the hummocky sedge-moss type. As discussed by Muc (this volume), these meadows are generally found in the older and better drained portions of the Lowland. The presence of fine textured soils, less standing water in early summer, and possible lower levels of nutrient input via surface water flow following snowmelt, probably account for the differences in plant cover and production, rather than a time-based successional sequence.

The dwarf shrub heath communities have a high species richness, total plant cover, and comparative high levels of net production. As with the sedge meadows, a large percentage of vascular plant production occurs belowground (Bliss et al. this volume). The standing crop of mosses is considerable, but net production is relatively low, due to their slower growth in these drier sites (Tables 2 and 3). Unlike the dominant species in the other major communities, *Cassiope* is seldom consumed by herbivores.

The raised beaches were divided ecologically into a cushion plant-lichen community on the crest and slope and a cushion plant-moss community on the lower slope (transition) with sedge meadow communities beyond. Both vascular plant and moss cover and production are significantly higher in the transition

Table 2. Plant standing crop (g m^{-2}) for the major plant communities at the peak of aboveground standing crop. The means are based upon 1 to 3 years of data (Bliss et al., Jefferies, Muc, Svoboda).

	Plant Community							
						Raised Beach		*Plateau*
Component	*Salt Marsh*	*Hummocky Sedge-moss Meadow*	*Wet Sedge-moss Meadow*	*Frost-boil Sedge Meadow (Total area)**	*Dwarf shrub heath (Total area)***	*Crest-Slope+ Cushion plant-lichen*	*Transition Cushion plant-moss*	*Moss-herb*
Lowland area (ha)	22	883	88	796	533	292	572	—
No. Vascular species	4	28	21	25	24	9	14	8
Vascular plant cover (%)		86	77	58	37	20	58	6
Cryptogam plant cover (%)	—	98	100	82	35	2	22	(5)
Vascular								
Aboveground								
Green	86	35	40	27	22	18	48	5
Non-green live	86	51	38	29	31	71	78	10
Dead	—	187	120	101	76	298	192	19
Belowground								
Rooting depth (cm)	—	25	25	25	12	20	20	10
Live	83	1085	691	353	260	57	50	4
Dead	—	938	604	313	87	—	—	—
Bryophytes								
Green	0	194	385	150	141	15	600	232
Non-green live	0	714	712	400	141	15	600	232
Lichens	0	0	0	0	16	49	23	0
Algae	—	4	2	1	—	—	—	—
Total	169	3,208	2,592	1,374	633	498	991	270
Vascular plant ratio Live above: Live below	1:1.0	1:2.6	1:8.9	1:6.3	1:4.9	1:0.6	1:0.4	1:0.3

* Vegetation occupies 51% of total area
** Vegetation occupies 32% of total area
+ Lichen barren on limestone assumed to have the same standing crop as on raised beach crests

zone. It is in this habitat that *Dicrostonyx* feed most intensively summer and winter (Fuller et al. this volume). A significant aspect of these communities is the reduction in root standing crop and net production. Most attached roots are alive. Generally, lichen and vascular plant cover and standing crop show little variation with increased age of the beach ridges (Svoboda this volume).

Other plant communities were studied in much less detail. The moss-herb community sampled 1 km from the edge of the plateau, within the Polar Desert, had much less cover and annual production (Tables 2 and 3). Further inland, species richness, plant cover, and production are significantly lower. The coastal salt marsh community is equally poor in species, but its carpet-like cover of grass has a higher level of aboveground production than what one might expect at this latitude (Jefferies 1977).

In the major plant communities (sedge meadow, cushion plant, and dwarf shrub heath) over 60% of the vascular plant production was contributed by one or two species, regardless of species richness. In the salt marsh and sedge meadow communities, one species contributed over 80% of total vascular plant production.

A striking feature of the Lowland is the contrast in plant structure, function, and net production, both above and belowground in the cushion

Table 3. Net plant production (g m^{-2}) for the major plant communities of Truelove Lowland. The mean values are based upon 1 to 3 years of data (Bliss et al., Jefferies, Muc, Svoboda).

Component	Salt Marsh	Hummocky Sedge-moss Meadow	Wet Sedge-moss Meadow	Frost-boil sedge Meadow: Vegetated Area	Frost-boil sedge Meadow: Total Area	Dwarf Shrub heath: Vegetated Area	Dwarf Shrub heath: Total Area	Raised Beach: Crest-Slope* Cushion plant-lichen	Raised Beach: Transition Cushion plant-moss	Raised Beach: Plateau Moss-herb
Lowland area (ha)	22	883	88	796		533		292	572	—
Vascular plants										
Aboveground										
Monototyledons	8.5	32.5	44.1	38.2	19.1	0.8	0.3	1.0	2.3	—
Forbs	—	4.5	1.5	3.2	1.6	1.5	0.5	4.9	5.0	2.0
Woody plants	0	7.7	0	16.4	8.2	15.6	5.0	9.3	20.0	0
Belowground	8.5	103.6	129.7	118.6	59.6	89.6	28.5	2.6	4.6	0.4
Total vascular plants	17.0	148.3	175.3	176.4	88.1	107.5	34.3	17.8	31.9	2.4
Bryophytes	0	33.0	102.6**	15.0	7.5	20.0	8.4+	2.0	20.0	5.0
Lichens	0	0	0	0	0	3.5	1.1	2.5	1.5	0
Algae	0	4.0	2.0	2.0	1.0	—	—	0.1	—	0
Total	17.0	185.3	279.9	193.4	96.6	131.0	43.8	22.4	53.4	7.4

* Lichen barren on limestone assumed to have the same net production as on the raised beach crest.

** Based on 15% streamside habitat with 293 g m^{-2} net production and 85% wet meadow with 69 g m^{-2} net production.

\+ Includes 2 g m^{-2} estimated moss production of *Rhacomitrium lanuginosum* upon rocks.

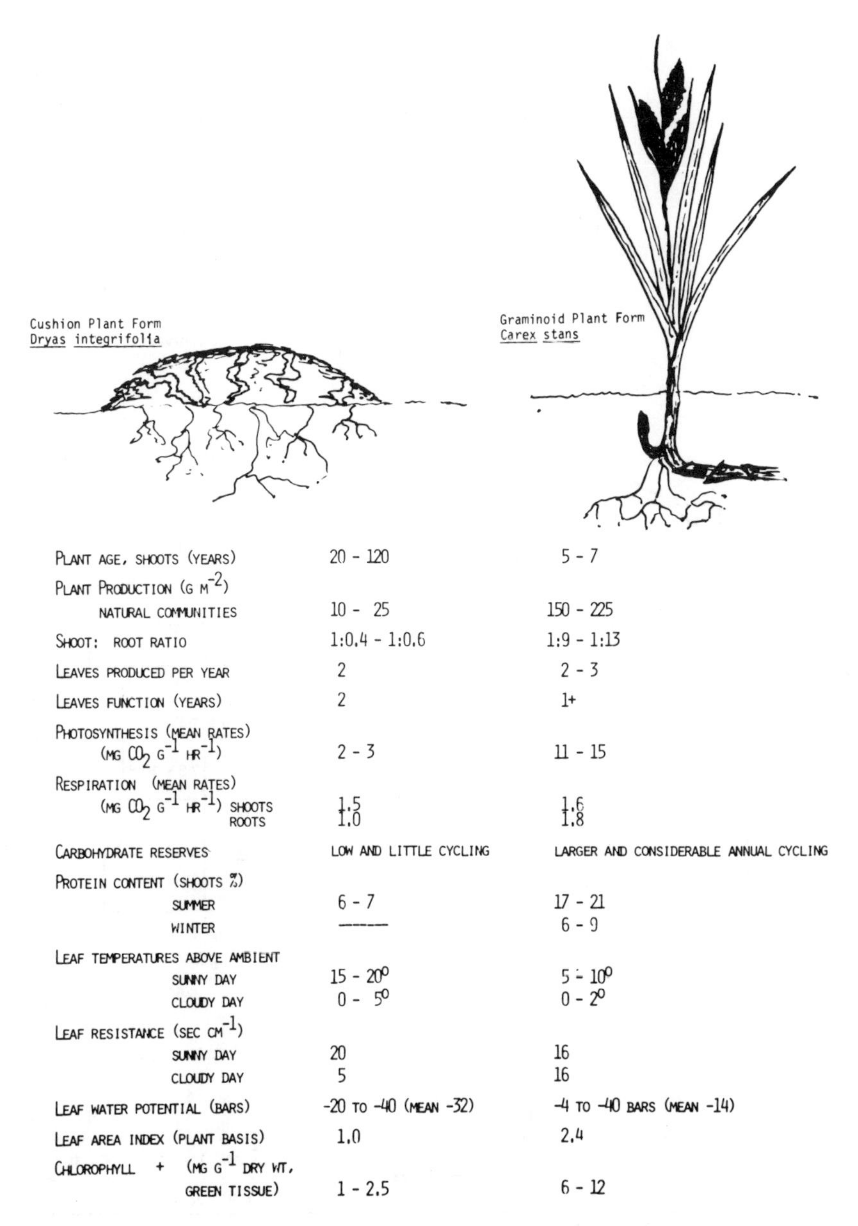

	Cushion Plant Form *Dryas integrifolia*	Graminoid Plant Form *Carex stans*
PLANT AGE, SHOOTS (YEARS)	20 - 120	5 - 7
PLANT PRODUCTION ($G\ M^{-2}$)		
NATURAL COMMUNITIES	10 - 25	150 - 225
SHOOT: ROOT RATIO	1:0.4 - 1:0.6	1:9 - 1:13
LEAVES PRODUCED PER YEAR	2	2 - 3
LEAVES FUNCTION (YEARS)	2	1+
PHOTOSYNTHESIS (MEAN RATES) ($MG\ CO_2\ G^{-1}\ HR^{-1}$)	2 - 3	11 - 15
RESPIRATION (MEAN RATES) ($MG\ CO_2\ G^{-1}\ HR^{-1}$) SHOOTS	1.5	1.6
ROOTS	1.0	1.8
CARBOHYDRATE RESERVES	LOW AND LITTLE CYCLING	LARGER AND CONSIDERABLE ANNUAL CYCLING
PROTEIN CONTENT (SHOOTS %)		
SUMMER	6 - 7	17 - 21
WINTER	------	6 - 9
LEAF TEMPERATURES ABOVE AMBIENT		
SUNNY DAY	15 - 20°	5 - 10°
CLOUDY DAY	0 - 5°	0 - 2°
LEAF RESISTANCE ($SEC\ CM^{-1}$)		
SUNNY DAY	20	16
CLOUDY DAY	5	16
LEAF WATER POTENTIAL (BARS)	-20 TO -40 (MEAN -32)	-4 TO -40 BARS (MEAN -14)
LEAF AREA INDEX (PLANT BASIS)	1.0	2.4
CHLOROPHYLL + ($MG\ G^{-1}$ DRY WT, GREEN TISSUE)	1 - 2.5	6 - 12

Fig. 3 Plant growth and physiological aspects of the cushion plant *Dryas integrifolia* and the upright sedge *Carex stans*. Data are based upon P.A. Addison, J.M. Mayo et al., M. Muc and J. Svoboda (this volume).

plants represented by *Dryas integrifolia* and graminoids represented by *Carex stans*. Their plant morphology, habitat selection, and physiological responses to environmental parameters are in sharp contrast. Since they are the dominant species in the cushion plant and meadow communities respectively, they were studied in greater detail. Much of the information is summarized in Fig. 3. *Dryas* is a conservative, slow growing, and physiologically less active plant than *Carex*. The former is the most characteristic species of the rolling upland of the Polar

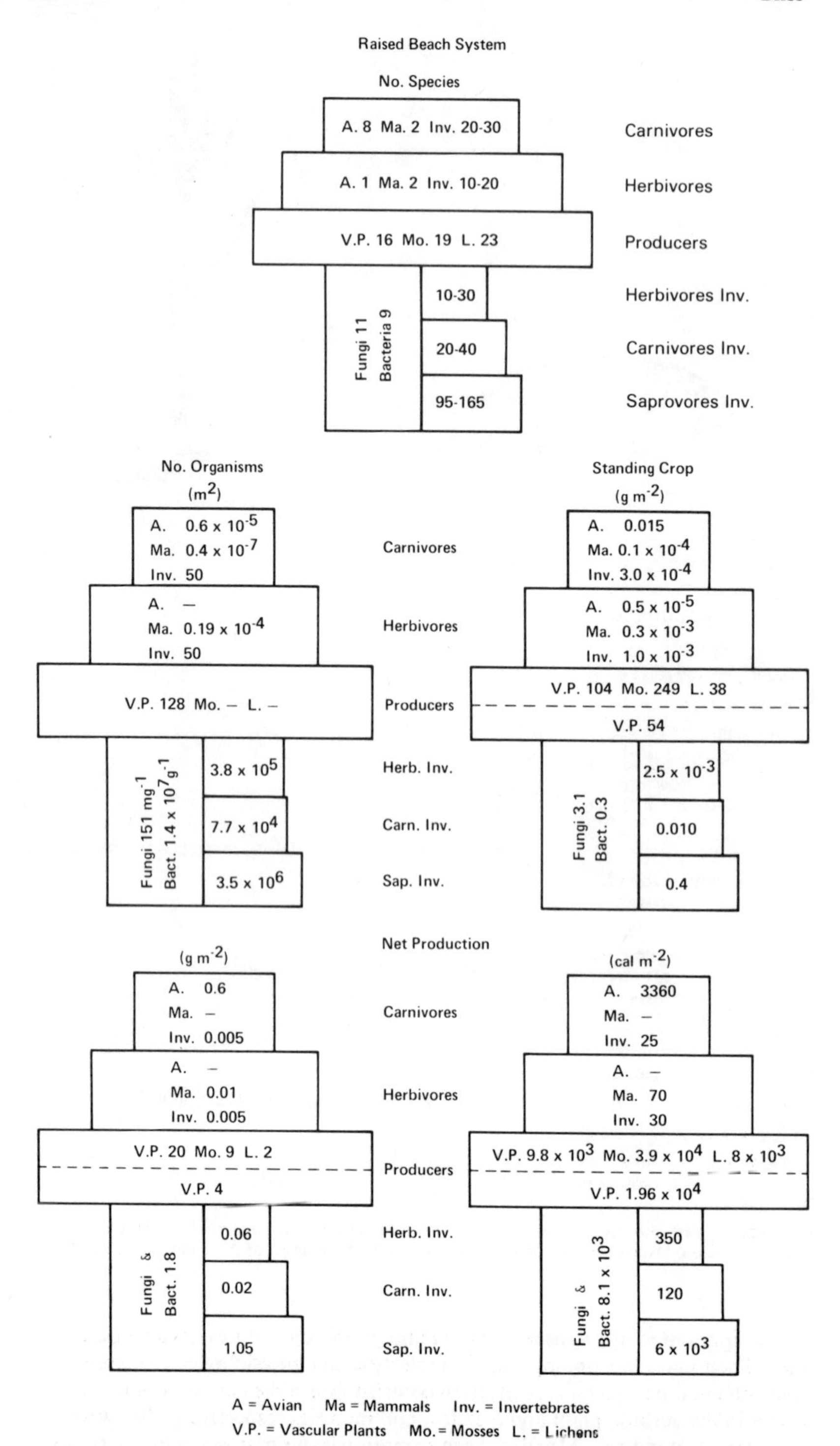

Fig. 4 Pyramids of species richness, number of organisms, standing crop (alive), and net production (g m^{-2} and cal m^{-2}) for the cushion plant-lichen/moss subsystem.

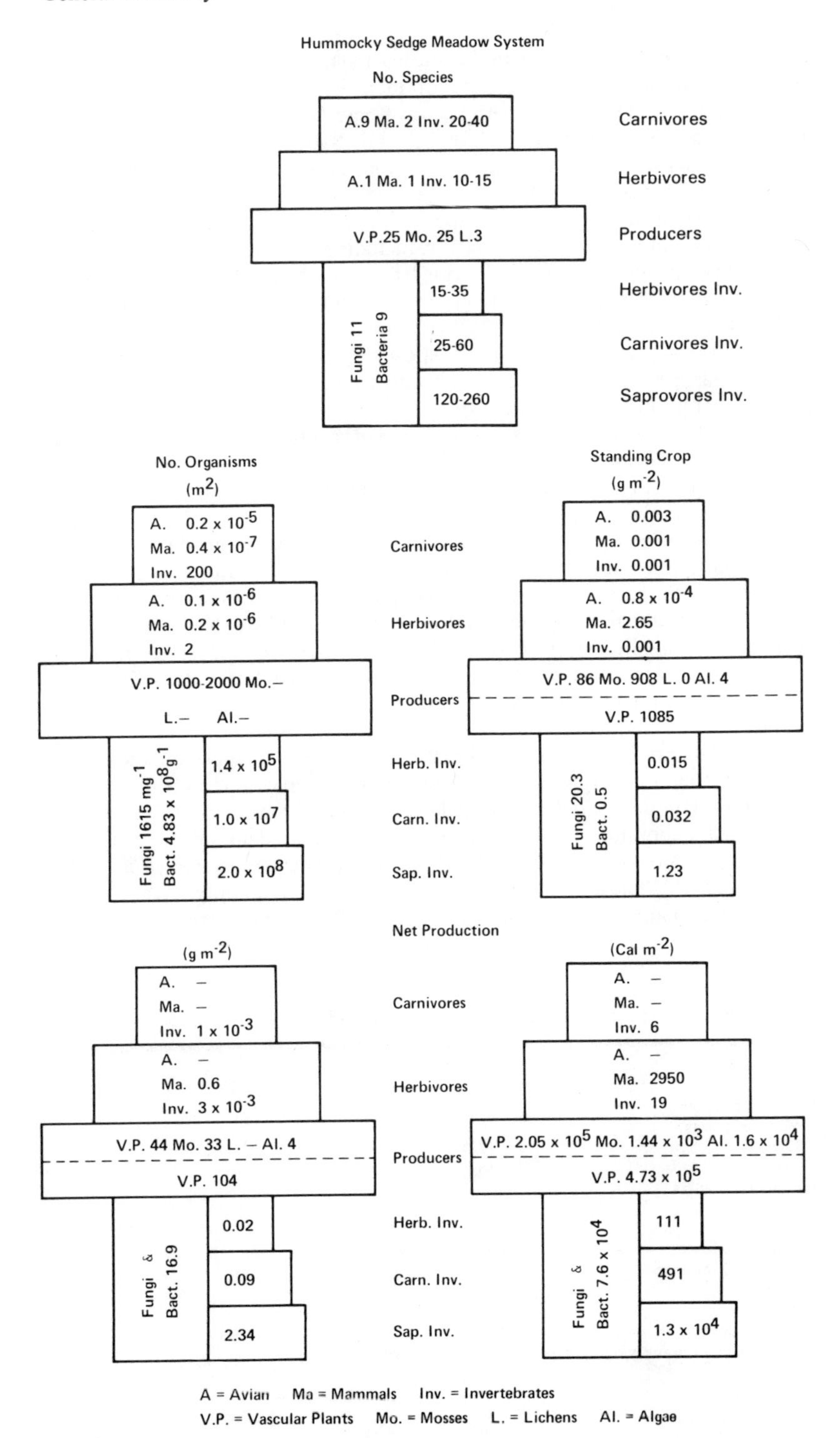

Fig. 5 Pyramids of species richness, number of organisms, standing crop (alive), and net production (g m^{-2} and cal m^{-2}) for the hummocky sedge-moss subsystem.

Semi-deserts that predominate in the southern islands of the Arctic Archipelago (Bliss and Svoboda 1977a). The same communities have a more limited distribution in the Queen Elizabeth Islands where *Luzula* and rosette plant dominated communities predominate (Bliss and Svoboda 1977b). The *Carex stans* meadows with associated graminoids and mosses are better represented on the southern islands and the Low Arctic of the continent.

The basic strategy of these two species (growth forms) is very different. *Dryas* because of its conservative nature, including an apparent mechanism for retaining its nutrient pool from decaying leaves (Svoboda, Widden—both this volume), can tolerate habitats with little winter snow cover, and soils that are warmer and drier in summer. *Dryas* does not fix nitrogen here although its roots are mycorrhizal (Stutz 1972). In contrast *Carex* grows in imperfectly to poorly drained soils, puts on more dry matter production below than aboveground, has much higher photosynthetic and respiratory rates, grows in soils where bluegreen algae and bacteria fix nitrogen, and in general is a far more productive plant (Fig. 3). Without the production efficiency of this species, including its rather high protein content in winter, there would not be the food base to support the relatively large population of muskox.

The quite close agreement in the estimates of dry matter production in *Carex stans* and *Dryas integrifolia* using harvest methods (Muc, Svoboda—both this volume) and expressing net production as a function of solar radiation and air temperature (Whitfield and Goodwin this volume) was very encouraging. In fact, the first results of the regression model showed the need to re-evaluate the calculations of root production in *Carex stans*. Thus this technique aided the final estimates of total plant production using the harvest method.

The plant production data for the cushion plant-lichen/moss and the sedge-moss communities are quite similar to data gathered on comparable communities on Banks and Victoria Islands (Bliss and Svoboda 1977a). The cushion plant community data are generally higher than data for the rush-forb-lichen communities (Polar Semi-desert) of the northwestern islands (Bliss and Svoboda 1977b). The data for the rosette plant-moss community on the plateau are higher than areas further inland and in other Polar Desert areas.

Table 4. Species richness within Truelove Lowland.

Species group	*Number*
Primary Producers	
Vascular plants	96
Lichens	182
Bryophytes	162
Primary consumers	
Mammals	3
Avian	2
Invertebrates	45-80
Secondary Consumers	
Mammals	3
Avian	14
Invertebrates	60-130
Tertiary and Quarternary consumers	
Avian	7
Decomposers	
Fungi	92
Bacteria (genera)	9
Invertebrates	175-365
Scavengers (avian)	4

Species richness

Floristically, this area is richer in species than other areas investigated to date at this latitude with the possible exception of the Lake Hazen area (82° N), Ellesmere Island. The latter area does not have as high a percentage of plant cover and habitat diversity as occurs on this and adjacent lowlands.

Six of the eight land mammals found in these arctic islands occur in the Lowland (Table 4) although wolves are rare and foxes are not common. Peary's caribou occurred here until recently when it was eliminated by excessive hunting. The avian fauna is unusually rich with 15 species regularly breeding and an additional 12 species occasionally breeding here or frequenting the area.

The greatest species richness is within the invertebrate groups. They probably total 500-600 species, although only *ca.* 235 species have been identified to date. The number of species of fungi and bacteria is probably at least double these preliminary values (Table 4).

Consumers and Decomposers

While vertebrates generally roam over diverse habitats, most species in this area, including lemming and muskox, are characteristic of either raised beaches or sedge-moss meadows. Most invertebrate groups are more easily assigned to one or the other habitat, although some groups are of equal importance in both habitats.

Since the Intensive Study Sites were located in these two habitats because of their areal extent and their known ecological importance to the total lowland system, the energy budgets (Whitfield this volume) and information discussed here refer to these two subsystems. We recognize that this is an over simplification in understanding total system function for the Lowland, yet this maximizes the available data. This method also permits comparisons of these two contrasting subsystems that represent important ecological systems elsewhere in the High Arctic.

Figures 4 and 5 summarize the biological data of the trophic components into aboveground and belowground pyramids for the two subsystems. These subsystems are very different in terms of microclimate, soils, plant communities, vertebrate herbivores, and dominant soil organism groups. They also differ greatly in species richness, standing crop, and net production, yet the percentage contribution of the various system components is very similar (Table 5). In terms of energy flow, the subsystems consist of solar radiation—plant producer—decomposer components. Primary producers account for 90-91% of total production in the two subsystems. The soil microflora (bacteria and fungi) is the second largest component.

Lemming, muskox, and birds are important to the subsystems in terms of species richness, their potential direct or indirect human use, and in landscape aesthetics, but they contribute very little to system function. An exception may be long term grazing of certain meadows by muskox where they may remove 15-20% of the total forage per year, although their overall grazing removes only 1-2% of the forage in all sedge meadows (Hubert this volume). Muskox also transfer nutrients via dung from meadows to raised beaches and the long term nutrient impact of this may be significant (Booth, Babb and Whitfield—both this volume).

Digestive efficiency was quite high in the various species studied. Pattie (this volume) reported efficiencies of 40% for seeds and 73% for arthropods in snow bunting. Values for other animals included 55% on Dicotyledonous plants

Table 5. Percentage contribution of the major components of the raised beach and sedge-moss subsystems. Data are based upon information in Figures 4 and 5.

		Biological summary		
System Component	*Species (No.)*	*Standing crop ($g\ m^{-2}$)*	*Net Production ($g\ m^{-2}$)*	*Net Production ($cal\ m^{-2}$)*
Herbivores + Carnivores aboveground				
Raised Beach	18	+	1.6	1.9
Sedge-moss Meadow	16	0.1	0.3	0.2
Primary Producers				
Raised Beach	23	98.7	90.9	90.5
Sedge-moss Meadow	18	98.9	90.3	90.0
Herbivores belowground				
Raised Beach	4	0.1	+	0.2
Sedge-moss Meadow	6	+	+	+
Carnivores belowground				
Raised Beach	8	+	+	0.1
Sedge-moss Meadow	9	+	+	0.1
Saprovores belowground				
Raised Beach	39	0.6	2.7	3.3
Sedge-moss Meadow	44	0.1	1.1	1.4
Microflora				
Raised Beach	8	0.7	4.8	4.4
Sedge-moss Meadow	7	0.9	8.3	8.3
Total Decomposer				
Raised Beach	59	1.3	7.5	8.0
Sedge-moss Meadow	66	1.0	9.4	9.7

for collared lemming (Fuller et al. this volume), 63% to 65% winter feeding (hay) and 75% summer feeding for muskox (Hubert this volume), 51% to 56% summer feeding for arctic hare (Smith and Wong this volume), and 86% for arctic fox and 74% for short-tailed weasel (ermine) using native wildlife (Riewe this volume).

The diversity of bird species and trophic levels (5) is high for this latitude and for systems that have such a low energy base for them. However, much of the food for some of the species (red throated loon, ducks, arctic tern, gulls) comes from the sea and there is a diversity of habitats for them to occupy. The land based birds often roam over much larger areas than this Lowland, especially the jaegers. All but two species are migratory. Nesting success of the various avian species is quite low, in part the result of predation (Pattie this volume).

In terms of standing crop and net production, vertebrate herbivores and carnivores constitute a larger component of the system than do belowground invertebrate herbivores and carnivores, but the saprovores are a larger component than are all of the herbivores and carnivores. In contrast with temperate grasslands and forests, aboveground invertebrates are a very minor component. This no doubt reflects the cold environment, slow growth rates, and long life cycles of these invertebrates (Ryan this volume).

Because of high turnover rates, both saprovores and the microflora (fungi and bacteria) are much more important (5 to 10 times) components of net production than of standing crop (Table 5). Bacterial numbers appear to be higher than in most tundra sites. High pH of the peat soils and a high correlation between amount of organic material and bacterial numbers may account for this (Nelson this volume).

Bacteria studied under laboratory conditions can be classified as facultative psychrophiles. They grew at $<5°$ C but *Bacillus* sp. would not grow below 0°. The three species studied were tolerant of low concentration of carbon in terms of growth and *Pseudomonas* sp. was resistant to starvation stress. *Bacillus* sp. was rarely isolated from the sedge-moss meadow soils and it was less successful at coping with low temperature, starvation, and freeze-thaw cycles (Nelson this volume).

The data on invertebrate respiration showed that some species have low Q_{10} values which may help maintain adequate metabolic rates at the low temperatures of both aquatic and terrestrial animals. Many of the species studied showed little adaptation of metabolic rates to low temperatures (Procter this volume). Similar findings have been reported for aquatic organisms in the Char Lake I.B.P. study (Rigler 1975).

Although the percentage contribution, of the number of species, standing crop, and net production, to the various trophic levels is very similar in the two subsystems (Table 5), the taxonomic components are very different. This is best exemplified by sedges in the wet sites and sub-shrubs or cushion plants in the dry sites, the dominance of Enchytraeidae and Protozoa in the wet sites and Nematoda, Enchytraeidae, and Collembola in the dry sites. Bacteria and fungi are much more abundant and more important in terms of energy flow in the sedge-moss meadows, yet the relative roles of the two groups remain similar in the two subsystems. It is unfortunate that at the time of preparation of this book, energy flow data were not available from the other arctic I.B.P. studies so that comparisons could be made.

Mineral cycling

Athough nutrient cycling was not studied in as great detail as was energy, sufficient data were gathered to permit certain generalizations. Available macronutrients (N, P, K) are low (Babb and Whitfield, Walker and Peters—this volume). The addition of fertilizer to field plots and to plants grown in the plant growth chamber experiments generally stimulated growth (Babb and Whitfield this volume). *Saxifraga oppositifolia* and *Cerastium alpinum* plants were flowering far more abundantly than controls three summers after treatment. Even the conservative *Dryas integrifolia* grew significantly more with the three macronutrients.

Both subsystems have considerable pools of organic matter, though the amount is much greater in the sedge-moss meadow where accumulation has probably gone on for *ca*. 6,000 years (Babb and Whitfield this volume). The frost-boil meadows have a shorter time period of lateral water movement than do the hummocky sedge-moss meadows. Since lateral movement of nutrients seems quite important in the Lowland, this may help to explain the lower production of the frost-boil meadow type. However where vascular plants occur, their levels of production are as high per unit area as they are in the hummocky sedge-moss type. The greatly reduced level of moss production in the frost-boil meadows may result from less nutrient absorption by moss leaves for it is known that moss production is greater where there is more surface runoff even though nutrient levels in the water are low (Pakarinen and Vitt 1974, Vitt and Pakarinen this volume). Immerk and Fish Lakes have a higher production (oxygen metabolism, mean summer chlorophyll *a* concentration, numbers of crustacean zooplankton, and early growth rates of arctic char) than does Loon Lake. The latter is fed by cold nutrient poor waters via Gully River from the plateau to the east, while the watersheds of the other two lakes consist of the surrounding meadows and beach ridges (Minns this volume). As a result the nutrient input via spring runoff is greater.

Nitrogen fixation is 5X greater than input of nitrogen via precipitation in the sedge-moss meadows but the reverse is true for the cushion plant communities of the raised beaches (Babb and Whitfield this volume). As with energy, most nitrogen and phosphorus mineralization occurs via the microflora and microbivores (Nematoda) in the raised beach subsystem and via the microflora and protozoa in the hummocky sedge-moss subsystem. The turnover time for N and P in living plant tissue is *ca*. 8 years in the sedge meadow and 10 and 12 years respectively in the cushion plant subsystem (Babb and Whitfield this volume). This is longer than the 4 years reported at Barrow, Alaska, for N and P and 5 years for N turnover at the Swedish mire (Bunnell et al. 1975, Rosswall et al. 1975 respectively). As in these other arctic and subarctic systems, only about 1% of the N and P is tied up in living tissue; the remainder is in the accumulated peat (Babb and Whitfield this volume).

System metabolism

The energy flow (Whitfield this volume) and nutrient cycle diagrams (Babb and Whitfield this volume) are excellent mechanisms for sufficiently presenting ecosystem function. Another more simplified approach is to discuss system production and respiration (Woodwell and Botkin 1970). Plant respiration was calculated by Mayo (pers. comm.) based upon our field measurements and literature values. The data were adjusted for mean temperature conditions of the two subsystems and the length of season in which respiration was estimated to have occurred. The time ranged from 320 hr in mosses and lichens on the drier raised beach to 1200 hr for all plant groups in the moist to wet meadow and 1440 hr for vascular plants on the raised beach. The following system properties were used:

Gross Primary Production	(GPP)	Gross photosynthesis
Net Primary Production	(NPP)	GPP-R_A
Autotrophic Respiration	(R_A)	GPP-NPP
Heterotrophic Respiration	(R_H)	Consumer and decomposer respiration
Ecosystem Respiration	(R_E)	$R_A + R_H$
Net Ecosystem Production	(NEP)	GPP-R_E

In this synthesis (Table 6) several important aspects of the subsystems are evident. Plant respiration accounts for a very high percentage of gross primary production (91-92%); net plant production (8-9%) and net ecosystem production (0.4-0.5%) are small in relation to gross primary production. The various ratios presented in Table 6 show how similar the two subsystems are.

In contrast with the summary data on carbon metabolism of two temperate deciduous forests, a tall grass prairie and the Barrow tundra ecosystem (Reichle 1975), our estimate of autotrophic respiration is a much larger percentage of the total system. The estimate of 5% for ecosystem productivity (NEP/GPP) at Barrow is an order of magnitude greater than for the cushion plant (0.5%) and sedge-moss (0.4%) subsystems in Truelove Lowland. If autotrophic respiration was assumed to be 2X net primary production rather than 10X as calculated, ecosystem production would still be only 2% of gross primary production.

Production efficiencies and high arctic systems

As discussed elsewhere in this book, the sedge-moss meadows are representative of only 1% of the landscape in the Queen Elizabeth Islands, yet these lands are essential to maintenance of muskox and a diversity of avian, mammalian, and invertebrate species. These sedge tundra communities are more characteristic of

Table 6. Carbon budget (g carbon m^{-2} yr^{-1}) for the cushion plant and hummocky sedge-moss meadow, Truelove Lowland.

Property	*Cushion plant*	*Sedge-moss meadow*
	Production and Respiration	
Gross Primary Production (GPP)	373	3,325
Autotrophic Respiration (R_A)	339	3,060
Net Primary Production (NPP)	34	265
Heterotrophic Respiration (R_H)	32	251
Ecosystem Respiration (RE)	371	3,311
Net Ecosystem Production (NEP)	2	14
	Efficiencies (%)	
Production Efficiency (R_A/GPP)	91	92
Effective Production (NPP/GPP)	9	8
Maintenance Efficiency (R_A/NPP)	10	12
Respiration Allocation (R_H/R_A)	9	8
Ecosystem Production (NEP/GPP)	0.5	0.4

Table 7. Net annual plant production and efficiency of annual plant production for the various communities and for the total lowland ecosystem. All data are based upon length of the growing season.

Component	*Growing Season (days)*	*Net Production (K cal m^{-2})*	*Total Radiation (K cal m^{-2})*	*Efficiency (%) Total Radiation*	*PAR**
Polar Desert (plateau)	45	33	$1.94x10^5$	0.02	0.04
Polar semi-desert					
cushion plant-lichen	60	114	$2.80x10^5$	0.04	0.09
cushion plant-moss	55	252	$2.45x10^5$	0.10	0.23
total subsystem	—	160	$2.68x10^5$	0.06	0.14
Sedge tundra					
hummocky sedge-moss	50	838	$2.16x10^5$	0.39	0.86
frost-boil sedge	50	442	$2.16x10^5$	0.20	0.45
total sedge-moss subsystem	50	680	$2.16x10^5$	0.31	0.70
Dwarf shrub tundra	50	657	$2.16x10^5$	0.30	0.68
Total lowland system	—	458	$2.30x10^5$	0.20	0.44

* PAR (photosynthetically active radiation) = 45% of total radiation.

the southern High Arctic and northern portions of the Low Arctic. Cushion plant-lichen/moss communities and the upland habitat they represent are well represented in the southern Arctic Archipelago as well as in the eastern Queen Elizabeth Islands. This is one of the two types of Polar Semi-desert communities. Finally the limited data collected on the plateau represent the upper or near upper limit of biological diversity and production of the true extreme Polar Deserts. Table 7 summarizes the data on net production and efficiency of production for these major landscape units as well as for the entire Lowland. The latter summation was based upon the percentage contribution of each major component within the total Lowland (raised beach types, meadow types, dwarf shrub heath, etc.). Efficiency of production in the sedge-moss and dwarf shrub heath tundra components (0.7-0.9%) is comparable to the values calculated for other tundra sites from the data in Rosswall and Heal (1975). The Semi-desert and Polar Desert systems on Devon Island have much lower levels of plant production and efficiency of production.

The ecological efficiencies calculated by Whitfield (this volume) speak for themselves. As he points out, herbivore consumption in Truelove Lowland is but a fraction of that in numerous systems, including Barrow, Alaska. Invertebrate herbivore consumption and efficiency are not very different from data in the literature from other biomes.

While the percentage contribution of some components of this system (invertebrate and vertebrate herbivory) may be different from more temperate region systems, the general functions of this northern one are very similar in spite of the severe climate limitations.

Ecosystem stability

In recent years much has been written regarding the fragility vs. stability of arctic ecosystems. Biological simplicity, low species richness, presence of ice-rich permafrost, wide fluctuations in animal populations, slow rates of succession and recovery from perturbations, and low rates of production and decomposition have been cited as characteristic of a fragile arctic system (see Bliss et al. 1973).

This study provides information on the subject. The permafrost system in the High Arctic is quite stable because of its low temperature and relatively low ice content in most areas compared with the Low Arctic. Mass wasting results more from water erosion during spring melt than from thermal erosion in summer. During the four years of intensive study, summer climatic conditions were very variable (lakes 50% to 6 weeks, ice-free end of summer, accumulated degree days 3 fold difference), yet plant production changed by only 8-10%. In most plant communities over 60% of net production and standing crop result from one species. One might expect that in such a severe environment, selection would be against single species dominance, as has occurred in the Polar Deserts. However these community types cover large areas and appear stable.

The lemming population ranged from a low in 1972 to a high the next year rather than following a cyclic pattern. Population dynamics in lemming and muskox appear to be controlled by climatic factors (storms that result in winter icings of vegetation or rapid spring melt and loss of nest sites before summer burrows thaw) rather than by available food per se.

Within the vertebrate fauna, species and trophic diversity occurs only in the birds where it might be least expected. This probably results from the diversity of feeding niches including many that are marine and the fact that most species are migratory. A great diversity of trophic pathways occurs in the decomposer component as in other ecosystems.

Results of this study point to high arctic systems of this diversity and productivity as being quite stable. The exceptions are crucial times in the year for vertebrate herbivores and carnivores when the food supply (icing of forage for herbivores, lack of food for weasel and fox, loss of winter nest habitat before summer burrows thaw for lemming) or habitat can be rapidly altered. Primary producers and the whole decomposer component appear to be more stable and conservative in growth and development. This is culminated in the life cycle strategy of *Dryas integrifolia* and the invertebrates studied to date, especially *Gynaephora* (Svoboda, Ryan and Hergert, both this volume).

Species can be eliminated as happened with excessive hunting of Peary's caribou. If key species of plants or animals were eliminated, there are few species able to fill the same trophic niche.

This system, like many of those studied in detail in temperate regions, is quite resiliant to severe natural perturbations and biologically some components respond rapidly to favorable climatic conditions. The overall stability should give reassurance to those concerned with system response to environmental problems and to long term management of land and its biological resources in the High Arctic.

Human use and land management

The research of Riewe (this volume) clearly shows the dramatic change in the hunting-fishing economy of the Grise Fiord Inuit as their travel changed from reliance on dog teams to snowmobiles. Mechanized travel costs more money, they extend their hunting range, are more effective in finding animals and thus have an impact on wildlife over a greater area, and they consume less of the food (energy) they harvest. Prior to snowmobiles (1965-66) only 10-15% of the animal harvest went to decomposers while in 1971-72, 67% went to decomposers and scavengers. This results in part because there are fewer dogs to feed and people are consuming more southern food. While these shifts in ecosystem use were taking place, their population greatly expanded, from 36 people in the early 1950s to 96 people in 1972. Here, as in other Inuit villages, the increase in human population has exceeded the ability of the natural ecosystems, both sea and land to support them. An exception to this pattern is the community at Sacks Harbour, Banks Island. Here, because of the extensive areas of sedge-moss meadows and rolling uplands of cushion plants (see Fig. 1, p. 2), much larger populations of muskox and Peary's caribou are available for human food. The abundance of lemming and waterfowl, especially snow geese, has enabled a high natural level of fox production and this animal, trapped in large numbers, provides the economic base for the community. Few if any other lands (possibly portions of Prince of Wales Island) in the Canadian High Arctic have this potential, certainly none in the Queen Elizabeth Islands.

At these latitudes there are relatively few areas with lakes. Regardless of whether the watersheds of the lakes are surrounded by barren lands (Polar Desert) as at Char Lake, Cornwallis Island (Rigler 1975) or by sedge-moss meadows as on the Truelove Lowland (Minns this volume), species diversity of invertebrates is low. As with most high arctic lakes, arctic char *(Salvelinus alpinus)* is the only fish species. Loon Lake, fed by cold, nutrient poor waters from the barren plateau, is much lower in production (similar to Char Lake) than are the other two lowland lakes studied, lakes fed by meltwater from the surrounding sedge-moss meadows. Benthic algae are the major primary producers.

These lakes contain land locked char and where similar conditions prevail, it can be predicted that the lakes have a very low level of sustained yield of fish compared with lakes open to sea run char. As discussed by Rigler (1975), invertebrates and vertebrates in these cold water arctic lakes appear to have no special adaptations for growth. Therefore production at all levels in these aquatic systems is low and their ability to produce fish for Inuit or commercial fishing camps is very limited.

Finally, the Arctic is a focal point in the 1970s because of mineral and petroleum discoveries. The Strathcona Sound, Baffin Island mine will begin production in 1977. Gas and now oil have been discovered in the western Queen Elizabeth Islands and plans are underway to bring these resources to market.

The Devon Island study included aspects of the impact of development on northern lands. From those and other studies (Babb this volume, Addison 1975, Bell 1975a, b) it can be concluded that surface erosion resulting from spring runoff is far more serious in the High than in the Low Arctic. Plant growth is slow, especially in the northwestern islands, because of cold soils and low air temperature. Reseeding with native species, even grasses from the Mackenzie Delta, holds little promise for site restoration.

Massive ice appears to be less of a problem than in the Low Arctic. Wet spots on scraped surfaces including runways, appear to result from ice rich soils *ca.* 1-2 m in depth. When this moisture evaporates in 2-3 summers the sites dry out and can be used throughout the summer (Price et al. 1974). An initial map showing the general susceptibility of landscapes to environmental impact (Babb and Bliss 1974) has been modified into a map showing sedge lands in the High Arctic in this volume, and into detailed land use maps prepared by the ALUR program within the Department of Indian and Northern Affairs.

Ecosystems as diverse and productive as these in northeastern Devon Island are very limited in the High Arctic and must receive protection if the wildlife they support is to be maintained. This becomes less problematic on Prince of Wales and Banks Islands where so much more of the landscape is occupied by these ecosystems.

Most of the High Arctic is a semi-barren or barren landscape (Polar Semi-desert, Polar Desert, ice caps) and thus supports relatively little biological diversity and production. Consequently industrial development including pipelines on land will have less of a biological impact than in the Low Arctic.

Although most of this research was confined to only 43 km^2 in the High Arctic, it is now evident that the results can be applied in general and sometimes in specific detail to large areas. This shows that beyond the specific scientific details on the structure and function of an arctic ecosystem, studies of this level provide information both basic and applied that can aid in land use decisions over vast areas of the High Arctic.

Acknowledgments

Grateful appreciation is extended to all of the researchers who not only provided information used in this summary but who have provided stimulating discussions through the years which culminate here. Special thanks are due to Jim Ryan for his help in the various aspects of invertebrate species and energetics that are presented here and to Doug Whitfield for his careful synthesis of energy flow and nutrient cycling so central to an ecosystem study.

References

Addison, P.A. 1975. Plant and surface responses to environmental conditions in the western High Arctic. pp. 3-20. *In:* Plant and Surface Responses to Environmental Conditions in the Western High Arctic. L.C. Bliss (ed.). ALUR 74-75-73. Ottawa. 72 pp.

Babb, T.A. and L.C. Bliss. 1974. Susceptibility to environmental impact in the Queen Elizabeth Islands. Arctic 27: 234-237 and map.

Bell K.L. 1975a. Root adaptations to a polar semi desert. pp. 21-61 (see Addison 1975).

———. 1975b. Aspects of seed production and germination in some high arctic plants. pp. 62-72 (see Addison 1975).

Bliss, L.C. 1975a. Devon Island, Canada. pp. 17-60. *In:* Structure and Function of Tundra Ecosystems. T. Rosswall and O.W. Heal (eds.). Ecol. Bull. (Stockholm) NFR. 20: 450 pp.

———. 1975b. Truelove Lowland: A high arctic ecosystem. pp. 51-85. *In:* Energy Flow — Its Biological Dimensions. T.W.M. Cameron and L.W. Billingsley (eds.). Royal Soc. Canada, Ottawa. 323 pp.

———. G.M. Courtin, D.L. Pattie, R.R. Riewe, D.W.A. Whitfield, and P. Widden. 1973. Arctic tundra ecosystems. Ann. Rev. Ecol. Syst. 4: 359-399.

———. and J. Svoboda. 1977a. Plant communities and plant production on Banks and Victoria Islands, N.W.T. (in preparation).

———. and ———. 1977b. Plant communities and plant production in the western Queen Elizabeth Islands, N.W.T. (in preparation).

Jefferies, R.L. 1977. Plant communities of muddy shores of arctic North America. J. Ecology. pp. 661-672.

Pakarinen, P. and D.H. Vitt. 1974. The major organic components and caloric contents of high arctic bryophytes. Can J. Bot. 52:1151-1161.

Price, L.W., L.C. Bliss, and J. Svoboda. 1974. Origin and significance of wet spots on scraped surfaces in the high arctic. Arctic 27: 304-306.

Reichle, D.W. 1975. Advances in ecosystem analysis. Bio Science 25: 257-264.

Rigler, F.H. 1975. The Char Lake project: an introduction to limnology in the Canadian Arctic. pp 171-198 (see Bliss 1975b).

Rosswall, T. et al. 1975. Stordalen (Abisko), Sweden. pp. 265-294 (see Bliss 1975).

Rosswall, T. and O.W. Heal (eds.). 1975. Structure and Function of Tundra Ecosystems. Ecol. Bull. (Stockholm) NFR. 20:450 pp.

Stutz, R.C. 1972. Survey of mycorrhizal plants. pp. 214-216. *In:* Devon Island I.B.P. Project: High Arctic Ecosystem, L.C. Bliss (ed.). Dept. Botany, Univ. Alberta, Edmonton. 413 pp.

Woodwell, G.M. and D.B. Botkin. 1970. Metabolism of terrestrial ecosystems by gas exchange techniques: The Brookhaven approach. pp. 73-85. *In:* Analysis of Temperate Forest Ecosystems. D.E. Reichle (ed.). Springer-Verlag. New York. 304 pp.

Appendices

Appendix 1

Administrative summary

L.C. Bliss

Projects with the size, diversity, and integration of those conducted within the I.B.P. have seldom been conducted in Canada or elsewhere in the world. Consequently a considerable knowledge was gained regarding scientific and administrative procedures. The purpose of this appendix is to briefly summarize aspects of these details that directly relate to this project.

Internationally, the first Tundra Biome meeting took place in September 1968, at Hardangervidda, Norway. At that time Canada had no I.B.P. tundra project, but a resolution was passed urging researchers to organize one. Largely through the efforts of R.T. Coupland, the Canadian Committee for the I.B.P. (C.C.I.B.P.) backed a request that a working group be struck to determine whether there was interest within the scientific community for such a project.

After much correspondence, a meeting attended by 16 people from across the country was held April 14-15, 1969, at the Kananaskis Environment Centre, University of Calgary. At that time the basic research diagram for nutrient cycling and energy flow approved in Norway for a tundra ecosystem study was agreed upon. It was agreed that the research should include all major components of an ecosystem (atmospheric, edaphic, producers, primary and secondary consumers, and decomposers including soil flora and fauna). Sampling procedures were agreed upon for plant sampling and many aspects of the animal population studies. It was agreed that research should concentrate on determining the kinds of organisms present, their numbers, standing crop, and production. The unifying aspects would be nutrient cycling within and energy flow through the system. Ten of the assumed 31 pathways within a theoretical system were felt most important to concentrate upon.

At that time, 20 locations, including 14 on the mainland and 6 within the Arctic Islands, were evaluated using 13 criteria ranging from logistics and previous research to biological diversity and applied value of the research. From this analysis, the group selected Cape Sparbo, Devon Island, Inuvik-Tuktoyaktuk, and St. Elias Range, Yukon, as the most favorable areas. The group endorsed the need for 3-4 major study sites. It was agreed that two people should visit various sites in 1969 to aid in final site selection. This was not accomplished.

Following further discussion and correspondence a second meeting was held in Edmonton, October 15-17, in connection with the International Conference on the Productivity and Conservation of Circumpolar Areas. At this time further discussion of Cape Sparbo, Mackenzie Delta, Baker Lake, and

Fort Churchill was held and the Truelove Lowland site on Devon Island was chosen for the following reasons:
1. Current facilities to support 10-15 people developed by the Arctic Institute of North America.
2. Previous and current research in biology and glacial geology.
3. Logistic nearness to Resolute, Cornwallis Island.
4. Diversity of plants and animals in a relatively small area with its own drainage basin and on adjacent upland with sparse plants and animals.
5. Herd of muskox which could provide a large herbivore.
6. No other country within the I.B.P. had developed a high arctic study and it was believed that Truelove Lowland was representive of such areas.

A considerable amount of credit for selecting a high arctic site is due several representatives from the other tundra projects who attended the Edmonton meeting. Dr. Paul Barrett was most helpful in the selection of the Truelove site and in providing details of the area prior to initiating the project.

The initial research proposal was submitted to the National Research Council of Canada's C.C.I.B.P. and to member companies of the Arctic Petroleum Operators Association in early 1970. An organizational meeting on research design was held in March and the field work began in May.

Each year a summary meeting was held with papers presented by each researcher. A major review of the project occurred October 9-10, 1971, with four people outside of the project making the evaluation. Their report was submitted to the federal funding agencies and the petroleum companies supporting the project. Following the January 24-26, 1972 research meeting, the first two years of research were summarized in a book (Bliss 1972) provided in limited distribution to researchers within the various tundra I.B.P. projects.

Because of the desire to include applied research in relation to petroleum exploration and the fact that monies from several federal departments and the petroleum industry, functioned throughout the project to review the research and proposed budgets, and discuss administrative details. Corresponding needs, and research progress.

A Steering Committee, chaired by J.C. Ritchie and with eight members representing the project, the Arctic Institute of North America, and the petroleum industry functioned throughout the project to review the research conducted, propose budgets, and discuss administrative details. Corresponding members represented the federal departments and petroleum companies that supported the project. The research was coordinated with a committee, each member representing one of the research areas. An advisory committee of three members and the project director made decisions on minor aspects.

Basic core funding for three years was provided by the National Research Council. In addition funds were provided by the Department of Indian and Northern Affairs, in part for the applied research, and the Department of Environment. Most of the logistic support from Resolute to Devon Island was provided by the Polar Continental Shelf Project of the Department of Energy, Mines and Resources. The Arctic Institute of North America managed the camp and provided some of the camp supplies. The 24 petroleum companies and consortia provided both funds and support in kind in the form of fuel and logistic support. Universities provided Graduate Teaching Assistantships for some researchers and paid the salaries of faculty members and some support staff (secretarial, expediting, etc.). Table 1 summarizes financial aspects. Of these totals, approximately 20% was spent on the transportation of people and freight, 53% on research, and 27% on support facilities and supplies. Direct administrative costs were less than 5% of the total. There were no overhead charges.

Table 1. Financial summary, Devon Island Project.

Agency	*1970*	*1971*	*1972*	*1973*	*1974*	*Total*
Arctic Institute	$ 9,000	$ 14,300	$ 10,000	$ 8,000	$ 3,000	$ 44,300
Government Agencies	144,700	275,000	250,000	197,500	3,000	870,200
Petroleum Industry	46,400	62,900	92,000	31,900	8,000	241,200
Universities	49,000	79,000	88,900	74,000	20,000	310,900
Total	$249,100	$431,200	$440,900	$311,400	$34,000	$1,466,600

Table 2. Personnel summary, Devon Island Project.

Category	*1970*	*1971*	*1972*	*1973*	*1974*	*Mean or Total*
	Personnel					
Graduate students	7	11	12	9	6	9.0
Post-doctoral fellows	2	5	5	2	—	2.8
Field Assistants	9	21	22	9	2	12.6
Faculty	9	13	13	15	16	13.2
Others	4	11	14	6	2	7.4
Total	31	61	66	41	26	45.0
	Man-Years					
Graduate students	3.5	6.5	7.7	3.4	3.2	24.3
Post-doctoral fellows	1.2	3.8	3.8	1.6	—	10.4
Field Assistants	3.1	7.0	8.2	3.8	0.3	22.4
Faculty	1.9	2.9	3.0	2.5	2.2	12.5
Others	1.5	4.9	7.0	1.1	0.3	14.8
Total	11.2	25.1	29.7	12.4	6.0	84.4

Most of the research was conducted by graduate students and post-doctoral fellows although several senior investigators conducted or assisted in the field research. The project resulted in the completion of five M.Sc. and seven Ph.D. theses. Table 2 summarizes man-years and related information on research.

Although the coordination of the project was made more difficult by the distant location of the field work, researchers spread across the country, and multi-source funding, there were numerous compensating aspects. The project was conducted on neutral ground where there was no established pattern of experienced researchers. Because of its distance, the aura of an expedition was a psychological aid. Monies were not available in large amounts when the project began, thus researchers took part because of their enthusiasm, rather than because of "a pot of gold." Most of the research was done by young, eager people who had the support of senior people, most of whom spent time in the field. A sense of cooperation and support was established early with the various funding groups, and thus competition for control was not a factor. Finally, the research was conducted during a high level of petroleum exploration in the High Arctic. This resulted in support that made the difference between a marginal and an adequately funded and logistically supported project.

The central role that D.W.A. Whitfield, our systems analyst, played cannot be overstated. He did not enter the project until the second year, but he rapidly became familiar with the research, largely because of his desire to spend time in the field with each researcher. We learned from this experience that projects of this complexity and with these objectives should incorporate systems analysis into the basic design.

A great deal of cooperation from all groups is essential for an integrated, multidisciplinary project of this kind to be successful. Fortunately this occurred. One of the most important aspects is the need to maintain internal communication between researchers, even though they are living and working together in a limited area. This was facilitated by weekly research meetings during the field season, the annual report meetings, and the frequent *ad hoc* meetings of researchers on this campus. Having researchers on different campuses makes communication more difficult but also more essential that it be treated as of central importance.

As a result of the broad arctic training that many young people obtained on the project, they have been able to enter the work force in a greater variety of capacities than would otherwise have been the case.

Reference

Bliss, L.C. (ed.) 1972. Devon Island I.B.P. Project, High Arctic Ecosystem. Dept. Botany. Univ. Alberta, Edmonton. 413 pp.

Appendix 2

Remote sensing and aerial photography of Truelove Lowland

G.M. Courtin

The intensive, quantitative biological research performed under the auspices of the International Biological Programme required photographic imagery of much higher quality than that available for this high arctic site prior to the start of the project in 1970. Since the major emphasis of the project was the ecosystem, where productivity was to be expressed on a per unit area basis, a detailed large-scale map was required. The concentration of expertise in both the biotic and abiotic aspects of ecology also made available a great quantity of accurate data on biotic, physiographic, and microclimatic parameters at the surface.

The foregoing elicited the interest and support of federal government agencies who either used the research site as an experimental area to test various types of imagery, or made aerial photographs to assist the project.

The main emphasis of this report is to describe the characteristics of the various types of imagery used, the conditions under which the work was done and where the information can be found should future researchers have use for the data.

Remote sensing

Three passes of Truelove Lowland were made on 28 July, 1971 by the Atmospheric Environment Service, using the ice reconnaissance aircraft. One pass covered a band about 1 km wide from the landward end of Rocky Point to the mouth of the Gully River. The other two passes were parallel to each other and overlapping to cover a band *ca*. 1.5 km wide from the end of Rocky Point to the mouth of the Truelove Valley. Each pass was filmed as a continuous scan simultaneously with both conventional black-and white-and infra-red film, sensitive to the 8-14 micron wavelength range. At the same time a laser profilometer scan with a resolution of 1 vertical foot in 100 horizontal feet (.3 m in 30 m) was recorded (Table 1). The latter was intended to give a topographic profile along the centre of the scan.

The two scans of greatest potential interest to the project were the infra-red and profilometer scans. The infra-red film used was sensitive to the principal thermal wavelengths and it was hoped to distinguish temperature differences across the Lowland. Since the aircraft passed from the limestone plateau, across the Lowland, to the sea ice, each end of the pass showed a very high reflectance in

Table 1. Technical data for the remote sensing and aerial photography conducted over Truelove Lowland, Devon Island, N.W.T. in July and August 1971.

	Remote sensing		*Aerial photography*		
	Infra-red	*Laser profilometer*	*Interpretive*	*Ortho-photography*	*False-color Infra-red*
Date	28 July 1971	28 July 1971	18 August 1971	18 August 1971	18 August 1971
Time	2236-2239	2236-2239	1600-1800	1600-1800	1600-1800
Flight height (Feet A.S.L.)	2,900	2,900	2,500	10,000	5,000
Scale	—	1 ft in 100 ft	1:5,000	1:20,000	1:10,000
Film	70 mm infra-red; sensitivity 8-14 microns	—	double x aerographic black & white	double x aerographic black & white	False-color infra-red
Roll No.	—	—	A-22541 35-267	A-22541 1-34	A-30370-I.R. 1-134
Agency	Atmospheric Environment Service		Inland Waters Branch, Environment Canada Polar Continental Shelf Project, E.M. & R.		
Availability	Atmospheric Environment Service		National Film Library, Ottawa		

comparison to the Lowland and the operator changed the sensitivity of the apparatus to accommodate these different substrates. The result was that the scan could only be interpreted in a relative rather than an absolute sense.

The profilometer recorder failed to advance sufficiently rapidly so that the scan was very crowded and therefore of little use.

It is the opinion of the author, however, that techniques of remote sensing temperature changes across little relieved terrain, and the means to describe that relief accurately, bear considerable promise for the future. Courtin and Labine (this volume) attempted to describe the spatial distribution of lowland temperature. Their efforts were only partially successful because even in relatively unrelieved terrain the network of sensors required must be at a much greater density than the 3 km^{-2} used in their study. Furthermore, the causative agents of temperature change, such as the sea, rock outcrops, valleys, and high cliffs, had to be inferred from experience rather than from any substantive data. A properly calibrated infra-red scan across the entire Lowland, even at a single point in time, would have given much greater confidence in the data that they presented.

Aerial photography

Total photographic coverage of Truelove Lowland was flown on 18 August, 1971 under clear skies. Black-and-white imagery was flown at 2,500 and 10,000 feet whereas false-color infra-red imagery was flown at 5,000 feet (Table 1). A photomosaic of the false-color infrared coverage was prepared by K. Arnold of the Inland Waters Branch as part of a much larger project to test the potential use of this film. This particular emulsion proved itself to be very responsive to water depth, to the point where one could not only determine the relative depth of the various lakes in the Lowland but also could make out the topography of the bottom of the shallower lakes. The photography was made too late in the growing season to show good tonal separation of the various vegetation types but it was possible to determine the most lush meadows even though these were becoming dormant.

The black-and-white photography, flown at 2,500 feet and at a scale of 1:5000, was used widely by project personnel as an interpretative tool, especially with respect to the field mapping of soils and vegetation. The 9×9 in format (230×230 mm) covered about 1 km on the ground, which together with very good quality photography, gave a high degree of resolution of surface features.

Stereo-orthophotography

The black-and-white photography flown from 10,000 feet was used to construct stero-orthophotographs of Truelove Lowland. The stereo-orthophotographs, through a special technique of photographic rectification (Blachut 1971), performed by Space Optic Limited, Ottawa, provided aerial coverage that was metrically accurate across the entire photo. To achieve this end, targets that were visible from the air were placed in a loose grid across the Lowland and the adjacent plateau, and subsequently surveyed. The triangulation survey provided the necessary control for the rectification of the original imagery to produce orthophotographs. The fourteen orthophoto "models" produced were used in two ways. The models were rephotographed as a photomosaic thus giving a "photographic map" of great accuracy. It was from this map that all areas were measured for use in the quantitative expression of the data used in this volume. Each model was used in the production of a "mate" photograph in which parallax was reintroduced in the x direction only thus permitting the photo pair to be viewed stereoscopically.

A special plotter has been developed for use with the stereo-orthophotographs that enables even those not versed in photogrammetry to determine heights, plot contours, and to interpret topographic detail. This is a radical departure from normal photogrammetric techniques that require very costly equipment and the services of an expert technician.

The orthophoto technique proved to be a vital aid in the intensive research that was conducted in an area with virtually no usable maps or photographs. It is to be hoped that further development will continue and that the cost of production of the models can be reduced to the point where the technique can be used on a wider scale.

Acknowledgments

The aerial photography and remote sensing resulted almost entirely from the efforts of individuals not directly connected with the Devon Island project. The author wishes to acknowledge the following for their participation: H. Hengeveld, Atmospheric Environment Service, work with the ice reconnaissance aircraft; K. Arnold, arctic aerial and scientific photography project — joint project between P.C.S.P. and Glaciology Division of the Inland Waters Branch, aerial photography; National Airphoto Library Reproduction Centre, infra-red, false color photomosaic; J. Saastamoinen, Photogrammetry Branch, field survey; T.J. Blachut, Photogrammetry Branch, for his enthusiasm and directorship; M. Van Wijk, Photogrammetry Branch, planning of the photography, aerial triangulation, and co-ordination of the orthophoto production; E. Tromanhauser, Space Optic Limited, for his personal interest in the project; and, C. Labine, I.B.P. Devon Island, placing and maintenance of targets for the aerial photography.

Reference

Blachut, T.J. 1971. Mapping and Photo Interpretation System Based on Stereo-Orthophotos. N.R.C. Pub. 12281. Ottawa. 147 pp.

Appendix 3

Lichens of Truelove Lowland

D.H.S. Richardson

Acarospora glaucocarpa (Wahlenb ex Ach.) Körb.
Alectoria chalybeiformis (L.) S. Gray
A. minuscuala Nyl.
A. nigricans (Ach.) Nyl.
A. nitidula (Th.Fr.) Vain.
A. ochroleuca (Hoffm.) Mass.
A. pubescens (L.) R.H. Howe
A. subdivergens Dahl
A. tenuis Dahl
Bacidia bagliettoana (Mass. & De Not.)
Baeomyces carneus (Retz.) Flörke
Buellia atrata (Sm.) Anzi.
B. papillata (Somm.) Tuck.
Caloplaca cinnamomea (Th. Fr.) Oliv.
C. holocarpa (Hoffm.) Wade
C. jungermanniae (Vahl.) Th. Fr.
C. stillicidiorum (Vahl.) Lynge
C. tetraspora (Nyl.) Oliv.
C. tiroliensis Zahlbr.
C. andelariella arctica (Körb.) Sant.
Candelariella terrigena Räs.
Cetraria cucullata (Bell.) Ach.
C. delisei (Bory *ex* Schaer.) Th. Fr.
C. erictorum Opiz.
C. islandica (L.) Ach.
C. nivalis (L.) Ach.
Cladonia amaurocraea (Flörke) Schaer.
C. bellidiflora (Ach.) Schaer.
C. coccifera (L.) Willd.
C. cornuta (L.) Hoffm.
C. gracilis (L.) Willd.
C. lepidota Nyl.
C. mitis (Sandst.) Hale & W. Culb.
C. pocillum (Ach.) O. Rich.
Collema ceraniscum Nyl.
C. undulatum Laur. *ex* Flot. var. *granulosum* Degel.
Coriscium viride (Ach.) Vain.
Cornicularia aculeata (Schreb.) Ach.
C. divergens Ach.
C. muricata (Ach.) Ach.
Dactylina arctica (Hook.) Nyl.
D. ramulosa (Hook.) Tuck.
Fulgensia bracteata (Hoffm.) Räs.
Gyalecta foveolaris (Ach.) Schaer.
G. peziza (Mont.) Anzi
Haematomma lapponicum Räs.
Hypogymnia physodes (L.) W Wats.
H. subobscura (Vain) Poelt.
Ionaspis euplotica (Ach.) Th. Fr. var. *arctica* (Lynge) Magn.
Lecanora atra (Huds.) Ach.
L. badia (Hoffm.) Ach.
L. behringii Nyl.
L. candida (Anzi.) Nyl.
L. castanea (Hepp) Th. Fr.
L. dispersa (Pers.) Somm.
L. epibryon (Ach.) Ach.
L. frustulosa (Dicks.) Ach.
L. intricata (Schrad.) Ach.
L. melanopthalma (Ram.) Ram.
L. cf. nordenskioldii Vain.
L. polytropa (Ehrh.) Rabenh.
L. proserpens Nyl.
L. rupicola (L.) Zahlbr.
L. cf. subrugosa Nyl.

L. urceolaria (Fr.) Wetm.
Lecidea armeniaca (DC.) Fr.
L. assimilata Nyl.
L. atrata (Ach.) Wahl.
L. auriculata Th. Fr.
L. crassipes (Th. Fr.) Nyl.
L. crustulata (Ach.) Spreng.
L. glaucophaea Körb.
L. lapicida (Ach.) Ach.
L. lulensis Hellb.
L. macrocarpa (DC.) Steud.
L. melinodes (Körb.) Magn.
L. micacea Körb.
L. pantherina (Hoffm.) Th. Fr.
L. purissima Darb.
L. ramulosa Th. Fr.
L. rubiformis (Wahlenb. *ex* Ach.) Wahlenb.
L. speira (Ach.) Ach.
L. tessellata (Ach.) Flörke
L. vernalis (L.) Ach.
L. vorticosa (Flörke) Körb.
Lecidella stigmatea (Ach.) Hert. & Leuck.
L. wulfenii (Hepp) Körb.
Leciophysma finmarkicum Th. Fr.
Lepraria neglecta (Nyl.) Lett.
Leptogium minutissimum (Flörke) Fr.
Lopadium pezizoideum (Ach.) Körb
Mycoblastus alpinus (Fr.) Kernst.
M. sanguinarius (L.) Norm.
Nephroma expallidum (Nyl.) Nyl.
Ochrolechia androgyna (Hoffm.) Arn.
O. frigida (Sw.) Lynge
O. geminipara (Th. Fr.) Vain.
O. gonatodes (Ach.) Räs.
O. inaequatula (Nyl.) Zahlbr.
O. upsaliensis (L.) Mass.
Omphalodiscus virginis (Schaer.) Schol.
Pannaria hookeri (Borr. *ex* Sm.) Nyl
Parmelia centrifuga (L.) Ach.
P. disjuncta Erichs.
P. exasperatula Nyl.
P. fraudans Nyl.
P. incurva (Pers.) Fr.
P. infumata Nyl.
P. omphalodes (L.) Ach.
P. saxatilis (L.) Ach.
P. separata Th. Fr.
P. stygia (L.) Ach.
P. sulcata Tayl.
Parmeliella praetermissa (Nyl.) P. James
Peltigera aphthosa (L.) Willd.
P. canina (L.) Willd.
P. canina (L.) Willd. var. *rufescens* (Weis.) Mudd.
P. malacea (Ach.) Funck
P. pulverulenta (Hook) Nyl.
Pertusaria bryontha (Ach.) Nyl.
P. coriacea (Th. Fr.) Th. Fr.
P. dactylina (Ach.) Nyl.
P. glomerata (Ach.) Schaer.
P. octomela (Norm.) Erichs.
P. panyrga (Ach.) Mass.
P. subobducens Nyl.
P. trochisea Norm.
Physcia caesia (Hoffm.) Hampe.
P. constipata (Nyl.) Norrl. & Nyl.
P. intermedia Vain.
P. muscigena (Ach.) Nyl.
P. muscigena f. squarrosa (Ach.) Lynge
P. sciastra (Ach.) Du Rietz
Placopsis gelida (L.) Linds.
Placynthium aspratile (Ach.) Henss.
P. nigrum (Huds.) S. Gray
Polyblastia bryophila Lönnr.
P. gelatinosa (Ach.) Th. Fr.
P. hyperborea Th. Fr.
P. integrascens (Nyl.) Vain.
P. theleodes (Somm.) Th. Fr.
Protoblastenia rupestris (Scop.) J. Stein.
Psoroma hypnorum (Vahl.) S. Gray
Pyrenopsis pulvinata (Schaer.) Th. Fr.
Racodium rupestre Pers.
Rhizocarpon copelandii (Körb.) Th. Fr.
R. crystalligenum Lynge
R. disporum (Naeg. *ex* Hepp) Mull. Arg.
R. geographicum (L.) DC.
R. jemtlandicum Malme.
R. polycarpum (Hepp) Th. Fr.
R. rittokense (Hellb.) Th. Fr.
Rinodina milvina (Wahlenb. ex Ach.) Th. Fr.
R. nimbosa (Fr.) Th. Fr.
R. occidentalis Lynge
R. roscida (Somm.) Arn.
R. turfacea (Wahlenb.) Körb.
Solorina bispora Nyl.
S. octospora Arn.

S. saccata (L.) Ach.
S. spongiosa (Sm.) Anzi.
Sphaerophorus globosus (Huds.) Vain.
Spilonema revertens Nyl.
Sporastatia testudinea (Ach.) Mass.
Stereocaulon alpinum Laur.
S. botryosum Ach.
S. rivulorum Magn.
Thamnolia vermicularis (Sw.) Ach. *ex* Schaer.
Toninia lobulata (Somm.) Lynge
Umbilicaria arctica (Ach.) Nyl.
U. havaasii Llano
U. hyperborea (Ach.) Ach.
U. lyngei (Ach.) Tuck.
U. proboscidea (L.) Schrad.
U. vellea (L.) Ach.
Verrucaria arctica Lynge
V. devergens Nyl.
V. deversa Vain.
V. nigrescens Pers.
Vestergrenopsis isidiata (Degel.) Dahl
Xanthoria candelaria (L.) Th. Fr.
X. elegans (Link) Th. Fr.

Appendix 4

Fungi of Truelove Lowland

T. Booth and P. Widden

Zoosporic Fungi

Catenophlyctis variabilis (Karling) Karling
Chytriomyces annulatus Dogma
C. polulatus Willoughby and Townley
Hyphochytrium catenoides Karling
Nowakowskiella elegans (Nowak.) Schroeter
N. macrospora Karling
N. ramosa Butler
Olpidium pendulum Zopf
Phlycotochytrium arcticum Barr
Phythium irregularae Buisman
Rhizidium verrucosum Karling
Rhizophlyctis harderi Uebelmesser
Rhizophydium angulosum Karling
R. coronum Hanson
R. elyensis Sparrow
R. karlingii Sparrow
R. pollinis-pini (Braun) Zopf
R. sphaerotheca Zopf
R. sp. *(nodulosum?)*
R. sp. *(patellarium?)*

Ascomycetes on dung

Ascobolus stictoideus Speg.
A. albidus Cr.
A. furfuraceous Pers. ex. Fr.
Cheilymenia stercorea (Pers. ex. Fr.) Boud.
Coprobia granulata (Bull. ex. Fr.) Boud.
Delitschia marchalii Berl. and Vogl.
Lilliputia refula (Berkeley and Broome) Hughes
Pleospora sp.
Podospora vesticola (Berk. and Broome) Mirza and Cain
Saccobolus quadrisporus Mass. and Salm.
S. versicolor (Karst.) Karst.
Sporormia ambigua Nies.
S. intermedia Aversw.
S. vexans Aversw.
Thelebolus stercoreus Tode per Fr.
Trichodelitschia bisporula (Crouan) Munk
Zopfiella biporosa n. sp.

Phycomycetes on dung

Mucor spp.
Pilobolus spp.

Fungi Imperfecti

Acremonium spp.
Alternaria alternata (Fr.) Keissl.
Arthrinium phaeospermum (Corda) M.B. Ellis
Aspergillus ustus (Bain.) Thom and Church
A. versicolor (Vuill.) Tiraboschi
Aureobasidium pullulans (de Bary) Arnaud

Beauveria bassiana (Bals.) Vuill.
Botrytis cinerea Pers. ex Fr.
Chaetophoma sp.
Chrysosporium pannorum (Link) Hughes
C. verrucosum Tubaki
Cladosporium cladosporoides (Fresen.) de Vries
C. herbarum (Pers.) Link
Cylindrocarpon sp.[1]
Dendrostilbella sp.
Epicoccum purpurascens Ehrenb. ex Schlecht.
Fusarium sp.
Geotrichum candidum Link
Harposporium sp.
Humicola fuscoatra Traaen
H. grisea Traaen
Paecilomyces elegans (Corda) Mason and Hughes
P. varioti Bain.
Penicillium atramentosum Thom.
P. corymbiferum Westling
P. decumbens Thom
P. implicatum Biourge
P. janthinellum Biourge
P. lanoso-viride Thom
P. lanosum Westling
P. miczynskii Zaleski
P. notatum Westling
P. oxalicum Currie and Thom
P. paxilli Bain
P. variabile Sopp
Penicillium spp.
Pestalotia sp.
Phialophora fastigiata (Lagerberg and Melin) Conant
Phialaphora spp.
Phoma herbarum Westend
Q. L.[2]
Rhinocladiella sp.
Sclerotium sp.
Sporobolomyces salmonicolor Kluy. and van Niel
Tolypocladium cylindrosporum Gams
T. inflatum Gams
Tolypocladium sp.[1]
Trichocladium canadense Hughes
Trichothecium roseum Link
Trimmatostroma sp.
Verticillium sp.

Zygomycetes

Mortierella sp.

Loculoascomycetes

Pleospora sp.

---

1. New, undescribed species
2. An unidentified Coelomycete

Appendix 5

Musci of the northern lowlands of Devon Island[1]

D.H. Vitt

Amphidium lapponicum (Hedw.) Schimp.
Andreaea rupestris Hedw.
Aplodon wormskjoldii (Horn.) R. Brown
Aulacomnium acuminatum (Lindb. & Arn.) Kindb.
A. palustre (Hedw.) Schwaegr.
A. turgidum (Wahlenb.) Schwaegr.
Barbula icmadophila Schimp. *ex* C. Muell.
Bartramia ithyphylla Brid.
Blindia acuta (Hedw.) B.S.G.
Brachythecium turgidum (C.J. Hartm.) Kindb.
Bryoerythrophyllum recurvirostre (Hedw.) Chen
Bryum arcticum (R. Brown) B.S.G.
B. argenteum Hedw.
B. cf capillare Hedw.
B. creberrimum Tayl.
B. cryophilum Mårt.
B. curvatum Kaur. & Arn.
B. neodamense var. *ovatum* Lindb. & Arn.
B. pseudotriquetrum (Hedw.) Gaertn., Meyer. & Scherb.
Calliergon giganteum (Schimp.) Kindb.
C. orbiculari-cordatum (Ren. & Card.) Broth.
C. sarmentosum (Wahlenb.) Kindb.
C. stramineum (Brid.) Kindb.
C. trifarium (Web. & Mohr) Kindb.
Campylium arcticum (Williams) Mitt.
C. stellatum (Hedw.) C. Jens.
Catoscopium nigritum (Hedw.) Brid.
Ceratodon purpureus (Hedw.) Brid.
Cinclidium arcticum (B.S.G.) Schimp.
C. latifolium Lindb.
Cirriphyllum cirrosum (Schwaegr. *ex* Schultes) Grout
Conostomum tetragonum (Hedw.) Lindb.
Cratoneuron arcticum Steere
C. filicinum (Hedw.) Spruce
Cnestrum alpestre (Wahlenb.) Nyh.
Cyrtomnium hymenophylloides (Hüb.) Kop.
C. hymenophyllum (B.S.G.) Holmen
Desmatodon heimii var. *arctica* (Lindb.) Crum
D. leucostoma (R. Brown) Berggr.
Dicranella crispa (Hedw.) Schimp.
Dicranoweisia crispula (Hedw.) Lindb. *ex* Milde
Dicranum angustum Lindb.

1. Based on Vitt, D.H. 1975. A key and annotated synopsis of the mosses of the northern lowlands of Devon Island, N.W.T., Canada, Can. J. Bot. 53:2158-2197.

D. elongatum Schleich. *ex* Schwaegr.
D. groenlandicum Brid.
D. scoparium Hedw.
Didymodon asperifolius (Mitt.) Crum, Steere, & Anderson
Distichium capillaceum (Hedw.) B.S.G.
D. hagenii Ryan *ex* Philib.
D. inclinatum (Hedw.) B.S.G.
D. flexicaule (Schwaegr.) Hampe
Drepanocladus aduncus (Hedw.) Warnst.
D. brevifolius (Lindb.) Warnst.
D. revolvens (Sw.) Warnst.
D. uncinatus (Hedw.) Warnst.
Encalypta alpina Sm.
E. brevicolla (B.S.G.) Bruch *ex* Ångstr.[2]
E. procera Bruch
E. rhaptocarpa Schwaegr.
E. vulgaris Hedw.
Eurhynchium pulchellum (Hedw.) Jenn.
Fissidens adiantoides Hedw.
F. arcticus Bryhn
F. bryoides Hedw.
F. osmundoides Hedw.
Funaria microstoma Bruch *ex* Schimp. (new var. to be described)
F. polaris Bryhn
Grimmia alpicola var. *rivularis* (Brid.) Wahlenb.
G. apocarpa Hedw. var. *apocarpa*
G. apocarpa var. *stricta* (Turn.) Hook. & Tayl.
G. ovalis (Hedw.) Lindb.[3]
G. tenera Zett.
G. torquata Hornsch. *ex* Grev.
Gymnostomum recurvirostrum Hedw.
Hygrohypnum luridium (Hedw.) Jenn.
H. polare (Lindb.) Loeske
Hylocomium splendens (Hedw.) B.S.G.
Hypnum bambergeri Schimp.
H. plicatulum (Lindb.) Jaeg. & Sauerb.
H. pratense Koch *ex* Brid.
H. procerrimum Mol.
H. revolutum (Mitt.) Lindb.
H. vaucheri Lesq.
Isopterygium pulchellum (Hedw.) Jaeg. & Sauerb.
Kiaeria blyttii (Schimp.) Broth.
Leptobryum pyriforme (Hedw.) Wils.
Meesia triquetra (Richt.) Ångstr.
M. uliginosa Hedw.
Mnium lycopodioides Schwaegr. (*coll.*)
Myurella julacea (Schwaegr.) B.S.G.
M. tenerrima (Brid.) Lindb.
Oncophorus wahlenbergii Brid.
Orthothecium chryseum (Schwaegr. *ex* Schultes) B.S.G.
O. strictum Lor.
Orthotrichum pylaisii Brid.
O. speciosum Nees *in* Sturm
Philonotis fontana var. *pumila* (Turn.) Brid.
Plagiobryum zierii (Hedw.) Lindb.
Plagiomnium ellipticum (Brid.) Kop.
Platydictya jungermannioides (Brid.) Crum
Pogonatum dentatum (Brid.) Brid.
Pohlia annotina (Hedw.) Loeske
P. cruda (Hedw.) Lindb.
P. nutans (Hedw.) Lindb.
Polytrichastrum alpinum (Hedw.) Smith
Polytrichum algidum Hag. & C. Jens.
P. juniperinum Hedw.
P. piliferum Hedw.
P. strictum Brid.
Pseudoleskeella tectorum (Funck *ex* Brid.) Kindb. *ex* Broth.
Psilopilum cavifolium (Wils.) Hag.
Rhacomitrium canescens (Hedw.) Brid. var. *canescens*
R. canescens var. *ericoides* (Hedw.) Hampe
R. lanuginosum (Hedw.) Brid.

---

2. This species was detected by D.G. Horton in 1976 and is an addition to the taxa listed by Vitt (1975).
3. Probably better considered under the concept of *Grimmia affinis* Hornsch.

R. sudeticum (Funck) B.S.G.[4] (presently being described as a new species of *Schistidium*)
Rhytidium rugosum (Hedw.) Kindb.
Saelania glaucescens (Hedw.) Broth. *in* Bomanss. & Broth.
Scorpidium scorpioides (Hedw.) Limpr.
S. turgescens (T. Jens.) Loeske
Seligeria polaris Berggr.
Sphagnum orientale L. Savicz.
Splachnum vasculosum var. *heterophyllum* (Hook. *in* Drumm.) Brassard
Stegonia latifolia (Schwaegr. *ex* Schultes) Vent. *ex* Broth.
Tetraplodon mnioides (Hedw.) B.S.G.
T. paradoxus (R. Brown) Hag.[5]
Thuidium abietinum (Hedw.) B.S.G.
Timmia austriaca Hedw.
T. bavarica Hessl.
T. norvegica Zett.
Tomenthypnum nitens (Hedw.) Loeske
Tortella arctica (Arn.) Crundw. & Nyh.
T. fragilis (Drumm.) Limpr.
Tortula mucronifolia Schwaegr.
T. ruralis (Hedw.) Gaertn., Meyer, & Scherb.
Voitia hyperborea Grev. & Arnott

Total: 71 genera
132 species
2 varieties

4. This material has recently been considered a new species *Schistidium holmenianum* Steere & Brassard.
5. If *Tetraplodon pallidus* Hag. is to be considered a distinct species, then the Devon Island material falls into the concept of that species.

Appendix 6

Vascular Plants of Truelove Lowland and adjacent areas including their relative importance[1]

(D—dominant, C—common, M—minor, R—rare)

L.C. Bliss

Cystopteris fragilis (L.) Bernh. M
Woodsia glabella R. Br. M
W. alpina (Boulton) S.F. Gray M
Equisetum arvense L. C
E. variegatum Schleich. C
Huperzia selago (L.) Bernh. ex. Schrank and Mart. According to Böcher et al. 1968. ssp. *arctica* (Grossh.) Löve and Löve. M
Hierochloe alpina (SW.) R. & S. M
H. pauciflora R. Br. M
Alopecurus alpinus L. M
Phippsia algida (Sol.) R. Br. M
Arctagrostis latifolia (R. Br.) Griseb. D
Trisetum spicatum (L.) Richt. M
Poa alpigena (Fr.) Lindm. var. *colpodea* (Fr.) Schol. M
P. arctica R. Br. C
P. glauca M. Vahl. R
P. abbreviata R. Br. M
P. hartzii Gand. M
Pleuropogon sabinei R. Br. M
Colpodium vahlianum (Liebm.) Nevski. M
Dupontia fisheri R. Br. M
Puccinellia vaginata (Lge.) Fern. and Weath. vap. *paradoxa* Th. S*ør.* M
P. phryganodes (Trin.) Scribn. & Merr. M
Festuca brachyphylla Schultes C
F. baffinensis Polunin M
Eriophorum angustifolium Honck. C
E. scheuchzeri Hoppe. M
E. triste (Th. Fr.) Hadac and Löve C
Kobresia myosuriodes (Vill.) Fiori and Paol. M
K. simpliciuscula (Wahlenb.) Mack. M
Carex nardina Fr. C
C. rupestris All. C
C. amblyorhyncha Krecz. R
C. stans Drej. D
C. atrofusca Schk. M
C. misandra R. Br. C
C. membranacea Hook. D
C. ursina Dew. R
Juncus biglumis L. C
Luzula nivalis (Least.) Beurl. M
L. confusa Lindeb. C
Tofieldia coccinea Richards. M
Salix reticulata L. R
S. arctica Pall. D
Oxyria digyna (L.) Hill C
Polygonum viviparum L. C
Stellaria humifusa Rottb. R
S. longipes Goldie s. lat. C
Cerastium alpinum L. C
C. arcticum Lge. R
C. regelii Ostf. R
Minuartia rubella (Wahlenb.) Hiern. C

1. Based mainly upon Barrett and Teeri (1973)

M. rossii (R. Br.) Graebn. M
Silene acaulis L. var. *exscapa* (All.) DC. C
Melandrium apetalum (L.) Fenzl. ssp. *arcticum* (Fr.) Hult. M
M. affine (J. Vahl.) Hartm. M
Ranunculus hyperboreus Rottb. M
R. sulphureus Sol. C
Papaver radicatum Rottb. C
Cochlearia officinalis L. ssp. *groenlandica* (L.) Porsild R
Eutrema edwardsii R. Br. R
Cardamine bellidifolia L. R
C. pratensis L. var. *angustifolia* Hook. C
Draba alpina L. C
D. corymbosa R. Br. ex. DC. C
D. nivalis Liljebl. C
D. lactea Adams C
D. subcapitata Simm. M
D. cinerea Adams M
D. oblongata R. Br. M
Braya purpurascens (R. Br.) Bunge M
Saxifraga caespitosa L. C
S. cernua L. C
S. hirculus L. var. *propinqua* (R. Br.) Simm. M
S. oppositifolia L. D
S. flagellaris Willd. ssp. *platysepala* (Trautv.) Porsild R
S. foliolosa R. Br. M
S. hieracifolia Waldst. & Kit. R. M
S. nivalis L. M
S. rivularis L. M
S. tenuis Sm. M
S. tricuspidata Rottb. M
Potentilla hyparctica Malte. M
P. rubricaulis Lehm. M
Dryas integrifolia M. Vahl. D
Epilobium latifolim L. M
E. arcticum Samuelss. R
Hippuris vulgaris L. R
Cassiope tetragona (L.) D. Don. D
Vaccinium uliginosum L. var. *alpinum* Big. R
Armeria maritima (Mill.) Willd. ssp. *labradorica* (Wallr.) Hult. R
Pedicularis capitata Adams C
P. lanata Cham. & Schlecht. C
P. hirsuta L. C
P. sudetica Willd. C
Campanula uniflora L. M
Taraxacum phymatocarpum J. Vahl. M

Appendix 7

Invertebrates of Truelove Lowland

J. Ryan

Protozoa

"It can be said that the protozoan fauna are predictable in species content—the ones here are common anywhere." (Comment by Heal while examining protozoa from meadow soil at Truelove.)

Mastigophora
  Bodonidae
    1 sp.
  "diflagellate"
    1 sp.
  Anisonemidae
    Entosiphon sulcata (Dujardin) Stein
Sarcodinea
  Amoebida
    Limax sp.
    ?*Astra* sp.
    Centropyxis aculeata (Ehernberg) Stein
    Cyphoderia aumpulla (Ehernberg) Leidy
    Euglypha sp.
    Nebela collaris (Ehernberg) Leidy
    Plagiopyxis sp.
    Trinema enchelys (Ehernberg) Leidy
    T. lineare Penard
Haplosporea
  Haplosporida
    1 sp. (host: *Branchinecta paludosa*)
Ciliophora
Ciliatea
  Holotrichia
    Hypotrichius sp.

Metazoa

Porifera
Demospongidae
  Spongillidae
    1 sp.
Platyhelminthes
Turbellaria
  Neorhabdocoela.
  Typhloplanidae
    1 sp.
Cestoda
    3 sp. (hosts: *Cepphus grylle* *Odobenus rosmarus* *Salvelinus alpinus* L.)
  Cyclophyllidia.
    1 sp. (host: *Dicrostonyx groenlandicus* Traill.)
Nematoda
    Cystidicola sp. (host: *Salvelinus alpinus*)
    Plectus sp.
Rotifera
    3 terrestrial sp.
    Filinia sp.
    Keratella cochlearis (Gosse)
    K. quadrata (Mueller)
    Polyarthra sp.

Annelida
Oligochaeta
Enchytraeidae
Bryodrilus parvus (Dash)
Cernovitoviella sp.
Henlea nasuta (Eisen)
H. perpusilla (Nielsen and Christensen)
H. sp. A
Marionina sp.
M. sp.
Tardigrada
Eutardigrada
Pseudechiniscidae
Pseudechiniscus sp.
Macrobiotidae
Macrobiotus sp.
Arthropoda
Crustacea
Anostraca
Branchinectidae
Branchinecta paludosa (Mueller)
Notostraca
Lepidurus arcticus Pallas
Cladocera
Daphnidae
Daphnia pulex Leydig
Bosminidae
Bosmina coregoni Baird
Copepoda
Temoridae
Eurytemora sp. prob. *composita*
Cyclopidae
Cyclops scutifer Sars.
C. vernalis Fischer
Harpacticoididae
Attheyella nordenskioldii (Lilljeborg)
Podocopa
Cypridae
Candona compressa Koch
C. willmani Staphlin
Eucypris crassa (Mueller)
Prionocypris glacialis (Sars)
Mysidacea
Mysidae
Mysis relicta Lovén
Arachnida
Araneida
Dictynidae
Dictyna borealis Pickard-Cambridge
Lycosidae
Tarentula exasperans Pickard-Cambridge
Linyphiidae
Acartauchenius pilifrons (Koch)
Collinsia spetsbergensis (Thorell)
C. thulensis (Jackson)
Diplocephalus barbatus (Koch)
Erigone psychrophila Thorell
Hilaira vexatrix (Pickard-Cambridge)
Minyriolus pamphia Chamberlin
Typhochraestus latithorax (Strand)
Acarina
Parasitidae
Parasitus sp. nov. near *fucorum* (DeGeer)
P. sp. nov.
Laelapidae
Laelaps alaskensis species-complex (host: *Dicrostonyx* groenlandicus)
Pneumolaelaps groenlandica Traegardh (host: *Bombus polaris*)
Haemogamasidae
Haemogamasus ambulans (Thorell)
Ascidae
Arctoseius multidentatus Evans
A. ornatus Evans
Cheiroseius groenlandicus (Haarlov)
Podapolipidae
Locustacaris buchneri Stammer
Rhagidiidae
Rhagidia sp. near *hamata* (Kramer & Neum.)
Eupodidae
Cocceupodes curviclava Thor
Bdellidae
Bdella muscorum Ewing
Neomolgus littoralis (L.)
Stigmaeidae
Stigmaeus parmatus Summers
Lebertiidae
Lebertia sp.
Acaridae
Kuzinia sp. nov. near *laevis* (Dujardin)
Hermanniidae
Hermannia subglabra Berlese
Camisiidae
Camisia horrida (Hermann)
Eremaeidae
Eremaeus sp. ?

Ceratozetidae
Iugoribates gracilis Sellnick
Trichoribates lucens (Koch)
T. polaris Hammer
Insecta
Collembola
Poduridae
Anurida granaria (Nicolet)
Ceratophysella armata ? (Nicolet)
Hypogastrura sp. near *trybomi* (Schoett)
H. tullbergi (Schaeffer)
H. sp. 1 near *tullbergi*
H. sp. 2 near *tullbergi*
H. sp. 3
Micranurida pygmaea Boerner
Morulina gigantea (Tullberg)
Podura aquatica L.
Willemia anophthalma Boerner
Onychiuridae
Onychiurus groenlandicus (Tullb.)
O. sp.
Tullbergia sp. nr. *simplex* Gisin
Isotomidae
Anurophorus? sp.
Folsomia bisetosa Gisin
F. duodecimsetosa Hammer
F. fimetaria (L.)
F. quadrioculata (Tullberg)
F. regularis Hammer
Isotoma notabilis Schaeffer var. *pallida* Agrell
I. olivacea Tullberg
I. violacea Tullberg
I. sp. n.
Isotomurus palustris (Mueller)
Proisotoma sp. n.
Entomobryidae
Entomobrya comparata Folsom
Sminthuridae
Megalothorax minimus (Willam)
Sminthurides malmgreni (Tullberg)
Sphaeridia sp. prob. *pumilis* (Kraus.)
Phthiraptera
Philopteridae
Anaticola rubromaculatus Rudow (host: *Sommateria mollissima*)
Anatoecus sp. (host: *Chen nivalis*)
Philopterus sp. (host: *Plectrophenax nivalis*)
Saemundssonia prob. *cephalus* (Denny) (prob. host: *Stercorarius parasiticus*)
Echinophthiriidae
Antarctophthirus trichechi Bohemann (host: *Odobenus rosmarus*)
Heteroptera
Pseudococcidae
Chorizococcus altoarcticus Richards
Coleoptera
Carabidae
Amara alpina Paykull
Dytiscidae
Hydroporus polaris Fall
Staphylinidae
Gynpeta sp.
Trichoptera
Limnephilidae
Apatania zonella Zett.
Lepidoptera
Nymphalidae
Boloria chariclea Schneid.
B. polaris Bdv.
Noctuidae
Anarta richardsoni Curtis
Lasiestra sp.
Sympistris labradoris Staud.
Lymantriidae
Gynaephora (=*Byrdia*) *groenlandica* (Homeyer)
G. (=*B.*) *rossii* (Curtis)
Geometridae
Dasyuris polata Dyp.
Psychophora sabini Kirby
Pyralidae
Udea torvalis Moesch.
Pterophoridae
Stenoptilia mengeli Fern.
Olethreutidae
Aphania frigidana Pack
Olethreutes inquietana Wlk.
O. mengelana Fern.
Diptera
Trichoceridae
Trichocera columbiana Alex.
T. borealis Lackschewitz

Tipulidae
Dactylolabis rhicnoptiloides (Alex.)
Nephrotoma lundbecki (Nielsen)
Tipula arctica Curtis
T. thulensis Alex.
Culicidae
Aedes impiger (Walker)
A. nigripes (Zetterstedt)
Ceratopogonidae
Ceratopogon (Isohelea) sp.
Chironomidae
Chaetocladius holmgreni (Jacobsen)
Corynoneura sp.
Cricotopus tibialis (Meigen)
C. sp.
Diamesa arctica (Boheman)
D. germinata Kieffer
D. simplex Kieffer
Diplocladius bilobatus Brundin
Heterotrissocladius subpilosus (Kieffer)
Limnophyes sp.
Metriocnemus sp.
Orthocladius (Eudactylocladius) mixtus (Holmgren)
O. (E.) sp.
Paracladius alpicola (Zett.)
Parakiefferiella sp.
Paraphaenocladius despectus (Kieffer)
Procladius culiciformis (Linne)
Psectrocladius sp. (Edw.)
Smittia velutina (Lundbeck)
Thienemanniella sp.
Trissocladius sp.
Mycetophilidae
Boletina sp. 21
Bolitophila n. sp.
Exechia frigida (Boh.)
Sciaridae
Bradysia spp. (about 4-5 spp.)
Cecidomyidae
Campylomyza sp.
Empididae
Clinocera n. sp.
Rhamphomyia hoeli Frey
R. nigrita Zett.
Dolichopodidae
Dolichopus humilis (Van Duzee)
Syrphidae
Melanostoma n. sp.
Piophilidae
Arctopiophila arctica Holmgren
Piophila fulviceps Holmgren
Heleomyzidae
Aecothea specus (Aldrich)
Neoleria prominens (Becker)
Scatophagidae
Allomyella unguiculata Mall.
Scatophaga apicalis Curtis
S. nigripalpis Becker
Muscidae
Spilogona almquisti (Holmgren)
S. deflorata (Holmgren)
S. dorsata (Zett.)
S. latilamina (Collin)
S. melanosoma (Huckett)
S. obsoleta (Mall.)
S. tundrae (Schnabl.)
Anthomyidae
Fucellia ariciiformis Holmgren
F. pictipennis Becker
Hylemya (Delia) fasciventris Ringdahl
Calliphoridae
Boreellus atriceps (Zett.)
Protocalliphora sapphira (Hall) (host: *Plectrophenax nivalis*)
Protophormia terraenovae (R.D.)
Tachinidae
Spoggosia gelida (Coq.) (hosts: *Gynaephora groenlandica* & *G. rossii*)
Hymenoptera
Tenthredinidae
Amauronematus nr. *abnormus*
A. nr. *alpacola*
A. nr. *anthracinus*
A. nr. *atratus*
A. nr. *coracinus*
A. nr. *histrio*
A. nr. *leucolaenus*
A. nr. *mcluckliei*
A. nr. *opacipleuris*
Pristiphora n. sp.
Braconidae
Apanteles n. sp. 130
A. n. sp. 133 (host: *Anarta richardsoni*)
Microplitis n. sp.
Oresbius tibialis Tow.
O. Trifasciatus Ashm.
O. n. sp.

Rogas n. sp. (host: *Gynaephora groenlandica & G. rossii*)

Ichneumonidae

Atractodes spp. (several spp.)

Bathythrix n. sp.

Campoletis spp. (2 spp.)

Cryptus leechi Mason

Diadegma sp.

Glypta sp.

Hyposoter sp.

Ichneumon lariae Curtis

I. n. sp.

Mesoleius n. sp.

Orthocentrus sp.

Plectiscus n. sp.

Stenomacrus femoralis Holmgren

S. sp. *U*

S. sp.

Trathala n. sp.

Encyrtidae

Copidosoma sp.

Pteromalidae

Seladerma n. sp.

Apidae

Bombus (Alpinobombus) polaris Curtis

I respectfully acknowledge the following authorities, who made the species determinations for this list: O.W. Heal (Protozoa), G. Daborn (Protozoa, Crustacea), J.O. Young (Planaria), R. Hobbs (Cestoda), J.C. Holmes (Nematoda), D.L.C. Procter (Nematoda), K. Minns (Rotifera, Crustacea), Mahdab Dash, Paul Skydt (Enchytraeidae), D. Nelson, R.P. Higgins (Tardigrada), H.C. Yeatman (Crustacea), L.D. Delorme (Ostracoda), R.E. Leech (Arachnida), L. Richards (Acarina), E.E. Lindquist (Acarina), I.M. Smith (Acarina), W.R. Richards (Collembola, Homoptera), J.A. Addison (Collembola), T.M. Clay (Phthiraptera), G.E. Ball (Carabidae), J.L. Carr (Dytiscidae), J.M. Campbell (Staphylinidae), G.B. Wiggins (Trichoptera), D. Pforr (Nymphalidae), E.W. Rockburne (Noctuidae), D.G. Ferguson (Lymantriidae), K. Bolte (Geometridae), G. Lewis (Lepidoptera), J.A. Downes (Lepidoptera), H.J. Teskey (Diptera), L. Forster (Diptera), D.R. Oliver (Chironomidae), J.R. Vockeroth (Diptera), B.V. Peterson (Diptera), J.F. McAlpine (Diptera), D.M. Wood (Diptera), B. Cooper (Calliphoridae), P. Graham (Calliphoridae), W.R. Wong (Tenthredinidae), W.R.M. Mason (Ichneumonoidea), C.M. Yoshimoto (Hymenoptera), K.W. Richards (Apidae).

Appendix 8

Birds and mammals of Truelove Lowland

D.L. Pattie

Class-Aves

Order Gaviiformes
  Family Gaviidae
    Gavia stellata (Pontoppidan)
Order Anseriformes
  Family Anatidae
    Branta bernicla (L.)
    Chen caerulescens (L.)
    Gythya affinis (Eyton)
    Clangula hyemalis (L.)
    Somateria mollissima (L.)
    S. spectabilis (L.)
Order Falconiformes
  Family Falconidae
    Falco rusticolus (L.)
    F. peregrinus Tunstall
Order Galliformes
  Family Tetraonidae
    Lagopus lagopus (L.)
Order Gruiformes
  Family Gruidae
    Grus candensis (L.)
Order Charadriiformes
  Family Charadriidae
    Charadrius hiaticula (L.)
    Pluvialis dominica (Müller)
    Squatarola squatarola (L.)
    Arenaria interpres (L.)
  Family Scolopacidae
    Calidris canutus (L.)
    Erolis maritima (Brunnich)
    E. fuscicollis (Vieillot)
    E. bairdii (Coues)
    E. alpina (L.)
    Crocethia alba (Pallas)
  Family Phalaropodidae
    Phalaropus fulicarius (L.)
  Family Stercorariidae
    Stercorarius longicaudus Vieillot
    S. paraciticus (L.)
  Family Laridae
    Larus hyperboreus Gunnerus
    L. thayeri Brooks
    Pagophila eburnea (Phipps)
    Sterna paradisaea Pontoppidan
  Family Alcidae
    Cepphus grylle (L.)
Order Cuculiformes
  Family Strigidae
    Nyctea scandiaca (L.)
Order Passeriformes
  Family Alaudidae
    Eremophila alpestris (L.)
  Family Corvidae
    Corvus corax (L.)
  Family Turdidae
    Oenanthe eonanthe (L.)
  Family Fringillidae
    Calcarius lapponicus (L.)
    Plectrophenax nivalis (L.)

Class-Mammalia

Order Lagomorpha
  Family Leporidae
    Lepus arcticus Nelson
Order Rodentia
  Family Muridae

Dicrostonyx groenlandicus Traill
Order Carnivora
Family Canidae
Alopex lagopus L.
Canis lupus Pocock
Family Mustelidae
Mustela erminea L.
Family Ursidae
Ursus maritimus Phipps
Order Artiodactyla
Family Bovidae
Ovibos moschatus (Zimmerman)

Authors' addresses

Addison, J.A.
Department of Zoology
CW312 Biological Sciences Centre
The University of Alberta
Edmonton, Alberta T6G 2E9

Addison, P.A.
Canadian Forestry Service
5320 - 122 Street
Edmonton, Alberta T6H 3S5

Babb, T.A.
General Delivery
Denman Island, British Columbia
V0R 1J0

Bliss, L.C.
Department of Botany
B414 Biological Sciences Centre
The University of Alberta
Edmonton, Alberta T6G 2E9

Booth, Tom
Department of Botany
The University of Manitoba
Winnipeg, Manitoba R3T 2N2

Brown, R.J.E.
Division of Building Research
National Research Council
Ottawa, Ontario K1A 0R6

Courtin, G.M.
Department of Biology
Laurentian University
Sudbury, Ontario P3E 2C6

Despain, Don G.
U.S. National Park Service
Yellowstone National Park
Wyoming, U.S.A.

Finnegan, E. (Mrs. Puckett)
c/o Dr. K.G. Puckett
Atmospheric Chemistry Division
Atmospheric Environment Service
4905 Dufferin Street
Downsview, Ontario M3H 5T4

Fuller, W.A.
Department of Zoology
CW312 Biological Sciences Centre
The University of Alberta
Edmonton, Alberta T6G 2E9

Goodwin, C.R.
Arctic Institute of North America
University Library Tower
2920 - 24 Avenue N.W.
Calgary, Alberta T2N 1N4

Hartgerink, A.P.
Department of Botany
The University of Illinois
Urbana, Illinois 61801 U.S.A.

Hergert, Colin R.
Alberta Hail & Crop Insurance
1110 - 1 Street S.W.
Calgary, Alberta T2M 2S2

Hubert, Ben A.
N.W.T. Fish & Wildlife Service
Department of Natural & Cultural Affairs
Yellowknife, N.W.T. X1A 2L9

Jankovska, Vlasta
Botanical Institute
Czechoslovak Academy of Sciences
Brno, Czechoslovakia

Judge, A.S.
Geothermal Studies Section
Seismology Division
Earth Physics Branch
Department of Energy, Mines
and Resources
Ottawa, Ontario

Kerik, J.
c/o Muttart Conservatory
9626 - 96A Street
Edmonton, Alberta T6C 3Z8

Krupicka, J.
Department of Geology
158 Agriculture Building
The University of Alberta
Edmonton, Alberta T6G 2E3

Labine, C.L.
Department of Geography
3-32 Henry Marshall Tory Building
The University of Alberta
Edmonton, Alberta T6G 2H4

Martell, A.M.
Department of Fisheries
and Environment
Canadian Wildlife Services
Sault Ste Marie, Ontario

Mayo, J.M.
Department of Botany
B414 Biological Sciences Centre
The University of Alberta
Edmonton, Alberta T6G 2E9

Minns, C.K.
Canadian Centre of Inland Waters
867 Lakeshore Drive
Burlington, Ontario L7S 1A1

Muc, Michael
Department of Biology
Camrose Lutheran College
Camrose, Alberta T4V 2R3

Nelson, Louise
Department of Microbiology
MacDonald Campus
McGill University
St. Anne de Bellevue
Montreal, Quebec

Nelson, Sherman D.
Department of Botany
B414 Biological Sciences Centre
The University of Alberta
Edmonton, Alberta T6G 2E9

Pakarinen, P.
Department of Botany
Toolonkatu 12 A 22
00100 Helsinki, Finland

Pattie, Donald L.
Department of Science
N.A.I.T.
11762 - 106 Street
Edmonton, Alberta T5G 2R1

Peters, T.W.
Soil Survey
Research Branch
Agriculture Canada
135 Agriculture Building
The University of Alberta
Edmonton, Alberta T6G 2E3

Peterson, W.
Department of Botany
B414 Biological Sciences Centre
The University of Alberta
Edmonton, Alberta T6G 2E9

Procter, Dennis L.C.
Department of Entomology
255 Agriculture Building
The University of Alberta
Edmonton, Alberta T6G 2E3

Richardson, D.H.S.
Department of Biology
Laurentian University
Sudbury, Ontario P3E 2C6

Riewe, R.R.
Department of Zoology
University of Manitoba
Winnipeg, Manitoba R3T 2N2

Ryan, James K.
Department of Entomology
255 Agriculture Building
The University of Alberta
Edmonton, Alberta T6G 2E3

Rydén, B.E.
Hydrological Division
Department of Physical Geography
The University of Uppsala
P.O.B. 554
S-751 22 Uppsala, Sweden

Smith, R.F.C.
Department of Biology
University of Brandon
Brandon, Manitoba R7A 6A9

Speller, W.
Chief, Wildlife Research
and Interpretation
Atlantic Region
Canadian Wildlife Service
Box 1590
Sackville
New Brunswick E0A 3C0

Stutz, R.C.
Renewable Resources Ltd.
11440 Kingsway Avenue
Edmonton, Alberta T5G 0X3

Svoboda, J.
Department of Botany
Erindale College
University of Toronto
3359 Mississauga Road
Mississauga, Ontario

Thompson, Robert G.
Department of Biology
Mount Allison University
Sackville, New Brunswick

van Zinderen Bakker, Edward M., Jr.
Kananaskis Environmental Centre
The University of Calgary
Calgary, Alberta T2N 1N4

Vitt, D.H.
Department of Botany
B414 Biological Sciences Centre
The University of Alberta
Edmonton, Alberta T6G 2E9

Walker, B.D.
Canadian Forestry Service
5320 - 122 Street
Edmonton, Alberta T6H 3S5

Wang, L.C.H.
Department of Zoology
CW312 Biological Sciences Centre
The University of Alberta
Edmonton, Alberta T6G 2E9

Whitfield, D.W.A.
Department of Botany
B414 Biological Sciences Centre
The University of Alberta
Edmonton, Alberta T6G 2E9

Widden, Paul
Department of Biology
Concordia University
7141 Sherbrooke Street W.
Montreal, Quebec H4B 1R6

Selective Index